Student Solutions Manual

D0224322

Finite Mathematics + Applied Calculus

FIFTH EDITION

Stefan Waner
Hofstra University

Steven R. Costenoble
Hofstra University

Prepared by

Stefan Waner
Hofstra University

Steven R. Costenoble
Hofstra University

BROOKS/COLE
CENGAGE Learning

Australia • Brazil • Japan • Korea • Mexico • Singapore • Spain • United Kingdom • United States

For product information and technology assistance, contact us at
Cengage Learning Customer & Sales Support, 1-800-354-9706

For permission to use material from this text or product, submit all requests online at
www.cengage.com/permissions
Further permissions questions can be emailed to
permissionrequest@cengage.com

ISBN-13: 978-0-538-73482-0
ISBN-10: 0-538-73482-5

Brooks/Cole
20 Channel Center Street
Boston, MA 02210
USA

Cengage Learning is a leading provider of customized learning solutions with office locations around the globe, including Singapore, the United Kingdom, Australia, Mexico, Brazil, and Japan. Locate your local office at:
www.cengage.com/global

Cengage Learning products are represented in Canada by Nelson Education, Ltd.

To learn more about Brooks/Cole, visit
www.cengage.com/brookscole

Purchase any of our products at your local college store or at our preferred online store
www.cengagebrain.com

Printed in the United States of America
1 2 3 4 5   14 13 12 11 10

ED363

Table of Contents

Chapter 0 Precalculus Review

0.1 Real Numbers

1. $2(4 + (-1))(2 \cdot -4)$
$= 2(3)(-8) = (6)(-8) = -48$

3. `20/(3*4)-1`
$= \dfrac{20}{12} - 1 = \dfrac{5}{3} - 1 = \dfrac{2}{3}$

5. $\dfrac{3 + ([3 + (-5)])}{3 - 2\times2} = \dfrac{3 + (-2)}{3 - 4}$
$= \dfrac{1}{-1} = -1$

7. `(2-5*(-1))/1-2*(-1)`
$= \dfrac{2 - 5\cdot(-1)}{1} - 2\cdot(-1)$
$= \dfrac{2+5}{1} + 2 = 7 + 2 = 9$

9. $2\cdot(-1)^2 / 2 = \dfrac{2\times(-1)^2}{2} = \dfrac{2\times1}{2} = \dfrac{2}{2} = 1$

11. $2\cdot4^2 + 1 = 2\times16 + 1 = 32 + 1 = 33$

13. `3^2+2^2+1`
$= 3^2 + 2^2 + 1 = 9 + 4 + 1 = 14$

15. $\dfrac{3 - 2(-3)^2}{-6(4 - 1)^2} = \dfrac{3 - 2\times9}{-6(3)^2} = \dfrac{3 - 18}{6\times9}$
$= \dfrac{-15}{54} = \dfrac{5}{18}$

17. `10*(1+1/10)^3`
$= 10\left(1 + \dfrac{1}{10}\right)^3 = 10(1.1)^3$
$= 10\times1.331 = 13.31$

19. $3\left[\dfrac{- 2\cdot3^2}{-(4 - 1)^2}\right] = 3\left[\dfrac{-2\times9}{-3^2}\right] = 3\left[\dfrac{-18}{-9}\right]$
$= 3\times2 = 6$

21. $3\left[1 - \left(-\dfrac{1}{2}\right)^2\right]^2 + 1 = 3\left[1 - \dfrac{1}{4}\right]^2 + 1$
$= 3\left[\dfrac{3}{4}\right]^2 = 3\left[\dfrac{9}{16}\right] + 1 = \dfrac{27}{16} + 1 = \dfrac{43}{16}$

23. `(1/2)^2-1/2^2`
$= \left[\dfrac{1}{2}\right]^2 - \dfrac{1}{2^2} = \dfrac{1}{4} - \dfrac{1}{4} = 0$

25. $3\times(2-5) =$ `3*(2-5)`

27. $\dfrac{3}{2-5} =$ `3/(2-5)`
Note `3/2-5` is wrong, since it corresponds to $\dfrac{3}{2} - 5$.

29. $\dfrac{3-1}{8+6} =$ `(3-1)/(8+6)`
Note `3-1/8-6` is wrong, since it corresponds to $3 -\dfrac{1}{8} - 6$.

31. $3 - \dfrac{4+7}{8} =$ `3-(4+7)/8`

33. $\dfrac{2}{3+x} - xy^2 =$ `2/(3+x)-x*y^2`

35. $3.1x^3 - 4x^{-2} - \dfrac{60}{x^2-1}$
$=$ `3.1x^3-4x^(-2)-60/(x^2-1)`

37. $\dfrac{\left[\dfrac{2}{3}\right]}{5} =$ `(2/3)/5`
Note that we use only (round) parentheses in technology formulas, and not brackets.

39. $3^{4-5}\times6 =$ `3^(4-5)*6`
Note that the entire exponent is in parentheses.

41. $3\left[1 + \dfrac{4}{100}\right]^{-3} =$ `3*(1+4/100)^(-3)`
Note that we use only (round) parentheses in technology formulas, and not brackets.

43. $3^{2x-1} + 4^x - 1 = $ `3^(2*x-1)+4^x-1`
Note that the entire exponent of 3 is in
parentheses.

45. $2^{2x^2-x+1} = $ `2^(2x^2-x+1)`
Note that the entire exponent is in parentheses.

47. $\dfrac{4e^{-2x}}{2-3e^{-2x}}$

= `4*e^(-2*x)/(2-3e^(-2*x))`
or `4(*e^(-2*x))/(2-3e^(-2*x))`
or `(4*e^(-2*x))/(2-3e^(-2*x))`

49. $3\left[1 - \left(-\dfrac{1}{2}\right)^2\right]^2 + 1$

= `3(1-(-1/2)^2)^2+1`
Note that we use only (round) parentheses in
technology formulas, and not brackets.

0.2 Exponents and Radicals

1. $3^3 = 27$

3. $-(2 \cdot 3)^2 = (2^2 \cdot 3^2) = -(4 \cdot 9) = -36$ or
$-(2 \cdot 3)^2 = -(6^2) = -36$

5. $\left(\dfrac{-2}{3}\right)^2 = \dfrac{(-2)^2}{3^2} = \dfrac{4}{9}$

7. $(-2)^{-3} = \dfrac{1}{(-2)^3} = \dfrac{1}{-8} = -\dfrac{1}{8}$

9. $\left(\dfrac{1}{4}\right)^{-2} = \dfrac{1}{(1/4)^2} = \dfrac{1}{1/4^2} = \dfrac{1}{1/16} = 16$

11. $2 \cdot 3^0 = 2 \cdot 1 = 2$

13. $2^3 2^2 = 2^{3+2} = 2^5 = 32$ or $2^3 2^2 = 8 \cdot 4 = 32$

15. $2^2 2^{-1} 2^4 2^{-4} = 2^{2-1+4-4} = 2^1 = 2$

17. $x^3 x^2 = x^{3+2} = x^5$

19. $-x^2 x^{-3} y = -x^{2-3} y = -x^{-1} y = -\dfrac{y}{x}$

21. $\dfrac{x^3}{x^4} = x^{3-4} = x^{-1} = \dfrac{1}{x}$

23. $\dfrac{x^2 y^2}{x^{-1} y} = x^{2-(-1)} y^{2-1} = x^3 y$

25. $\dfrac{(xy^{-1}z^3)^2}{x^2 yz^2} = \dfrac{x^2(y^{-1})^2(z^3)^2}{x^2 yz^2} = x^{2-2} y^{-2-1} z^{6-2} =$
$y^{-3} z^4 = \dfrac{z^4}{y^3}$

27. $\left(\dfrac{xy^{-2}z}{x^{-1}z}\right)^3 = \dfrac{(xy^{-2}z)^3}{(x^{-1}z)^3} = \dfrac{x^3 y^{-6} z^3}{x^{-3} z^3} = x^{3-(-3)} y^{-6} z^{3-3}$
$= x^6 y^{-6} = \dfrac{x^6}{y^6}$

29. $\left(\dfrac{x^{-1}y^{-2}z^2}{xy}\right)^{-2} = (x^{-1-1}y^{-2-1}z^2)^{-2} = (x^{-2}y^{-3}z^2)^{-2}$
$= x^4 y^6 z^{-4} = \dfrac{x^4 y^6}{z^4}$

31. $3x^{-4} = \dfrac{3}{x^4}$

33. $\dfrac{3}{4} x^{-2/3} = \dfrac{3}{4x^{2/3}}$

35. $1 - \dfrac{0.3}{x^{-2}} - \dfrac{6}{5} x^{-1} = 1 - 0.3x^2 - \dfrac{6}{5x}$

37. $\sqrt{4} = 2$

39. $\sqrt{\dfrac{1}{4}} = \dfrac{\sqrt{1}}{\sqrt{4}} = \dfrac{1}{2}$

41. $\sqrt{\dfrac{16}{9}} = \dfrac{\sqrt{16}}{\sqrt{9}} = \dfrac{4}{3}$

43. $\dfrac{\sqrt{4}}{5} = \dfrac{2}{5}$

45. $\sqrt{9} + \sqrt{16} = 3 + 4 = 7$

47. $\sqrt{9 + 16} = \sqrt{25} = 5$

49. $\sqrt[3]{8 - 27} = \sqrt[3]{-19} \approx -2.668$

51. $\sqrt[3]{\dfrac{27}{8}} = \dfrac{\sqrt[3]{27}}{\sqrt[3]{8}} = \dfrac{3}{2}$

53. $\sqrt{(-2)^2} = \sqrt{4} = 2$

55. $\sqrt{\dfrac{1}{4}(1 + 15)} = \sqrt{\dfrac{16}{4}} = \dfrac{\sqrt{16}}{\sqrt{4}} = \dfrac{4}{2} = 2$

57. $\sqrt{a^2 b^2} = \sqrt{a}\,\sqrt{b} = ab$

3

59. $\sqrt{(x+9)^2} = x + 9$ ($x + 9 > 0$ because x is positive)

61. $\sqrt[3]{x^3(a^3 + b^3)} = \sqrt[3]{x^3}\sqrt[3]{a^3 + b^3} = x\sqrt[3]{a^3 + b^3}$
(Notice: *Not* $x(a + b)$.)

63. $\sqrt{\dfrac{4xy^3}{x^2 y}} = \sqrt{\dfrac{4y^2}{x}} = \dfrac{\sqrt{4}\sqrt{y^2}}{\sqrt{x}} = \dfrac{2y}{\sqrt{x}}$

65. $\sqrt{3} = 3^{1/2}$

67. $\sqrt{x^3} = x^{3/2}$

69. $\sqrt[3]{xy^2} = (xy^2)^{1/3}$

71. $\dfrac{x^2}{\sqrt{x}} = \dfrac{x^2}{x^{1/2}} = x^{2-1/2} = x^{3/2}$

73. $\dfrac{3}{5x^2} = \dfrac{3}{5}x^{-2}$

75. $\dfrac{3x^{-1.2}}{2} - \dfrac{1}{3x^{2.1}} = \dfrac{3}{2}x^{-1.2} - \dfrac{1}{3}x^{-2.1}$

77. $\dfrac{2x}{3} - \dfrac{x^{0.1}}{2} + \dfrac{4}{3x^{1.1}} = \dfrac{2}{3}x - \dfrac{1}{2}x^{0.1} + \dfrac{4}{3}x^{-1.1}$

79. $\dfrac{3\sqrt{x}}{4} - \dfrac{5}{3\sqrt{x}} + \dfrac{4}{3x\sqrt{x}} = \dfrac{3x^{1/2}}{4} - \dfrac{5}{3x^{1/2}} + \dfrac{4}{3xx^{1/2}} = \dfrac{3}{4}x^{1/2} - \dfrac{5}{3}x^{-1/2} + \dfrac{4}{3}x^{-3/2}$

81. $\dfrac{3\sqrt[5]{x^2}}{4} - \dfrac{7}{2\sqrt{x^3}}$
$= \dfrac{3x^{2/5}}{4} - \dfrac{7}{2x^{3/2}} = \dfrac{3}{4}x^{2/5} - \dfrac{7}{2}x^{-3/2}$

83. $\dfrac{1}{(x^2 + 1)^3} - \dfrac{3}{4\sqrt[3]{(x^2 + 1)}} =$
$\dfrac{1}{(x^2 + 1)^3} - \dfrac{3}{4(x^2 + 1)^{1/3}} =$
$(x^2 + 1)^{-3} - \dfrac{3}{4}(x^2 + 1)^{-1/3}$

85. $2^{2/3} = \sqrt[3]{2^2}$

87. $x^{4/3} = \sqrt[3]{x^4}$

89. $(x^{1/2}y^{1/3})^{1/5} = \sqrt[5]{\sqrt{x}\sqrt[3]{y}}$

91. $-\dfrac{3}{2}x^{-1/4} = -\dfrac{3}{2x^{1/4}} = -\dfrac{3}{2\sqrt[4]{x}}$

93. $0.2x^{-2/3} + \dfrac{3}{7x^{-1/2}} = \dfrac{0.2}{x^{2/3}} + \dfrac{3x^{1/2}}{7} = \dfrac{0.2}{\sqrt[3]{x^2}} + $
$\dfrac{3\sqrt{x}}{7}$

95. $\dfrac{3}{4(1 - x)^{5/2}} = \dfrac{3}{4\sqrt{(1 - x)^5}}$

97. $4^{-1/2}4^{7/2} = 4^{-1/2+7/2} = 4^3 = 64$

99. $3^{2/3}3^{-1/6} = 3^{2/3-1/6} = 3^{1/2} = \sqrt{3}$

101. $\dfrac{x^{3/2}}{x^{5/2}} = x^{3/2-5/2} = x^{-1} = \dfrac{1}{x}$

103. $\dfrac{x^{1/2}y^2}{x^{-1/2}y} = x^{1/2+1/2}y^{2-1} = xy$

105. $\left(\dfrac{x}{y}\right)^{1/3}\left(\dfrac{y}{x}\right)^{2/3} = \left(\dfrac{y}{x}\right)^{-1/3}\left(\dfrac{y}{x}\right)^{2/3} = \left(\dfrac{y}{x}\right)^{1/3}$

107. $x^2 - 16 = 0$, $x^2 = 16$, $x = \pm\sqrt{16} = \pm4$

109. $x^2 - \dfrac{4}{9} = 0$, $x^2 = \dfrac{4}{9}$, $x = \pm\sqrt{\dfrac{4}{9}} = \pm\dfrac{2}{3}$

4

111. $x^2 - (1 + 2x)^2 = 0$, $x^2 = (1 + 2x)^2$, $x = \pm(1 + 2x)$; if $x = 1 + 2x$ then $-x = 1$, $x = -1$; if $x = -(1 + 2x)$ then $3x = -1$, $x = -1/3$. So, $x = -1$ or $-1/3$.

113. $x^5 + 32 = 0$, $x^5 = -32$, $x = \sqrt[5]{-32} = -2$

115. $x^{1/2} - 4 = 0$, $x^{1/2} = 4$, $x = 4^2 = 16$

117. $1 - \dfrac{1}{x^2} = 0$, $1 = \dfrac{1}{x^2}$, $x^2 = 1$, $x = \pm\sqrt{1} = \pm1$

119. $(x - 4)^{-1/3} = 2$, $x - 4 = 2^{-3} = \dfrac{1}{8}$,

$x = 4 + \dfrac{1}{8} = \dfrac{33}{8}$

0.3 Multiplying and Factoring Algebraic Expressions

1. $x(4x + 6) = 4x^2 + 6x$

3. $(2x - y)y = 2xy - y^2$

5. $(x + 1)(x - 3) = x^2 + x - 3x - 3 = x^2 - 2x - 3$

7. $(2y + 3)(y + 5) = 2y^2 + 3y + 10y + 15 = 2y^2 + 13y + 15$

9. $(2x - 3)^2 = 4x^2 - 12x + 9$

11. $\left(x + \dfrac{1}{x}\right)^2 = x^2 + 2 + \dfrac{1}{x^2}$

13. $(2x - 3)(2x + 3) = (2x)^2 - 3^2 = 4x^2 - 9$

15. $\left(y - \dfrac{1}{y}\right)\left(y + \dfrac{1}{y}\right) = y^2 - \left(\dfrac{1}{y}\right)^2 = y^2 - \dfrac{1}{y^2}$

17. $(x^2 + x - 1)(2x + 4) = (x^2 + x - 1)2x + (x^2 + x - 1)4 = 2x^3 + 2x^2 - 2x + 4x^2 + 4x - 4 = 2x^3 + 6x^2 + 2x - 4$

19. $(x^2 - 2x + 1)^2 = (x^2 - 2x + 1)(x^2 - 2x + 1) = x^2(x^2 - 2x + 1) - 2x(x^2 - 2x + 1) + (x^2 - 2x + 1) = x^4 - 2x^3 + x^2 - 2x^3 + 4x^2 - 2x + x^2 - 2x + 1 = x^4 - 4x^3 + 6x^2 - 4x + 1$

21. $(y^3 + 2y^2 + y)(y^2 + 2y - 1) = y^3(y^2 + 2y - 1) + 2y^2(y^2 + 2y - 1) + y(y^2 + 2y - 1) = y^5 + 2y^4 - y^3 + 2y^4 + 4y^3 - 2y^2 + y^3 + 2y^2 - y = y^5 + 4y^4 + 4y^3 - y$

23. $(x + 1)(x + 2) + (x + 1)(x + 3) = (x + 1)(x + 2 + x + 3) = (x + 1)(2x + 5)$

25. $(x^2 + 1)^5(x + 3)^4 + (x^2 + 1)^6(x + 3)^3 = (x^2 + 1)^5(x + 3)^3(x + 3 + x^2 + 1) = (x^2 + 1)^5(x + 3)^3(x^2 + x + 4)$

27. $(x^3 + 1)\sqrt{x + 1} - (x^3 + 1)^2\sqrt{x + 1} = (x^3 + 1)\sqrt{x + 1}\,[1 - (x^3 + 1)] = -x^3(x^3 + 1)\sqrt{x + 1}$

29. $\sqrt{(x + 1)^3} + \sqrt{(x + 1)^5} = \sqrt{(x + 1)^3} \cdot [1 + \sqrt{(x + 1)^2}\,] = \sqrt{(x + 1)^3}\,(1 + x + 1) = (x + 2)\sqrt{(x + 1)^3}$

31. (a) $2x + 3x^2 = x(2 + 3x)$ **(b)** $x(2 + 3x) = 0$; $x = 0$ or $2 + 3x = 0$; $x = 0$ or $-2/3$

33. (a) $6x^3 - 2x^2 = 2x^2(3x - 1)$
(b) $2x^2(3x - 1) = 0$; $x = 0$ or $3x - 1 = 0$; $x = 0$ or $1/3$

35. (a) $x^2 - 8x + 7 = (x - 1)(x - 7)$
(b) $(x - 1)(x - 7) = 0$; $x - 1 = 0$ or $x - 7 = 0$; $x = 1$ or 7

37. (a) $x^2 + x - 12 = (x - 3)(x + 4)$
(b) $(x - 3)(x + 4) = 0$; $x - 3 = 0$ or $x + 4 = 0$; $x = 3$ or -4

39. (a) $2x^2 - 3x - 2 = (2x + 1)(x - 2)$
(b) $(2x + 1)(x - 2) = 0$; $2x + 1 = 0$ or $x - 2 = 0$; $x = -1/2$ or 2

41. (a) $6x^2 + 13x + 6 = (2x + 3)(3x + 2)$
(b) $(2x + 3)(3x + 2) = 0$; $2x + 3 = 0$ or $3x + 2 = 0$; $x = -3/2$ or $-2/3$

43. (a) $12x^2 + x - 6 = (3x - 2)(4x + 3)$
(b) $(3x - 2)(4x + 3) = 0$; $3x - 2 = 0$ or $4x + 3 = 0$; $x = 2/3$ or $-3/4$

45. (a) $x^2 + 4xy + 4y^2 = (x + 2y)^2$
(b) $(x + 2y)^2 = 0$; $x + 2y = 0$; $x = -2y$

47. (a) $x^4 - 5x^2 + 4 = (x^2 - 1)(x^2 - 4) = (x - 1)(x + 1)(x - 2)(x + 2)$
(b) $(x - 1)(x + 1)(x - 2)(x + 2) = 0$; $x - 1 = 0$ or $x + 1 = 0$ or $x - 2 = 0$ or $x + 2 = 0$; $x = \pm 1$ or ± 2

0.4 Rational Expressions

1. $\dfrac{x-4}{x+1} \cdot \dfrac{2x+1}{x-1} = \dfrac{(x-4)(2x+1)}{(x+1)(x-1)} =$

$\dfrac{2x^2 - 7x - 4}{x^2 - 1}$

3. $\dfrac{x-4}{x+1} + \dfrac{2x+1}{x-1} =$

$\dfrac{(x-4)(x-1) + (x+1)(2x+1)}{(x+1)(x-1)} =$

$\dfrac{3x^2 - 2x + 5}{x^2 - 1}$

5. $\dfrac{x^2}{x+1} - \dfrac{x-1}{x+1} = \dfrac{x^2 - (x-1)}{x+1} = \dfrac{x^2 - x + 1}{x+1}$

7. $\dfrac{1}{\left(\dfrac{x}{x-1}\right)} + x - 1 = \dfrac{x-1}{x} + x - 1 =$

$\dfrac{x - 1 + x(x-1)}{x} = \dfrac{x^2 - 1}{x}$

9. $\dfrac{1}{x}\left(\dfrac{x-3}{xy} + \dfrac{1}{y}\right) = \dfrac{1}{x}\left(\dfrac{x-3+x}{xy}\right) = \dfrac{2x-3}{x^2 y}$

11. $\dfrac{(x+1)^2(x+2)^3 - (x+1)^3(x+2)^2}{(x+2)^6} =$

$\dfrac{(x+1)^2(x+2)^2[(x+2) - (x+1)]}{(x+2)^6} =$

$\dfrac{(x+1)^2}{(x+2)^4}$

13. $\dfrac{(x^2-1)\sqrt{x^2+1} - \dfrac{x^4}{\sqrt{x^2+1}}}{x^2 + 1} =$

$\dfrac{(x^2-1)(x^2+1) - x^4}{(x^2+1)\sqrt{x^2+1}} = \dfrac{-1}{\sqrt{(x^2+1)^3}}$

15. $\dfrac{\dfrac{1}{(x+y)^2} - \dfrac{1}{x^2}}{y} = \dfrac{x^2 - (x+y)^2}{yx^2(x+y)^2} =$

$\dfrac{x^2 - x^2 - 2xy - y^2}{yx^2(x+y)^2} = \dfrac{-y(2x+y)}{yx^2(x+y)^2} = \dfrac{-(2x+y)}{x^2(x+y)^2}$

7

0.5 Solving Polynomial Equations

1. $x + 1 = 0$, $x = -1$

3. $-x + 5 = 0$, $x = 5$

5. $4x - 5 = 8$, $4x = 13$, $x = 13/4$

7. $7x + 55 = 98$, $7x = 43$, $x = 43/7$

9. $x + 1 = 2x + 2$, $-x = 1$, $x = -1$

11. $ax + b = c$, $ax = c - b$, $x = (c - b)/a$

13. $2x^2 + 7x - 4 = 0$, $(2x - 1)(x + 4) = 0$, $x = -4, \dfrac{1}{2}$

15. $x^2 - x + 1 = 0$, $\Delta = -3 < 0$, so this equation has no real solutions

17. $2x^2 - 5 = 0$, $x^2 = \dfrac{5}{2}$, $x = \pm\sqrt{\dfrac{5}{2}}$

19. $-x^2 - 2x - 1 = 0$, $-(x + 1)^2 = 0$, $x = -1$

21. $\dfrac{1}{2}x^2 - x - \dfrac{3}{2} = 0$, $x^2 - 2x - 3 = 0$, $(x + 1)(x - 3) = 0$, $x = -1, 3$

23. $x^2 - x = 1$, $x^2 - x - 1 = 0$, $x = \dfrac{1 \pm \sqrt{5}}{2}$ by the quadratic formula

25. $x = 2 - \dfrac{1}{x}$, $x^2 = 2x - 1$, $x^2 - 2x + 1 = 0$, $(x - 1)^2 = 0$, $x = 1$

27. $x^4 - 10x^2 + 9 = 0$, $(x^2 - 1)(x^2 - 9) = 0$, $x^2 = 1$ or $x^2 = 0$, $x = \pm1, \pm3$

29. $x^4 + x^2 - 1 = 0$, $x^2 = \dfrac{-1 \pm \sqrt{5}}{2}$ by the quadratic formula, $x = \pm\sqrt{\dfrac{-1 \pm \sqrt{5}}{2}}$

31. $x^3 + 6x^2 + 11x + 6 = 0$,

$(x + 1)(x + 2)(x + 3) = 0$, $x = -1, -2, -3$

33. $x^3 + 4x^2 + 4x + 3 = 0$, $(x + 3)(x^2 + x + 1) = 0$, $x = -3$ (For $x^2 + x + 1 = 0$, $\Delta = -3 < 0$, so there are no real solutions to this quadratic equation.)

35. $x^3 - 1 = 0$, $x^3 = 1$, $x = \sqrt[3]{1} = 1$

37. $y^3 + 3y^2 + 3y + 2 = 0$, $(y + 2)(y^2 + y + 1) = 0$, $y = -2$ (For $y^2 + y + 1 = 0$, $\Delta = -3 < 0$, so there are no real solutions to this quadratic equation.)

39. $x^3 - x^2 - 5x + 5 = 0$, $(x - 1)(x^2 - 5) = 0$, $x = 1, \pm\sqrt{5}$

41. $2x^6 - x^4 - 2x^2 + 1 = 0$, $(2x^2 - 1)(x^4 - 1) = 0$, [or $(2x^2 - 1)(x^2 - 1)(x^2 + 1) = 0$; in any case, think of the cubic you get by substituting y for x^2], $x = \pm1, \pm\dfrac{1}{\sqrt{2}}$

43. $(x^2 + 3x + 2)(x^2 - 5x + 6) = 0$, $(x + 2)(x + 1)(x - 2)(x - 3) = 0$, $x = -2, -1, 2, 3$

0.6 Solving Miscellaneous Equations

1. $x^4 - 3x^3 = 0$, $x^3(x - 3) = 0$, $x = 0, 3$

3. $x^4 - 4x^2 = -4$, $x^4 - 4x^2 + 4 = 0$, $(x^2 - 2)^2 = 0$, $x = \pm\sqrt{2}$

5. $(x + 1)(x + 2) + (x + 1)(x + 3) = 0$, $(x + 1)(x + 2 + x + 3) = 0$, $(x + 1)(2x + 5) = 0$, $x = -1, -5/2$

7. $(x^2 + 1)^5(x + 3)^4 + (x^2 + 1)^6(x + 3)^3 = 0$, $(x^2 + 1)^5(x + 3)^3(x + 3 + x^2 + 1) = 0$, $(x^2 + 1)^5(x + 3)^3(x^2 + x + 4) = 0$, $x = -3$ (Neither $x^2 + 1 = 0$ nor $x^2 + x + 4 = 0$ has a real solution.)

9. $(x^3 + 1)\sqrt{x + 1} - (x^3 + 1)^2\sqrt{x + 1} = 0$, $(x^3 + 1)\sqrt{x + 1}\,[1 - (x^3 + 1)] = 0$, $-x^3(x^3 + 1)\sqrt{x + 1} = 0$, $x = 0, -1$

11. $\sqrt{(x + 1)^3} + \sqrt{(x + 1)^5} = 0$, $\sqrt{(x + 1)^3}\,(1 + x + 1) = 0$, $(x + 2)\sqrt{(x + 1)^3} = 0$, $x = -1$ ($x = -2$ is not a solution because $\sqrt{(x + 1)^3}$ is not defined for $x = -2$.)

13. $(x + 1)^2(2x + 3) - (x + 1)(2x + 3)^2 = 0$, $(x + 1)(2x + 3)(x + 1 - 2x - 3) = 0$, $(x + 1)(2x + 3)(-x - 2) = 0$, $x = -2, -3/2, -1$

15. $\dfrac{(x + 1)^2(x + 2)^3 - (x + 1)^3(x + 2)^2}{(x + 2)^6} = 0$, $\dfrac{(x + 1)^2(x + 2)^2[(x + 2) - (x + 1)]}{(x + 2)^6} = 0$, $\dfrac{(x + 1)^2}{(x + 2)^4} = 0$, $(x + 1)^2 = 0$, $x = -1$

17. $\dfrac{2(x^2 - 1)\sqrt{x^2 + 1} - \dfrac{x^4}{\sqrt{x^2 + 1}}}{x^2 + 1} = 0$, $\dfrac{2(x^2 - 1)(x^2 + 1) - x^4}{(x^2 + 1)\sqrt{x^2 + 1}} = 0$, $\dfrac{x^4 - 2}{(x^2 + 1)\sqrt{x^2 + 1}} = 0$, $x^4 - 2 = 0$, $x = \pm\sqrt[4]{2}$

19. $x - \dfrac{1}{x} = 0$, $x^2 - 1 = 0$, $x = \pm 1$

21. $\dfrac{1}{x} - \dfrac{9}{x^3} = 0$, $x^2 - 9 = 0$, $x = \pm 3$

23. $\dfrac{x - 4}{x + 1} - \dfrac{x}{x - 1} = 0$, $\dfrac{(x - 4)(x - 1) - x(x + 1)}{(x + 1)(x - 1)} = 0$, $\dfrac{-6x + 4}{(x + 1)(x - 1)} = 0$, $-6x + 4 = 0$, $x = 2/3$

25. $\dfrac{x + 4}{x + 1} + \dfrac{x + 4}{3x} = 0$, $\dfrac{3x(x + 4) + (x + 1)(x + 4)}{3x(x + 1)} = 0$, $\dfrac{(x + 4)(3x + x + 1)}{3x(x + 1)} = 0$, $\dfrac{(x + 4)(4x + 1)}{3x(x + 1)} = 0$, $(x + 4)(4x + 1) = 0$, $x = -4, -1/4$

0.7 The Coordinate Plane

1. $P(0, 2)$, $Q(4, -2)$, $R(-2, 3)$, $S(-3.5, -1.5)$, $T(-2.5, 0)$, $U(2, 2.5)$

3.

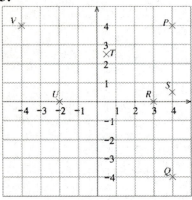

5. Solve the equation $x + y = 1$ for y to get $y = 1-x$. Then plot some points:

x	y = 1−x
−2	3
−1	2
0	1
1	0
2	−1

Graph:

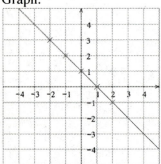

7. Solve the equation $2y - x^2 = 1$ for y to get $y = (1+x^2)/2$. Then plot some points:

x	y = (1+x²)/2
−2	2.5
−1	1
0	0.5
1	1
2	2.5

Graph:

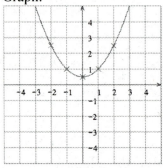

9. Solve the equation $xy = 4$ for y to get $y = 4/x$. Then plot some points:

x	y = 4/x
−3	−1.333
−2	−2
−1	−4
1	4
2	2
3	1.3333

Graph:

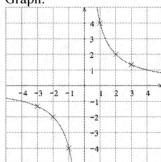

11. Solve the equation $xy = x^2 + 1$ for y to get $y = x + 1/x$. Then plot some points:

x	$y = x + 1/x$
-2	-2.5
-1	2
$-1/2$	-2.5
$1/2$	2.5
1	2
2	2.5

Graph:

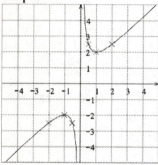

13. $\sqrt{(2-1)^2 + (-2+1)^2} = \sqrt{2}$

15. $\sqrt{(0-a)^2 + (b-0)^2} = \sqrt{a^2 + b^2}$

17. Set the two distances equal and solve:
$\sqrt{(1-0)^2 + (k-0)^2} = \sqrt{(1-2)^2 + (k-1)^2}$;
$\sqrt{1 + k^2} = \sqrt{2 - 2k + k^2}$;
$1 + k^2 = 2 - 2k + k^2$; $2k = 1$; $k = \frac{1}{2}$

19. Circle with center $(0,0)$ and radius 3

11

Chapter 1 Functions and Linear Models

1.1 Functions from the Numerical and Algebraic Viewpoints

1. Using the table,
(a) $f(0) = 2$ **(b)** $f(2) = 0.5$.

3. Using the table,
(a) $f(2) - f(-2) = 0.5 - 2 = -1.5$
(b) $f(-1)f(-2) = (4)(2) = 8$
(c) $-2f(-1) = -2(4) = -8$

5. From the graph, we find
(a) $f(1) = 20$ **(b)** $f(2) = 30$

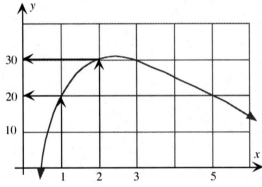

In a similar way, we find

(c) $f(3) = 30$ **(d)** $f(5) = 20$

(e) $f(3) - f(2) = 30 - 30 = 0$

7. From the graph,

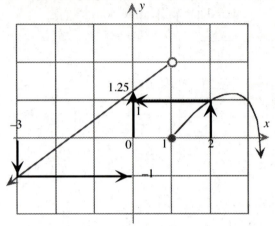

(a) $f(-3) = -1$ **(b)** $f(0) = 1.25$

(c) $f(1) = 0$ since the solid dot is on $(1, 0)$
($f(1) \neq 2$ since the hollow dot at $(1, 2)$ indicates that $(1, 2)$ is not a point of the graph.)

(d) $f(2) = 1$
(e) Since $f(3) = 1$ and $f(2) - 1$,
$$\frac{f(3) - f(2)}{3 - 2} = \frac{1 - 1}{3 - 2} = 0$$

9. $f(x) = x - \dfrac{1}{x^2}$, with domain $(0, +\infty)$
(a) Since 4 is in $(0, +\infty)$, $f(4)$ is defined.
$$f(4) = 4 - \frac{1}{4^2} = 4 - \frac{1}{16} = \frac{63}{16}$$
(b) Since 0 is not in $(0, +\infty)$, $f(0)$ is not defined.
(c) Since -1 is not in $(0, +\infty)$, $f(-1)$ is not defined.

11. $f(x) = \sqrt{x+10}$, with domain $[-10, 0)$
(a) Since 0 is not in $[-10, 0)$, $f(0)$ is not defined.
(b) Since 9 is not in $[-10, 0)$, $f(9)$ is not defined.
(c) Since -10 is in $[-10, 0)$, $f(-10)$ is defined.
$$f(-10) = \sqrt{-10+10} = \sqrt{0} = 0$$

13. $f(x) = 4x - 3$
(a) $f(-1) = 4(-1) - 3 = -4 - 3 = -7$
(b) $f(0) = 4(0) - 3 = 0 - 3 = -3$
(c) $f(1) = 4(1) - 3 = 4 - 3 = 1$

(**d**) Substitute y for x to obtain
$$f(y) = 4y - 3$$
(**e**) Substitute $(a+b)$ for x to obtain
$$f(a+b) = 4(a+b) - 3$$

15. $f(x) = x^2 + 2x + 3$
(**a**) $f(0) = (0)^2 + 2(0) + 3$
$\qquad = 0 + 0 + 3 = 3$
(**b**) $f(1) = 1^2 + 2(1) + 3$
$\qquad = 1 + 2 + 3 = 6$
(**c**) $f(-1) = (-1)^2 + 2(-1) + 3$
$= 1 - 2 + 3 = 2$
(**d**) $f(-3) = (-3)^2 + 2(-3) + 3$
$= 9 - 6 + 3 = 6$
(**e**) Substitute a for x to obtain
$f(a) = a^2 + 2a + 3$
(**f**) Substitute $(x+h)$ for x to obtain
$f(x+h) = (x+h)^2 + 2(x+h) + 3$

17. $g(s) = s^2 + \dfrac{1}{s}$

(**a**) $g(1) = 1^2 + \dfrac{1}{1} = 1 + 1 = 2$

(**b**) $g(-1) = (-1)^2 + \dfrac{1}{(-1)} = 1 - 1 = 0$

(**c**) $g(4) = 4^2 + \dfrac{1}{4} = 16 + \dfrac{1}{4}$

$\qquad = 16\frac{1}{4}$ or $\dfrac{65}{4}$ or 16.25

(**d**) Substitute x for s to obtain

$$g(x) = x^2 + \dfrac{1}{x}$$

(**e**) Substitute $(s+h)$ for s to obtain

$$g(s+h) = (s+h)^2 + \dfrac{1}{s+h}$$

(**f**) $g(s+h) - g(s)$
= Answer to part (e) – Original function

$$= \left((s+h)^2 + \dfrac{1}{s+h}\right) - \left(s^2 + \dfrac{1}{s}\right)$$

19. $f(x) = -x^3$ (domain $(-\infty, +\infty)$)
Technology formula: $\texttt{-(x\^{}3)}$

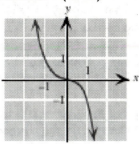

21. $f(x) = x^4$ (domain $(-\infty, +\infty)$)
Technology formula: $\texttt{x\^{}4}$

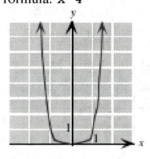

23. $f(x) = \dfrac{1}{x^2}$ $(x \ne 0)$

Technology formula: $\texttt{1/x\^{}2}$

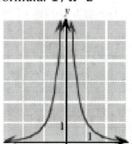

25. (**a**) $f(x) = x$ $(-1 \le x \le 1)$
Since the graph of $f(x) = x$ is a diagonal 45° line through the origin inclining up from left to right, the correct graph is (I)
(**b**) $f(x) = -x$ $(-1 \le x \le 1)$
Since the graph of $f(x) = -x$ is a diagonal 45° line through the origin inclining down from left to right, the correct graph is (IV)
(**c**) $f(x) = \sqrt{x}$ $(0 < x < 4)$
Since the graph of $f(x) = \sqrt{x}$ is the top half of a sideways parabola, the correct graph is (V)

(d) $f(x) = x + \dfrac{1}{x} - 2 \ (0 < x < 4)$

If we plot a few points like $x = 0.1, 1, 2$, and 3 we find that the correct graph is (VI).

(e) $f(x) = |x| \ (-1 \le x \le 1)$

Since the graph of $f(x) = |x|$ is a "V"-shape with its vertex at the origin, the correct graph is (III).

(f) $f(x) = x - 1 \ (-1 \le x \le 1)$

Since the graph of $f(x) = x-1$ is a straight line through $(0, -1)$ and $(1, 0)$, the correct graph is (II).

27. $f(x) = 0.1x^2 - 4x + 5$

Technology formula: `0.1*x^2-4*x+5`

Table of Values:

x	0	1	2	3
$f(x)$	5	1.1	−2.6	−6.1
x	4	5	6	7
$f(x)$	−9.4	−12.5	−15.4	−18.1
x	8	9	10	
$f(x)$	−20.6	−22.9	−25	

29. $h(x) = \dfrac{x^2-1}{x^2+1}$

Technology formula: `(x^2-1)/(x^2+1)`

Table of Values (rounded to four decimal places):

x	0.5	1.5	2.5	3.5
$h(x)$	−0.6000	0.3846	0.7241	0.8491
x	4.5	5.5	6.5	7.5
$h(x)$	0.9059	0.9360	0.9538	0.9651
x	8.5	9.5	10.5	
$h(x)$	0.9727	0.9781	0.9820	

31. $f(x) = \begin{cases} x & \text{if } -4 \le x < 0 \\ 2 & \text{if } 0 \le x \le 4 \end{cases}$

Technology formula: `x*(x<0)+2*(x>=0)`
(For a graphing calculator, use ≥ instead of >=.)

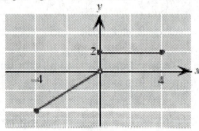

(a) $f(-1) = -1$. We used the first formula, since −1 is in $[-4, 0)$.

(b) $f(0) = 2$. We used the second formula, since 0 is in $[0, 4]$.

(c) $f(1) = 2$. We used the second formula, since 1 is in $[0, 4]$.

33. $f(x) = \begin{cases} x^2 & \text{if } -2 < x \le 0 \\ 1/x & \text{if } 0 < x \le 4 \end{cases}$

Technology formula:
`(x^2)*(x<=0)+(1/x)*(0<x)`
(For a graphing calculator, use ≤ instead of <=.)

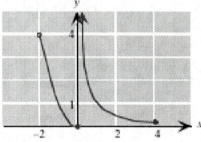

(a) $f(-1) = 1^2 = 1$. We used the first formula, since −1 is in $(-2, 0]$.

(b) $f(0) = 0^2 = 0$. We used the first formula, since 0 is in $(-2, 0]$.

(c) $f(1) = 1/1 = 1$. We used the second formula, since 1 is in $(0, 4]$.

35. $f(x) = \begin{cases} x & \text{if } -1 < x \le 0 \\ x+1 & \text{if } 0 < x \le 2 \\ x & \text{if } 2 < x \le 4 \end{cases}$

Technology formula:
$$x*(x<=0)+(x+1)*(0<x)*(x<=2)+$$
$$x*(2<x)$$
(For a graphing calculator, use ≤ instead of <=.)

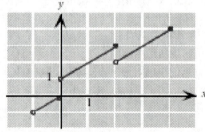

(a) $f(0) = 0$. We used the first formula, since 0 is in $(-1, 0]$.

(b) $f(1) = 1+1 = 2$. We used the second formula, since 1 is in $(0, 2]$.

(c) $f(2) = 2+1 = 3$. We used the second formula, since 2 is in $(0, 2]$.

14

(d) $f(3) = 3$. We used the third formula, since 3 is in $(2, 4]$.

37. $f(x) = x^2$

(a) $f(x+h) = (x+h)^2$ Therefore,

$$\begin{aligned}
f(x+h) - f(x) &= (x+h)^2 - x^2 \\
&= x^2 + 2xh + h^2 - x^2 \\
&= 2xh + h^2 \\
&= h(2x + h)
\end{aligned}$$

(b) Using the answer to part (a)

$$\frac{f(x+h) - f(x)}{h} = \frac{h(2x + h)}{h}$$

$$= 2x + h$$

39. $f(x) = 2 - x^2$

(a) $f(x+h) = 2 - (x+h)^2$

Therefore,

$$f(x+h) - f(x)$$
$$= 2 - (x+h)^2 - (2 - x^2)$$
$$= 2 - x^2 - 2xh - h^2 - 2 + x^2$$
$$= -2xh - h^2$$
$$= -h(2x + h)$$

(b) Using the answer to part (a)

$$\frac{f(x+h) - f(x)}{h} = -\frac{h(2x + h)}{h}$$

$$= -(2x + h)$$

41. (a) $I(3)$ = value of I at time 3 = 1.55.

   $I(5)$ = value of I at time 5 = 1.5.

   $I(6)$ = value of I at time 6 = 1.5.

Interpretation:

   $I(3) = 1.55$. In 2003 the U.S. imported 1.55 million gallons/day.

   $I(5) = 1.5$. In 2005 the U.S. imported 1.5 million gallons/day.

   $I(6) = 1.5$. In 2006 the U.S. imported 1.5 million gallons/day.

(b) Since I is specified only for $1 \le t \le 6$, the domain of I is $[1, 6]$.

(c) Graph:

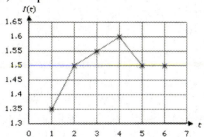

From the graph we can estimate $4(4.5) = 1.55$ as shown:

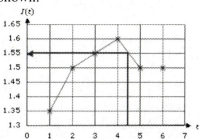

To interpret the data, notice that $I(4)$ is the amount of oil imported in 2004; that is, the year ending Dec 31, 2004, while $I(5)$ is the amount of oil imported in the year ending Dec 31, 2005. Thus, $I(4.5)$ is amount of oil is the amount of oil imported in the year ending June 30, 2005 (or the year starting July 1, 2004). Thus the U.S. imported 1.55 million gallons/day in the year ending June 30, 2005.

43. Reading from the graph, $f(4) \approx 1600$, $f(5) \approx 1700$, and $f(6.5) \approx 1300$. Since f is the number of thousands of housing starts in year, t we interpret the results as follows:

$f(4) \approx 1600$: There were 1.6 million housing starts in 2004.

$f(5) \approx 1700$: There were 1.7 million housing starts in 2005.

$f(6.5) \approx 1300$: There were 1.3 million housing starts in the year beginning July, 2006.

45. $f(5) - f(0) = 1700 - 1200 = 500$

$f(7) - f(5) = 1100 - 1700 = -600$

Thus, the magnitude of $f(7) - f(5)$ is the larger.

Interpretation:

$f(5) - f(0)$ is the change in the number of housing starts from 2000 to 2005, and $f(7) - f(5)$ is the

15

change in the number of housing starts from 2005 to 2007. The change from 2005 to 2007 was larger in magnitude.

47. (a) The model is valid for the range 1994 (t = 0) through 2004 (t = 10). Thus, an appropriate domain is [0, 10]. $t \geq 0$ is not an appropriate domain because it would predict U.S. trade with China into the indefinite future with no basis.

(b) $C(t) = 3t^2 - 7t + 50$

$C(10) = 3(10)^2 - 7(10) + 50 = 300 - 70 + 50 = \$280$ billion.

Since t = 10 represents 2004, the answer tells us that U.S. trade with China in 2004 was valued at approximately \$280 billion.

49. (a) Since the graph specifies the values of $L(t)$ from t = 1 to t = 6, the domain of the graph is [1, 6].

(b) Reading from the graph, $L(2) \approx 400$; $L(5) \approx 900$, $L(6) \approx 1100$. Since $L(t)$ specifies the loss in millions of dollars, we have the following interpretation:

$L(2) \approx 400$: The loss in 2002 was about \$400 million.

$L(5) \approx 900$: The loss in 2005 was about \$900 million.

$L(6) \approx 1100$: The loss in 2006 was about \$1100 million, or \$1.1 billion.

(c) $L(t)$ increasing most rapidly at the point where the graph is steepest: around t = 4.5 (midway through 2004). Thus, Sirius' losses were increasing fastest approximately midway through 2004.

51. $P(t) = \begin{cases} 75t + 200 & \text{if } 0 \leq t \leq 4 \\ 600t - 1900 & \text{if } 4 < t \leq 9 \end{cases}$

(a) $P(0) = 75(0) + 200 = 200$ We used the first formula, since 0 is in [0, 4].

$P(4) = 75(4) + 200 = = 300 + 200 = 500$ We used the first formula, since 4 is in [0, 4].

$P(5) = 600(5) - 1900 = 3000 - 1900 = 1100$ We used the second formula, since 5 is in (4, 9].

Interpretation:

Since $P(t)$ gives the speed of Intel processors at the start of year 1995+t, we have:

$P(0) = 200$: At the start of 1995 the processor speed was 200 megahertz.

$P(4) = 500$: At the start of 1999 the processor speed was 500 megahertz.

$P(5) = 1100$: At the start of 2000 the processor speed was 1100 megahertz.

(b) Technology formula:

`(75*t+200)*(t<=4)+(600*t-1900)*(t>4)`

(For a graphing calculator, use x instead of t, and $\geq$ instead of >=) Graph:

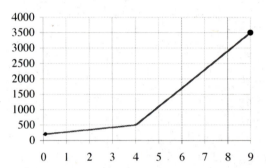

To answer the question, we need to find the value of t such that $P(t) = 2000$ (because 2.0 gigahertz = 2000 megahertz). From the graph, $P(6.5) \approx 2000$, and 6.5 years since the start of 1995 was midway through 2001. Thus, processor speeds first reached 2.0 gigahertz midway through 2001.

(c) We use the technology formula from part (b) to obtain the following table of values:

t	P(t)
0	200
1	275
2	350
3	425
4	500
5	1100
6	1700
7	2300
8	2900
9	3500

53. A taxable income of \$26,000 falls in the category "Over \$7,825 but not over \$31,850" and so we use the formula "\$782.50 + 15% of the amount over \$7,825."

$T(26,000) - \$782.50 + 0.15(26,000 - 7,825)$

16

= \$3508.75

A taxable income of \$65,000 falls in the category "Over \$31,850 but not over \$77,100" and so we use the formula "\$4,386.25 + 25% of the amount over \$31,850."

$T(65,000) = \$4386.25 + 0.25(65,000-31,850)$
$= \$12,673.75$

55. $p(t) = 100\left(1 - \dfrac{12,200}{t^{4.48}}\right)$ $\qquad (t \geq 8.5)$

(a) Technology formula:
$$100*(1-12200/t\char94 4.48)$$

(b) Graph:

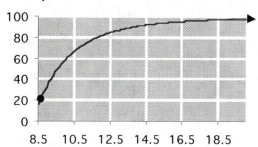

(c) Table of values:

t	9	10	11	12	13	14
$p(t)$	35.2	59.6	73.6	82.2	87.5	91.1
t	15	16	17	18	19	20
$p(t)$	93.4	95.1	96.3	97.1	97.7	98.2

(d) From the table, $p(12) = 82.2$, so that 82.2% of children are able to speak in at least single words by the age of 12 months.

(e) We seek the first value of t such that $p(t)$ is at least 90. Since $t = 14$ has this property ($p(14) = 91.1$) we conclude that, at 14 months, 90% or more children are able to speak in at least single words.

57. The dependent variable is a function of the independent variable. Here, the market price of gold m is a function of time t. Thus, the independent variable is t and the dependent variable is m.

59. To obtain the function notation, write the dependent as a function of the independent variable. Thus $y = 4x^2 - 2$ can be written as $f(x) = 4x^2 - 2$ or $y(x) = 4x^2 - 2$

61. True. A graphically specified function is specified by a graph. Given a graph, we can read off a set of values to construct a table, and hence specify the function numerically. (The more accurate the graph is, the more accurate the numerical values are.)

63. False. In a numerically specified function, only certain values of the function are specified so we cannot know its value on every real number in [0, 10], whereas an algebraically specified function would give values for every real number in [0, 10].

65. Functions with infinitely many points in their domain (such as $f(x) = x^2$) cannot be specified numerically. So, the assertion is false.

67. As the text reminds us: to evaluate f of a quantity (such as $x+h$) replace x everywhere by the *whole quantity $x+h$*:
$$f(x) = x^2 - 1$$
$$f(x+h) = (x+h)^2 - 1.$$

69. If two functions are specified by the same formula $f(x)$ say, their graphs must follow the same curve $y = f(x)$. However, it is the domain of the function that specifies what portion of the curve appears on the graph. Thus, if the functions have different domains, their graphs will be different portions of the curve $y = f(x)$.

71. Suppose we already have the graph of f and want to construct the graph of g. We can plot a point of the graph of g as follows: Choose a value for x ($x = 7$, say) and then "look back" 5 units to read off $f(x-5)$ ($f(2)$ in this instance). This value gives the y-coordinate we want. In other words, points on the graph of g are obtained by "looking back 5 units" to the graph of f and then copying that portion of the curve. Put another way, the graph of g is the same as the graph of f, but shifted 5 units to the right.

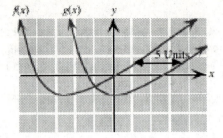

1.2 Functions and Models

1. Number of sound files
= Starting number + New files
= 200 + 10×Number of days
So, $N(t) = 200 + 10t$
(N = number of sound files, t = time in days)

3. Take y to be the width. Since the length is twice the width,
$$x = 2y$$
so $y = x/2$.
The area is therefore
$$A = xy = x(x/2) = x^2/2$$

5. Since the patch is square the width and length are both equal to x. The costs are:
East and West sides: $4x + 4x = 8x$.
North and South Sides: $2x + 2x = 4x$.
Total cost $C(x) = 8x + 4x = 12x$

7. For a linear cost function,
$$C(x) = mx + b$$
m = marginal cost = $1500 per piano
b = fixed cost = $1200
Thus, the daily cost function is
$$C(x) = 1500x + 1200.$$
(a) The cost of manufacturing 3 pianos is
$$C(3) = 1500(3) + 1200$$
$$= 4500 + 1200 = \$5700$$
(b) The cost of manufacturing each additional piano (such as the third one or the 11th one) is the marginal cost, m = $1500.
(c) Same answer as (b).
(d) Variable cost = part of the cost function that depends on x = $1500x$
Fixed cost = constant summand of the cost function = $1200
Marginal cost = slope of the cost function = $1500 per piano.

(e) Graph:

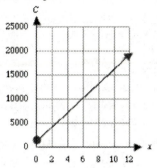

9. (a) For a linear cost function,
$$C(x) = mx + b.$$
m = marginal cost = $0.40 per copy
b = fixed cost = $70
Thus, the cost function is
$$C(x) = 0.4x + 70.$$
The revenue function is
$$R(x) = 0.50x \quad (x \text{ copies @ 50¢ per copy})$$
The profit function is
$$P(x) = R(x) - C(x)$$
$$= 0.5x - (0.4x + 70)$$
$$= 0.5x - 0.4x - 70$$
$$= 0.1x - 70$$
(b) $P(500) = 0.1(500) - 70$
$$= 50 - 70 = -20$$
Since P is negative, this represents a loss of $20.
(c) For break-even,
$$P(x) = 0$$
$$0.1x - 70 = 0$$
$$0.1x = 70$$
$$x = \frac{70}{0.1} = 700 \text{ copies}$$

11. The revenue per jersey is $100. Therefore,
Revenue $R(x) = \$100x$
Profit = Revenue – Cost
$P(x) = R(x) - C(x)$
$$= 100x - (2000 + 10x + 0.2x^2)$$
$$= -2000 + 90x - 0.2x^2$$
To break even, $P(x) = 0$, so
$$-2000 + 90x - 0.2x^2 = 0$$
This is a quadratic equation with $a = -0.2$, $b = 90$, $c = -2000$ and solution

19

$$x = \frac{-b \pm \sqrt{b^2 - 4ac}}{2a}$$

$$= \frac{0.02 \pm \sqrt{(90)^2 - 4(-0.2)(-2000)}}{2(-0.2)}$$

$$\approx 23.44 \text{ or } 426.56 \text{ jerseys.}$$

Since the second value is outside the domain, we use the first: $x = 23.22$ jerseys. To make a profit, x should be larger than this value: at least 24 jerseys.

13. The revenue from one thousand square feet ($x = 1$) is $0.1 million. Therefore,

Revenue $R(x) = \$0.1x$

Profit = Revenue − Cost

$P(x) = R(x) − C(x)$

$\quad = 0.1x − (1.7 + 0.12x − 0.0001x^2)$

$\quad = −1.7 − 0.02x + 0.0001x^2$

To break even, $P(x) = 0$, so

$−1.7 − 0.02x + 0.0001x^2 = 0$

This is a quadratic equation with $a = 0.0001$, $b = −0.02$, $c = -1.7$ and solution

$$x = \frac{-b \pm \sqrt{b^2 - 4ac}}{2a}$$

$$= \frac{0.02 \pm \sqrt{(-0.02)^2 - 4(0.0001)(-1.7)}}{2(0.0001)}$$

$$\approx \frac{0.02 \pm 0.03286}{2(0.0001)} \approx 264 \text{ thousand square}$$

feet

15. The hourly profit function is given by

Profit = Revenue − Cost

$P(x) = R(x) − C(x)$

(Hourly) cost function: This is a fixed cost of $5132 only:

$C(x) = 5132$

(Hourly) revenue function: This is a variable of $100 per passenger cost only:

$R(x) = 100x$

Thus, the profit function is

$P(x) = R(x) − C(x)$

$P(x) = 100x − 5132$

For the domain of $P(x)$, the number of passengers x cannot exceed the capacity: 405. Also, x cannot be negative. Thus, the domain is given by $0 \le x \le 405$, or $[0, 405]$.

For break-even, $P(x) = 0$

$100x − 5132 = 0$

$100x = 5132$, or $x = \dfrac{5132}{100} = 51.32$

If x is larger than this, then the profit function is positive, and so there should be at least 52 passengers; $x \ge 52$, for a profit.

17. To compute the break-even point, we use the profit function:

Profit = Revenue − Cost

$P(x) = R(x) − C(x)$

$R(x) = 2x$ \$2 per unit

$C(x) = $ Variable Cost + Fixed Cost

$\quad = 40\% \text{ of Revenue} + 6000$

$\quad = 0.4(2x) + 6000$

$\quad = 0.8x + 6000$

Thus,

$P(x) = R(x) − C(x)$

$P(x) = 2x − (0.8x + 6000)$

$P(x) = 1.2x − 6000$

For break-even, $P(x) = 0$

$1.2x − 6000 = 0$

$1.2x = 6000$

$x = \dfrac{6000}{1.2} = 5000$

Therefore, 5000 units should be made to break even.

19. To compute the break-even point, we use the revenue and cost functions:

$R(x) = $ Selling price $\times$ Number of units

$\quad = SPx$

$C(x) = $ Variable Cost + Fixed Cost

$\quad = VCx + FC$

(Note that "variable cost per unit" is marginal cost.) For break-even

$R(x) = C(x)$

$SPx = VCx + FC$

$SPx − VCx = FC$

$x(SP − VC) = FC$

$x = \dfrac{FC}{SP − VC}$

21. Take x to be the number of grams of perfume he buys and sells. The profit function is given by

Profit = Revenue − Cost

$P(x) = R(x) − C(x)$

Cost function $C(x)$:

Fixed costs: 20,000

Cheap perfume @ \$20 per g: $20x$

Transportation @ \$30 per 100 g: $0.3x$

Thus the cost function is

$C(x) = 20x + 0.3x + 20,000$

$C(x) = 20.3x + 20,000$

Revenue function $R(x)$

$R(x) = 600x$ \$600 per gram

Thus, the profit function is

$P(x) = R(x) - C(x)$

$P(x) = 600x - (20.3x + 20,000)$

$P(x) = 579.7x - 20,000$,

with domain $x \geq 0$.

For break-even, $P(x) = 0$

$579.7x - 20,000 = 0$

$579.7x = 20,000$

$x = \dfrac{20,000}{579.7} \approx 34.50$

Thus, he should buy and sell 34.50 grams of perfume per day to break even.

23. (a) To graph the demand function we use technology with the formula

    `64*x^(-0.76)`

with xMin = 3 and xMax = 5.

Graph:

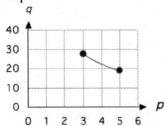

(b) $q(3) = 64(3)^{-0.76} \approx 27.77$

$q(3.5) = 64(3.5)^{-0.76} \approx 24.70$

So the change in demand is $24.70 - 27.77 = -3.07$
Thus, ridership drops by about 3070 rides per day.

25. $q(p) = 361,201 - (p+1)^2$

(a) Note p is expressed in cents, not dollars, so 50¢ is represented by $p = 50$, not 0.50.

$q(50) = 361,201 - (50+1)^2$

$= 361,201 - 2601$

$= 358,600$ brownie dishes

(b) If they give them away, $p = 0$, so

$q(0) = 361,201 - (0+1)^2$

$= 361,201 - 1$

$= 361,200$ brownie dishes

(c) If they sell no dishes, $q = 0$, and so

$0 = 361,201 - (p+1)^2$

$(p+1)^2 = 361,201$

$p+1 = \sqrt{361,201} = 601$

$p = 601 - 1 = 600$¢, or \$6.00.

27. The price at which there is neither a shortage nor surplus is the equilibrium price, which occurs when demand = supply:

$-3p + 700 = 2p - 500$

$5p = 1200$

$p = \$240$ per skateboard.

29. (a) The equilibrium price occurs when demand = supply:

$-p + 156 = 4p - 394$

$5p = 550$

$p = \$110$ per phone.

(b) Since \$105 is below the equilibrium price, there would be a shortage at that price. To calculate it, compute demand and supply:

Demand: $q = -105 + 156 = 51$ million phones

Supply: $q = 4(105) - 394 = 26$ million phones

Shortage = Demand – Supply = $51 - 26 = 25$ million phones

31. (a) Graph:

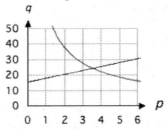

Technology formulas:

Demand: `y = 64*x^-0.76`

Supply: `y = 2.5x+15.5`

The graphs cross where $p \approx 3.5$, so the equilibrium price is about \$3.50 per ride.

(b) Since \$5 is above the equilibrium price, there would be a surplus at that price. To calculate it, compute demand and supply:

21

Demand: $q = 64(5)^{-0.76} = 18.83$ thousand rides

Supply: $q = 2.5(5)+15.5 = 28$ thousand rides

Surplus = Supply − Demand = $28 - 18.83 \approx 9.17$ thousand rides, or 9170 rides.

33. $C(q) = 2000 + 100q^2$

(a) $C(10) = 2000+100(10)^2 = 2000+10{,}000$
$= \$12{,}000$

(b) Net Cost = Total Cost − Subsidy
$$N(q) = C(q) - S(q)$$
$$= 2000 + 100q^2 - 500q$$
$N(20) = 2000 + 100(20)^2 - 500(20)$
$= 2000 + 40{,}000 - 10{,}000$
$= \$32{,}000$

35. The technology formulas are:

(A) `-0.2*t^2+t+16`

(B) `0.2*t^2+t+16`

(C) `t+16`

The following table shows the values predicted by the three models.

t	0	2	4	6	7
S	16	18	22	28	30
(A)	16	17.2	16.8	14.8	13.2
(B)	16	18.8	23.2	29.2	32.8
(C)	16	18	20	22	23

As shown in the table, the values predicted by model (B) are much closer to the observed values S than those predicted by the other models.

(b) Since 1998 corresponds to $t = 8$,
$$S(t) = 0.2\,t^2 + t + 16$$
$$S(8) = 0.2(8)^2 + 8 + 16 = 36.8$$
So the spending on corrections in 1998 was predicted to be \$36.8 billion.

37. (a) Following are the graphs of the four models using the following technology formulas (in the online grapher):

Model (A): `y = 98*1.2^x`

Model (B): `y = 4.6*x^2+1.2*x+109`

Model (C): `y = 47*x+48`

Model (D): `y = 98*1.2^-x`

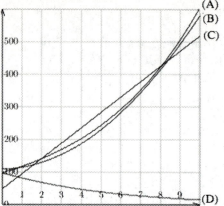

It is clear from the graph that models (A) and (B) give the best fit.

(b) Comparing the predictions of models (A) and (B) for 2020 ($t = 20$):

Model (A): $f(20) = 98(1.2^{20}) \approx 3757$ tons of freon

Model (B): $f(20) = 4.6(20)^2 + 1.2(20) + 109$ ≈ 1973 tons of freon

The first model predicts the largest quantity of freon in 2020: around 3757 tons.

39. (a) The technology formulas are:

(A): `0.005*x+2.75`

(B): `0.01*x+20+25/x`

(C): `0.0005*x^2-0.07*x+23.25`

(D): `25.5*1.08^(x-5)`

The following table shows the values predicted by the four models.

x	5	25	40	100	125
$A(x)$	22.91	21.81	21.25	21.25	22.31
(A)	2.775	2.875	2.95	3.25	3.375
(B)	25.05	21.25	21.025	21.25	21.45
(C)	22.913	21.813	21.25	21.25	22.313
(D)	25.5	118.85	377.03	38177	261451

Model (C) fits the data almost perfectly — more closely that any of the other models.

(b) Graph of model (C):

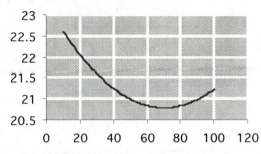

$$0.0005*x^2-0.07*x+23.25$$

The lowest point on the graph occurs at $x = 70$ with a y-coordinate of 20.8. Thus, the lowest cost per shirt is $20.80, which the team can obtain by buying 70 shirts.

41. A plot of the given data suggests a curve with a low point somewhere between $t = 0$ and $t = 8$. A linear model would predict perpetually increasing or decreasing value of the euro (depending on whether the slope is positive or negative) and an exponential model $n(t) = Ab^t$ would also be perpetually increasing or decreasing (depending whether b is larger than 1 or less than 1). This leaves a quadratic model as the only possible choice. In fact, a quadratic can always be found that passes through any three points not on the same straight line with different x-coordinates. Therefore, a quadratic model would give an exact fit.

43. Solution: Apply the formula

$$A(t) = P\left(1+\frac{r}{m}\right)^{mt}$$

with $P = 5000$, $r = 0.0494$, and $m = 12$. We get the model

$$A(t) = 5000(1+0.0494/12)^{12t}$$

At the start of 2014 ($t = 7$), the deposit would be worth $5000(1+0.0494/12)^{12(7)} \approx \$7061$

45. From the answer to Exercise 43, the value of the investment after t years is

$$A(t) = 5000(1+0.0494/12)^{12t}.$$

TI 83/84 Plus: Enter

    Y₁ = 5000*(1+0.0494/12)^(12*X),

Press [2nd] [TBLSET], and set Indpnt to Ask.

(You do this once and for all; it will permit you to specify values for x in the table screen.) Then, press [2nd] [TABLE], and you will be able to evaluate the function at several values of x. Here are some values of x and the resulting values of Y_1.

x	Y₁
1	5252.67
2	5518.11
3	5796.96
4	6089.9
5	6397.65
6	6720.95
7	7060.59
8	7417.39
9	7792.22
10	8185.99

Notice that Y_1 first exceeds 7500 when $x = 9$. Since $x = 0$ represents the beginning of 2007, $x = 9$ represents the start of 2016, so the investment will first exceed $7500 at the beginning of 2016.

47. $C(t) = 104(0.999879)^t$, so $C(10,000) \approx 31.0$ grams, $C(20,000) \approx 9.25$ grams, and $C(30,000) \approx 2.76$ grams.

49. We are looking for t such that $4.06 = 46(0.999879)^t$. Among the values suggested we find that $C(15,000) \approx 7.5$, $C(20,000) \approx 4.09$ and $C(25,000) \approx 2.23$. Thus, the answer is 20,000 years to the nearest 5000 years.

51. (a) Amount left after 1000 years:
$$C(1000) = A(0.999\,567)^{1000} \approx 0.6485A,$$
or about 65% of the original amount.
Amount left after 2000 years:
$$C(2000) = A(0.999\,567)^{2000} \approx 0.4206A,$$
or about 42% of the original amount.
Amount left after 3000 years:
$$C(3000) = A(0.999\,567)^{3000} \approx 0.2727A,$$
or about 27% of the original amount.
(b) For a sample of 100g,
$$C(t) = 100(0.999\,567)^t.$$
Here is the graph, together with the line $y = 50$ (one half the original sample):

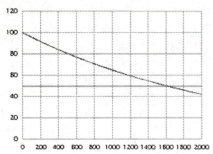

Technology: `100*(0.999567)^x`
Since the graphs intersect close to $x = 1600$, we
conclude that half the sample will have decayed
after about 1600 years.

53. $P(0) = 300$, $P(1) = 330$, $P(2) = 260$,... and so
on, Thus, the population is increasing by 30 per
year.

55. Curve-fitting. The model is based on fitting a
curve to a given set of observed data.

57. The given model is $c(t) = 4 - 0.2t$. This tells
us that c is $4 at time $t = 0$ (January) and is
decreasing by $0.20 per month. So, the cost of
downloading a movie was $4 in January and is
decreasing by 20¢ per month.

59. In a linear cost function, the variable cost is x
times the marginal cost.

61. Yes, as long as the supply is going up at a
faster rate, as illustrated by the following graph:

63. Extrapolate both models and choose the one
that gives the most reasonable predictions.

1.3 Linear Functions and Models

1.

x	-1	0	1
y	5	8	

We calculate the slope m first. The first two points shown give a changes in x and y of

$$\Delta x = 0 - (-1) = 1$$
$$\Delta y = 8 - 5 = 3$$

This gives a slope of

$$m = \frac{\Delta y}{\Delta x} = \frac{3}{1} = 3.$$

Now look at the second and third points: The change in x is again

$$\Delta x = 1 - 0 = 1$$

and so Δy must be given by the formula

$$\Delta y = m\Delta x$$
$$\Delta y = 3(1) = 3.$$

This means that the missing value of y is

$$8 + \Delta y = 8 + 3 = 11.$$

3.

x	2	3	5
y	-1	-2	

We calculate the slope m first. The first two points shown give a changes in x and y of

$$\Delta x = 3 - 2 = 1$$
$$\Delta y = -2 - (-1) = -1$$

This gives a slope of

$$m = \frac{\Delta y}{\Delta x} = \frac{-1}{1} = -1.$$

Now look at the second and third points: The change in x is

$$\Delta x = 5 - 3 = 2$$

and so Δy must be given by the formula

$$\Delta y = m\Delta x$$
$$\Delta y = (-1)(2) = -2.$$

This means that the missing value of y is

$$-2 + \Delta y = -2 + (-2) = -4.$$

5.

x	-2	0	2
y	4		10

We calculate the slope m first. The first and third points shown give a changes in x and y of

$$\Delta x = 2 - (-2) = 4$$

$$\Delta y = 10 - 4 = 6$$

This gives a slope of

$$m = \frac{\Delta y}{\Delta x} = \frac{6}{4} = \frac{3}{2}.$$

Now look at the first and second points: The change in x is

$$\Delta x = 0 - (-2) = 2$$

and so Δy must be given by the formula

$$\Delta y = m\Delta x$$

$$\Delta y = (\frac{3}{2})(2) = 3.$$

This means that the missing value of y is

$$4 + \Delta y = 4 + 3 = 7.$$

7. From the table,

$$b = f(0) = -2.$$

The slope (using the first two points) is

$$m = \frac{y_2 - y_2}{x_2 - x_1} = \frac{-2 - (-1)}{0 - (-2)} = \frac{-1}{2} = -\frac{1}{2}.$$

Thus, the linear equation is

$$f(x) = mx + b = -\frac{1}{2}x - 2,$$

or $f(x) = -\frac{x}{2} - 2.$

9. The slope (using the first two points) is

$$m = \frac{y_2 - y_2}{x_2 - x_1} = \frac{-2 - (-1)}{-3 - (-4)} = \frac{-1}{1} = -1.$$

To obtain $f(0) = b$, notice that, since the slope is -1, y decreases by 1 for every one-unit increase in x. Thus,

$$f(0) = f(-1) + m = -4 - 1 = -5.$$

This gives

$$f(x) = mx + b = -x - 5.$$

11. In the table, x increases in steps of 1 and f increases in steps of 4, showing that f is linear with slope

$$m = \frac{\Delta y}{\Delta x} = \frac{4}{1} = 4$$

and intercept

$$b = f(0) = 6$$

giving

$$f(x) = mx + b = 4x + 6.$$

The function g does not increase in equal steps, so g is not linear.

13. In the first three points listed in the table, x increases in steps of 3, but f does not increase in equal steps, whereas g increases in steps of 6. Thus, based on the first three points, only g could possibly be linear, with slope

$$m = \frac{\Delta y}{\Delta x} = \frac{6}{3} = 2$$

and intercept

$$b = g(0) = -1$$

giving

$$g(x) = mx + b = 2x - 1.$$

We can now check that the remaining points in the table fit the formula $g(x) = 2x-1$, showing that g is indeed linear.

15. Slope = coefficient of $x = -\dfrac{3}{2}$

17. Write the equation as

$$y = \frac{x}{6} + \frac{1}{6}$$

Slope = coefficient of $x = \dfrac{1}{6}$

19. If we solve for x we find that the given equation represents the vertical line $x = -1/3$, and so its slope is infinite (undefined).

21. $3y + 1 = 0$
Solving for y:

$$3y = -1$$

$$y = -\frac{1}{3}$$

Slope = coefficient of $x = 0$

23. $4x + 3y = 7$
Solve for y:

$$3y = -4x + 7$$

$$y = -\frac{4}{3}x + \frac{7}{3}$$

Slope = coefficient of $x = -\dfrac{4}{3}$

25. $y = 2x - 1$
y-intercept $= -1$, slope $= 2$

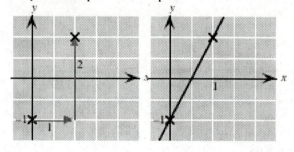

27. $y = -\dfrac{2}{3}x + 2$

y-intercept $= 2$, slope $= -\dfrac{2}{3}$

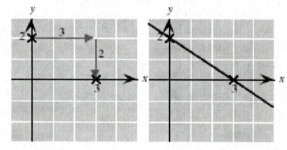

29. $y + \dfrac{1}{4}x = -4$

Solve for y to obtain $y = -\dfrac{1}{4}x - 4$

y-intercept $= -4$, slope $= -\dfrac{1}{4}$

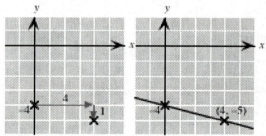

31. $7x - 2y = 7$
Solve for y:

$$-2y = -7x + 7$$

$$y = \frac{7}{2}x - \frac{7}{2}$$

y–intercept $= -\dfrac{7}{2} = -3.5$, slope $= \dfrac{7}{2} = 3.5$

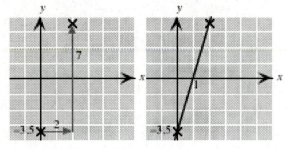

33. $3x = 8$

Solve for x to obtain $x = \dfrac{8}{3}$.

The graph is a vertical line:

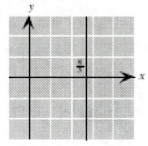

35. $6y = 9$

Solve for y to obtain $y = \dfrac{9}{6} = \dfrac{3}{2} = 1.5$

y–intercept $= \dfrac{3}{2} = 1.5$, slope $= 0$

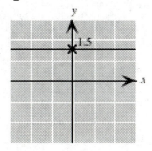

37. $2x = 3y$

Solve for y to obtain $y = \dfrac{2}{3} x$

y–intercept $= 0$, slope $= \dfrac{2}{3}$

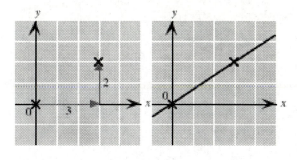

39. $(0, 0)$ and $(1, 2)$

$$m = \frac{y_2 - y_1}{x_2 - x_1} = \frac{2 - 0}{1 - 0} = 2$$

41. $(-1, -2)$ and $(0, 0)$

$$m = \frac{y_2 - y_1}{x_2 - x_1} = \frac{0 - (-2)}{0 - (-1)} = \frac{2}{1} = 2$$

43. $(4, 3)$ and $(5, 1)$

$$m = \frac{y_2 - y_1}{x_2 - x_1} = \frac{1 - 3}{5 - 4} = \frac{-2}{1} = -2$$

45. $(1, -1)$ and $(1, -2)$

$$m = \frac{y_2 - y_1}{x_2 - x_1} = \frac{-2 - (-1)}{1 - 1} \quad \text{Undefined}$$

47. $(2, 3.5)$ and $(4, 6.5)$

$$m = \frac{y_2 - y_1}{x_2 - x_1} = \frac{6.5 - 3.5}{4 - 2} = \frac{3}{2} = 1.5$$

49. $(300, 20.2)$ and $(400, 11.2)$

$$m = \frac{y_2 - y_1}{x_2 - x_1} = \frac{11.2 - 20.2}{400 - 300}$$

$$= \frac{-9}{100} = -0.09$$

51. $(0, 1)$ and $\left(-\frac{1}{2}, \frac{3}{4}\right)$

$$m = \frac{y_2 - y_1}{x_2 - x_1} = \frac{\frac{3}{4} - 1}{-\frac{1}{2} - 0}$$

$$= \frac{-\frac{1}{4}}{-\frac{1}{2}} = \frac{1}{4} \cdot 2 = \frac{1}{2}$$

27

53. (a, b) and (c, d) $(a \neq c)$

$$m = \frac{y_2 - y_1}{x_2 - x_1} = \frac{d - b}{c - a}$$

55.

(a)

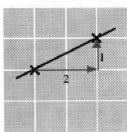

$m = \dfrac{\Delta y}{\Delta x} = \dfrac{1}{1}$
$= 1$

(b)

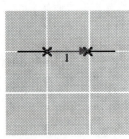

$m = \dfrac{\Delta y}{\Delta x} = \dfrac{1}{2}$

(c)

$m = \dfrac{\Delta y}{\Delta x} = \dfrac{0}{1}$
$= 0$

(d)

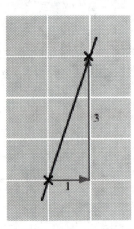

$m = \dfrac{\Delta y}{\Delta x} = \dfrac{3}{1}$
$= 3$

(e)

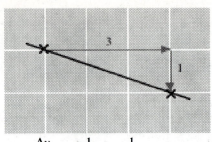

$m = \dfrac{\Delta y}{\Delta x} = \dfrac{-1}{3} = -\dfrac{1}{3}$

(f)

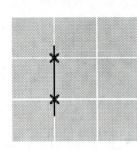

$m = \dfrac{\Delta y}{\Delta x} = \dfrac{-1}{1}$
$= -1$

(g)

Vertical line; undefined slope

(h)

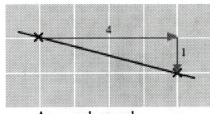

$m = \dfrac{\Delta y}{\Delta x} = \dfrac{-1}{4} = -\dfrac{1}{4}$

(i)

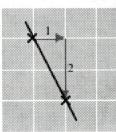

$m = \dfrac{\Delta y}{\Delta x} = \dfrac{-2}{1}$
$= -2$

57. Through $(1, 3)$ with slope 3
Point: $(1, 3)$ **Slope:** $m = 3$
$$b = y_1 - mx_1$$
$$= 3 - 3 \cdot 1 = 0$$
Thus, the equation is
$$y = mx + b$$
$$y = 3x + 0, \quad y = 3x$$

59. Through $(1, -\frac{3}{4})$ with slope $\frac{1}{4}$
Point: $(1, -\frac{3}{4})$ **Slope:** $m = \frac{1}{4}$
$$b = y_1 - mx_1$$
$$= -\frac{3}{4} - \frac{1}{4} \cdot 1 = -1$$
Thus, the equation is
$$y = mx + b$$
$$y = \frac{1}{4}x - 1$$

61. Through $(20, -3.5)$ and increasing at a rate of 10 units of y per unit of x
Point: $(20, -3.5)$
Slope: $m = \dfrac{\Delta y}{\Delta x} = \dfrac{10}{1} = 10$
$$b = y_1 - mx_1$$
$$= -3.5 - (10)(20)$$
$$= -3.5 - 200 = -203.5$$
Thus, the equation is
$$y = mx + b$$
$$y = 10x - 203.5$$

63. Through $(2, -4)$ and $(1, 1)$
Point: $(2, -4)$
Slope: $m = \dfrac{y_2 - y_1}{x_2 - x_1} = \dfrac{1 - (-4)}{1 - 2} = \dfrac{5}{-1} = -5$
$$b = y_1 - mx_1$$
$$= -4 - (-5)(2) = 6$$
Thus, the equation is
$$y = mx + b$$
$$y = -5x + 6$$

65. Through $(1, -0.75)$ and $(0.5, 0.75)$
Point: $(1, -0.75)$
Slope: $m = \dfrac{y_2 - y_1}{x_2 - x_1} = \dfrac{0.75 - (-0.75)}{0.5 - 1}$

$$= \frac{1.5}{-0.5} = -3$$
$$b = y_1 - mx_1$$
$$= -0.75 - (-3)(1) = -0.75 + 3 = 2.25$$
Thus, the equation is
$$y = mx + b$$
$$y = -3x + 2.25$$

67. Through $(6, 6)$ and parallel to the line $x + y = 4$
Point: $(6, 6)$
Slope: Same as slope of $x + y = 4$. To find the slope, solve for y, getting
$$y = -x + 4$$
Thus, $m = -1$.
$$b = y_1 - mx_1$$
$$= 6 - (-1)(6) = 6 + 6 = 12$$
Thus, the equation is
$$y = mx + b$$
$$y = -x + 12$$

69. Through $(0.5, 5)$ and parallel to the line $4x - 2y = 11$
Point: $(0.5, 5)$
Slope: Same as slope of $4x - 2y = 11$. To find the slope, solve for y, getting
$$2y = 4x - 11$$
$$y = 2x - \frac{11}{2}$$
Thus, $m = 2$.
$$b = y_1 - mx_1$$
$$= 5 - (2)(0.5) = 5 - 1 = 4$$
Thus, the equation is
$$y = mx + b$$
$$y = 2x + 4 \quad .$$

71. Slope: $m = \dfrac{y_2 - y_1}{x_2 - x_1} = \dfrac{q - 0}{p - 0} = \dfrac{q}{p}$

Intercept: $b = y_1 - mx_1 = 0 - \dfrac{q}{p}(0) = 0$

Equation: $y = mx + b$

$$y = \frac{q}{p}x$$

73. Slope: $m = \dfrac{y_2 - y_1}{x_2 - x_1} = \dfrac{q-q}{r-p} = 0$

Intercept: $b = y_1 - mx_1 = q - (0)p = q$

Equation: $y = mx + b$

$y = q$

75. We are given two points on the graph of the linear cost function: (100, 10,500) and (120, 11,0000) (x is the number of items, and y is the cost C).

Marginal cost:

$m = \dfrac{C_2 - C_1}{x_2 - x_1} = \dfrac{11,000 - 10,500}{120 - 100}$

$= \dfrac{500}{20} = \$25$ per bicycle.

Fixed cost:

$b = C_1 - mx_1 = 10,500 - (25)(100)$

$= 10,500 - 2500 = \$8000$

77. We are given two points on the graph of the linear cost function: $(5, 800)$ and $(25, 3700)$ (x is the number of iPods, and y is the cost C).

Marginal cost:

$m = \dfrac{C_2 - C_1}{x_2 - x_1} = \dfrac{3700 - 800}{25 - 5}$

$= \dfrac{2900}{20} = \$145$ per iPod.

Fixed cost:

$b = C_1 - mx_1 = 800 - (145)(5)$

$= 800 - 725 = \$75$

Cost function:

$C = mx + b$

$C = 145x + 75$

the cost to manufacture each iPod is the marginal cost: \$145.

The cost to manufacture 100 ipods is obtained by setting $x = 100$ in the demand equation:

$C(100) = 145(100) + 75 = \$14,575.$

79. A linear demand function has the form

$q = mp + b.$

(x is the price p, and y is the demand q). We are given two points on its graph: (1, 1960) and (5, 1800).

Slope:

$m = \dfrac{q_2 - q_1}{p_2 - p_1} = \dfrac{1800 - 1960}{5 - 1} = \dfrac{-160}{4} = -40$

Intercept:

$b = q_1 - mp_1 = 1960 - (-40)(1)$

$= 1960 + 40 = 2000$

Thus, the demand equation is

$q = mp + b$

$q = -40p + 2000$

81. (a) A linear demand function has the form

$q = mp + b.$

(x is the price p, and y is the demand q). We are given two points on its graph:

2004 second quarter data: (111, 45.4)

2004 fourth quarter data: (105, 51.4)

Slope:

$m = \dfrac{q_2 - q_1}{p_2 - p_1} = \dfrac{51.4 - 45.4}{105 - 111} = \dfrac{6}{-6} = -1$

Intercept:

$b = q_1 - mp_1 = 45.4 - (-1)(111) = 156.4$

Thus, the demand equation is

$q = mp + b$

$q = -p + 156.4$

If $p = \$103$, then

$q = -p + 156.4 = -103 + 156.4$

$= 53.4$ million phones

(b) Since the slope is -1 million phones per unit increase in price, we interpret of the slope as follows: For every __\$1__ increase in price, sales of cellphones decrease by __1 million__ units.

83. (a) A linear demand function has the form

$q = mp + b.$

(x is the price p, and y is the demand q). We are given two points on its graph: (3, 28,000) and (5, 19,000).

Slope:

$m = \dfrac{q_2 - q_1}{p_2 - p_1} = \dfrac{19,000 - 28,000}{5 - 3} = \dfrac{-9000}{2} = -4500$

Intercept:

$b = q_1 - mp_1 = 28000 - (-4500)(3)$

$= 28,000 + 13,500 = 41,500$

Thus, the demand equation is

$q = mp + b$

$q = -4500p + 41{,}500$

(b) The units of measeurement of the slope are generally units of y per unit of x. In this case:

Units of q per unit of p

That is,

Rides/day per \$1 increase in the fare.

Since the slope is -4500 rides/day per \$1 increase in the price, we interpret it as saying that ridership decreases by 4500 rides per day for every \$1 increase in the fare.

(c) From part (a), the demand equation is

$q = -4500p + 41{,}500$

If the fare is \$6, we have $p = 6$, so

$q = -4500(6) + 41{,}500$

$\quad = -27{,}000 + 41{,}500$

$\quad = 14{,}500$ rides/day.

85. (a) Demand Function: The given points are

$(p, q) = (1, 90)$ and $(2, 30)$

Slope:

$m = \dfrac{q_2 - q_1}{p_2 - p_1} = \dfrac{30 - 90}{2 - 1} = -60$

Intercept:

$b = q_1 - mp_1 = 90 - (-60(1) = 150$

Thus, the demand equation is

$q = mp + b$

$q = -60p + 150$

Supply Function: The given points are

$(p, q) = (1, 20)$ and $(2, 100)$

Slope:

$m = \dfrac{q_2 - q_1}{p_2 - p_1} = \dfrac{100 - 20}{2 - 1} = 80$

Intercept:

$b = q_1 - mp_1 = 20 - (80)1 = -60$

Thus, the supply equation is

$q = mp + b$

$q = 80p - 60$

(b) For equilibrium,

Supply = Demand

$80p - 60 = -60p + 150$

$140p = 210$

$p = \dfrac{210}{140} = 1.5$

Thus, the chias should be marked at \$1.50 each.

87. (a) We are given the q-intercept as 290 and the slope as 40. Thus,

$q = 40t + 290$ million pounds of pasta

(b) In 2005, $t = 15$, and so

$q(15) = 40(15) + 290 = 890$ million pounds.

89. (a) Linear Model: $N = mt + n$

Points: $(3, 0.3)$ and $(5, 3.2)$

Slope: $m = \dfrac{N_2 - N_1}{t_2 - t_1}$

$\quad = \dfrac{3.2 - 0.3}{5 - 3} = 1.45$

Intercept: $b = N_1 - mt_1$

$\quad = 0.3 - (1.45)(3) = -4.05$

Model: $N = mt + b = 1.45t - 4.05$

(b) The units of measurement of the slope are units of N per unit of t; that is, millions of subscribers per year. The number of Sirius Satellite Radio subscribers grew at a rate of 1.45 million subscribers per year.

(c) The year 2006 corresponds to $t = 6$, and so $N = 1.45(6) - 4.05 = 4.65$ million subscribers which is considerably less than the actual number.

91. $s(t) = 2.5t + 10$

(a) Velocity = slope = 2.5 feet/sec.

(b) After 4 seconds, $t = 4$, so

$s(4) = 2.5(4) + 10 = 10 + 10 = 20$

Thus the model train has moved 20 feet along the track.

(c) The train will be 25 feet along the track when $s = 25$. Substituting gives

$25 = 2.5t + 10$

Solving for time t gives

$2.5t = 25 - 10 = 15$

$t = \dfrac{15}{2.5} = 6$ seconds

93. (a) Take s to be displacement from Jones Beach, and t to be time in hours. We are given two points

$(t, s) = (10, 0)$ $s = 0$ for Jones Beach.

$(t, s) = (10.1, 13)$ 6 minutes = 0.1 hours

We are asked for the speed, which equals the magnitude of the slope.

$$m = \frac{s_2 - s_1}{t_2 - t_1} = \frac{13 - 0}{10.1 - 10} = \frac{13}{0.1} = 130$$

Units of slope = units of s per unit of t
= miles per hour

Thus, the police car was traveling at 130 mph.
(b) For the displacement from Jones Beach at
time t, we want to express s as a linear function of
t; namely, $s = mt + b$. We already know $m =$
130 from part (a). For the intercept, use

$$b = s_1 - mt_1 = 0 - 130(10) = -1300$$

Therefore, the displacement at time t is

$s = mt + b$
$s = 130t - 1300$

95. F = Fahrenheit temperature,
C = Celsius temperature,
and we want F as a linear function of C. That is,

$F = mC + b$

(F plays the role of y and C plays the role of x.)
We are given two points:

$(C, F) = (0. 32)$ Freezing point
$(C, F) = (100. 212)$ Boiling point

Slope:

$$m = \frac{F_2 - F_1}{C_2 - C_1} = \frac{212 - 32}{100 - 0} = \frac{180}{100} = 1.8$$

Intercept:

$$b = F_1 - mC_1 = 32 - 1.8(0) = 32.$$

Thus, the linear relation is

$F = mC + b$
$F = 1.8C + 32$

When $C = 30°$

$F = 1.8(30) + 32 = 54 + 32 = 86°$

When $C = 22°$

$F = 1.8(22) + 32 = 39.6 + 32 = 71.6°$

Rounding to the nearest degree gives 72°F.
When $C = -10°$

$F = 1.8(-10) + 32 = -18 + 32 = 14°$

When $C = -14°$

$F = 1.8(-14) + 32 = -25.2 + 32 = 6.8°$

Rounding to the nearest degree gives 7°F.

97. Income = royalties + screen rights
I = 5% of net profits + 50,000
$I = 0.05N + 50,000$ Equation notation
$I(N) = 0.05N + 50,000$ Function notation
For an income of $100,000,

$100,000 = 0.05N + 50,000$
$0.05N = 50,000$
$$N = \frac{50,000}{0.05} = \$1,000,000$$

Her marginal income is her increase in income
per $1 increase in net profit. This is the slope, m
= 0.05 dollars of income per dollar of net profit,
or 5¢ per dollar of net profit.

99. We want the temperature T as a linear
function of the rate r of chirping. That is,

$T(r) = mr + b$.

Thus, T plays the role of y and r plays the role of
x. We are given two points:

$(r, T) = (140, 80)$ and $(120, 75)$

Slope:

$$m = \frac{T_2 - T_1}{r_2 - r_1} = \frac{75 - 80}{120 - 140} = \frac{-5}{-20} = \frac{1}{4}$$

Intercept:

$$b = T_1 - mr_1 = 80 - \frac{1}{4}(140) = 80 - 35 = 45$$

Thus, the linear function is

$T(r) = mr + b$
$T(r) = \frac{1}{4}r + 45$

When the chirping rate is 100 chirps per minute, r
= 100, and so the temperature is

$$T(100) = \frac{1}{4}(100) + 45 = 25 + 45 = 70°F$$

101. The year 2006 corresponds to $t = 16$, which
is in the range $8 \le t \le 17$, so we use the second
equation: $C(t) = 0.13t + 0.20$. The slope is 0.13
million dollars/year, telling us that the cost of an
ad was increasing by $0.13 = $130,000/year.

103. The data is

t	0	5	12
y	200	50	470

(a) 1995–2000 (first two data points):

Slope: $m = \dfrac{y_2 - y_1}{t_2 - t_1} = \dfrac{50 - 200}{5 - 0} = -30$

Intercept: $b = 200$ Specified in first data point
Thus, the linear model is

$y = mt + b$
$y = -30t + 200$

32

(b) 2000–2004 (second and third data points):

Slope: $m = \dfrac{y_2 - y_1}{t_2 - t_1} = \dfrac{470-50}{12-5} = 60$

Intercept: $b = y_1 - mt_1 = 50 - 60(5) = -250$

Thus, the linear model is

$$y = mt + b$$
$$y = 60t - 250$$

(c) Since the first model is valid for $0 \le t \le 5$ and the second one for $5 \le t \le 9$, we put them together as

$$y = \begin{cases} -30t + 200 & \text{if } 0 \le t \le 5 \\ 60t - 250 & \text{if } 5 < t \le 12 \end{cases}$$

Notice that, since both formulas agree at $t = 5$, we can also say

$$y = \begin{cases} -30t + 200 & \text{if } 0 \le t < 5 \\ 60t - 250 & \text{if } 5 \le t \le 12 \end{cases} \cdot$$

(d) Since 2002 is represented by $t = 7$, we use the second formula to obtain

$$y = 60(7) - 250 = 170$$

105. 1995–2000:

Points: $(t,C) = (0,3)$ 1995 data

$\qquad\qquad (t,C) = (5,4.1)$ 2000 data

Slope: $m = \dfrac{N_2 - N_1}{t_2 - t_1} = \dfrac{4.1-3}{5-0} = 0.22$

Intercept: $b = N_1 - mt_1 = 3 - (0.22)(0) = 3$

Equation: $\begin{array}{l} N = mt + b \\ N = 0.22t + 3 \end{array}$

2000–2004:

Points: $(t,C) = (5,4.1)$ 2000 data

$\qquad\qquad (t,C) = (9,3.5)$ 2004 data

Slope: $m = \dfrac{N_2 - N_1}{t_2 - t_1} = \dfrac{3.5-4.1}{9-5} = -0.15$

Intercept: $\begin{array}{l} b = N_1 - mt_1 = 4.1 - (-0.15)(5) \\ \quad = 4.1 + 0.75 = 4.85 \end{array}$

Equation: $\begin{array}{l} N = mt + b \\ N = -0.15t + 4.85 \end{array}$

Putting them together gives

$$N = \begin{cases} 0.22t + 3 & \text{if } 0 \le t \le 5 \\ -0.15t + 4.85 & \text{if } 5 < t \le 9. \end{cases}$$

The number of manufacturing jobs in Mexico in 2002 is $N(7)$, so we use the second formula to obtain

$$N(7) = -0.15(7) + 4.85$$
$$= -1.05 + 4.85 = 3.8 \text{ million jobs}$$

107. Compute the corresponding successive changes Δx in x and Δy in y, and compute the ratios $\Delta y / \Delta x$. If the answer is always the same number, then the values in the table come from a linear function.

109. To find the linear function, solve the equation $ax + by = c$ for y:

$$by = -ax + c$$
$$y = -\frac{a}{b}x + \frac{c}{b}$$

Thus, the desired function is $f(x) = -\dfrac{a}{b}x + \dfrac{c}{b}$.

If $b = 0$, then $\dfrac{a}{b}$ and $\dfrac{c}{b}$ are undefined, and y cannot be specified as a function of x. (The graph of the resulting equation would be a vertical line.)

111. The slope of the line is $m = \dfrac{\Delta y}{\Delta x} = \dfrac{3}{1} = 3$.

Therefore, if, in a straight line, y is increasing three times as fast as x, then its <u>slope</u> is 3 .

113. If m is positive then y will increase as x increases; if m is negative then y will decrease as x increases; if m is zero then y will not change as x changes.

115.

	A	B	C	D
1	x	y	m	b
2		1	2	=(B3-B2)/(A3-A2)	=B2-C2*A2
3		3	-1	Slope	Intercept

The slope computed in cell C2 is given by

$$m = \frac{y_2 - y_1}{x_2 - x_1} = \frac{-1 - 2}{3 - 1} = -1.5$$

If we increase the y-coordinate in cell B3, this increases y_2, and thus increases the numerator $\Delta y = y_2 - y_1$ without effecting the denominator Δx. Thus the slope will increase.

117. The units of the slope m are units of y (bootlags) per unit of x (zonars). The intercept b is on the y-axis, and is thus measured in units of y (bootlags). Thus, m is measured in <u>bootlags per zonar</u> and b is measured in <u>bootlags</u>.

119. If a quantity changes linearly with time, it must change by the same amount for every unit change in time. Thus, since it increases by 10 units in the first day, it must increase by 10 units each day, including the third.

121. $v = 0.1t + 20$ m/sec
Since the slope is 0.1, the velocity is increasing at a rate of 0.1 m/sec every second. Since the velocity is increasing, the object is accelerating (choice B).

123. Increasing the number of items from the breakeven results in a profit: Because the slope of the revenue graph is larger than the slope of the cost graph, it is higher than the cost graph to the right of the point of intersection, and hence corresponds to a profit.

34

1.4 Linear Regression

1. $(1, 1)$, $(2, 2)$, $(3, 4)$; $y = x-1$

		Predicted $\hat{y} = x-1$	Residual $y - \hat{y}$	Residual^2 $(y - \hat{y})^2$
x	y			
1	1	0	1	1
2	2	1	1	1
3	4	2	2	4

SSE = Sum of squares of residuals
   $= 4+1+1 = 6$

3. $(0,-1)$, $(1,3)$, $(4,6)$, $(5,0)$; $y = -x+2$

		Predicted $\hat{y} = -x+2$	Residual $y - \hat{y}$	Residual^2 $(y - \hat{y})^2$
x	y			
0	-1	2	-3	9
1	3	1	2	4
4	6	-2	8	64
5	0	-3	3	9

SSE = Sum of squares of residuals
   $= 9 + 4 + 64 + 9 = 86$

5. $(1, 1)$, $(2, 2)$, $(3, 4)$
(a) $y = 1.5x-1$

x	y	$\hat{y}$	$y - \hat{y}$	$(y - \hat{y})^2$
1	1	0.5	0.5	0.25
2	2	2	0	0
3	4	3.5	0.5	0.25

SSE = Sum of squares of residuals = 0.5
(b) $y = 2x - 1.5$

x	y	$\hat{y}$	$y - \hat{y}$	$(y - \hat{y})^2$
1	1	0.5	0.5	0.25
2	2	2.5	-0.5	0.25
3	4	4.5	-0.5	0.25

SSE = Sum of squares of residuals = 0.75
The model that gives the better fit is (a) because it gives the smaller value of SSE.

7. $(0, -1)$, $(1, 3)$, $(4, 6)$, $(5, 0)$
(a) $y = 0.3x + 1.1$

x	y	$\hat{y}$	$y - \hat{y}$	$(y - \hat{y})^2$
0	-1	1.1	-2.1	4.41
1	3	1.4	1.6	2.56
4	6	2.3	3.7	13.69
5	0	2.6	-2.6	6.76

SSE = Sum of squares of residuals = 27.42
(b) $y = 0.4x+0.9$

x	y	$\hat{y}$	$y - \hat{y}$	$(y - \hat{y})^2$
0	-1	0.9	-1.9	3.61
1	3	1.3	1.7	2.89
4	6	2.5	3.5	12.25
5	0	2.9	-2.9	8.41

SSE = Sum of squares of residuals = 27.16
The model that gives the better fit is (b) because it gives the smaller value of SSE.

9. $(1,1)$, $(2,2)$, $(3,4)$

x	y	xy	x^2
1	1	1	1
2	2	4	4
3	4	12	9
Σ **(Sum)** 6	7	17	14

$n = 3$ (number of data points)

Slope: $m = \dfrac{n(\Sigma xy) - (\Sigma x)(\Sigma y)}{n(\Sigma x^2) - (\Sigma x)^2}$

$\qquad = \dfrac{3(17) - (6)(7)}{3(14) - 6^2} = \dfrac{9}{6} = 1.5$

Intercept: $b = \dfrac{\Sigma y - m(\Sigma x)}{n}$

$\qquad = \dfrac{7 - 1.5(6)}{3} = -\dfrac{2}{3} \approx -0.6667$

Thus, the regression line is
$\quad y = mx + b$
$\quad y = 1.5x - 0.6667$

Graph:

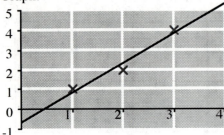

$n = 3$ (number of data points)

$$r = \frac{n(\Sigma xy) - (\Sigma x)(\Sigma y)}{\sqrt{n(\Sigma x^2) - (\Sigma x)^2} \cdot \sqrt{n(\Sigma y^2) - (\Sigma y)^2}}$$

$$= \frac{3(41) - (8)(13)}{\sqrt{3(30)-(8)^2} \sqrt{3(61)-(13)^2}}$$

$$\approx \frac{19}{19.078784} \approx 0.9959$$

(b) $\{(0, -1), (2, 1), (3, 4)\}$

x	y	xy	x^2	y^2
0	−1	0	0	1
2	1	2	4	1
3	4	12	9	16
Σ 5	4	14	13	18

$n = 3$ (number of data points)

$$r = \frac{n(\Sigma xy) - (\Sigma x)(\Sigma y)}{\sqrt{n(\Sigma x^2) - (\Sigma x)^2} \cdot \sqrt{n(\Sigma y^2) - (\Sigma y)^2}}$$

$$= \frac{3(14) - (5)(4)}{\sqrt{3(13)-(5)^2} \sqrt{3(18)-(4)^2}}$$

$$\approx \frac{22}{23.0651252} \approx 0.9538$$

(c) $\{(4, -3), (5, 5), (0, 0)\}$

x	y	xy	x^2	y^2
4	−3	−12	16	9
5	5	25	25	25
0	0	0	0	0
Σ 9	2	13	41	34

$n = 3$ (number of data points)

$$r = \frac{n(\Sigma xy) - (\Sigma x)(\Sigma y)}{\sqrt{n(\Sigma x^2) - (\Sigma x)^2} \cdot \sqrt{n(\Sigma y^2) - (\Sigma y)^2}}$$

$$= \frac{3(13) - (9)(2)}{\sqrt{3(41)-(9)^2} \sqrt{3(34)-(2)^2}}$$

$$\approx \frac{21}{64.1560597} \approx 0.3273$$

The value of r in part (a) has the largest absolute value. Therefore, the regression line for the data in part (a) is the best fit.

The value of r in part (c) has the smallest absolute value. Therefore, the regression line for the data in part (c) is the worst fit.

Since r is not ± 1 for any of these lines, none of them is a perfect fit.

11. $(0, -1), (1, 3), (4, 6), (5, 0)$

x	y	xy	x^2
0	−1	0	0
1	3	3	1
4	6	24	16
5	0	0	25
Σ **(Sum)** 10	8	27	42

$n = 4$ (number of data points)

Slope: $m = \dfrac{n(\Sigma xy) - (\Sigma x)(\Sigma y)}{n(\Sigma x^2) - (\Sigma x)^2}$

$$= \frac{4(27) - (10)(8)}{4(42) - 10^2} = \frac{28}{68} \approx 0.4118$$

Intercept: $b = \dfrac{\Sigma y - m(\Sigma x)}{n}$

$$= \frac{8 - \left(\dfrac{28}{68}\right)(10)}{4} \approx 0.9706$$

Thus, the regression line is

$y = mx + b$

$y = 0.4118x + 0.9706$

Graph:

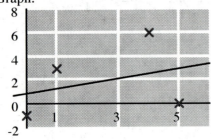

13. (a) $\{(1, 3), (2, 4), (5, 6)\}$

x	y	xy	x^2	y^2
1	3	3	1	9
2	4	8	4	16
5	6	30	25	36
Σ 8	13	41	30	61

15. The entries in the xy column are obtained by multiplying the entries in the x column by the corresponding entries in the y column. The entries in the x^2 column are the squares of the entries in the x column.

x	y	xy	x^2
3	500	1500	9
5	600	3000	25
7	800	5600	49
Σ **(Sum)** 15	1900	10100	83

$n = 3$ (number of data points)

Slope: $m = \dfrac{n(\Sigma xy) - (\Sigma x)(\Sigma y)}{n(\Sigma x^2) - (\Sigma x)^2}$

$= \dfrac{3(10100) - (15)(1900)}{3(83) - (15)^2}$

$= \dfrac{1800}{24} = 75$

Intercept: $b = \dfrac{\Sigma y - m(\Sigma x)}{n}$

$= \dfrac{1900 - (75)(15)}{3}$

$= \dfrac{775}{3} \approx 258.33$ (to 2 decimal places)

Thus, the regression line is

$y = mx + b$

$y = 75x + 258.33$

To estimate the 2008 sales we put $x = 8$:

$y = 75(8) + 258.33 \approx 858.33$ million

17. Calculation of the regression line:

x	y	xy	x^2
0	6	0	0
4	16	64	16
7	32	224	49
Σ **(Sum)** 11	54	288	65

$n = 3$ (number of data points)

Slope: $m = \dfrac{n(\Sigma xy) - (\Sigma x)(\Sigma y)}{n(\Sigma x^2) - (\Sigma x)^2}$

$= \dfrac{3(288) - (11)(54)}{3(65) - (11)^2}$

$= \dfrac{270}{74} = 3.65$

Intercept: $b = \dfrac{\Sigma y - m(\Sigma x)}{n}$

$= \dfrac{54 - (3.65)(11)}{3}$

≈ 4.62 (to 2 decimal places)

Thus, the regression line is

$y = mt + b$ Independent variable is called t

$y = 3.65t + 4.62$

$y(6) \approx 3.65(6) + 4.62 = \$26.52$ billion

19. Calculation of the regression line:

x	y	xy	x^2
20	3	60	400
40	6	240	1600
80	9	720	6400
100	15	1500	10000
Σ **(Sum)** 240	33	2520	18400

$n = 4$ (number of data points)

Slope: $m = \dfrac{n(\Sigma xy) - (\Sigma x)(\Sigma y)}{n(\Sigma x^2) - (\Sigma x)^2}$

$= \dfrac{4(2520) - (240)(33)}{4(18400) - (240)^2}$

$= \dfrac{2160}{16,000} = 0.135$

Intercept: $b = \dfrac{\Sigma y - m(\Sigma x)}{n}$

$= \dfrac{33 - (0.135)(240)}{4}$

$= \dfrac{0.6}{4} = 0.15$

Thus, the regression line is

$y = mx + b$

$y = 0.135x + 0.15$

$y(50) = 0.135(50) + 0.15 = 6.9$ million jobs

21. (a) Since production is a function of cultivated area, we take x as cultivated area, and y as production:

x	25	30	32	40	52
y	15	25	30	40	60

Using the method of Example 3, we obtain the following regression line and plot (coefficients rounded to two decimal places):
$$y = 1.62x - 23.87$$

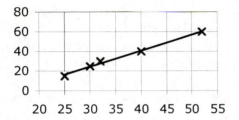

(b) To interpret the slope $m = 1.62$, recall that units of m are units of y per unit of x; That is, millions of tons of production of soybeans per million acres of cultivated land. Thus, production increases by 1.62 million tons of soybeans per million acres of cultivated land. More simply, each acre of cultivated land produces about 1.62 tons of soybeans.

23. (a) Using x = Number of automobiles (in millions) and y = Number of motorcycles (in millions) gives the following table of values:

x	y
4	0.28
5.3	0.25
6.6	0.25
7.5	0.13
10.2	0.29
14.7	0.61

Using the technology method of Example 2, we obtain the following regression line and plot (coefficients rounded to two significant digits):
$$y = 0.032x + 0.042$$
Graph:

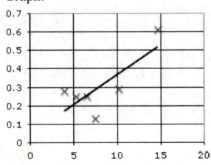

(b) To interpret the slope, recall that units of the slope are units of y (millions of motorcycles) per unit of x (millions of automobiles). Thus,

$$m \approx 0.032 \text{ million motorcycles per million automobiles, or just motorcycles per automobile}$$

indicating that the number of motorcycles increases by 0.032 million per 1 million increase in the number of automobiles. That is, there are about 0.032 additional motorcycles per additional automobile, or 32 motorcycles per 1000 automobiles.

(c) Using the technology method of Example 3, we can use technology to show the value of r^2:
$$r^2 \approx 0.5989$$
so $r = \sqrt{r^2} \approx -\sqrt{0.5989} \approx 0.7739$.

Since r is not close to 1, the correlation between x and y is not a strong one.

25. (a) Using the method of Example 3, we obtain the following regression line and plot (coefficients rounded to two significant digits):
$$p = 0.13t + 0.22$$
Graph:

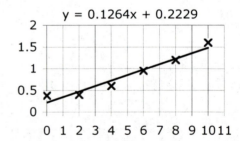

(b) The first and last points lie above the regression line, while the central points lie below it, suggesting a curve.
(c) Here is an Excel worksheet showing the computation of the residuals (based on Example 1 in the text):

	A	B	C	D
1	t	p (Observed)	p (predicted)	Residual
2	0	0.38	=0.13*A2+0.22	=B2-C2
3	2	0.4		
4	4	0.6		
5	6	0.95		
6	8	1.2		
7	10	1.6		

$\downarrow$

38

◇	A	B	C	D
1	t	p (Observed)	p (predicted)	Residual
2	0	0.38	0.22	0.16
3	2	0.4	0.48	-0.08
4	4	0.6	0.74	-0.14
5	6	0.95	1	-0.05
6	8	1.2	1.26	-0.06
7	10	1.6	1.52	0.08

Notice that the residuals are positive at first, become negative, and then become positive, confirming the impression from the graph.

27. The regression line is defined to be the line that gives the lowest sum-of-squares error, SSE. If we are given two points, (a, b) and (c, d) with $a \neq c$ then there is a line that passes through these two points, giving SSE = 0. Since 0 is the smallest value possible, this line must be the regression line.

29. If the points (x_1, y_1), (x_2, y_2), . . . , (x_n, y_n) lie on a straight line, then the sum-of-squares error, SSE, for this line is zero. Since 0 is the smallest value possible, this line must be the regression line.

31. Calculation of the regression line:

x	y	xy	x^2
0	0	0	0
$-a$	a	$-a^2$	a^2
a	a	a^2	a^2

	x	y	xy	x^2
Σ	0	$2a$	0	$2a^2$

n = 3 (number of data points)

Slope: $m = \dfrac{n(\Sigma xy) - (\Sigma x)(\Sigma y)}{n(\Sigma x^2) - (\Sigma x)^2}$

$= \dfrac{3(0) - (0)(2a)}{3(2a^2) - 0^2} = 0$

Regression Coefficient:

$r = \dfrac{n(\Sigma xy) - (\Sigma x)(\Sigma y)}{\sqrt{n(\Sigma x^2) - (\Sigma x)^2} \cdot \sqrt{n(\Sigma y^2) - (\Sigma y)^2}}$

has the same numerator as m, and we have just seen that this numerator is zero, Hence, $r = 0$.

33. No. The regression line through $(-1, 1)$, $(0, 0)$, and $(1, 1)$ passes through none of these points.

Chapter 1 Review Exercises

1.

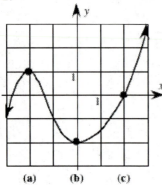

(a) 1 **(b)** −2 **(c)** 0
(d) $f(2) - f(-2) = 0 - 1 = -1$

3.

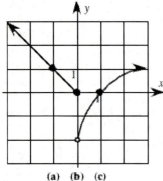

(a) 1 **(b)** 0 **(c)** 0
(d) $f(1) - f(-1) = 0 - 1 = -1$

5. $y = -2x + 5$
y−intercept = 5 slope = −2

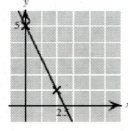

7. $y = \begin{cases} \frac{1}{2}x & \text{if } -1 \le x \le 1 \\ x - 1 & \text{if } 1 < x \le 3 \end{cases}$

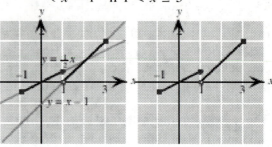

Here are the graphs of the functions, with the points connected:

9.

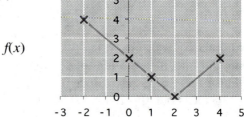

$f(x)$

V-shape indicates **absolute value** model.

11.

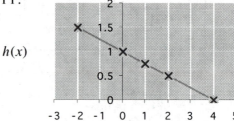

$h(x)$

Linear

13.

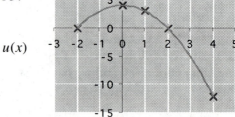

$u(x)$

Parabolic shape of graph indicates a **quadratic** model.

40

15. Line through $(3, 2)$ with slope -3
Point: $(3, 2)$
Slope: $m = -3$
Intercept: $b = y_1 - mx_1$
$$= 2-(-3)(3) = 2+9 = 11$$
Thus, the equation is
$$y = mx + b$$
$$y = -3x + 11$$

17. Through $(1, 2)$ parallel to $x - 2y = 2$
Point: $(1, 2)$
Slope: Parallel to the line $x - 2y = 2$, and so has the same slope. Solve for y to obtain
$$-2y = -x - 2$$
$$y = \tfrac{1}{2}x + 1$$
Thus the slope is $m = \tfrac{1}{2}$

Intercept: $b = y_1 - mx_1 = 2-\tfrac{1}{2}(1) = \dfrac{3}{2}$
Thus, the equation is
$$y = mx + b$$
$$y = \tfrac{1}{2}x + \tfrac{3}{2}$$

19. $y = -0.5x + 1$:

x	observed y	predicted y	residual^2
-1	1	1.5	0.25
1	1	0.5	0.25
2	0	0	0
		SSE:	0.5

$y = -x/4 + 1$:

x	observed y	predicted y	residual^2
-1	1	1.25	0.0625
1	1	0.75	0.0625
2	0	0.5	0.25
		SSE:	0.375

The second line, $y = -x/4 + 1$, is a better fit.

21.

x	y	xy	x^2	y^2
-1	1	-1	1	1
1	2	2	1	4
2	0	0	4	0
Σ **(Sum)**	2	3	1	6	5

$n = 3$ (number of data points)
Slope: $m = \dfrac{n(\Sigma xy) - (\Sigma x)(\Sigma y)}{n(\Sigma x^2) - (\Sigma x)^2}$
$$= \dfrac{3(1) - (2)(3)}{3(6) - 2^2} = \dfrac{-3}{14} \approx -0.214$$

Intercept: $b = \dfrac{\Sigma y - m(\Sigma x)}{n}$
$$= \dfrac{3 - (-0.214)(2)}{3} \approx 1.14$$

Thus, the regression line is
$$y = mx + b$$
$$y = -0.214x + 1.14$$
The correlation coefficient is
$$r = \dfrac{n(\Sigma xy) - (\Sigma x)(\Sigma y)}{\sqrt{n(\Sigma x^2) - (\Sigma x)^2} \cdot \sqrt{n(\Sigma y^2) - (\Sigma y)^2}}$$
$$= \dfrac{3(1) - (2)(3)}{\sqrt{3(6)-(2)^2} \sqrt{3(5)-(3)^2}}$$
$$\approx -0.33$$

23. (a) Graph:

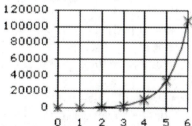

Since the data definitely suggests a curve, we rule out a linear function, leaving us with a choice of quadratic or exponential. Of the two,, an exponential function would fit best, given the leveling-off we see on the left; the graph of a quadratic function would not flatten out, but instead form a low point and begin rising again toward the left.

(b) The ratios (rounded to 1 decimal place) are:

$V(1)/V(0)$	$V(2)/V(1)$	$V(3)/V(2)$
300/100	1000/300	3300/1000
= 3	≈ 3.3	= 3.3

$V(4)/V(3)$	$V(5)/V(4)$	$V(6)/V(5)$
10,500/3300	33,600/10,500	107,400/33,600
≈ 3.2	≈ 3.2	≈ 3.2

They are close to 3.2.

(c) The data suggest that website traffic is increasing by a factor of around 3.2 per year, so the prediction for next year (year 6) would be around 3.2×107,400 ≈ 343,700 visits/day.

25.

t	1	2	3	4	5	6
$S(t)$	12.5	37.5	62.5	72.0	74.5	75.0

(a) Technology formulas:

    (A): `300/(4+100*5^(-t))`
    (B): `13.3*t+8.0`
    (C): `-2.3*t^2+30.0*t-3.3`
    (D): `7*3^(0.5*t)`

Here are the values for the three given models (rounded to 1 decimal place):

t	1	2	3	4	5	6
(A)	12.5	37.5	62.5	72.1	74.4	74.9
(B)	21.3	34.6	47.9	61.2	74.5	87.8
(C)	24.4	47.5	66.0	79.9	89.2	93.9
(D)	12.1	21.0	36.4	63.0	109.1	189.0

Model (A) gives an almost perfect fit, whereas the other models are not even close.

(b) Looking at the table, we see the following behavior as t increases:
(A) Leveling off; (B) Rising (C) Rising (they begin to fall after 7 months, however) (D) Rising

27. $h = -0.000005c^2 + 0.085c + 1750$

(a) Currently, $c = \$6000$, so

$h = -0.000005(6000)^2 + 0.085(6000) + 1750$
   $= -180 + 510 + 1750 = 2080$ hits per day

(b) Here is a portion of the graph of h:

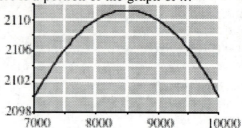

Technology formula:
   `-0.000005*x^2+0.085*x+1750`

For $c = 8500$ or larger, we see that Web site traffic is projected to decrease as advertising increases, and then drop toward zero. Thus, the model does not appear to give a reasonable prediction of traffic at expenditures larger than $8,500 per month.

29. (a) Demand: We are given two points:
   $(p, q) = (10, 350)$ and $(5.5, 620)$ Slope:

$$m = \frac{q_2 - q_1}{p_2 - p_1} = \frac{620-350}{5.5-10} = \frac{270}{-4.5} = -60$$

Intercept:

$b = q_1 - mp_1 = 350-(-60)(10)$
   $= 350 + 600 = 950$

Thus, the demand equation is

   $q = mp + b$
   $q = -60p + 950$

(b) When $p = \$15$, the demand is

   $q = -60(15) + 950 = -900 + 950$
   $= 50$ novels per month

(c) From Question 8, the cost function is

   $C = 4q + 900$

We use q for the monthly sales rather than x

   $= 4(-60p + 950) + 900$

We want everything expressed in terms of p, so we used the demand equation.

   $= -240p + 3800 + 900$
   $C = -240p + 4700$

To compute the profit in terms of price, we need the revenue as well:

   $R = pq = p(-60p + 950)$
   $= -60p^2 + 950p$

Profit: $P = R - C$

$$P = -60p^2 + 950p - (-240p + 4700)$$
$$= -60p^2 + 1190p - 4700$$

Now we compare profits for the three prices:

$$P(5.50) = -60(5.5)^2 + 1190(5.5) - 4700$$
$$= \$30$$

$$P(10) = -60(10)^2 + 1190(10) - 4700$$
$$= \$1200$$

$$P(15) = -60(15)^2 + 1190(15) - 4700$$
$$= -\$350 \text{ (loss)}$$

Thus, charging \$10 will result in the largest monthly profit of \$1200.

Chapter 2 Systems of Linear Equations and Matrices

2.1 Systems of Two Equations in Two Unknowns

1. $x - y = 0$
$x + y = 4$
Adding gives
$\quad 2x = 4$
$\quad x = 2$
Substituting $x=2$ in first equation:
$\quad 2 - y = 0$
$\quad y = 2$
Solution: $(2, 2)$
Graph: $y = x$; $y = 4 - x$

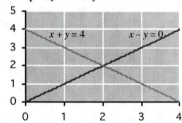

3. $x + y = 4$
$x - y = 2$
Adding gives
$\quad 2x = 6$
$\quad x = 3$
Substituting $x = 3$ in first equation:
$\quad 3 + y = 4$
$\quad y = 4 - 3 = 1$
Solution: $(3, 1)$
Graph: $y = 4 - x$; $y = x - 2$

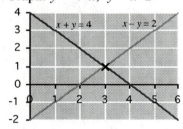

5. $3x - 2y = 6$
$2x - 3y = -6$
Multiply first equation by 2 and second by -3:

$6x - 4y = 12$
$-6x + 9y = 18$
Adding gives
$\quad 5y = 30$
$\quad y = 6$
Substituting $y=6$ in first equation gives
$\quad 3x - 12 = 6$
$\quad 3x = 18$
$\quad x = 6$
Solution: $(6, 6)$
Graph: $y = \dfrac{3}{2}x - 3$; $y = \dfrac{2}{3}x + 2$

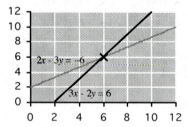

7. $0.5x + 0.1y = 0.7$
$0.2x - 0.2y = 0.6$
Multiply both equations by 10:
$\quad 5x + y = 7$
$\quad 2x - 2y = 6$
Divide second equation by 2:
$\quad 5x + y = 7$
$\quad x - y = 3$
Adding gives
$\quad 6x = 10$
$\quad x = \dfrac{10}{6} = \dfrac{5}{3}$
Substituting $x=\dfrac{5}{3}$ in $x - y = 3$ gives
$\quad \dfrac{5}{3} - y = 3$
$\quad y = \dfrac{5}{3} - 3 = -\dfrac{4}{3}$
Solution: $\left(\dfrac{5}{3}, -\dfrac{4}{3} \right)$

Graph: $y = -5x+7$; $y = x-3$

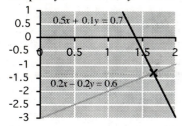

9. $\dfrac{x}{3} - \dfrac{y}{2} = 1$

$\dfrac{x}{4} + y = -2$

Multiply first equation by 6 and second by 4:

$2x - 3y = 6$

$x + 4y = -8$

Multiply the second equation by -2:

$2x - 3y = 6$

$-2x - 8y = 16$

Adding gives

$-11y = 22$

$y = \dfrac{22}{-11} = -2$

Substituting $y=-2$ into $x+4y = -8$ gives

$x + 4(-2) = -8$

$x - 8 = -8$

$x = 0$

Solution: $(0, -2)$

Graph: $y = -x/4-2$; $y = 2(x/3-1)$

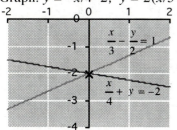

11. $2x + 3y = 1$

$-x - \dfrac{3y}{2} = -\dfrac{1}{2}$

Multiply the second equation by 2

$2x + 3y = 1$

$-2x - 3y = -1$

Adding gives $0 = 0$, indicating that the system is redundant: The graphs are the same, so there are infinitely many solutions.

We obtain the solutions by solving either equation for y:

$2x + 3y = 1$

$3y = -2x+1$

$y = (-2x+1)/3$

Solution: $(x, [-2x+1]3)$; x arbitrary

Graph: $y = (-2x+1)/3$

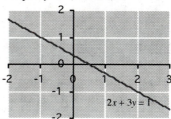

Every point on the line represents a solution.

13. $2x + 3y = 2$

$-x - \dfrac{3y}{2} = -\dfrac{1}{2}$

Multiply the second equation by 2:

$2x + 3y = 2$

$-2x - 3y = -1$

Adding gives $0 = 1$, indicating that the system is inconsistent. There is no solution; the lines are parallel.

Graph: $y = (-2x+2)/3$; $y = (-2x+1)/3$

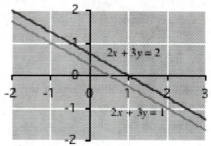

Parallel lines; no solution

15. $2x + 8y = 10$

$x + y = 5$

To graph these, solve for y:

$y = (-2x + 10)/8$

$y = -x + 5$

45

Graph, with vertical and horizontal scales of 0.1:

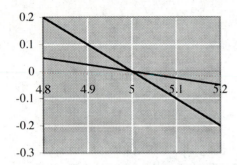

The grid point closest to the intersection of the lines gives the approximate solution (in this case the exact solution)
Solution: (5, 0)

17. $3.1x - 4.5y = 6$
$4.5x + 1.1y = 0$
To graph these, solve for y:
$y = (3.1x - 6)/4.5$
$y = -4.5x/1.1$
Graph, with vertical and horizontal scales of 0.1:

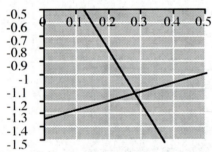

The grid point closest to the intersection of the lines gives the approximate solution.
Approximate solution: (0.3, −1.1)

19. $10.2x + 14y = 213$
$4.5x + 1.1y = 448$
To graph these, solve for y:
$y = (-10.2x + 213)/14$
$y = (-4.5x + 448)/1.1$

Graph, with vertical and horizontal scales of 0.1:

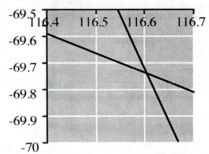

The grid point closest to the intersection of the lines gives the approximate solution.
Approximate solution: (116.6, −69.7)

21. Line through (0, 1) and (4.2, 2)
Point: (0, 1)

Slope: $m = \dfrac{y_2 - y_1}{x_2 - x_1} = \dfrac{2 - 1}{4.2 - 0} = \dfrac{1}{4.2}$

Intercept: $b = 1$ Given
Thus, the equation is
$y = mx + b$
$y = \dfrac{1}{4.2}x + 1$

Line through (2.1, 3) and (5.2, 0)
Point: (5.2, 0)

Slope: $m = \dfrac{y_2 - y_1}{x_2 - x_1} = \dfrac{0 - 3}{5.2 - 2.1} = -\dfrac{3}{3.1}$

Intercept:
$b = y_1 - mx_1$
$= 0 - \left(-\dfrac{3}{3.1}\right)5.2 = \dfrac{15.6}{3.1}$

Thus, the equation is
$y = mx + b$
$y = -\dfrac{3}{3.1}x + \dfrac{15.6}{3.1}$

Graph, with vertical and horizontal scales of 0.1:

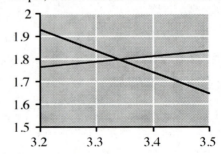

The grid point closest to the intersection of the lines gives the approximate solution.
Approximate solution: (3.3, 1.8)

23. Line through $(0, 0)$ and $(5.5, 3)$
Point: $(0, 0)$
Slope: $m = \dfrac{y_2 - y_1}{x_2 - x_1} = \dfrac{3 - 0}{5.5 - 0} = \dfrac{3}{5.5}$
Intercept: 0   Given
Thus, the equation is
$$y = mx + b$$
$$y = \dfrac{3}{5.5}x$$

Line through $(5, 0)$ and $(0, 6)$.
Point: $(0, 6)$
Slope: $m = \dfrac{y_2 - y_1}{x_2 - x_1} = \dfrac{6 - 0}{0 - 5} = -\dfrac{6}{5}$
Intercept: 6   Given
Thus, the equation is
$$y = mx + b$$
$$y = -\dfrac{6}{5}x + 6$$

Graph, with vertical and horizontal scales of 0.1:

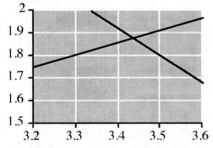

The grid point closest to the intersection of the lines gives the approximate solution.
Approximate solution: (3.4, 1.9)

25. Unknowns:
x = the number of quarts of Creamy Vanilla
y = the number of quarts of Continental Mocha
Arrange the given information in a table with unknowns across the top:

	Vanilla (x)	Mocha (y)	Available
Eggs	2	1	500
Cream	3	3	900

We can now set up an equation for each of the items listed on the left:
Eggs: $2x + y = 500$
Cream: $3x + 3y = 900$
Multiply the first equation by –1 and divide the second by 3:
$$-2x - y = -500$$
$$x + y = 300$$
Adding gives
$$-x = -200$$
$$x = 200 \text{ quarts of vanilla}$$
Substituting $x = 200$ in the first equation gives
$$2(200) + y = 500$$
$$400 + y = 500$$
$$y = 500 - 400 = 100 \text{ quarts of mocha}$$
Solution: Make 200 quarts of vanilla and 100 quarts of mocha.

27. Unknowns:
x = the number of servings of Mixed Cereal
y = the number of servings of Mango Tropical Fruit
Arrange the given information in a table with unknowns across the top:

	Cereal (x)	Mango (y)	Desired
Calories	60	80	200
Carbs.	11	21	43

We can now set up an equation for each of the items listed on the left:
Calories: $60x + 80y = 200$
Carbs: $11x + 21y = 43$
Divide first equation by 20:
$$3x + 4y = 10$$
$$11x + 21y = 43$$
Multiply first equation by –11 and second by 3:
$$-33x - 44y = -110$$
$$33x + 63y = 129$$
Adding gives
$$19y = 19$$
$$y = 1 \text{ serving of Mango Tropical Fruit}$$
Substituting $y = 1$ in the equation $3x + 4y = 10$ gives
$$3x + 4(1) = 10$$
$$3x = 6$$
$$x = 2 \text{ servings of Mixed Cereal}$$
Solution: Use 1 serving of Mango Tropical Fruit and 2 servings of Mixed Cereal.

47

29. (a) one half of the US RDA for protein, is 30g.

Unknowns:

x = the number of servings of Pork & Beans

y = the number of slices of white bread

Arrange the given information in a table with unknowns across the top:

	Pork & Beans (x)	Bread (y)	Desired
Protein	5	2	30
Carbs.	21	11	139

We can now set up an equation for each of the items listed on the left:

Protein $\qquad 5x + 2y = 30$

Carbs: $\qquad 21x + 11y = 139$

Multiply first equation by 11 and second by –2:

$\qquad 55x + 22y = 330$

$\qquad -42x - 22y = -278$

Adding gives

$\qquad 13x = 52$

$\qquad x = \dfrac{52}{13}$ = 4 servings of Pork & Beans

Substituting $x=4$ in the equation $5x+2y = 30$ gives

$\qquad 5(4) + 2y = 30$

$\qquad 20 + 2y = 30$

$\qquad 2y = 10$

$\qquad y = 5$ slices of bread

Solution: Prepare 4 servings of Pork & Beans and 5 slices of bread.

(b) Decreasing the Carbohydrate total to 100 gives:

Protein $\qquad 5x + 2y = 30$

Carbs: $\qquad 21x + 11y = 100$

Multiply first equation by 11 and second by –2:

$\qquad 55x + 22y = 330$

$\qquad -42x - 22y = -200$

Adding gives

$\qquad 13x = 130$

$\qquad x = \dfrac{130}{13}$ = 10 servings of Pork & Beans

Substituting $x=10$ in the equation $5x+2y = 30$ gives

$\qquad 5(10) + 2y = 30$

$\qquad 50 + 2y = 30$

$\qquad 2y = 30-50 = -20$

$y = -10$ slices of bread

Since it is impossible to prepare a negative number of slices of bread, we conclude that it is not possible to prepare such a meal.

31. Unknowns:

x = number of servings of Designer Whey

y = number of servings of Muscle Milk

Arrange the given information in a table with unknowns across the top:

	Designer Whey	Muscle Milk	Desired
Protein	18	32	280
Carbs	2	16	56
Cost	$0.50	$1.60	

We can now set up an equation for each nutrient listed on the left:

Protein: $18x + 32y=280$

Carbs: $\qquad 2x + 16y=56$

Multiply the second equation by -2 and add to the first:

$\qquad 14x = 168$

$\qquad x = 12$ servings of Designer Whey

Substitute into the second equation to obtain

$\qquad 24+16y = 56$

$\qquad 16y = 32$

$\qquad y = 2$ servings of Muscle Milk

Cost $= 12(0.50) + 2(1.60) = \$9.20$

Solution: Mix 12 servings of Designer Whey and 2 servings of Muscle Milk for a cost of $9.20.

33. We first determine how many servings of each kind of supplement it contains:

Unknowns:

x = number of servings of Designer Whey

y = number of servings of Pure Whey Protein Stack

Arrange the given information in a table with unknowns across the top:

	Designer Whey	Pure Whey Stack	Total
Protein	18	24	540
Cost	0.50	0.60	14
Carbs	2	3	

We can now set up an equation for protein content and for cost:

Protein: $18x + 24y = 540$
Cost: $\quad\quad\quad 0.5x + 0.6y = 14$
Divide the first equation by 6 and multiply the second by 10:
$\quad 3x + 4y = 90$
$\quad 5x + 6y = 140$
2 times the second minus 3 times the first gives
$\quad x = 10$ servings of Designer Whey
Substituting in the first gives
$\quad 4y = 60$
$\quad y = 15$ Pure Whey Protein Stack
From the table,
Total amount of carbohydrates = $10(2) + 15(3) = 65$g

35. Unknowns:
x = number of shares of AAPL
y = number of shares of MSFT
Arrange the given information in a table with unknowns across the top:

	AAPL (x)	MSFT (y)	Total
October 31	105	21	3570
November 30	95	20	3250

We can now set up equations for the start and end of November:
Oct 31: $105x + 21y = 3570$
Nov 30: $95x + 20y = 3250$
Divide the first equation by 21 (or first by 3 and then by 7) and the second by 5:
$\quad 5x + y = 170$
$\quad 19x + 4y = 650$
Multiply the first by –4:
$\quad -20x - 4y = -680$
$\quad 19x + 4y = 650$
Adding gives
$\quad -x = -30$
$\quad x = 30$ shares of AAPL
Substituting $x = 30$ into $5x + y = 170$ gives
$\quad 5(30) + y = 170$
$\quad y = 20$ shares of MSFT
Solution: You bought 30 shares of AAPL and 20 shares of MSFT.

37. Unknowns:
x = number of shares of BAC
y = number of shares of JPM

The total investment was $15,200:
Investment in BAC = x shares @ $16 = $16x$
Investment in JPM = y shares @ $30 = $30y$
Thus,
$\quad 16x + 30y = 15,200$
You earned $696 in dividends:
BAC dividend = 3% of $16x$ invested
$\quad = 0.03(16x) = 0.48x$
JPM dividend = 5% of $30y$ invested
$\quad = 0.05(30y) = 1.5y$
Thus,
$0.48x + 1.5y = 696$
We therefore have the following system:
$\quad 16x + 30y = 15,200$
$\quad 0.48x + 1.5y = 696$
Eliminate decimals: Multiply the second equation by 100:
$\quad 16x + 30y = 15,200$
$\quad 48x + 150y = 69,600$
Simplify: Divide the second equation by 3:
$\quad 16x + 30y = 15,200$
$\quad 16x + 50y = 23,200$
Eliminate x: Multiply the first equation by –1 and add to the second:
$\quad 20y = 8000$
$\quad y = 400$ shares of JPM
Substituting $y = 400$ in $16x + 30y = 15,200$ gives
$\quad 16x + 12,000 = 15,200$
$\quad 16x = 3200$
$\quad x = 200$ shares of BAC
Solution: you purchased 200 shares of BAC and 400 shares of JPM.

39. Unknowns:
x = number of members voting in favor
y = number of members voting against
We are given two pieces of information:
(1) Total number of votes is 435
$\quad x + y = 435$
(2) 49 more members voted in favor than against.
Rephrasing this gives:
The number of members voting in favor exceeded the number of members voting against by 49.
$\quad x - y = 49$
Thus we have two equations:

$x + y = 435$
$x - y = 49$
Adding gives
$\quad 2x = 484$
$\quad x = 242$
Substituting $x = 242$ in the first equation gives
$\quad 242 + y = 435$
$\quad y = 435 - 242 = 193$
Solution: 242 voted in favor and 193 voted against.

41. Unknowns:
x = number of soccer games won
y = number of football games won
We are given two pieces of information:
(1) The total number of games was 12
$\quad x + y = 12$
(2) Total number of points earned was 38
$\quad 2x + 4y = 38$
(Two points per soccer game and 4 per football game)
Thus we have two equations:
$\quad x + y = 12$
$\quad 2x + 4y = 38$
Dividing the second by –2:
$\quad x + y = 12$
$\quad -x - 2y = -19$
Adding gives
$\quad -y = -7$
$\quad y = 7$ football games
Substituting $y = 7$ in the first equation gives
$\quad x + 7 = 12$
$\quad x = 5$ soccer games
Solution: Lombardi House won 5 soccer games and 7 football games.

43. Unknowns:
x = number of brand X pens
y = number of brand Y pens
We are given two pieces of information:
(1) The total number of pens is 12
$\quad x + y = 12$
(2) The total amount spent was $42.
$\quad 4x + 2.8y = 42$
($4 per band X pen and $2.80 per brand Y pen)
Thus we have two equations:
$\quad x + y = 12$

$4x + 2.8y = 42$
Multiply the second by 10:
$\quad x + y = 12$
$\quad 40x + 28y = 420$
Multiply the first by 10 and divide the second by –4:
$\quad 10x + 10y = 120$
$\quad -10x - 7y = -105$
Adding,
$\quad 3y = 15$
$\quad y = 5$ brand Y pens
Substituting this value in the first equation $x+y = 12$ gives
$\quad x + 5 = 12$
$\quad x = 7$ brand X pens
Solution: Elena purchased 7 brand X pens and 5 brand Y pens

45. Rewrite the demand and supply equations in standard form:
Demand: $q + 60p = 150$
Supply: $q - 80p = -60$
For neither a shortage nor surplus, both equations must be satisfied. Multiplying the second equation by –1 and adding gives
$\quad 140p = 210$
$\quad p = \dfrac{210}{140} = \$1.50$ per pet chia
(We do not need to solve for q.)

47. Demand: $D = 85 - 5P$
Supply: $S = 25 + 5P$
For the equilibrium, we can equate the supply and demand:
$\quad$ Demand = Supply
$\quad 85 - 5P = 25 + 5P$
$\quad -10P = -60$
$\quad P = \$6$ per widget
Substituting $P = \$6$ in the demand curve gives
$\quad D = 85 - 5(6) = 85 - 30 = 55$ widgets

49. Demand: The given points are
$\quad (p, q) = (8, 15)$ and $(11, 3)$
$\quad m = \dfrac{q_2 - q_1}{p_2 - p_1} = \dfrac{3 - 15}{11 - 8} = \dfrac{-12}{3} = -4$
$\quad b = q_1 - mp_1 = 15 - (-4)(8) = 15+32 = 47$

Thus, the demand equation is

$$q = mp + b$$
$$q = -4p + 47$$

Supply: The given points are

$$(p, q) = (8, 3) \text{ and } (11, 15)$$
$$m = \frac{q_2 - q_1}{p_2 - p_1} = \frac{15 - 3}{11 - 8} = \frac{12}{3} = 4$$
$$b = q_1 - mp_1 = 3 - (4)(8) = 3 - 32 = -29$$

Thus, the supply equation is

$$q = mp + b$$
$$q = 4p - 29$$

For the equilibrium price, we can equate the supply and demand:

$$\text{Supply} = \text{Demand}$$
$$-4p + 47 = 4p - 29$$
$$-8p = -76$$
$$p = \frac{76}{8} = \$9.50$$

51. Unknowns:

x = number of pairs of dirty socks
y = number of T shirts

We are given two pieces of information:
(1) A total of 44 items were washed:

$$x + y = 44$$

(2) There were three times as many pairs of dirty socks as T shirts, Rephrasing this gives:

The number of pairs of dirty socks was three times the number of T shirts.

$$x = 3y$$

Thus we have two equations:

$$x + y = 44$$
$$x - 3y = 0$$

Subtracting (or multiplying the second by -1 and adding) gives

$$4y = 44$$
$$y = 11 \text{ T shirts}$$

Substituting $y = 11$ in the first equation gives

$$x + 11 = 44$$
$$x = 44 - 11 = 33 \text{ pairs of dirty socks.}$$

Solution: Joe's roommate threw out 33 pairs of dirty socks and 11 T shirts.

53. Unknowns:

x = size of raise for each full-time employee
y = size of raise for each part-time employee

We are given two pieces of information:
(1) Total budget = \$6000. There are 4 full-time employees each getting a raise of x and 2 part-time employees each getting a raise of y. Thus,

$$4x + 2y = 6000$$

(2) The raise received by each full-time employee is twice the raise that each of the part-time employees receives.

$$x = 2y$$

Thus we have two equations:

$$4x + 2y = 6000$$
$$x - 2y = 0$$

Adding gives

$$5x = 6000$$
$$x = \$1200 \text{ per full time employee}$$

(We are not asked for the value of y.)

55. The three lines in a plane must intersect in a single point for there to be a unique solution. This can happen in two ways: (1) the three lines intersect in a single point, or (2) two of the lines are the same, and the third line intersects it in a single point.

57. The equilibrium price occurs at the point where the demand and supply lines cross. Even if two lines have negative slope, they will still intersect if the slopes differ. Therefore, there can be an equilibrium price.

59. You cannot round both of them up, since there will not be sufficient eggs and cream. Rounding both answers down will ensure that you will not run out of ingredients. It may be possible to round one answer down and the other up, and still have sufficient eggs and cream. and this should be tried.

61. Since multiplying both sides of an equation by a non-zero number has no effect on its solutions, the graph (which represents the set of all solutions) is unchanged: (B).

63. The associated system has no solutions, and so the lines do not intersect. Thus, they must be parallel: (B).

65. Answers will vary.

67. Choosing two lines at random gives two random slopes (which are numbers). Since two randomly selected numbers are unlikely to be the same, it follows that two randomly chosen straight lines are very unlikely to be parallel (or the same line). Thus, the two lines are very likely to intersect in a point, giving a unique solution.

2.2 Using Matrices to Solve Systems of Equations

1. $x + y = 4$
$x - y = 2$

$\begin{bmatrix} 1 & 1 & 4 \\ 1 & -1 & 2 \end{bmatrix} R_2 - R_1$

$\begin{bmatrix} 1 & 1 & 4 \\ 0 & -2 & -2 \end{bmatrix} (1/2)R_2$

$\begin{bmatrix} 1 & 1 & 4 \\ 0 & -1 & -1 \end{bmatrix} R_1 + R_2$

$\begin{bmatrix} 1 & 0 & 3 \\ 0 & -1 & -1 \end{bmatrix} -R_2 \rightarrow \begin{bmatrix} 1 & 0 & 3 \\ 0 & 1 & 1 \end{bmatrix}$

Solution: $x = 3$, $y = 1$, or $(3, 1)$

3. $3x - 2y = 6$
$2x - 3y = -6$

$\begin{bmatrix} 3 & -2 & 6 \\ 2 & -3 & -6 \end{bmatrix} 3R_2 - 2R_1$

$\begin{bmatrix} 3 & -2 & 6 \\ 0 & -5 & -30 \end{bmatrix} (1/5)R_2$

$\begin{bmatrix} 3 & -2 & 6 \\ 0 & -1 & -6 \end{bmatrix} R_1 - 2R_2$

$\begin{bmatrix} 3 & 0 & 18 \\ 0 & -1 & -6 \end{bmatrix} (1/3)R_1$

$\begin{bmatrix} 1 & 0 & 6 \\ 0 & -1 & -6 \end{bmatrix} -R_2 \rightarrow \begin{bmatrix} 1 & 0 & 6 \\ 0 & 1 & 6 \end{bmatrix}$

Solution: $x = 6$, $y = 6$, or $(6, 6)$

5. $2x + 3y = 1$
$-x - \dfrac{3y}{2} = -\dfrac{1}{2}$

$\begin{bmatrix} 2 & 3 & 1 \\ -1 & -3/2 & -1/2 \end{bmatrix} 2R_2$

$\begin{bmatrix} 2 & 3 & 1 \\ -2 & -3 & -1 \end{bmatrix} R_2 + R_1$

$\begin{bmatrix} 2 & 3 & 1 \\ 0 & 0 & 0 \end{bmatrix} \begin{matrix}(1/2)R_1 \\ (1/2)R_2\end{matrix} \rightarrow \begin{bmatrix} 1 & 3/2 & 1/2 \\ 0 & 0 & 0 \end{bmatrix}$

Converting back to equations:
$x + \dfrac{3}{2} y = \dfrac{1}{2}$

Solve for x:
$x = \dfrac{1}{2} - \dfrac{3}{2} y$

General Solution:
$x = \dfrac{1}{2} - \dfrac{3}{2} y = \dfrac{1}{2}(1 - 3y);$

y is arbitrary

or $(\dfrac{1}{2}[1 - 3y], y);$ y arbitrary

7. $2x + 3y = 2$
$-x - \dfrac{3y}{2} = -\dfrac{1}{2}$

$\begin{bmatrix} 2 & 3 & 2 \\ -1 & -3/2 & -1/2 \end{bmatrix} 2R_2$

$\begin{bmatrix} 2 & 3 & 2 \\ -2 & -3 & -1 \end{bmatrix} R_2 + R_1 \rightarrow \begin{bmatrix} 2 & 3 & 2 \\ 0 & 0 & 1 \end{bmatrix}$

Since the bottom row translates to the false statement $0 = 1$, there is no solution.

9. $x + y = 1$
$3x - y = 0$
$x - 3y = -2$

$\begin{bmatrix} 1 & 1 & 1 \\ 3 & -1 & 0 \\ 1 & -3 & -2 \end{bmatrix} \begin{matrix} \\ R_2 - 3R_1 \\ R_3 - R_1 \end{matrix}$

$\begin{bmatrix} 1 & 1 & 1 \\ 0 & -4 & -3 \\ 0 & -4 & -3 \end{bmatrix} \begin{matrix} 4R_1 + R_2 \\ \\ R_3 - R_2 \end{matrix}$

$\begin{bmatrix} 4 & 0 & 1 \\ 0 & -4 & -3 \\ 0 & 0 & 0 \end{bmatrix} \begin{matrix} (1/4)R_1 \\ -(1/4)R_2 \\ -(1/4)R_3 \end{matrix}$

$\begin{bmatrix} 1 & 0 & 1/4 \\ 0 & 1 & 3/4 \\ 0 & 0 & 0 \end{bmatrix}$

53

© 2011 Cengage Learning. All Rights Reserved. May not be scanned, copied or duplicated, or posted to a publicly accessible website, in whole or in part.

Solution: $x = 1/4$, $y = 3/4$, or $(1/4. 3/4)$

11. $x + y = 0$
$3x - y = 1$
$x - y = -1$

$$\begin{bmatrix} 1 & 1 & 0 \\ 3 & -1 & 1 \\ 1 & -1 & -1 \end{bmatrix} \begin{matrix} \\ R_2 - 3R_1 \\ R_3 - R_1 \end{matrix}$$

$$\begin{bmatrix} 1 & 1 & 0 \\ 0 & -4 & 1 \\ 0 & -2 & -1 \end{bmatrix} \begin{matrix} 4R_1 + R_2 \\ \\ 2R_3 - R_2 \end{matrix}$$

$$\begin{bmatrix} 4 & 0 & 1 \\ 0 & -4 & 1 \\ 0 & 0 & -3 \end{bmatrix}$$

Since the bottom row translates to the false statement $0 = -3$, there is no solution.

13. $0.5x + 0.1y = 1.7$
$0.1x - 0.1y = 0.3$
$x + y = \dfrac{11}{3}$

$$\begin{bmatrix} 0.5 & 0.1 & 1.7 \\ 0.1 & -0.1 & 0.3 \\ 1 & 1 & 11/3 \end{bmatrix} \begin{matrix} 10R_1 \\ 10R_2 \\ 3R_3 \end{matrix}$$

$$\begin{bmatrix} 5 & 1 & 17 \\ 1 & -1 & 3 \\ 3 & 3 & 11 \end{bmatrix} \begin{matrix} \\ 5R_2 - R_1 \\ 5R_3 - 3R_1 \end{matrix}$$

$$\begin{bmatrix} 5 & 1 & 17 \\ 0 & -6 & -2 \\ 0 & 12 & 4 \end{bmatrix} \begin{matrix} \\ (1/2)R_2 \\ (1/4)R_3 \end{matrix} \rightarrow$$

$$\begin{bmatrix} 5 & 1 & 17 \\ 0 & -3 & -1 \\ 0 & 3 & 1 \end{bmatrix} \begin{matrix} 3R_1 + R_2 \\ \\ R_3 + R_2 \end{matrix} \rightarrow$$

$$\begin{bmatrix} 15 & 0 & 50 \\ 0 & -3 & -1 \\ 0 & 0 & 0 \end{bmatrix} \begin{matrix} (1/15)R_1 \\ -(1/3)R_2 \\ \end{matrix} \rightarrow$$

$$\begin{bmatrix} 1 & 0 & 10/3 \\ 0 & 1 & 1/3 \\ 0 & 0 & 0 \end{bmatrix}$$

Solution: $x = 10/3$, $y = 1/3$

15. $-x + 2y - z = 0$
$-x - y + 2z = 0$
$2x - z = 4$

$$\begin{bmatrix} -1 & 2 & -1 & 0 \\ -1 & -1 & 2 & 0 \\ 2 & 0 & -1 & 4 \end{bmatrix} \begin{matrix} \\ R_2 - R_1 \\ R_3 + 2R_1 \end{matrix}$$

$$\begin{bmatrix} -1 & 2 & -1 & 0 \\ 0 & -3 & 3 & 0 \\ 0 & 4 & -3 & 4 \end{bmatrix} \begin{matrix} \\ (1/3)R_2 \\ \end{matrix}$$

$$\begin{bmatrix} -1 & 2 & -1 & 0 \\ 0 & -1 & 1 & 0 \\ 0 & 4 & -3 & 4 \end{bmatrix} \begin{matrix} R_1 + 2R_2 \\ \\ R_3 + 4R_2 \end{matrix}$$

$$\begin{bmatrix} -1 & 0 & 1 & 0 \\ 0 & -1 & 1 & 0 \\ 0 & 0 & 1 & 4 \end{bmatrix} \begin{matrix} R_1 - R_3 \\ R_2 - R_3 \\ \end{matrix}$$

$$\begin{bmatrix} -1 & 0 & 0 & -4 \\ 0 & -1 & 0 & -4 \\ 0 & 0 & 1 & 4 \end{bmatrix} \begin{matrix} -R_1 \\ -R_2 \\ \end{matrix}$$

$$\begin{bmatrix} 1 & 0 & 0 & 4 \\ 0 & 1 & 0 & 4 \\ 0 & 0 & 1 & 4 \end{bmatrix}$$

Solution: $x = 4$, $y = 4$, $z = 4$

17. $x + y + 6z = -1$
$\dfrac{1}{3}x - \dfrac{1}{3}y + \dfrac{2}{3}z = 1$
$\dfrac{1}{2}x \phantom{- \frac{1}{3}y} + z = 0$

$$\begin{bmatrix} 1 & 1 & 6 & -1 \\ 1/3 & -1/3 & 2/3 & 1 \\ 1/2 & 0 & 1 & 0 \end{bmatrix} \begin{matrix} \\ 3R_2 \\ 2R_3 \end{matrix}$$

$$\begin{bmatrix} 1 & 1 & 6 & -1 \\ 0 & -2 & -4 & 4 \\ 0 & -1 & -4 & 1 \end{bmatrix} \begin{matrix} \\ (1/2)R_2 \\ \end{matrix}$$

$$\begin{bmatrix} 1 & 1 & 6 & -1 \\ 0 & -1 & -2 & 2 \\ 0 & -1 & -4 & 1 \end{bmatrix} \begin{matrix} R_1 + R_2 \\ \\ R_3 - R_2 \end{matrix}$$

$$\begin{bmatrix} 1 & 0 & 4 & 1 \\ 0 & -1 & -2 & 2 \\ 0 & 0 & -2 & -1 \end{bmatrix} \begin{matrix} R_1 + 2R_3 \\ R_2 - R_3 \\ \end{matrix}$$

$$\begin{bmatrix} 1 & 0 & 0 & -1 \\ 0 & -1 & 0 & 3 \\ 0 & 0 & -2 & -1 \end{bmatrix} \begin{matrix} \\ -R_2 \\ -(1/2)R_3 \end{matrix}$$

$$\begin{bmatrix} 1 & 0 & 0 & -1 \\ 0 & 1 & 0 & -3 \\ 0 & 0 & 1 & 1/2 \end{bmatrix}$$

Solution: $x = -1$, $y = -3$, $z = 1/2$

19. $-\frac{1}{2}x + y - \frac{1}{2}z = 0$

$-\frac{1}{2}x - \frac{1}{2}y + z = 0$

$x - \frac{1}{2}y - \frac{1}{2}z = 0$

$$\begin{bmatrix} -1/2 & 1 & -1/2 & 0 \\ -1/2 & -1/2 & 1 & 0 \\ 1 & -1/2 & -1/2 & 0 \end{bmatrix} \begin{matrix} 2R_1 \\ 2R_2 \\ 2R_3 \end{matrix}$$

$$\begin{bmatrix} -1 & 2 & -1 & 0 \\ -1 & -1 & 2 & 0 \\ 2 & -1 & -1 & 0 \end{bmatrix} \begin{matrix} \\ R_2 - R_1 \\ R_3 + 2R_1 \end{matrix}$$

$$\begin{bmatrix} -1 & 2 & -1 & 0 \\ 0 & -3 & 3 & 0 \\ 0 & 3 & -3 & 0 \end{bmatrix} \begin{matrix} \\ (1/3)R_2 \\ (1/3)R_3 \end{matrix}$$

$$\begin{bmatrix} -1 & 2 & -1 & 0 \\ 0 & -1 & 1 & 0 \\ 0 & 1 & -1 & 0 \end{bmatrix} \begin{matrix} R_1 + 2R_2 \\ \\ R_3 + R_2 \end{matrix}$$

$$\begin{bmatrix} -1 & 0 & 1 & 0 \\ 0 & -1 & 1 & 0 \\ 0 & 0 & 0 & 0 \end{bmatrix} \begin{matrix} -R_1 \\ -R_2 \\ \end{matrix}$$

$$\begin{bmatrix} 1 & 0 & -1 & 0 \\ 0 & 1 & -1 & 0 \\ 0 & 0 & 0 & 0 \end{bmatrix}$$

Translating back to equations gives
  $x - z = 0$
  $y - z = 0$
Thus, the general solution is
  $x = z$, $y = z$, z arbitrary,
or (z, z, z), z arbitrary

21. $x + y + 2z = -1$

$2x + 2y + 2z = 2$

$\frac{3}{5}x + \frac{3}{5}y + \frac{3}{5}z = \frac{2}{5}$

$$\begin{bmatrix} 1 & 1 & 2 & -1 \\ 2 & 2 & 2 & 2 \\ 3/5 & 3/5 & 3/5 & 2/5 \end{bmatrix} \begin{matrix} \\ (1/2)R_2 \\ 5R_3 \end{matrix}$$

$$\begin{bmatrix} 1 & 1 & 2 & -1 \\ 1 & 1 & 1 & 1 \\ 3 & 3 & 3 & 2 \end{bmatrix} \begin{matrix} \\ R_2 - R_1 \\ R_3 - 3R_1 \end{matrix}$$

$$\begin{bmatrix} 1 & 1 & 2 & -1 \\ 0 & 0 & -1 & 2 \\ 0 & 0 & -3 & 5 \end{bmatrix} \begin{matrix} R_1 + 2R_2 \\ \\ R_3 - 3R_2 \end{matrix}$$

$$\begin{bmatrix} 1 & 1 & 0 & 3 \\ 0 & 0 & -1 & 2 \\ 0 & 0 & 0 & -1 \end{bmatrix}$$

Since the bottom row translates to the false statement $0 = -1$, there is no solution.

23. $-0.5x + 0.5y + 0.5z = 1.5$

$4.2x + 2.1y + 2.1z = 0$

$0.2x \qquad + 0.2z = 0$

$$\begin{bmatrix} -.5 & .5 & .5 & 1.5 \\ 4.2 & 2.1 & 2.1 & 0 \\ .2 & 0 & .2 & 0 \end{bmatrix} \begin{matrix} 2R_1 \\ 10R_2 \\ 5R_3 \end{matrix}$$

$$\begin{bmatrix} -1 & 1 & 1 & 3 \\ 42 & 21 & 21 & 0 \\ 1 & 0 & 1 & 0 \end{bmatrix} (1/21)R_2$$

$$\begin{bmatrix} -1 & 1 & 1 & 3 \\ 2 & 1 & 1 & 0 \\ 1 & 0 & 1 & 0 \end{bmatrix} \begin{matrix} \\ R_2 + 2R_1 \\ R_3 + R_1 \end{matrix}$$

$$\begin{bmatrix} -1 & 1 & 1 & 3 \\ 0 & 3 & 3 & 6 \\ 0 & 1 & 2 & 3 \end{bmatrix} (1/3)R_2$$

$$\begin{bmatrix} -1 & 1 & 1 & 3 \\ 0 & 1 & 1 & 2 \\ 0 & 1 & 2 & 3 \end{bmatrix} \begin{matrix} R_1 - R_2 \\ \\ R_3 - R_2 \end{matrix}$$

$$\begin{bmatrix} -1 & 0 & 0 & 1 \\ 0 & 1 & 1 & 2 \\ 0 & 0 & 1 & 1 \end{bmatrix} \begin{matrix} \\ R_2 - R_3 \\ \end{matrix}$$

$$\begin{bmatrix} -1 & 0 & 0 & 1 \\ 0 & 1 & 0 & 1 \\ 0 & 0 & 1 & 1 \end{bmatrix} -R_1$$

$$\begin{bmatrix} 1 & 0 & 0 & -1 \\ 0 & 1 & 0 & 1 \\ 0 & 0 & 1 & 1 \end{bmatrix}$$

Solution: $x = -1$, $y = 1$, $z = 1$

25. $2x - y + z = 4$
  $3x - y + z = 5$

$$\begin{bmatrix} 2 & -1 & 1 & 4 \\ 3 & -1 & 1 & 5 \end{bmatrix} 2R_2 - 3R_1$$

$$\begin{bmatrix} 2 & -1 & 1 & 4 \\ 0 & 1 & -1 & -2 \end{bmatrix} R_1 + R_2$$

$$\begin{bmatrix} 2 & 0 & 0 & 2 \\ 0 & 1 & -1 & -2 \end{bmatrix} (1/2)R_1$$

$$\begin{bmatrix} 1 & 0 & 0 & 1 \\ 0 & 1 & -1 & -2 \end{bmatrix}$$

Translating back to equations gives
  $x = 1$

$y - z = -2$
Thus, the general solution is
  $x = 1$
  $y = z - 2$
  z arbitrary,
or
  $(1, z-2, z)$. z arbitrary

27. $0.75x - 0.75y - z = 4$
  $x - y + 4z = 0$

$$\begin{bmatrix} 0.75 & -.075 & -1 & 4 \\ 1 & -1 & 4 & 0 \end{bmatrix} 4R_1$$

$$\begin{bmatrix} 3 & -3 & -4 & 16 \\ 1 & -1 & 4 & 0 \end{bmatrix} 3R_2 - R_1$$

$$\begin{bmatrix} 3 & -3 & -4 & 16 \\ 0 & 0 & 16 & -16 \end{bmatrix} (1/16)R_2$$

$$\begin{bmatrix} 3 & -3 & -4 & 16 \\ 0 & 0 & 1 & -1 \end{bmatrix} R_1 + 4R_2$$

$$\begin{bmatrix} 3 & -3 & 0 & 12 \\ 0 & 0 & 1 & -1 \end{bmatrix} (1/3)R_1$$

$$\begin{bmatrix} 1 & -1 & 0 & 4 \\ 0 & 0 & 1 & -1 \end{bmatrix}$$

Translating back to equations gives
  $x - y = 4$
  $z = -1$
Thus, the general solution is
  $x = y + 4$
  y arbitrary
  $z = -1$,
or $(y + 4, y, -1)$; y arbitrary

29. $3x + y - z = 12$

$$[\ 3 \quad 1 \quad -1 \quad 12\ ] (1/3)R_1$$

$$[\ 1 \quad 1/3 \quad -1/3 \quad 4\ ]$$

Translating back to equations gives
$x + y/3 - z/3 = 4$
General Solution:
$x = 4 - y/3 + z/3$
y arbitrary
z arbitrary,
or $(4 - y/3 + z/3, y, z)$; y, z arbitrary

56

31. $x + y + 2z = -1$
  $2x + 2y + 2z = 2$
  $0.75x + 0.75y + z = 0.25$
  $-x - 2z = 21$

$$\begin{bmatrix} 1 & 1 & 2 & -1 \\ 2 & 2 & 2 & 2 \\ .75 & .75 & 1 & .25 \\ -1 & 0 & -2 & 21 \end{bmatrix} \begin{matrix} \\ (1/2)R_2 \\ 4R_3 \\ \\ \end{matrix}$$

$$\begin{bmatrix} 1 & 1 & 2 & -1 \\ 1 & 1 & 1 & 1 \\ 3 & 3 & 4 & 1 \\ -1 & 0 & -2 & 21 \end{bmatrix} \begin{matrix} \\ R_2 - R_1 \\ R_3 - 3R_1 \\ R_4 + R_1 \end{matrix}$$

$$\begin{bmatrix} 1 & 1 & 2 & -1 \\ 0 & 0 & -1 & 2 \\ 0 & 0 & -2 & 4 \\ 0 & 1 & 0 & 20 \end{bmatrix} \begin{matrix} \\ \\ (1/2)R_3 \\ \\ \end{matrix}$$

$$\begin{bmatrix} 1 & 1 & 2 & -1 \\ 0 & 0 & -1 & 2 \\ 0 & 0 & -1 & 2 \\ 0 & 1 & 0 & 20 \end{bmatrix} \begin{matrix} R_1 + 2R_2 \\ \\ R_3 - R_2 \\ \\ \end{matrix}$$

$$\begin{bmatrix} 1 & 1 & 0 & 3 \\ 0 & 0 & -1 & 2 \\ 0 & 0 & 0 & 0 \\ 0 & 1 & 0 & 20 \end{bmatrix} \begin{matrix} R_1 - R_4 \\ \\ \\ \\ \end{matrix}$$

$$\begin{bmatrix} 1 & 0 & 0 & -17 \\ 0 & 0 & -1 & 2 \\ 0 & 0 & 0 & 0 \\ 0 & 1 & 0 & 20 \end{bmatrix} \begin{matrix} \\ -R_2 \\ \\ \\ \end{matrix}$$

$$\begin{bmatrix} 1 & 0 & 0 & -17 \\ 0 & 0 & 1 & -2 \\ 0 & 0 & 0 & 0 \\ 0 & 1 & 0 & 20 \end{bmatrix} \textit{Rearrange Rows}$$

$$\begin{bmatrix} 1 & 0 & 0 & -17 \\ 0 & 1 & 0 & 20 \\ 0 & 0 & 1 & -2 \\ 0 & 0 & 0 & 0 \end{bmatrix}$$

Solution: $x = -17$, $y = 20$, $z = -2$

33. $x + y + 5z = 1$
  $y + 2z + w = 1$
  $x + 3y + 7z + 2w = 2$
  $x + y + 5z + w = 1$

$$\begin{bmatrix} 1 & 1 & 5 & 0 & 1 \\ 0 & 1 & 2 & 1 & 1 \\ 1 & 3 & 7 & 2 & 2 \\ 1 & 1 & 5 & 1 & 1 \end{bmatrix} \begin{matrix} \\ \\ R_3 - R_1 \\ R_4 - R_1 \end{matrix}$$

$$\begin{bmatrix} 1 & 1 & 5 & 0 & 1 \\ 0 & 1 & 2 & 1 & 1 \\ 0 & 2 & 2 & 2 & 1 \\ 0 & 0 & 0 & 1 & 0 \end{bmatrix} \begin{matrix} R_1 - R_2 \\ \\ R_3 - 2R_2 \\ \\ \end{matrix}$$

$$\begin{bmatrix} 1 & 0 & 3 & -1 & 0 \\ 0 & 1 & 2 & 1 & 1 \\ 0 & 0 & -2 & 0 & -1 \\ 0 & 0 & 0 & 1 & 0 \end{bmatrix} \begin{matrix} 2R_1 + 3R_3 \\ R_2 + R_3 \\ \\ \\ \end{matrix}$$

$$\begin{bmatrix} 2 & 0 & 0 & -2 & -3 \\ 0 & 1 & 0 & 1 & 0 \\ 0 & 0 & -2 & 0 & -1 \\ 0 & 0 & 0 & 1 & 0 \end{bmatrix} \begin{matrix} R_1 + 2R_4 \\ R_2 - R_4 \\ \\ \\ \end{matrix}$$

$$\begin{bmatrix} 2 & 0 & 0 & 0 & -3 \\ 0 & 1 & 0 & 0 & 0 \\ 0 & 0 & -2 & 0 & -1 \\ 0 & 0 & 0 & 1 & 0 \end{bmatrix} \begin{matrix} (1/2)R_1 \\ \\ -(1/2)R_3 \\ \\ \end{matrix}$$

$$\begin{bmatrix} 1 & 0 & 0 & 0 & -3/2 \\ 0 & 1 & 0 & 0 & 0 \\ 0 & 0 & 1 & 0 & 1/2 \\ 0 & 0 & 0 & 1 & 0 \end{bmatrix}$$

Solution: $x = -3/2$, $y = 0$, $z = 1/2$, $w = 0$

35. $x + y + 5z = 1$
   $y + 2z + w = 1$
   $x + y + 5z + w = 1$
   $x + 2y + 7z + 2w = 2$

$$\begin{bmatrix} 1 & 1 & 5 & 0 & 1 \\ 0 & 1 & 2 & 1 & 1 \\ 1 & 1 & 5 & 1 & 1 \\ 1 & 2 & 7 & 2 & 2 \end{bmatrix} \begin{matrix} \\ \\ R_3 - R_1 \\ R_4 - R_1 \end{matrix}$$

$$\begin{bmatrix} 1 & 1 & 5 & 0 & 1 \\ 0 & 1 & 2 & 1 & 1 \\ 0 & 0 & 0 & 1 & 0 \\ 0 & 1 & 2 & 2 & 1 \end{bmatrix} \begin{matrix} R_1 - R_2 \\ \\ \\ R_4 - R_2 \end{matrix}$$

$$\begin{bmatrix} 1 & 0 & 3 & -1 & 0 \\ 0 & 1 & 2 & 1 & 1 \\ 0 & 0 & 0 & 1 & 0 \\ 0 & 0 & 0 & 1 & 0 \end{bmatrix} \begin{matrix} R_1 + R_3 \\ R_2 - R_3 \\ \\ R_4 - R_3 \end{matrix}$$

$$\begin{bmatrix} 1 & 0 & 3 & 0 & 0 \\ 0 & 1 & 2 & 0 & 1 \\ 0 & 0 & 0 & 1 & 0 \\ 0 & 0 & 0 & 0 & 0 \end{bmatrix}$$

Translating back to equations gives
$x + 3z = 0$
$y + 2z = 1$
$w = 0$
General solution:
$x = -3z$
$y = 1 - 2z$
z arbitrary
$w = 0$,
or $(-3z, 1-2z. z. 0)$; z arbitrary

37. $x - 2y + z - 4w = 1$
   $x + 3y + 7z + 2w = 2$

$2x + y + 8z - 2w = 3$

$$\begin{bmatrix} 1 & -2 & 1 & -4 & 1 \\ 1 & 3 & 7 & 2 & 2 \\ 2 & 1 & 8 & -2 & 3 \end{bmatrix} \begin{matrix} \\ R_2 - R_1 \\ R_3 - 2R_1 \end{matrix}$$

$$\begin{bmatrix} 1 & -2 & 1 & -4 & 1 \\ 0 & 5 & 6 & 6 & 1 \\ 0 & 5 & 6 & 6 & 1 \end{bmatrix} \begin{matrix} 5R_1 + 2R_2 \\ \\ R_3 - R_2 \end{matrix}$$

$$\begin{bmatrix} 5 & 0 & 17 & -8 & 7 \\ 0 & 5 & 6 & 6 & 1 \\ 0 & 0 & 0 & 0 & 0 \end{bmatrix} \begin{matrix} (1/5)R_1 \\ (1/5)R_2 \\ \end{matrix}$$

$$\begin{bmatrix} 1 & 0 & 17/5 & -8/5 & 7/5 \\ 0 & 1 & 6/5 & 6/5 & 1/5 \\ 0 & 0 & 0 & 0 & 0 \end{bmatrix}$$

$x + 17z/5 - 8w/5 = 7/5$
$y + 6z/5 + 6w/5 = 1/5$
General solution:
$x = 7/5 - 17z/5 + 8w/5$
$y = 1/5 - 6z/5 - 6w/5$
z, w arbitrary,
or $(7/5 - 17z/5 + 8w/5, \ 1/5 - 6z/5 - 6w/5, \ z, \ w)$
z, w arbitrary
or $(\frac{1}{5}(7 - 17z + 8w), \frac{1}{5}(1 - 6z - 6w), z, w)$, z, w
arbitrary

39. $x + y + z + u + v = 15$
   $y - z + u - v = -2$
   $z + u + v = 12$
   $u - v = -1$
   $v = 5$

$$\begin{bmatrix} 1 & 1 & 1 & 1 & 1 & 15 \\ 0 & 1 & -1 & 1 & -1 & -2 \\ 0 & 0 & 1 & 1 & 1 & 12 \\ 0 & 0 & 0 & 1 & -1 & -1 \\ 0 & 0 & 0 & 0 & 1 & 5 \end{bmatrix} \begin{matrix} R_1 - R_2 \\ \\ \\ \\ \\ \end{matrix}$$

58

$$\begin{bmatrix} 1 & 0 & 2 & 0 & 2 & 17 \\ 0 & 1 & -1 & 1 & -1 & -2 \\ 0 & 0 & 1 & 1 & 1 & 12 \\ 0 & 0 & 0 & 1 & -1 & -1 \\ 0 & 0 & 0 & 0 & 1 & 5 \end{bmatrix} \begin{matrix} R_1 - 2R_3 \\ R_2 + R_3 \\ \\ \\ \\ \end{matrix}$$

$$\begin{bmatrix} 1 & 0 & 0 & -2 & 0 & -7 \\ 0 & 1 & 0 & 2 & 0 & 10 \\ 0 & 0 & 1 & 1 & 1 & 12 \\ 0 & 0 & 0 & 1 & -1 & -1 \\ 0 & 0 & 0 & 0 & 1 & 5 \end{bmatrix} \begin{matrix} R_1 + 2R_4 \\ R_2 - 2R_4 \\ R_3 - R_4 \\ \\ \\ \end{matrix}$$

$$\begin{bmatrix} 1 & 0 & 0 & 0 & -2 & -9 \\ 0 & 1 & 0 & 0 & 2 & 12 \\ 0 & 0 & 1 & 0 & 2 & 13 \\ 0 & 0 & 0 & 1 & -1 & -1 \\ 0 & 0 & 0 & 0 & 1 & 5 \end{bmatrix} \begin{matrix} R_1 + 2R_5 \\ R_2 - 2R_5 \\ R_3 - 2R_5 \\ R_4 + R_5 \\ \\ \end{matrix}$$

$$\begin{bmatrix} 1 & 0 & 0 & 0 & 0 & 1 \\ 0 & 1 & 0 & 0 & 0 & 2 \\ 0 & 0 & 1 & 0 & 0 & 3 \\ 0 & 0 & 0 & 1 & 0 & 4 \\ 0 & 0 & 0 & 0 & 1 & 5 \end{bmatrix}$$

Solution: $x = 1, y = 2, z = 3, u = 4, v = 5$

41. $x - y + z - u + v = 0$
$\quad\quad y - z + u - v = -2$
$\quad x \quad\quad\quad\quad\quad - 2v = -2$
$\quad 2x - y + z - u - 3v = -2$
$\quad 4x - y + z - u - 7v = -6$

$$\begin{bmatrix} 1 & -1 & 1 & -1 & 1 & 0 \\ 0 & 1 & -1 & 1 & -1 & -2 \\ 1 & 0 & 0 & 0 & -2 & -2 \\ 2 & -1 & 1 & -1 & -3 & -2 \\ 4 & -1 & 1 & -1 & -7 & -6 \end{bmatrix} \begin{matrix} \\ \\ R_3 - R_1 \\ R_4 - 2R_1 \\ R_5 - 4R_1 \end{matrix}$$

$$\begin{bmatrix} 1 & -1 & 1 & -1 & 1 & 0 \\ 0 & 1 & -1 & 1 & -1 & -2 \\ 0 & 1 & -1 & 1 & -3 & -2 \\ 0 & 1 & -1 & 1 & -5 & -2 \\ 0 & 3 & -3 & 3 & -11 & -6 \end{bmatrix} \begin{matrix} R_1 + R_2 \\ \\ R_3 - R_2 \\ R_4 - R_2 \\ R_5 - 3R_2 \end{matrix}$$

$$\begin{bmatrix} 1 & 0 & 0 & 0 & 0 & -2 \\ 0 & 1 & -1 & 1 & -1 & -2 \\ 0 & 0 & 0 & 0 & -2 & 0 \\ 0 & 0 & 0 & 0 & -4 & 0 \\ 0 & 0 & 0 & 0 & -8 & 0 \end{bmatrix} \begin{matrix} \\ \\ (1/2)R_3 \\ (1/4)R_4 \\ (1/8)R_5 \end{matrix}$$

$$\begin{bmatrix} 1 & 0 & 0 & 0 & 0 & -2 \\ 0 & 1 & -1 & 1 & -1 & -2 \\ 0 & 0 & 0 & 0 & -1 & 0 \\ 0 & 0 & 0 & 0 & -1 & 0 \\ 0 & 0 & 0 & 0 & -1 & 0 \end{bmatrix} \begin{matrix} \\ R_2 - R_3 \\ \\ R_4 - R_3 \\ R_5 - R_3 \end{matrix}$$

$$\begin{bmatrix} 1 & 0 & 0 & 0 & 0 & -2 \\ 0 & 1 & -1 & 1 & 0 & -2 \\ 0 & 0 & 0 & 0 & -1 & 0 \\ 0 & 0 & 0 & 0 & 0 & 0 \\ 0 & 0 & 0 & 0 & 0 & 0 \end{bmatrix} \begin{matrix} \\ \\ -R_3 \\ \\ \end{matrix}$$

$$\begin{bmatrix} 1 & 0 & 0 & 0 & 0 & -2 \\ 0 & 1 & -1 & 1 & 0 & -2 \\ 0 & 0 & 0 & 0 & 1 & 0 \\ 0 & 0 & 0 & 0 & 0 & 0 \\ 0 & 0 & 0 & 0 & 0 & 0 \end{bmatrix}$$

$x = -2$
$y - z + u = -2$
$v = 0$

General solution:
$x = -2$
$y = -2 + z - u$
z is arbitrary
u is arbitrary
$v = 0$
or
$(-2, -2+z-u. z. u. 0)$; z, u arbitrary

43. $x + 2y - z + w = 30$
$\quad 2x \quad - z + 2w = 30$
$\quad x + 3y + 3z - 4w = 2$
$\quad 2x - 9y \quad + w = 4$

Using technology:

Matrix #1

x	y	z	w	
1	2	-1	1	30
2	0	-1	2	30
1	3	3	-4	2
2	-9	0	1	4

Matrix #2

x	y	z	w	
1	2	-1	1	30
0	-4	1	0	-30
0	1	4	-5	-28
0	-13	2	-1	-56

Matrix #3

x	y	z	w	
2	0	-1	2	30
0	-4	1	0	-30
0	0	17	-20	-142
0	0	-5	-4	166

Matrix #4

x	y	z	w	
34	0	0	14	368
0	-68	0	20	-368
0	0	17	-20	-142
0	0	0	-168	2112

Matrix #5

x	y	z	w	
17	0	0	7	184
0	-17	0	5	-92
0	0	17	-20	-142
0	0	0	-7	88

Matrix #6

x	y	z	w	
17	0	0	0	272
0	-119	0	0	-204
0	0	119	0	-2754
0	0	0	-7	88

Matrix #7

x	y	z	w	
1	0	0	0	16
0	1	0	0	12/7
0	0	1	0	-162/7
0	0	0	1	-88/7

Solution: $(16, 12/7, -162/7, -88/7)$

45. $x + 2y + 3z + 4w + 5t = 6$
$\quad 2x + 3y + 4z + 5w + t = 5$
$\quad 3x + 4y + 5z + w + 2t = 4$
$\quad 4x + 5y + z + 2w + 3t = 3$
$\quad 5x + y + 2z + 3w + 4t = 2$

Using technology:

Matrix #1

x	y	z	w	t	
1	2	3	4	5	6
2	3	4	5	1	5
3	4	5	1	2	4
4	5	1	2	3	3
5	1	2	3	4	2

Matrix #2

x	y	z	w	t	
1	2	3	4	5	6
0	-1	-2	-3	-9	-7
0	-2	-4	-11	-13	-14
0	-3	-11	-14	-17	-21
0	-9	-13	-17	-21	-28

Matrix #3

x	y	z	w	t	
1	0	-1	-2	-13	-8
0	-1	-2	-3	-9	-7
0	0	0	-5	5	0
0	0	-5	-5	10	0
0	0	5	10	60	35

Matrix #4

x	y	z	w	t	
1	0	-1	-2	-13	-8
0	-1	-2	-3	-9	-7
0	0	0	-1	1	0
0	0	-1	-1	2	0
0	0	1	2	12	7

Matrix #5

x	y	z	w	t	
1	0	-1	0	-15	-8
0	-1	-2	0	-12	-7
0	0	0	-1	1	0
0	0	-1	0	1	0
0	0	1	0	14	7

Matrix #6

x	y	z	w	t	
1	0	0	0	-16	-8
0	-1	0	0	-14	-7
0	0	0	-1	1	0
0	0	-1	0	1	0
0	0	0	0	15	7

Matrix #7

x	y	z	w	t	
15	0	0	0	0	-8
0	-15	0	0	0	-7
0	0	0	-15	0	-7
0	0	-15	0	0	-7
0	0	0	0	15	7

Matrix #8

x	y	z	w	t	
1	0	0	0	0	-8/15
0	1	0	0	0	7/15
0	0	0	1	0	7/15
0	0	1	0	0	7/15
0	0	0	0	1	7/15

Solution: $(-8/15, 7/15, 7/15, 7/15, 7/15)$

47. $1.6x + 2.4y - 3.2z = 4.4$
$5.1x - 6.3y + 0.6z = -3.2$
$4.2x + 3.5y + 4.9z = 10.1$

We use the Excel Matrix Pivot Tool (on the web site):

x	y	z	
1.6	2.4	-3.2	4.4
5.1	-6.3	0.6	-3.2
4.2	3.5	4.9	10.1

x	y	z	
1	1.5	-2	2.75
0	-13.95	10.8	-17.225
0	-2.8	13.3	-1.45

x	y	z	
1	0	-0.8387097	0.89784946
0	1	-0.7741935	1.23476703
0	0	11.1322581	2.00734767

x	y	z	
1	0	0	1.049084
0	1	0	1.37436814
0	0	1	0.1803181

Solution (rounded to 1 decimal place):
$(1.0, 1.4, 0.2)$

49. $-0.2x + 0.3y + 0.4z - t = 4.5$
$2.2x + 1.1y - 4.7z + 2t = 8.3$
$9.2y - 1.3t = 0$
$3.4x + 0.5z - 3.4t = 0.1$

We use the Excel Matrix Pivot Tool (on the web site):

x	y	z	t	
-0.2	0.3	0.4	-1	4.5
2.2	1.1	-4.7	2	8.3
0	9.2	0	-1.3	0
3.4	0	0.5	-3.4	0.1

x	y	z	t	
1	-1.5	-2	5	-22.5
0	4.4	-0.3	-9	57.8
0	9.2	0	-1.3	0
0	5.1	7.3	-20.4	76.6

x	y	z	t	
1	0	-2.1022727	1.93181818	-2.7954545
0	1	-0.0681818	-2.0454545	13.1363636
0	0	0.62727273	17.5181818	-120.85455
0	0	7.64772727	-9.9681818	9.60454545

x	y	z	t	
1	0	0	60.6431159	-407.83333
0	1	0	-0.1413043	1.7764E-15
0	0	1	27.9275362	-192.66667
0	0	0	-223.55036	1483.06667

x	y	z	t	
1	0	0	0	-5.5177974
0	1	0	0	-0.9374343
0	0	1	0	-7.3911984
0	0	0	1	-6.6341501

Solution (rounded to 1 decimal place):
$(-5.5, -0.9, -7.4, -6.6)$

51. A pivot is an entry in a matrix that is selected to "clear a column;" that is, use the row operations of a certain type to obtain zeros everywhere above and below it. "Pivoting" is the procedure of clearing a column using a designated pivot.

53. $2R_1 + 5R_4$, or $6R_1 + 15R_4$ (which is less desirable, since it will produce a row in which every entry is divisible by 3)

55. It will include a row of zeros. (Subtracting the two rows produces a row of zeros.)

57. The claim is wrong. If there are more equations than unknowns, there can be a unique solution as well as row(s) of zeros in the reduced matrix, as in Example 6.

59. Since there are 5 columns, there are 4 unknowns. (The last column is for the answers.) Since there are 5 rows of which 3 are zero, that leaves 2 rows with pivots. Thus, there are 2 unknowns that are *not* parameters. The remaining 2 unknowns are arbitrary (parameters).

61. The number of pivots must equal the number of variables, since no variable will be used as a parameter.

63. A simple example is:
$x = 1; y - z = 1; x + y - z = 2$

2.3 Applications of Systems of Linear Equations

1. Unknowns:

x = the number of batches of vanilla
y = the number of batches of mocha
z = the number of batches of strawberry

Arrange the given information in a table with unknowns across the top:

	Vanilla (x)	Mocha (y)	Strawberry (z)	Avail.
Eggs	2	1	1	350
Milk	1	1	2	350
Cream	2	2	1	400

We can now set up an equation for each of the items listed on the left:

Eggs: $2x + y + z = 350$
Milk: $x + y + 2z = 350$
Cream: $2x + 2y + z = 400$

$$\begin{bmatrix} 2 & 1 & 1 & 350 \\ 1 & 1 & 2 & 350 \\ 2 & 2 & 1 & 400 \end{bmatrix} \begin{matrix} \\ 2R_2 - R_1 \\ R_3 - R_1 \end{matrix}$$

$$\begin{bmatrix} 2 & 1 & 1 & 350 \\ 0 & 1 & 3 & 350 \\ 0 & 1 & 0 & 50 \end{bmatrix} \begin{matrix} R_1 - R_2 \\ \\ R_3 - R_2 \end{matrix}$$

$$\begin{bmatrix} 2 & 0 & -2 & 0 \\ 0 & 1 & 3 & 350 \\ 0 & 0 & -3 & -300 \end{bmatrix} \begin{matrix} (1/2)R_1 \\ \\ (1/3)R_3 \end{matrix}$$

$$\begin{bmatrix} 1 & 0 & -1 & 0 \\ 0 & 1 & 3 & 350 \\ 0 & 0 & -1 & -100 \end{bmatrix} \begin{matrix} R_1 - R_3 \\ R_2 + 3R_3 \\ \\ \end{matrix}$$

$$\begin{bmatrix} 1 & 0 & 0 & 100 \\ 0 & 1 & 0 & 50 \\ 0 & 0 & -1 & -100 \end{bmatrix} \begin{matrix} \\ \\ -R_3 \end{matrix}$$

$$\begin{bmatrix} 1 & 0 & 0 & 100 \\ 0 & 1 & 0 & 50 \\ 0 & 0 & 1 & 100 \end{bmatrix}$$

$x = 100, y = 50, z = 50$

Solution: Make 100 batches of vanilla, 50 batches of mocha, and 100 batches of strawberry

3. Unknowns:

x = the number of sections of Finite Math
y = the number of sections of Applied Calculus
z = the number of sections of Computer Methods.

We are given three pieces of information:

(1) There are a total of 6 sections
   $x + y + z = 6$
(2) The total number of students is 210:
   $40x + 40y + 10z = 210$
(3) The total revenue is $260,000:
   $40,000x + 60,000y + 20,000z = 260,000$
or, working in thousands of dollars,
   $40x + 60y + 20z = 260$

Thus, we have a system of 3 equations in 3 unknowns:

$x + y + z = 6$
$40x + 40y + 10z = 210$
$40x + 60y + 20z = 260$

$$\begin{bmatrix} 1 & 1 & 1 & 6 \\ 40 & 40 & 10 & 210 \\ 40 & 60 & 20 & 260 \end{bmatrix} \begin{matrix} \\ (1/10)R_2 \\ (1/20)R_3 \end{matrix}$$

$$\begin{bmatrix} 1 & 1 & 1 & 6 \\ 4 & 4 & 1 & 21 \\ 2 & 3 & 1 & 13 \end{bmatrix} \begin{matrix} \\ R_2 - 4R_1 \\ R_3 - 2R_1 \end{matrix}$$

$$\begin{bmatrix} 1 & 1 & 1 & 6 \\ 0 & 0 & -3 & -3 \\ 0 & 1 & -1 & 1 \end{bmatrix} \begin{matrix} \\ (1/3)R_2 \\ \\ \end{matrix}$$

$$\begin{bmatrix} 1 & 1 & 1 & 6 \\ 0 & 0 & -1 & -1 \\ 0 & 1 & -1 & 1 \end{bmatrix} \begin{matrix} R_1 + R_2 \\ \\ R_3 - R_2 \end{matrix}$$

$$\begin{bmatrix} 1 & 1 & 0 & 5 \\ 0 & 0 & -1 & -1 \\ 0 & 1 & 0 & 2 \end{bmatrix} \begin{matrix} R_1 - R_3 \\ \\ \\ \end{matrix}$$

$$\begin{bmatrix} 1 & 0 & 0 & 3 \\ 0 & 0 & -1 & -1 \\ 0 & 1 & 0 & 2 \end{bmatrix} \begin{matrix} \\ -R_2 \\ \\ \end{matrix}$$

$$\begin{bmatrix} 1 & 0 & 0 & 3 \\ 0 & 0 & 1 & 1 \\ 0 & 1 & 0 & 2 \end{bmatrix} R_2 \leftrightarrow R_3$$

$$\begin{bmatrix} 1 & 0 & 0 & 3 \\ 0 & 1 & 0 & 2 \\ 0 & 0 & 1 & 1 \end{bmatrix}$$

$x = 3, y = 2, z = 1$

Solution: Offer 3 sections of Finite Math, 2 sections of Applied Calculus and 1 section of Computer Methods

5. Unknowns:

x = revenue (in billions) earned from rock music
y = revenue (in billions) earned from religious music
z = revenue (in billions) earned from classical music

Total revenues were 4 billion:

$\quad x + y + z = 4$

Religious music brought in twice as much classical music. Reword this follows:
The revenue earned from religious music was twice the revenue earned from classical music:

$\quad y = 2z$, or
$\quad y - 2z = 0$

Rock music brought in $3 billion more than religious music. Reword this follows:
The revenue earned from rock music was $3 billion more than the revenue earned from religious music:

$\quad x = y + 3$, or
$\quad x - y = 3$

We thus solve the system

$\quad x+y+z = 4$
$\quad y-2z = 0$
$\quad x-y = 3$

$$\begin{bmatrix} 1 & 1 & 1 & 4 \\ 0 & 1 & -2 & 0 \\ 1 & -1 & 0 & 3 \end{bmatrix} R_3 - R_1$$

$$\begin{bmatrix} 1 & 1 & 1 & 4 \\ 0 & 1 & -2 & 0 \\ 0 & -2 & -1 & -1 \end{bmatrix} \begin{matrix} R_1 - R_2 \\ \\ R_3 + 2R_2 \end{matrix}$$

$$\begin{bmatrix} 1 & 0 & 3 & 4 \\ 0 & 1 & -2 & 0 \\ 0 & 0 & -5 & -1 \end{bmatrix} \begin{matrix} 5R_1 + 3R_3 \\ 5R_2 - 2R_3 \end{matrix}$$

$$\begin{bmatrix} 5 & 0 & 0 & 17 \\ 0 & 5 & 0 & 2 \\ 0 & 0 & -5 & -1 \end{bmatrix} \begin{matrix} (1/5)R_1 \\ (1/5)R_2 \\ -(1/5)R_3 \end{matrix}$$

$$\begin{bmatrix} 1 & 0 & 0 & 17/5 \\ 0 & 1 & 0 & 2/5 \\ 0 & 0 & 1 & 1/5 \end{bmatrix}$$

$x = 17/5 = 3.4, y = 2/5 = 0.4, z = 1/5 = 0.2$
Solution: Revenues were $3.4 billion for rock music, $0.4 billion for religious music, and $0.2 billion for classical music

7. Unknowns:

x = the number of Airbus A330-300s
y = the number of Boeing 767-200ERs
z = the number of Boeing Dreamliner 787-9s
Passengers: $320x + 250y + 275z = 4480$
Cost: $200x + 125y + 200z = 2900$
The number of Dreamliners is twice the number of Airbus : $z = 2x$, or

$\quad 2x - z = 0$
Solving:

$$\begin{bmatrix} 2 & 0 & -1 & 0 \\ 320 & 250 & 275 & 4480 \\ 200 & 125 & 200 & 2900 \end{bmatrix} \begin{matrix} (1/5)R_2 \\ (1/25)R_3 \end{matrix}$$

$$\begin{bmatrix} 2 & 0 & -1 & 0 \\ 64 & 50 & 55 & 896 \\ 8 & 5 & 8 & 116 \end{bmatrix} \begin{matrix} R_2 - 32R_1 \\ R_3 - 4R_1 \end{matrix}$$

$$\begin{bmatrix} 2 & 0 & -1 & 0 \\ 0 & 50 & 87 & 896 \\ 0 & 5 & 12 & 116 \end{bmatrix} 10R_3 - R_2$$

$$\begin{bmatrix} 2 & 0 & -1 & 0 \\ 0 & 50 & 87 & 896 \\ 0 & 0 & 33 & 264 \end{bmatrix} (1/33)R_3$$

$$\begin{bmatrix} 2 & 0 & -1 & 0 \\ 0 & 50 & 87 & 896 \\ 0 & 0 & 1 & 8 \end{bmatrix} \begin{matrix} R_1 + R_3 \\ R_2 - 87R_3 \end{matrix}$$

$$\begin{bmatrix} 2 & 0 & 0 & 8 \\ 0 & 50 & 0 & 200 \\ 0 & 0 & 1 & 8 \end{bmatrix} \begin{matrix} (1/2)R_1 \\ (1/50)R_2 \end{matrix}$$

$$\begin{bmatrix} 1 & 0 & 0 & 4 \\ 0 & 1 & 0 & 4 \\ 0 & 0 & 1 & 8 \end{bmatrix}$$

$x = 4, y = 4, z = 8$

Solution: Order 4 Airbus A330-300s, 4 Boeing 767-200ERs and 8 Dreamliners

9. Unknowns:

x = the number of tons from CCC

y = the number of tons from SSS

z = the number of tons from BBF

Total order of cheese is 100 tons:

$\quad x + y + z = 100$

Total cost = \$5990:

$\quad 80x + 50y + 65z = 5990$

Same amount from CCC and BBF:

$\quad x = z$, or

$\quad x - z = 0$

Solving:

$$\begin{bmatrix} 1 & 1 & 1 & 100 \\ 80 & 50 & 65 & 5990 \\ 1 & 0 & -1 & 0 \end{bmatrix} (1/5)R_2$$

$$\begin{bmatrix} 1 & 1 & 1 & 100 \\ 16 & 10 & 13 & 1198 \\ 1 & 0 & -1 & 0 \end{bmatrix} \begin{matrix} R_2 - 16R_1 \\ R_3 - R_1 \end{matrix}$$

$$\begin{bmatrix} 1 & 1 & 1 & 100 \\ 0 & -6 & -3 & -402 \\ 0 & -1 & -2 & -100 \end{bmatrix} (1/3)R_2$$

$$\begin{bmatrix} 1 & 1 & 1 & 100 \\ 0 & -2 & -1 & -134 \\ 0 & -1 & -2 & -100 \end{bmatrix} \begin{matrix} 2R_1 + R_2 \\ \\ 2R_3 - R_2 \end{matrix}$$

$$\begin{bmatrix} 2 & 0 & 1 & 66 \\ 0 & -2 & -1 & -134 \\ 0 & 0 & -3 & -66 \end{bmatrix} (1/3)R_3$$

$$\begin{bmatrix} 2 & 0 & 1 & 66 \\ 0 & -2 & -1 & -134 \\ 0 & 0 & -1 & -22 \end{bmatrix} \begin{matrix} R_1 + R_3 \\ R_2 - R_3 \end{matrix}$$

$$\begin{bmatrix} 2 & 0 & 0 & 44 \\ 0 & -2 & 0 & -112 \\ 0 & 0 & -1 & -22 \end{bmatrix} \begin{matrix} (1/2)R_1 \\ -(1/2)R_2 \\ -R_3 \end{matrix}$$

$$\begin{bmatrix} 1 & 0 & 0 & 22 \\ 0 & 1 & 0 & 56 \\ 0 & 0 & 1 & 22 \end{bmatrix}$$

$x = 22, y = 56, z = 22$

Solution: The store ordered 22 tons from Cheesy Cream, 56 tons from Super Smooth & Sons, and 22 tons from Bagel's Best Friend.

11. Unknowns:

x = the number of evil sorcerers slain

y = the number of trolls slain

z = the number of orcs slain

Total number slain was 560:

$\quad x + y + z = 560$

Total number of sword thrusts was 620:

$\quad 2x + 2y + z = 620$

The number of trolls slain is five times the number of evil sorcerers slain:

$\quad y = 5x$, or

$\quad -5x + y = 0$

Solving:

$$\begin{bmatrix} 1 & 1 & 1 & 560 \\ 2 & 2 & 1 & 620 \\ -5 & 1 & 0 & 0 \end{bmatrix} \begin{matrix} \\ R_2 - 2R_1 \\ R_3 + 5R_1 \end{matrix}$$

$$\begin{bmatrix} 1 & 1 & 1 & 560 \\ 0 & 0 & -1 & -500 \\ 0 & 6 & 5 & 2800 \end{bmatrix} \begin{matrix} R_1 + R_2 \\ \\ R_3 + 5R_2 \end{matrix}$$

$$\begin{bmatrix} 1 & 1 & 0 & 60 \\ 0 & 0 & -1 & -500 \\ 0 & 6 & 0 & 300 \end{bmatrix} \begin{matrix} \\ \\ (1/6)R_3 \end{matrix}$$

$$\begin{bmatrix} 1 & 1 & 0 & 60 \\ 0 & 0 & -1 & -500 \\ 0 & 1 & 0 & 50 \end{bmatrix} \begin{matrix} R_1 - R_3 \\ \\ \end{matrix}$$

$$\begin{bmatrix} 1 & 0 & 0 & 10 \\ 0 & 0 & -1 & -500 \\ 0 & 1 & 0 & 50 \end{bmatrix} \begin{matrix} \\ -R_2 \\ \end{matrix}$$

$$\begin{bmatrix} 1 & 0 & 0 & 10 \\ 0 & 0 & 1 & 500 \\ 0 & 1 & 0 & 50 \end{bmatrix} \begin{matrix} \\ R_2 \leftrightarrow R_3 \\ \end{matrix}$$

$$\begin{bmatrix} 1 & 0 & 0 & 10 \\ 0 & 1 & 0 & 50 \\ 0 & 0 & 1 & 500 \end{bmatrix}$$

$x = 10, y = 50, z = 500$

Solution: Conan has slain 10 evil sorcerers, 50 trolls and 500 orcs.

13. Unknowns:

x = amount of money donated to the MPBF

y = amount of money donated to the SCN

z = amount of money donated to the NY Jets

Given information:

(1) The society donated twice as much to the NY Jets as to the MPBF. Rephrase this as follows: The amount of money donated to the NY Jets was equal to twice the amount of money donated to the MPBF:

$z = 2x$, or $2x - z = 0$

(2) The society donated equal amounts to the first two funds:

$x = y$, or $x - y \ 0$

(3) Money donated back to the society:

$x + 2y + 2z = 4200$

Solving:

$$\begin{bmatrix} 2 & 0 & -1 & 0 \\ 1 & -1 & 0 & 0 \\ 1 & 2 & 2 & 4200 \end{bmatrix} \begin{matrix} \\ 2R_2 - R_1 \\ 2R_3 - R_1 \end{matrix}$$

$$\begin{bmatrix} 2 & 0 & -1 & 0 \\ 0 & -2 & 1 & 0 \\ 0 & 4 & 5 & 8400 \end{bmatrix} \begin{matrix} \\ \\ R_3 + 2R_2 \end{matrix}$$

$$\begin{bmatrix} 2 & 0 & -1 & 0 \\ 0 & -2 & 1 & 0 \\ 0 & 0 & 7 & 8400 \end{bmatrix} \begin{matrix} \\ \\ (1/7)R_3 \end{matrix}$$

$$\begin{bmatrix} 2 & 0 & -1 & 0 \\ 0 & -2 & 1 & 0 \\ 0 & 0 & 1 & 1200 \end{bmatrix} \begin{matrix} R_1 + R_3 \\ R_2 - R_3 \\ \end{matrix}$$

$$\begin{bmatrix} 2 & 0 & 0 & 1200 \\ 0 & -2 & 0 & -1200 \\ 0 & 0 & 1 & 1200 \end{bmatrix} \begin{matrix} (1/2)R_1 \\ -(1/2)R_2 \\ \end{matrix}$$

$$\begin{bmatrix} 1 & 0 & 0 & 600 \\ 0 & 1 & 0 & 600 \\ 0 & 0 & 1 & 1200 \end{bmatrix}$$

Solution: It donated $600 to each of the MPBF and the SCN, and $1200 to the Jets

15. Unknowns:

x = the number of empty seats on United

y = the number of empty seats on American

z = the number of empty seats on Southwest

We are given the following information:

(1) United, American and Southwest flew a total of 210 empty seats:

$x + y + z = 210$

(2) The total cost of these seats was $86,010. (note that the figures in the table are for one mile, but the trip was 3000 miles):

$(3000)0.148x + (3000)0.139y + (3000)0.107z = 86,010$

That is,

$444x + 417y + 321z = 86,010$

(3) United had three times as many empty seats as American

$x = 3y$, or $x - 3y = 0$

We use the Excel Matrix Pivot Tool (on the web site):

x1	x2	x3	x4
1	1	1	210
444	417	321	86010
1	-3	0	0

x1	x2	x3	x4
1	1	1	210
0	-27	-123	-7230
0	-4	-1	-210

x1	x2	x3	x4
1	0	-3.5555556	-57.777778
0	1	4.55555556	267.777778
0	0	17.2222222	861.111111

x1	x2	x3	x4
1	0	0	120
0	1	0	40
0	0	1	50

Solution: United: 120; American: 40; Southwest: 50

17. Unknowns:

x = amount invested in FCTFX

y = amount invested in FSAZX

z = amount invested in FUSFX

The total investment was $9000:

$x + y + z = 9000$

You invested an equal amount in FSAZX and FUSFX:

$y = z$, or

$y - z = 0$

Loss for the year from the first two funds was $400:

$0.06x + 0.05y = 400$

Solving:

$$\begin{bmatrix} 1 & 1 & 1 & 9000 \\ 0 & 1 & -1 & 0 \\ 0.06 & 0.05 & 0 & 400 \end{bmatrix} \begin{matrix} \\ \\ 100R_3 \end{matrix}$$

$$\begin{bmatrix} 1 & 1 & 1 & 9000 \\ 0 & 1 & -1 & 0 \\ 6 & 5 & 0 & 40000 \end{bmatrix} \begin{matrix} \\ \\ R_3 - 6R_1 \end{matrix}$$

$$\begin{bmatrix} 1 & 1 & 1 & 9000 \\ 0 & 1 & -1 & 0 \\ 0 & -1 & -6 & -14000 \end{bmatrix} \begin{matrix} R_1 - R_2 \\ \\ R_3 + R_2 \end{matrix}$$

$$\begin{bmatrix} 1 & 0 & 2 & 9000 \\ 0 & 1 & -1 & 0 \\ 0 & 0 & -7 & -14000 \end{bmatrix} (1/7)R_3$$

$$\begin{bmatrix} 1 & 0 & 2 & 9000 \\ 0 & 1 & -1 & 0 \\ 0 & 0 & -1 & -2000 \end{bmatrix} \begin{matrix} R_1 + 2R_3 \\ R_2 - R_3 \\ \\ \end{matrix}$$

$$\begin{bmatrix} 1 & 0 & 0 & 5000 \\ 0 & 1 & 0 & 2000 \\ 0 & 0 & -1 & -2000 \end{bmatrix} -R_3$$

$$\begin{bmatrix} 1 & 0 & 0 & 5000 \\ 0 & 1 & 0 & 2000 \\ 0 & 0 & 1 & 2000 \end{bmatrix}$$

$x = 5000, y = 2000, z = 2000$

Solution: You invested $5000 in FCTFX, $2000 in FSAZX, $2000 in FUSFX.

19. Unknowns:

x = the number of shares of GE

y = the number of shares of WMT

z = the number of shares of XOM

The total investment was $8400

Investment in GE = x shares @ $16 = $16x$

Investment in WMT = y shares @ $56 = $56y$

Investment in XOM = z shares @ $80 = $80z$

Thus,

$16x + 56y + 80z = 8400$

You expected to earn $248 in dividends:

GE dividend = 7% of $16x$ invested

$= 0.07(16x) = 1.12x$

WMT dividend = 2% of $56y$ invested

$= 0.02(56y) = 1.12y$

XOM dividend = 2% of $80y$ invested

$= 0.02(80y) = 1.6y$

Thus,

$1.12x + 1.12y + 1.6z = 248$

You purchased a total of 200 shares:

$x + y + z = 200$

We therefore have the following system:

$x + y + z = 200$

$16x + 56y + 80z = 8400$

$1.12x + 1.12y + 1.6z = 248$

Solving:

$$\left[\begin{array}{ccc|c} 1 & 1 & 1 & 200 \\ 16 & 56 & 80 & 8400 \\ 1.12 & 1.12 & 1.6 & 248 \end{array}\right] \begin{array}{l} \\ \\ 25R_3 \end{array}$$

$$\left[\begin{array}{ccc|c} 1 & 1 & 1 & 200 \\ 16 & 56 & 80 & 8400 \\ 28 & 28 & 40 & 6200 \end{array}\right] \begin{array}{l} \\ (1/8)R_2 \\ (1/4)R_3 \end{array}$$

$$\left[\begin{array}{ccc|c} 1 & 1 & 1 & 200 \\ 2 & 7 & 10 & 1050 \\ 7 & 7 & 10 & 1550 \end{array}\right] \begin{array}{l} \\ R_2 - 2R_1 \\ R_3 - 7R_1 \end{array}$$

$$\left[\begin{array}{ccc|c} 1 & 1 & 1 & 200 \\ 0 & 5 & 8 & 650 \\ 0 & 0 & 3 & 150 \end{array}\right] \begin{array}{l} \\ \\ (1/3)R_3 \end{array}$$

$$\left[\begin{array}{ccc|c} 1 & 1 & 1 & 200 \\ 0 & 5 & 8 & 650 \\ 0 & 0 & 1 & 50 \end{array}\right] \begin{array}{l} 5R_1 - R_2 \\ \\ \end{array}$$

$$\left[\begin{array}{ccc|c} 5 & 0 & -3 & 350 \\ 0 & 5 & 8 & 650 \\ 0 & 0 & 1 & 50 \end{array}\right] \begin{array}{l} R_1 + 3R_3 \\ R_2 - 8R_3 \\ \end{array}$$

$$\left[\begin{array}{ccc|c} 5 & 0 & 0 & 500 \\ 0 & 5 & 0 & 250 \\ 0 & 0 & 1 & 50 \end{array}\right] \begin{array}{l} (1/5)R_1 \\ (1/5)R_2 \\ \end{array}$$

$$\left[\begin{array}{ccc|c} 1 & 0 & 0 & 100 \\ 0 & 1 & 0 & 50 \\ 0 & 0 & 1 & 50 \end{array}\right]$$

$x = 100, y = 50, z = 50$

Solution: You purchased 100 shares of GE, 50 shares of WMT, and 50 shares of XOM.

21. Solution: With x, y, z, and u as indicated, the first piece of information we hve is that
$x + y + z + u = 284$
The remaining equations must be written in standard form:
$-3x + 3y + z - u = 6$
$x + y - z - u = 50$
$x - y + z - u = 42$
Using pivoting technology, we obtain:

Matrix #1

x	y	z	u	
1	1	1	1	284
-3	3	1	-1	6
1	1	-1	-1	50
1	-1	1	-1	42

Matrix #2

x	y	z	u	
1	1	1	1	284
0	6	4	2	858
0	0	-2	-2	-234
0	-2	0	-2	-242

Matrix #3

x	y	z	u	
1	1	1	1	284
0	3	2	1	429
0	0	-1	-1	-117
0	-1	0	-1	-121

Matrix #4

x	y	z	u	
3	0	1	2	423
0	3	2	1	429
0	0	-1	-1	-117
0	0	2	-2	66

Matrix #5

x	y	z	u	
3	0	1	2	423
0	3	2	1	429
0	0	-1	-1	-117
0	0	1	-1	33

Matrix #6

x	y	z	u	
3	0	0	1	306
0	3	0	-1	195
0	0	-1	-1	-117
0	0	0	-2	-84

Matrix #7

x	y	z	u	
3	0	0	1	306
0	3	0	-1	195
0	0	-1	-1	-117
0	0	0	-1	-42

Matrix #8

x	y	z	u	
3	0	0	0	264
0	3	0	0	237
0	0	-1	0	-75
0	0	0	-1	-42

Matrix #9

x	y	z	u	
1	0	0	0	88
0	1	0	0	79
0	0	1	0	75
0	0	0	1	42

Solution: $x = 88$, $y = 79$, $z = 75$, $u = 42$
Microsoft: 88 million, Time Warner: 79 million,
Yahoo: 75 million, Google: 42 million

23. Solution: Since the percentage market shares
add up to 100, the third equation is
$x + y + z + w = 100$
If we rewrite the given equations in standard
form, we get the second and third equations:
$x - y - z = 1$
$z + 0.2w = 16$
Row reduction:

$$\begin{bmatrix} 1 & 1 & 1 & 1 & 100 \\ 1 & -1 & -1 & 0 & 1 \\ 0 & 0 & 1 & 0.2 & 16 \end{bmatrix} 5R_3$$

$$\begin{bmatrix} 1 & 1 & 1 & 1 & 100 \\ 1 & -1 & -1 & 0 & 1 \\ 0 & 0 & 5 & 1 & 80 \end{bmatrix} R_2 - R_1$$

$$\begin{bmatrix} 1 & 1 & 1 & 1 & 100 \\ 0 & -2 & -2 & -1 & -99 \\ 0 & 0 & 5 & 1 & 80 \end{bmatrix} 2R_1 + R_2$$

$$\begin{bmatrix} 2 & 0 & 0 & 1 & 101 \\ 0 & -2 & -2 & -1 & -99 \\ 0 & 0 & 5 & 1 & 80 \end{bmatrix} 5R_2 + 2R_3$$

$$\begin{bmatrix} 2 & 0 & 0 & 1 & 101 \\ 0 & -10 & 0 & -3 & -335 \\ 0 & 0 & 5 & 1 & 80 \end{bmatrix} \begin{matrix}(1/2)R_1 \\ -(1/10)R_2 \\ (1/5)R_3\end{matrix}$$

$$\begin{bmatrix} 1 & 0 & 0 & 0.5 & 50.5 \\ 0 & 1 & 0 & 0.3 & 33.5 \\ 0 & 0 & 1 & 0.2 & 16 \end{bmatrix}$$

Translating back to equations (and rounding to
two decimal places) gives
$x + 0.5w = 50.5$
$y + 0.3w = 33.5$
$z + 0.2w = 16$
Solving for x, y, and z in terms of w gives the
general solution:
$x = 50.5 - 0.5w$
$y = 33.5 - 0.3w$
$z = 16 - 0.2w$
w arbitrary
We now answer the question: Which of the three
companies' market share is most impacted by the
share held by Other? Since Other is represented
by w, we determine which of the four unknowns
has the coefficient of w with the greatest absolute
value—namely, x, representing State Farm. Thus,
State Farm is most impacted by Other.

25. Unknowns:
x = the number of books sent from Brooklyn to
Long Island
y = the number of books sent from Queens to
Long Island

69

z = the number of books sent from Brooklyn to Manhattan

w = the number of books sent from Queens to Manhattan

We represent the given information in a diagram:

Note that, since a total of 3000 books are ordered and there are a total of 3000 in stock, both warehouses need to clear all their stocks.

Books to Long Island: $x + y = 1500$

Books to Manhattan Order: $z + w = 1500$

Books from Brooklyn: $x + z = 1000$

Books from Queens: $y + w = 2000$

(a) Transportation budget:

$5x + 4y + z + 2w = 9000$

We have five equations in 4 unknowns. Solving:

$$\begin{bmatrix} 1 & 1 & 0 & 0 & 1500 \\ 0 & 0 & 1 & 1 & 1500 \\ 1 & 0 & 1 & 0 & 1000 \\ 0 & 1 & 0 & 1 & 2000 \\ 5 & 4 & 1 & 2 & 9000 \end{bmatrix} \begin{matrix} \\ \\ R_3 - R_1 \\ \\ R_5 - 5R_1 \end{matrix}$$

$$\begin{bmatrix} 1 & 1 & 0 & 0 & 1500 \\ 0 & 0 & 1 & 1 & 1500 \\ 0 & -1 & 1 & 0 & -500 \\ 0 & 1 & 0 & 1 & 2000 \\ 0 & -1 & 1 & 2 & 1500 \end{bmatrix} \begin{matrix} \\ \\ R_3 - R_2 \\ \\ R_5 - R_2 \end{matrix}$$

$$\begin{bmatrix} 1 & 1 & 0 & 0 & 1500 \\ 0 & 0 & 1 & 1 & 1500 \\ 0 & -1 & 0 & -1 & -2000 \\ 0 & 1 & 0 & 1 & 2000 \\ 0 & -1 & 0 & 1 & 0 \end{bmatrix} \begin{matrix} R_1 + R_3 \\ \\ \\ R_4 + R_3 \\ R_5 - R_3 \end{matrix}$$

$$\begin{bmatrix} 1 & 0 & 0 & -1 & -500 \\ 0 & 0 & 1 & 1 & 1500 \\ 0 & -1 & 0 & -1 & -2000 \\ 0 & 0 & 0 & 0 & 0 \\ 0 & 0 & 0 & 2 & 2000 \end{bmatrix} \begin{matrix} \\ \\ \\ \\ (1/2)R_5 \end{matrix}$$

$$\begin{bmatrix} 1 & 0 & 0 & -1 & -500 \\ 0 & 0 & 1 & 1 & 1500 \\ 0 & -1 & 0 & -1 & -2000 \\ 0 & 0 & 0 & 0 & 0 \\ 0 & 0 & 0 & 1 & 1000 \end{bmatrix} \begin{matrix} R_1 + R_5 \\ R_2 - R_5 \\ R_3 + R_5 \\ \\ \end{matrix}$$

$$\begin{bmatrix} 1 & 0 & 0 & 0 & 500 \\ 0 & 0 & 1 & 0 & 500 \\ 0 & -1 & 0 & 0 & -1000 \\ 0 & 0 & 0 & 0 & 0 \\ 0 & 0 & 0 & 1 & 1000 \end{bmatrix} \begin{matrix} \\ \\ -R_3 \\ \\ \end{matrix}$$

$$\begin{bmatrix} 1 & 0 & 0 & 0 & 500 \\ 0 & 0 & 1 & 0 & 500 \\ 0 & 1 & 0 & 0 & 1000 \\ 0 & 0 & 0 & 0 & 0 \\ 0 & 0 & 0 & 1 & 1000 \end{bmatrix} \text{Rearrange rows}$$

$$\begin{bmatrix} 1 & 0 & 0 & 0 & 500 \\ 0 & 1 & 0 & 0 & 1000 \\ 0 & 0 & 1 & 0 & 500 \\ 0 & 0 & 0 & 1 & 1000 \\ 0 & 0 & 0 & 0 & 0 \end{bmatrix}$$

Solution: Brooklyn to Long Island: 500 books; Queens to Long Island: 1000 books; Brooklyn to Manhattan: 500 books; Queens to Manhattan: 1000 books

(b) We try setting $x = 0$ to avoid the expensive \$5 per book cost. This reduces the system to:

$0 + y = 1500$, so $y = 1500$

$z + w = 1500$

$0 + z = 1000$, so $z = 1000$

$y + w = 2000$

Substituting the values of y and z in the second equation gives $w = 1500 - 1000 = 500$. The total transportation cost is

$5x + 4y + z + 2w$
$= 5(0) + 4(1500) + 1000 + 2(500)$
$= \$8000,$

which is less than the $9000 budget. Thus, a solution is:

Brooklyn to Long Island: no books; Queens to Long Island: 1500 books; Brooklyn to Manhattan: 1000 books; Queens to Manhattan: 500 books, for a total cost of $8000.

27. Unknowns:

x = the number of tourists from North America to Australia
y = the number of tourists from North America to South Africa
z = the number of tourists from Europe to Australia
w = the number of tourists from Europe to South Africa

We represent the given information in a diagram:

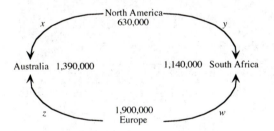

(a)
North America: $x + y = 630{,}000$
Europe: $z + w = 1{,}900{,}000$
Australia: $x + z = 1{,}390{,}000$
South Africa: $y + w = 1{,}140{,}000$
Reducing the matrix form:

$$\begin{bmatrix} 1 & 1 & 0 & 0 & 630000 \\ 0 & 0 & 1 & 1 & 1900000 \\ 1 & 0 & 1 & 0 & 1390000 \\ 0 & 1 & 0 & 1 & 1140000 \end{bmatrix} \begin{matrix} \\ \\ R_3 - R_1 \\ \\ \end{matrix}$$

$$\begin{bmatrix} 1 & 1 & 0 & 0 & 630000 \\ 0 & 0 & 1 & 1 & 1900000 \\ 0 & -1 & 1 & 0 & 760000 \\ 0 & 1 & 0 & 1 & 1140000 \end{bmatrix} \begin{matrix} \\ \\ R_3 - R_2 \\ \\ \end{matrix}$$

$$\begin{bmatrix} 1 & 1 & 0 & 0 & 630000 \\ 0 & 0 & 1 & 1 & 1900000 \\ 0 & -1 & 0 & -1 & -1140000 \\ 0 & 1 & 0 & 1 & 1140000 \end{bmatrix} \begin{matrix} R_1 + R_3 \\ \\ \\ R_4 + R_3 \end{matrix}$$

$$\begin{bmatrix} 1 & 0 & 0 & -1 & -510000 \\ 0 & 0 & 1 & 1 & 1900000 \\ 0 & -1 & 0 & -1 & -1140000 \\ 0 & 0 & 0 & 0 & 0 \end{bmatrix} \begin{matrix} \\ \\ -R_3 \\ \\ \end{matrix}$$

$$\begin{bmatrix} 1 & 0 & 0 & -1 & -510000 \\ 0 & 0 & 1 & 1 & 1900000 \\ 0 & 1 & 0 & 1 & 1140000 \\ 0 & 0 & 0 & 0 & 0 \end{bmatrix}$$

This system has infinitely many solutions, which is why the given information is not sufficient to determine the number of tourists from each region to each destination.

(b) We are told that that

$x + y + z + y = 2{,}530{,}000$

However, this equation can be obtained by adding the first two equations in part (a). Thus, the new equation gives us no additional information, and so the associated system of linear equations will have the same infinite solution set as part (a).

(c) The additional information is, rephrased:

The number of people from Europe to Australia = the number of people from Europe to South Africa: $z = w$, or

$z - w = 0.$

Adding this to our list of equations and solving gives:

$$\begin{bmatrix} 1 & 1 & 0 & 0 & 630000 \\ 0 & 0 & 1 & 1 & 1900000 \\ 1 & 0 & 1 & 0 & 1390000 \\ 0 & 1 & 0 & 1 & 1140000 \\ 0 & 0 & 1 & -1 & 0 \end{bmatrix} R_3 - R_1$$

$$\begin{bmatrix} 1 & 1 & 0 & 0 & 630000 \\ 0 & 0 & 1 & 1 & 1900000 \\ 0 & -1 & 1 & 0 & 760000 \\ 0 & 1 & 0 & 1 & 1140000 \\ 0 & 0 & 1 & -1 & 0 \end{bmatrix} \begin{matrix} \\ \\ R_3 - R_2 \\ \\ R_5 - R_2 \end{matrix}$$

$$\begin{bmatrix} 1 & 1 & 0 & 0 & 630000 \\ 0 & 0 & 1 & 1 & 1900000 \\ 0 & -1 & 0 & -1 & -1140000 \\ 0 & 1 & 0 & 1 & 1140000 \\ 0 & 0 & 0 & -2 & -1900000 \end{bmatrix} (1/2)R_5$$

$$\begin{bmatrix} 1 & 1 & 0 & 0 & 630000 \\ 0 & 0 & 1 & 1 & 1900000 \\ 0 & -1 & 0 & -1 & -1140000 \\ 0 & 1 & 0 & 1 & 1140000 \\ 0 & 0 & 0 & -1 & -950000 \end{bmatrix} \begin{matrix} R_1 + R_3 \\ \\ \\ R_4 + R_3 \\ \end{matrix}$$

$$\begin{bmatrix} 1 & 0 & 0 & -1 & -510000 \\ 0 & 0 & 1 & 1 & 1900000 \\ 0 & -1 & 0 & -1 & -1140000 \\ 0 & 0 & 0 & 0 & 0 \\ 0 & 0 & 0 & -1 & -950000 \end{bmatrix} \begin{matrix} R_1 - R_5 \\ R_2 + R_5 \\ R_3 - R_5 \\ \\ \end{matrix}$$

$$\begin{bmatrix} 1 & 0 & 0 & 0 & 440000 \\ 0 & 0 & 1 & 0 & 950000 \\ 0 & -1 & 0 & 0 & -190000 \\ 0 & 0 & 0 & 0 & 0 \\ 0 & 0 & 0 & -1 & -950000 \end{bmatrix} \begin{matrix} \\ \\ -R_3 \\ \\ -R_5 \end{matrix}$$

$$\begin{bmatrix} 1 & 0 & 0 & 0 & 440000 \\ 0 & 0 & 1 & 0 & 950000 \\ 0 & 1 & 0 & 0 & 190000 \\ 0 & 0 & 0 & 0 & 0 \\ 0 & 0 & 0 & 1 & 950000 \end{bmatrix} \text{Rearrange rows}$$

$$\begin{bmatrix} 1 & 0 & 0 & 0 & 440000 \\ 0 & 1 & 0 & 0 & 190000 \\ 0 & 0 & 1 & 0 & 950000 \\ 0 & 0 & 0 & 1 & 950000 \end{bmatrix}$$

$x = 440{,}000$, $y = 190{,}000$, $z = 950{,}000$, $w = 950{,}000$

Since this is a unique solution, we can determine the numbers from each country to each destination:

North America to Australia: 440,000, North America to South Africa: 190,000, Europe to Australia: 950,000, Europe to South Africa: 950,000

29.

	Used alcohol	Alcohol-free	Totals
US	x	y	14,000
Europe	z	w	95,000
Totals	63,550	45,450	

(a) Using the row and column totals, we get four equations:

U.S.: $x + y = 14{,}000$
Europe: $z + w = 95{,}000$
Used Alcohol: $x + z = 63{,}550$
Alcohol-free: $y + w = 45{,}450$
Row-reducing the augmented matrix gives:

$$\begin{bmatrix} 1 & 1 & 0 & 0 & 14000 \\ 0 & 0 & 1 & 1 & 95000 \\ 1 & 0 & 1 & 0 & 63550 \\ 0 & 1 & 0 & 1 & 45450 \end{bmatrix} R_3 - R_1$$

$$\begin{bmatrix} 1 & 1 & 0 & 0 & 14000 \\ 0 & 0 & 1 & 1 & 95000 \\ 0 & -1 & 1 & 0 & 49550 \\ 0 & 1 & 0 & 1 & 45450 \end{bmatrix} R_3 - R_2$$

$$\begin{bmatrix} 1 & 1 & 0 & 0 & 14000 \\ 0 & 0 & 1 & 1 & 95000 \\ 0 & -1 & 0 & -1 & -45450 \\ 0 & 1 & 0 & 1 & 45450 \end{bmatrix} \begin{matrix} R_1+R_3 \\ \\ \\ R_4+R_3 \end{matrix}$$

$$\begin{bmatrix} 1 & 0 & 0 & -1 & -31450 \\ 0 & 0 & 1 & 1 & 95000 \\ 0 & -1 & 0 & -1 & -45450 \\ 0 & 0 & 0 & 0 & 0 \end{bmatrix} -R_3$$

$$\begin{bmatrix} 1 & 0 & 0 & -1 & -31450 \\ 0 & 0 & 1 & 1 & 95000 \\ 0 & 1 & 0 & 1 & 45450 \\ 0 & 0 & 0 & 0 & 0 \end{bmatrix}$$

The row-reduced matrix shows that there are infinitely many solutions, and hence no unique solution. Thus, the given data are insufficient to obtain the missing data (x, y, z and w).

(b) The number of US 10th graders who were alcohol-free was 50% more than the number who had used alcohol:

$y = x + 0.50x = 1.50x$, or
$-1.50x + y = 0$

If we include this additional equation, we get the system

$x + y = 14.000$
$z + w = 95.000$
$x + z = 63.550$
$y + w = 45.450$
$-1.50x + y = 0$

Solving:

$$\begin{bmatrix} 1 & 1 & 0 & 0 & 14000 \\ 0 & 0 & 1 & 1 & 95000 \\ 1 & 0 & 1 & 0 & 63550 \\ 0 & 1 & 0 & 1 & 45450 \\ -1.5 & 1 & 0 & 0 & 0 \end{bmatrix} 2R_5$$

$$\begin{bmatrix} 1 & 1 & 0 & 0 & 14000 \\ 0 & 0 & 1 & 1 & 95000 \\ 1 & 0 & 1 & 0 & 63550 \\ 0 & 1 & 0 & 1 & 45450 \\ -3 & 2 & 0 & 0 & 0 \end{bmatrix} \begin{matrix} \\ \\ R_3-R_1 \\ \\ R_5+3R_1 \end{matrix}$$

$$\begin{bmatrix} 1 & 1 & 0 & 0 & 14000 \\ 0 & 0 & 1 & 1 & 95000 \\ 0 & -1 & 1 & 0 & 49550 \\ 0 & 1 & 0 & 1 & 45450 \\ 0 & 5 & 0 & 0 & 42000 \end{bmatrix} (1/5)R_5$$

$$\begin{bmatrix} 1 & 1 & 0 & 0 & 14000 \\ 0 & 0 & 1 & 1 & 95000 \\ 0 & -1 & 1 & 0 & 49550 \\ 0 & 1 & 0 & 1 & 45450 \\ 0 & 1 & 0 & 0 & 8400 \end{bmatrix} R_3-R_2$$

$$\begin{bmatrix} 1 & 1 & 0 & 0 & 14000 \\ 0 & 0 & 1 & 1 & 95000 \\ 0 & -1 & 0 & -1 & -45450 \\ 0 & 1 & 0 & 1 & 45450 \\ 0 & 1 & 0 & 0 & 8400 \end{bmatrix} \begin{matrix} R_1+R_3 \\ \\ \\ R_4+R_3 \\ R_5+R_3 \end{matrix}$$

$$\begin{bmatrix} 1 & 0 & 0 & -1 & -31450 \\ 0 & 0 & 1 & 1 & 95000 \\ 0 & -1 & 0 & -1 & -45450 \\ 0 & 0 & 0 & 0 & 0 \\ 0 & 0 & 0 & -1 & -37050 \end{bmatrix} \begin{matrix} R_1-R_5 \\ R_2+R_5 \\ R_3-R_5 \\ \\ \end{matrix}$$

$$\begin{bmatrix} 1 & 0 & 0 & 0 & 5600 \\ 0 & 0 & 1 & 0 & 57950 \\ 0 & -1 & 0 & 0 & -8400 \\ 0 & 0 & 0 & 0 & 0 \\ 0 & 0 & 0 & -1 & -37050 \end{bmatrix} \begin{matrix} \\ \\ -R_3 \\ \\ -R_5 \end{matrix}$$

$$\begin{bmatrix} 1 & 0 & 0 & 0 & 5600 \\ 0 & 0 & 1 & 0 & 57950 \\ 0 & 1 & 0 & 0 & 8400 \\ 0 & 0 & 0 & 0 & 0 \\ 0 & 0 & 0 & 1 & 37050 \end{bmatrix}$$ Rearrange Rows

$$\begin{bmatrix} 1 & 0 & 0 & 0 & 5600 \\ 0 & 1 & 0 & 0 & 8400 \\ 0 & 0 & 1 & 0 & 57950 \\ 0 & 0 & 0 & 1 & 37050 \\ 0 & 0 & 0 & 0 & 0 \end{bmatrix}$$

Thus, the missing data is:
$x = 5600, y = 8400, z = 57,950, w = 37,050$

31. $x =$ daily traffic flow along Eastward Blvd.
$y =$ daily traffic flow along Northwest La.
$z =$ daily traffic flow along Southwest La.

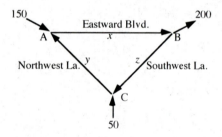

(a) Intersection A: Traffic in = Traffic out:
  $150 + y = x$, or $x - y = 150$
Intersection B: Traffic in = Traffic out:
  $x = 200 + z$, or $x - z = 200$
Intersection C: Traffic in = Traffic out:
  $50 + z = y$, or $y - z = 50$
This gives us a system of 3 linear equations:
  $x - y = 150$
  $x - z = 200$
  $y - z = 50$
Solving:

$$\begin{bmatrix} 1 & -1 & 0 & 150 \\ 1 & 0 & -1 & 200 \\ 0 & 1 & -1 & 50 \end{bmatrix} R_2 - R_1$$

$$\begin{bmatrix} 1 & -1 & 0 & 150 \\ 0 & 1 & -1 & 50 \\ 0 & 1 & -1 & 50 \end{bmatrix} \begin{matrix} R_1 + R_2 \\ \\ R_3 - R_2 \end{matrix}$$

$$\begin{bmatrix} 1 & 0 & -1 & 200 \\ 0 & 1 & -1 & 50 \\ 0 & 0 & 0 & 0 \end{bmatrix}$$

Since there are infinitely many solutions, it is not possible to determine the daily flow of traffic along each of the three streets from the information given.
Translating the row-reduced matrix back into equations gives
  $x - z = 200$, or $x = 200 + z$
  $y - z = 50$, or $y = 50 + z$
thus, the general solution is:
  $x = 200 + z$
  $y = 50 + z$
  $z \geq 0$ arbitrary,
where z is traffic along Southwest Lane. Thus, it would suffice to know the traffic along Southwest to obtain the other traffic flows.
(b) If we set $z = 60$, we obtain, from the general solution in part (a),
$x = 200 + 60 = 260$ vehicles along Eastward Blvd., $y = 50 + 60 = 110$ vehicles along Northwest La., $z = 60$ vehicles along Southwest La.
(c) From the general solution in part (a), the traffic flow along Northwest La. is given by
  $y = 50 + z$
Since $z \geq 0$, the value of y must be at least 50 vehicles per day.

33. (a) Unknowns:
$x =$ traffic on middle section of Bree
$y =$ traffic on middle section of Jeppe
$z =$ traffic on middle section of Simmons
$w =$ traffic on middle section of Harrison

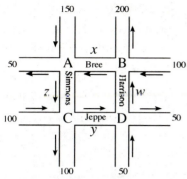

Intersection A: Traffic in = Traffic out:
$150 + x = 50 + z$ or $x - z = -100$
Intersection B: Traffic in = Traffic out:
$100 + w = 200 + x$ or $x - w = -100$
Intersection C: Traffic in = Traffic out:
$100 + z = 100 + y$ or $y - z = 0$
Intersection D: Traffic in = Traffic out:
$50 + y = 50 + w$ or $y - w = 0$
This gives us a system of 4 linear equations:
$x - z = -100$
$x - w = -100$
$y - z = 0$
$y - w = 0$
Solving:

$$\begin{bmatrix} 1 & 0 & -1 & 0 & -100 \\ 1 & 0 & 0 & -1 & -100 \\ 0 & 1 & -1 & 0 & 0 \\ 0 & 1 & 0 & -1 & 0 \end{bmatrix} R_2 - R_1$$

$$\begin{bmatrix} 1 & 0 & -1 & 0 & -100 \\ 0 & 0 & 1 & -1 & 0 \\ 0 & 1 & -1 & 0 & 0 \\ 0 & 1 & 0 & -1 & 0 \end{bmatrix} \begin{matrix} R_1 + R_2 \\ \\ R_3 + R_2 \end{matrix}$$

$$\begin{bmatrix} 1 & 0 & 0 & -1 & -100 \\ 0 & 0 & 1 & -1 & 0 \\ 0 & 1 & 0 & -1 & 0 \\ 0 & 1 & 0 & -1 & 0 \end{bmatrix} R_4 - R_3$$

$$\begin{bmatrix} 1 & 0 & 0 & -1 & -100 \\ 0 & 0 & 1 & -1 & 0 \\ 0 & 1 & 0 & -1 & 0 \\ 0 & 0 & 0 & 0 & 0 \end{bmatrix} R_2 \leftrightarrow R_3$$

$$\begin{bmatrix} 1 & 0 & 0 & -1 & -100 \\ 0 & 1 & 0 & -1 & 0 \\ 0 & 0 & 1 & -1 & 0 \\ 0 & 0 & 0 & 0 & 0 \end{bmatrix}$$

For the general solution:
Translating the row-reduced matrix back into equations gives
$x - w = -100$, or $x = w - 100$
$y - w = 0$, or $y = w$
$z - w = 0$, or $z = w$
thus, the general solution is:
$x = w - 100$
$y = w$
$z = w$
w arbitrary,
where w is traffic along middle section of Harrison.
(b) Since $y = w$ has more than one possible value, it is not possible to determine the traffic flow along the middle section of Jeppe.
(c) Given $x = 400$, the first equation of the general solution says that
$400 = w - 100$
so $w = 500$. We need Simmons (z). But the third equation says $z = w$. So $w = 500$ cars down the middle section of Simmons.
(d) If w is less than 100, x would become negative, by the first equation of the solution. Thus, $w \geq 100$.

75

(e) Simmons is $z = w$, which can be as large as we like without making any of the variables negative (see the solution). Thus, there is no upper limit to the traffic on the middle section of Simmons. We can visualize this by imagining thousands of cars going around and around the center block without effecting the recorded numbers in the diagram.

35. Let us take the unknowns to be the net traffic flow going east on the three stretches of Broadway as shown in the figure. (If any of the unknowns is negative, it indicates a net positive flow in the opposite direction.)

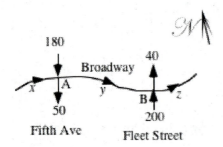

(a) Intersection A: Traffic in = Traffic out:
$180 + x = 50 + y$, or $x - y = -130$
Intersection B: Traffic in = Traffic out:
$200 + y = 40 + z$, or $y - z = -160$
Solving:

$$\begin{bmatrix} 1 & -1 & 0 & -130 \\ 0 & 1 & -1 & -160 \end{bmatrix} R_1 + R_2$$

$$\begin{bmatrix} 1 & 0 & -1 & -290 \\ 0 & 1 & -1 & -160 \end{bmatrix}$$

The general solution is:
$x - z = -290$, or $x = z - 290$
$y - z = -160$, or $y = z - 160$
z is arbitrary

Since there are infinitely many solutions, we cannot determine the traffic flow along each stretch of Broadway. Knowing any one of z, y, or z would enable us to solve for the other two unknowns uniquely.

(b) The general solution from part (a) is:
$x = z - 290$
$y = z - 160$
z is arbitrary

East of Fleet Street, the traffic flow is z. If z is smaller than 160, the above solution shows that values of z and y are negative, indicating a net flow to the west on those stretches.

37. We rewrite each given equation in standard form, using the given information that $M = 120$ billion dollars:

$$\begin{aligned} 120 &= C + D & &\rightarrow C + D = 120 \\ C &= 0.2D & &\rightarrow C - 0.2D = 0 \\ R &= 0.1D & &\rightarrow R - 0.1D = 0 \\ H &= R + C & &\rightarrow H - R - C = 0 \end{aligned}$$

The matrices below are set up with the unknowns in the order: C, R, D, H

$$\begin{bmatrix} 1 & 0 & 1 & 0 & 120 \\ 1 & 0 & -.2 & 0 & 0 \\ 0 & 1 & -.1 & 0 & 0 \\ -1 & -1 & 0 & 1 & 0 \end{bmatrix} \begin{matrix} \\ 5R_2 \\ 10R_3 \\ \\ \end{matrix}$$

$$\begin{bmatrix} 1 & 0 & 1 & 0 & 120 \\ 5 & 0 & -1 & 0 & 0 \\ 0 & 10 & -1 & 0 & 0 \\ -1 & -1 & 0 & 1 & 0 \end{bmatrix} \begin{matrix} \\ R_2 - 5R_1 \\ \\ R_4 + R_1 \end{matrix}$$

$$\begin{bmatrix} 1 & 0 & 1 & 0 & 120 \\ 0 & 0 & -6 & 0 & -600 \\ 0 & 10 & -1 & 0 & 0 \\ 0 & -1 & 1 & 1 & 120 \end{bmatrix} \begin{matrix} \\ (1/6)R_2 \\ \\ \\ \end{matrix}$$

$$\begin{bmatrix} 1 & 0 & 1 & 0 & 120 \\ 0 & 0 & -1 & 0 & -100 \\ 0 & 10 & -1 & 0 & 0 \\ 0 & -1 & 1 & 1 & 120 \end{bmatrix} \begin{matrix} R_1 + R_2 \\ \\ R_3 - R_2 \\ R_4 + R_2 \end{matrix}$$

$$\begin{bmatrix} 1 & 0 & 0 & 0 & 20 \\ 0 & 0 & -1 & 0 & -100 \\ 0 & 10 & 0 & 0 & 100 \\ 0 & -1 & 0 & 1 & 20 \end{bmatrix} \begin{matrix} \\ \\ (1/10)R_3 \\ \\ \end{matrix}$$

$$\begin{bmatrix} 1 & 0 & 0 & 0 & 20 \\ 0 & 0 & -1 & 0 & -100 \\ 0 & 1 & 0 & 0 & 10 \\ 0 & -1 & 0 & 1 & 20 \end{bmatrix} R_4 + R_3$$

$$\begin{bmatrix} 1 & 0 & 0 & 0 & 20 \\ 0 & 0 & -1 & 0 & -100 \\ 0 & 1 & 0 & 0 & 10 \\ 0 & 0 & 0 & 1 & 30 \end{bmatrix} -R_2$$

$$\begin{bmatrix} 1 & 0 & 0 & 0 & 20 \\ 0 & 0 & 1 & 0 & 100 \\ 0 & 1 & 0 & 0 & 10 \\ 0 & 0 & 0 & 1 & 30 \end{bmatrix} R_2 \leftrightarrow R_3$$

$$\begin{bmatrix} 1 & 0 & 0 & 0 & 20 \\ 0 & 1 & 0 & 0 & 10 \\ 0 & 0 & 1 & 0 & 100 \\ 0 & 0 & 0 & 1 & 30 \end{bmatrix}$$

$C = 20, R = 10, D = 100, H = 30$

We are asked for bank reserves R, which are therefore $10 billion.

39.

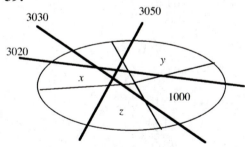

Adding the values along each beam gives

$x + y + z = 3050$

$x + z + 1000 = 3030$, or $x + z = 2030$

$x + y + 1000 = 3020$, or $x + y = 2020$

$$\begin{bmatrix} 1 & 1 & 1 & 3050 \\ 1 & 0 & 1 & 2030 \\ 1 & 1 & 0 & 2020 \end{bmatrix} \begin{matrix} \\ R_2 - R_1 \\ R_3 - R_1 \end{matrix}$$

$$\begin{bmatrix} 1 & 1 & 1 & 3050 \\ 0 & -1 & 0 & -1020 \\ 0 & 0 & -1 & -1030 \end{bmatrix} R_1 + R_2$$

$$\begin{bmatrix} 1 & 0 & 1 & 2030 \\ 0 & -1 & 0 & -1020 \\ 0 & 0 & -1 & -1030 \end{bmatrix} R_1 + R_3$$

$$\begin{bmatrix} 1 & 0 & 0 & 1000 \\ 0 & -1 & 0 & -1020 \\ 0 & 0 & -1 & -1030 \end{bmatrix} \begin{matrix} \\ -R_2 \\ -R_3 \end{matrix}$$

$$\begin{bmatrix} 1 & 0 & 0 & 1000 \\ 0 & 1 & 0 & 1020 \\ 0 & 0 & 1 & 1030 \end{bmatrix}$$

Solution: $x = 1000$, $y = 1020$, $z = 1030$

From the table, we find the corresponding components:

$x =$ water, $y =$ gray matter, $z =$ tumor

41.

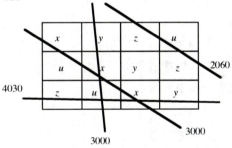

Adding the values along each beam gives

$x + y + z + u = 4030$

$x + y + u = 3000$

$3x + 2u = 3000$

$2z + u = 2060$

Solving:

$$\begin{bmatrix} 1 & 1 & 1 & 1 & 4030 \\ 1 & 1 & 0 & 1 & 3000 \\ 3 & 0 & 0 & 2 & 3000 \\ 0 & 0 & 2 & 1 & 2060 \end{bmatrix} \begin{matrix} \\ R_2 - R_1 \\ R_3 - 3R_1 \\ \\ \end{matrix}$$

$$\begin{bmatrix} 1 & 1 & 1 & 1 & 4030 \\ 0 & 0 & -1 & 0 & -1030 \\ 0 & -3 & -3 & -1 & -9090 \\ 0 & 0 & 2 & 1 & 2060 \end{bmatrix} \begin{matrix} R_1 + R_2 \\ \\ R_3 - 3R_2 \\ R_4 + 2R_2 \end{matrix}$$

$$\begin{bmatrix} 1 & 1 & 0 & 1 & 3000 \\ 0 & 0 & -1 & 0 & -1030 \\ 0 & -3 & 0 & -1 & -6000 \\ 0 & 0 & 0 & 1 & 0 \end{bmatrix} 3R_1 + R_3$$

$$\begin{bmatrix} 3 & 0 & 0 & 2 & 3000 \\ 0 & 0 & -1 & 0 & -1030 \\ 0 & -3 & 0 & -1 & -6000 \\ 0 & 0 & 0 & 1 & 0 \end{bmatrix} \begin{matrix} R_1 - 2R_4 \\ \\ R_3 + R_4 \end{matrix}$$

$$\begin{bmatrix} 3 & 0 & 0 & 0 & 3000 \\ 0 & 0 & -1 & 0 & -1030 \\ 0 & -3 & 0 & 0 & -6000 \\ 0 & 0 & 0 & 1 & 0 \end{bmatrix} \begin{matrix} (1/3)R_1 \\ -R_2 \\ -(1/3)R_3 \end{matrix}$$

$$\begin{bmatrix} 1 & 0 & 0 & 0 & 1000 \\ 0 & 0 & 1 & 0 & 1030 \\ 0 & 1 & 0 & 0 & 2000 \\ 0 & 0 & 0 & 1 & 0 \end{bmatrix} R_2 \leftrightarrow R_3$$

$$\begin{bmatrix} 1 & 0 & 0 & 0 & 1000 \\ 0 & 1 & 0 & 0 & 2000 \\ 0 & 0 & 1 & 0 & 1030 \\ 0 & 0 & 0 & 1 & 0 \end{bmatrix}$$

$x = 1000, y = 2000, z = 1030, u = 0$
From the table, we find the corresponding components:
$x =$ water, $y =$ bone, $z =$ tumor, $u =$ air

43. Since the three beams extending from left to right pass though the same four squares on the left, let us label those squares y, The vertical beam passes through an additional two squares, which we will label as z and u as shown:

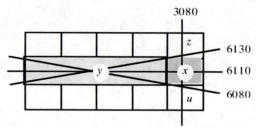

Adding the values along each beam gives
$$x + z + u = 3080$$
$$y + z = 6130$$
$$x + y = 6110$$
$$y + u = 6080$$
Solving:

$$\begin{bmatrix} 1 & 0 & 1 & 1 & 3080 \\ 0 & 1 & 1 & 0 & 6130 \\ 1 & 1 & 0 & 0 & 6110 \\ 0 & 1 & 0 & 1 & 6080 \end{bmatrix} R_3 - R_1$$

$$\begin{bmatrix} 1 & 0 & 1 & 1 & 3080 \\ 0 & 1 & 1 & 0 & 6130 \\ 0 & 1 & -1 & -1 & 3030 \\ 0 & 1 & 0 & 1 & 6080 \end{bmatrix} \begin{matrix} \\ \\ R_3 - R_2 \\ R_4 - R_2 \end{matrix}$$

$$\begin{bmatrix} 1 & 0 & 1 & 1 & 3080 \\ 0 & 1 & 1 & 0 & 6130 \\ 0 & 0 & -2 & -1 & -3100 \\ 0 & 0 & -1 & 1 & -50 \end{bmatrix} \begin{matrix} 2R_1 + R_3 \\ 2R_2 + R_3 \\ \\ 2R_4 - R_3 \end{matrix}$$

$$\begin{bmatrix} 2 & 0 & 0 & 1 & 3060 \\ 0 & 2 & 0 & -1 & 9160 \\ 0 & 0 & -2 & -1 & -3100 \\ 0 & 0 & 0 & 3 & 3000 \end{bmatrix} (1/3)R_4$$

$$\begin{bmatrix} 2 & 0 & 0 & 1 & 3060 \\ 0 & 2 & 0 & -1 & 9160 \\ 0 & 0 & -2 & -1 & -3100 \\ 0 & 0 & 0 & 1 & 1000 \end{bmatrix} \begin{matrix} R_1 - R_4 \\ R_2 + R_4 \\ R_3 + R_4 \end{matrix}$$

$$\begin{bmatrix} 2 & 0 & 0 & 0 & 2060 \\ 0 & 2 & 0 & 0 & 10160 \\ 0 & 0 & -2 & 0 & -2100 \\ 0 & 0 & 0 & 1 & 1000 \end{bmatrix} \begin{matrix} (1/2)R_1 \\ (1/2)R_2 \\ -(1/2)R_3 \end{matrix}$$

$$\begin{bmatrix} 1 & 0 & 0 & 0 & 1030 \\ 0 & 1 & 0 & 0 & 5080 \\ 0 & 0 & 1 & 0 & 1050 \\ 0 & 0 & 0 & 1 & 1000 \end{bmatrix}$$

All we need is the value of x: 1030, which corresponds to tumor

45. Unknowns:

x = the number of Democrats who voted in favor

y = the number of Republicans who voted in favor

z = the number of Others who voted in favor

Note:

   $333-x$ Democrats voted against,

   $89-y$ Republicans voted against, and

   $13-z$ Others voted against.

Given information:

(1) There were 31 more votes in favor than against:

   The total number of votes in favor $= x+y+z$

   Total number of votes against

     $= (333-x) + (89-y) + (13-z)$

     $= 435 - x - y - z$

Thus,

   $x + y + z - (435 - x - y - z) = 31$, or

   $2x + 2y + 2z = 466$, or

   $x + y + z = 233$

(2) 10 times as many Democrats voted for the bill as Republicans. Rephrasing this:
The number of Democrats voting for the bill was 10 times the number of Republicans voting for the bill:

   $x = 10y$, or

   $x - 10y = 0$

(3) 36 more non-Democrats voted against the bill than for it. Rephrasing this: The number of non-Democrats voting against the bill exceeded the number of non-Democrats voting for by 36

   $(89-y) + (13-z) - (y+z) = 36$, or

   $-2y - 2z = -66$, or

   $y + z = 33$

Thus, we have 3 equations with 3 unknowns:

   $x + y + z = 233$

   $x - 10y = 0$

   $y + z = 33$

Solving:

$$\begin{bmatrix} 1 & 1 & 1 & 233 \\ 1 & -10 & 0 & 0 \\ 0 & 1 & 1 & 33 \end{bmatrix} R_2 - R_1$$

$$\begin{bmatrix} 1 & 1 & 1 & 233 \\ 0 & -11 & -1 & -233 \\ 0 & 1 & 1 & 33 \end{bmatrix} \begin{matrix} 11R_1 + R_2 \\ \\ 11R_3 + R_2 \end{matrix}$$

$$\begin{bmatrix} 11 & 0 & 10 & 2330 \\ 0 & -11 & -1 & -233 \\ 0 & 0 & 10 & 130 \end{bmatrix} (1/10)R_3$$

$$\begin{bmatrix} 11 & 0 & 10 & 2330 \\ 0 & -11 & -1 & -233 \\ 0 & 0 & 1 & 13 \end{bmatrix} \begin{matrix} R_1 - 10R_3 \\ R_2 + R_3 \end{matrix}$$

$$\begin{bmatrix} 11 & 0 & 0 & 2200 \\ 0 & -11 & 0 & -220 \\ 0 & 0 & 1 & 13 \end{bmatrix} \begin{matrix} (1/11)R_1 \\ -(1/11)R_2 \end{matrix}$$

$$\begin{bmatrix} 1 & 0 & 0 & 200 \\ 0 & 1 & 0 & 20 \\ 0 & 0 & 1 & 13 \end{bmatrix}$$

Solution: 200 Democrats, 20 Republicans, 13 of other parties voted for the bill.

47. Unknowns:

x = amount invested in company X

y = amount invested in company Y

z = amount invested in company Z

w = amount invested in company W

Investments totaled $65 million:

   $x + y + z + w = 65$

Total return on investments was $8:

   $0.15x - 0.20y + 0.20w = 8$

Colossal invested twice as much in company X as in company Z. Rewording this: The amount invested in company X was twice the amount invested in company Z:

$x = 2z$, or

$x - 2z = 0$

The amount invested in company W was 3 times the amount invested in company Z:

$w = 3z$, or

$-3z + w = 0$

$$\begin{bmatrix} 1 & 1 & 1 & 1 & 65 \\ 0.15 & -.2 & 0 & 0.2 & 8 \\ 1 & 0 & -2 & 0 & 0 \\ 0 & 0 & -3 & 1 & 0 \end{bmatrix} \begin{matrix} \\ 20R_2 \\ \\ \end{matrix}$$

$$\begin{bmatrix} 1 & 1 & 1 & 1 & 65 \\ 3 & -4 & 0 & 4 & 160 \\ 1 & 0 & -2 & 0 & 0 \\ 0 & 0 & -3 & 1 & 0 \end{bmatrix} \begin{matrix} \\ R_2 - 3R_1 \\ R_3 - R_1 \\ \end{matrix}$$

$$\begin{bmatrix} 1 & 1 & 1 & 1 & 65 \\ 0 & -7 & -3 & 1 & -35 \\ 0 & -1 & -3 & -1 & -65 \\ 0 & 0 & -3 & 1 & 0 \end{bmatrix} \begin{matrix} 7R_1 + R_2 \\ \\ 7R_3 - R_2 \\ \end{matrix}$$

$$\begin{bmatrix} 7 & 0 & 4 & 8 & 420 \\ 0 & -7 & -3 & 1 & -35 \\ 0 & 0 & -18 & -8 & -420 \\ 0 & 0 & -3 & 1 & 0 \end{bmatrix} \begin{matrix} \\ \\ (1/2)R_3 \\ \end{matrix}$$

$$\begin{bmatrix} 7 & 0 & 4 & 8 & 420 \\ 0 & -7 & -3 & 1 & -35 \\ 0 & 0 & -9 & -4 & -210 \\ 0 & 0 & -3 & 1 & 0 \end{bmatrix} \begin{matrix} 9R_1 + 4R_3 \\ 3R_2 - R_3 \\ \\ 3R_4 - R_3 \end{matrix}$$

$$\begin{bmatrix} 63 & 0 & 0 & 56 & 2940 \\ 0 & -21 & 0 & 7 & 105 \\ 0 & 0 & -9 & -4 & -210 \\ 0 & 0 & 0 & 7 & 210 \end{bmatrix} \begin{matrix} (1/7)R_1 \\ (1/7)R_2 \\ \\ (1/7)R_4 \end{matrix}$$

$$\begin{bmatrix} 9 & 0 & 0 & 8 & 420 \\ 0 & -3 & 0 & 1 & 15 \\ 0 & 0 & -9 & -4 & -210 \\ 0 & 0 & 0 & 1 & 30 \end{bmatrix} \begin{matrix} R_1 - 8R_4 \\ R_2 - R_4 \\ R_3 + 4R_4 \\ \end{matrix}$$

$$\begin{bmatrix} 9 & 0 & 0 & 0 & 180 \\ 0 & -3 & 0 & 0 & -15 \\ 0 & 0 & -9 & 0 & -90 \\ 0 & 0 & 0 & 1 & 30 \end{bmatrix} \begin{matrix} (1/9)R_1 \\ -(1/3)R_2 \\ -(1/9)R_3 \\ \end{matrix}$$

$$\begin{bmatrix} 1 & 0 & 0 & 0 & 20 \\ 0 & 1 & 0 & 0 & 5 \\ 0 & 0 & 1 & 0 & 10 \\ 0 & 0 & 0 & 1 & 30 \end{bmatrix}$$

$x = 20, y = 5, z = 10, w = 30$

Since this is a unique solution, Smiley has sufficient information to piece together Colossal's investment portfolio: Colossal invested $20m in company X; $5m in company Y, $10m in company Z, and $30m in company W.

49. It is not realistic to expect to use exactly all of the ingredients. Solutions of the associated system may involve negative numbers or not exist. Only solutions with nonnegative values for all the unknowns correspond to being able to use up all of the ingredients.

51. The blend consists of 100 pounds of ingredient X. This says that $x = 100$, which is a linear equation.

53. The blend contains 30% ingredient Y by weight. Rephrasing: The weight of ingredient Y is 30% of the combined weights of X, Y, and Z:

$y = 0.30(x + y + z)$

$y = 0.30x + 0.30y + 0.30z$

$0.3x - 0.7y + 0.3z = 0$,

which is a linear equation.

55. There is at least 30% ingredient Y by weight. Rephrasing: the weight of ingredient Y is at least 30% of the combined weights of X, Y, and Z:

$y \geq 0.30(x + y + z)$

This gives a linear inequality—not an equation.

Chapter 2 Review Exercises

1. To graph with technology, solve the equations for y:

First line: $y = -x/2 + 2$
Second line: $y = 2x - 1$

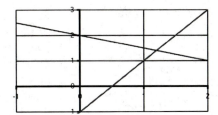

One solution

3. To graph with technology, solve the equations for y:

First line: $y = 2x/3$
Second line: $y = 2x/3$

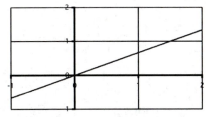

Infinitely many solutions

5. To graph with technology, solve the equations for y:

First line: $y = -x + 1$
Second line: $y = -2x + 0.3$
Third line: $y = -3x/2 + 13/20$

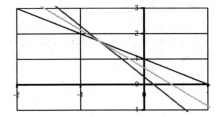

One solution

7. $x + 2y = 4$
    $2x - y = 1$
Multiply the second equation by 2:

$x + 2y = 4$
$4x - 2y = 2$
Adding gives

$5x = 6$, so $x = \dfrac{6}{5}$

Substituting $x = 6/5$ in the first equation gives

$\dfrac{6}{5} + 2y = 4$

$2y = 4 - \dfrac{6}{5} = \dfrac{14}{5}$

$y = \dfrac{7}{5}$

Solution: $x = \dfrac{6}{5}$, $y = \dfrac{7}{5}$

9. $\frac{1}{2}x - \frac{3}{4}y = 0$
     $6x - 9y = 0$

Multiply the first equation by 4 and divide the second by -3:

$2x - 3y = 0$
$-2x + 3y = 0$

Adding gives $0 = 0$, so the system is redundant (the equations give the same lines).
To get the general solution, we solve for x:

$2x = 3y$

$x = \dfrac{3y}{2}$

General solution:

$x = \dfrac{3y}{2}$, y arbitrary, or

$\left(\dfrac{3y}{2}, y\right)$; y arbitrary

11. $x + y = 1$
      $2x + y = 0.3$
      $3x + 2y = \dfrac{13}{10}$

$$\begin{bmatrix} 1 & 1 & 1 \\ 2 & 1 & 3/10 \\ 3 & 2 & 13/10 \end{bmatrix} \begin{matrix} \\ 10R_2 \\ 10R_3 \end{matrix}$$

$$\begin{bmatrix} 1 & 1 & 1 \\ 20 & 10 & 3 \\ 30 & 20 & 13 \end{bmatrix} \begin{matrix} \\ R_2 - 20R_1 \\ R_3 - 30R_1 \end{matrix}$$

$$\begin{bmatrix} 1 & 1 & 1 \\ 0 & -10 & -17 \\ 0 & -10 & -17 \end{bmatrix} \begin{matrix} 10R_1 + R_2 \\ \\ R_3 - R_2 \end{matrix}$$

$$\begin{bmatrix} 10 & 0 & -7 \\ 0 & -10 & -17 \\ 0 & 0 & 0 \end{bmatrix} \begin{matrix} (1/10)R_1 \\ -(1/10)R_2 \\ \\ \end{matrix}$$

$$\begin{bmatrix} 1 & 0 & -7/10 \\ 0 & 1 & 17/10 \\ 0 & 0 & 0 \end{bmatrix}$$

Solution: $x = -0.7$, $y = 1.7$

13. $\begin{aligned} x + 2y \quad &= -3 \\ x \quad - z &= 0 \\ x + 3y - 2z &= -2 \end{aligned}$

$$\begin{bmatrix} 1 & 2 & 0 & -3 \\ 1 & 0 & -1 & 0 \\ 1 & 3 & -2 & -2 \end{bmatrix} \begin{matrix} \\ R_2 - R_1 \\ R_3 - R_1 \end{matrix}$$

$$\begin{bmatrix} 1 & 2 & 0 & -3 \\ 0 & -2 & -1 & 3 \\ 0 & 1 & -2 & 1 \end{bmatrix} \begin{matrix} R_1 + R_2 \\ \\ 2R_3 + R_2 \end{matrix}$$

$$\begin{bmatrix} 1 & 0 & -1 & 0 \\ 0 & -2 & -1 & 3 \\ 0 & 0 & -5 & 5 \end{bmatrix} \begin{matrix} \\ \\ (1/5)R_3 \end{matrix}$$

$$\begin{bmatrix} 1 & 0 & -1 & 0 \\ 0 & -2 & -1 & 3 \\ 0 & 0 & -1 & 1 \end{bmatrix} \begin{matrix} R_1 - R_3 \\ R_2 - R_3 \\ \\ \end{matrix}$$

$$\begin{bmatrix} 1 & 0 & 0 & -1 \\ 0 & -2 & 0 & 2 \\ 0 & 0 & -1 & 1 \end{bmatrix} \begin{matrix} \\ -(1/2)R_2 \\ -R_3 \end{matrix}$$

$$\begin{bmatrix} 1 & 0 & 0 & -1 \\ 0 & 1 & 0 & -1 \\ 0 & 0 & 1 & -1 \end{bmatrix}$$

Solution: $x = -1$, $y = -1$, $z = -1$

15. $\begin{aligned} x - \tfrac{1}{2}y + z &= 0 \\ \tfrac{1}{2}x \quad - \tfrac{1}{2}z &= -1 \\ \tfrac{3}{2}x - \tfrac{1}{2}y + \tfrac{1}{2}z &= -1 \end{aligned}$

$$\begin{bmatrix} 1 & -1/2 & 1 & 0 \\ 1/2 & 0 & -1/2 & -1 \\ 3/2 & -1/2 & 1/2 & -1 \end{bmatrix} \begin{matrix} 2R_1 \\ 2R_2 \\ 2R_3 \end{matrix}$$

$$\begin{bmatrix} 2 & -1 & 2 & 0 \\ 1 & 0 & -1 & -2 \\ 3 & -1 & 1 & -2 \end{bmatrix} \begin{matrix} \\ 2R_2 - R_1 \\ 2R_3 - 3R_1 \end{matrix}$$

$$\begin{bmatrix} 2 & -1 & 2 & 0 \\ 0 & 1 & -4 & -4 \\ 0 & 1 & -4 & -4 \end{bmatrix} \begin{matrix} R_1 + R_2 \\ \\ R_3 - R_2 \end{matrix}$$

$$\begin{bmatrix} 2 & 0 & -2 & -4 \\ 0 & 1 & -4 & -4 \\ 0 & 0 & 0 & 0 \end{bmatrix} \begin{matrix} (1/2)R_1 \\ \\ \\ \end{matrix}$$

$$\begin{bmatrix} 1 & 0 & -1 & -2 \\ 0 & 1 & -4 & -4 \\ 0 & 0 & 0 & 0 \end{bmatrix}$$

Translating back to equations:
$$\begin{aligned} x - z &= -2 \\ y - 4z &= -4 \end{aligned}$$
General solution:
$$\begin{aligned} x &= z - 2 \\ y &= 4z - 4 = 4(z-1) \\ z &\text{ arbitrary.} \end{aligned}$$
or
$$(z-2, 4(z-1), z); z \text{ arbitrary}$$

17. Rewrite the given equations in standard form:
$$\begin{aligned} x - \tfrac{1}{2}y &= 0 \\ \tfrac{1}{2}x + \tfrac{1}{2}z &= 2 \\ 3x - y + z &= 0 \end{aligned}$$

$$\begin{bmatrix} 1 & -1/2 & 0 & 0 \\ 1/2 & 0 & 1/2 & 2 \\ 3 & -1 & 1 & 0 \end{bmatrix} \begin{matrix} 2R_1 \\ 2R_2 \\ \\ \end{matrix}$$

$$\begin{bmatrix} 2 & -1 & 0 & 0 \\ 1 & 0 & 1 & 4 \\ 3 & -1 & 1 & 0 \end{bmatrix} \begin{array}{l} \\ 2R_2 - R_1 \\ 2R_3 - 3R_1 \end{array}$$

$$\begin{bmatrix} 2 & -1 & 0 & 0 \\ 0 & 1 & 2 & 8 \\ 0 & 1 & 2 & 0 \end{bmatrix} \begin{array}{l} R_1 + R_2 \\ \\ R_3 - R_2 \end{array}$$

$$\begin{bmatrix} 2 & 0 & 2 & 8 \\ 0 & 1 & 2 & 8 \\ 0 & 0 & 0 & -8 \end{bmatrix}$$

The last row translates to the false statement $0 = -8$, showing that there is no solution for the system.

19. $5F - 9C = 160$
We are given the additional information
$\qquad F = C$, or $F - C = 0$
giving us a second linear equation.
We can solve the resulting system of 2 equations in 2 unknowns by multiplying the second equation by -5:
$\qquad 5F - 9C = 160$,
$\qquad -5F + 5C = 0$
adding gives
$\qquad -4C = 160$
$\qquad C = -40°$
Since $C = F$, $F = -40°$ also.

21. $5F - 9C = 160$
We are told that the Fahrenheit temperature is 1.8 times the Celsius temperature: $F = 1.8C$, or $F - 1.8C = 0$, giving us the system
$\qquad 5F - 9C = 160$
$\qquad F - 1.8C = 0$
Multiplying the second equation by -5 gives
$\qquad 5F - 9C = 160$
$\qquad -5F + 9C = 0$
Adding gives the false statement $0 = 160$, showing that the given system is inconsistent (has no solution). Thus, it is not possible for the Fahrenheit temperature of an object to be 1.8 times its Celsius temperature.

In Exercises 23–28, use the following unknowns:
$\qquad x =$ the population (in millions) of city A
$\qquad y =$ the population (in millions) of city B
$\qquad z =$ the population (in millions) of city C
$\qquad w =$ the population (in millions) of city D

23. The total population of the four cities is 10 million people
$\qquad x + y + z + w = 10$
This is a linear equation.

25. there are no people living in city D:
$\qquad w = 0$
This is a linear equation.

27. City C has 30% more people than City B.
Rephrasing: the population of City C is 130% of the population of City B:
$\qquad z = 1.30y$, or
$\qquad -1.30y + z = 0$
This is a linear equation.

29. Unknowns:
$x =$ the number of packages from Duffin House
$y =$ the number of packages from Higgins Press
Arrange the given information in a table:

	Duffin (x)	Higgins (y)	Desired totals
Horror	5	5	4500
Romance	5	11	6600

Horror: $5x + 5y = 4500$
Romance: $5x + 11y = 6600$
To solve, multiply the first equation by -1 and add, to obtain
$\qquad 6y = 2100$
$\qquad y = 350$
The first equation can be divided by 5 and rewritten as $x + y = 900$ Substituting $y = 350$ gives us
$\qquad x + 350 = 900$
$\qquad x = 550$
Solution: Purchase 550 packages from Duffin House, 350 from Higgins Press

31. Unknowns:
$x =$ the number of packages from Duffin House
$y =$ the number of packages from Higgins Press

Cost: $50x + 150y = 90{,}000$
Also, you have promised to spend twice as much money for books from Duffin as from Higgins:
Amount spent on Duffin = 2× amount spent on Higgins

$50x = 2(150)y$, or
$50x - 300y = 0$

Dividing each of these equations by 50 gives:

$x + 3y = 1800$
$x - 6y = 0$

Muiltiplying the first equation by 2 and adding gives:

$3x = 3600$
$x = 1200$

Substituting $x = 1200$ in the second equation gives

$1200 - 6y = 0$
$6y = 1200$
$y = 200$

Solution: Purchase 1200 packages from Duffin House, 200 from Higgins Press.

33. Demand: $q = -1000p + 140{,}000$
   Supply: $q = 2000p + 20{,}000$
For the equilibrium price, we can equate the supply and demand:

Demand = Supply
$-1000p + 140{,}000 = 2000p + 20{,}000$
$-3000p = -120{,}000$
$p = \dfrac{120{,}000}{3000}$ = \$40 per book

35. Unknowns:
x = the number of baby sharks
y = the number of piranhas
z = the number of squids
Arrange the given data in a table with the unknowns across the top:

	Sharks (x)	Piranha (y)	Squid (z)	Total Consumed
Goldfish	1	1	1	21
Angelfish	2	0	1	21
Butterfly fish	2	3	0	35

Goldfish: $x + y + z = 21$
Angelfish: $2x + z = 21$

Butterfly fish: $2x + 3y = 35$

$$\left[\begin{array}{ccc|c} 1 & 1 & 1 & 21 \\ 2 & 0 & 1 & 21 \\ 2 & 3 & 0 & 35 \end{array}\right] \begin{array}{l} \\ R_2 - 2R_1 \\ R_3 - 2R_1 \end{array}$$

$$\left[\begin{array}{ccc|c} 1 & 1 & 1 & 21 \\ 0 & -2 & -1 & -21 \\ 0 & 1 & -2 & -7 \end{array}\right] \begin{array}{l} 2R_1 + R_2 \\ \\ 2R_3 + R_2 \end{array}$$

$$\left[\begin{array}{ccc|c} 2 & 0 & 1 & 21 \\ 0 & -2 & -1 & -21 \\ 0 & 0 & -5 & -35 \end{array}\right] \begin{array}{l} \\ \\ (1/5)R_3 \end{array}$$

$$\left[\begin{array}{ccc|c} 2 & 0 & 1 & 21 \\ 0 & -2 & -1 & -21 \\ 0 & 0 & -1 & -7 \end{array}\right] \begin{array}{l} R_1 + R_3 \\ R_2 - R_3 \\ \\ \end{array}$$

$$\left[\begin{array}{ccc|c} 2 & 0 & 0 & 14 \\ 0 & -2 & 0 & -14 \\ 0 & 0 & -1 & -7 \end{array}\right] \begin{array}{l} (1/2)R_1 \\ -(1/2)R_2 \\ -R_3 \end{array}$$

$$\left[\begin{array}{ccc|c} 1 & 0 & 0 & 7 \\ 0 & 1 & 0 & 7 \\ 0 & 0 & 1 & 7 \end{array}\right]$$

$x = 7, y = 7, z = 7$
Solution: You have 7 of each type of carnivorous creature.

37. Unknowns:
x = The number of hits at OHaganBooks.com
y = The number of hits at JungleBooks.com
z = The number of hits at FarmerBooks.com
We are given the following information:
(1) Combined web site traffic at the three sites is estimated at 10,000 hits per day.
   $x + y + z = 10{,}000$
(2) The total number of orders is 1500 per day:
   $0.10x + 0.20y + 0.20z = 1500$
(3) FarmerBooks.com gets as many book orders as the other two combined.
   $0.20z = 0.10x + 0.20y$, or
   $0.10x + 0.20y - 0.20z = 0$
Solving this system of 3 linear equations in 2 unknowns:

$$\begin{bmatrix} 1 & 1 & 1 & 10000 \\ 0.1 & 0.2 & 0.2 & 1500 \\ 0.1 & 0.2 & -.2 & 0 \end{bmatrix} \begin{matrix} \\ 10R_2 \\ 10R_3 \end{matrix}$$

$$\begin{bmatrix} 1 & 1 & 1 & 10000 \\ 1 & 2 & 2 & 15000 \\ 1 & 2 & -2 & 0 \end{bmatrix} \begin{matrix} \\ R_2 - R_1 \\ R_3 - R_1 \end{matrix}$$

$$\begin{bmatrix} 1 & 1 & 1 & 10000 \\ 0 & 1 & 1 & 5000 \\ 0 & 1 & -3 & -10000 \end{bmatrix} \begin{matrix} R_1 - R_2 \\ \\ R_3 - R_2 \end{matrix}$$

$$\begin{bmatrix} 1 & 0 & 0 & 5000 \\ 0 & 1 & 1 & 5000 \\ 0 & 0 & -4 & -15000 \end{bmatrix} \begin{matrix} \\ \\ (1/4)R_3 \end{matrix}$$

$$\begin{bmatrix} 1 & 0 & 0 & 5000 \\ 0 & 1 & 1 & 5000 \\ 0 & 0 & -1 & -3750 \end{bmatrix} \begin{matrix} \\ R_2 + R_3 \\ \end{matrix}$$

$$\begin{bmatrix} 1 & 0 & 0 & 5000 \\ 0 & 1 & 0 & 1250 \\ 0 & 0 & -1 & -3750 \end{bmatrix} \begin{matrix} \\ \\ -R_3 \end{matrix}$$

$$\begin{bmatrix} 1 & 0 & 0 & 5000 \\ 0 & 1 & 0 & 1250 \\ 0 & 0 & 1 & 3750 \end{bmatrix}$$

Solution: 5000 hits per day at OHaganBooks.com, 1250 at JungleBooks.com, 3750 at FarmerBooks.com

39. Unknowns:
x = the number of shares of HAL
y = the number of shares of POM
z = the number of shares of WELL
The total investment was $12,400:
  Investment in HAL = x shares @ $100 = $100x$
  Investment in POM = y shares @ $20 = $20y$
  Investment in WELL = z shares @ $25 = $25z$
Thus,
  $100x + 20y + 25z = 12,400$
You earned $56 in dividends:
  HAL dividend = 0.5% of $100x$ invested
    = $0.005(100x)$ = $0.5x$

POM dividend = 1.5% of $20y$ invested
  = $0.015(20y)$ = $0.3y$
WELL dividend = 0
Thus,
$0.5x + 0.3y = 56$
You purchased a total of 200 shares:
  $x + y + z = 200$
We therefore have the following system:
  $100x + 20y + 25z = 12,400$
  $0.5x + 0.3y = 56$
  $x + y + z = 200$
Solving:

$$\begin{bmatrix} 100 & 20 & 25 & 12400 \\ 0.5 & 0.3 & 0 & 56 \\ 1 & 1 & 1 & 200 \end{bmatrix} \begin{matrix} \\ 10R_2 \\ \end{matrix}$$

$$\begin{bmatrix} 100 & 20 & 25 & 12400 \\ 5 & 3 & 0 & 560 \\ 1 & 1 & 1 & 200 \end{bmatrix} \begin{matrix} (1/5)R_1 \\ \\ \end{matrix}$$

$$\begin{bmatrix} 20 & 4 & 5 & 2480 \\ 5 & 3 & 0 & 560 \\ 1 & 1 & 1 & 200 \end{bmatrix} \begin{matrix} \\ 4R_2 - R_1 \\ 20R_3 - R_1 \end{matrix}$$

$$\begin{bmatrix} 20 & 4 & 5 & 2480 \\ 0 & 8 & -5 & -240 \\ 0 & 16 & 15 & 1520 \end{bmatrix} \begin{matrix} 2R_1 - R_2 \\ \\ R_3 - 2R_2 \end{matrix}$$

$$\begin{bmatrix} 40 & 0 & 15 & 5200 \\ 0 & 8 & -5 & -240 \\ 0 & 0 & 25 & 2000 \end{bmatrix} \begin{matrix} (1/5)R_1 \\ \\ (1/25)R_3 \end{matrix}$$

$$\begin{bmatrix} 8 & 0 & 3 & 1040 \\ 0 & 8 & -5 & -240 \\ 0 & 0 & 1 & 80 \end{bmatrix} \begin{matrix} R_1 - 3R_3 \\ R_2 + 5R_3 \\ \end{matrix}$$

$$\begin{bmatrix} 8 & 0 & 0 & 800 \\ 0 & 8 & 0 & 160 \\ 0 & 0 & 1 & 80 \end{bmatrix} \begin{matrix} (1/8)R_1 \\ (1/8)R_2 \\ \end{matrix}$$

$$\begin{bmatrix} 1 & 0 & 0 & 100 \\ 0 & 1 & 0 & 20 \\ 0 & 0 & 1 & 80 \end{bmatrix}$$

$x = 100, y = 20, z = 80$

Solution: He purchased 100 shares of HAL, 20 shares of POM, and 80 shares of WELL.

41. Unknowns:

x = the number of credits of Liberal Arts

y = the number of credits of Sciences

z = the number of credits of Fine Arts

w = the number of credits of Mathematics

Given information:

(1) The total number of credits is 124:

$x + y + z + w = 124$

(2) An equal number of Science and Fine Arts credit:

$y = z$, or $y - z = 0$

(3) Twice as many Mathematics credits as Science credits and Fine Arts credits combined. Rephrasing: The number of Mathematics credits is twice the sum of the numbers of Science and Fine Arts credits:

$w = 2(y + z)$, or

$2y + 2z - w = 0$

(4) Liberal Arts credits exceed Mathematics credits by one third of the number of Fine Arts credits. Rephrasing: the number of Liberal Arts credits minus the number of Mathematics credits is one third of the number of Fine Arts credits:

$x - w = \frac{1}{3}z$, or

$x - \frac{1}{3}z - w = 0$

Solving:

$$\begin{bmatrix} 1 & 1 & 1 & 1 & 124 \\ 0 & 1 & -1 & 0 & 0 \\ 0 & 2 & 2 & -1 & 0 \\ 1 & 0 & -1/3 & -1 & 0 \end{bmatrix} \begin{matrix} \\ \\ \\ 3R_4 \end{matrix}$$

$$\begin{bmatrix} 1 & 1 & 1 & 1 & 124 \\ 0 & 1 & -1 & 0 & 0 \\ 0 & 2 & 2 & -1 & 0 \\ 3 & 0 & -1 & -3 & 0 \end{bmatrix} \begin{matrix} \\ \\ \\ R_4 - 3R_1 \end{matrix}$$

$$\begin{bmatrix} 1 & 1 & 1 & 1 & 124 \\ 0 & 1 & -1 & 0 & 0 \\ 0 & 2 & 2 & -1 & 0 \\ 0 & -3 & -4 & -6 & -372 \end{bmatrix} \begin{matrix} R_1 - R_2 \\ \\ R_3 - 2R_2 \\ R_4 + 3R_2 \end{matrix}$$

$$\begin{bmatrix} 1 & 0 & 2 & 1 & 124 \\ 0 & 1 & -1 & 0 & 0 \\ 0 & 0 & 4 & -1 & 0 \\ 0 & 0 & -7 & -6 & -372 \end{bmatrix} \begin{matrix} 2R_1 - R_3 \\ 4R_2 + R_3 \\ \\ 4R_4 + 7R_3 \end{matrix}$$

$$\begin{bmatrix} 2 & 0 & 0 & 3 & 248 \\ 0 & 4 & 0 & -1 & 0 \\ 0 & 0 & 4 & -1 & 0 \\ 0 & 0 & 0 & -31 & -1488 \end{bmatrix} \begin{matrix} \\ \\ \\ (1/31)R_4 \end{matrix}$$

$$\begin{bmatrix} 2 & 0 & 0 & 3 & 248 \\ 0 & 4 & 0 & -1 & 0 \\ 0 & 0 & 4 & -1 & 0 \\ 0 & 0 & 0 & -1 & -48 \end{bmatrix} \begin{matrix} R_1 + 3R_4 \\ R_2 - R_4 \\ R_3 - R_4 \\ \end{matrix}$$

$$\begin{bmatrix} 2 & 0 & 0 & 0 & 104 \\ 0 & 4 & 0 & 0 & 48 \\ 0 & 0 & 4 & 0 & 48 \\ 0 & 0 & 0 & -1 & -48 \end{bmatrix} \begin{matrix} (1/2)R_1 \\ (1/4)R_2 \\ (1/4)R_3 \\ -R_4 \end{matrix}$$

$$\begin{bmatrix} 1 & 0 & 0 & 0 & 52 \\ 0 & 1 & 0 & 0 & 12 \\ 0 & 0 & 1 & 0 & 12 \\ 0 & 0 & 0 & 1 & 48 \end{bmatrix}$$

Solution: Billy Sean is forced to take exactly the following combination: Liberal Arts: 52 credits, Sciences: 12 credits, Fine Arts: 12 credits, Mathematics: 48 credits.

43.

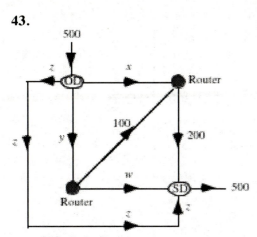

OD = Order Department
SD = Shipping Department

(a) Order Department: Traffic in = Traffic out
$$500 = x + y + z$$
Top right router: Traffic in = Traffic out
$$100 + x = 200, \text{ or } x = 100$$
Bottom left router: Traffic in = Traffic out
$$y = 100 + w, \text{ or } y - w = 100$$
Shipping Department: Traffic in = Traffic out
$$z + w + 200 = 500, \text{ or } z + w = 300$$
Solving:

$$\begin{bmatrix} 1 & 1 & 1 & 0 & 500 \\ 1 & 0 & 0 & 0 & 100 \\ 0 & 1 & 0 & -1 & 100 \\ 0 & 0 & 1 & 1 & 300 \end{bmatrix} \begin{matrix} \\ R_2 - R_1 \\ \\ \\ \end{matrix}$$

$$\begin{bmatrix} 1 & 1 & 1 & 0 & 500 \\ 0 & -1 & -1 & 0 & -400 \\ 0 & 1 & 0 & -1 & 100 \\ 0 & 0 & 1 & 1 & 300 \end{bmatrix} \begin{matrix} R_1 + R_2 \\ \\ R_3 + R_2 \\ \\ \end{matrix}$$

$$\begin{bmatrix} 1 & 0 & 0 & 0 & 100 \\ 0 & -1 & -1 & 0 & -400 \\ 0 & 0 & -1 & -1 & -300 \\ 0 & 0 & 1 & 1 & 300 \end{bmatrix} \begin{matrix} \\ R_2 - R_3 \\ \\ R_4 + R_3 \end{matrix}$$

$$\begin{bmatrix} 1 & 0 & 0 & 0 & 100 \\ 0 & -1 & 0 & 1 & -100 \\ 0 & 0 & -1 & -1 & -300 \\ 0 & 0 & 0 & 0 & 0 \end{bmatrix} \begin{matrix} \\ -R_2 \\ -R_3 \\ \end{matrix}$$

$$\begin{bmatrix} 1 & 0 & 0 & 0 & 100 \\ 0 & 1 & 0 & -1 & 100 \\ 0 & 0 & 1 & 1 & 300 \\ 0 & 0 & 0 & 0 & 0 \end{bmatrix}$$

Translating back to equations:
$$x = 100$$
$$y - w = 100$$
$$z + w = 300$$
General solution:
$$x = 100$$
$$y = 100 + w$$
$$z = 300 - w$$
$$w \text{ arbitrary}$$

(b) Since $y = 100 + w$, and w can be any number ≥ 0 (w cannot be negative because it represents the number of book orders along any route). Therefore, the smallest possible value of y is 100 books per day.

(c) The equation $z = 300 - w$ tells us that w cannot exceed 300 books per day., or else x would become negative.

(d) If there is no traffic along z, then $z = 0$, giving:
$$x = 100$$
$$y = 100 + w$$
$$0 = 300 - w$$
Thus, from the third equation, $w = 300$, giving us the particular solution
$$x = 100, y = 100+300 = 400, z = 0, w = 300$$

(e) If there is the same volume of traffic along y and z, then $y = z$, and so
$$100 + w = 300 - w$$
Thus,
$$2w = 200,$$
so $w = 100$ books per day.

45. Unknowns:

x = the number of packages from New York to Texas

y = the number of packages from New York to California

z = the number of packages from Illinois to Texas

w = the number of packages from Illinois to California

Books to Texas: $x + z = 600$

Books to California: $y + w = 200$

Books from Illinois: $z + w = 300$

Budget: $20x + 50y + 30z + 40w = 22{,}000$

Solving:

$$\begin{bmatrix} 1 & 0 & 1 & 0 & 600 \\ 0 & 1 & 0 & 1 & 200 \\ 0 & 0 & 1 & 1 & 300 \\ 20 & 50 & 30 & 40 & 22000 \end{bmatrix} \begin{matrix} \\ \\ \\ (1/10)R_4 \end{matrix}$$

$$\begin{bmatrix} 1 & 0 & 1 & 0 & 600 \\ 0 & 1 & 0 & 1 & 200 \\ 0 & 0 & 1 & 1 & 300 \\ 2 & 5 & 3 & 4 & 2200 \end{bmatrix} \begin{matrix} \\ \\ \\ R_4 - 2R_1 \end{matrix}$$

$$\begin{bmatrix} 1 & 0 & 1 & 0 & 600 \\ 0 & 1 & 0 & 1 & 200 \\ 0 & 0 & 1 & 1 & 300 \\ 0 & 5 & 1 & 4 & 1000 \end{bmatrix} \begin{matrix} \\ \\ \\ R_4 - 5R_2 \end{matrix}$$

$$\begin{bmatrix} 1 & 0 & 1 & 0 & 600 \\ 0 & 1 & 0 & 1 & 200 \\ 0 & 0 & 1 & 1 & 300 \\ 0 & 0 & 1 & -1 & 0 \end{bmatrix} \begin{matrix} R_1 - R_3 \\ \\ \\ R_4 - R_3 \end{matrix}$$

$$\begin{bmatrix} 1 & 0 & 0 & -1 & 300 \\ 0 & 1 & 0 & 1 & 200 \\ 0 & 0 & 1 & 1 & 300 \\ 0 & 0 & 0 & -2 & -300 \end{bmatrix} \begin{matrix} \\ \\ \\ (1/2)R_4 \end{matrix}$$

$$\begin{bmatrix} 1 & 0 & 0 & -1 & 300 \\ 0 & 1 & 0 & 1 & 200 \\ 0 & 0 & 1 & 1 & 300 \\ 0 & 0 & 0 & -1 & -150 \end{bmatrix} \begin{matrix} R_1 - R_4 \\ R_2 + R_4 \\ R_3 + R_4 \\ \end{matrix}$$

$$\begin{bmatrix} 1 & 0 & 0 & 0 & 450 \\ 0 & 1 & 0 & 0 & 50 \\ 0 & 0 & 1 & 0 & 150 \\ 0 & 0 & 0 & -1 & -150 \end{bmatrix} \begin{matrix} \\ \\ \\ -R_4 \end{matrix}$$

$$\begin{bmatrix} 1 & 0 & 0 & 0 & 450 \\ 0 & 1 & 0 & 0 & 50 \\ 0 & 0 & 1 & 0 & 150 \\ 0 & 0 & 0 & 1 & 150 \end{bmatrix}$$

Solution: New York to Texas: 450 packages, New York to California: 50 packages, Illinois to Texas: 150 packages, Illinois to California: 150 packages.

Chapter 3 Matrix Algebra and Applications

3.1 Matrix Addition and Scalar Multiplication

1. $A = \begin{bmatrix} 1 & 5 & 0 & \frac{1}{4} \end{bmatrix}$

A has 1 row and 4 columns. Therefore, it is a 1×4 matrix. a_{13} is the entry in row 1 and column 3, so $a_{13} = 0$.

3. $C = \begin{bmatrix} \frac{5}{2} \\ 1 \\ -2 \\ 8 \end{bmatrix}$

C has 4 rows and 1 column. Therefore, it is a 4×1 matrix. C_{11} is the entry in row 1 and column 1, so $C_{11} = \frac{5}{2}$.

5. $E = \begin{bmatrix} e_{11} & e_{12} & e_{13} & \cdots & e_{1q} \\ e_{21} & e_{22} & e_{23} & \cdots & e_{2q} \\ \vdots & \vdots & \vdots & \ddots & \vdots \\ e_{p1} & e_{p2} & e_{p3} & \cdots & e_{pq} \end{bmatrix}$

E has p rows and q columns. Therefore, it is a $p \times q$ matrix. E_{22} is the entry in row 2 and column 2, so $E_{22} = c_{22}$.

7. $B = \begin{bmatrix} 1 & 3 \\ 5 & -6 \end{bmatrix}$

B has 2 rows and 2 columns. Therefore, it is a 2×2 matrix. b_{12} is the entry in row 1 and column 2, so $b_{12} = 3$.

9. $D = \begin{bmatrix} d_1 & d_2 & \cdots & d_n \end{bmatrix}$

D has 1 row and n columns. Therefore, it is a $1 \times n$ matrix. D_{1r} is the entry in row 1 and column r, so $D_{1r} = d_r$.

11. $\begin{bmatrix} x+y & x+z \\ y+z & w \end{bmatrix} = \begin{bmatrix} 3 & 4 \\ 5 & 4 \end{bmatrix}$

Equating corresponding entries gives

$x + y = 3$

$x + z = 4$

$y + z = 5$

$w = 4$

The first three equations can be solved in various ways. Let us use row reduction:

$\begin{bmatrix} 1 & 1 & 0 & 3 \\ 1 & 0 & 1 & 4 \\ 0 & 1 & 1 & 5 \end{bmatrix} R_2 - R_1$

$\begin{bmatrix} 1 & 1 & 0 & 3 \\ 0 & -1 & 1 & 1 \\ 0 & 1 & 1 & 5 \end{bmatrix} \begin{matrix} R_1 + R_2 \\ \\ R_3 + R_2 \end{matrix}$

$\begin{bmatrix} 1 & 0 & 1 & 4 \\ 0 & -1 & 1 & 1 \\ 0 & 0 & 2 & 6 \end{bmatrix} (1/2)R_3$

$\begin{bmatrix} 1 & 0 & 1 & 4 \\ 0 & -1 & 1 & 1 \\ 0 & 0 & 1 & 3 \end{bmatrix} \begin{matrix} R_1 - R_3 \\ R_2 - R_3 \end{matrix}$

$\begin{bmatrix} 1 & 0 & 0 & 1 \\ 0 & -1 & 0 & -2 \\ 0 & 0 & 1 & 3 \end{bmatrix} -R_2$

$\begin{bmatrix} 1 & 0 & 0 & 1 \\ 0 & 1 & 0 & 2 \\ 0 & 0 & 1 & 3 \end{bmatrix}$

Thus, $x = 1$, $y = 2$, $z = 3$, $w = 4$.

13. We obtain $A+B$ by adding corresponding entries:

$A + B = \begin{bmatrix} 0 & -1 \\ 1 & 0 \\ -1 & 2 \end{bmatrix} + \begin{bmatrix} 0.25 & -1 \\ 0 & 0.5 \\ -1 & 3 \end{bmatrix}$

$= \begin{bmatrix} 0.25 & -2 \\ 1 & 0.5 \\ -2 & 5 \end{bmatrix}$

15. To obtain $A+B-C$, add corresponding entries of A and B and subtract those of C:

$$A+B-C = \begin{bmatrix} 0 & -1 \\ 1 & 0 \\ -1 & 2 \end{bmatrix} + \begin{bmatrix} 0.25 & -1 \\ 0 & 0.5 \\ -1 & 3 \end{bmatrix} -$$

$$\begin{bmatrix} 1 & -1 \\ 1 & 1 \\ -1 & -1 \end{bmatrix}$$

$$= \begin{bmatrix} -0.75 & -1 \\ 0 & -0.5 \\ -1 & 6 \end{bmatrix}$$

17. $2A-C = 2\begin{bmatrix} 0 & -1 \\ 1 & 0 \\ -1 & 2 \end{bmatrix} - \begin{bmatrix} 1 & -1 \\ 1 & 1 \\ -1 & -1 \end{bmatrix}$

$$= \begin{bmatrix} -1 & -1 \\ 1 & -1 \\ -1 & 5 \end{bmatrix}$$

19. The transpose, A^T, of A is obtained by writing the rows of A as columns:

$$A = \begin{bmatrix} 0 & -1 \\ 1 & 0 \\ -1 & 2 \end{bmatrix}; A^T = \begin{bmatrix} 0 & 1 & -1 \\ -1 & 0 & 2 \end{bmatrix}.$$

Thus, $2A^T = 2\begin{bmatrix} 0 & 1 & -1 \\ -1 & 0 & 2 \end{bmatrix} = \begin{bmatrix} 0 & 2 & -2 \\ -2 & 0 & 4 \end{bmatrix}$

21. $A+B = \begin{bmatrix} 1 & -1 & 0 \\ 0 & 2 & -1 \end{bmatrix} + \begin{bmatrix} 3 & 0 & -1 \\ 5 & -1 & 1 \end{bmatrix}$

$$= \begin{bmatrix} 4 & -1 & -1 \\ 5 & 1 & 0 \end{bmatrix}$$

23. $A-B+C = \begin{bmatrix} 1 & -1 & 0 \\ 0 & 2 & -1 \end{bmatrix} - \begin{bmatrix} 3 & 0 & -1 \\ 5 & -1 & 1 \end{bmatrix}$

$+ \begin{bmatrix} x & 1 & w \\ z & r & 4 \end{bmatrix}$

$$= \begin{bmatrix} -2+x & 0 & 1+w \\ -5+z & 3+r & 2 \end{bmatrix}$$

25. $2A-B = 2\begin{bmatrix} 1 & -1 & 0 \\ 0 & 2 & -1 \end{bmatrix} - \begin{bmatrix} 3 & 0 & -1 \\ 5 & -1 & 1 \end{bmatrix}$

$$= \begin{bmatrix} -1 & -2 & 1 \\ -5 & 5 & -3 \end{bmatrix}$$

27. $B = \begin{bmatrix} 3 & 0 & -1 \\ 5 & -1 & 1 \end{bmatrix}; B^T = \begin{bmatrix} 3 & 5 \\ 0 & -1 \\ -1 & 1 \end{bmatrix}$

Therefore, $3B^T = 3\begin{bmatrix} 3 & 5 \\ 0 & -1 \\ -1 & 1 \end{bmatrix} = \begin{bmatrix} 9 & 15 \\ 0 & -3 \\ -3 & 3 \end{bmatrix}$

Exercises 29–36 can be answered using either a graphing calculator or the on-line Matrix Algebra Tool at
FiniteMath.org →Student Web Site → On-Line Utilities → Matrix Algebra Tool

29. $A - C = \begin{bmatrix} 1.5 & -2.35 & 5.6 \\ 44.2 & 0 & 12.2 \end{bmatrix} -$

$\begin{bmatrix} 10 & 20 & 30 \\ -10 & -20 & -30 \end{bmatrix}$

$$= \begin{bmatrix} -8.5 & -22.35 & -24.4 \\ 54.2 & 20 & 42.2 \end{bmatrix}$$

TI-83/84 Plus format: `[A]-[C]`
On-line Matrix Algebra Tool format: `A-C`

31. $1.1B = 1.1\begin{bmatrix} 1.4 & 7.8 \\ 5.4 & 0 \\ 5.6 & 6.6 \end{bmatrix} = \begin{bmatrix} 1.54 & 8.58 \\ 5.94 & 0 \\ 6.16 & 7.26 \end{bmatrix}$

TI-83/84 Plus format: `1.1[B]`
On-line Matrix Algebra Tool format: `1.1*B`

33. $A^T+4.2B = \begin{bmatrix} 1.5 & 44.2 \\ -2.35 & 0 \\ 5.6 & 12.2 \end{bmatrix} + 4.2\begin{bmatrix} 1.4 & 7.8 \\ 5.4 & 0 \\ 5.6 & 6.6 \end{bmatrix}$

$$= \begin{bmatrix} 7.38 & 76.96 \\ 20.33 & 0 \\ 29.12 & 39.92 \end{bmatrix}$$

TI-83/84 Plus format: `[A]T+4.2[B]`
On-line Matrix Algebra Tool format:
`A^T+4.2*B`

35. $(2.1A-2.3C)^T = \begin{bmatrix} -19.85 & 115.82 \\ -50.935 & 46 \\ -57.24 & 94.62 \end{bmatrix}$

TI-83/84 Plus format: `(2.1[A]-2.3[C])T`
On-line Matrix Algebra Tool format: `(2.1*A-2.3*C)^T`

37.

	Macs	**iPods**	**iPhones***
Q4 2006	1.6	8.8	0
Change in 2007	0.6	1.5	1.1
Change in 2008	0.4	0.9	5.8

Regard each row of the table as a matrix:

Sales in 2006 Q4 = [1.6 8.8 0]

Increase in 2007 = [0.6 0.5 1.1]

Increase in 2008 = [0.4 0.9 5.8]

Sales in 2007 Q4 = Sales in 2006 Q4 + Increase in 2007

= [1.6 8.8 0] + [0.6 0.5 1.1]

= [2.2 10.3 1.1]

Sales in 2008 Q4 = Sales in 2007 Q4 + Increase in 2008

= [2.2 10.3 1.1] + [0.4 0.9 5.8]

= [2.6 11.2 6.9]

39. Write the given inventory table as a 2×3 matrix:

$$\text{Inventory} = \begin{bmatrix} 1000 & 2000 & 5000 \\ 1000 & 5000 & 2000 \end{bmatrix}$$

Write the given sales figures as a similar matrix:

$$\text{Sales} = \begin{bmatrix} 700 & 1300 & 2000 \\ 400 & 300 & 500 \end{bmatrix}$$

We then compute the remaining inventory by subtracting the sales:

Remaining Inventory = Inventory – Sales

$$= \begin{bmatrix} 1000 & 2000 & 5000 \\ 1000 & 5000 & 2000 \end{bmatrix} -$$

$$\begin{bmatrix} 700 & 1300 & 2000 \\ 400 & 300 & 500 \end{bmatrix}$$

$$= \begin{bmatrix} 300 & 700 & 3000 \\ 600 & 4700 & 1500 \end{bmatrix}$$

41.

◇	A	B	C	D
1	**Revenue**			
2		**2004**	**2004**	**2006**
3	**Full Boots**	$10,000	$9,000	$11,000
4	**Half Boots**	$8,000	$7,200	$8,800
5	**Sandals**	$4,000	$5,000	$6,000
6				
7	**Production Costs**			
8		**2004**	**2004**	**2006**
9	**Full Boots**	$2,000	$1,800	$2,200
10	**Half Boots**	$2,400	$1,440	$1,760
11	**Sandals**	$1,200	$1,500	$2,000

Arrange the revenue and costs in two 3×3 matrices

$$\text{Revenue} = \begin{bmatrix} 10,000 & 9000 & 11,000 \\ 8000 & 7200 & 8800 \\ 4000 & 5000 & 6000 \end{bmatrix}$$

$$\text{Cost} = \begin{bmatrix} 2000 & 1800 & 2200 \\ 2400 & 1440 & 1760 \\ 1200 & 1500 & 2000 \end{bmatrix}$$

Profit = Revenue – Cost

$$= \begin{bmatrix} 10,000 & 9000 & 11,000 \\ 8000 & 7200 & 8800 \\ 4000 & 5000 & 6000 \end{bmatrix} - \begin{bmatrix} 2000 & 1800 & 2200 \\ 2400 & 1440 & 1760 \\ 1200 & 1500 & 2000 \end{bmatrix}$$

$$= \begin{bmatrix} 8000 & 7200 & 8800 \\ 5600 & 5760 & 7040 \\ 2800 & 3500 & 4000 \end{bmatrix}$$

43. Row vectors for the populations, in millions:

1980 distribution = A = [49.1 58.9 75.4 43.2]

1990 distribution = B = [50.8 59.7 85.4 52.8]

Net change 1980 to 1990 = $B - A$ = [1.7 0.8 10 9.6]

Since all the net changes are positive, they represent net increases in the population.

45.

	Aug	**Sept**	**Oct**	**Nov**	**Dec**
California	7100	3300	5700	5000	4300
Florida	3500	900	2400	2800	2300
Texas	1900	1000	1900	2100	2100

Regard each row in the table as a matrix.

Foreclosures in California
= [7100 3300 5700 5000 4300]

Foreclosures in Florida
= [3500 900 2400 2800 2300]

Foreclosures in Texas

91

$= [1900\ 1000\ 1900\ 2100\ 2100]$
Total Foreclosures = Foreclosures in California
          + Foreclosures in Florida + Foreclosures
in Texas
$= [7100\ 3300\ 5700\ 5000\ 4300]$
     $+ [3500\ 900\ 2400\ 2800\ 2300]$
     $+ [1900\ 1000\ 1900\ 2100\ 2100]$
$= [12500\ 5200\ 10000\ 9900\ 8700]$

47. (Refer to the solution to Exercise 45.) The difference between the number of foreclosures in California and in Florida is given by
  Difference = Foreclosures in California
    − Foreclosures in Florida
$= [7100\ 3300\ 5700\ 5000\ 4300]$
  $- [3500\ 900\ 2400\ 2800\ 2300]$
$= [3600\ 2400\ 3300\ 2200\ 2000]$
The difference was greatest in August 2008.

49. First, organize the starting inventory as a 2×3 matrix:

$$\begin{array}{c} \text{Proc Mem Tubes} \\ \begin{array}{c}\text{Pom II}\\\text{Pom Classic}\end{array}\left[\begin{array}{ccc}500 & 5000 & 10{,}000\\200 & 2000 & 20{,}000\end{array}\right]\end{array}$$

Thus,
$$\text{Inventory} = \left[\begin{array}{ccc}500 & 5000 & 10{,}000\\200 & 2000 & 20{,}000\end{array}\right]$$
The parts used in making one of each computer can also be arranged in a matrix:
$$\text{Use} = \left[\begin{array}{ccc}2 & 16 & 20\\1 & 4 & 40\end{array}\right]$$
(a) After two months, the company has made 100 of each computer, so the inventory remaining is
  Inventory Remaining = Inventory − 100·Use
$$= \left[\begin{array}{ccc}500 & 5000 & 10{,}000\\200 & 2000 & 20{,}000\end{array}\right]$$
$$-100\left[\begin{array}{ccc}2 & 16 & 20\\1 & 4 & 40\end{array}\right]$$
$$= \left[\begin{array}{ccc}300 & 3400 & 8000\\100 & 1600 & 16{,}000\end{array}\right]$$
(b) After n months, the company has made $50n$ of each computer. Therefore, the inventory remaining is
$$= \left[\begin{array}{ccc}500 & 5000 & 10{,}000\\200 & 2000 & 20{,}000\end{array}\right]$$
$$-50n\left[\begin{array}{ccc}2 & 16 & 20\\1 & 4 & 40\end{array}\right]$$

$$= \left[\begin{array}{ccc}500-100n & 5000-800n & 10{,}000-1000n\\200-50n & 2000-200n & 20{,}000-2000n\end{array}\right]$$
The 1,1 entry is zero after $n = 5$ months
The 1,2 entry is zero after $n = 6.25$ months
The 1,3 entry is zero after $n = 10$ months
The 2,1 entry is zero after $n = 4$ months
The 2,2 entry is zero after $n = 10$ months
The 2,3 entry is zero after $n = 10$ months
Thus, after 4 months, the first entry becomes zero, meaning that the company will run out of Pom Classic processor chips after 4 months.

51.

TO	Australia	South Africa
FROM		
North America	440	190
Europe	950	950
Asia	1790	200

(a) Arrange the above 1998 tourism figures in a matrix
$$A = \left[\begin{array}{cc}440 & 190\\950 & 950\\1790 & 200\end{array}\right]$$
Then, arrange the predicted changes from 1998 to 2008 in a new matrix:
$$D = \left[\begin{array}{cc}-20 & 40\\50 & 50\\0 & 100\end{array}\right]$$
Tourism in 2008 = Tourism in 1998 + Change
$$= A + D = \left[\begin{array}{cc}420 & 230\\1000 & 1000\\1790 & 300\end{array}\right]$$
(b) The average pf the tourism figures is obtained by taking the average of the entries in A and B. To obtain the average, add and divide by 2:
  Average $= \frac{1}{2}(A+B)$
$$A = 1998 \text{ figures} = \left[\begin{array}{cc}440 & 190\\950 & 950\\1790 & 200\end{array}\right]$$
$$B = 2008 \text{ figures} = \left[\begin{array}{cc}420 & 230\\1000 & 1000\\1790 & 300\end{array}\right]$$
(From part (a))

$$\text{Average} = \tfrac{1}{2}(A+B) = \tfrac{1}{2}\begin{bmatrix} 860 & 420 \\ 1950 & 1950 \\ 3580 & 500 \end{bmatrix}$$

$$= \begin{bmatrix} 430 & 210 \\ 975 & 975 \\ 1790 & 250 \end{bmatrix}$$

53. No; for two matrices to be equal, they must have the same dimension.

55. $(A+B)_{ij} = A_{ij} + B_{ij}$
The left-hand side is the ijth entry of $A+B$. The right-hand side is the sum of the ijth entries of A and B. Thus, the equation tells us that the ij th entry of the sum $A+B$ is obtained by adding the ij th entries of A and B.

57. If A is any matrix, then $A_{11}, A_{22}, ..., A_{ii}, ...$ denote the entries going down the main diagonal (top left to bottom right). Thus, if $A_{ii} = 0$ for every i, then A would have zeros down the diagonal:

$$A = \begin{bmatrix} 0 & \# & \# & \# & \# \\ \# & 0 & \# & \# & \# \\ \# & \# & 0 & \# & \# \\ \# & \# & \# & 0 & \# \\ \# & \# & \# & \# & 0 \end{bmatrix}$$

(The symbols # indicate arbitrary numbers.)

59. The transpose of an $m\times n$ matrix is the $n\times m$ matrix obtained by writing its rows as columns. Thus, the entry originally in Row j and Column i winds up in Row i and Column j when a matrix is transposed. In other words,
  ijth entry of the transpose $= ji$th entry of the original matrix
  $(A^T)_{ij} = A_{ji}$

61. In a skew-symmetric matrix, the entry in the ij position is the negative of that in the ji position. In other words, each entry is the negative of its mirror image in the diagonal.
  What about the diagonal entries? Each diagonal entry stays the same when the matrix is transposed. So, in a skew-symmetric matrix, each diagonal entry must equal its own negative. The

only way this can happen is if the diagonal entries are zero.
  Examples will vary.

(a) $\begin{bmatrix} 0 & -4 \\ 4 & 0 \end{bmatrix}$ **(b)** $\begin{bmatrix} 0 & -4 & 5 \\ 4 & 0 & 1 \\ -5 & -1 & 0 \end{bmatrix}$

63. The associativity of matrix addition is a consequence of the associativity of addition of numbers, since we add matrices by adding the corresponding entries (which are real numbers).

3.2 Matrix Multiplication

1. $[1 \quad 3 \quad -1] \begin{bmatrix} 9 \\ 1 \\ -1 \end{bmatrix} = [9+3+1] = [13]$

3. $[-1 \quad \frac{1}{2}] \begin{bmatrix} -\frac{1}{3} \\ 1 \end{bmatrix} = [\frac{1}{3}+\frac{1}{2}] = [\frac{5}{6}]$

5. $[0 \quad -2 \quad 1] \begin{bmatrix} x \\ y \\ z \end{bmatrix} = [-2y+z]$

7. $[1 \quad 3 \quad 2] \begin{bmatrix} 1 \\ -1 \end{bmatrix}$ is undefined.

(The dimensions are 1×3 and 2×1; the 3 and the 2 do not match.)

9. $[-1 \quad 1] \begin{bmatrix} -3 & 1 & 4 & 3 \\ 0 & 1 & -2 & 1 \end{bmatrix}$

$= [(3+0) \quad (-1+1) \quad (-4-2) \quad (-3+1)]$

$= [3 \quad 0 \quad -6 \quad -2]$

11. $[1 \quad -1 \quad 2 \quad 3] \begin{bmatrix} -1 & 2 & 0 \\ 2 & -1 & 0 \\ 0 & 5 & 2 \\ -1 & 8 & 1 \end{bmatrix}$

$= [(-1-2+0-3) \quad (2+1+10+24) \quad (0+0+4+3)]$

$= [-6 \quad 37 \quad 7]$

13. $\begin{bmatrix} 1 & 0 & -1 \\ 1 & 1 & 2 \end{bmatrix} \begin{bmatrix} 0 & 1 & -1 \\ 1 & 0 & 1 \\ 4 & 8 & 0 \end{bmatrix}$

$= \begin{bmatrix} (0+0-4) & (1+0-8) & (-1+0+0) \\ (0+1+8) & (1+0+16) & (-1+1+0) \end{bmatrix}$

$= \begin{bmatrix} -4 & -7 & -1 \\ 9 & 17 & 0 \end{bmatrix}$

15. $\begin{bmatrix} 1 & 0 \\ 1 & -1 \end{bmatrix} \begin{bmatrix} 0 & 1 \\ 0 & 1 \end{bmatrix}$

$= \begin{bmatrix} (0+0) & (1+0) \\ (0+0) & (1-1) \end{bmatrix} = \begin{bmatrix} 0 & 1 \\ 0 & 0 \end{bmatrix}$

17. $\begin{bmatrix} 0 & 1 \\ 0 & 1 \end{bmatrix} \begin{bmatrix} 1 & 0 \\ 1 & -1 \end{bmatrix}$

$= \begin{bmatrix} (0+1) & (0-1) \\ (0+1) & (0-1) \end{bmatrix} = \begin{bmatrix} 1 & -1 \\ 1 & -1 \end{bmatrix}$

19. $\begin{bmatrix} 1 & -1 \\ 1 & -1 \end{bmatrix} \begin{bmatrix} 2 & 3 \\ 2 & 3 \end{bmatrix}$

$= \begin{bmatrix} (2-2) & (3-3) \\ (2-2) & (3-3) \end{bmatrix} = \begin{bmatrix} 0 & 0 \\ 0 & 0 \end{bmatrix}$

21. $\begin{bmatrix} 1 & -1 \\ -1 & 1 \end{bmatrix} \begin{bmatrix} 2 & 3 \\ 2 & 3 \\ 1 & 1 \end{bmatrix}$ is undefined.

(The dimensions are 2×2 and 3×2; the 2 and the 3 do not match.)

23. $\begin{bmatrix} 1 & 0 & -1 \\ 2 & -2 & 1 \\ 0 & 0 & 1 \end{bmatrix} \begin{bmatrix} 1 & -1 & 4 \\ 1 & 1 & 0 \\ 0 & 4 & 1 \end{bmatrix}$

$= \begin{bmatrix} (1+0+0) & (-1+0-4) & (4+0-1) \\ (2-2+0) & (-2-2+4) & (8+0+1) \\ (0+0+0) & (0+0+4) & (0+0+1) \end{bmatrix}$

$= \begin{bmatrix} 1 & -5 & 3 \\ 0 & 0 & 9 \\ 0 & 4 & 1 \end{bmatrix}$

25. $\begin{bmatrix} 1 & 0 & 1 & 0 \\ -1 & 1 & 0 & 1 \\ -2 & 0 & 1 & 4 \\ 0 & -1 & 0 & 1 \end{bmatrix} \begin{bmatrix} 1 \\ -3 \\ 2 \\ 0 \end{bmatrix} = \begin{bmatrix} (1+0+2+0) \\ (-1-3+0+0) \\ (-2+0+2+0) \\ (0+3+0+0) \end{bmatrix}$

$= \begin{bmatrix} 3 \\ -4 \\ 0 \\ 3 \end{bmatrix}$

27. An on-line matrix algebra utility is available at

FiniteMath.org $\rightarrow$ Student Web Site $\rightarrow$ On-Line Utilities $\rightarrow$ Matrix Algebra Tool

$\begin{bmatrix} 1.1 & 2.3 & 3.4 & -1.2 \\ 3.4 & 4.4 & 2.3 & 1.1 \\ 2.3 & 0 & -2.2 & 1.1 \\ 1.2 & 1.3 & 1.1 & 1.1 \end{bmatrix} \begin{bmatrix} -2.1 & 0 & -3.3 \\ -3.4 & -4.8 & -4.2 \\ 3.4 & 5.6 & 1 \\ 1 & 2.2 & 9.8 \end{bmatrix}$

$= \begin{bmatrix} 0.23 & 5.36 & -21.65 \\ -13.18 & -5.82 & -16.62 \\ -11.21 & -9.9 & 0.99 \\ -2.1 & 2.34 & 2.46 \end{bmatrix}$

TI-83/84 Plus format: `[A]*[B]`

On-line Matrix Algebra Tool format: `A*B`

29. $A = \begin{bmatrix} 0 & 1 & 1 & 1 \\ 0 & 0 & 1 & 1 \\ 0 & 0 & 0 & 1 \\ 0 & 0 & 0 & 0 \end{bmatrix}$

$A^2 = A \cdot A = \begin{bmatrix} 0 & 1 & 1 & 1 \\ 0 & 0 & 1 & 1 \\ 0 & 0 & 0 & 1 \\ 0 & 0 & 0 & 0 \end{bmatrix} \begin{bmatrix} 0 & 1 & 1 & 1 \\ 0 & 0 & 1 & 1 \\ 0 & 0 & 0 & 1 \\ 0 & 0 & 0 & 0 \end{bmatrix}$

$=$

$\begin{bmatrix} (0+0+0+0) & (0+0+0+0) & (0+1+0+0) & (0+1+1+0) \\ (0+0+0+0) & (0+0+0+0) & (0+0+0+0) & (0+0+1+0) \\ (0+0+0+0) & (0+0+0+0) & (0+0+0+0) & (0+0+0+0) \\ (0+0+0+0) & (0+0+0+0) & (0+0+0+0) & (0+0+0+0) \\ (0+0+0+0) & (0+0+0+0) & (0+0+0+0) & (0+0+0+0) \end{bmatrix}$

$= \begin{bmatrix} 0 & 0 & 1 & 2 \\ 0 & 0 & 0 & 1 \\ 0 & 0 & 0 & 0 \\ 0 & 0 & 0 & 0 \end{bmatrix}$

$A^3 = A \cdot A^2 = \begin{bmatrix} 0 & 1 & 1 & 1 \\ 0 & 0 & 1 & 1 \\ 0 & 0 & 0 & 1 \\ 0 & 0 & 0 & 0 \end{bmatrix} \begin{bmatrix} 0 & 0 & 1 & 2 \\ 0 & 0 & 0 & 1 \\ 0 & 0 & 0 & 0 \\ 0 & 0 & 0 & 0 \end{bmatrix}$

$=$

$\begin{bmatrix} (0+0+0+0) & (0+0+0+0) & (0+0+0+0) & (0+1+0+0) \\ (0+0+0+0) & (0+0+0+0) & (0+0+0+0) & (0+0+0+0) \\ (0+0+0+0) & (0+0+0+0) & (0+0+0+0) & (0+0+0+0) \\ (0+0+0+0) & (0+0+0+0) & (0+0+0+0) & (0+0+0+0) \\ (0+0+0+0) & (0+0+0+0) & (0+0+0+0) & (0+0+0+0) \end{bmatrix}$

$= \begin{bmatrix} 0 & 0 & 0 & 1 \\ 0 & 0 & 0 & 0 \\ 0 & 0 & 0 & 0 \\ 0 & 0 & 0 & 0 \end{bmatrix}$

$A^4 = A \cdot A^3 = \begin{bmatrix} 0 & 1 & 1 & 1 \\ 0 & 0 & 1 & 1 \\ 0 & 0 & 0 & 1 \\ 0 & 0 & 0 & 0 \end{bmatrix} \begin{bmatrix} 0 & 0 & 0 & 1 \\ 0 & 0 & 0 & 0 \\ 0 & 0 & 0 & 0 \\ 0 & 0 & 0 & 0 \end{bmatrix}$

$= \begin{bmatrix} 0 & 0 & 0 & 0 \\ 0 & 0 & 0 & 0 \\ 0 & 0 & 0 & 0 \\ 0 & 0 & 0 & 0 \end{bmatrix}.$

Continuing to multiply by A continues to yield zero, so $A^{100} = 0$.

31. $AB = \begin{bmatrix} 0 & -1 & 0 & 1 \\ 10 & 0 & 1 & 0 \end{bmatrix} \begin{bmatrix} 0 & -1 \\ 1 & 1 \\ -1 & 3 \\ 5 & 0 \end{bmatrix}$

$= \begin{bmatrix} (0-1+0+5) & (0-1+0+0) \\ (0+0-1+0) & (-10+0+3+0) \end{bmatrix}$

$= \begin{bmatrix} 4 & -1 \\ -1 & -7 \end{bmatrix}$

TI-83/84 Plus format: `[A]*[B]`
On-line Matrix Algebra Tool format: `A*B`

33. $A(B-C)$

$= \begin{bmatrix} 0 & -1 & 0 & 1 \\ 10 & 0 & 1 & 0 \end{bmatrix} \left(\begin{bmatrix} 0 & -1 \\ 1 & 1 \\ -1 & 3 \\ 5 & 0 \end{bmatrix} - \begin{bmatrix} 1 & -1 \\ 1 & 1 \\ 1 & 1 \\ 1 & 1 \end{bmatrix} \right)$

$= \begin{bmatrix} 0 & -1 & 0 & 1 \\ 10 & 0 & 1 & 0 \end{bmatrix} \begin{bmatrix} -1 & 0 \\ 0 & 0 \\ -2 & 2 \\ 4 & -1 \end{bmatrix} = \begin{bmatrix} 4 & -1 \\ -12 & 2 \end{bmatrix}$

TI-83/84 Plus format: `[A]*([B]-[C])`
On-line Matrix Algebra Tool format: `A*(B-C)`

35. $AB = \begin{bmatrix} 1 & -1 \\ 0 & 2 \\ 0 & -2 \end{bmatrix} \begin{bmatrix} 3 & 0 & -1 \\ 5 & -1 & 1 \end{bmatrix}$

$= \begin{bmatrix} -2 & 1 & -2 \\ 10 & -2 & 2 \\ -10 & 2 & -2 \end{bmatrix}$

TI-83/84 Plus format: `[A]*[B]`
On-line Matrix Algebra Tool format: `A*B`

37. $A(B+C)$

$= \begin{bmatrix} 1 & -1 \\ 0 & 2 \\ 0 & -2 \end{bmatrix} \left(\begin{bmatrix} 3 & 0 & -1 \\ 5 & -1 & 1 \end{bmatrix} + \begin{bmatrix} x & 1 & w \\ z & r & 4 \end{bmatrix} \right)$

$= \begin{bmatrix} 1 & -1 \\ 0 & 2 \\ 0 & -2 \end{bmatrix} \begin{bmatrix} 3+x & 1 & -1+w \\ 5+z & -1+r & 5 \end{bmatrix}$

$= \begin{bmatrix} -2+x-z & 2-r & -6+w \\ 10+2z & -2+2r & 10 \\ -10-2z & 2-2r & -10 \end{bmatrix}$

39.

(a) $P^2 = P \cdot P = \begin{bmatrix} 0.2 & 0.8 \\ 0.2 & 0.8 \end{bmatrix} \begin{bmatrix} 0.2 & 0.8 \\ 0.2 & 0.8 \end{bmatrix}$

95

$$= \begin{bmatrix} (0.04+0.16) & (0.16+0.64) \\ (0.04+0.16) & (0.16+0.64) \end{bmatrix}$$

$$= \begin{bmatrix} 0.2 & 0.8 \\ 0.2 & 0.8 \end{bmatrix}$$

(b) $P^4 = P^2 \cdot P^2 = \begin{bmatrix} 0.2 & 0.8 \\ 0.2 & 0.8 \end{bmatrix} \begin{bmatrix} 0.2 & 0.8 \\ 0.2 & 0.8 \end{bmatrix}$

$$= \begin{bmatrix} 0.2 & 0.8 \\ 0.2 & 0.8 \end{bmatrix} \text{ again.}$$

(c) $P^8 = P^4 \cdot P^4 = \begin{bmatrix} 0.2 & 0.8 \\ 0.2 & 0.8 \end{bmatrix} \begin{bmatrix} 0.2 & 0.8 \\ 0.2 & 0.8 \end{bmatrix}$

$$= \begin{bmatrix} 0.2 & 0.8 \\ 0.2 & 0.8 \end{bmatrix} \text{ again.}$$

(d) $P^{1000} = P^8 \cdot P^8 \cdot P^8 \cdot \ldots P^8$ (125 times)

$$= \begin{bmatrix} 0.2 & 0.8 \\ 0.2 & 0.8 \end{bmatrix}.$$

41.

(a) $P^2 = P \cdot P = \begin{bmatrix} 0.1 & 0.9 \\ 0 & 1 \end{bmatrix} \begin{bmatrix} 0.1 & 0.9 \\ 0 & 1 \end{bmatrix}$

$$= \begin{bmatrix} 0.01 & 0.99 \\ 0 & 1 \end{bmatrix}$$

(b) $P^4 = P^2 \cdot P^2$

$$= \begin{bmatrix} 0.01 & 0.99 \\ 0 & 1 \end{bmatrix} \begin{bmatrix} 0.01 & 0.99 \\ 0 & 1 \end{bmatrix}$$

$$= \begin{bmatrix} 0.0001 & 0.9999 \\ 0 & 1 \end{bmatrix}$$

(c) $P^8 = P^4 \cdot P^4$

$$= \begin{bmatrix} 0.0001 & 0.9999 \\ 0 & 1 \end{bmatrix} \begin{bmatrix} 0.0001 & 0.9999 \\ 0 & 1 \end{bmatrix}$$

$$\approx \begin{bmatrix} 0 & 1 \\ 0 & 1 \end{bmatrix} \text{ when we round to 4 decimal places.}$$

(d) $P^{1000} = P^8 \cdot P^8 \cdot P^8 \cdot \ldots P^8$ (125 times)

$$\approx \begin{bmatrix} 0 & 1 \\ 0 & 1 \end{bmatrix} \begin{bmatrix} 0 & 1 \\ 0 & 1 \end{bmatrix} \begin{bmatrix} 0 & 1 \\ 0 & 1 \end{bmatrix} \cdots \begin{bmatrix} 0 & 1 \\ 0 & 1 \end{bmatrix}$$

$$= \begin{bmatrix} 0 & 1 \\ 0 & 1 \end{bmatrix}$$

43.

(a) $P^2 = \begin{bmatrix} 0.25 & 0.25 & 0.50 \\ 0.25 & 0.25 & 0.50 \\ 0.25 & 0.25 & 0.50 \end{bmatrix} \begin{bmatrix} 0.25 & 0.25 & 0.50 \\ 0.25 & 0.25 & 0.50 \\ 0.25 & 0.25 & 0.50 \end{bmatrix}$

=

$$\begin{bmatrix} .0625+.0625+.125 & .0625+.0625+.125 & .125+.125+.25 \\ .0625+.0625+.125 & .0625+.0625+.125 & .125+.125+.25 \\ .0625+.0625+.125 & .0625+.0625+.125 & .125+.125+.25 \end{bmatrix}$$

$$= \begin{bmatrix} 0.25 & 0.25 & 0.50 \\ 0.25 & 0.25 & 0.50 \\ 0.25 & 0.25 & 0.50 \end{bmatrix}$$

(b) $P^4 = P^2 \cdot P^2$

$$= \begin{bmatrix} 0.25 & 0.25 & 0.50 \\ 0.25 & 0.25 & 0.50 \\ 0.25 & 0.25 & 0.50 \end{bmatrix} \begin{bmatrix} 0.25 & 0.25 & 0.50 \\ 0.25 & 0.25 & 0.50 \\ 0.25 & 0.25 & 0.50 \end{bmatrix}$$

$$= \begin{bmatrix} 0.25 & 0.25 & 0.50 \\ 0.25 & 0.25 & 0.50 \\ 0.25 & 0.25 & 0.50 \end{bmatrix} \text{ again}$$

(c) $P^8 = P^4 \cdot P^4$

$$= \begin{bmatrix} 0.25 & 0.25 & 0.50 \\ 0.25 & 0.25 & 0.50 \\ 0.25 & 0.25 & 0.50 \end{bmatrix} \begin{bmatrix} 0.25 & 0.25 & 0.50 \\ 0.25 & 0.25 & 0.50 \\ 0.25 & 0.25 & 0.50 \end{bmatrix}$$

$$= \begin{bmatrix} 0.25 & 0.25 & 0.50 \\ 0.25 & 0.25 & 0.50 \\ 0.25 & 0.25 & 0.50 \end{bmatrix} \text{ again.}$$

(d) $P^{1000} = P^8 \cdot P^8 \cdot P^8 \cdot \ldots P^8$ (125 times)

$$= \begin{bmatrix} 0.25 & 0.25 & 0.50 \\ 0.25 & 0.25 & 0.50 \\ 0.25 & 0.25 & 0.50 \end{bmatrix}.$$

45. $\begin{bmatrix} 2 & -1 & 4 \\ -4 & \frac{3}{4} & \frac{1}{3} \\ -3 & 0 & 0 \end{bmatrix} \begin{bmatrix} x \\ y \\ z \end{bmatrix} = \begin{bmatrix} 3 \\ -1 \\ 0 \end{bmatrix}$

$$\begin{bmatrix} 2x-y+4z \\ -4x+\frac{3}{4}y+\frac{1}{3}z \\ -3x \end{bmatrix} = \begin{bmatrix} 3 \\ -1 \\ 0 \end{bmatrix}$$

Equating entries gives the following system of linear equations:

$$2x - y + 4z = 3$$
$$-4x + \tfrac{3}{4}y + \tfrac{1}{3}z = -1$$
$$-3x = 0$$

47. $\begin{bmatrix} 1 & -1 & 0 & 1 \\ 1 & 1 & 2 & 4 \end{bmatrix} \begin{bmatrix} x \\ y \\ z \\ w \end{bmatrix} = \begin{bmatrix} -1 \\ 2 \end{bmatrix}$

$$\begin{bmatrix} x-y+w \\ x+y+2z+4w \end{bmatrix} = \begin{bmatrix} -1 \\ 2 \end{bmatrix}$$

Equating entries gives the following system of linear equations:

$$x - y + w = -1$$
$$x + y + 2z + 4w = 2$$

49. $x - y = 4$
$2x - y = 0$

The matrix form is $AX = B$, where A is the matrix of coefficients, X is the column matrix of unknowns, and B is the column matrix of right-hand sides.

$$A = \begin{bmatrix} 1 & -1 \\ 2 & -1 \end{bmatrix}, X = \begin{bmatrix} x \\ y \end{bmatrix}, B = \begin{bmatrix} 4 \\ 0 \end{bmatrix}$$

Thus, the matrix system is

$$\begin{bmatrix} 1 & -1 \\ 2 & -1 \end{bmatrix}\begin{bmatrix} x \\ y \end{bmatrix} = \begin{bmatrix} 4 \\ 0 \end{bmatrix}$$

51. $x + y - z = 8$
$2x + y + z = 4$
$\dfrac{3x}{4} + \dfrac{z}{2} = 1$

The matrix form is $AX = B$, where A is the matrix of coefficients, X is the column matrix of unknowns, and B is the column matrix of right-hand sides.

$$A = \begin{bmatrix} 1 & 1 & -1 \\ 2 & 1 & 1 \\ \frac{3}{4} & 0 & \frac{1}{2} \end{bmatrix} X = \begin{bmatrix} x \\ y \\ z \end{bmatrix}, B = \begin{bmatrix} 8 \\ 4 \\ 1 \end{bmatrix}$$

Thus, the matrix system is

$$\begin{bmatrix} 1 & 1 & -1 \\ 2 & 1 & 1 \\ \frac{3}{4} & 0 & \frac{1}{2} \end{bmatrix}\begin{bmatrix} x \\ y \\ z \end{bmatrix} = \begin{bmatrix} 8 \\ 4 \\ 1 \end{bmatrix}$$

53. We have three prices and three quantities. Arrange the prices as a row matrix and quantities as a column matrix (so that we can multiply them):

$$\text{Price} = \begin{bmatrix} 15 & 10 & 12 \end{bmatrix} \quad \text{Quantity} = \begin{bmatrix} 50 \\ 40 \\ 30 \end{bmatrix}$$

Revenue = Price × Quantity $= \begin{bmatrix} 15 & 10 & 12 \end{bmatrix}\begin{bmatrix} 50 \\ 40 \\ 30 \end{bmatrix}$

$= [750+400+360] = [1510]$

55. We are asked to write the prices as a column matrix:

$$\text{Price:} \begin{matrix} \text{Hard} \\ \text{Soft} \\ \text{Plastic} \end{matrix} \begin{bmatrix} 30 \\ 10 \\ 15 \end{bmatrix}$$

We are asked to find the revenue, Since the prices are given as a column matrix, we expect it to go on the right when we multiply rows by columns. So we write the formula for revenue with prices on the right:

Revenue = Quantity × Price Note the reversal of the usual order

This means that the quantities (hard, soft, plastic) should appear as rows in the quantity matrix:

$$\text{Quantity:} \begin{bmatrix} 700 & 1300 & 2000 \\ 400 & 300 & 500 \end{bmatrix}, \text{Price} = \begin{bmatrix} 30 \\ 10 \\ 15 \end{bmatrix}$$

Revenue = Quantity × Price

$$= \begin{bmatrix} 700 & 1300 & 2000 \\ 400 & 300 & 500 \end{bmatrix}\begin{bmatrix} 30 \\ 10 \\ 15 \end{bmatrix} = \begin{bmatrix} \$64,000 \\ \$22,500 \end{bmatrix}$$

57. Total Cost = Cost per sq. foot × Number of sq. ft

$$= \begin{bmatrix} 180 & 160 & 100 \end{bmatrix}\begin{bmatrix} 200 \\ 800 \\ 1000 \end{bmatrix} = [264,000]$$

Therefore, the total cost is $264,000

59. We are given the mean income per person for each age group as well as the number people.

Total income = Income per person × Number of females in 2020

Thus, we could write the income per person as a row matrix and the number of females in 2020 as a column matrix:

Income per person = $\begin{bmatrix} 14 & 42 & 48 & 29 \end{bmatrix}$ thousand dollars

$$\text{Number of females in 2020} = \begin{bmatrix} 23 \\ 43 \\ 43 \\ 31 \end{bmatrix} \text{million}$$

Total income = Income per person × Number of females in 2020

$$= \begin{bmatrix} 14 & 42 & 48 & 29 \end{bmatrix}\begin{bmatrix} 23 \\ 43 \\ 43 \\ 31 \end{bmatrix}$$

$= \$5091$ thousand million
$= \$5091$ billion

The total income, rounded to two significant digits, is $5100 billion (or $5.1 trillion).

97

61. Take N to be the income per person

$N = [14 \; 42 \; 48 \; 29]$,

take F and M to be, respectively, the female and male populations in 2020:

$$F = \begin{bmatrix} 23 \\ 43 \\ 43 \\ 31 \end{bmatrix} \quad M = \begin{bmatrix} 24 \\ 43 \\ 41 \\ 20 \end{bmatrix}$$

Then the female income is NF and the male income is NM. The difference is therefore

$D = NF - NM$

$= N(F - M)$,

which is the single formula required. Computing,

$N(F - M) = [14 \; 42 \; 48 \; 29]$

$$\left(\begin{bmatrix} 23 \\ 43 \\ 43 \\ 31 \end{bmatrix} - \begin{bmatrix} 24 \\ 43 \\ 41 \\ 20 \end{bmatrix} \right)$$

$$= [14 \; 42 \; 48 \; 29] \begin{bmatrix} -1 \\ 0 \\ 2 \\ 11 \end{bmatrix}$$

$= \$401$ thousand million (or billion) $\approx \$400$ billion

63. $P = \begin{bmatrix} 2.7 & 3.0 \\ 3.9 & 4.0 \end{bmatrix}$

$[-1 \; 1] P = [-1 \; 1] \begin{bmatrix} 2.7 & 3.0 \\ 3.9 & 4.0 \end{bmatrix} = [(-2.7+3.9) \; (-3.0+4.0)] = [1.2 \; 1.0]$

To interpret this, look at what we are doing when we multiply the matrices: Multiplying $[-1, 1]$ by the first column of P computes the north central states total for 1999 – the western states total for 1999, giving the the amount, in billions of pounds, by which cheese production in north central states exceeded that in western states in 1999. Multiplying by the second columns gives a similar calculation for 2000. Thus, $[-1 \; 1] P$ represents the amount, in billions of pounds, by which cheese production in north central states exceeded that in western states.

65. Take A to be the matrix given by the table:

A = Total number of filings

$$= \begin{bmatrix} 5700 & 5000 & 4300 \\ 2400 & 2800 & 2300 \\ 1900 & 2100 & 2100 \end{bmatrix}$$

The percentages for the three states handled by the firm can be represented by a row matrix:

Percentage handled by firm = $[0.10 \; 0.05 \; 0.20]$

(We used a row so that the three percentages match the three states down each column of A.)

Number of foreclosures filings handled by firm

= Percentage handled by firm $\times$ Total number

$$= [0.10 \; 0.05 \; 0.20] \begin{bmatrix} 5700 & 5000 & 4300 \\ 2400 & 2800 & 2300 \\ 1900 & 2100 & 2100 \end{bmatrix}$$

$= [1070 \; 1060 \; 965]$

67. $B = [1 \; 1 \; 0]$, $A = \begin{bmatrix} 5700 & 5000 & 4300 \\ 2400 & 2800 & 2300 \\ 1900 & 2100 & 2100 \end{bmatrix}$

When we multiply B by a column of A, we are adding the foreclosures in California and Florida for the corresponding month. Therefore, BA gives number of foreclosures in California and Florida combined in each of the months shown.

69. To compute the amount by which the combined foreclosures in California and Texas exceeded the foreclosures in Florida each month, we add the California and Texas entries and subtract the Florida entry in each column. This may be accomplished by multiplying each column by the row matrix $[1 \; -1 \; 1]$:

$$[1 \; -1 \; 1] \begin{bmatrix} 5700 & 5000 & 4300 \\ 2400 & 2800 & 2300 \\ 1900 & 2100 & 2100 \end{bmatrix}$$

$= [5200 \; 4300 \; 4100]$

This product gives the result for each of the three months. To add them up, we can multiply on the right by a column matrix whose entries are all 1:

$$[5200 \; 4300 \; 4100] \begin{bmatrix} 1 \\ 1 \\ 1 \end{bmatrix} = [13600]$$

Therefore, the matrix product we used was

$$[1 \; -1 \; 1] \begin{bmatrix} 5700 & 5000 & 4300 \\ 2400 & 2800 & 2300 \\ 1900 & 2100 & 2100 \end{bmatrix} \begin{bmatrix} 1 \\ 1 \\ 1 \end{bmatrix}$$

$= [13600]$

71. We are asked to organize the parts required data in a matrix, and the prices per part from each supplier in another,

Quantity of parts:

Proc Mem Tubes
$$\begin{matrix} \text{Pom II} \\ \text{Pom Classic} \end{matrix} \begin{bmatrix} 2 & 16 & 20 \\ 1 & 4 & 40 \end{bmatrix}$$

$= Q$

We multiply this by the prices per part matrix to obtain the total costs. Thus, we should write the price per parts matrix with the prices for each component as columns (to match the rows above):

Motorel Intola
$$\text{Prices:}\ \begin{matrix} \text{Proc} \\ \text{Mem} \\ \text{Tubes} \end{matrix} \begin{bmatrix} 100 & 150 \\ 50 & 40 \\ 10 & 15 \end{bmatrix} = P$$

Total cost = Quantity × Price = QP

$$= \begin{bmatrix} 2 & 16 & 20 \\ 1 & 4 & 40 \end{bmatrix} \begin{bmatrix} 100 & 150 \\ 50 & 40 \\ 10 & 15 \end{bmatrix}$$

$$= \begin{bmatrix} 1200 & 1240 \\ 700 & 910 \end{bmatrix}$$

How are these costs organized? The clue is that the rows of the product QP correspond to the rows of the left-hand matrix, Q. Here Q has rows corresponding to Pom II and Classic. On the other hand, the *columns* of the product QP correspond to the columns of P, which are Motorel and Intola. Thus, the prices are organized as follows:

Motorel Intola
$$\begin{matrix} \text{Pom II} \\ \text{Pom Classic} \end{matrix} \begin{bmatrix} \$1200 & \$1240 \\ \$700 & \$910 \end{bmatrix}$$

Solution:

Motorel parts on the Pom II: $1200
Intola parts on the Pom II: $1240
Motorel parts on the Pom Classic: $700
Intola parts on the Pom Classic: $910

73. $A = \begin{bmatrix} 440 & 190 \\ 950 & 950 \\ 1790 & 200 \end{bmatrix}$

$AB = \begin{bmatrix} 440 & 190 \\ 950 & 950 \\ 1790 & 200 \end{bmatrix} \begin{bmatrix} 0.05 \\ 0.04 \end{bmatrix} = \begin{bmatrix} 29.6 \\ 85.5 \\ 97.5 \end{bmatrix}$

In computing this product, we are adding 5% of the number of people visiting Australia and 4% of the number visiting South Africa for each of the three region. These are exactly the percentages that decide to settle there. Thus, the entries of AB give the number of people from each of the three regions who settle in Australia or South Africa.

$AC = \begin{bmatrix} 440 & 190 \\ 950 & 950 \\ 1790 & 200 \end{bmatrix} \begin{bmatrix} 0.05 & 0 \\ 0 & 0.04 \end{bmatrix}$

$= \begin{bmatrix} 22 & 7.6 \\ 47.5 & 38 \\ 89.5 & 8 \end{bmatrix}$

Each row of the product is computed by taking 5% of all the tourists to Australia and 4% of those to South Africa separately, Thus, the entries of AC give the number of people from each of the three regions who settle in Australia, and the number that settle in South Africa.

75. $P = \begin{bmatrix} 0.9892 & 0.0017 & 0.0073 & 0.0018 \\ 0.0010 & 0.9920 & 0.0048 & 0.0022 \\ 0.0018 & 0.0024 & 0.9934 & 0.0024 \\ 0.0008 & 0.0033 & 0.0045 & 0.9914 \end{bmatrix}$

2006 Distribution = $A = [55.1\ \ 66.2\ \ 110.0\ \ 70.0]$
To compute the population distribution in 2007, compute $AP \approx [54.8\ \ 66.3\ \ 110.3\ \ 69.9]$

77. Answers will vary.
One example: $A = [1\ \ 2]\ B = \begin{bmatrix} 1 & 2 & 3 \\ 4 & 5 & 6 \end{bmatrix}$
Another example: $A = [1]\ \ B = [1\ \ 2]$

79. We find that the addition and multiplication of 1×1 matrices is identical to the addition and multiplication of numbers.

81. The claim is correct. Every matrix equation represents the equality of two matrices. When two matrices are equal, each of their corresponding entries must be equal. Equating the corresponding entries gives a system of equations.

83. Here is a possible scenario:
Costs of items A, B and C in 1995 = $[10\ \ 20\ \ 30]$

99

Percentage increases in these costs in 1996
  $= [0.5 \quad 0.1 \quad 0.20]$
Actual increases in costs
  $= [10 \times 0.5 \quad 20 \times 0.1 \quad 30 \times 0.20]$

85. It produces a matrix whose *ij* entry is the product of the *ij* entries of the two matrices.

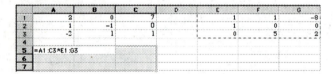

	A	B	C	D	E	F	G
1	2	0	7		1	1	-8
2	1	-1	0		1	0	0
3	-2	1	1		0	5	2
4							
5	=A1:C3*E1:G3						
6							
7							

3.3 Matrix Inversion

1. $A = \begin{bmatrix} 0 & 1 \\ 1 & 0 \end{bmatrix}$, $B = \begin{bmatrix} 0 & 1 \\ 1 & 0 \end{bmatrix}$

$AB = \begin{bmatrix} 0 & 1 \\ 1 & 0 \end{bmatrix} \begin{bmatrix} 0 & 1 \\ 1 & 0 \end{bmatrix} = \begin{bmatrix} 1 & 0 \\ 0 & 1 \end{bmatrix} = I$

Since A and B are square matrices with $AB = I$, A and B are inverses. (We do not have to check that $BA = I$ as well—see the note in the text after the definition of inverse.)

3. $A = \begin{bmatrix} 2 & 1 & 1 \\ 0 & 1 & 1 \\ 0 & 0 & 1 \end{bmatrix}$, $B = \begin{bmatrix} \frac{1}{2} & -\frac{1}{2} & 0 \\ 0 & 1 & -1 \\ 0 & 0 & 1 \end{bmatrix}$

$AB = \begin{bmatrix} 2 & 1 & 1 \\ 0 & 1 & 1 \\ 0 & 0 & 1 \end{bmatrix} \begin{bmatrix} \frac{1}{2} & -\frac{1}{2} & 0 \\ 0 & 1 & -1 \\ 0 & 0 & 1 \end{bmatrix}$

$= \begin{bmatrix} 1 & 0 & 0 \\ 0 & 1 & 0 \\ 0 & 0 & 1 \end{bmatrix} = I$

Since A and B are square matrices with $AB = I$, A and B are inverses. (We do not have to check that $BA = I$ as well—see the note in the text after the definition of inverse.)

5. $A = \begin{bmatrix} a & 0 & 0 \\ 0 & b & 0 \\ 0 & 0 & 0 \end{bmatrix}$, $B = \begin{bmatrix} a^{-1} & 0 & 0 \\ 0 & b^{-1} & 0 \\ 0 & 0 & 0 \end{bmatrix}$ $(a, b \neq 0)$

$AB = \begin{bmatrix} a & 0 & 0 \\ 0 & b & 0 \\ 0 & 0 & 0 \end{bmatrix} \begin{bmatrix} a^{-1} & 0 & 0 \\ 0 & b^{-1} & 0 \\ 0 & 0 & 0 \end{bmatrix}$

$= \begin{bmatrix} 1 & 0 & 0 \\ 0 & 1 & 0 \\ 0 & 0 & 0 \end{bmatrix} \neq I$

Since $AB \neq I$, A and B are not inverses.

7. To find the inverse of $\begin{bmatrix} 1 & 1 \\ 2 & 1 \end{bmatrix}$, we augment with the 2×2 identity matrix and row-reduce:

$\begin{bmatrix} 1 & 1 & 1 & 0 \\ 2 & 1 & 0 & 1 \end{bmatrix}_{R_2 - 2R_1}$

$\begin{bmatrix} 1 & 1 & 1 & 0 \\ 0 & -1 & -2 & 1 \end{bmatrix}_{R_1 + R_2}$

$\begin{bmatrix} 1 & 0 & -1 & 1 \\ 0 & -1 & -2 & 1 \end{bmatrix}_{-R_2}$

$\begin{bmatrix} 1 & 0 & -1 & 1 \\ 0 & 1 & 2 & -1 \end{bmatrix}$

The right-hand 2×2 block is the desired inverse:

$\begin{bmatrix} 1 & 1 \\ 2 & 1 \end{bmatrix}^{-1} = \begin{bmatrix} -1 & 1 \\ 2 & -1 \end{bmatrix}$

9. To find the inverse of $\begin{bmatrix} 0 & 1 \\ 1 & 0 \end{bmatrix}$, we augment with the 2×2 identity matrix and row-reduce:

$\begin{bmatrix} 0 & 1 & 1 & 0 \\ 1 & 0 & 0 & 1 \end{bmatrix}_{R_1 \leftrightarrow R_2} \rightarrow \begin{bmatrix} 1 & 0 & 0 & 1 \\ 0 & 1 & 1 & 0 \end{bmatrix}$

The right-hand 2×2 block is the desired inverse:

$\begin{bmatrix} 0 & 1 \\ 1 & 0 \end{bmatrix}^{-1} = \begin{bmatrix} 0 & 1 \\ 1 & 0 \end{bmatrix}$

11. $\begin{bmatrix} 2 & 1 & 1 & 0 \\ 1 & 1 & 0 & 1 \end{bmatrix}_{2R_2 - R_1}$

$\begin{bmatrix} 2 & 1 & 1 & 0 \\ 0 & 1 & -1 & 2 \end{bmatrix}_{R_1 - R_2}$

$\begin{bmatrix} 2 & 0 & 2 & -2 \\ 0 & 1 & -1 & 2 \end{bmatrix}_{(1/2)R_1}$

$\begin{bmatrix} 1 & 0 & 1 & -1 \\ 0 & 1 & -1 & 2 \end{bmatrix}$

The right-hand 2×2 block is the desired inverse:

$\begin{bmatrix} 2 & 1 \\ 1 & 1 \end{bmatrix}^{-1} = \begin{bmatrix} 1 & -1 \\ -1 & 2 \end{bmatrix}$

13.

$\begin{bmatrix} 2 & 1 & 1 & 0 \\ 4 & 2 & 0 & 1 \end{bmatrix}_{R_2 - 2R_1} \rightarrow \begin{bmatrix} 2 & 1 & 1 & 0 \\ 0 & 0 & -2 & 1 \end{bmatrix}$

Since the left-hand 2×2 block has a row of zeros, we cannot reduce the matrix to obtain the 2×2 identity on the left. Therefore, the matrix is singular.

15. To find the inverse of a 3×3 matrix, we augment it with the 3×3 identity matrix and row-reduce:

$$\begin{bmatrix} 1 & 1 & 1 & 1 & 0 & 0 \\ 0 & 1 & 1 & 0 & 1 & 0 \\ 0 & 0 & 1 & 0 & 0 & 1 \end{bmatrix} \begin{matrix} R_1 - R_2 \\ \\ \\ \end{matrix}$$

$$\begin{bmatrix} 1 & 0 & 0 & 1 & -1 & 0 \\ 0 & 1 & 1 & 0 & 1 & 0 \\ 0 & 0 & 1 & 0 & 0 & 1 \end{bmatrix} \begin{matrix} \\ R_2 - R_3 \\ \end{matrix}$$

$$\begin{bmatrix} 1 & 0 & 0 & 1 & -1 & 0 \\ 0 & 1 & 0 & 0 & 1 & -1 \\ 0 & 0 & 1 & 0 & 0 & 1 \end{bmatrix}$$

The right-hand 3×3 block is the desired inverse:

$$\begin{bmatrix} 1 & 1 & 1 \\ 0 & 1 & 1 \\ 0 & 0 & 1 \end{bmatrix}^{-1} = \begin{bmatrix} 1 & -1 & 0 \\ 0 & 1 & -1 \\ 0 & 0 & 1 \end{bmatrix}$$

17.

$$\begin{bmatrix} 1 & 1 & 1 & 1 & 0 & 0 \\ 1 & 0 & 2 & 0 & 1 & 0 \\ 1 & -1 & 1 & 0 & 0 & 1 \end{bmatrix} \begin{matrix} \\ R_2 - R_1 \\ R_3 - R_1 \end{matrix}$$

$$\begin{bmatrix} 1 & 1 & 1 & 1 & 0 & 0 \\ 0 & -1 & 1 & -1 & 1 & 0 \\ 0 & -2 & 0 & -1 & 0 & 1 \end{bmatrix} \begin{matrix} R_1 + R_2 \\ \\ R_3 - 2R_2 \end{matrix}$$

$$\begin{bmatrix} 1 & 0 & 2 & 0 & 1 & 0 \\ 0 & -1 & 1 & -1 & 1 & 0 \\ 0 & 0 & -2 & 1 & -2 & 1 \end{bmatrix} \begin{matrix} R_1 + R_3 \\ 2R_2 + R_3 \\ \end{matrix}$$

$$\begin{bmatrix} 1 & 0 & 0 & 1 & -1 & 1 \\ 0 & -2 & 0 & -1 & 0 & 1 \\ 0 & 0 & -2 & 1 & -2 & 1 \end{bmatrix} \begin{matrix} \\ -(1/2)R_2 \\ -(1/2)R_3 \end{matrix}$$

$$\begin{bmatrix} 1 & 0 & 0 & 1 & -1 & 1 \\ 0 & 1 & 0 & 1/2 & 0 & -1/2 \\ 0 & 0 & 1 & -1/2 & 1 & -1/2 \end{bmatrix}$$

The right-hand 3×3 block is the desired inverse:

$$\begin{bmatrix} 1 & 1 & 1 \\ 1 & 0 & 2 \\ 1 & -1 & 1 \end{bmatrix}^{-1} = \begin{bmatrix} 1 & -1 & 1 \\ 1/2 & 0 & -1/2 \\ -1/2 & 1 & -1/2 \end{bmatrix}$$

19.

$$\begin{bmatrix} 1 & 1 & 1 & 1 & 0 & 0 \\ 1 & -1 & 0 & 0 & 1 & 0 \\ 1 & 2 & 3 & 0 & 0 & 1 \end{bmatrix} \begin{matrix} \\ R_2 - R_1 \\ R_3 - R_1 \end{matrix}$$

$$\begin{bmatrix} 1 & 1 & 1 & 1 & 0 & 0 \\ 0 & -2 & -1 & -1 & 1 & 0 \\ 0 & 1 & 2 & -1 & 0 & 1 \end{bmatrix} \begin{matrix} 2R_1 + R_2 \\ \\ 2R_3 + R_2 \end{matrix}$$

$$\begin{bmatrix} 2 & 0 & 1 & 1 & 1 & 0 \\ 0 & -2 & -1 & -1 & 1 & 0 \\ 0 & 0 & 3 & -3 & 1 & 2 \end{bmatrix} \begin{matrix} 3R_1 - R_3 \\ 3R_2 + R_3 \\ \end{matrix}$$

$$\begin{bmatrix} 6 & 0 & 0 & 6 & 2 & -2 \\ 0 & -6 & 0 & -6 & 4 & 2 \\ 0 & 0 & 3 & -3 & 1 & 2 \end{bmatrix} \begin{matrix} (1/6)R_1 \\ -(1/6)R_2 \\ (1/3)R_3 \end{matrix}$$

$$\begin{bmatrix} 1 & 0 & 0 & 1 & 1/3 & -1/3 \\ 0 & 1 & 0 & 1 & -2/3 & -1/3 \\ 0 & 0 & 1 & -1 & 1/3 & 2/3 \end{bmatrix}$$

The right-hand 3×3 block is the desired inverse:

$$\begin{bmatrix} 1 & 1 & 1 \\ 1 & -1 & 0 \\ 1 & 2 & 3 \end{bmatrix}^{-1} = \begin{bmatrix} 1 & 1/3 & -1/3 \\ 1 & -2/3 & -1/3 \\ -1 & 1/3 & 2/3 \end{bmatrix}$$

21.

$$\begin{bmatrix} 1 & 1 & 1 & 1 & 0 & 0 \\ 1 & 0 & 1 & 0 & 1 & 0 \\ 1 & -1 & 1 & 0 & 0 & 1 \end{bmatrix} \begin{matrix} \\ R_2 - R_1 \\ R_3 - R_1 \end{matrix}$$

$$\begin{bmatrix} 1 & 1 & 1 & 1 & 0 & 0 \\ 0 & -1 & 0 & -1 & 1 & 0 \\ 0 & -2 & 0 & -1 & 0 & 1 \end{bmatrix} \begin{matrix} R_1 + R_2 \\ \\ R_3 - 2R_2 \end{matrix}$$

$$\begin{bmatrix} 1 & 0 & 1 & 0 & 1 & 0 \\ 0 & -1 & 0 & -1 & 1 & 0 \\ 0 & 0 & 0 & 1 & -2 & 1 \end{bmatrix}$$

Since the left-hand 3×3 block has a row of zeros, we cannot reduce the matrix to obtain the 3×3 identity on the left. Therefore, the matrix is singular.

23. To find the inverse of a 4×4 matrix, we augment it with the 4×4 identity matrix and row-reduce:

$$\begin{bmatrix} 1 & 0 & 1 & 0 & 1 & 0 & 0 & 0 \\ -1 & 1 & 0 & 1 & 0 & 1 & 0 & 0 \\ -1 & 0 & 0 & 1 & 0 & 0 & 1 & 0 \\ 0 & 1 & 0 & 1 & 0 & 0 & 0 & 1 \end{bmatrix} \begin{matrix} \\ R_2 + R_1 \\ R_3 + R_1 \\ \\ \end{matrix}$$

$$\begin{bmatrix} 1 & 0 & 1 & 0 & 1 & 0 & 0 & 0 \\ 0 & 1 & 1 & 1 & 1 & 1 & 0 & 0 \\ 0 & 0 & 1 & 1 & 1 & 0 & 1 & 0 \\ 0 & -1 & 0 & 1 & 0 & 0 & 0 & 1 \end{bmatrix} \begin{matrix} \\ \\ \\ R_4 + R_2 \end{matrix}$$

$$\begin{bmatrix} 1 & 0 & 1 & 0 & 1 & 0 & 0 & 0 \\ 0 & 1 & 1 & 1 & 1 & 1 & 0 & 0 \\ 0 & 0 & 1 & 1 & 1 & 0 & 1 & 0 \\ 0 & 0 & 1 & 2 & 1 & 1 & 0 & 1 \end{bmatrix} \begin{matrix} R_1 - R_3 \\ R_2 - R_3 \\ \\ R_4 - R_3 \end{matrix}$$

$$\begin{bmatrix} 1 & 0 & 0 & -1 & 0 & 0 & -1 & 0 \\ 0 & 1 & 0 & 0 & 0 & 1 & -1 & 0 \\ 0 & 0 & 1 & 1 & 1 & 0 & 1 & 0 \\ 0 & 0 & 0 & 1 & 0 & 1 & -1 & 1 \end{bmatrix} \begin{matrix} R_1 + R_4 \\ \\ R_3 - R_4 \\ \\ \end{matrix}$$

$$\begin{bmatrix} 1 & 0 & 0 & 0 & 0 & 1 & -2 & 1 \\ 0 & 1 & 0 & 0 & 0 & 1 & -1 & 0 \\ 0 & 0 & 1 & 0 & 1 & -1 & 2 & -1 \\ 0 & 0 & 0 & 1 & 0 & 1 & -1 & 1 \end{bmatrix}$$

The right-hand 4×4 block is the desired inverse:

$$\begin{bmatrix} 1 & 0 & 1 & 0 \\ -1 & 1 & 0 & 1 \\ -1 & 0 & 0 & 1 \\ 0 & -1 & 0 & 1 \end{bmatrix}^{-1} = \begin{bmatrix} 0 & 1 & -2 & 1 \\ 0 & 1 & -1 & 0 \\ 1 & -1 & 2 & -1 \\ 0 & 1 & -1 & 1 \end{bmatrix}$$

25.

$$\begin{bmatrix} 1 & 2 & 3 & 4 & 1 & 0 & 0 & 0 \\ 0 & 1 & 2 & 3 & 0 & 1 & 0 & 0 \\ 0 & 0 & 1 & 2 & 0 & 0 & 1 & 0 \\ 0 & 0 & 0 & 1 & 0 & 0 & 0 & 1 \end{bmatrix} \begin{matrix} R_1 - 2R_2 \\ \\ \\ \\ \end{matrix}$$

$$\begin{bmatrix} 1 & 0 & -1 & -2 & 1 & -2 & 0 & 0 \\ 0 & 1 & 2 & 3 & 0 & 1 & 0 & 0 \\ 0 & 0 & 1 & 2 & 0 & 0 & 1 & 0 \\ 0 & 0 & 0 & 1 & 0 & 0 & 0 & 1 \end{bmatrix} \begin{matrix} R_1 + R_3 \\ R_2 - 2R_3 \\ \\ \\ \end{matrix}$$

$$\begin{bmatrix} 1 & 0 & 0 & 0 & 1 & -2 & 1 & 0 \\ 0 & 1 & 0 & -1 & 0 & 1 & -2 & 0 \\ 0 & 0 & 1 & 2 & 0 & 0 & 1 & 0 \\ 0 & 0 & 0 & 1 & 0 & 0 & 0 & 1 \end{bmatrix} \begin{matrix} \\ R_2 + R_4 \\ R_3 - 2R_4 \\ \\ \end{matrix}$$

$$\begin{bmatrix} 1 & 0 & 0 & 0 & 1 & -2 & 1 & 0 \\ 0 & 1 & 0 & 0 & 0 & 1 & -2 & 1 \\ 0 & 0 & 1 & 0 & 0 & 0 & 1 & -2 \\ 0 & 0 & 0 & 1 & 0 & 0 & 0 & 1 \end{bmatrix}$$

The right-hand 4×4 block is the desired inverse:

$$\begin{bmatrix} 1 & 2 & 3 & 4 \\ 0 & 1 & 2 & 3 \\ 0 & 0 & 1 & 2 \\ 0 & 0 & 0 & 1 \end{bmatrix}^{-1} = \begin{bmatrix} 1 & -2 & 1 & 0 \\ 0 & 1 & -2 & 1 \\ 0 & 0 & 1 & -2 \\ 0 & 0 & 0 & 1 \end{bmatrix}$$

27. $\begin{bmatrix} a & b \\ c & d \end{bmatrix} = \begin{bmatrix} 1 & 1 \\ 1 & -1 \end{bmatrix}$

$\det \begin{bmatrix} 1 & 1 \\ 1 & -1 \end{bmatrix} = ad - bc$

$= (1)(-1) - (1)(1) = -2$

$\begin{bmatrix} 1 & 1 \\ 1 & -1 \end{bmatrix}^{-1} = \dfrac{1}{ad - bc} \begin{bmatrix} d & -b \\ -c & a \end{bmatrix}$

$= \dfrac{1}{-2} \begin{bmatrix} -1 & -1 \\ -1 & 1 \end{bmatrix} = \begin{bmatrix} 1/2 & 1/2 \\ 1/2 & -1/2 \end{bmatrix}$

29. $\begin{bmatrix} a & b \\ c & d \end{bmatrix} = \begin{bmatrix} 1 & 2 \\ 3 & 4 \end{bmatrix}$

$\det \begin{bmatrix} 1 & 2 \\ 3 & 4 \end{bmatrix} = ad - bc$

$= (1)(4) - (2)(3) = -2$

$\begin{bmatrix} 1 & 2 \\ 3 & 4 \end{bmatrix}^{-1} = \dfrac{1}{ad - bc} \begin{bmatrix} d & -b \\ -c & a \end{bmatrix}$

$= \dfrac{1}{-2} \begin{bmatrix} 4 & -2 \\ -3 & 1 \end{bmatrix} = \begin{bmatrix} -2 & 1 \\ 3/2 & -1/2 \end{bmatrix}$

31. $\begin{bmatrix} a & b \\ c & d \end{bmatrix} = \begin{bmatrix} 1/6 & -1/6 \\ 0 & 1/6 \end{bmatrix}$

$\det \begin{bmatrix} 1/6 & -1/6 \\ 0 & 1/6 \end{bmatrix} = ad - bc$

$= (1/6)(1/6) - (-1/6)(0) = 1/36$

$\begin{bmatrix} 1/6 & -1/6 \\ 0 & 1/6 \end{bmatrix}^{-1} = \dfrac{1}{ad - bc} \begin{bmatrix} d & -b \\ -c & a \end{bmatrix}$

$= 36 \begin{bmatrix} 1/6 & 1/6 \\ 0 & 1/6 \end{bmatrix} = \begin{bmatrix} 6 & 6 \\ 0 & 6 \end{bmatrix}$

33. $\begin{bmatrix} a & b \\ c & d \end{bmatrix} = \begin{bmatrix} 1 & 0 \\ 3/4 & 0 \end{bmatrix}$

$\det\begin{bmatrix} 1 & 0 \\ 3/4 & 0 \end{bmatrix} = ad - bc$

$= (1)(0) - (0)(3/4) = 0$

Singular matrix

Exercises 35–42 can be answered using a graphing calculator, Excel, or the on-line Matrix Algebra Tool at
FiniteMath.org → Student Web Site → On-Line Utilities → Matrix Algebra Tool

35. $\begin{bmatrix} 1.1 & 1.2 \\ 1.3 & -1 \end{bmatrix}^{-1} = \begin{bmatrix} 0.38 & 0.45 \\ 0.49 & -0.41 \end{bmatrix}$

TI-83/84 Plus format: `[A]`$^{-1}$
On-line Matrix Algebra Tool format: `A^(-1)`
Excel: `=MINVERSE(A1:B2)`
(Assuming the matrix is in cells A1→B2)

37. $\begin{bmatrix} 3.56 & 1.23 \\ -1.01 & 0 \end{bmatrix}^{-1} = \begin{bmatrix} 0.00 & -0.99 \\ 0.81 & 2.87 \end{bmatrix}$

TI-83/84 Plus format: `[A]`$^{-1}$
On-line Matrix Algebra Tool format: `A^(-1)`
Excel: `=MINVERSE(A1:B2)`
(Assuming the matrix is in cells A1→B2)

39. $\begin{bmatrix} 1.1 & 3.1 & 2.4 \\ 1.7 & 2.4 & 2.3 \\ 0.6 & -0.7 & -0.1 \end{bmatrix}$ is singular

TI-83/84 Plus format: `[A]`$^{-1}$
On-line Matrix Algebra Tool format: `A^(-1)`
Excel: `=MINVERSE(A1:C3)`
(Assuming the matrix is in cells A1→C3)

41. $\begin{bmatrix} 0.01 & 0.32 & 0 & 0.04 \\ -0.01 & 0 & 0 & 0.34 \\ 0 & 0.32 & -0.23 & 0.23 \\ 0 & 0.41 & 0 & 0.01 \end{bmatrix}^{-1}$

$= \begin{bmatrix} 91.35 & -8.65 & 0 & -71.30 \\ -0.07 & -0.07 & 0 & 2.49 \\ 2.60 & 2.60 & -4.35 & 1.37 \\ 2.69 & 2.69 & 0 & -2.10 \end{bmatrix}$

TI-83/84 Plus format: `[A]`$^{-1}$
On-line Matrix Algebra Tool format: `A^(-1)`

Excel: `=MINVERSE(A1:D4)`
(Assuming the matrix is in cells A1→D4)

43. $x + y = 4$
$x - y = 1$
Matrix form:
$\begin{bmatrix} 1 & 1 \\ 1 & -1 \end{bmatrix}\begin{bmatrix} x \\ y \end{bmatrix} = \begin{bmatrix} 4 \\ 1 \end{bmatrix}$
$AX = B$
$X = A^{-1}B$
$A^{-1} = \begin{bmatrix} 1 & 1 \\ 1 & -1 \end{bmatrix}^{-1} = \begin{bmatrix} 1/2 & 1/2 \\ 1/2 & -1/2 \end{bmatrix}$
See Exercise 27
Thus,
$X = A^{-1}B = \begin{bmatrix} 1/2 & 1/2 \\ 1/2 & -1/2 \end{bmatrix}\begin{bmatrix} 4 \\ 1 \end{bmatrix} = \begin{bmatrix} 5/2 \\ 3/2 \end{bmatrix}$
So, $(x, y) = (5/2, 3/2)$

45. $\frac{x}{3} + \frac{y}{2} = 0$
$\frac{x}{2} + y = -1$
Matrix form:
$\begin{bmatrix} 1/3 & 1/2 \\ 1/2 & 1 \end{bmatrix}\begin{bmatrix} x \\ y \end{bmatrix} = \begin{bmatrix} 0 \\ -1 \end{bmatrix}$
$AX = B$
$X = A^{-1}B$
$A^{-1} = \begin{bmatrix} 1/3 & 1/2 \\ 1/2 & 1 \end{bmatrix}^{-1} = \begin{bmatrix} 12 & -6 \\ -6 & 4 \end{bmatrix}$
Thus,
$X = A^{-1}B = \begin{bmatrix} 12 & -6 \\ -6 & 4 \end{bmatrix}\begin{bmatrix} 0 \\ -1 \end{bmatrix} = \begin{bmatrix} 6 \\ -4 \end{bmatrix}$
So, $(x, y) = (6, -4)$

47. $-x + 2y - z = 0$
$-x - y + 2z = 0$
$2x \quad - z = 6$
Matrix form:
$\begin{bmatrix} -1 & 2 & -1 \\ -1 & -1 & 2 \\ 2 & 0 & -1 \end{bmatrix}\begin{bmatrix} x \\ y \\ z \end{bmatrix} = \begin{bmatrix} 0 \\ 0 \\ 6 \end{bmatrix}$
$AX = B$
$X = A^{-1}B$
$A^{-1} = \begin{bmatrix} -1 & 2 & -1 \\ -1 & -1 & 2 \\ 2 & 0 & -1 \end{bmatrix}^{-1} = \begin{bmatrix} 1/3 & 2/3 & 1 \\ 1 & 1 & 1 \\ 2/3 & 4/3 & 1 \end{bmatrix}$

Thus,

$$X = A^{-1}B = \begin{bmatrix} 1/3 & 2/3 & 1 \\ 1 & 1 & 1 \\ 2/3 & 4/3 & 1 \end{bmatrix} \begin{bmatrix} 0 \\ 0 \\ 6 \end{bmatrix} = \begin{bmatrix} 6 \\ 6 \\ 6 \end{bmatrix}$$

So, $(x, y) = (6, 6, 6)$

49. The three systems of equations have the matrix form $AX = B$ where

$$A = \begin{bmatrix} -1 & -4 & 2 \\ 1 & 2 & -1 \\ 1 & 1 & -1 \end{bmatrix}; A^{-1} = \begin{bmatrix} 1 & 2 & 0 \\ 0 & 1 & -1 \\ 1 & 3 & -2 \end{bmatrix}$$

(a) $B = \begin{bmatrix} 4 \\ 3 \\ 8 \end{bmatrix}$

$$X = A^{-1}B = \begin{bmatrix} 1 & 2 & 0 \\ 0 & 1 & -1 \\ 1 & 3 & -2 \end{bmatrix} \begin{bmatrix} 4 \\ 3 \\ 8 \end{bmatrix} = \begin{bmatrix} 10 \\ -5 \\ -3 \end{bmatrix}$$

$(x, y, z) = (10, -5, -3)$

(b) $B = \begin{bmatrix} 0 \\ 3 \\ 2 \end{bmatrix}$

$$X = A^{-1}B = \begin{bmatrix} 1 & 2 & 0 \\ 0 & 1 & -1 \\ 1 & 3 & -2 \end{bmatrix} \begin{bmatrix} 0 \\ 3 \\ 2 \end{bmatrix} = \begin{bmatrix} 6 \\ 1 \\ 5 \end{bmatrix}$$

$(x, y, z) = (6, 1, 5)$

(c) $B = \begin{bmatrix} 0 \\ 0 \\ 0 \end{bmatrix}$

$$X = A^{-1}B = \begin{bmatrix} 1 & 2 & 0 \\ 0 & 1 & -1 \\ 1 & 3 & -2 \end{bmatrix} \begin{bmatrix} 0 \\ 0 \\ 0 \end{bmatrix} = \begin{bmatrix} 0 \\ 0 \\ 0 \end{bmatrix}$$

$(x, y, z) = (0, 0, 0)$

51. Unknowns:

x = the number of servings of Pork & Beans
y = the number of slices of bread

(a) Arrange the given information in a table with unknowns across the top:

	Pork & Beans (x)	Bread (y)	Desired
Protein	5	4	20
Carbs.	21	12	80

We can now set up an equation for each of the items listed on the left:

Protein $\qquad 5x + 4y = 20$

Carbs: $\qquad 21x + 12y = 80$

We put this system in matrix form:

$$\begin{bmatrix} 5 & 4 \\ 21 & 12 \end{bmatrix} \begin{bmatrix} x \\ y \end{bmatrix} = \begin{bmatrix} 20 \\ 80 \end{bmatrix}$$

$$AX = B$$
$$X = A^{-1}B$$

$$A^{-1} = \begin{bmatrix} 5 & 4 \\ 21 & 12 \end{bmatrix}^{-1} = \begin{bmatrix} -1/2 & 1/6 \\ 7/8 & -5/24 \end{bmatrix}$$

Thus,

$$X = A^{-1}B = \begin{bmatrix} -1/2 & 1/6 \\ 7/8 & -5/24 \end{bmatrix} \begin{bmatrix} 20 \\ 80 \end{bmatrix}$$

$$= \begin{bmatrix} 10/3 \\ 5/6 \end{bmatrix}$$

So, $(x, y) = (10/3, 5/6)$

Solution: Prepare 10/3 servings of beans, and 5/6 slices of bread.

(b) We must solve the following system:

$5x + 4y = A$
$21x + 12y = B$

We put this system in matrix form:

$$\begin{bmatrix} 5 & 4 \\ 21 & 12 \end{bmatrix} \begin{bmatrix} x \\ y \end{bmatrix} = \begin{bmatrix} A \\ B \end{bmatrix}$$

Solving gives

$$\begin{bmatrix} x \\ y \end{bmatrix} = \begin{bmatrix} 5 & 4 \\ 21 & 12 \end{bmatrix}^{-1} \begin{bmatrix} A \\ B \end{bmatrix}$$

$$= \begin{bmatrix} -1/2 & 1/6 \\ 7/8 & -5/24 \end{bmatrix} \begin{bmatrix} A \\ B \end{bmatrix}$$

$$= \begin{bmatrix} -A/2 + B/6 \\ 7A/8 - 5B/24 \end{bmatrix}$$

Solution: Prepare $-A/2 + B/6$ servings of beans and $7A/8 - 5B/24$ slices of bread.

53. Unknowns:

x = the number of batches of vanilla
y = the number of batches of mocha
z = the number of batches of strawberry

Arrange the given information in a table with unknowns across the top:

	Vanilla (x)	Mocha (y)	Strawberry (z)
Eggs	2	1	1
Milk	1	1	2
Cream	2	2	1

(a)

Eggs: $\quad 2x + y + z = 350$

Milk: $\quad x + y + 2z = 350$

Cream: $2x + 2y + z = 400$
Matrix form:

$$\begin{bmatrix} 2 & 1 & 1 \\ 1 & 1 & 2 \\ 2 & 2 & 1 \end{bmatrix} \begin{bmatrix} x \\ y \\ z \end{bmatrix} = \begin{bmatrix} 350 \\ 350 \\ 400 \end{bmatrix}$$

$AX = B \Rightarrow X = A^{-1}B$

$$X = \begin{bmatrix} 2 & 1 & 1 \\ 1 & 1 & 2 \\ 2 & 2 & 1 \end{bmatrix}^{-1} \begin{bmatrix} 350 \\ 350 \\ 400 \end{bmatrix}$$

$$= \begin{bmatrix} 1 & -1/3 & -1/3 \\ -1 & 0 & 1 \\ 0 & 2/3 & -1/3 \end{bmatrix} \begin{bmatrix} 350 \\ 350 \\ 400 \end{bmatrix} = \begin{bmatrix} 100 \\ 50 \\ 100 \end{bmatrix}$$

Solution: Use 100 batches of vanilla, 50 batches of mocha, and 100 batches of strawberry.

(b) The requirements lead to the following matrix equation:

$$\begin{bmatrix} 2 & 1 & 1 \\ 1 & 1 & 2 \\ 2 & 2 & 1 \end{bmatrix} \begin{bmatrix} x \\ y \\ z \end{bmatrix} = \begin{bmatrix} 400 \\ 500 \\ 400 \end{bmatrix}$$

$AX = B \Rightarrow X = A^{-1}B$

$$X = \begin{bmatrix} 2 & 1 & 1 \\ 1 & 1 & 2 \\ 2 & 2 & 1 \end{bmatrix}^{-1} \begin{bmatrix} 400 \\ 500 \\ 400 \end{bmatrix}$$

$$= \begin{bmatrix} 1 & -1/3 & -1/3 \\ -1 & 0 & 1 \\ 0 & 2/3 & -1/3 \end{bmatrix} \begin{bmatrix} 400 \\ 500 \\ 400 \end{bmatrix} = \begin{bmatrix} 100 \\ 0 \\ 200 \end{bmatrix}$$

Solution: Use 100 batches of vanilla, no mocha, and 200 batches of strawberry.

(c) The requirements lead to the following matrix equation:

$$\begin{bmatrix} 2 & 1 & 1 \\ 1 & 1 & 2 \\ 2 & 2 & 1 \end{bmatrix} \begin{bmatrix} x \\ y \\ z \end{bmatrix} = \begin{bmatrix} A \\ B \\ C \end{bmatrix}$$

$AX = B \Rightarrow X = A^{-1}B$

$$X = \begin{bmatrix} 2 & 1 & 1 \\ 1 & 1 & 2 \\ 2 & 2 & 1 \end{bmatrix}^{-1} \begin{bmatrix} A \\ B \\ C \end{bmatrix} =$$

$$= \begin{bmatrix} 1 & -1/3 & -1/3 \\ -1 & 0 & 1 \\ 0 & 2/3 & -1/3 \end{bmatrix} \begin{bmatrix} A \\ B \\ C \end{bmatrix} =$$

$$\begin{bmatrix} A-B/3-C/3 \\ -A+C \\ 2B/3-C/3 \end{bmatrix}$$

Solution: Use $A-B/3-C/3$ batches of Vanilla, $-A+C$ batches of mocha, and $2B/3-C/3$ batches of strawberry.

55. Unknowns:
x = amount invested in FCTFX
y = amount invested in FSAZX
z = amount invested in FUSFX
The total investment was $9000:
   $x + y + z = 9000$
You invested an equal amount in FSAZX and FUSFX:
   $y = z$, or
   $y - z = 0$
Loss for the year from the first two funds was $400:
   $.06x + .05y = 400$
Matrix form:

$$\begin{bmatrix} 1 & 1 & 1 \\ 0 & 1 & -1 \\ 0.06 & 0.05 & 0 \end{bmatrix} \begin{bmatrix} x \\ y \\ z \end{bmatrix} = \begin{bmatrix} 9000 \\ 0 \\ 400 \end{bmatrix}$$

$AX = B \Rightarrow X = A^{-1}B$

$$X = \begin{bmatrix} 1 & 1 & 1 \\ 0 & 1 & -1 \\ 0.06 & 0.05 & 0 \end{bmatrix}^{-1} \begin{bmatrix} 9000 \\ 0 \\ 400 \end{bmatrix}$$

$$= \begin{bmatrix} -5/7 & -5/7 & 200/7 \\ 6/7 & 6/7 & -100/7 \\ 6/7 & -1/7 & -100/7 \end{bmatrix} \begin{bmatrix} 9000 \\ 0 \\ 400 \end{bmatrix} = \begin{bmatrix} 5000 \\ 2000 \\ 2000 \end{bmatrix}$$

Solution: You invested $5000 in FCTFX, $2000 in FSAZX, $2000 in FUSFX

57. Unknowns:
x = the number of shares of GE
y = the number of shares of WMT
z = the number of shares of XOM
The total investment was $8400
   Investment in GE = x shares @ $16 = $16x$
   Investment in WMT = y shares @ $56 = $56y$
   Investment in XOM = z shares @ $80 = $80z$
   Thus,
   $16x + 56y + 80z = 8400$
You expected to earn $248 in dividends:
   GE dividend = 7% of $16x$ invested
       $= .07(16x) = 1.12x$
   WMT dividend = 2% of $56y$ invested
       $= .02(56y) = 1.12y$

XOM dividend = 2% of 80y invested
    = .02(80y) = 1.6y
Thus,
1.12x +1.12y+1.6z = 248
You purchased a total of 200 shares:
x + y + z = 200
We therefore have the following system:
x + y + z = 200
16x+56y+80z = 8400
1.12x +1.12y+1.6z = 248
Matrix form:

$$\begin{bmatrix} 1 & 1 & 1 \\ 16 & 56 & 80 \\ 28/25 & 28/25 & 8/5 \end{bmatrix} \begin{bmatrix} x \\ y \\ z \end{bmatrix} = \begin{bmatrix} 200 \\ 8400 \\ 248 \end{bmatrix}$$

$$AX = B \Rightarrow X = A^{-1}B$$

$$X = \begin{bmatrix} 1 & 1 & 1 \\ 16 & 56 & 80 \\ 28/25 & 28/25 & 8/5 \end{bmatrix}^{-1} \begin{bmatrix} 200 \\ 8400 \\ 248 \end{bmatrix}$$

$$= \begin{bmatrix} 0 & -1/40 & 5/4 \\ 10/3 & 1/40 & -10/3 \\ -7/3 & 0 & 25/12 \end{bmatrix} \begin{bmatrix} 200 \\ 8400 \\ 248 \end{bmatrix} = \begin{bmatrix} 100 \\ 50 \\ 50 \end{bmatrix}$$

Solution: You purchased 100 shares of GE, 50 shares of WMT, 50 shares of XOM.

59. $P = \begin{bmatrix} 0.9892 & 0.0017 & 0.0073 & 0.0018 \\ 0.0010 & 0.9920 & 0.0048 & 0.0022 \\ 0.0018 & 0.0024 & 0.9934 & 0.0024 \\ 0.0008 & 0.0033 & 0.0045 & 0.9914 \end{bmatrix}$

2006 Distribution = A = [55.1 66.2 110.0 70.0]
To compute the population distribution in 2005,
compute $A \cdot P^{-1} \approx$ [55.4 66.1 109.7 70.1]

61. $R = \begin{bmatrix} \sqrt{1/2} & -\sqrt{1/2} \\ \sqrt{1/2} & \sqrt{1/2} \end{bmatrix} \approx$

$\begin{bmatrix} 0.7071 & -0.7071 \\ 0.7071 & 0.7071 \end{bmatrix}$

(a) The coordinates of a rotated point are given by

$$\begin{bmatrix} x' \\ y' \end{bmatrix} = R \begin{bmatrix} x \\ y \end{bmatrix} \approx \begin{bmatrix} 0.7071 & -0.7071 \\ 0.7071 & 0.7071 \end{bmatrix}$$

$$\begin{bmatrix} 2 \\ 3 \end{bmatrix}$$

$$= \begin{bmatrix} -0.7071 \\ 3.5355 \end{bmatrix}$$

Thus, $(x', y') \approx (-0.7071, 3.5355)$

(b) Multiplication by R rotates points through 45°. To rotate through 90°, multiply by R again, obtaining

$$R\left(R\begin{bmatrix} x \\ y \end{bmatrix}\right) = R^2 \begin{bmatrix} x \\ y \end{bmatrix}$$

In other words, multiplying by R^2 will result in a counterclockwise rotation of 90°.

To rotate by 135°, multiply yet again by R. This amounts to multiplying the original column vector by R^3.

(c) Let S be the matrix representing a clockwise rotation through 45°. Since a rotation of 45° clockwise followed by a rotation of 45° counterclockwise results in no change,

$$R\left(S\begin{bmatrix} x \\ y \end{bmatrix}\right) = \begin{bmatrix} x \\ y \end{bmatrix} = I\begin{bmatrix} x \\ y \end{bmatrix}$$

In other words, multiplication by RS should have the same effect as multiplication by the 2×2 identity matrix. This will happen if $S = R^{-1}$.

63. First, write the uncoded message as a string of numbers using A = 1, B = 2, C = 3, and so on:
"GO TO PLAN B" = [7 15 0 20 15 0 16 12 1 14 0 2] .
First, arrange the numbers in a matrix with 2 rows:
Uncoded message =
$\begin{bmatrix} 7 & 0 & 15 & 16 & 1 & 0 \\ 15 & 20 & 0 & 12 & 14 & 2 \end{bmatrix}$
Then encode by multiplying by $A = \begin{bmatrix} 1 & 2 \\ 3 & 4 \end{bmatrix}$:

Coded message

$= \begin{bmatrix} 1 & 2 \\ 3 & 4 \end{bmatrix} \begin{bmatrix} 7 & 0 & 15 & 16 & 1 & 0 \\ 15 & 20 & 0 & 12 & 14 & 2 \end{bmatrix}$

$= \begin{bmatrix} 37 & 40 & 15 & 40 & 29 & 4 \\ 81 & 80 & 45 & 96 & 59 & 8 \end{bmatrix}$

Arrange as a single row:
[37 81 40 80 15 45 40 96 29 59 4 8]

65. Coded message = [33 69 54 126 11 27 20 60 29 59 65 149 41 87]
First, arrange the numbers in a matrix with 2 rows:

Coded message = $\begin{bmatrix} 33 & 54 & 11 & 20 & 29 & 65 & 41 \\ 69 & 126 & 27 & 60 & 59 & 149 & 87 \end{bmatrix}$

To decode the message, multiply by A^{-1}

107

$$A^{-1} = \begin{bmatrix} 1 & 2 \\ 3 & 4 \end{bmatrix}^{-1} = \begin{bmatrix} -2 & 1 \\ 1.5 & -0.5 \end{bmatrix}$$

Decoded message

$$\begin{bmatrix} -2 & 1 \\ 1.5 & -0.5 \end{bmatrix} \begin{bmatrix} 33 & 54 & 11 & 20 & 29 & 65 & 41 \\ 69 & 126 & 27 & 60 & 59 & 149 & 87 \end{bmatrix}$$

$$= \begin{bmatrix} 3 & 18 & 5 & 20 & 1 & 19 & 5 \\ 15 & 18 & 3 & 0 & 14 & 23 & 18 \end{bmatrix}$$

Arrange as a single row:
[3 15 18 18 5 3 20 0 1 14 19 23 5 18]
Translate to letters using A = 1, B = 2, C = 3, and so on:
Decoded message = "CORRECT ANSWER"

67. If $AB = I$ and $BA = I$, then A and B are inverse matrices. (See the definition of the inverse matrix in the text.) Thus, choice (A) is the correct choice.

69. The given matrix has two identical rows. If two rows of a matrix are the same, then row reducing it will lead to a row of zeros, and so it cannot be reduced to the identity. Thus, the inverse of the given matrix does not exist—the matrix is singular.

71. We check that the given matrix is the inverse of $\begin{bmatrix} a & b \\ c & d \end{bmatrix}$ by multiplying the two matrices:

$$\left(\frac{1}{ad - bc} \begin{bmatrix} d & -b \\ -c & a \end{bmatrix} \right) \begin{bmatrix} a & b \\ c & d \end{bmatrix}$$

$$= \frac{1}{ad - bc} \begin{bmatrix} ad-bc & db-bd \\ -ac+ca & -bc+ad \end{bmatrix}$$

$$= \frac{1}{ad - bc} \begin{bmatrix} ad-bc & 0 \\ 0 & ad-bc \end{bmatrix}$$

$$= \begin{bmatrix} 0 & 1 \\ 1 & 0 \end{bmatrix}.$$

Since the product of the two matrices is the identity, they must be inverse matrices as claimed. In other words,

$$\begin{bmatrix} a & b \\ c & d \end{bmatrix}^{-1} = \frac{1}{ad - bc} \begin{bmatrix} d & -b \\ -c & a \end{bmatrix} \text{ (provided } ad$$
$$- bc \neq 0).$$

73. D is singular when one or more of the d_i are zero. If that is the case, then the matrix $[D \mid I]$ easily reduces to a matrix that has a row of zeros on the left-hand portion, so that D is singular,

Conversely, if none of the d_i are zero, then $[D \mid I]$ easily reduces to a matrix of the form $[I \mid E]$, showing that D is invertible.

75. To check that $B^{-1}A^{-1}$ is the inverse of AB, we multiply these two matrices and check that we obtain the identity matrix:
$$(AB)(B^{-1}A^{-1}) = A(BB^{-1})A^{-1} \text{ Associative law}$$
$$= AIA^{-1} \text{ Since } BB^{-1} = I$$
$$= AA^{-1} = I$$
Since the product of the given matrices is the identity, they must be inverses as claimed.

77. If a square matrix A reduces to one with a row of zeros, then it cannot have an inverse. The reason is that, if A has an inverse, then every system of equations $AX = B$ has a unique solution, namely $X = A^{-1}B$. But if A reduces to a matrix with a row of zeros, then such a system has either infinitely many solutions or no solution at all.

3.4 Game Theory

1. $e = RPC$

$$= [0 \ 1 \ 0 \ 0] \begin{bmatrix} 2 & 0 & -1 & 2 \\ -1 & 0 & 0 & -2 \\ -2 & 0 & 0 & 1 \\ 3 & 1 & -1 & 1 \end{bmatrix} \begin{bmatrix} 1 \\ 0 \\ 0 \\ 0 \end{bmatrix}$$

$$= -1$$

3. $e = RPC$

$$= [0.5 \ 0.5 \ 0 \ 0] \begin{bmatrix} 2 & 0 & -1 & 2 \\ -1 & 0 & 0 & -2 \\ -2 & 0 & 0 & 1 \\ 3 & 1 & -1 & 1 \end{bmatrix} \begin{bmatrix} 0 \\ 0 \\ 0.5 \\ 0.5 \end{bmatrix}$$

$$= -0.25$$

5. Since we are given a column matrix, we take our row strategy to be

$R = [x \ y \ z \ t]$.

$e = RPC$

$$= [x \ y \ z \ t] \begin{bmatrix} 0 & -1 & 5 \\ 2 & -2 & 4 \\ 0 & 3 & 0 \\ 1 & 0 & -5 \end{bmatrix} \begin{bmatrix} 0.25 \\ 0.75 \\ 0 \end{bmatrix}$$

$$= [x \ y \ z \ t] \begin{bmatrix} -0.75 \\ -1 \\ 2.25 \\ 0.25 \end{bmatrix}$$

$$= -0.74x - y + 2.25z + 0.25t.$$

The greatest coefficient is the coefficient of z, so we take $z = 1$ and $x = y = t = 0$. This gives the strategy

$R = [0 \ 0 \ 1 \ 0]$

and resulting expected value of the game

$e = -0.74(0) - 0 + 2.25(1) + 0.25(0)$

$= 2.25$.

7. Since we are given a row matrix, we take our column strategy to be

$C = [x \ y \ z]^T$.

$e - RPC$

$$= [1/2 \ 0 \ 1/4 \ 1/4] \begin{bmatrix} 0 & -1 & 5 \\ 2 & -2 & 4 \\ 0 & 3 & 0 \\ 1 & 0 & -5 \end{bmatrix} \begin{bmatrix} x \\ y \\ z \end{bmatrix}$$

$$= [1/4 \ 1/4 \ 5/4] \begin{bmatrix} x \\ y \\ z \end{bmatrix}$$

$$= (1/4)x + (1/4)y + (5/4)z$$

The lowest coefficients are the coefficients of x and y, so we can either take $x = 1$ and $y = z = 0$, giving the strategy $C = [1 \ 0 \ 0]^T$, or we could take $y = 1$ and $x = z = 0$, giving the strategy $C = [0 \ 1 \ 0]^T$, and resulting expected value of the game

$e = (1/4)(1) + (1/4)(0) + (5/4)(0)$

$= 1/4$

9. Checking the rows: Neither of rows a and b dominate the other: 1 is worse than 2, but 10 is better than –4.

Checking the columns: Column p dominates column q because each of the payoffs in column p is lower than or equal to the corresponding payoff in column q. We therefore eliminate column q to obtain

$$\begin{array}{cc} & \mathbf{B} \\ & \begin{array}{cc} p & r \end{array} \\ \mathbf{A} \begin{array}{c} a \\ b \end{array} & \begin{bmatrix} 1 & 10 \\ 2 & -4 \end{bmatrix} \end{array}$$

This matrix cannot be reduced further.

11. Checking the rows: Row 3 dominates both rows 1 and 2 because each of its payoffs is greater than or equal to the corresponding payoff in row 1 and 2 so we eliminate rows 1 and 2 to obtain

$$\begin{array}{c} \mathbf{B} \\ \begin{array}{ccc} a & b & c \end{array} \\ \mathbf{A} \ 3 \begin{array}{ccc} [5 & 0 & -1] \end{array} \end{array}$$

Checking the columns: Column c dominates columns a and b because –1 is less than 0 and $5p$. We therefore eliminate columns a and b to obtain

$$\begin{array}{c} \mathbf{B} \\ c \\ \mathbf{A} \ 3 \ [-1] \end{array}$$

13. Checking the rows: Row q dominates rows p and s because each of its payoffs is greater than or equal to the corresponding payoff in row p and s so we eliminate rows p and s to obtain

$$\begin{array}{cc} & \mathbf{B} \\ & \begin{array}{ccc} a & b & c \end{array} \\ \mathbf{A} \begin{array}{c} q \\ r \end{array} & \left[\begin{array}{ccc} 4 & 0 & 2 \\ 3 & -3 & 10 \end{array} \right] \end{array}$$

Checking the columns: Column b dominates the other two columns (each of its payoffs is less than or equal to the corresponding payoff in the other two columns). We therefore eliminate columns a and c to obtain

$$\begin{array}{cc} & \mathbf{B} \\ & b \\ \mathbf{A} \quad q & [\, 0 \,] \end{array}$$

15. Circle the row minima:

$$\begin{array}{cc} & \mathbf{B} \\ & \begin{array}{cc} p & q \end{array} \\ \mathbf{A} \begin{array}{c} a \\ b \end{array} & \left[\begin{array}{cc} ① & ① \\ 2 & ④ \end{array} \right] \end{array}$$

The largest row minimum is the 1 occurring in row a.
Box the column maxima:

$$\begin{array}{cc} & \mathbf{B} \\ & \begin{array}{cc} p & q \end{array} \\ \mathbf{A} \begin{array}{c} a \\ b \end{array} & \left[\begin{array}{cc} 1 & \boxed{1} \\ \boxed{2} & -4 \end{array} \right] \end{array}$$

The smallest column maximum is the 1 occurring in column q.
Since the a,q-entry 1 was both circled and boxed, it is a saddle point, and the game is strictly determined with A's optimal strategy being a, B's optimal strategy being q, and the value = saddle point = 1.

17. Circle the row minima:

$$\begin{array}{cc} & \mathbf{B} \\ & \begin{array}{ccc} p & q & r \end{array} \\ \mathbf{A} \begin{array}{c} a \\ b \end{array} & \left[\begin{array}{ccc} 2 & 0 & ②\\ ① & 3 & 0 \end{array} \right] \end{array}$$

The largest row minimum is the -1 occurring in row b. Box the column maxima:

$$\begin{array}{cc} & \mathbf{B} \\ & \begin{array}{ccc} p & q & r \end{array} \\ \mathbf{A} \begin{array}{c} a \\ b \end{array} & \left[\begin{array}{ccc} \boxed{2} & 0 & -2 \\ -1 & \boxed{3} & \boxed{0} \end{array} \right] \end{array}$$

The smallest column maximum is the 0 occurring in column r.
Since no entry is both circled and boxed, there are no saddle points, and so the game is not strictly determined.

19. Circle the row minima:

$$\begin{array}{cc} & \mathbf{B} \\ & \begin{array}{ccc} a & b & c \end{array} \\ \mathbf{A} \begin{array}{c} P \\ Q \\ R \\ S \end{array} & \left[\begin{array}{ccc} 1 & -1 & ⑤ \\ 4 & ④ & 2 \\ 3 & -3 & ⑩ \\ 5 & ⑤ & -4 \end{array} \right] \end{array}$$

The largest row minimum is the -4 occurring in row Q.
Box the column maxima:

$$\begin{array}{cc} & \mathbf{B} \\ & \begin{array}{ccc} a & b & c \end{array} \\ \mathbf{A} \begin{array}{c} P \\ Q \\ R \\ S \end{array} & \left[\begin{array}{ccc} 1 & \boxed{-1} & -5 \\ 4 & -4 & \boxed{2} \\ 3 & -3 & -10 \\ \boxed{5} & -5 & -4 \end{array} \right] \end{array}$$

The smallest column maximum is the -1 occurring in column b.
Since no entry is both circled and boxed, there are no saddle points, and so the game is not strictly determined.

21. (a) Row strategy: Take $R = [x,\ 1-x]$, $C = [1\ \ 0]^T$

$$e = RPC = [x\ \ 1-x] \left[\begin{array}{cc} -1 & 2 \\ 0 & -1 \end{array} \right] \left[\begin{array}{c} 1 \\ 0 \end{array} \right]$$

$$= -x$$

Take $R = [x\ \ 1-x]$, $C = [0\ \ 1]^T$

$$f = RPC = [x\ \ 1-x] \left[\begin{array}{cc} -1 & 2 \\ 0 & -1 \end{array} \right] \left[\begin{array}{c} 0 \\ 1 \end{array} \right]$$

$$= 2x - (1-x) = 3x - 1$$

Graphs of e and f:

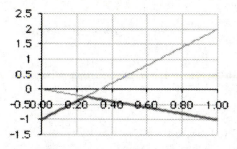

Lower (heavy) portion has its highest point at the intersection of the two lines. The x-coordinate of the intersection is given when $e = f$:

$$-x = 3x - 1$$
$$4x = 1$$
$$x = 1/4$$

So the optimal row strategy is

$$R = [x \quad 1-x]$$
$$= [1/4 \quad 3/4]$$

(b) Column Strategy:

Take $R = [1 \quad 0], C = [x \quad 1-x]^T$

$$e = RPC = [1 \ 0] \begin{bmatrix} -1 & 2 \\ 0 & -1 \end{bmatrix} \begin{bmatrix} x \\ 1-x \end{bmatrix}$$

$$= -x + 2(1-x)$$
$$= -3x + 2$$

Take $R = [0 \quad 1], C = [x \quad 1-x]^T$

$$f = RPC = [0 \ 1] \begin{bmatrix} -1 & 2 \\ 0 & -1 \end{bmatrix} \begin{bmatrix} x \\ 1-x \end{bmatrix}$$

$$= x - 1$$

Graphs of e and f:

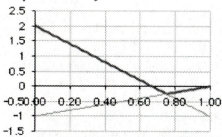

The upper (heavy) portion has its lowest point at the intersection of the two lines. The x-coordinate of the intersection is given when $e = f$:

$$-3x + 2 = x - 1$$
$$4x = 3$$
$$x = 3/4$$

So the optimal column strategy is

$$C = [x \quad 1-x]^T$$

$$= [3/4 \quad 1/4]^T$$

(c) To compute the expected value of the game, compute the y-coordinate of the point on either graph with the given x-coordinate. For instance, using the second graph (part (b)),

$$e = x - 1 = 3/4 - 1 = -1/4.$$

Solution: $R = [1/4 \ \ 3/4], C = [3/4 \ \ 1/4]^T,$

$$e = -1/4$$

23. (a) Row strategy: Take $R = [x, \ 1-x], C = [1 \ \ 0]^T$

$$e = RPC = [x \ 1-x] \begin{bmatrix} -1 & -2 \\ -2 & 1 \end{bmatrix} \begin{bmatrix} 1 \\ 0 \end{bmatrix}$$

$$= -x - 2(1-x) = x - 2$$

Take $R = [x \ \ 1-x], C = [0 \ \ 1]^T$

$$f = RPC = [x \ 1-x] \begin{bmatrix} -1 & -2 \\ -2 & 1 \end{bmatrix} \begin{bmatrix} 0 \\ 1 \end{bmatrix}$$

$$= -2x + (1-x) = -3x + 1$$

Graphs of e and f:

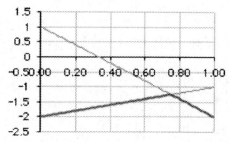

Lower (heavy) portion has its highest point at the intersection of the two lines. The x-coordinate of the intersection is given when $e = f$:

$$x - 2 = -3x + 1$$
$$4x = 3$$
$$x = 3/4$$

So the optimal row strategy is

$$R = [x \quad 1-x]$$
$$= [3/4 \quad 1/4]$$

(b) Column Strategy:

Take $R = [1 \quad 0], C = [x \quad 1-x]^T$

$$e = RPC = [1 \ 0] \begin{bmatrix} -1 & -2 \\ -2 & 1 \end{bmatrix} \begin{bmatrix} x \\ 1-x \end{bmatrix}$$

$$= -x - 2(1-x) = x - 2$$

Take $R = [0 \quad 1], C = [x \quad 1-x]^T$

$$f = RPC = [0 \ 1] \begin{bmatrix} -1 & -2 \\ -2 & 1 \end{bmatrix} \begin{bmatrix} x \\ 1-x \end{bmatrix}$$

$$= -2x + (1-x) = -3x + 1$$

Note that *e* and *f* are the same as for part (a). Therefore, their graphs are the same, except that this time we are interested in the lowest point of the upper region:

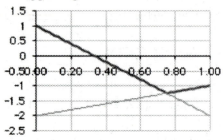

This is the same intersection point as in part (a), so $x = 3/4$ and the optimal column strategy is

$$C = [x \quad 1-x]^T$$
$$= [3/4 \quad 1/4]^T$$

(c) To compute the expected value of the game, compute the *y*-coordinate of the point on either graph with the given *x*-coordinate. For instance, using the second graph (part (b)),

$$e = x - 2 = 3/4 - 2 = -5/4.$$

Solution: $R = [3/4 \quad 1/4]$, $C = [3/4 \quad 1/4]^T$, $e = -5/4$

25. The possible outcomes are:
HH: You lose 1 point. Payoff: -1
HT: You win 1 point. Payoff: 1
TH: You win 1 point. Payoff: 1
TT: You lose 1 point. Payoff: -1
Payoff matrix:

$$\text{You} \quad \begin{array}{c} \\ H \\ T \end{array} \begin{array}{cc} \multicolumn{2}{c}{\text{Friend}} \\ H \quad T \\ \left[\begin{array}{cc} -1 & 1 \\ 1 & -1 \end{array} \right] \end{array}$$

27. Your strategies are to attack by sea or air. Your oppoenent's strategies are to defend by sea, air, or both. The possible outcomes are:

(Sea, Sea) or (Air, Air): Your attack is met by an all-out defence: you lose 200 points.
Payoff: -200

(Sea, Air) or (Air, Sea): Your attack is met by no defence: you win 100 points.
Payoff: 100

(Sea, Both) or (Air, Both): Your attack is met by a shared defence: you win 50 points.
Payoff: 50

Payoff matrix:

		Your Opponent Defends		
		Sea	Air	Both
You Attack	Sea	-200	100	50
	Air	100	-200	50

29. Take B = Brakpan; N = Nigel; S = Springs. Your strategies are to locate in one of B, N, S, and your opponent's strategies are the same. The possible outcomes are:
BB, SS, or SS: You share the total business. No net gain or loss of sales. Payoff: 0
BN or NB: One of you locates at B, the other at N. These two cities provide the same potential market, so again there is no net gain or loss. Payoff: 0
SB or SN: You locate at S (total market: 1000 bugers/day), your opponent locates at N or B (total market: 2000 burgers/day). Your net loss is then 1000 burgers/day. Payoff: -1000.
BS or NS: You locate at N or B (total market: 2000 burgers/day), your opponent locates at. S (total market: 1000 bugers/day) Your net gain is then 1000 burgers/day. Payoff: 1000.
Payoff matrix:

		Your Opponent		
		B	N	S
	B	0	0	1000
You	N	0	0	1000
	S	-1000	-1000	0

31. Let P = PleasantTap; T = Thunder Rumble; S = Strike the Gold, N = None. Your strategies: P, T, S. Your "opponent" is Nature, which decides the winner. Opponent's strategies: P, T, S, N. The possible outcomes are:
PP: You bet on P and P wins. Odds: 5-2. $2 wins you $5. Therefore $10 wins you 5×$5 = $25.
Payoff: 25
TT: You bet on T and T wins. Odds: 7-2. $2 wins you $7. Therefore $10 wins you 5×$7 = $35.
Payoff: 35
SS: You bet on S and S wins. Odds: 4-1. $1 wins you $4. Therefore $10 wins you 10×$4 = $40.
Payoff: 40

All other outsomes: You lose your $10 because your horse fails to win. Payoff: –10
Payoff matrix:

$$\begin{array}{c} & \textbf{Winner} \\ & \begin{array}{cccc} \text{P} & \text{T} & \text{S} & \text{N} \end{array} \\ \textbf{You Bet} \begin{array}{c} \text{P} \\ \text{T} \\ \text{S} \end{array} & \left[\begin{array}{cccc} 25 & -10 & -10 & -10 \\ -10 & 35 & -10 & -10 \\ -10 & -10 & 40 & -10 \end{array}\right] \end{array}$$

33. $P = \begin{bmatrix} -60 & -40 \\ 30 & -50 \end{bmatrix}$, $R = [.50 \quad .50]$, $C = \begin{bmatrix} 0.20 \\ 0.80 \end{bmatrix}$

$e = RPC$

$= [.50 \quad .50] \begin{bmatrix} -60 & -40 \\ 30 & -50 \end{bmatrix} \begin{bmatrix} 0.20 \\ 0.80 \end{bmatrix}$

$= [-15 \quad -45] \begin{bmatrix} 0.20 \\ 0.80 \end{bmatrix} = -39$

You can expect to lose 39 customers.

35. $P = \begin{bmatrix} 200 & 150 & 140 \\ 130 & 220 & 130 \\ 110 & 110 & 220 \end{bmatrix}$, $R = [x \ y \ z]$,

$C = \begin{bmatrix} 0.2 \\ 0.5 \\ 0.3 \end{bmatrix}$

$e = RPC$

$= [x \ y \ z] \begin{bmatrix} 200 & 150 & 140 \\ 130 & 220 & 130 \\ 110 & 110 & 220 \end{bmatrix} \begin{bmatrix} 0.2 \\ 0.5 \\ 0.3 \end{bmatrix}$

$= [x \ y \ z] \begin{bmatrix} 157 \\ 175 \\ 143 \end{bmatrix} = 157x + 175y + 143z$

The largest coefficient is the coefficient of y, so we take
$R = [0 \quad 1 \quad 0]$
meaning that Option 2: Move to the suburbs (corresponding to y) is the best option.

37. (a) $P = \begin{bmatrix} 90 & 70 & 70 \\ 40 & 90 & 40 \\ 60 & 40 & 90 \end{bmatrix}$,

$R = [0.25 \quad 0.50 \quad 0.25]$, $C = \begin{bmatrix} 0.25 \\ 0.50 \\ 0.25 \end{bmatrix}$

$e = RPC$

$= [0.25 \quad 0.50 \quad 0.25] \begin{bmatrix} 90 & 70 & 70 \\ 40 & 90 & 40 \\ 60 & 40 & 90 \end{bmatrix} \begin{bmatrix} 0.25 \\ 0.50 \\ 0.25 \end{bmatrix}$

$= [0.25 \quad 0.50 \quad 0.25] \begin{bmatrix} 75 \\ 65 \\ 57.5 \end{bmatrix}$

$= [65.625]$

You can expect to get about 66% on the test.

(b) $P = \begin{bmatrix} 90 & 70 & 70 \\ 40 & 90 & 40 \\ 60 & 40 & 90 \end{bmatrix}$, $R = [x \ y \ z]$,

$C = \begin{bmatrix} 0.25 \\ 0.50 \\ 0.25 \end{bmatrix}$

$e = RPC$

$= [x \ y \ z] \begin{bmatrix} 90 & 70 & 70 \\ 40 & 90 & 40 \\ 60 & 40 & 90 \end{bmatrix} \begin{bmatrix} 0.25 \\ 0.50 \\ 0.25 \end{bmatrix}$

$= [x \ y \ z] \begin{bmatrix} 75 \\ 65 \\ 57.5 \end{bmatrix} = 75x + 65y + 57.5z$

The largest coefficient is the coefficient of x, so we take
$R = [1 \quad 0 \quad 0]$
meaning that Game Theory (coresponding to x) is the best option. Your expected score is then
$75(1) + 65(0) + 57.5(0) = 75\%$

(c) $P = \begin{bmatrix} 90 & 70 & 70 \\ 40 & 90 & 40 \\ 60 & 40 & 90 \end{bmatrix}$, $R = [0.25 \quad 0.50 \quad 0.25]$,

$C = \begin{bmatrix} x \\ y \\ z \end{bmatrix}$

$e = RPC$

$= [0.25 \quad 0.50 \quad 0.25] \begin{bmatrix} 90 & 70 & 70 \\ 40 & 90 & 40 \\ 60 & 40 & 90 \end{bmatrix} \begin{bmatrix} x \\ y \\ z \end{bmatrix}$

$= [57.5 \quad 72.5 \quad 60] \begin{bmatrix} x \\ y \\ z \end{bmatrix} = 57.5x + 72.5y + 60z$

The lowest coefficient is the coefficient of x, so we take
$C = [1 \quad 0 \quad 0]^T$

113

meaning that game theory would be worst for you, and you could expect to earn
$57.5(1) + 72.5(0) + 60(0) = 57.5\%$
for the test.

39. (a)

$$P = \begin{bmatrix} -500,000 & -200,000 & 10,000 & 200,000 \\ -200,000 & 0 & 0 & 0 \\ -100,000 & 10,000 & -200,000 & -300,000 \end{bmatrix}$$

$$R = [x \quad y \quad z], C = \begin{bmatrix} 0.2 \\ 0.2 \\ 0.3 \\ 0.3 \end{bmatrix},$$

(based on the past 10 years)
$e = RPC$
$= [x \quad y \quad z]$
$\times$

$$\begin{bmatrix} -500,000 & -200,000 & 10,000 & 200,000 \\ -200,000 & 0 & 0 & 0 \\ -100,000 & 10,000 & -200,000 & -300,000 \end{bmatrix}$$

$$\times \begin{bmatrix} 0.2 \\ 0.2 \\ 0.3 \\ 0.3 \end{bmatrix}$$

$$= [x \quad y \quad z] \begin{bmatrix} -77,000 \\ -40,000 \\ -168,000 \end{bmatrix}$$

$= -77,000x - 40,000y - 168,000z$
The largest coefficient is the coefficient of y, so we take
$R = [0 \quad 1 \quad 0]$
meaning that laying off 10 workers (coresponding to y) is the best option. The expected payoff is
$-77,000(0) - 40,000(1) - 168,000(0) = -40,000,$
corresponding to a cost of \$40,000.

(b)
$P =$

$$\begin{bmatrix} -500,000 & -200,000 & 10,000 & 200,000 \\ -200,000 & 0 & 0 & 0 \\ -100,000 & 10,000 & -200,000 & -300,000 \end{bmatrix}$$

$$R = [0.50 \quad 0.50 \quad 0], C = \begin{bmatrix} x \\ y \\ z \\ t \end{bmatrix}$$

$e = RPC$

$= [0.50 \quad 0.50 \quad 0] \times$

$$\begin{bmatrix} -500,000 & -200,000 & 10,000 & 200,000 \\ -200,000 & 0 & 0 & 0 \\ -100,000 & 10,000 & -200,000 & -300,000 \end{bmatrix}$$

$$\begin{bmatrix} x \\ y \\ z \\ t \end{bmatrix}$$

$$= [-350,000 \quad -100,000 \quad 5000 \quad 100,000] \begin{bmatrix} x \\ y \\ z \\ t \end{bmatrix}$$

$= -350,000x - 100,000y + 5000z + 100,000t.$
The lowest coefficient is the coefficient of x, so we take
$C = [1 \quad 0 \quad 0 \quad 0]^T$
meaning that 60 inches of snow would be worst for him, costing him \$350,000.

(c) The strategy from part (a) is
$R = [0 \quad 1 \quad 0];$
laying off 10 workers. If he does so, the worst that could happen is 0 inches of snow, (which would cost him \$20,000). Since he feels that the Gods of Chaos are planning on 0 inches of snow, his best option would be to lay off 15 workers (according to the playoff matrix) and cut his losses to \$100,000.

41. (a) The payoff matrix is

$$P = \begin{bmatrix} 15 & 60 & 80 \\ 15 & 60 & 60 \\ 10 & 20 & 40 \end{bmatrix}$$

Checking the rows: Row 1 dominates each of the other rows, so we eliminate rows 2 and 3, leaving
$[15 \quad 60 \quad 80]$
Checking the columns: Column 1 dominates each of the other columns, so we eliminate columns 2 and 3, leaving the 1×1 game
$[15],$
which corresponds to the first CE strategy (charge \$1000) and the first GCS strategy (charge \$900). So, CE should charge \$1000 and GCS should charge \$900. Since the payoff is 15, a 15% gain in market share for CE results.

114

(b) Look at the original (unreduced) payoff matrix. CE is aware that GCS is planning to charge $900. For the best market share, CE should charge either $1000 or $1200 because either will result in the best market share (15% gain) under the circumstances. Thus, in terms of market share, the added information has no effect. However, in terms of revenue, the better of the two options would be to charge the larger price: $1200. (the more CE can charge for the same market, the better!).

43. Take CCC as the row player and MMA as the column player. Thus, CCC's strategies are Pablo, Sal and Edison, while MMA's strategies are Carlos, Marcus and Noto. To set up the payoff matrix, enter 1 for every combination in which the CCC wrestler beats the MMA wrestler, –1 for every combination in which the MMA wrestler beats the CCC wrestler, and 0 in all the evenly matched combinations. The resulting payoff matrix is

MMA

	Carlos	Marcus	Noto
Pablo	1	1	0
Sal	0	–1	0
Edison	0	–1	–1

CCC

Comparing rows, we see that Row 1 dominates the other two rows, so we eliminate Rows 2 and 3 leaving

	Carlos	Marcus	Noto
Pablo	1	1	0

Comparing columns, we see that Column 3 dominates the other two columns, so we eliminate Columns 1 and 2, leaving us with

	Noto
Pablo	0

Thus the game is reduced to Pablo vs. Noto. Since the payoff is 0, the game is evenly matched.

45. We first reduce by dominance (following the "FAQ" in the textbook):
Comparing the rows, we find that neither row dominates the other.
Comparing the columns, we see than Column 1 dominates Column 2. So we eliminate Column 2, leaving

	Northern Route
Northern Route	2
Southern Route	1

Comparing the rows, we now see that Row 1 dominates Row 2, so we eliminate Row 2, leaving us with

	Northern Route
Northern Route	1

Thus, both commanders should use the northern route, resulting in an estimate of one day's bombing time.

47. Take C = Confess and N = Do not confess. If we take the (negative) payoffs as the amount of time Slim faces behind bars, we get:

Joe

	C	N
Slim C	–2	0
N	–10	–5

Since Row 1 dominates Row 2, Slim's optimal strategy is to confess.

49. (a) Let F represent a visit to Florida and O a visit to Ohio. The outcomes are:
FF and OO: Both candidates visit the same state, so McCain still has a 24% chance of winning, so the payoff is 24 in both cases.
FO: McCain visits Florida, giving him a 60+10 = 70% chance in that state, and Obama visits Ohio, reducing McCain's chances there to 40–10 = 30%. Thus, the probability of McCain winning both states is 0.70×0.30 = 0.21

OF: McCain visits Ohio, giving him a 40+10 = 50% chance in that state, and Obama visits Florida, reducing McCain's chances there to 60–10 = 50%. Thus, the probability of McCain winning both states is 0.50×0.50 = 0.25
This gives the following payoff matrix:

$$\begin{array}{c} \textbf{Obama} \\ \begin{array}{cc} F & O \end{array} \\ \textbf{McCain} \begin{array}{c} F \\ O \end{array} \left[\begin{array}{cc} 24 & 21 \\ 25 & 24 \end{array} \right] \end{array}$$

(b) In the payoff matrix in part (a), Row 2 dominates Row 1 (meaning that McCain should visit Ohio. in that case):

$$\begin{array}{c} \textbf{Obama} \\ \begin{array}{cc} F & O \end{array} \\ \textbf{McCain} \begin{array}{c} O \end{array} \left[\begin{array}{cc} 25 & 24 \end{array} \right] \end{array}$$

Column 2 now dominates Column 1, leaving us with the following solution

$$\begin{array}{c} \textbf{Obama} \\ O \\ \textbf{McCain} \begin{array}{c} O \end{array} \left[\begin{array}{c} 24 \end{array} \right] \end{array}$$

meaning that both candidates should visit Ohio, leaving McCain with a 21% chance of winning the election.

51. The payoff matrix (payoffs in thousands of dollars) is

$$\begin{array}{c} \textbf{Splish} \\ \begin{array}{cc} \text{WISH} & \text{WASH} \end{array} \\ \textbf{Softex} \begin{array}{c} \text{WISH} \\ \text{WASH} \end{array} \left[\begin{array}{cc} -20 & 100 \\ 0 & -20 \end{array} \right] \end{array}$$

We are aksed to provide the optimal row strategy.
Take $R = [x, 1-x], C = [1 \ 0]^T$

$$e = RPC = [x \ 1-x] \left[\begin{array}{cc} -20 & 100 \\ 0 & -20 \end{array} \right] \left[\begin{array}{c} 1 \\ 0 \end{array} \right]$$

$$= -20x + 0(1-x) = -20x$$

Take $R = [x \ 1-x], C = [0 \ 1]^T$

$$f - RPC = [x \ 1-x] \left[\begin{array}{cc} -20 & 100 \\ 0 & 20 \end{array} \right] \left[\begin{array}{c} 0 \\ 1 \end{array} \right]$$

$$= 100x - 20(1-x) = 120x - 20$$

Graphs of e and f:

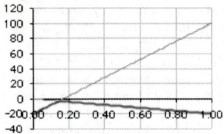

Lower (heavy) portion has its highest point at the intersection of the two lines. The x-coordinate of the intersection is given when $e = f$:

$$-20x = 120x - 20$$
$$140x = 20$$
$$x = 1/7$$

So the optimal row strategy is

$$R = [x \ 1-x]$$
$$= [1/7 \ 6/7]$$

This solution corresponds to allocating 1/7 of its advertising budget to WISH and the rest (6/7) to WASH.
The value of the game is then

$$e = -20x = -20/7 \approx -2.86$$

So Softex can expect to lose approximately $2,860.

53. Like a saddle point in a payoff matrix, the center of a saddle is a low point (minimum height) in one direction and a high point (maximum) in a perpendicular direction.

55. Although there is a saddle point in the 2,4 position, you would be wrong to use saddle points (based on the minimax criterion) to reach the conclusion that row strategy 2 is best. One reason is that the entries in the matrix do not represent payoffs, since high numbers of employees in an area do not necessarily represent benefit to the row player. Another reason for this is that there is no opponent deciding what your job will be in such a way as to force you into the least populated job.

57. If you strictly alternate the two strategies the column player will know which pure strategy you will play on each move, and can choose a pure strategy accordingly. For example, consider the game

$$A \begin{bmatrix} \overset{a}{1} & \overset{b}{0} \\ 0 & 1 \end{bmatrix}.$$

By the analysis of Example 3 (or the symmetry of the game), the best strategy for the row player is [0.5 0.5] and the best strategy for the column player is [0.5 0.5]^T. This gives an expected value of 0.5 for the game. However, suppose that the row player alternates A and B strictly and that the column player catches on to this. Then, whenever the row player plays A the column player will play b and whenever the row player plays B the column player will play a. This gives a payoff of 0 each time, worse for the row player than the expected value of 0.5.

3.5 Input-Output Models

1. $A = \begin{bmatrix} 0.2 & 0.05 \\ 0.8 & 0.01 \end{bmatrix}$

Sector 1 = Paper; Sector 2 = Wood

(a) Wood → Paper = Sector 2→Sector 1
The corresponding entry of the technology matrix is

$a_{21} = 0.8$

(b) Paper → Paper = Sector 1→Sector 1
The corresponding entry of the technology matrix is

$a_{11} = 0.2$

(c) Paper → Wood = Sector 1→Sector 2
The corresponding entry of the technology matrix is

$a_{12} = 0.05$

3. Sector 1 = Television news, Sector 2 = Radio news
a_{11} = units of Television news needed to produce one unit of Television news = 0.2
a_{12} = units of Television news needed to produce one unit of Radio news = 0.1
a_{21} = units of Radio news needed to produce one unit of Television news = 0.5
a_{22} = units of Radio news needed to produce one unit of Radio news = 0

Thus, $A = \begin{bmatrix} 0.2 & 0.1 \\ 0.5 & 0 \end{bmatrix}$

5. $X = (I-A)^{-1}D$

$= \left(\begin{bmatrix} 1 & 0 \\ 0 & 1 \end{bmatrix} - \begin{bmatrix} 0.5 & 0.4 \\ 0 & 0.5 \end{bmatrix} \right)^{-1} \begin{bmatrix} 10,000 \\ 20,000 \end{bmatrix}$

$= \begin{bmatrix} 0.5 & -0.4 \\ 0 & 0.5 \end{bmatrix}^{-1} \begin{bmatrix} 10,000 \\ 20,000 \end{bmatrix}$

$= \begin{bmatrix} 2 & 1.6 \\ 0 & 2 \end{bmatrix} \begin{bmatrix} 10,000 \\ 20,000 \end{bmatrix} = \begin{bmatrix} 52,000 \\ 40,000 \end{bmatrix}$

7. $X = (I-A)^{-1}D$

$= \left(\begin{bmatrix} 1 & 0 \\ 0 & 1 \end{bmatrix} - \begin{bmatrix} 0.1 & 0.4 \\ 0.2 & 0.5 \end{bmatrix} \right)^{-1} \begin{bmatrix} 25,000 \\ 15,000 \end{bmatrix}$

$= \begin{bmatrix} 0.9 & -0.4 \\ -0.2 & 0.5 \end{bmatrix}^{-1} \begin{bmatrix} 25,000 \\ 15,000 \end{bmatrix}$

$= \begin{bmatrix} 50/37 & 40/37 \\ 20/37 & 90/37 \end{bmatrix} \begin{bmatrix} 25,000 \\ 15,000 \end{bmatrix}$

$= \begin{bmatrix} 50,000 \\ 50,000 \end{bmatrix}$

9. $X = (I-A)^{-1}D$

$= \left(\begin{bmatrix} 1 & 0 & 0 \\ 0 & 1 & 0 \\ 0 & 0 & 1 \end{bmatrix} - \begin{bmatrix} 0.5 & 0.1 & 0 \\ 0 & 0.5 & 0.1 \\ 0 & 0 & 0.5 \end{bmatrix} \right)^{-1} \begin{bmatrix} 1000 \\ 1000 \\ 2000 \end{bmatrix}$

$= \begin{bmatrix} 0.5 & -.1 & 0 \\ 0 & 0.5 & -.1 \\ 0 & 0 & 0.5 \end{bmatrix}^{-1} \begin{bmatrix} 1000 \\ 1000 \\ 2000 \end{bmatrix}$

$= \begin{bmatrix} 2 & 0.4 & 0.08 \\ 0 & 2 & 0.4 \\ 0 & 0 & 2 \end{bmatrix} \begin{bmatrix} 1000 \\ 1000 \\ 2000 \end{bmatrix} = \begin{bmatrix} 2560 \\ 2800 \\ 4000 \end{bmatrix}$

11. $X = (I-A)^{-1}D$

$= \left(\begin{bmatrix} 1 & 0 & 0 \\ 0 & 1 & 0 \\ 0 & 0 & 1 \end{bmatrix} - \begin{bmatrix} 0.2 & 0.2 & 0 \\ 0.2 & 0.4 & 0.2 \\ 0 & 0.2 & 0.2 \end{bmatrix} \right)^{-1}$
$\begin{bmatrix} 16,000 \\ 8000 \\ 8000 \end{bmatrix}$

$= \begin{bmatrix} 0.8 & -.2 & 0 \\ -.2 & 0.6 & -.2 \\ 0 & -.2 & 0.8 \end{bmatrix}^{-1} \begin{bmatrix} 16,000 \\ 8000 \\ 8000 \end{bmatrix}$

$= \begin{bmatrix} 11/8 & 1/2 & 1/8 \\ 1/2 & 2 & 1/2 \\ 1/8 & 1/2 & 11/8 \end{bmatrix} \begin{bmatrix} 16,000 \\ 8000 \\ 8000 \end{bmatrix} = \begin{bmatrix} 27,000 \\ 28,000 \\ 17,000 \end{bmatrix}$

13. Change in Production
$= (I-A)^{-1} \times$ Change in Demand

$= \left(\begin{bmatrix} 1 & 0 \\ 0 & 1 \end{bmatrix} - \begin{bmatrix} 0.1 & 0.4 \\ 0.2 & 0.5 \end{bmatrix} \right)^{-1} \begin{bmatrix} 50 \\ 30 \end{bmatrix}$

$= \begin{bmatrix} 0.9 & -0.4 \\ -0.2 & 0.5 \end{bmatrix}^{-1} \begin{bmatrix} 50 \\ 30 \end{bmatrix}$

$= \begin{bmatrix} 50/37 & 40/37 \\ 20/37 & 90/37 \end{bmatrix} \begin{bmatrix} 50 \\ 30 \end{bmatrix} = \begin{bmatrix} 100 \\ 100 \end{bmatrix}$

15. Change in Production
$= (I-A)^{-1} \times$ Change in Demand

$$= \begin{bmatrix} 1.5 & 0.1 & 0 \\ 0.2 & 1.2 & 0.1 \\ 0.1 & 0.7 & 1.6 \end{bmatrix} \begin{bmatrix} 1 \\ 0 \\ 0 \end{bmatrix} = \begin{bmatrix} 1.5 \\ 0.2 \\ 0.1 \end{bmatrix}$$

Note that this is the first column of $(I-A)^{-1}$. In general, the ith column of $(I-A)^{-1}$ gives the change in production necessary to meet an increase in external demand of one unit for the product of Sector i.

17.

	To	A	B	C
From	**A**	1000	2000	3000
	B	0	4000	0
	C	0	1000	3000
Total Output		5000	5000	6000

We obtain the technology matrix from the input-output table by dividing each column by its total:

$$A = \begin{bmatrix} 1000/5000 & 2000/5000 & 3000/6000 \\ 0 & 4000/5000 & 0 \\ 0 & 1000/5000 & 3000/6000 \end{bmatrix}$$

$$= \begin{bmatrix} 0.2 & 0.4 & 0.5 \\ 0 & 0.8 & 0 \\ 0 & 0.2 & 0.5 \end{bmatrix}$$

19. First obtain the technology matrix from the input-output table by dividing each column by its total:

$$A = \begin{bmatrix} 10,000/50,000 & 20,000/40,000 \\ 5000/50,000 & 0 \end{bmatrix}$$

$$= \begin{bmatrix} 0.2 & 0.5 \\ 0.1 & 0 \end{bmatrix}$$

Production = $(I-A)^{-1} \times$ Demand

$$= \left(\begin{bmatrix} 1 & 0 \\ 0 & 1 \end{bmatrix} - \begin{bmatrix} 0.2 & 0.5 \\ 0.1 & 0 \end{bmatrix} \right)^{-1} \begin{bmatrix} 45,000 \\ 30,000 \end{bmatrix}$$

$$= \begin{bmatrix} 0.8 & -.5 \\ -.1 & 1 \end{bmatrix}^{-1} \begin{bmatrix} 45,000 \\ 30,000 \end{bmatrix}$$

$$= \begin{bmatrix} 4/3 & 2/3 \\ 2/15 & 16/15 \end{bmatrix} \begin{bmatrix} 45,000 \\ 30,000 \end{bmatrix}$$

$$= \begin{bmatrix} 80,000 \\ 38,000 \end{bmatrix}$$

Solution: The Main DR had to produce $80,000 worth of food, while Bits & Bytes had to produce $38,000 worth of food.

21. First obtain the technology matrix from the input-output table by dividing each column by its total:

$$A = \begin{bmatrix} 6000/90,000 & 500/140,000 \\ 24,000/90,000 & 30,000/140,000 \end{bmatrix} = \begin{bmatrix} 1/15 & 1/280 \\ 4/15 & 3/14 \end{bmatrix}$$

Production = $(I-A)^{-1} \times$ Demand

$$= \left(\begin{bmatrix} 1 & 0 \\ 0 & 1 \end{bmatrix} - \begin{bmatrix} 1/15 & 1/280 \\ 4/15 & 3/14 \end{bmatrix} \right)^{-1} \begin{bmatrix} 80,000 \\ 90,000 \end{bmatrix}$$

$$= \begin{bmatrix} 14/15 & -1/280 \\ -4/15 & 11/14 \end{bmatrix}^{-1} \begin{bmatrix} 80,000 \\ 90,000 \end{bmatrix}$$

$$\approx \begin{bmatrix} 1.07282 & 0.00488 \\ 0.36411 & 1.27438 \end{bmatrix} \begin{bmatrix} 80,000 \\ 90,000 \end{bmatrix}$$

$$\approx \begin{bmatrix} 86265 \\ 143823 \end{bmatrix}$$

Solution: Equipment Sector production approximately $86,000 million, Components Sector production approximately $140,000 million.

23. $(I-A)^{-1} = \begin{bmatrix} 1.228 & 0.182 \\ 0.006 & 1.1676 \end{bmatrix}$

Sector 1 = Textiles, Sector 2 = Clothing & footwear

(a) The missing term is the number of units of Sector 2 needed to produce one additional unit of Sector 1. This quantity is given by the 2,1-entry of $(I-A)^{-1}$, or 0.006.

(b) The missing terms refer to the 1,2-entry of $(I-A)^{-1}$, which gives the number of units of Sector 1 needed to produce one additional unit of Sector 2—in other words, the additional dollars worth of textiles that must be produced one-dollar increase in the demand for clothing and footwear.

In Exercises 25–28, the technology we show in the solutions is the on-line Excel tutorial found at

FiniteMath.org → Student Web Site → Everything for Finite Math → Chapter 3 → Excel Tutorials for Chapter 3 → Section 3.4

25.

To	1	2	3	4
From 1	11,937	9	109	855
2	26,649	4,285	0	4,744
3	0	0	439	61
4	5,423	10,952	3,002	216
Total	97,795	120,594	14,642	47,473

To determine how each sector would need to react to an increase in demand, we compute the columns of $(I-A)^{-1}$. First, we obtain the technology matrix A by dividing each column by its total. (This process is automated in the Excel tutorial):

0.12206	0.00007	0.00744	0.01801
0.27250	0.03553	0.00000	0.09993
0.00000	0.00000	0.02998	0.00128
0.05545	0.09082	0.20503	0.00455

Then compute the matrix $(I-A)^{-1}$

   TI-83/84 Plus Format:

   `(Identity(4)-[A])`$^{-1}$

   Matrix Algebra Tool: `(I-A)^-1`

1.14099	0.00205	0.01317	0.02087
0.33210	1.04734	0.02605	0.11118
0.00012	0.00013	1.03119	0.00135
0.09388	0.09569	0.21550	1.01615

We are given external demand increases of $1000 million in each sector, so we multiply $(I-A)^{-1}$ by the column matrix

$$D = \begin{bmatrix} 1000 \\ 1000 \\ 1000 \\ 1000 \end{bmatrix}$$

and obtain

1140.99	2.05	13.17	20.87
332.10	1047.34	26.05	111.18
0.12	0.13	1031.19	1.35
93.88	95.69	215.50	1016.15

The columns of this matrix show the amounts, in millions of dollars, by which each sector needs to increase production to meet the additional demand.

27.

To	1	2	3	4
From 1	678.4	3.7	3341.5	1023.5
2	15.5	6.9	17.1	124.5
3	47.3	4.3	893.1	145.8
4	312.5	22.1	83.2	693.5
Total	9401.3	685.8	6997.3	4818.3

To determine how each sector would need to react to an increase in demand, we compute the columns of $(I-A)^{-1}$. First, we obtain the technology matrix A by dividing each column by its total. (This process is automated in the Excel tutorial):

0.07216	0.00540	0.47754	0.21242
0.00165	0.01006	0.00244	0.02584
0.00503	0.00627	0.12763	0.03026
0.03324	0.03223	0.01189	0.14393

Then compute the matrix $(I-A)^{-1}$

   TI-83/84 Plus Format:

   `(Identity(4)-[A])`$^{-1}$

   Matrix Algebra Tool: `(I-A)^-1`

1.09155	0.01929	0.60157	0.29270
0.00295	1.01123	0.00488	0.03143
0.00779	0.00873	1.15118	0.04289
0.04260	0.03894	0.03953	1.18127

(a) We are given an external demand increase of $100 in Sector 1, so we multiply $(I-A)^{-1}$ by the column matrix

$$D = \begin{bmatrix} 100 \\ 0 \\ 0 \\ 0 \end{bmatrix}$$

and obtain

$$X = \begin{bmatrix} 109.155 \\ 0.295 \\ 0.779 \\ 4.260 \end{bmatrix}$$

The additional production required from the meat and milk sector (Sector 3) is the 1,3 entry: $0.78 (rounded to the nearest 1¢).

(b) The diagonal entries in $(I–A)^{-1}$ show the additional production required from that sector to meet a $1 increase for the product of that sector. Since the largest diagonal entry is the 4,4-entry: 1.18127, we conclude that Sector 4 (other food products) requires the most of its own product in order to meet a $1 increase in external demand for that product.

29. It would mean that all of the sectors require neither their own product or the product of any other sector.

31. The sum of the entries in a row of an input-output table gives the total internal demand for that sector's products. If that total was equal to the total output for that sector, it would mean that all of the output of that sector was used internally in the economy. Thus, none of the output was available for export and no importing was necessary.

33. If an entry in the matrix $(I–A)^{-1}$ is zero, then an increase in demand for one sector (the column sector) has no effect on the production of another sector (the row sector).

35. The off-diagonal entries in $(I–A)^{-1}$ show the additional production required by each sector to meet a one-unit increase for the product of some other sector. Usually, to produce one unit of one sector requires less than one unit of input from another. We would expect then that an increase in demand of one unit for one sector would require a smaller increase in production in another sector.

Chapter 3 Review Exercises

1. The sum of two matrices is defined only when they have the same dimensions. Since A is 2×3 and B is 2×2, their dimensions differ, so the sum $A+B$ is undefined.

3. $A^{\mathrm{T}} = \begin{bmatrix} 1 & 4 \\ 2 & 5 \\ 3 & 6 \end{bmatrix}$, so

$$2A^T+C = 2\begin{bmatrix} 1 & 4 \\ 2 & 5 \\ 3 & 6 \end{bmatrix} + \begin{bmatrix} -1 & 0 \\ 1 & 1 \\ 0 & 1 \end{bmatrix}$$

$$= \begin{bmatrix} 1 & 8 \\ 5 & 11 \\ 6 & 13 \end{bmatrix}$$

5. $A^T B = \begin{bmatrix} 1 & 4 \\ 2 & 5 \\ 3 & 6 \end{bmatrix}\begin{bmatrix} 1 & -1 \\ 0 & 1 \end{bmatrix}$

$$= \begin{bmatrix} (1+0) & (-1+4) \\ (2+0) & (-2+5) \\ (3+0) & (-3+6) \end{bmatrix} = \begin{bmatrix} 1 & 3 \\ 2 & 3 \\ 3 & 3 \end{bmatrix}$$

7. $B^2 = \begin{bmatrix} 1 & -1 \\ 0 & 1 \end{bmatrix}\begin{bmatrix} 1 & -1 \\ 0 & 1 \end{bmatrix}$

$$= \begin{bmatrix} (1+0) & (-1-1) \\ (0+0) & (0+1) \end{bmatrix} = \begin{bmatrix} 1 & -2 \\ 0 & 1 \end{bmatrix}$$

9. $AC + B$

$$= \begin{bmatrix} 1 & 2 & 3 \\ 4 & 5 & 6 \end{bmatrix}\begin{bmatrix} -1 & 0 \\ 1 & 1 \\ 0 & 1 \end{bmatrix} + \begin{bmatrix} 1 & -1 \\ 0 & 1 \end{bmatrix}$$

$$= \begin{bmatrix} 1 & 5 \\ 1 & 11 \end{bmatrix} + \begin{bmatrix} 1 & -1 \\ 0 & 1 \end{bmatrix} = \begin{bmatrix} 2 & 4 \\ 1 & 12 \end{bmatrix}$$

11. To find the inverse of $\begin{bmatrix} 1 & -1 \\ 0 & 1 \end{bmatrix}$, we augment with the 2×2 identity matrix and row-reduce:

$$\begin{bmatrix} 1 & -1 & 1 & 0 \\ 0 & 1 & 0 & 1 \end{bmatrix}\!\!\begin{array}{l} R_1 + R_2 \\ \\ \end{array} \rightarrow \begin{bmatrix} 1 & 0 & 1 & 1 \\ 0 & 1 & 0 & 1 \end{bmatrix}$$

The right-hand 2×2 block is the desired inverse:

$$\begin{bmatrix} 1 & -1 \\ 0 & 1 \end{bmatrix}^{-1} = \begin{bmatrix} 1 & 1 \\ 0 & 1 \end{bmatrix}$$

13. To find the inverse of a 3×3 matrix, we augment it with the 3×3 identity matrix and row-reduce:

$$\begin{bmatrix} 1 & 2 & 3 & 1 & 0 & 0 \\ 0 & 4 & 1 & 0 & 1 & 0 \\ 0 & 0 & 1 & 0 & 0 & 1 \end{bmatrix}\!\!\begin{array}{l} 2R_1 - R_2 \\ \\ \end{array}$$

$$\begin{bmatrix} 2 & 0 & 5 & 2 & -1 & 0 \\ 0 & 4 & 1 & 0 & 1 & 0 \\ 0 & 0 & 1 & 0 & 0 & 1 \end{bmatrix}\!\!\begin{array}{l} R_1 - 5R_3 \\ R_2 - R_3 \\ \end{array}$$

$$\begin{bmatrix} 2 & 0 & 0 & 2 & -1 & -5 \\ 0 & 4 & 0 & 0 & 1 & -1 \\ 0 & 0 & 1 & 0 & 0 & 1 \end{bmatrix}\!\!\begin{array}{l} (1/2)R_1 \\ (1/4)R_2 \\ \end{array}$$

$$\begin{bmatrix} 1 & 0 & 0 & 1 & -1/2 & -5/2 \\ 0 & 1 & 0 & 0 & 1/4 & -1/4 \\ 0 & 0 & 1 & 0 & 0 & 1 \end{bmatrix}$$

The right-hand 3×3 block is the desired inverse:

$$\begin{bmatrix} 1 & 2 & 3 \\ 0 & 4 & 1 \\ 0 & 0 & 1 \end{bmatrix}^{-1} = \begin{bmatrix} 1 & -1/2 & -5/2 \\ 0 & 1/4 & -1/4 \\ 0 & 0 & 1 \end{bmatrix}$$

15. $\begin{bmatrix} 1 & 2 & 3 & 4 & 1 & 0 & 0 & 0 \\ 2 & 3 & 3 & 3 & 0 & 1 & 0 & 0 \\ 0 & 1 & 2 & 3 & 0 & 0 & 1 & 0 \\ 0 & 0 & 1 & 2 & 0 & 0 & 0 & 1 \end{bmatrix}\!\!\begin{array}{l} \\ R_2 - 2R_1 \\ \\ \\ \end{array}$

$$\begin{bmatrix} 1 & 2 & 3 & 4 & 1 & 0 & 0 & 0 \\ 0 & -1 & -3 & -5 & -2 & 1 & 0 & 0 \\ 0 & 1 & 2 & 3 & 0 & 0 & 1 & 0 \\ 0 & 0 & 1 & 2 & 0 & 0 & 0 & 1 \end{bmatrix}\!\!\begin{array}{l} R_1 + 2R_2 \\ \\ R_3 + R_2 \\ \\ \end{array}$$

$$\begin{bmatrix} 1 & 0 & -3 & -6 & -3 & 2 & 0 & 0 \\ 0 & -1 & -3 & -5 & -2 & 1 & 0 & 0 \\ 0 & 0 & -1 & -2 & -2 & 1 & 1 & 0 \\ 0 & 0 & 1 & 2 & 0 & 0 & 0 & 1 \end{bmatrix}\!\!\begin{array}{l} R_1 - 3R_3 \\ R_2 - 3R_3 \\ \\ R_4 + R_3 \\ \end{array}$$

$$\begin{bmatrix} 1 & 0 & 0 & 0 & 3 & -1 & -3 & 0 \\ 0 & -1 & 0 & 1 & 4 & -2 & -3 & 0 \\ 0 & 0 & -1 & -2 & -2 & 1 & 1 & 0 \\ 0 & 0 & 0 & 0 & -2 & 1 & 1 & 1 \end{bmatrix}$$

122

Since the left-hand 4×4 block has a row of zeros, we cannot reduce the matrix to obtain the 4×4 identity on the left. Therefore, the matrix is singular.

17. $x + 2y = 0$
     $3x + 4y = 2$
Matrix form:
$$\begin{bmatrix} 1 & 2 \\ 3 & 4 \end{bmatrix} \begin{bmatrix} x \\ y \end{bmatrix} = \begin{bmatrix} 0 \\ 2 \end{bmatrix}$$
Solving gives
$$\begin{bmatrix} x \\ y \end{bmatrix} = \begin{bmatrix} 1 & 2 \\ 3 & 4 \end{bmatrix}^{-1} \begin{bmatrix} 0 \\ 2 \end{bmatrix}$$
$$= \begin{bmatrix} -2 & 1 \\ 3/2 & -1/2 \end{bmatrix} \begin{bmatrix} 0 \\ 2 \end{bmatrix}$$
$$= \begin{bmatrix} 2 \\ -1 \end{bmatrix}$$
Solution: $(x, y) = (2, -1)$

19. $x + y + z = 2$
     $x + 2y + z = 3$
     $x + y + 2z = 1$
Matrix form:
$$\begin{bmatrix} 1 & 1 & 1 \\ 1 & 2 & 1 \\ 1 & 1 & 2 \end{bmatrix} \begin{bmatrix} x \\ y \\ z \end{bmatrix} = \begin{bmatrix} 2 \\ 3 \\ 1 \end{bmatrix}$$
Solving gives
$$\begin{bmatrix} x \\ y \\ z \end{bmatrix} = \begin{bmatrix} 1 & 1 & 1 \\ 1 & 2 & 1 \\ 1 & 1 & 2 \end{bmatrix}^{-1} \begin{bmatrix} 2 \\ 3 \\ 1 \end{bmatrix}$$
$$= \begin{bmatrix} 3 & -1 & -1 \\ -1 & 1 & 0 \\ -1 & 0 & 1 \end{bmatrix} \begin{bmatrix} 2 \\ 3 \\ 1 \end{bmatrix}$$
$$= \begin{bmatrix} 2 \\ 1 \\ -1 \end{bmatrix}$$
Solution: $(x, y, z) = (2, 1, -1)$

21. Reduce by dominance: Row 1 dominates row 2, so eliminate row 2:
$$\begin{bmatrix} 2 & 1 & 3 & 2 \\ 2 & 0 & 1 & 3 \end{bmatrix}$$
Now column 2 dominates all the other columns, so eliminate all but column 2:
$$\begin{bmatrix} 1 \\ 0 \end{bmatrix}$$

Row 1 dominates row 2, so we eliminate row 2, leaving just the single entry 1. This means that the optimal strategies are pure strategies, corresponding to the first row and second column: $R = [1, 0, 0]$, $C = [0, 1, 0, 0]^{T}$. The corresponding expected value is the entry in the first row, second column, $e = 1$.

23. We begin by reducing by dominance. The second row dominates the first, so eliminate the first:
$$\begin{bmatrix} -1 & 3 & 0 \\ 3 & 3 & -1 \end{bmatrix}$$
Now, the first column dominates the second, so eliminate the second:
$$\begin{bmatrix} -1 & 0 \\ 3 & -1 \end{bmatrix}$$
We can't reduce any further, so now we look for the players' optimal mixed strategies. We begin with the row player. Take $R = [x \ \ 1-x]$ and $C = [1 \ \ 0]^{T}$. $e = RPC = -4x + 3$. If we take $R = [x \ \ 1-x]$ and $C = [0 \ \ 1]^{T}$, we get $f = RPC = x - 1$. Here are the graphs of e and f:

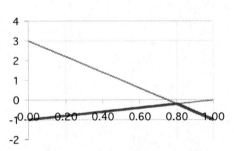

The lower (heavy) portion has its highest point at the intersection of the two lines:
$$-4x + 3 = x - 1$$
$$-5x = -4$$
$$x = 4/5 = 0.8$$
So the optimal row strategy is $R = [0.8 \ \ 0.2]$ for the 2×2 game, which corresponds to $[0 \ \ 0.8 \ \ 0.2]$ for the original game.

For the column player we take $R = [1 \ \ 0]$ and $C = [x \ \ 1-x]$, getting $e = RPC = -x$. Taking $R = [0 \ \ 1]$ and $C = [x \ \ 1-x]$ we get $f = RPC = 4x - 1$. Here are the graphs of e and f:

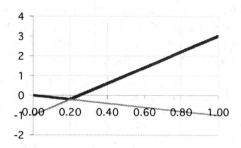

The upper (heavy) portion has its lowest point at the intersection of the two lines:
$$-x = 4x - 1$$
$$-5x = -1$$
$$x = 1/5 = 0.2$$
So, the optimal column strategy is $C = [0.2 \quad 0.8]$ for the 2×2 game, which corresponds to $[0.2 \quad 0 \quad 0.8]$ for the original game.

The expected value is the 2nd coordinate of either intersection point above, or the product RPC with the optimal strategies. In any case, it is $e = -0.2$.

25. $(I - A)^{-1} = \begin{bmatrix} 0.7 & -0.1 \\ 0 & 0.7 \end{bmatrix}^{-1}$

$= \begin{bmatrix} 10/7 & 10/49 \\ 0 & 10/7 \end{bmatrix}$

$X = (I - A)^{-1}D = \begin{bmatrix} 1100 \\ 700 \end{bmatrix}$

27. $(I - A)^{-1} = \begin{bmatrix} 0.8 & -0.2 & -0.2 \\ 0 & 0.8 & -0.2 \\ 0 & 0 & 0.8 \end{bmatrix}^{-1}$

$= \begin{bmatrix} 5/4 & 5/16 & 25/64 \\ 0 & 5/4 & 5/16 \\ 0 & 0 & 5/4 \end{bmatrix}$

$X = (I - A)^{-1}D = \begin{bmatrix} 48,125 \\ 22,500 \\ 10,000 \end{bmatrix}$

29. Inventory $= \begin{bmatrix} 2500 & 4000 & 3000 \\ 1500 & 3000 & 1000 \end{bmatrix}$

Sales $= \begin{bmatrix} 300 & 500 & 100 \\ 100 & 600 & 200 \end{bmatrix}$

Purchases $= \begin{bmatrix} 400 & 400 & 300 \\ 200 & 400 & 300 \end{bmatrix}$

To obtain the inventory in the warehouses at the end of June, we subtract the June sales from the

June 1 stock (inventory) and add the new books purchased:

Inventory − Sales + Purchases
$= \begin{bmatrix} 2500 & 4000 & 3000 \\ 1500 & 3000 & 1000 \end{bmatrix}$

$- \begin{bmatrix} 300 & 500 & 100 \\ 100 & 600 & 200 \end{bmatrix} + \begin{bmatrix} 400 & 400 & 300 \\ 200 & 400 & 300 \end{bmatrix} =$

$\begin{bmatrix} 2600 & 3900 & 3200 \\ 1600 & 2800 & 1100 \end{bmatrix}$

31. Inventory on July 1
$= \begin{bmatrix} 2600 & 3900 & 3200 \\ 1600 & 2800 & 1100 \end{bmatrix}$ from Exercise 29

Sales each month $= \begin{bmatrix} 280 & 550 & 100 \\ 50 & 500 & 120 \end{bmatrix}$

Purchases each month $= \begin{bmatrix} 400 & 400 & 300 \\ 200 & 400 & 300 \end{bmatrix}$

To obtain the inventory in the warehouses x months after July 1, we subtract x times the monthly sales from the July 1 stock and add x times the new books purchased:

Inventory $+ x$Purchases $- x$Sales
$= \begin{bmatrix} 2600 & 3900 & 3200 \\ 1600 & 2800 & 1100 \end{bmatrix}$

$- x\begin{bmatrix} 280 & 550 & 100 \\ 50 & 500 & 120 \end{bmatrix}$

$+ x\begin{bmatrix} 400 & 400 & 300 \\ 200 & 400 & 300 \end{bmatrix}$

$= \begin{bmatrix} 2600 & 3900 & 3200 \\ 1600 & 2800 & 1100 \end{bmatrix}$

$+ x\begin{bmatrix} 120 & -150 & 200 \\ 150 & -100 & 180 \end{bmatrix}$

Nevada Sci Fi inventory is given by the 2,2 entry:
$2800 + x(-500 + 400) = 2950 - 100x$,
which is zero when $x = 28$ months from July 1, or December 1 of next year.

33. Projected July sales figures are as $\begin{bmatrix} 280 & 550 & 100 \\ 50 & 500 & 120 \end{bmatrix}$.

Revenue = Quantity × Price

$= \begin{bmatrix} 280 & 550 & 100 \\ 50 & 500 & 120 \end{bmatrix}\begin{bmatrix} 5 \\ 6 \\ 5.5 \end{bmatrix}$

$= \begin{bmatrix} 5250 \\ 3910 \end{bmatrix}\begin{matrix} \text{Texas} \\ \text{Nevada} \end{matrix}$

35. Unknowns
x = Number of shares purchased on July 1
y = Number of shares purchased on August 1
z = Number of shares purchased on September 1

Given information:
Total of 5000 shares purchased:
  $x + y + z = 5000$
Total of \$50,000 invested:
  $20x + 10y + 5z = 50,000$
Total dividends of \$300 on shares held as of
August 15:
  $0.10(x + y) = 300$
  $0.10x + 0.10y = 300$
We have three equations in three unknowns:
  $x + y + z = 5000$
  $20x + 10y + 5z = 50,000$
  $0.10x + 0.10y = 300$
Matrix form:

$$\begin{bmatrix} 1 & 1 & 1 \\ 20 & 10 & 5 \\ 0.1 & 0.1 & 0 \end{bmatrix}\begin{bmatrix} x \\ y \\ z \end{bmatrix} = \begin{bmatrix} 5000 \\ 50,000 \\ 300 \end{bmatrix}$$

Solving gives

$$\begin{bmatrix} x \\ y \\ z \end{bmatrix} = \begin{bmatrix} 1 & 1 & 1 \\ 20 & 10 & 5 \\ 0.1 & 0.1 & 0 \end{bmatrix}^{-1}\begin{bmatrix} 5000 \\ 50,000 \\ 300 \end{bmatrix}$$

$$= \begin{bmatrix} -.5 & 0.1 & -5 \\ 0.5 & -.1 & 15 \\ 1 & 0 & -10 \end{bmatrix}\begin{bmatrix} 5000 \\ 50,000 \\ 300 \end{bmatrix}$$

$$= \begin{bmatrix} 1000 \\ 2000 \\ 2000 \end{bmatrix}$$

Solution: The company made the following
investments: July 1: 1000 shares, August 1: 2000
shares, September 1: 2000 shares

37. We first need to solve #35:
Unknowns
x = Number of shares purchased on July 1
y = Number of shares purchased on August 1
z = Number of shares purchased on September 1
Given information:
Total of 5000 shares purchased:
  $x + y + z = 5000$
Total of \$50,000 invested:
  $20x + 10y + 5z = 50,000$

Total dividends of \$300 on shares held as of
August 15:
  $0.10(x + y) = 300$
  $0.10x + 0.10y = 300$
We have three equations in three unknowns:
  $x + y + z = 5000$
  $20x + 10y + 5z = 50,000$
  $0.10x + 0.10y = 300$
Matrix form:

$$\begin{bmatrix} 1 & 1 & 1 \\ 20 & 10 & 5 \\ 0.1 & 0.1 & 0 \end{bmatrix}\begin{bmatrix} x \\ y \\ z \end{bmatrix} = \begin{bmatrix} 5000 \\ 50,000 \\ 300 \end{bmatrix}$$

Solving gives

$$\begin{bmatrix} x \\ y \\ z \end{bmatrix} = \begin{bmatrix} 1 & 1 & 1 \\ 20 & 10 & 5 \\ 0.1 & 0.1 & 0 \end{bmatrix}^{-1}\begin{bmatrix} 5000 \\ 50,000 \\ 300 \end{bmatrix}$$

$$= \begin{bmatrix} -.5 & 0.1 & -5 \\ 0.5 & -.1 & 15 \\ 1 & 0 & -10 \end{bmatrix}\begin{bmatrix} 5000 \\ 50,000 \\ 300 \end{bmatrix}$$

$$= \begin{bmatrix} 1000 \\ 2000 \\ 2000 \end{bmatrix}$$

To answer the current question, let us write the
number of shares purchased (calculated in
Exercise 35) as a row matrix:
Shares Purchased = [1000 2000 2000]
Then we can calculate the loss by computing the
total purchase cost and subtracting the total
proceeds (dividends plus selling price):
Loss = Number of shares×(Purchase price –
Dividends – Selling price)
= [1000 2000 2000]

$$\left(\begin{bmatrix} 20 \\ 10 \\ 5 \end{bmatrix} - \begin{bmatrix} 0.10 \\ 0.10 \\ 0 \end{bmatrix} - \begin{bmatrix} 3 \\ 1 \\ 1 \end{bmatrix}\right) = [42,700]$$

39. July 1 customers = [2000 4000 4000]
Customers at the end of July

$$= [2000 \quad 4000 \quad 4000]\begin{bmatrix} 0.8 & 0.1 & 0.1 \\ 0.4 & 0.6 & 0 \\ 0.2 & 0 & 0.8 \end{bmatrix}$$

$$= [4000 \quad 2600 \quad 3400]$$

41. The matrix shows that no JungleBooks
customers switched directly to FarmerBooks, so
the only way to get to FarmerBooks is via
OHaganBooks.

43. $P = \begin{bmatrix} 0 & -60 & -40 \\ 30 & 20 & 10 \\ 20 & 0 & 15 \end{bmatrix}$

$R = [x \quad y \quad z], C = \begin{bmatrix} 0.4 \\ 0.2 \\ 0.4 \end{bmatrix}$,

$e = RPC$

$= [x \quad y \quad z] \begin{bmatrix} 0 & -60 & -40 \\ 30 & 20 & 10 \\ 20 & 0 & 15 \end{bmatrix} \begin{bmatrix} 0.4 \\ 0.2 \\ 0.4 \end{bmatrix}$

$= [x \quad y \quad z] \begin{bmatrix} -28 \\ 20 \\ 14 \end{bmatrix}$

$= -28 + 20y + 14z$

The largest coefficient is the coefficient of y, so we take

$R = [0 \quad 1 \quad 0]$

meaning that you should go with the "3 for 1" promotion. The resulting effect on your customer base is then

$e = -28(0) + 20(1) + 14(0) = 20$,

so you will gain 20,000 customers from JungleBooks.

45. Since JungleBooks now knows that OHaganBooks is assuming it will use the mixed column strategy of Exercise 43, it also knows that OHaganBooks must logically respond by going with the "3 for 1" promotion to counter this. Therefore, its best response will be to use its third strategy "3 for 2" and thus cut its losses to 10,000 customers. But, having seen the e-mail, O'Hagan knows this as well, and so its logical move will be to go with the Finite Math promo and thereby gain 15,000 customers!

47. Since JungleBooks now has full information about the game, the fundamental principle of game theory comes into effect and you must solve the game to find your minimax optimal strategy. You start by reducing by dominance, which leads to throwing out the "no promotion" options for both players, leaving the following 2×2 game:

$P = \begin{bmatrix} 20 & 10 \\ 0 & 15 \end{bmatrix}$

We need only find your optimal strategy as row player. Take $R = [x \quad 1-x]$ and $C = [1 \quad 0]^T$. $e = RPC = 20x$. If we take $R = [x \quad 1-x]$ and $C = [0 \quad 1]^T$, we get $f = RPC = -5x + 15$. Here are the graphs of e and f:

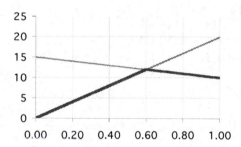

The lower (heavy) portion has its highest point at the intersection of the two lines:

$20x = -5x + 15$

$25x = 15$

$x = 3/5 = 0.6$

So the optimal row strategy is $R = [0.6 \quad 0.4]$ for the 2×2 game, which corresponds to $[0 \quad 0.6 \quad 0.4]$ for the original game. The expected value is the height of the point of intersection, $e = 20(0.6) = 12$, corresponding to a gain of 12,000 customers.

49. We obtain the technology matrix A by dividing each entry in the inoput-output table by the total in the last row:

$A = \begin{bmatrix} 0.1 & 0.5 \\ 0.01 & 0.05 \end{bmatrix}$

51. First, compute

$(I - A)^{-1} = \left(\begin{bmatrix} 1 & 0 \\ 0 & 1 \end{bmatrix} - \begin{bmatrix} 0.1 & 0.5 \\ 0.01 & 0.05 \end{bmatrix} \right)^{-1}$

$= \begin{bmatrix} 0.9 & -.5 \\ -0.01 & 0.95 \end{bmatrix}^{-1} = \begin{bmatrix} 19/17 & 10/17 \\ 1/85 & 18/17 \end{bmatrix}$

We are told that the total (external) demand for Bruno Mills' products is

$\text{Demand} = \begin{bmatrix} 170 \\ 1700 \end{bmatrix}$

$\text{Production} = (I - A)^{-1} \times \text{Demand}$

$$= \begin{bmatrix} 19/17 & 10/17 \\ 1/85 & 18/17 \end{bmatrix} \begin{bmatrix} 170 \\ 1700 \end{bmatrix}$$

$$= \begin{bmatrix} 1190 \\ 1802 \end{bmatrix}$$

Thus, $1190 worth of paper and $1802 worth of books must be produced.

Chapter 4 Linear Programming

4.1 Graphing Linear Inequalities

1. $2x + y \le 10$

First sketch the graph of
$$2x + y = 10$$
(graph on the left).

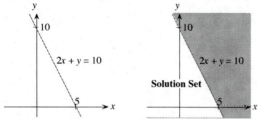

Choose $(0, 0)$ as a test point:
$$2(0) + (0) \le 10 \checkmark$$

Since $(0, 0)$ is in the solution set, we block out the region on the other side of the line as shown above on the right. Since the solution set is not completely enclosed, it is unbounded.

3. $-x - 2y \le 8$

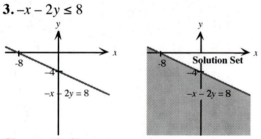

Choose $(0, 0)$ as a test point:
$$-0 - 2(0) \le 8 \checkmark$$

Since $(0, 0)$ is in the solution set, we block out the region on the other side of the line as shown above on the right. Since the solution set is not completely enclosed, it is unbounded.

5. $3x + 2y \ge 5$

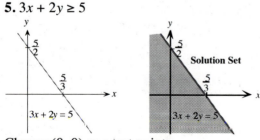

Choose $(0, 0)$ as a test point:

$$3(0) + 2(0) \ge 5 \; \textbf{X}$$

Since $(0, 0)$ is not in the solution set, we block out the region on the same side of the line as shown above on the right. Since the solution set is not completely enclosed, it is unbounded.

7. $x \le 3y$

To sketch $x = 3y$, solve for y to obtain $y = \frac{1}{3}x$.

This is a line of slope $\frac{1}{3}$ passing through the origin.

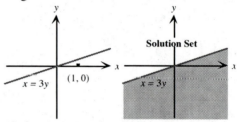

Since $(0, 0)$ is on the line, we choose another test point—say, $(1, 0)$:
$$1 \le 3(0) \; \textbf{X}$$

Since $(1, 0)$ is not in the solution set, we block out the region on the same side of the line as shown above on the right. Since the solution set is not completely enclosed, it is unbounded.

9. $\dfrac{3x}{4} - \dfrac{y}{4} \le 1$

To sketch the associated line, replace the inequality by equality and solve for y solve for y:
$$\frac{3x}{4} - \frac{y}{4} = 1 \Rightarrow \frac{y}{4} = \frac{3x}{4} - 1$$
$$\Rightarrow y = 3x - 4$$

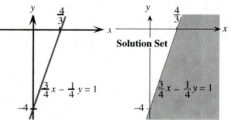

Choose $(0, 0)$ as a test point:
$$\frac{3(0)}{4} - \frac{0}{4} \le 1 \checkmark$$

Since $(0, 0)$ is in the solution set, we block out the region on the other side of the line as shown above on the right. Since the solution set is not completely enclosed, it is unbounded.

128

11. $x \geq -5$

The line $x = -5$ is a vertical line as shown on the left:

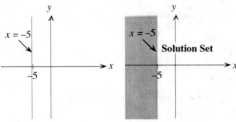

Choose $(0, 0)$ as a test point:

$0 \geq -5$ ✓

Since $(0, 0)$ is in the solution set, we block out the region on the other side of the line as shown above on the right. Since the solution set is not completely enclosed, it is unbounded.

13. $4x - y \leq 8$
$\quad x + 2y \leq 2$

The two associated lines are shown below on the left:

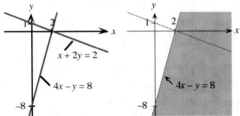

Choose $(0, 0)$ as a test point for the region

$4x - y \leq 8$:

$4(0) - 0 \leq 8$ ✓

Therefore we shade to the right of the line $4x - y = 8$ as shown above on the right.

Choose $(0, 0)$ as a test point for the region

$x + 2y \leq 2$:

$0 + 2(0) \leq 2$ ✓

Therefore we shade above the line $x + 2y = 2$ as shown below.

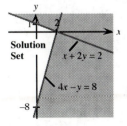

The white region shown is the solution set, which, being not entirely enclosed, is unbounded.

For the corner point, we solve the system

$\quad 4x - y = 8$
$\quad x + 2y = 2$

Multiplying the first equation by 2 and adding gives

$\quad 9x = 18$, so $x = 2$

Substituting $x = 2$ in the first equation gives

$\quad 8 - y = 8$, so $y = 0$

Therefore, the corner point is $(2, 0)$.

15. $3x + 2y \geq 6$
$\quad 3x - 2y \leq 6$
$\quad x \geq 0$

The three associated lines are shown below on the left:

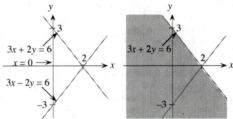

Choose $(0, 0)$ as a test point for the region

$3x + 2y \geq 6$:

$3(0) + 2(0) \geq 6$ ✗

Therefore we shade to the left of the line $3x + 2y = 6$ as shown above on the right.

Now choose $(0, 0)$ as a test point for the region

$3x - 2y \leq 6$:

$3(0) - 2(0) \leq 6$ ✓

Therefore we shade below the line $3x - 2y = 6$ as shown below on the left:

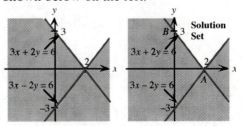

Now choose $(1, 0)$ as a test point for the region $x \geq 0$:

$\quad 1 \geq 0$ ✓

Therefore we shade to the left of $x = 0$ as shown above on the right, leaving the solution set as the unshaded area. Again, the solution set is unbounded, since it is no entirely enclosed.

Corner points:

Point	Lines through point	Coordinates
A	$3x + 2y = 6$ $3x - 2y = 6$	$(2, 0)$
B	$x = 0$ $3x + 2y = 6$	$(0, 3)$

17. $x + y \geq 5$
$\quad x \leq 10$
$\quad y \leq 8$
$\quad x \geq 0, y \geq 0$

The last two inequalities $x \geq 0$, $y \geq 0$ tell us that the solution set is in the first quadrant, so we block out everything else, as shown on the left below:

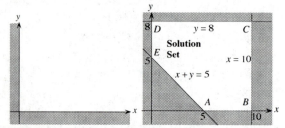

We then add the lines $x + y = 5$, $x = 10$, and $y = 8$ and then shade the appropriate regions, as shown on the right above. The solution set is bounded, since it is entirely enclosed.

Corner points: We can easily read these off the graph:
A: $(5, 0)$, B: $(10, 0)$, C: $(10. 8)$, D: $(0. 8)$, E: $(0, 5)$

19. $20x + 10y \leq 100$
$\quad 10x + 20y \leq 100$
$\quad 10x + 10y \leq 60$
$\quad x \geq 0, y \geq 0$

The last two inequalities $x \geq 0$, $y \geq 0$ tell us that the solution set is in the first quadrant, so we bock out everything else. We then add the lines $20x + 10y = 100$, $10x + 20y = 100$, and $10x + 10y = 60$

and then shade the appropriate regions, as shown below.

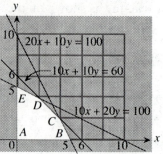

The solution set is the bounded white region.

Corner points:

Point	Lines through point	Coordinates
A	$x = 0, y = 0$	$(0, 0)$
B	$y = 0$ $20x + 10y = 100$	$(5, 0)$
C	$20x + 10y = 100$ $10x + 10y = 60$	$(4, 2)$
D	$10x + 10y = 60$ $10x + 20y = 100$	$(2, 4)$
E	$x = 0$ $10x + 20y = 100$	$(0, 5)$

21. $20x + 10y \geq 100$
$\quad 10x + 20y \geq 100$
$\quad 10x + 10y \geq 80$
$\quad x \geq 0, \quad y \geq 0$

Proceeding as before, we obtain the unbounded solution set shown below:

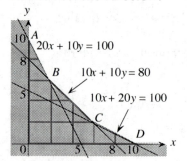

Corner points:

Point	Lines through point	Coordinates
A	$x = 0$ $20x + 10y = 100$	$(0, 10)$
B	$20x + 10y = 100$ $10x + 10y = 80$	$(2, 6)$
C	$10x + 10y = 80$ $10x + 20y = 100$	$(6, 2)$
D	$10x + 20y = 100$ $y = 0$	$(10, 0)$

23. $-3x + 2y \le 5$

$3x - 2y \le 6$

$x \le 2y$

$x \ge 0, y \ge 0$

Solution set (unbounded):

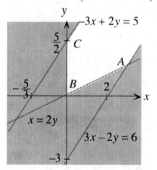

Corner points: We can read two of the corner points, B: $(0, 0)$ and C: $(0, \frac{5}{2})$ directly from the graph. The remaining corner point is A, at the intersection of the lines $x = 2y$ and $3x - 2y = 6$. Solving this system of 2 equations gives the point A as A: $(3, \frac{3}{2})$.

25. $2x - y \ge 0$

$x - 3y \le 0$

$x \ge 0, y \ge 0$

To sketch the lines, solve for y in each case:

$2x - y = 0 \Rightarrow y = 2x$

$x - 3y = 0 \Rightarrow y = \frac{1}{3}x$

Solution set (unbounded):

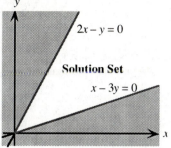

The only corner point is the origin: $(0, 0)$

27. $2.1x - 4.3y \ge 9.7$

To draw this region using technology, solve the associated equation for y:

$4.3y = 2.1x - 9.7$

$y = \frac{2.1}{4.3}x - \frac{8.7}{4.3}$

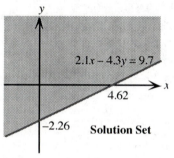

29. $-0.2x + 0.7y \ge 3.3$

$1.1x + 3.4y \ge 0$

To draw this region using technology, solve the associated equations for y:

$y = \frac{0.2}{0.7}x + \frac{3.3}{0.7} = \frac{2}{7}x + \frac{33}{7}$

$y = -\frac{1.1}{3.4}x = -\frac{11}{34}x$

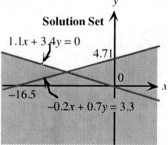

To obtain the coordinates of (the only) corner point, zoom in to it until you can read off the coordinates to two decimal places: $(-7.74, 2.50)$

131

31. $4.1x - 4.3y \le 4.4$
$7.5x - 4.4y \le 5.7$
$4.3x + 8.5y \le 10$

To obtain the associated lines, solve for y:

$$y = \frac{4.1}{4.3}x - \frac{4.4}{4.3}$$

$$y = \frac{7.5}{4.4}x - \frac{5.7}{4.4}$$

$$= -\frac{4.3}{8.5}x + \frac{10}{4.3}$$

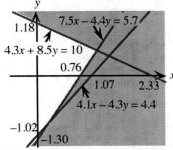

To obtain the coordinates of the corner points, zoom in to it until you can read off the coordinates to two decimal places:
A: $(0.36, -0.68)$, B: $(1.12, 0.61)$

33. Unknowns:
$x =$ # quarts of Creamy Vanilla
$y =$ # quarts of Continental Mocha
Arrange the given information in a table with unknowns across the top:

	Vanilla (x)	Mocha (y)	Available
Eggs	2	1	500
Cream	3	3	900

We can now set up an inequality for each of the items listed on the left:
Eggs: $2x + y \le 500$
Cream: $3x + 3y \le 900$
  or $x + y \le 300$
Since the factory cannot manufacture negative amounts, we also have
  $x \ge 0, y \ge 0$
The solution set is the unshaded region shown below:

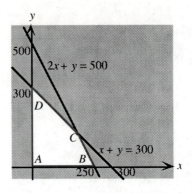

Corner points:

Point	Lines through point	Coordinates
A	$x = 0$ $y = 0$	$(0, 0)$
B	$y = 0$ $2x + y = 500$	$(250, 0)$
C	$2x + y = 500$ $x + y = 300$	$(200, 100)$
D	$x + y = 300$ $x = 0$	$(0, 300)$

35. Unknowns:
$x =$ # ounces of chicken
$y =$ # ounces of grain
Arrange the given information in a table with unknowns across the top:

	Chicken (x)	Grain (y)	Required
Protein	10	2	200
Fat	5	2	150

We can now set up an inequality for each of the items listed on the left (Note that "at least" is represented by "$\ge$"):
Protein: $10x + 2y \ge 200$,
  or $5x + y \ge 100$
Fat: $5x + 2y \ge 150$
The amounts of ingredients cannot be negative:
  $x \ge 0, y \ge 0$
The solution set is the unshaded region shown below:

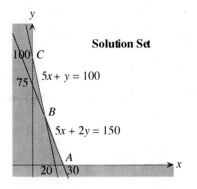

Solution Set

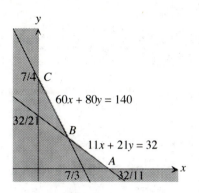

Corner points:

Point	Lines through point	Coordinates
A	$y = 0$ $5x + 2y = 150$	(30, 0)
B	$5x + 2y = 150$ $5x + y = 100$	(10, 50)
C	$5x + y = 100$ $x = 0$	(0, 100)

Corner points:

Point	Lines through point	Coordinates
A	$y = 0$ $11x + 21y = 32$	(32/11, 0)
B	$11x + 21y = 32$ $3x + 4y = 7$	(1, 1)
C	$3x + 4y = 7$ $x = 0$	(0, 7/4)

37. Unknowns:

x = # servings of Mixed Cereal

y = # servings of Mango Tropical Fruit

Arrange the given information in a table with unknowns across the top:

	Cereal (x)	Mango (y)	Required
Calories	60	80	140
Carbs.	11	21	32

We can now set up an inequality for each of the items listed on the left (Note that "at least" is represented by "≥"):

Calories: $\quad 60x + 80y \geq 140,$

$\quad$ or $\quad 3x + 4y \geq 7$

Carbs: $\quad 11x + 21y \geq 32$

The values of the unknowns cannot be negative:

$\quad x \geq 0, y \geq 0$

The solution set is the unshaded region shown below:

39. Unknowns:

x = # dollars in PNF

y = # dollars in FDMMX

Total invested is up to $80,000:

$\quad x + y \leq 80,000$

Interest earned is at least $4200:

$\quad 0.06x + 0.05y \geq 4200,$

or $\quad 6x + 5y \geq 420,000$

Unknowns nonnegative:

$\quad x \geq 0, y \geq 0$

The solution set is the thin unshaded triangle shown below:

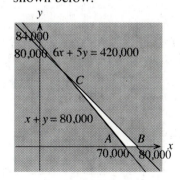

Corner points:

Point	Lines through point	Coordinates
A	$y = 0$ $6x + 5y = 420{,}000$	$(70{,}000, 0)$
B	$y = 0$ $x + y = 80{,}000$	$(80{,}000, 0)$
C	$x + y = 80{,}000$ $6x + 5y = 420{,}000$	$(20{,}000, 60{,}000)$

41. Unknowns:

x = # shares in ED

y = # shares in GE

Total invested is up to \$10,000:

$40x + 16y \le 10{,}000$

At least \$600 in dividends

$0.06(40x) + 0.075(16y) \ge 600$,

or $2.4x + 1.2y \ge 600$

Unknowns nonnegative:

$x \ge 0, \, y \ge 0$

The solution set is the thin unshaded triangle shown below:

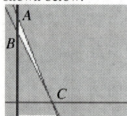

Corner points:

Point	Lines through point	Coordinates
A	$x = 0$ $40x + 16y = 10{,}000$	$(0, 625)$
B	$x = 0$ $2.4x + 1.2y = 600$	$(0, 500)$
C	$40x + 16y = 10{,}000$ $2.4x + 1.2y = 600$	$(250, 0)$

43. Unknowns:

x = # full-page ads in Sports Illustrated

y = # full-page ads of ads in GQ

Readership: $0.65x + 0.15y \ge 3$

At least 3 full-page ads in each magazine:

$x \ge 3, \, y \ge 3$

The solution set is the unshaded region shown below:

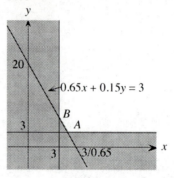

Corner points (coordinates rounded to the nearest whole number):

Point	Lines through point	Coordinates
A	$y = 3$ $0.65x + 0.15y = 3$	$\left(\dfrac{2.55}{0.65}, 3\right)$ $\approx (4, 3)$ (rounded)
B	$x = 3$ $0.65x + 0.15y = 3$	$\left(3, \dfrac{1.05}{0.15}\right)$ $= (3, 7)$

45. Many of the systems of inequalities in the earlier exercises have unbounded solution sets. Another example is: $x \ge 0, \, y \ge 0, \, x+y \ge 1$.

47. The given triangle is the region enclosed by the lines $x = 0$, $y = 0$, and $x + 2y = 2$ (see figure).

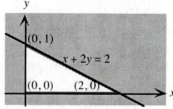

Thus, the region can be described as the solution set of the system $x \ge 0, \, y \ge 0, \, x + 2y \le 2$.

49. (Answers may vary.) One limitation is that the method is only suitable for situations with two unknown quantities. Another limitation is accuracy, which is never perfect when graphing.

51. There should be at least 3 more grams of ingredient A than ingredient B. Rephrasing this statement gives:
The number of grams of ingredient A exceeds the grams of ingredient B by 3:

$x - y \geq 3$

Choice (C)

53. There should be at least 3 parts (by weight) of ingredient A to 2 parts of ingredient B. That is, $3/2 = 1.5$ parts of ingredient A to 1 part of ingredient B. Rephrasing this statement gives:
The number of grams of ingredient A is 1.5 times the number of grams of ingredient B:

$x = 1.5y$
$2x = 3y$
$2x - 3y = 0$

Choice (B)

55. There are no feasible solutions; that is, it is impossible to satisfy all the constraints.

4.2 Solving Linear Programming Problems Graphically

1. Maximize $p = x + y$ subject to

$x + 2y \le 9$

$2x + y \le 9$

$x \ge 0, y \ge 0$

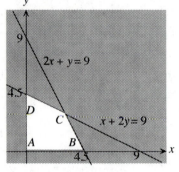

Point	Lines	Coords.	$p = x + y$
A	$x = 0, y = 0$	$(0,0)$	0
B	$y = 0$ $2x + y = 9$	$(4.5, 0)$	4.5
C	$2x + y = 9$ $x + 2y = 9$	$(3,3)$	**6**
D	$x + 2y = 9$ $x = 0$	$(0, 4.5)$	4.5

Maximum value occurs at point C: $p = 6$ for $x = 3, y = 3$.

3. Minimize $c = x + y$ subject to

$x + 2y \ge 6$

$2x + y \ge 6$

$x \ge 0, \ y \ge 0$

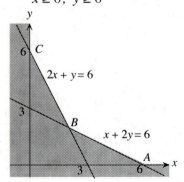

Although the feasible region is unbounded, there is no need to add a bounding rectangle since, by the *FAQ* at the end of the section in the text:

If you are minimizing $c = ax + by$ with a and b nonnegative, $x \ge 0$, and $y \ge 0$, then optimal solutions always exist.

Point	Lines	Coords.	$c = x + y$
A	$y = 0$ $x + 2y = 6$	$(6,0)$	6
B	$x + 2y = 6$ $2x + y = 6$	$(2,2)$	**4**
C	$2x + y = 6$ $x = 0$	$(0,6)$	6

Minimum value occurs at point B: $c = 4$, $x = 2$, $y = 2$.

5. Maximize $p = 3x + y$ subject to

$3x - 7y \le 0$

$7x - 3y \ge 0$

$x + y \le 10$

$x \ge 0, y \ge 0$

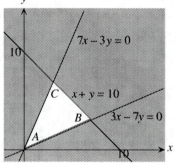

Point	Lines	Coords.	$p = 3x + y$
A	$x = 0, y = 0$	$(0,0)$	0
B	$3x - 7y = 0$ $x + y = 10$	$(7,3)$	**24**
C	$7x - 3y = 0$ $x + y = 10$	$(3,7)$	16

Maximum value occurs at point B: $p = 24$, $x = 7$, $y = 3$.

7. Maximize $p = 3x + 2y$ subject to

$0.2x + 0.1y \le 1$

$0.15x + 0.3y \le 1.5$

$10x + 10y \le 60$

$x \ge 0, \ y \ge 0$

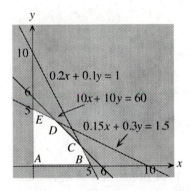

Point	Lines	Coords.	$p =$ $3x+2y$
A	$x = 0$, $y = 0$	$(0, 0)$	0
B	$y = 0$ $0.2x+0.1y = 1$	$(5, 0)$	15
C	$0.2x+0.1y = 1$ $10x+10y = 60$	$(4, 2)$	**16**
D	$10x+10y = 60$ $0.15x+0.3y = 1.5$	$(2, 4)$	14
E	$0.15x+0.3y = 1.5$ $x = 0$	$(0, 5)$	10

Maximum value occurs at point C: $p = 16$, $x = 4$, $y = 2$.

9. Minimize $c = 0.2x + 0.3y$ subject to
$$0.2x + 0.1y \geq 1$$
$$0.15x + 0.3y \geq 1.5$$
$$10x + 10y \geq 80$$
$$x \geq 0, \quad y \geq 0$$

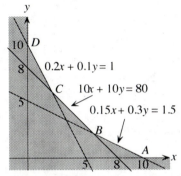

Although the feasible region is unbounded, there is no need to add a bounding rectangle since the coefficients of c are nonnegative.

Point	Lines	Coords	$c =$ $0.2x+0.3y$
A	$y = 0$ $0.15x+0.3y = 1.5$	$(10, 0)$	2
B	$0.15x+0.3y = 1.5$ $10x+10y = 80$	$(6, 2)$	**1.8**
C	$10x+10y = 80$ $0.2x+0.1y = 1$	$(2, 6)$	2.2
D	$0.2x+0.1y = 1$ $x = 0$	$(0, 10)$	3

Minimum value occurs at point B: $c = 1.8$, $x = 6$, $y = 2$.

11. Maximize and minimize $p = x + 2y$ subject to
$$x + y \geq 2$$
$$x + y \leq 10$$
$$x - y \leq 2$$
$$x - y \geq -2$$

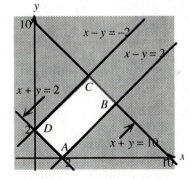

Point	Lines	Coords.	$p = x + 2y$
A	$x + y = 2$ $x - y = 2$	$(2, 0)$	**2** **Min**
B	$x - y = 2$ $x + y = 10$	$(6, 4)$	14
C	$x + y = 10$ $x - y = -2$	$(4, 6)$	**16** **Max**
D	$x - y = -2$ $x + y = 2$	$(0, 2)$	4

Maximum value occurs at point C: $p = 16$, $x = 4$, $y = 6$; minimum value occurs at point A: $p = 2$, $x = 2$, $y = 0$.

13. Maximize $p = 2x + 3y$ subject to
$$0.1x + 0.2y \geq 1$$
$$2x + y \geq 10$$
$$x \geq 0, \ y \geq 0$$

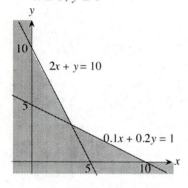

This is an unbounded region, and we wish to maximize $p = 2x + 3y$. According to the *FAQ* at the end of the Section 4.2:

If you are maximizing $p = ax + by$ with a and b nonnegative, then there is no optimal solution unless the feasible region is bounded.

Thus, there is no optimal solution.

15. Minimize $c = 2x + 4y$ subject to
$$0.1x + 0.1y \geq 1$$
$$x + 2y \geq 14$$
$$x \geq 0, \ y \geq 0$$

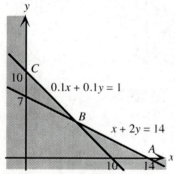

Although the feasible region is unbounded, there is no need to add a bounding rectangle since the coefficients of c are nonnegative

Point	Lines	Coords.	$c = 2x+4y$
A	$y = 0$ $x + 2y = 14$	(14, 0)	**28**
B	$x + 2y = 14$ $0.1x + 0.1y = 1$	(6, 4)	**28**
C	$0.1x + 0.1y = 1$ $x = 0$	(0, 10)	40

Minimum value occurs at points A and B: $c = 28$; $(x, y) = (14, 0)$ and $(6, 4)$, and the line connecting them.

17. Minimize $c = 3x - 3y$ subject to
$$\frac{x}{4} \leq y, \quad y \leq \frac{2x}{3}$$
$$x + y \geq 5$$
$$x + 2y \leq 10$$
$$x \geq 0, \ y \geq 0$$

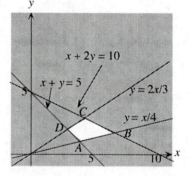

Point	Lines	Coords.	$c = 3x-3y$
A	$y = x/4$ $x + y = 5$	(4, 1)	9
B	$y = x/4$ $x + 2y = 10$	(20/3, 5/3)	15
C	$x + 2y = 10$ $y = 2x/3$	(30/7, 20/7)	30/7
D	$y = 2x/3$ $x + y = 5$	(3, 2)	**3**

Minimum value occurs at point D: $c = 3$, $x = 3$, $y = 2$.

19. Maximize $p = x + y$ subject to
$$x + 2y \geq 10$$
$$2x + 2y \leq 10$$
$$2x + y \geq 10$$

$x \geq 0, \ y \geq 0$

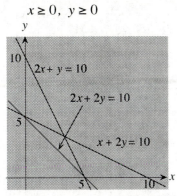

Feasible region is empty—no solutions.

21. (See # 33 in Section 4.1) Unknowns:
x = # quarts of Creamy Vanilla
y = # quarts of Continental Mocha
Maximize $p = 3x + 2y$ subject to
$\qquad 2x + y \leq 500$
$\qquad 3x + 3y \leq 900$ or $x + y \leq 300$
$\qquad x \geq 0, y \geq 0.$

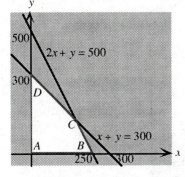

Corner points:

Point	Lines	Coords.	$p =$ $3x+2y$
A	$x = 0$ $y = 0$	$(0, 0)$	0
B	$y = 0$ $2x + y = 500$	$(250, 0)$	750
C	$2x + y = 500$ $x + y = 300$	$(200, 100)$	**800**
D	$x + y = 300$ $x = 0$	$(0, 300)$	600

Maximum value occurs at the point C: $p = 800$; $x = 200, y = 100.$

Solution: You should make 200 quarts of vanilla and 100 quarts of mocha.

23. (See # 35 in Section 4.1) Unknowns:
x = # ounces of chicken
y = # ounces of grain
Minimize $c = 10x + y$ subject to
$\qquad 10x + 2y \geq 200$
$\qquad 5x + 2y \geq 150$
$\qquad x \geq 0, y \geq 0.$

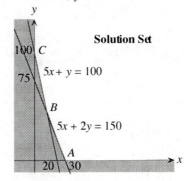

Corner points:

Point	Lines	Coords.	$c =$ $10x+y$
A	$y = 0$ $5x+2y = 150$	$(30, 0)$	300
B	$5x+2y = 150$ $5x+y = 100$	$(10, 50)$	150
C	$5x+y = 100$ $x = 0$	$(0, 100)$	**100**

Minimum value occurs at the point C: $c = 800$; $x = 0, y = 100.$

Solution: Ruff, Inc., should use 100 oz of grain and no chicken.

25. (See # 37 in Section 4.1.) Unknowns:
x = # servings of Mixed Cereal
y = # servings of Mango Tropical Fruit
Minimize $c = 30x + 50y$ subject to
$\qquad 11x + 21y \geq 32$
$\qquad 3x + 4y \geq 7$
$\qquad x \geq 0, y \geq 0.$

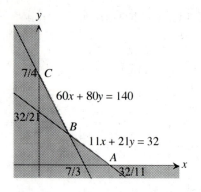

Corner points:

Point	Lines	Coords.	$c = 30x+50y$
A	$y = 0$ $11x+21y = 32$	(32/11, 0)	87.3
B	$11x+21y = 32$ $3x+4y = 7$	(1, 1)	**80**
C	$3x+4y = 7$ $x = 0$	(0, 7/4)	87.5

Minimum value occurs at the point B: $c = 80$; $x = 1$, $y = 1$.

Solution: Feed your child 1 serving of cereal and 1 serving of dessert.

27. Unknowns:
x = # compact fluorescent light bulbs
y = # square ft of insulation
Maximize $p = 2x + 0.2y$ subject to
$$4x + y \le 1200$$
$$x \le 60$$
$$y \le 1100$$
$$x \ge 0, y \ge 0.$$

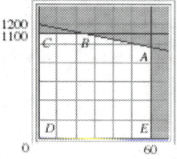

Corner points:

Point	Lines	Coords.	$p = 2x + 0.2y$
A	$4x + y = 1200$ $x = 60$	(60,960)	**312**
B	$y = 1100$ $4x + y = 1200$	(25,1100)	270
C	$y = 1100$ $x = 0$	(0, 1100)	220
D	$y = 0$ $x = 0$	(0, 0)	0
E	$x = 60$ $y = 0$	(60, 0)	120

Maximum value occurs at the point A: $p = 312$; $x = 60$, $y = 960$.

Solution: Purchase 60 compact fluorescent lightbulbs and 960 square feet of insulation for a total saving of $312 per year in energy costs.

29. Unknowns:
x = # of servings of Cell-Tech
y = # servings of Riboforce HP.
Minimize $c = 2.20x + 1.60y$ subject to
$$10x + 5y \ge 80$$
$$2x + y \ge 10$$
$$75x + 15y \le 750$$
$$200x \le 1000$$
$$x \ge 0, y \ge 0$$

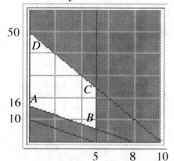

Corner points:

Point	Lines	Coords.	$c = 2.20x+1.60y$
A	$10x+5y = 80$ $x = 0$	(0, 16)	25.6
B	$10x+5y = 80$ $200x = 1000$	(5, 6)	**20.6**
C	$75x+15y = 750$ $200x = 1000$	(5, 25)	51

D	$75x+15y = 750$ $x = 0$	(0, 50)	80

Minimum value occurs at the point B: $c = 20.6$; $x = 5$, $y = 6$.

Solution: Mix 5 servings of Cell-Tech and 6 servings of Riboforce HP for a cost of $20.60.

31. Unknowns:
x = # Dracula salamis
y = # Frankenstein sausages
Maximize $p = x + 3y$ subject to
$x + 2y \le 1000$
$3x + 2y \le 2400$
$y \le 2x$
$x \ge 0$, $y \ge 0$

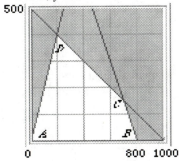

Corner points:

Point	Lines	Coords.	$p = x+3y$
A	$x = 0$ $y = 0$	(0, 0)	0
B	$3x+2y = 2400$ $y = 0$	(800, 0)	800
C	$x+2y = 1000$ $3x+2y = 2400$	(700, 150)	1150
D	$x+2y = 1000$ $y-2x = 0$	(200, 400)	**1400**

Maximum value occurs at the point D: $p = 1400$; $x = 200$, $y = 400$.

Solution: You should make 200 Dracula Salamis and 400 Frankenstein Sausages, for a profit of $1400.

33. Unknowns:

x = # spots on "American Idol"
y = # spots on "Back to You"
Maximize $V = 28.5x + 12.3y$, subject to
$x + y \ge 30$
$3x + y \le 120$
$-x + y \le 0$
$x \ge 0$, $y \ge 0$

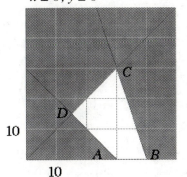

Corner points:

Point	Lines	Coords.	$V = 28.5x + 12.3y$
A	$x+y = 30$ $y = 0$	(30, 0)	612
B	$3x+y = 120$ $y = 0$	(40, 0)	1140
C	$3x+y = 120$ $-x+y = 0$	(30, 30)	**1224**
D	$x+y = 30$ $-x+y = 0$	(15, 15)	612

Maximum value occurs at the point C: $V = 316$; $x = 20$, $y = 20$.

Solution: You should purchase 30 spots on "American Idol" and 30 spots on "Back to You."

35. Unknowns:
x = # ATRI shares
y = # BBY shares
Maximize $p = 7.5x + 3.2y$ subject to
$90x + 20y \le 10,000$
or $9x + 2y \le 1000$
and $(0.01)(90x) + (0.02)(20y) \ge 120$
or $0.9x + 0.4y \ge 120$
or $9x + 4y \ge 1200$
$x \ge 0$, $y \ge 0$

141

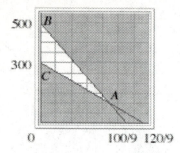

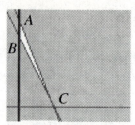

Corner points:

Point	Lines	Coords.	$p = 7.5x + 3.2y$
A	$9x+2y = 1000$ $9x+4y = 1200$	$= (88.889, 100)$	986.67
B	$9x+2y = 1000$ $x = 0$	$(0, 500)$	**1600**
C	$9x+4y = 1200$ $x = 0$	$(0, 300)$	360

Maximum value occurs at the point B: $p = 1600$; $x = 0$, $y = 500$.

Solution: Buy no shares of ATRI and 500 shares of BBY for maximum company earnings of $1600.

37. Unknowns:
x = # shares in ED
y = # shares in GE
Total invested is up to $10,000:
  $40x + 16y \le 10,000$
At least $600 in dividends
  $0.06(40x) + 0.075(16y) \ge 600$,
or
Unknowns nonnegative:
  $x \ge 0$, $y \ge 0$
LP problem: Minimize $c = 2x + 3y$ subject to
  $40x + 16y \le 10,000$
  $2.4x + 1.2y \ge 600$
The solution set is the thin unshaded triangle shown below:

Corner points:

Point	Lines through point	Coords.	$c = 2x + 3y$
A	$x = 0$ $40x + 16y = 10,000$	$(0, 625)$	1875
B	$x = 0$ $2.4x + 1.2y = 600$	$(0, 500)$	1500
C	$40x + 16y = 10,000$ $2.4x + 1.2y = 600$	$(250, 0)$	**500**

Solution: Purchase 250 shares of ED and no shares of GE for a total risk index of 500.

39. Unknowns:
x = # hours spent in battle instruction per week
y = # hours spent per week in diplomacy instruction
Maximize $p = 50x + 40y$ subject to
  $x + y \le 50$
  $x \ge 2y$
  $y \ge 10$
  $10x + 5y \ge 400$
  $x \ge 0$, $y \ge 0$

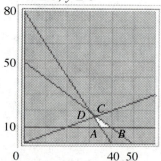

Corner points:

Point	Lines	Coords.	$p = 50x+40y$

A	$y = 10$ $10x+5y = 400$	$(35, 10)$	2150
B	$x+y = 50$ $y = 10$	$(40, 10)$	**2400**
C	$x+y = 50$ $x-2y = 0$	$(100/3, 50/3)$	7000/3
D	$x-2y = 0$ $10x+5y = 400$	$(32, 16)$	2240

Maximum value occurs at the point B: $p = 2400$; $x = 40$, $y = 10$.

Solution: He should instruct in diplomacy for 10 hours per week and in battle for 40 hours per week, giving a weekly profit of 2400 ducats.

41. Unknowns:
x = # sleep spells
y = # shock spells
Minimize $c = 500x + 750y$ subject to
$$2x+3y \geq 1440$$
$$3x+2y \geq 1200$$
$$x-3y \leq 0$$
$$-x + 2y \leq 0$$
$$x \geq 0, y \geq 0$$

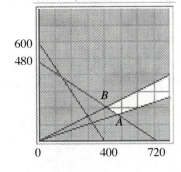

Corner points:

Point	Lines	Coords.	$c = 500x + 750y$
A	$2x+3y = 1440$ $x-3y = 0$	$(480, 160)$	360,000
B	$2x+3y = 1440$ $-x+2y = 0$	$(2880/7, 1440/7)$	360,000

Minimum value occurs at points A and B: $c = 360,000$; $x = 480$, $y = 160$ and $x = 2880/7$, $y = 1440/7$.

Solution: Gillian could expend a minimum of 360,000 pico-shirleys of energy by using 480 sleep spells and 160 shock spells. (There is actually a whole line of solutions joining the one above with $x = 2880/7$, $y = 1440/7$.)

43. Unknowns:
x = # hours for new customers
y = # hours for old customers
Maximize $p = 10x + 30y$ subject to
$$10x + 30y \geq 1200$$
$$x + y \leq 160$$
$$x \geq 100$$
$$x \geq 0, y \geq 0$$

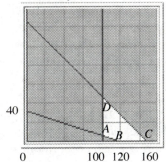

Corner points:

Point	Lines	Coords.	$p = 10x+30y$
A	$10x+30y = 1200$ $x = 100$	$(100, 20/3)$	1200
B	$10x+30y = 1200$ $y = 0$	$(120, 0)$	1200
C	$x+y = 160$ $y = 0$	$(160, 0)$	1600
D	$x+y = 160$ $x = 100$	$(100, 60)$	**2800**

Maximum value occurs at the point D: $p = 2800$; $x = 100$, $y = 60$.

Solution: Allocate 100 hours per week for new customers and 60 hours per week for old customers.

45. By the Fundamental Theorem of linear programming, linear programming problems with bounded, non-empty feasible regions always have optimal solutions (Choice (A)).

47. Every point along the line connecting them is also an optimal solution.

49. Here are two simple examples:
(1) (Empty feasible region) Maximize $p = x + y$ subject to $x + y \leq 10$; $x + y \geq 11$, $x \geq 0$, $y \geq 0$.
(2) (Unbounded feasible region and no optimal solutions) Minimize $c = x - y$ subject to $x + y \geq 10$, $x \geq 0$, $y \geq 0$.

51. Answers will vary.

53. A simple example is the following: Maximize profit $= p = x + y$ subject to $x \geq 0$, $y \geq 0$. Then p can be made as large as we like by choosing large values of x and/or y. Thus there is no optimal solution to the problem.

55. Mathematically, this means that there are infinitely many possible solutions: one for each point along the line joining the two corner points in question. In practice, select those points with integer solutions (since x and y must be whole numbers in this problem) that are in the feasible region and close to this line, and choose the one that gives the largest profit.

57. The corner points are shown in the following figure:

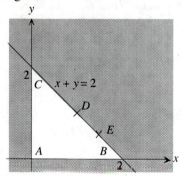

The corner point have coordinates A: $(0, 0)$, B: $(2, 0)$, C:$(0, 2)$, and so $p = xy$ is zero at each of these points. However, at infinitely many points along the diagonal such as D: $(1, 1)$ and E: $(1.5, 0.5)$, the value of p is positive. (For example, at D, $p = (1)(1) = 1$, and at E, $p = (1.5)(0.5) = 0.75$.)

4.3 The Simplex Method: Solving Standard Maximization Problems

1. Introduce slack variables and rewrite the constraints and objective function in standard form, and set up the first tableau.

$$x + 2y + s = 6$$
$$-x + y + t = 4$$
$$x + y + u = 4$$
$$-2x - y + p = 0$$

	x	y	s	t	u	p	
s	1	2	1	0	0	0	6
t	−1	1	0	1	0	0	4
u	1	1	0	0	1	0	4
p	−2	−1	0	0	0	1	0

We search for the most negative number in the bottom row (if any). The most negative entry in the bottom row is the −2 in the x-column. So we use this column as the pivot column. The test ratios are: s: 6/1, u: 4/1. The smallest test ratio is u: 4/1. Thus, we pivot on the 1 in the u-row.

	x	y	s	t	u	p		
s	1	2	1	0	0	0	6	$R_1 - R_3$
t	−1	1	0	1	0	0	4	$R_2 + R_3$
u	[1]	1	0	0	1	0	4	
p	−2	−1	0	0	0	1	0	$R_4 + 2R_3$

	x	y	s	t	u	p	
s	0	1	1	0	−1	0	2
t	0	2	0	1	1	0	8
x	1	1	0	0	1	0	4
p	0	1	0	0	2	1	8

Since there are no more negative numbers in the bottom row, we are done, and read off the solution.

Optimal Solution: p = 8/1 = 8; x = 4/1 = 4, y = 0.

3. Introduce slack variables and rewrite the constraints and objective function in standard form, and set up the first tableau.

$$5x - 5y + s = 20$$
$$2x - 10y + t = 40$$
$$-x + y + p = 0$$

	x	y	s	t	p	
s	5	−5	1	0	0	20
t	2	−10	0	1	0	40
p	−1	1	0	0	1	0

We search for the most negative number in the bottom row (if any). The most negative entry in the bottom row is the −1 in the x–column. So we use this column as the pivot column. The test ratios are: s: 20/5, t: 40/2. The smallest test ratio is s: 20/5. Thus, we pivot on the 5 in the s–row.

	x	y	s	t	p		
s	[5]	−5	1	0	0	20	
t	2	−10	0	1	0	40	$5R_2 - 2R_1$
p	−1	1	0	0	1	0	$5R_3 + R_1$

	x	y	s	t	p	
x	5	−5	1	0	0	20
t	0	−40	−2	5	0	160
p	0	0	1	0	5	20

Since there are no more negative numbers in the bottom row, we are done, and read off the solution: Optimal Solution: $p = 20/5 = 4$; $x = 20/5 = 4$, $y = 0$.

5. Introduce slack variables and rewrite the constraints and objective function in standard form, and set up the first tableau.

$$5x + 5z + s = 100$$
$$5y - 5z + t = 50$$
$$5x - 5y + u = 50$$
$$-5x + 4y - 3z + p = 0$$

	x	y	z	s	t	u	p	
s	5	0	5	1	0	0	0	100
t	0	5	−5	0	1	0	0	50
u	5	−5	0	0	0	1	0	50
p	−5	4	−3	0	0	0	1	0

The most negative entry in the bottom row is the −5 in the x–column. So we use this column as the pivot column. The test ratios are: s: 100/5, u: 50/5. The smallest test ratio is u: 50/5. Thus, we pivot on the 5 in the u–row.

	x	y	z	s	t	u	p		
s	5	0	5	1	0	0	0	100	$R_1 - R_3$
t	0	5	-5	0	1	0	0	50	
u	[5]	-5	0	0	0	1	0	50	
p	-5	4	-3	0	0	0	1	0	$R_4 + R_3$

	x	y	z	s	t	u	p	
s	0	5	5	1	0	-1	0	50
t	0	5	-5	0	1	0	0	50
x	5	-5	0	0	0	1	0	50
p	0	-1	-3	0	0	1	1	50

The most negative entry in the bottom row is the −3 in the z–column. So we use this column as the pivot column. The only positive entry in this column is the 5 in the s–row, so we pivot on this entry.

	x	y	z	s	t	u	p		
s	0	5	[5]	1	0	-1	0	50	
t	0	5	-5	0	1	0	0	50	$R_2 + R_1$
x	5	-5	0	0	0	1	0	50	
p	0	-1	-3	0	0	1	1	50	$5R_4 + 3R_1$

	x	y	z	s	t	u	p	
z	0	5	5	1	0	-1	0	50
t	0	10	0	1	1	-1	0	100
x	5	-5	0	0	0	1	0	50
p	0	10	0	3	0	2	5	400

Since there are no more negative numbers in the bottom row, we are done, and read off the solution:
Optimal Solution: $p = 400/5 = 80$; $x = 50/5 = 10$, $y = 0$, $z = 50/5 = 10$.

7. Introduce slack variables and rewrite the constraints and objective function in standard form, and set up the first tableau.

$x + y - z + s = 3$
$x + 2y + z + t = 8$
$x + y + u = 5$
$-7x - 5y - 6z + p = 0$

	x	y	z	s	t	u	p	
s	1	1	-1	1	0	0	0	3
t	1	2	1	0	1	0	0	8
u	1	1	0	0	0	1	0	5
p	-7	-5	-6	0	0	0	1	0

147

The most negative entry in the bottom row is the -7 in the x–column. So we use this column as the pivot column. The test ratios are: s: $3/1$, t: $8/1$, u: $5/1$. The smallest test ratio is s: $3/1$. Thus, we pivot on the 1 in the s–row.

	x	y	z	s	t	u	p	
s	$\boxed{1}$	1	-1	1	0	0	0	3
t	1	2	1	0	1	0	0	8
u	1	1	0	0	0	1	0	5
p	-7	-5	-6	0	0	0	1	0

$R_2 - R_1$
$R_3 - R_1$
$R_4 + 7R_1$

	x	y	z	s	t	u	p	
x	1	1	-1	1	0	0	0	3
t	0	1	2	-1	1	0	0	5
u	0	0	1	-1	0	1	0	2
p	0	2	-13	7	0	0	1	21

The most negative entry in the bottom row is the -13 in the z–column. So we use this column as the pivot column. The test ratios are: t: $5/2$, u: $2/1$. The smallest test ratio is u: $2/1$. Thus, we pivot on the 1 in the u–row.

	x	y	z	s	t	u	p	
x	1	1	-1	1	0	0	0	3
t	0	1	2	-1	1	0	0	5
u	0	0	$\boxed{1}$	-1	0	1	0	2
p	0	2	-13	7	0	0	1	21

$R_1 + R_3$
$R_2 - 2R_3$

$R_4 + 13R_3$

	x	y	z	s	t	u	p	
x	1	1	0	0	0	1	0	5
t	0	1	0	1	1	-2	0	1
z	0	0	1	-1	0	1	0	2
p	0	2	0	-6	0	13	1	47

The most negative entry in the bottom row is the -6 in the s–column. So we use this column as the pivot column. The only positive entry in this column is the 1 in the t–row, so we pivot on this entry.

	x	y	z	s	t	u	p	
x	1	1	0	0	0	1	0	5
t	0	1	0	$\boxed{1}$	1	-2	0	1
z	0	0	1	-1	0	1	0	2
p	0	2	0	-6	0	13	1	47

$R_3 + R_2$
$R_4 + 6R_2$

	x	y	z	s	t	u	p	
x	1	1	0	0	0	1	0	5
s	0	1	0	1	1	-2	0	1
z	0	1	1	0	1	-1	0	3
p	0	8	0	0	6	1	1	53

Since there are no more negative numbers in the bottom row, we are done, and read off the solution:
Optimal Solution: $p = 53/1 = 53$; $x = 5/1 = 5$, $y = 0$, $z = 3/1 = 3$.

9. Introduce slack variables and rewrite the constraints and objective function in standard form, and set up the first tableau.

$$5x_1 - x_2 + x_3 + s = 1500$$
$$2x_1 + 2x_2 + x_3 + t = 2500$$
$$4x_1 + 2x_2 + x_3 + u = 2000$$
$$-3x_1 - 7x_2 - 8x_3 + z = 0$$

	x_1	x_2	x_3	s	t	u	z	
s	5	-1	1	1	0	0	0	1500
t	2	2	1	0	1	0	0	2500
u	4	2	1	0	0	1	0	2000
z	-3	-7	-8	0	0	0	1	0

The most negative entry in the bottom row is the -8 in the x_3–column. So we use this column as the pivot column. The test ratios are: s: 1500/1, t: 2500/1, u: 2000/1. The smallest test ratio is s: 1500/1. Thus, we pivot on the 1 in the s–row.

	x_1	x_2	x_3	s	t	u	z	
s	5	-1	$\boxed{1}$	1	0	0	0	1500
t	2	2	1	0	1	0	0	2500
u	4	2	1	0	0	1	0	2000
z	-3	-7	-8	0	0	0	1	0

$R_2 - R_1$
$R_3 - R_1$
$R_4 + 8R_1$

	x_1	x_2	x_3	s	t	u	z	
x_3	5	−1	1	1	0	0	0	1500
t	−3	3	0	−1	1	0	0	1000
u	−1	3	0	−1	0	1	0	500
z	37	−15	0	8	0	0	1	12000

The most negative entry in the bottom row is the −15 in the x_2–column. So we use this column as the pivot column. The test ratios are: t: 1000/3, u: 500/3. The smallest test ratio is u: 500/3. Thus, we pivot on the 3 in the u–row.

	x_1	x_2	x_3	s	t	u	z		
x_3	5	−1	1	1	0	0	0	1500	$3R_1 + R_3$
t	−3	3	0	−1	1	0	0	1000	$R_2 - R_3$
u	−1	$\boxed{3}$	0	−1	0	1	0	500	
z	37	−15	0	8	0	0	1	12000	$R_4 + 5R_3$

	x_1	x_2	x_3	s	t	u	z	
x_3	14	0	3	2	0	1	0	5000
t	−2	0	0	0	1	−1	0	500
x_2	−1	3	0	−1	0	1	0	500
z	32	0	0	3	0	5	1	14500

Since there are no more negative numbers in the bottom row, we are done, and read off the solution:
Optimal Solution: $z = 14500/1 = 14500$; $x_1 = 0$, $x_2 = 500/3$, $x_3 = 5000/3$.

11. Introduce slack variables and rewrite the constraints and objective function in standard form, and set up the first tableau.

$$x + y + z + s = 3$$
$$y + z + w + t = 4$$
$$x + z + w + u = 5$$
$$x + y + w + v = 6$$
$$-x + y - z - w = 0$$

	x	y	z	w	s	t	u	v	p	
s	1	1	1	0	1	0	0	0	0	3
t	0	1	1	1	0	1	0	0	0	4
u	1	0	1	1	0	0	1	0	0	5
v	1	1	0	1	0	0	0	1	0	6
p	−1	−1	−1	−1	0	0	0	0	1	0

The (first) most negative entry in the bottom row is the -1 in the x–column. So we use this column as the pivot column. The test ratios are: s: 3/1, u: 5/1, v: 6/1. The smallest test ratio is s: 3/1. Thus, we pivot on the 1 in the s–row.

	x	y	z	w	s	t	u	v	p	
s	☐1	1	1	0	1	0	0	0	0	3
t	0	1	1	1	0	1	0	0	0	4
u	1	0	1	1	0	0	1	0	0	5
v	1	1	0	1	0	0	0	1	0	6
p	-1	-1	-1	-1	0	0	0	0	1	0

$R_3 - R_1$
$R_4 - R_1$
$R_5 + R_1$

	x	y	z	w	s	t	u	v	p	
x	1	1	1	0	1	0	0	0	0	3
t	0	1	1	1	0	1	0	0	0	4
u	0	-1	0	1	-1	0	1	0	0	2
v	0	0	-1	1	-1	0	0	1	0	3
p	0	0	0	-1	1	0	0	0	1	3

The most negative entry in the bottom row is the -1 in the w–column. So we use this column as the pivot column. The test ratios are: t: 4/1, u: 2/1, v: 3/1. The smallest test ratio is u: 2/1. Thus, we pivot on the 1 in the u–row.

	x	y	z	w	s	t	u	v	p	
x	1	1	1	0	1	0	0	0	0	3
t	0	1	1	1	0	1	0	0	0	4
u	0	-1	0	☐1	-1	0	1	0	0	2
v	0	0	-1	1	-1	0	0	1	0	3
p	0	0	0	-1	1	0	0	0	1	3

$R_2 - R_3$

$R_4 - R_3$
$R_5 + R_3$

	x	y	z	w	s	t	u	v	p	
x	1	1	1	0	1	0	0	0	0	3
t	0	2	1	0	1	1	-1	0	0	2
w	0	-1	0	1	-1	0	1	0	0	2
v	0	1	-1	0	0	0	-1	1	0	1
p	0	-1	0	0	0	0	1	0	1	5

The most negative entry in the bottom row is the -1 in the y–column. So we use this column as the pivot column. The test ratios are: x: 3/1, t: 2/2, v: 1/1. The smallest test ratio is v: 1/1. Thus, we pivot on the 1 in the v–row.

	x	y	z	w	s	t	u	v	p		
x	1	1	1	0	1	0	0	0	0	3	$R_1 - R_4$
t	0	2	1	0	1	1	−1	0	0	2	$R_2 - 2R_4$
w	0	−1	0	1	−1	0	1	0	0	2	$R_3 + R_4$
v	0	$\boxed{1}$	−1	0	0	0	−1	1	0	1	
p	0	−1	0	0	0	0	1	0	1	5	$R_5 + R_4$

	x	y	z	w	s	t	u	v	p	
x	1	0	2	0	1	0	1	−1	0	2
t	0	0	3	0	1	1	1	−2	0	0
w	0	0	−1	1	−1	0	0	1	0	3
y	0	1	−1	0	0	0	−1	1	0	1
p	0	0	−1	0	0	0	0	1	1	6

The most negative entry in the bottom row is the −1 in the z–column. So we use this column as the pivot column. The test ratios are: x: 2/2, t: 0/3. The smallest test ratio is t: 0/3. Thus, we pivot on the 3 in the t–row.

	x	y	z	w	s	t	u	v	p		
x	1	0	2	0	1	0	1	−1	0	2	$3R_1 - 2R_2$
t	0	0	$\boxed{3}$	0	1	1	1	−2	0	0	
w	0	0	−1	1	−1	0	0	1	0	3	$3R_3 + R_2$
y	0	1	−1	0	0	0	−1	1	0	1	$3R_4 + R_2$
p	0	0	−1	0	0	0	0	1	1	6	$3R_5 + R_2$

	x	y	z	w	s	t	u	v	p	
x	3	0	0	0	1	−2	1	1	0	6
z	0	0	3	0	1	1	1	−2	0	0
w	0	0	0	3	−2	1	1	1	0	9
y	0	3	0	0	1	1	−2	1	0	3
p	0	0	0	0	1	1	1	1	3	18

Since there are no more negative numbers in the bottom row, we are done, and read off the solution: Optimal Solution: $p = 6$; $x = 6/3 = 2$, $y = 3/3 = 1$, $z = 0/3 = 0$, $w = 9/3 = 3$.

13. Introduce slack variables and rewrite the constraints and objective function in standard form, and set up the first tableau.

$$x + y + s = 1$$
$$y + z + t = 2$$
$$z + w + u = 3$$
$$w + v + r = 4$$

$$-x - y - z - w - v + p = 0$$

	x	y	z	w	v	s	t	u	r	p		
s	☐1	1	0	0	0	1	0	0	0	0	1	
t	0	1	1	0	0	0	1	0	0	0	2	
u	0	0	1	1	0	0	0	1	0	0	3	
r	0	0	0	1	1	0	0	0	1	0	4	
p	−1	−1	−1	−1	−1	0	0	0	0	1	0	$R_5 + R_1$

The most negative entry in the bottom row is the −1 in the x–column. So we use this column as the pivot column. The only positive entry in this column is the 1 in the s–row, so we pivot on it and obtain the second tableau:

	x	y	z	w	v	s	t	u	r	p		
x	1	1	0	0	0	1	0	0	0	0	1	
t	0	1	☐1	0	0	0	1	0	0	0	2	
u	0	0	1	1	0	0	0	1	0	0	3	$R_3 - R_2$
r	0	0	0	1	1	0	0	0	1	0	4	
p	0	0	−1	−1	−1	1	0	0	0	1	1	$R_5 + R_2$

The most negative entry in the bottom row is the −1 in the z–column. So we use this column as the pivot column. The test ratios are: t: 2/1, u: 3/1. The smallest test ratio is t: 2/1. Thus, we pivot on the 1 in the t–row, and obtain the third tableau:

	x	y	z	w	v	s	t	u	r	p		
x	1	1	0	0	0	1	0	0	0	0	1	
z	0	1	1	0	0	0	1	0	0	0	2	
u	0	−1	0	☐1	0	0	−1	1	0	0	1	
r	0	0	0	1	1	0	0	0	1	0	4	$R_4 - R_3$
p	0	1	0	−1	−1	1	1	0	0	1	3	$R_5 + R_3$

The most negative entry in the bottom row is the −1 in the w–column. So we use this column as the pivot column. The test ratios are: u: 1/1, r: 4/1. The smallest test ratio is u: 1/1. Thus, we pivot on the 1 in the u–row, and obtain the fourth tableau:

	x	y	z	w	v	s	t	u	r	p	
x	1	1	0	0	0	1	0	0	0	0	1
z	0	1	1	0	0	0	1	0	0	0	2
w	0	−1	0	1	0	0	−1	1	0	0	1
r	0	1	0	0	☐1	0	1	−1	1	0	3
p	0	0	0	0	−1	1	0	1	0	1	4

$R_5 + R_4$

The most negative entry in the bottom row is the −1 in the v–column. So we use this column as the pivot column. The only positive entry in this column is the 1 in the r–row, so we picot on this entry and obtain the fifth tableau:

	x	y	z	w	v	s	t	u	r	p	
x	1	1	0	0	0	1	0	0	0	0	1
z	0	1	1	0	0	0	1	0	0	0	2
w	0	−1	0	1	0	0	−1	1	0	0	1
v	0	1	0	0	1	0	1	−1	1	0	3
p	0	1	0	0	0	1	1	0	1	1	7

Since there are no more negative numbers in the bottom row, we are done, and read off the solution:
Optimal Solution: $p = 7$; $x = 1/1 = 1$, $y = 0$, $z = 2/1 = 2$, $w = 1/1 = 1$, $v = 3/1 = 3$

Note: Making different choices for the pivot columns results in the following alternate solution: $p = 7$; $x = 1$, $y = 0$, $z = 2$, $w = 0$, $v = 4$

Note: In Exercises 15–20, we used the online Simplex Method Tool. When entering the problems, there is no need to enter the inequalities x ≥ 0, y ≥ 0, *etc.*

15. Maximize $p = 2.5x + 4.2y + 2z$ subject to
$0.1x + y - 2.2z \le 4.5$
$2.1x + y + z \le 8$
$x + 2.2y \le 5$
$x \ge 0, y \ge 0, z \ge 0$

```
Tableau #1
x         y         z         s1        s2        s3        p
0.1       1         -2.2      1         0         0         0         4.5
2.1       1         1         0         1         0         0         8
1         2.2       0         0         0         1         0         5
-2.5      -4.2      -2        0         0         0         1         0
```

```
Tableau #2
x         y        z        s1       s2       s3        p
-0.355    0       -2.2      1        0       -0.455     0        2.23
1.65      0        1        0        1       -0.455     0        5.73
0.455     1        0        0        0        0.455     0        2.27
-0.591    0       -2        0        0        1.91      1        9.55


Tableau #3
x         y        z        s1       s2       s3        p
3.27      0        0        1        2.2     -1.45      0        14.8
1.65      0        1        0        1       -0.455     0        5.73
0.455     1        0        0        0        0.455     0        2.27
2.7       0        0        0        2        1         1        21
```

Optimal Solution: p = 21; x = 0, y = 2.27, z = 5.73

17. Maximize $p = x + 2y + 3z + w$ subject to

$x + 2y + 3z \le 3$

$y + z + 2.2w \le 4$

$x + z + 2.2w \le 5$

$x + y + 2.2w \le 6$

$x \ge 0, y \ge 0, z \ge 0, w \ge 0$

```
Tableau #1
x         y        z        w        s1       s2       s3       s4       p
1         2        3        0        1        0        0        0        0        3
0         1        1        2.2      0        1        0        0        0        4
1         0        1        2.2      0        0        1        0        0        5
1         1        0        2.2      0        0        0        1        0        6
-1       -2       -3       -1        0        0        0        0        1        0


Tableau #2
x         y        z        w        s1       s2       s3       s4       p
0.333     0.667    1        0        0.333    0        0        0        0        1
-0.333    0.333    0        2.2     -0.333    1        0        0        0        3
0.667    -0.667    0        2.2     -0.333    0        1        0        0        4
1         1        0        2.2      0        0        0        1        0        6
0         0        0       -1        1        0        0        0        1        3


Tableau #3
x         y        z        w        s1       s2       s3       s4       p
0.333     0.667    1        0        0.333    0        0        0        0        1
-0.152    0.152    0        1       -0.152    0.455    0        0        0        1.36
1        -1        0        0        0       -1        1        0        0        1
1.33      0.667    0        0        0.333   -1        0        1        0        3
-0.152    0.152    0        0        0.848    0.455    0        0        1        4.36
```

Tableau #4

x	y	z	w	s1	s2	s3	s4	p	
0	1	1	0	0.333	0.333	-0.333	0	0	0.667
0	0	0	1	-0.152	0.303	0.152	0	0	1.52
1	-1	0	0	0	-1	1	0	0	1
0	2	0	0	0.333	0.333	-1.33	1	0	1.67
0	0	0	0	0.848	0.303	0.152	0	1	4.52

Optimal Solution: p = 4.52; x = 1, y = 0, z = 0.67, w = 1.52

19. Maximize $p = x - y + z - w + v$ subject to

$x + y \le 1.1$
$y + z \le 2.2$
$z + w \le 3.3$
$w + v \le 4.4$
$x \ge 0, y \ge 0, z \ge 0, w \ge 0, v \ge 0$

Tableau #1

x	y	z	w	v	s1	s2	s3	s4	p	
1	1	0	0	0	1	0	0	0	0	1.1
0	1	1	0	0	0	1	0	0	0	2.2
0	0	1	1	0	0	0	1	0	0	3.3
0	0	0	1	1	0	0	0	1	0	4.4
-1	1	-1	1	-1	0	0	0	0	1	0

Tableau #2

x	y	z	w	v	s1	s2	s3	s4	p	
1	1	0	0	0	1	0	0	0	0	1.1
0	1	1	0	0	0	1	0	0	0	2.2
0	0	1	1	0	0	0	1	0	0	3.3
0	0	0	1	1	0	0	0	1	0	4.4
0	2	-1	1	-1	1	0	0	0	1	1.1

Tableau #3

x	y	z	w	v	s1	s2	s3	s4	p	
1	1	0	0	0	1	0	0	0	0	1.1
0	1	1	0	0	0	1	0	0	0	2.2
0	-1	0	1	0	0	-1	1	0	0	1.1
0	0	0	1	1	0	0	0	1	0	4.4
0	3	0	1	-1	1	1	0	0	1	3.3

Tableau #4

x	y	z	w	v	s1	s2	s3	s4	p	
1	1	0	0	0	1	0	0	0	0	1.1
0	1	1	0	0	0	1	0	0	0	2.2
0	-1	0	1	0	0	-1	1	0	0	1.1
0	0	0	1	1	0	0	0	1	0	4.4
0	3	0	2	0	1	1	0	1	1	7.7

Optimal Solution: p = 7.7; x = 1.1, y = 0, z = 2.2, w = 0, v = 4.4

21. Unknowns:

x = # calculus texts

y = # history texts

z = # marketing texts

Maximize $p = 10x + 4y + 8z$ subject to

$$x + y + z \le 650$$
$$2x + y + 3z \le 1000$$
$$x \ge 0, y \ge 0, z \ge 0.$$

	x	y	z	s	t	p	
s	1	1	1	1	0	0	650
t	[2]	1	3	0	1	0	1000
p	−10	−4	−8	0	0	1	0

$2R_1 - R_2$

$R_3 + 5R_2$

	x	y	z	s	t	p	
s	0	1	−1	2	−1	0	300
x	2	1	3	0	1	0	1000
p	0	1	7	0	5	1	5000

Optimal Solution: $p = 5000$; $x = 1000/2 = 500$, $y = 0$, $z = 0$. You should purchase 500 calculus texts, no history texts and no marketing texts. The maximum profit is $5000 per semester.

23. Unknowns:

x = # gallons of PineOrange

y = # gallons of PineKiwi

z = # gallons of OrangeKiwi

Maximize $p = x + 2y + z$ subject to

$$2x + 3y \le 800$$
$$2x + 3z \le 650$$
$$y + z \le 350$$
$$x \ge 0, y \ge 0, z \ge 0.$$

	x	y	z	s	t	u	p	
s	2	[3]	0	1	0	0	0	800
t	2	0	3	0	1	0	0	650
u	0	1	1	0	0	1	0	350
p	−1	−2	−1	0	0	0	1	0

$3R_3 - R_1$

$3R_4 + 2R_1$

	x	y	z	s	t	u	p	
y	2	3	0	1	0	0	0	800
t	2	0	3	0	1	0	0	650
u	−2	0	[3]	−1	0	3	0	250
p	1	0	−3	2	0	0	3	1600

$R_2 - R_3$

$R_4 + R_3$

	x	y	z	s	t	u	p	
y	2	3	0	1	0	0	0	800
t	[4]	0	0	1	1	−3	0	400
z	−2	0	3	−1	0	3	0	250
p	−1	0	0	1	0	3	3	1850

$2R_1 - R_2$

$2R_3 + R_2$

$4R_4 + R_2$

	x	y	z	s	t	u	p	
y	0	6	0	1	−1	3	0	1200
x	4	0	0	1	1	−3	0	400
z	0	0	6	−1	1	3	0	900
p	0	0	0	5	1	9	12	7800

Optimal Solution: $p = 650$; $x = 400/4 = 100$, $y = 1200/6 = 200$, $z = 900/6 = 150$. The company makes a maximum profit of \$650 by making 100 gallons of PineOrange, 200 gallons of PineKiwi, and 150 gallons of OrangeKiwi.

25. Unknowns:
x = # sections of Ancient History
y = # sections of Medieval History
z = # sections of Modern History
Maximize $p = x + 2y + 3z$ (in tens of thousands of dollars) subject to
$$x + y + z \le 45$$
$$100x + 50y + 200z \le 5000$$
$$x + y + 2z \le 60$$
$$x \ge 0, y \ge 0, z \ge 0.$$

	x	y	z	s	t	u	p	
s	1	1	1	1	0	0	0	45
t	100	50	[200]	0	1	0	0	5000
u	1	1	2	0	0	1	0	60
p	−1	−2	−3	0	0	0	1	0

$200R_1 - R_2$

$100R_3 - R_2$

$200R_4 + 3R_2$

	x	y	z	s	t	u	p		
s	100	150	0	200	−1	0	0	4000	$R_1 - 3R_3$
z	100	50	200	0	1	0	0	5000	$R_2 - R_3$
u	0	$\boxed{50}$	0	0	−1	100	0	1000	
p	100	−250	0	0	3	0	200	15000	$R_4 + 5R_3$

	x	y	z	s	t	u	p		
s	100	0	0	200	$\boxed{2}$	−300	0	1000	
z	100	0	200	0	2	−100	0	4000	$R_2 - R_1$
y	0	50	0	0	−1	100	0	1000	$2R_3 + R_1$
p	100	0	0	0	−2	500	200	20000	$R_4 + R_1$

	x	y	z	s	t	u	p	
t	100	0	0	200	2	−300	0	1000
z	0	0	200	−200	0	200	0	3000
y	100	100	0	200	0	−100	0	3000
p	200	0	0	200	0	200	200	21000

Optimal Solution: $p = 105$; $x = 0$, $y = 3000/100 = 30$, $z = 3000/200 = 15$. The department should offer no Ancient History, 30 sections of Medieval History, and 15 sections of Modern History, for a profit of $\$105 \times 10{,}000 = \$1{,}050{,}000$.

Answers to additional question:
The values of the slack variables are:
$t = 1000/2 = 500$, meaning that there will be 500 students without classes.
$s = u = 0$, meaning that all sections and professors are used.

27. Unknowns:
$x = $ # acres of tomatoes
$y = $ # acres of lettuce
$z = $ # acres of carrots
Maximize $p = 20x + 15y + 5z$ (in hundreds of dollars) subject to
$$x + y + z \leq 100$$
$$5x + 4y + 2z \leq 400$$
$$4x + 2y + 2z \leq 500$$
$$x \geq 0, y \geq 0, z \geq 0.$$

	x	y	z	s	t	u	p		
s	1	1	1	1	0	0	0	100	$5R_1 - R_2$
t	[5]	4	2	0	1	0	0	400	
u	4	2	2	0	0	1	0	500	$5R_3 - 4R_2$
p	−20	−15	−5	0	0	0	1	0	$R_4 + 4R_2$

	x	y	z	s	t	u	p	
s	0	1	3	5	−1	0	0	100
x	5	4	2	0	1	0	0	400
u	0	−6	2	0	−4	5	0	900
p	0	1	3	0	4	0	1	1600

Optimal Solution: $p = 1600$; $x = 400/5 = 80$, $y = 0$, $z = 0$. Plant 80 acres of tomatoes and no lettuce or carrots.

The slack variable corresponding to the number of acres available is s. The value of s is $s = 1200/5 = 20$, meaning that you will leave 20 acres unplanted.

29. Unknowns:
x = # servings of granola
y = # servings of nutty granola
z = # servings of nuttiest granola
Maximize $p = 6x + 8y + 3z$ subject to
$x + y + 5z \le 1500$
$4x + 8y + 8z \le 10{,}000$
$2x + 4y + 8z \le 4000$
$x \ge 0, y \ge 0, z \ge 0$.

	x	y	z	s	t	u	p		
s	1	1	5	1	0	0	0	1500	$4R_1 - R_3$
t	4	8	8	0	1	0	0	10000	$R_2 - 2R_3$
u	2	[4]	8	0	0	1	0	4000	
p	−6	−8	−3	0	0	0	1	0	$R_4 + 2R_3$

	x	y	z	s	t	u	p		
s	[2]	0	12	4	0	−1	0	2000	
t	0	0	−8	0	1	−2	0	2000	
y	2	4	8	0	0	1	0	4000	$R_3 - R_1$
p	−2	0	13	0	0	2	1	8000	$R_4 + R_1$

	x	y	z	s	t	u	p	
x	2	0	12	4	0	−1	0	2000
t	0	0	−8	0	1	−2	0	2000
y	0	4	−4	−4	0	2	0	2000
p	0	0	25	4	0	1	1	10,000

Optimal Solution: $p = 10,000$; $x = 2000/2 = 1000$, $y = 2000/4 = 500$, $z = 0$. The Choral Society can make a profit of $10,000 by selling 1000 servings of Granola, 500 servings of Nutty Granola and no Nuttiest Granola.

To obtain the ingredients left over, look at the values of the slack variables in the final tableau:

$s = 0$, so there are no toasted oats left over.

$t = 2000/1 = 2000$, so there are 2000 oz. of almonds left over.

$u = 0$, so there are no raisins left over.

31. Unknowns:

x = # oil allocated to process A

y = # oil allocated to process B

z = # oil allocated to process C

Maximize $p = 4x + 4y + 4z$ subject to

$x + y + z \le 50$

$15x + 10y + 5z \le 300$, or $3x + 2y + z \le 60$

$60x + 55y + 50z \le 5000$, or $12x + 11y + 10z \le 1000$

$x \ge 0, y \ge 0. z \ge 0.$

	x	y	z	s	t	u	p		
s	1	1	1	1	0	0	0	50	$3R_1 - R_2$
t	[3]	2	1	0	1	0	0	60	
u	12	11	10	0	0	1	0	1000	$R_3 - 4R_2$
p	−4	−4	−4	0	0	0	1	0	$3R_4 + 4R_2$

	x	y	z	s	t	u	p		
s	0	1	[2]	3	−1	0	0	90	
x	3	2	1	0	1	0	0	60	$2R_2 - R_1$
u	0	3	6	0	−4	1	0	760	$R_3 - 3R_1$
p	0	−4	−8	0	4	0	3	240	$R_4 + 4R_1$

	x	y	z	s	t	u	p	
z	0	1	2	3	−1	0	0	90
x	6	3	0	−3	3	0	0	30
u	0	0	0	−9	−1	1	0	490
p	0	0	0	12	0	0	3	600

Optimal Solution: $p = 200$; $x = 30/6 = 5$, $y = 0$, $z = 90/2 = 45$. Allocate 5 million gals to process A and 45 million gals. to process C.

Another solution (obtained by making different choices for pivoting): Allocate 10 million gals to process B and 40 million gals. to process C.

33. Unknowns:
x = # of servings of Cell-Tech
y = # servings of Riboforce HP
z = # servings of Creatine Transport.
Maximize $p = 10x + 5y + 5z$ subject to
$\quad\quad 200x + 100z \le 1000$, or $2x + z \le 10$
$\quad\quad 75x + 15y + 35z \le 225$, or $15x + 3y + 7z \le 45$
$\quad\quad x \ge 0, y \ge 0, z \ge 0.$

	x	y	z	s	t	p		
s	2	0	1	1	0	0	10	$15R_1 - 2R_2$
t	15	3	7	0	1	0	45	
p	-10	-5	-5	0	0	1	0	$3R_3 + 2R_2$

	x	y	z	s	t	p		
s	0	-6	1	15	-2	0	60	$R_1 + 2R_2$
x	15	3	7	0	1	0	45	
p	0	-9	-1	0	2	3	90	$R_3 + 3R_2$

	x	y	z	s	t	p	
s	30	0	15	15	0	0	150
y	15	3	7	0	1	0	45
p	45	0	20	0	5	3	225

Optimal Solution: $p = 75$; $x = 0$, $y = 45/3 = 15$, $z = 0$. Use 15 servings of RiboForce HP and none of the others for a maximum of 75g creatine.

35. Unknowns:
x = # MON shares
y = # SNE shares
z = # IMCL shares
Maximize $p = 4x + 3y - z$ subject to $80x + 20y + 70z \le 10,000$,
$(0.01)(80x) + (0.02)(20y) \le 200$, or $8x + 4y \le 2000$

	x	y	z	s	t	p	
s	80	20	70	1	0	0	10000
t	8	4	0	0	1	0	2000
p	−4	−3	1	0	0	1	0

The most negative entry in the bottom row is the −4 in the x-column. So we use this column as the pivot column. The test ratios are: s: 10000/80, t: 2000/8. The smallest test ratio is s: 10000/80. Thus, we pivot on the 80 in the s-row.

	x	y	z	s	t	p	
s	80	20	70	1	0	0	10000
t	8	4	0	0	1	0	2000
p	−4	−3	1	0	0	1	0

$10R_2 - R_1$

$20R_3 + R_1$

	x	y	z	s	t	p	
x	80	20	70	1	0	0	10000
t	0	20	-70	-1	10	0	10000
p	0	−40	90	1	0	20	10000

The most negative entry in the bottom row is the -40 in the y-column. So we use this column as the pivot column. The test ratios are: x: 10000/20, t: 10000/20. These are equal, but we've already pivoted on the first row, so we pivot on the 20 in the t-row.

	x	y	z	s	t	p	
x	80	20	70	1	0	0	10000
t	0	20	−70	−1	10	0	10000
p	0	−40	90	1	0	20	10000

$R_1 - R_2$

$R_3 + 2R_2$

	x	y	z	s	t	p	
x	80	0	140	2	−10	0	0
y	0	20	−70	−1	10	0	10000
p	0	0	−50	−1	20	20	30000

The most negative entry in the bottom row is the −50 in the z-column. So we use this column as the pivot column. The only usable test ratio is: x: 0/140. Thus, we pivot on the 140 in the x-row.

	x	y	z	s	t	p	
x	80	0	$\boxed{140}$	2	-10	0	0
y	0	20	-70	-1	10	0	10000
p	0	0	-50	-1	20	20	30000

$2R_2 + R_1$

$140R_3 + 50R_1$

	x	y	z	s	t	p	
z	80	0	140	2	-10	0	0
y	80	40	0	0	10	0	20000
p	4000	0	0	-40	2300	2800	4200000

The most negative entry in the bottom row is the -40 in the s-column. So we use this column as the pivot column. The only usable test ratio is: z: 0/2. Thus, we pivot on the 2 in the z-row.

	x	y	z	s	t	p	
z	80	0	140	$\boxed{2}$	-10	0	0
y	80	40	0	0	10	0	20000
p	4000	0	0	-40	2300	2800	4200000

$R_3 + 20R_1$

	x	y	z	s	t	p	
s	80	0	140	2	-10	0	0
y	80	40	0	0	10	0	20000
p	5600	0	2800	0	2100	2800	4200000

Since there are no more negative numbers in the bottom row, we are done, and read off the solution:
Optimal Solution: $p = 4{,}200{,}000/2800 = 1500$; $x = 0$, $y = 20{,}000/40 = 500$, $z = 0$
Buy 500 shares of SNE and no others. The broker is wrong.

37. Unknowns:
x = amount allocated to automobile loans (in \$ millions)
y = amount allocated to furniture loans (in \$ millions)
z = amount allocated to signature loans (in \$ millions)
w = amount allocated to other secured loans
Maximize $p = 8x + 10y + 12z + 10w$ (100 times the return) subject to
$\quad x + y + z + w \le 5$
$\quad -x - y + 9z - w \le 0$
$\quad -x + y + w \le 0$
$\quad -2x + w \le 0$
$\quad x \ge 0,\ y \ge 0,\ z \ge 0,\ w \ge 0.$

	x	y	z	w	s	t	u	v	p		
s	1	1	1	1	1	0	0	0	0	5	$9R_1 - R_2$
t	-1	-1	[9]	-1	0	1	0	0	0	0	
u	-1	1	0	1	0	0	1	0	0	0	
v	-2	0	0	1	0	0	0	1	0	0	
p	-8	-10	-12	-10	0	0	0	0	1	0	$9R_5 + 12R_2$

	x	y	z	w	s	t	u	v	p		
s	10	10	0	10	9	-1	0	0	0	45	$R_1 - 10R_3$
z	-1	-1	9	-1	0	1	0	0	0	0	$R_2 + R_3$
u	-1	[1]	0	1	0	0	1	0	0	0	
v	-2	0	0	1	0	0	0	1	0	0	
p	-84	-102	0	-102	0	12	0	0	9	0	$R_5 + 102R_3$

	x	y	z	w	s	t	u	v	p		
s	[20]	0	0	0	9	-1	-10	0	0	45	
z	-2	0	9	0	0	1	1	0	0	0	$10R_2 + R_1$
y	-1	1	0	1	0	0	1	0	0	0	$20R_3 + R_1$
v	-2	0	0	1	0	0	0	1	0	0	$10R_4 + R_1$
p	-186	0	0	0	0	12	102	0	9	0	$20R_5 + 186R_1$

	x	y	z	w	s	t	u	v	p	
x	20	0	0	0	9	-1	-10	0	0	45
z	0	0	90	0	9	9	0	0	0	45
y	0	20	0	20	9	-1	10	0	0	45
v	0	0	0	10	9	-1	-10	10	0	45
p	0	0	0	0	1674	54	180	0	180	8370

Optimal Solution: $p = 8370/180 = 46.5$; $x = 45/20 = 2.25$, $y = 45/20 = 2.25$, $z = 45/90 = 0.5$, $w = 0$. Allocate \$2,250,000 to automobile loans, \$2,250,000 to furniture loans, and \$500,000 to signature loans. Another optimal solution (pivot in the w-column): $p = 8370/180 = 46.5$; $x = 45/20 = 2.25$, $y = 0$, $z = 45/90 = 0.5$, $w = 45/20 = 2.25$. Allocate \$2,250,000 to automobile loans, \$500,000 to signature loans, and \$2,250,000 to other secured loans.

In general, Allocate \$2,250,000 to automobile loans, \$500,000 to signature loans, and \$2,250,000 to any combination of furniture loans and other secured loans.

39. Unknowns:

x = amount invested in Warner
y = amount invested in Universal
z = amount invested in Sony
w = amount invested in EMI

Maximize $p = 8x + 20y + 10z + 15w$ subject to

$0.12x + 0.20y + 0.20z + 0.15w \le 15,000$, or $12x + 20y + 20z + 15w \le 1,500,000$
$x - 4y + z + w \le 0$
$x + y + z + w \le 100,000$
$x \ge 0, y \ge 0, z \ge 0, w \ge 0$.

	x	y	z	w	s	t	u	p	
s	12	[20]	20	15	1	0	0	0	1500000
t	1	−4	1	1	0	1	0	0	0
u	1	1	1	1	0	0	1	0	100000
p	−8	−20	−10	−15	0	0	0	1	0

$5R_2 + R_1$
$20R_3 - R_1$
$R_4 + R_1$

	x	y	z	w	s	t	u	p	
y	12	20	20	15	1	0	0	0	1500000
t	17	0	25	20	1	5	0	0	1500000
u	8	0	0	5	−1	0	20	0	500000
p	4	0	10	0	1	0	0	1	1500000

Optimal Solution: $p = 1,500,000$; $x = 0$, $y = 1,500,000/20 = 75,000$, $z = 0$, $w = 0$. Invest \$75,000 in Universal, none in the rest. Another optimal solution (pivot in the w-column) is: Invest \$18,750 in Universal, and \$75,000 in EMI.

41. Unknowns:

x = # boards from Tucson to Honolulu
y = # boards from Tucson to Venice Beach
z = # boards from Toronto to Honolulu
w = # boards from Toronto to Venice Beach

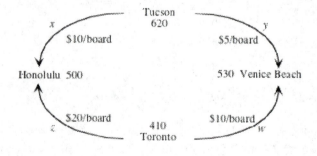

Maximize $p = x + y + z + w$ subject to

$x + y \le 620$
$z + w \le 410$
$x + z \le 500$
$y + w \le 530$
$10x + 5y + 20z + 10w \le 6550$
$x \ge 0, y \ge 0, z \ge 0$.

	x	y	z	w	s	t	u	v	r	p		
s	1	1	0	0	1	0	0	0	0	0	620	$R_1 - R_3$
t	0	0	1	1	0	1	0	0	0	0	410	
u	[1]	0	1	0	0	0	1	0	0	0	500	
v	0	1	0	1	0	0	0	1	0	0	530	
r	10	5	20	10	0	0	0	0	1	0	6550	$R_5 - 10R_3$
p	-1	-1	-1	-1	0	0	0	0	0	1	0	$R_6 + R_3$

	x	y	z	w	s	t	u	v	r	p		
s	0	[1]	-1	0	1	0	-1	0	0	0	120	
t	0	0	1	1	0	1	0	0	0	0	410	
x	1	0	1	0	0	0	1	0	0	0	500	
v	0	1	0	1	0	0	0	1	0	0	530	$R_4 - R_1$
r	0	5	10	10	0	0	-10	0	1	0	1550	$R_5 - 5R_1$
p	0	-1	0	-1	0	0	1	0	0	1	500	$R_6 + R_1$

	x	y	z	w	s	t	u	v	r	p		
y	0	1	-1	0	1	0	-1	0	0	0	120	$15R_1 + R_5$
t	0	0	1	1	0	1	0	0	0	0	410	$15R_2 - R_5$
x	1	0	1	0	0	0	1	0	0	0	500	$15R_3 - R_5$
v	0	0	1	1	-1	0	1	1	0	0	410	$15R_4 - R_5$
r	0	0	[15]	10	-5	0	-5	0	1	0	950	
p	0	0	-1	-1	1	0	0	0	0	1	620	$15R_6 + R_5$

	x	y	z	w	s	t	u	v	r	p		
y	0	15	0	10	10	0	-20	0	1	0	2750	$R_1 - R_5$
t	0	0	0	5	5	15	5	0	-1	0	5200	$2R_2 - R_5$
x	15	0	0	-10	5	0	20	0	-1	0	6550	$R_3 + R_5$
v	0	0	0	5	-10	0	20	15	-1	0	5200	$2R_4 - R_5$
z	0	0	15	10	-5	0	-5	0	1	0	950	
p	0	0	0	-5	10	0	-5	0	1	15	10250	$2R_6 + R_5$

	x	y	z	w	s	t	u	v	r	p		
y	0	15	−15	0	15	0	−15	0	0	0	1800	$3R_1 + R_4$
t	0	0	−15	0	15	30	15	0	−3	0	9450	$3R_2 - R_4$
x	15	0	15	0	0	0	15	0	0	0	7500	$3R_3 - R_4$
v	0	0	−15	0	−15	0	45	30	−3	0	9450	
w	0	0	15	10	−5	0	−5	0	1	0	950	$9R_5 + R_4$
p	0	0	15	0	15	0	−15	0	3	30	21450	$3R_6 + R_4$

	x	y	z	w	s	t	u	v	r	p	
y	0	45	−60	0	30	0	0	30	−3	0	14850
t	0	0	−30	0	60	90	0	−30	−6	0	18900
x	45	0	60	0	15	0	0	−30	3	0	13050
u	0	0	−15	0	−15	0	45	30	−3	0	9450
w	0	0	120	90	−60	0	0	30	6	0	18000
p	0	0	30	0	30	0	0	30	6	90	73800

Optimal Solution: $p = 73,800/90 = 820$; $x = 13,050/45 = 290$, $y = 14,850/45 = 330$, $z = 0$, $w = 18,000/90 = 200$. Make the following shipments: Tucson to Honolulu: 290 boards; Tucson to Venice Beach: 330 boards; Toronto to Honolulu: 0 boards; Toronto to Venice Beach: 200 boards, giving 820 boards shipped.

43. Unknowns:
x = # people you fly from Chicago to Los Angeles
y = # people you fly from Chicago to New York
z = # people you fly from Denver to Los Angeles
w = # people you fly from Denver to New York
Maximize $p = x + y + z + w$ subject to
$$195x + 182y + 395z + 166w \le 4520$$
$$x + y \le 20$$
$$z + w \le 10$$
$$x + z \le 10$$
$$y + w \le 15$$
$$x \ge 0, y \ge 0, z \ge 0, w \ge 0.$$

To solve, we used the online Simplex Method Tool. When entering the problems, there is no need to enter the inequalities $x \ge 0$, $y \ge 0$, etc.

Tableau #1

x	y	z	w	s1	s2	s3	s4	s5	p	
195	182	395	166	1	0	0	0	0	0	4520
1	1	0	0	0	1	0	0	0	0	20
0	0	1	1	0	0	1	0	0	0	10
1	0	1	0	0	0	0	1	0	0	10
0	1	0	1	0	0	0	0	1	0	15

-1	-1	-1	-1	0	0	0	0	0	1	0

Tableau #2

x	y	z	w	s1	s2	s3	s4	s5	p	
0	182	200	166	1	0	0	-195	0	0	2570
0	1	-1	0	0	1	0	-1	0	0	10
0	0	1	1	0	0	1	0	0	0	10
1	0	1	0	0	0	0	1	0	0	10
0	1	0	1	0	0	0	0	1	0	15
0	-1	0	-1	0	0	0	1	0	1	10

Tableau #3

x	y	z	w	s1	s2	s3	s4	s5	p	
0	0	382	166	1	-182	0	-13	0	0	750
0	1	-1	0	0	1	0	-1	0	0	10
0	0	1	1	0	0	1	0	0	0	10
1	0	1	0	0	0	0	1	0	0	10
0	0	1	1	0	-1	0	1	1	0	5
0	0	-1	-1	0	1	0	0	0	1	20

Tableau #4

x	y	z	w	s1	s2	s3	s4	s5	p	
0	0	1	0.43	0	-0.48	0	-0.03	0	0	1.96
0	1	0	0.43	0	0.52	0	-1.03	0	0	11.96
0	0	0	0.57	0	0.48	1	0.03	0	0	8.04
1	0	0	-0.43	0	0.48	0	1.03	0	0	8.04
0	0	0	0.57	0	-0.52	0	1.03	1	0	3.04
0	0	0	-0.57	0	0.52	0	-0.03	0	1	21.96

Tableau #5

x	y	z	w	s1	s2	s3	s4	s5	p	
0	0	2.3	1	0.01	-1.1	0	-0.08	0	0	4.52
0	1	-1	0	0	1	0	-1	0	0	10
0	0	-1.3	0	-0.01	1.1	1	0.08	0	0	5.48
1	0	1	0	0	0	0	1	0	0	10
0	0	-1.3	0	-0.01	0.1	0	1.08	1	0	0.48
0	0	1.3	0	0.01	-0.1	0	-0.08	0	1	

24.52

Tableau #6

x	y	z	w	s1	s2	s3	s4	s5	p	
0	0	1	1	0	0	1	0	0	0	10
0	1	0.19	0	0.01	0	-0.91	-1.07	0	0	5
0	0	-1.19	0	-0.01	1	0.91	0.07	0	0	5
1	0	1	0	0	0	0	1	0	0	10
0	0	-1.19	0	-0.01	0	-0.09	1.07	1	0	0
0	0	1.19	0	0.01	0	0.09	-0.07	0	1	25

```
Tableau #7
```

x	y	z	w	s1	s2	s3	s4	s5	p	
0	0	1	1	0	0	1	0	0	0	10
0	1	-1	0	0	0	-1	0	1	0	5
0	0	-1.11	0	-0.01	1	0.92	0	-0.07	0	5
1	0	2.11	0	0.01	0	0.08	0	-0.93	0	10
0	0	-1.11	0	-0.01	0	-0.08	1	0.93	0	0
0	0	1.11	0	0.01	0	0.08	0	0.07		25

Optimal Solution: $p = 25$; $x = 10$, $y = 5$, $z = 0$, $w = 10$. Fly 10 people from Chicago to Los Angeles, 5 people from Chicago to New York, and 10 people from Denver to New York.

45. The only difficulty with the stated problem is the "≥" in the first constraint: $x - y + z \geq 0$. We can reverse the inequality by multiplying both sides by -1: $-x + y - z \leq 0$
Thus, the given problem can be stated as a standard maximization problem:
Maximize $p = 3x - 2y$ subject to
$$-x + y - z \leq 0$$
$$x - y - z \leq 6$$
$$x \geq 0,\ y \geq 0,\ z \geq 0$$

47. The graphical method applies only to LP problems in two unknowns, whereas the simplex method can be used to solve LP problems with any number of unknowns.

49. She is correct. Since there are only two constraints, there can only be two active variables, giving two or fewer non-zero values for the unknowns at each stage.

51. A basic solution to a system of linear equations is a solution in which all the non-pivotal variables are taken to be zero; that is, all variables whose value is arbitrary are assigned the value zero. To obtain a basic solution for a given system of linear equations, one can row reduce the associated augmented matrix, write down the general solution, and then set all the parameters (variables with "arbitrary" values) equal to zero.

53. No. Let us assume for the sake of simplicity that all the pivots are 1's. (They may certainly be changed to 1's without affecting the value of any of the variables.) Since the entry at the bottom of the pivot column is negative, the bottom row gets replaced by itself plus a positive multiple of the pivot row. The value of the objective function (bottom-right entry) is thus replaced by itself plus a positive multiple of the nonnegative rightmost entry of the pivot row. Therefore, it cannot decrease.

4.4 The Simplex Method: Solving General Linear Programming Problems

1. Introduce slack and surplus variables, rewrite the constraints and objective function in standard form, set up the first tableau, and star rows coming from "≥" constraints.

$$x + 2y - s = 6$$
$$-x + y + t = 4$$
$$2x + y + u = 8$$
$$-x - y + p = 0$$

	x	y	s	t	u	p	
*s	1	2	−1	0	0	0	6
t	−1	1	0	1	0	0	4
u	2	1	0	0	1	0	8
p	−1	−1	0	0	0	1	0

The first (only) starred row is the s–row, and its largest positive entry is the 2 in the y–column. So we use this column as the pivot column. The test ratios are: s: 6/2, t: 4/1, u: 8/1. The smallest test ratio is s: 6/2. Thus, we pivot on the 2 in the s–row.

	x	y	s	t	u	p		
*s	1	2	−1	0	0	0	6	
t	−1	1	0	1	0	0	4	$2R_2 - R_1$
u	2	1	0	0	1	0	8	$2R_3 - R_1$
p	−1	−1	0	0	0	1	0	$2R_4 + R_1$

	x	y	s	t	u	p	
y	1	2	−1	0	0	0	6
t	−3	0	1	2	0	0	2
u	3	0	1	0	2	0	10
p	−1	0	−1	0	0	2	6

Since there are no more starred rows, we go to Phase 2, and do the standard simplex method:

	x	y	s	t	u	p		
y	1	2	−1	0	0	0	6	$3R_1 - R_3$
t	−3	0	1	2	0	0	2	$R_2 + R_3$
u	3	0	1	0	2	0	10	
p	−1	0	−1	0	0	2	6	$3R_4 + R_3$

171

	x	y	s	t	u	p		
y	0	6	−4	0	−2	0	8	$R_1 + 2R_2$
t	0	0	2	2	2	0	12	
x	3	0	1	0	2	0	10	$2R_3 - R_2$
p	0	0	−2	0	2	6	28	$R_4 + R_2$

	x	y	s	t	u	p	
y	0	6	0	4	2	0	32
s	0	0	2	2	2	0	12
x	6	0	0	−2	2	0	8
p	0	0	0	2	4	6	40

Optimal Solution: $p = 40/6 = 20/3$; $x = 8/6 = 4/3$, $y = 32/6 = 16/3$.

3. Introduce slack and surplus variables, rewrite the constraints and objective function in standard form, set up the first tableau, and star rows coming from "≥" constraints.

$$x + y + s = 25$$
$$x \quad\quad - t = 10$$
$$-x + 2y - u = 0$$
$$-12x - 10y + p = 0$$

	x	y	s	t	u	p		
s	1	1	1	0	0	0	25	$R_1 - R_2$
*t	1	0	0	−1	0	0	10	
*u	−1	2	0	0	−1	0	0	$R_3 + R_2$
p	−12	−10	0	0	0	1	0	$R_4 + 12R_2$

The first starred row above is the t-row, and its largest positive entry is the 1 in the x-column. So we use this column as the pivot column. The test ratios are: s: 25/1, t: 10/1. The smallest test ratio is t: 10/1. Thus, we pivot on the 1 in the t-row.

	x	y	s	t	u	p		
s	0	1	1	1	0	0	15	$2R_1 - R_3$
x	1	0	0	−1	0	0	10	
*u	0	2	0	−1	−1	0	10	
p	0	−10	0	−12	0	1	120	$R_4 + 5R_3$

The only remaining starred row above is the u–row, and its largest positive entry is the 2 in the y–column. So we use this column as the pivot column. The test ratios are: s: 15/1, u: 10/2. The smallest test ratio is u: 10/2. Thus, we pivot on the 2 in the u–row.

172

	x	y	s	t	u	p	
s	0	0	2	3	1	0	20
x	1	0	0	−1	0	0	10
y	0	2	0	−1	−1	0	10
p	0	0	0	−17	−5	1	170

Since there are no more starred rows, we go to Phase 2, and do the usual simplex method:

	x	y	s	t	u	p		
s	0	0	2	$\boxed{3}$	1	0	20	
x	1	0	0	−1	0	0	10	$3R_2 + R_1$
y	0	2	0	−1	−1	0	10	$3R_3 + R_1$
p	0	0	0	−17	−5	1	170	$3R_4 + 17R_1$

	x	y	s	t	u	p	
t	0	0	2	3	1	0	20
x	3	0	2	0	1	0	50
y	0	6	2	0	−2	0	50
p	0	0	34	0	2	3	850

Optimal Solution: $p = 850/3$; $x = 50/3 = 50/3$, $y = 50/6 = 25/3$.

5. Introduce slack and surplus variables, rewrite the constraints and objective function in standard form, and set up the first tableau.

$$x + y + z + s = 150$$
$$x + y + z \geq -t = 100$$
$$-2x - 5y - 3z + p = 0$$

	x	y	z	s	t	p		
s	1	1	1	1	0	0	150	$R_1 - R_2$
*t	$\boxed{1}$	1	1	0	−1	0	100	
p	−2	−5	−3	0	0	1	0	$R_3 + 2R_2$

The first starred row above is the t–row, and its largest positive entry is the 1 in the x–column. So we use this column as the pivot column. The test ratios are: s: 150/1, t: 100/1. The smallest test ratio is t: 100/1. Thus, we pivot on the 1 in the t–row.

	x	y	z	s	t	p	
s	0	0	0	1	1	0	50
x	1	1	1	0	−1	0	100
p	0	−3	−1	0	−2	1	200

Since there are no more starred rows, we go to Phase 2.

	x	y	z	s	t	p	
s	0	0	0	1	1	0	50
x	1	[1]	1	0	–1	0	100
p	0	–3	–1	0	–2	1	200

$R_3 + 3R_2$

	x	y	z	s	t	p	
s	0	0	0	1	[1]	0	50
y	1	1	1	0	–1	0	100
p	3	0	2	0	–5	1	500

$R_2 + R_1$

$R_3 + 5R_1$

	x	y	z	s	t	p	
t	0	0	0	1	1	0	50
y	1	1	1	1	0	0	150
p	3	0	2	5	0	1	750

Optimal Solution: $p = 750$; $x = 0$, $y = 150/1 = 150$, $z = 0$.

7. $x + y + z + w + s = 40$
$2x + y - z - w - t = 10$
$x + y + z + w - u = 10$
$-2x - 3y - z - 4w + p = 0$

	x	y	z	w	s	t	u	p	
s	1	1	1	1	1	0	0	0	40
*t	[2]	1	–1	–1	0	–1	0	0	10
*u	1	1	1	1	0	0	–1	0	10
p	–2	–3	–1	–4	0	0	0	1	0

$2R_1 - R_2$

$2R_3 - R_2$

$R_4 + R_2$

The first starred row above is the t–row, and its largest positive entry is the 2 in the x–column. So we use this column as the pivot column. The test ratios are: s: 40/1, t: 10/2, u: 10/1. The smallest test ratio is t: 10/2. Thus, we pivot on the 2 in the t–row.

	x	y	z	w	s	t	u	p	
s	0	1	3	3	2	1	0	0	70
x	2	1	–1	–1	0	–1	0	0	10
*u	0	1	3	[3]	0	1	–2	0	10
p	0	–2	–2	–5	0	–1	0	1	10

$R_1 - R_3$

$3R_2 + R_3$

$3R_4 + 2R_3$

The first starred row above is the u–row, and its largest positive entry is the 3 in the z–column. So we use this column as the pivot column. The test ratios are: s: 70/3, u: 10/3. The smallest test ratio is u: 10/3. Thus, we pivot on the 3 in the u–row. This will take us to Phase 2:

	x	y	z	w	s	t	u	p		
s	0	0	0	0	2	0	2	0	60	
x	6	4	0	0	0	−2	−2	0	40	
z	0	1	3	[3]	0	1	−2	0	10	
p	0	−4	0	−9	0	−1	−4	3	50	$R_4 + 3R_3$

	x	y	z	w	s	t	u	p		
s	0	0	0	0	2	0	[2]	0	60	
x	6	4	0	0	0	−2	−2	0	40	$R_2 + R_1$
w	0	1	3	3	0	1	−2	0	10	$R_3 + R_1$
p	0	−1	9	0	0	2	−10	3	80	$R_4 + 5R_1$

	x	y	z	w	s	t	u	p		
u	0	0	0	0	2	0	2	0	60	
x	6	[4]	0	0	2	−2	0	0	100	
w	0	1	3	3	2	1	0	0	70	$4R_3 - R_2$
p	0	−1	9	0	10	2	0	3	380	$4R_4 + R_2$

	x	y	z	w	s	t	u	p	
u	0	0	0	0	2	0	2	0	60
y	6	4	0	0	2	−2	0	0	100
w	−6	0	12	12	6	6	0	0	180
p	6	0	36	0	42	6	0	12	1620

Optimal Solution: $p = 1620/12 = 135$; $x = 0$, $y = 100/4 = 25$, $z = 0$, $w = 180/12 = 15$.

9. Convert the given problem into a maximization problem:

Maximize $p = -6x - 6y$ subject to $x + 2y \geq 20$, $2x + y \geq 20$, $x \geq 0$, $y \geq 0$.

Introduce slack and surplus variables, rewrite the constraints and objective function in standard form, and set up the first tableau.

$$x + 2y - s = 20$$
$$2x + y - t = 20$$
$$6x + 6y + p = 0$$

	x	y	s	t	p		
*s	1	[2]	–1	0	0	20	
*t	2	1	0	–1	0	20	$2R_2 - R_1$
p	6	6	0	0	1	0	$R_3 - 3R_1$

The first starred row above is the s–row, and its largest positive entry is the 2 in the y–column. So we use this column as the pivot column. The test ratios are: s: 20/2, t: 20/1. The smallest test ratio is s: 20/2. Thus, we pivot on the 2 in the s–row.

	x	y	s	t	p		
y	1	2	–1	0	0	20	$3R_1 - R_2$
*t	[3]	0	1	–2	0	20	
p	3	0	3	0	1	–60	$R_3 - R_2$

The first starred row above is the t–row, and its largest positive entry is the 3 in the x–column. So we use this column as the pivot column. The test ratios are: y: 20/1, t: 20/3. The smallest test ratio is t: 20/3. Thus, we pivot on the 3 in the t–row.

	x	y	s	t	p	
y	0	6	–4	2	0	40
x	3	0	1	–2	0	20
p	0	0	2	2	1	–80

Since there are no more starred rows, we go to Phase 2, and search for the most negative entry in the bottom row (if any). Since there are no negative entries in the bottom row, we are done, and read off the optimal solution: $p = -80$; $x = 20/3$, $y = 40/6 = 20/3$. Since $c = -p$, the minimum value of c is $c = 80$.

11. Convert the given problem into a maximization problem:
Maximize $p = -2x - y - 3z$ subject to $x + y + z \geq 100$, $2x + y \geq 50$, $y + z \geq 50$, $x \geq 0$, $y \geq 0$, $z \geq 0$.
Introduce slack and surplus variables, rewrite the constraints and objective function in standard form, and solve this LP problem:.

$$x + y + z - s = 100$$
$$2x + y = t - 50$$
$$y + z - u = 50$$

	x	y	z	s	t	u	p		
*s	1	1	1	–1	0	0	0	100	$2R_1 - R_2$
*t	[2]	1	0	0	–1	0	0	50	
*u	0	1	1	0	0	–1	0	50	
p	2	1	3	0	0	0	1	0	$R_4 - R_2$

	x	y	z	s	t	u	p		
*s	0	1	2	–2	1	0	0	150	$R_1 - 2R_3$
x	2	1	0	0	–1	0	0	50	
*u	0	1	$\boxed{1}$	0	0	–1	0	50	
p	0	0	3	0	1	0	1	–50	$R_4 - 3R_3$

	x	y	z	s	t	u	p		
*s	0	–1	0	–2	1	$\boxed{2}$	0	50	
x	2	1	0	0	–1	0	0	50	
z	0	1	1	0	0	–1	0	50	$2R_3 + R_1$
p	0	–3	0	0	1	3	1	–200	$2R_4 - 3R_1$

Phase 2:

	x	y	z	s	t	u	p		
u	0	–1	0	–2	1	2	0	50	$R_1 + R_2$
x	2	$\boxed{1}$	0	0	–1	0	0	50	
z	0	1	2	–2	1	0	0	150	$R_3 - R_2$
p	0	–3	0	6	–1	0	2	–550	$R_4 + 3R_2$

	x	y	z	s	t	u	p		
u	2	0	0	–2	0	2	0	100	
y	2	1	0	0	–1	0	0	50	$2R_2 + R_3$
z	–2	0	2	–2	$\boxed{2}$	0	0	100	
p	6	0	0	6	–4	0	2	–400	$R_4 + 2R_3$

	x	y	z	s	t	u	p		
u	2	0	0	–2	0	2	0	100	
y	2	2	2	–2	0	0	0	200	
t	–2	0	2	–2	2	0	0	100	
p	2	0	4	2	0	0	2	–200	

Optimal Solution: $c = -p = 200/2 = 100$; $x = 0$, $y = 200/2 = 100$, $z = 0$.

13. Convert the given problem into a maximization problem: Maximize $p = -50x - 50y - 11z$ subject to $2x + z \geq 3$, $2x + y - z \geq 2$, $3x + y - z \leq 3$, $x \geq 0$, $y \geq 0$, $z \geq 0$

	x	y	z	s	t	u	p		
*s	2	0	1	-1	0	0	0	3	$3R_1 - 2R_3$
*t	2	1	-1	0	-1	0	0	2	$3R_2 - 2R_3$
u	[3]	1	-1	0	0	1	0	3	
p	50	50	11	0	0	0	1	0	$3R_4 - 50R_3$

	x	y	z	s	t	u	p		
*s	0	-2	[5]	-3	0	-2	0	3	
*t	0	1	-1	0	-3	-2	0	0	$5R_2 + R_1$
x	3	1	-1	0	0	1	0	3	$5R_3 + R_1$
p	0	100	83	0	0	-50	3	-150	$5R_4 - 83R_1$

	x	y	z	s	t	u	p		
z	0	-2	5	-3	0	-2	0	3	$3R_1 + 2R_2$
*t	0	[3]	0	-3	-15	-12	0	3	
x	15	3	0	-3	0	3	0	18	$R_3 - R_2$
p	0	666	0	249	0	-84	15	-999	$R_4 - 222R_2$

	x	y	z	s	t	u	p	
z	0	0	15	-15	-30	-30	0	15
y	0	3	0	-3	-15	-12	0	3
x	15	0	0	0	15	15	0	15
p	0	0	0	915	3330	2580	15	-1665

Optimal Solution: $c = -p = 1665/15 = 111$; $x = 15/15 = 1$, $y = 3/3 = 1$, $z = 15/15 = 1$.

15. Convert the given problem into a maximization problem: Maximize $p = -x - y - z - w$ subject to $5x - y + w \geq 1000$, $z + w \leq 2000$, $x + y \leq 500$, $x \geq 0$, $y \geq 0$, $z \geq 0$, $w \geq 0$.

	x	y	z	w	s	t	u	p		
*s	[5]	-1	0	1	-1	0	0	0	1000	
t	0	0	1	1	0	1	0	0	2000	
u	1	1	0	0	0	0	1	0	500	$5R_3 - R_1$
p	1	1	1	1	0	0	0	1	0	$5R_4 - R_1$

	x	y	z	w	s	t	u	p	
x	5	−1	0	1	−1	0	0	0	1000
t	0	0	1	1	0	1	0	0	2000
u	0	6	0	−1	1	0	5	0	1500
p	0	6	5	4	1	0	0	5	−1000

Optimal Solution: $p = -c = 1000/5 = 200$; $x = 1000/5 = 200$, $y = 0$, $z = 0$, $w = 0$.

Note: *In Exercises 17–22, we used the online Simplex Method Tool. When entering the problems, there is no need to enter the inequalities* x ≥ 0, y ≥ 0, *etc.*

17. Maximize $p = 2x + 3y + 1.1z + 4w$ subject to
$1.2x + y + z + w \le 40.5$
$2.2x + y - z - w \ge 10$
$1.2x + y + z + 1.2w \ge 10.5$
$x \ge 0, y \ge 0, z \ge 0, w \ge 0.$

```
Tableau #1
x        y        z        w        s1       s2       s3       p
1.2      1        1        1        1        0        0        0        40.5
2.2      1        -1       -1       0        -1       0        0        10
1.2      1        1        1.2      0        0        -1       0        10.5
-2       -3       -1.1     -4       0        0        0        1        0

Tableau #2
x        y        z        w        s1       s2       s3       p
0        0.45     1.55     1.55     1        0.55     0        0        35.05
1        0.45     -0.45    -0.45    0        -0.45    0        0        4.55
0        0.45     1.55     1.75     0        0.55     -1       0        5.05
0        -2.09    -2.01    -4.91    0        -0.91    0        1        9.09

Tableau #3
x        y        z        w        s1       s2       s3       p
0        0.05     0.18     0        1        0.06     0.89     0        30.58
1        0.57     -0.05    0        0        -0.31    -0.26    0        5.86
0        0.26     0.89     1        0        0.31     -0.57    0        2.89
0        -0.81    2.34     0        0        0.63     -2.81    1        23.28

Tableau #4
x        y        z        w        s1       s2       s3       p
0        0.06     0.2      0        1.13     0.07     1        0        34.54
1        0.59     0        0        0.29     -0.29    0        0        14.85
0        0.29     1        1        0.65     0.35     0        0        22.68
0        -0.65    2.9      0        3.18     0.82     0        1        120.41
```

Tableau #5

x	y	z	w	s1	s2	s3	p	
-0.1	0	0.2	0	1.1	0.1	1	0	33.05
1.7	1	0	0	0.5	-0.5	0	0	25.25
-0.5	0	1	1	0.5	0.5	0	0	15.25
1.1	0	2.9	0	3.5	0.5	0	1	136.75

Optimal Solution: $p = 136.75$; $x = 0$, $y = 25.25$, $z = 0$, $w = 15.25$.

19. Minimize $c = 2.2x + y + 3.3z$ subject to

$x + 1.5y + 1.2z \geq 100$
$2x + 1.5y \geq 50$
$1.5y + 1.1z \geq 50$
$x \geq 0, y \geq 0, z \geq 0$.

Tableau #1

x	y	z	s1	s2	s3	-c	
1	1.5	1.2	-1	0	0	0	100
2	1.5	0	0	-1	0	0	50
0	1.5	1.1	0	0	-1	0	50
2.2	1	3.3	0	0	0	1	0

Tableau #2

x	y	z	s1	s2	s3	-c	
1	0	0.1	-1	0	1	0	50
2	0	-1.1	0	-1	1	0	0
0	1	0.73	0	0	-0.67	0	33.33
2.2	0	2.57	0	0	0.67	1	-33.33

Tableau #3

x	y	z	s1	s2	s3	-c	
0	0	0.65	-1	0.5	0.5	0	50
1	0	-0.55	0	-0.5	0.5	0	0
0	1	0.73	0	0	-0.67	0	33.33
0	0	3.78	0	1.1	-0.43	1	-33.33

Tableau #4

x	y	z	s1	s2	s3	-c	
0	-0.89	0	-1	0.5	1.09	0	20.45
1	0.75	0	0	-0.5	0	0	25
0	1.36	1	0	0	-0.91	0	45.45
0	-5.15	0	0	1.1	3	1	-205

```
Tableau #5
x         y         z         s1        s2        s3        -c
0         -0.81     0         -0.92     0.46      1         0         18.75
1         0.75      0         0         -0.5      0         0         25
0         0.63      1         -0.83     0.42      0         0         62.5
0         -2.71     0         2.75      -0.27     0         1         -261.25

Tableau #6
x         y         z         s1        s2        s3        -c
1.08      0         0         -0.92     -0.08     1         0         45.83
1.33      1         0         0         -0.67     0         0         33.33
-0.83     0         1         -0.83     0.83      0         0         41.67
3.62      0         0         2.75      -2.08     0         1         -170.83

Tableau #7
x         y         z         s1        s2        s3        -c
1         0         0.1       -1        0         1         0         50
0.67      1         0.8       -0.67     0         0         0         66.67
-1        0         1.2       -1        1         0         0         50
1.53      0         2.5       0.67      0         0         1         -66.67
```

Optimal Solution: $c = 66.67$; $x = 0$, $y = 66.67$, $z = 0$.

21. Minimize $c = 1.1x + y + 1.5z - w$ subject to

$$5.12x - y + w \le 1000$$
$$z + w \ge 2000$$
$$1.22x + y \le 500$$
$$x \ge 0, y \ge 0, z \ge 0, w \ge 0.$$

```
Tableau #1
x         y         z         w         s1        s2        s3        -c
5.12      -1        0         1         1         0         0         0         1000
0         0         1         1         0         -1        0         0         2000
1.22      1         0         0         0         0         1         0         500
1.1       1         1.5       -1        0         0         0         1         0

Tableau #2
x         y         z         w         s1        s2        s3        -c
5.12      -1        0         1         1         0         0         0         1000
0         0         1         1         0         -1        0         0         2000
1.22      1         0         0         0         0         1         0         500
1.1       1         0         -2.5      0         1.5       0         1         -3000
```

Tableau #3

x	y	z	w	s1	s2	s3	-c	
5.12	-1	0	1	1	0	0	0	1000
-5.12	1	1	0	-1	-1	0	0	1000
1.22	1	0	0	0	0	1	0	500
13.9	-1.5	0	0	2.5	1.5	0	1	-500

Tableau #4

x	y	z	w	s1	s2	s3	-c	
6.34	0	0	1	1	0	1	0	1500
-6.34	0	1	0	-1	-1	-1	0	500
1.22	1	0	0	0	0	1	0	500
15.73	0	0	0	2.5	1.5	1.5	1	250

Optimal Solution: $c = -250$; $x = 0$, $y = 500$, $z = 500$, $w = 1500$.

23. Unknowns:

x = # acres of tomatoes

y = # acres of lettuce

z = # acres of carrots

Maximize $p = 20x + 15y + 5z$ (in hundreds of dollars) subject to

$x + y + z \le 100$

$5x + 4y + 2z \ge 400$

$4x + 2y + 2z \le 500$

$x \ge 0$, $y \ge 0$, $z \ge 0$.

	x	y	z	s	t	u	p		
s	1	1	1	1	0	0	0	100	$5R_1 - R_2$
*t	[5]	4	2	0	-1	0	0	400	
u	4	2	2	0	0	1	0	500	$5R_3 - 4R_2$
p	-20	-15	-5	0	0	0	1	0	$R_4 + 4R_2$

	x	y	z	s	t	u	p		
s	0	1	3	5	[1]	0	0	100	
x	5	4	2	0	-1	0	0	400	$R_2 + R_1$
u	0	-6	2	0	4	5	0	900	$R_3 - 4R_1$
p	0	1	3	0	-4	0	1	1600	$R_4 + 4R_1$

	x	y	z	s	t	u	p	
t	0	1	3	5	1	0	0	100
x	5	5	5	5	0	0	0	500
u	0	–10	–10	–20	0	5	0	500
p	0	5	15	20	0	0	1	2000

Optimal Solution: $p = 2000/1 = 2000$; $x = 500/5 = 100$, $y = 0$, $z = 0$
Plant 100 acres of tomatoes and no other crops. This will give you a profit of $200,000. (You will be using all 100 acres of your farm.)

25. Unknowns: x = # mailings to the East Coast, y = # mailings to the Midwest, z = # mailings to the West Coast
Minimize $c = 40x + 60y + 50z$ subject to
$$100x + 100y + 50z \geq 1500$$
$$50x + 100y + 100z \geq 1500$$
$$x \geq 0, y \geq 0, z \geq 0.$$
Associated maximization problem: Maximize $p = -40x - 60y - 50z$ subject to $100x + 100y + 50z \geq 1500$, $50x + 100y + 100z \geq 1500$, $x \geq 0, y \geq 0, z \geq 0$.

	x	y	z	s	t	p		
*s	100	100	50	–1	0	0	1500	
*t	50	100	100	0	–1	0	1500	$2R_2 - R_1$
p	40	60	50	0	0	1	0	$5R_3 - 2R_1$

	x	y	z	s	t	p		
x	100	100	50	–1	0	0	1500	$3R_1 - R_2$
*t	0	100	150	1	–2	0	1500	
p	0	100	150	2	0	5	–3000	$R_3 - R_2$

	x	y	z	s	t	p	
x	300	200	0	–4	2	0	3000
z	0	100	150	1	–2	0	1500
p	0	0	0	1	2	5	–4500

Optimal Solution: $p = -c = 4500/5 = 900$; $x = 3000/300 = 10$, $y = 0$, $z = 1500/150 = 10$. Send 10 mailings to the East coast, none to the Midwest, 10 to the West Coast. Cost: $900. Another Solution resulting in the same cost (pivot on the y-column in the last tableau above) is no mailings to the East coast, 15 to the Midwest, none to the West Coast.

27. Unknowns: x = # quarts orange juice, y = # quarts orange concentrate
Minimize $c = 0.5x + 2y$ subject to
$$x \geq 10,000$$
$$y \geq 1000$$

$$10x + 50y \geq 200,000$$
$$x \geq 0, y \geq 0.$$

Convert the given problem into a maximization problem: Maximize $p = -0.5x - 2y$ subject to $x \geq 10,000$, $y \geq 1000$, $10x + 50y \geq 200,000$, $x \geq 0$, $y \geq 0$.

	x	y	s	t	u	p	
*s	[1]	0	−1	0	0	0	10,000
*t	0	1	0	−1	0	0	1000
*u	10	50	0	0	−1	0	200,000
p	1	4	0	0	0	2	0

$R_3 - 10R_1$
$R_4 - R_1$

(Note that we cleared fractions in the last row of the first tableau before starting.)

	x	y	s	t	u	p	
x	1	0	−1	0	0	0	10,000
*t	0	[1]	0	−1	0	0	1000
*u	0	50	10	0	−1	0	100,000
p	0	4	1	0	0	2	−10,000

$R_3 - 50R_2$
$R_4 - 4R_2$

	x	y	s	t	u	p	
x	1	0	−1	0	0	0	10,000
y	0	1	0	−1	0	0	1000
*u	0	0	10	[50]	−1	0	50,000
p	0	0	1	4	0	2	−14,000

$50R_2 + R_3$

$50R_4 - 4R_3$

	x	y	s	t	u	p	
x	1	0	−1	0	0	0	10,000
y	0	50	10	0	−1	0	100,000
t	0	0	10	50	−1	0	50,000
p	0	0	10	0	4	100	−900,000

Optimal Solution: $c = -p = 900,000/100 = 9000$; $x = 10,000/1 = 10,000$, $y = 100,000/50 = 2000$.
Succulent Citrus should produce 10,000 quarts of orange juice and 2000 quarts of orange concentrate.

29. Unknowns: $x = \#$ rock music CDs, $y = \#$ rap CDs, and $z = \#$ classical CDs.
Maximize $p = 12x + 15y + 12z$ subject to $x - 2y \geq 0$, $x + y + z \leq 20000$, $z \geq 5000$, $x \geq 0$, $y \geq 0$, $z \geq 0$.

	x	y	z	s	t	u	p		
*s	$\boxed{1}$	-2	0	-1	0	0	0	0	
t	1	1	1	0	1	0	0	$20{,}000$	$R_2 - R_1$
*u	0	0	1	0	0	-1	0	5000	
p	-12	-15	-12	0	0	0	1	0	$R_4 + 12R_1$

	x	y	z	s	t	u	p		
x	1	-2	0	-1	0	0	0	0	
t	0	3	1	1	1	0	0	$20{,}000$	$R_2 - R_3$
*u	0	0	$\boxed{1}$	0	0	-1	0	5000	
p	0	-39	-12	-12	0	0	1	0	$R_4 + 12R_3$

	x	y	z	s	t	u	p		
x	1	-2	0	-1	0	0	0	0	$3R_1 + 2R_2$
t	0	$\boxed{3}$	0	1	1	1	0	$15{,}000$	
z	0	0	1	0	0	-1	0	5000	
p	0	-39	0	-12	0	-12	1	$60{,}000$	$R_4 + 13R_2$

	x	y	z	s	t	u	p	
x	3	0	0	-1	2	2	0	$30{,}000$
y	0	3	0	1	1	1	0	$15{,}000$
z	0	0	1	0	0	-1	0	5000
p	0	0	0	1	13	1	1	$255{,}000$

Optimal Solution: $p = 255{,}000$; $x = 30{,}000/3 = 10{,}000$, $y = 15{,}000/3 = 5000$, $z = 5000/1 = 5000$.
Stock 10,000 rock CDs, 5000 rap CDs and 5000 classical CDs for a maximum retail value of \$255,000.

31. Unknowns: $x = $ # servings of cereal, $y = $ # servings of dessert, $z = $ # servings of juice.
Minimize $c = 10x + 53y + 27z$ subject to
  $60x + 80y + 60z \geq 120$, or $3x + 4y + 3z \geq 6$
  $45y + 120z \geq 120$, or $3y + 8z \geq 8$
  $x \geq 0, y \geq 0, z \geq 0$.
Associated maximization problem: Maximize $p = -10x - 53y - 27z$ subject to $3x + 4y + 3z \geq 6$, $3y + 8z \geq 8$, $x \geq 0, y \geq 0, z \geq 0$.

	x	y	z	s	t	p		
*s	3	$\boxed{4}$	3	-1	0	0	6	
*t	0	3	8	0	-1	0	8	$4R_2 - 3R_1$
p	10	53	27	0	0	1	0	$4R_3 - 53R_1$

	x	y	z	s	t	p		
y	3	4	3	−1	0	0	6	$23R_1 - 3R_2$
*t	−9	0	[23]	3	−4	0	14	
p	−119	0	−51	53	0	4	−318	$23R_3 + 51R_2$

	x	y	z	s	t	p		
y	[96]	92	0	−32	12	0	96	
z	−9	0	23	3	−4	0	14	$96R_2 + 9R_1$
p	−3196	0	0	1372	−204	92	−6600	$96R_3 + 3196R_1$

	x	y	z	s	t	p	
x	96	92	0	−32	12	0	96
z	0	828	2208	0	−276	0	2208
p	0	294032	0	29440	18768	8832	−326784

Optimal Solution: $c = -p = 326784/8832 = 37$; $x = 96/96 = 1$, $y = 0$, $z = 2208/2208 = 1$. Provide one serving of cereal, one serving of juice, and no dessert!

33. Unknowns: x = # bundles from Nadir, y = # bundles from Blunt, z = # bundles from Sonny.
Minimize $c = 3x + 4y + 5z$ (in thousands of dollars) subject to $5x + 10y + 15z \geq 150$, $10x + 10y + 10z \geq 200$, $15x + 10y + 10z \geq 150$, $x \geq 0$, $y \geq 0$, $z \geq 0$.
Associated maximization problem: Maximize $p = -3x - 4y - 5z$ subject to $5x + 10y + 15z \geq 150$, $10x + 10y + 10z \geq 200$, $15x + 10y + 10z \geq 150$, $x \geq 0$, $y \geq 0$, $z \geq 0$.

	x	y	z	s	t	u	p		
*s	5	10	[15]	−1	0	0	0	150	
*t	10	10	10	0	−1	0	0	200	$3R_2 - 2R_1$
*u	15	10	10	0	0	−1	0	150	$3R_3 - 2R_1$
p	3	4	5	0	0	0	1	0	$3R_4 - R_1$

	x	y	z	s	t	u	p		
z	5	10	15	−1	0	0	0	150	$7R_1 - R_3$
*t	20	10	0	2	−3	0	0	300	$3.5R_2 - 2R_3$
*u	[35]	10	0	2	0	−3	0	150	
p	4	2	0	1	0	0	3	−150	$35R_4 - 4R_3$

	x	y	z	s	t	u	p		
z	0	60	105	-9	0	3	0	900	$R_1 - 6R_3$
*t	0	30	0	6	-21	12	0	1500	$2R_2 - 3R_3$
x	35	[10]	0	2	0	-3	0	150	
p	0	30	0	27	0	12	105	-5850	$R_4 - 3R_3$

	x	y	z	s	t	u	p		
z	-210	0	105	-21	0	[21]	0	0	
*t	-105	0	0	0	-21	21	0	1050	$R_2 - R_1$
y	35	10	0	2	0	-3	0	150	$7R_3 + R_1$
p	-105	0	0	21	0	21	105	-6300	$R_4 - R_1$

	x	y	z	s	t	u	p		
u	-210	0	105	-21	0	21	0	0	$R_1 + 2R_2$
*t	[105]	0	-105	21	-21	0	0	1050	
y	35	70	105	-7	0	0	0	1050	$3R_3 - R_2$
p	105	0	-105	42	0	0	105	-6300	$R_4 - R_2$

	x	y	z	s	t	u	p	
u	0	0	-105	21	-42	21	0	2100
x	105	0	-105	21	-21	0	0	1050
y	0	210	420	-42	21	0	0	2100
p	0	0	0	21	21	0	105	-7350

Optimal Solution: $p = -c = 7350/105 = 70$; $x = 1050/105 = 10$, $y = 2100/210 = 10$, $z = 0$. Buy 10 bundles from Nadir, 10 from Blunt, and none from Sony. Cost: $70,000. Another solution resulting in the same cost (pivot on the z-column in the last tableau) is 15 bundles from Nadir, none from Blunt, and 5 from Sonny.

Note: In the Exercises marked for technology, we used the online Simplex Method Tool. When entering the problems, there is no need to enter the inequalities x ≥ 0, y ≥ 0, *etc.*

35. Unknowns: x = # servings of Cell–Tech, y = # servings of Riboforce HP, z = # servings of Creatine Transport.
Minimize $c = 2.20x + 1.60y + 0.60z$ subject to $10x + 5y + 5z \geq 80$, $2x + y + z \geq 10$, $75x + 15y + 35z \leq 750$, $200x + 100z \leq 1000$, $x \geq 0$, $y \geq 0$, $z \geq 0$.
Associated maximization problem: Maximize $p = -2.20x - 1.60y - 0.60z$ subject to $10x + 5y + 5z \geq 80$, $2x + y + z \geq 10$, $75x + 15y + 35z < 750$, $200x + 100z \leq 1000$, $x \geq 0$, $y \geq 0$, $z \geq 0$.

Tableau #1

x	y	z	s1	s2	s3	s4	p	
10	5	5	-1	0	0	0	0	80
2	1	1	0	-1	0	0	0	10
75	15	35	0	0	1	0	0	750
200	0	100	0	0	0	1	0	1000
2.2	1.6	0.6	0	0	0	0	1	0

Tableau #2

x	y	z	s1	s2	s3	s4	p	
0	5	0	-1	0	0	-0.05	0	30
0	1	0	0	-1	0	-0.01	0	0
0	15	-2.5	0	0	1	-0.37	0	375
1	0	0.5	0	0	0	0.01	0	5
0	1.6	-0.5	0	0	0	-0.01	1	-11

Tableau #3

x	y	z	s1	s2	s3	s4	p	
0	0	0	-1	5	0	0	0	30
0	1	0	0	-1	0	-0.01	0	0
0	0	-2.5	0	15	1	-0.22	0	375
1	0	0.5	0	0	0	0.01	0	5
0	0	-0.5	0	1.6	0	0.01	1	-11

Tableau #4

x	y	z	s1	s2	s3	s4	p	
0	0	0	-0.2	1	0	0	0	6
0	1	0	-0.2	0	0	-0.01	0	6
0	0	-2.5	3	0	1	-0.22	0	285
1	0	0.5	0	0	0	0.01	0	5
0	0	-0.5	0.32	0	0	0.01	1	-20.6

Tableau #5

x	y	z	s1	s2	s3	s4	p	
0	0	0	-0.2	1	0	0	0	6
0	1	0	-0.2	0	0	-0.01	0	6
5	0	0	3	0	1	-0.2	0	310
2	0	1	0	0	0	0.01	0	10
1	0	0	0.32	0	0	0.01	1	-15.6

Optimal Solution: $c = p = 15.6$; $x = 0$, $y = 6$, $z = 10$. Mix 6 servings of Riboforce HP and 10 servings of Creatine Transport for a cost of \$15.60.

37. Unknowns: x = # convention–style hotels, y = # vacation–style hotels, and z = # motels.
(a) Minimize $C = 100x + 20y + 4z$ subject to $500x + 200y + 50z \geq 1400$, $x \geq 1$, $z \leq 2$, $x \geq 0$, $y \geq 0$, $z \geq 0$.

Associated maximization problem: Maximize $p = -100x - 20y - 4z$ subject to $500x + 200y + 50z \geq 1400$,
$x \geq 1, z \leq 2, x \geq 0, y \geq 0, z \geq 0$.

	x	y	z	s	t	u	p		
*s	500	200	50	-1	0	0	0	1400	$R_1 - 500R_2$
*t	1	0	0	0	-1	0	0	1	
u	0	0	1	0	0	1	0	2	
p	100	20	4	0	0	0	1	0	$R_4 - 100R_2$

	x	y	z	s	t	u	p		
*s	0	200	50	-1	500	0	0	900	
x	1	0	0	0	-1	0	0	1	$500R_2 + R_1$
u	0	0	1	0	0	1	0	2	
p	0	20	4	0	100	0	1	-100	$5R_4 - R_1$

	x	y	z	s	t	u	p		
t	0	200	50	-1	500	0	0	900	
x	500	200	50	-1	0	0	0	1400	$R_2 - R_1$
u	0	0	1	0	0	1	0	2	
p	0	-100	-30	1	0	0	5	-1400	$2R_4 + R_1$

	x	y	z	s	t	u	p		
y	0	200	50	-1	500	0	0	900	$R_1 - 50R_3$
x	500	0	0	0	-500	0	0	500	
u	0	0	1	0	0	1	0	2	
p	0	0	-10	1	500	0	10	-1900	$R_4 + 10R_3$

	x	y	z	s	t	u	p	
y	0	200	0	-1	500	-50	0	800
x	500	0	0	0	-500	0	0	500
z	0	0	1	0	0	1	0	2
p	0	0	0	1	500	10	10	-1880

Optimal Solution: $c = -p = 1880/10 = 188$; $x = 500/500 = 1$, $y = 800/200 = 4$, $z = 2/1 = 2$. Build 1 convention–style hotel, 4 vacation style hotels and 2 small motels. The total cost will amount to $188 million.

(b) Since 20% of the \$188 million cost is \$37.6 million, you will still be covered by the \$50 million subsidy.

39. x = # boards sent from Tucson to Honolulu, y = # boards sent from Tucson to Venice Beach, z = # boards sent from Toronto to Honolulu, w = # boards sent from Toronto to Venice Beach.

Minimize $c = 10x + 5y + 20z + 10w$ subject to

$x + y \le 620$

$z + w \le 410$

$x + z \ge 500$

$y + w \ge 530$

$x \ge 0, y \ge 0, z \ge 0, w \ge 0.$

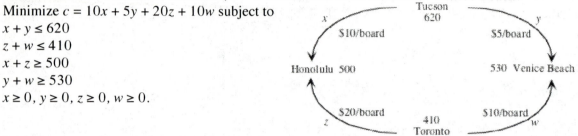

Convert to a minimization problem and solve:

	x	y	z	w	s	t	u	v	p		
s	1	1	0	0	1	0	0	0	0	620	$R_1 - R_3$
t	0	0	1	1	0	1	0	0	0	410	
$*u$	[1]	0	1	0	0	0	-1	0	0	500	
$*v$	0	1	0	1	0	0	0	-1	0	530	
p	10	5	20	10	0	0	0	0	1	0	$R_5 - 10R_3$

	x	y	z	w	s	t	u	v	p		
s	0	[1]	-1	0	1	0	1	0	0	120	
t	0	0	1	1	0	1	0	0	0	410	
x	1	0	1	0	0	0	-1	0	0	500	
$*v$	0	1	0	1	0	0	0	-1	0	530	$R_4 - R_1$
p	0	5	10	10	0	0	10	0	1	-5000	$R_5 - 5R_1$

	x	y	z	w	s	t	u	v	p		
y	0	1	-1	0	1	0	1	0	0	120	$R_1 + R_4$
t	0	0	1	1	0	1	0	0	0	410	$R_2 - R_4$
x	1	0	1	0	0	0	-1	0	0	500	$R_3 - R_4$
$*v$	0	0	[1]	1	-1	0	-1	-1	0	410	
p	0	0	15	10	-5	0	5	0	1	-5600	$R_5 - 15R_4$

	x	y	z	w	s	t	u	v	p		
y	0	1	0	1	0	0	0	–1	0	530	$R_1 - R_4$
t	0	0	0	0	1	1	1	1	0	0	
x	1	0	0	–1	1	0	0	1	0	90	$R_3 + R_4$
z	0	0	1	☐1	–1	0	–1	–1	0	410	
p	0	0	0	–5	10	0	20	15	1	–11750	$R_5 + 5R_4$

	x	y	z	w	s	t	u	v	p	
y	0	1	–1	0	1	0	1	0	0	120
t	0	0	0	0	1	1	1	1	0	0
x	1	0	1	0	0	0	–1	0	0	500
w	0	0	1	1	–1	0	–1	–1	0	410
p	0	0	5	0	5	0	15	10	1	–9700

Optimal Solution: $c = -p = 9700/1 = 9700$; $x = 500/1 = 500$, $y = 120/1 = 120$, $z = 0$, $w = 410/1 = 410$.
Make the following shipments: Tucson to Honolulu: 500 boards per week; Tucson to Venice Beach: 120
boards per week; Toronto to Honolulu: 0 boards per week; Toronto to Venice Beach: 410 boards per
week. Minimum weekly cost is \$9,700.

41. Unknowns: x = amount overdrawn from Congressional Integrity Bank, y = amount from Citizens'
Trust, z = amount from Checks R Us.

Maximize $p = x$ subject to $x + y + z \leq 10{,}000$, $x + y \leq \frac{1}{4}(x + y + z)$, or $3x + 3y - z \leq 0$, $x \geq 2500$, $x \geq 0$, $y \geq 0$, $z \geq 0$.

	x	y	z	s	t	u	p		
s	1	1	1	1	0	0	0	10,000	$3R_1 - R_2$
t	☐3	3	–1	0	1	0	0	0	
*u	1	0	0	0	0	–1	0	2500	$3R_3 - R_2$
p	–1	0	0	0	0	0	1	0	$3R_4 + R_2$

	x	y	z	s	t	u	p		
s	0	0	4	3	–1	0	0	30,000	$R_1 - 4R_3$
x	3	3	–1	0	1	0	0	0	$R_2 + R_3$
*u	0	–3	☐1	0	–1	–3	0	7500	
p	0	3	–1	0	1	0	3	0	$R_4 + R_3$

	x	y	z	s	t	u	p		
s	0	12	0	3	3	[12]	0	0	
x	3	0	0	0	0	-3	0	7500	$4R_2 + R_1$
z	0	-3	1	0	-1	-3	0	7500	$4R_3 + R_1$
p	0	0	0	0	0	-3	3	7500	$4R_4 + R_1$

	x	y	z	s	t	u	p	
u	0	12	0	3	3	12	0	0
x	12	12	0	3	3	0	0	30,000
z	0	0	4	3	-1	0	0	30,000
p	0	12	0	3	3	0	12	30,000

Optimal Solution: $p = 30{,}000/12 = 2500$; $x = 30{,}000/12 = 2500$, $y = 0$, $z = 30{,}000/4 = 7500$. Withdraw $2500 from Congressional Integrity Bank, $0 from Citizens' Trust, $7500 from Checks R Us.

43. Unknowns: x = # people you fly from Chicago to Los Angeles, y = # people you fly from Chicago to New York, z = # people you fly from Denver to Los Angeles, w = # people you fly from Denver to New York

Minimize $c = 195x + 182y + 395z + 166w$ subject to $x + y \le 20$, $z + w \le 10$, $x + z \ge 10$, $y + w \ge 15$ $x \ge 0$, $y \ge 0$, $z \ge 0$, $w \ge 0$.

Tableau #1

x	y	z	w	s1	s2	s3	s4	-c	
1	1	0	0	1	0	0	0	0	20
0	0	1	1	0	1	0	0	0	10
1	0	1	0	0	0	-1	0	0	10
0	1	0	1	0	0	0	-1	0	15
195	182	395	166	0	0	0	0	1	0

Tableau #2

x	y	z	w	s1	s2	s3	s4	-c	
0	1	-1	0	1	0	1	0	0	10
0	0	1	1	0	1	0	0	0	10
1	0	1	0	0	0	-1	0	0	10
0	1	0	1	0	0	0	-1	0	15
0	182	200	166	0	0	195	0	1	-1950

Tableau #3

x	y	z	w	s1	s2	s3	s4	-c	
0	1	-1	0	1	0	1	0	0	10
0	0	1	1	0	1	0	0	0	10
1	0	1	0	0	0	-1	0	0	10
0	0	1	1	-1	0	-1	-1	0	5
0	0	382	166	-182	0	13	0	1	-3770

Tableau #4

x	y	z	w	s1	s2	s3	s4	-c	
0	1	0	1	0	0	0	-1	0	15
0	0	0	0	1	1	1	1	0	5
1	0	0	-1	1	0	0	1	0	5
0	0	1	1	-1	0	-1	-1	0	5
0	0	0	-216	200	0	395	382	1	-5680

Tableau #5

x	y	z	w	s1	s2	s3	s4	-c	
0	1	-1	0	1	0	1	0	0	10
0	0	0	0	1	1	1	1	0	5
1	0	1	0	0	0	-1	0	0	10
0	0	1	1	-1	0	-1	-1	0	5
0	0	216	0	-16	0	179	166	1	-4600

Tableau #6

x	y	z	w	s1	s2	s3	s4	-c	
0	1	-1	0	0	-1	0	-1	0	5
0	0	0	0	1	1	1	1	0	5
1	0	1	0	0	0	-1	0	0	10
0	0	1	1	0	1	0	0	0	10
0	0	216	0	0	16	195	182	1	-4520

Optimal Solution: $c = 4520$; $x = 10$, $y = 5$, $z = 0$, $w = 10$. Fly 10 people from Chicago to LA, 5 from Chicago to New York, none from Denver to LA, 10 from Denver to NY at a total cost of $4520.

45. Unknowns: x = # cardiologists hired, y = # rehabilitation specialists hired, z = # infectious disease specialists hired.

Maximize $r = 12x + 19y + 14z$ ($10,000) subject to $x + y + 3z \geq 27$, $10x + 10y + 10z \leq 200 - 30 = 170$, $x \geq 0$,

$y \geq 0$, $z \geq 0$.

Tableau #1

x	y	z	s1	s2	r	
1	1	3	-1	0	0	27
1	1	1	0	1	0	17
-12	-19	-14	0	0	1	0

Tableau #2

x	y	z	s1	s2	r	
1/3	1/3	1	-1/3	0	0	9
2/3	2/3	0	1/3	1	0	8
-22/3	-43/3	0	-14/3	0	1	126

Tableau #3

x	y	z	s1	s2	r	
0	0	1	-1/2	-1/2	0	5
1	1	0	1/2	3/2	0	12
7	0	0	5/2	43/2	1	298

Optimal Solution: $r = 298$; $x = 0$, $y = 12$, $z = 5$. Hire no more cardiologists, 12 rehabilitation specialists, and 5 infectious disease specialists.

47. In a general linear programming problem, the solution $x = 0$, $y = 0$, ...) represented by the initial tableau may not be feasible (feasible solutions must satisfy all the constraints)., and this shows up as a basic solution with some negative values for surplus variables. In order for a basic solution to be feasible, none of the basic variables (including surplus and slack variables) can be negative. In phase I we use pivoting to arrive at a basic solution where no basic variables are negative

49. The basic solution corresponding to the initial tableau has all the unknowns equal to zero, and this is not a feasible solution, since it does not satisfy the given inequality.

51. We can rewrite the given problem as a standard LP problem by multiplying both sides of the first constraint by -1: Maximize $p = x + y$ subject to $-x + 2y \le 0$, $2x + y \le 10$, $x \ge 0$, $y \ge 0$. Therefore, this problem can be solved using the techniques of either section (choice (C)).

53. Examples are Exercises 1 and 2.

55. A simple example is: Maximize $p = x + y$ subject to $x + y \le 10$, $x + y \ge 20$, $x \ge 0$, $y \ge 0$.

	x	y	s	t	p	
s	1	[1]	1	0	0	10
*t	1	1	0	-1	0	20
p	-1	-1	0	0	1	0

$R_2 - R_1$
$R_3 + R_1$

	x	y	s	t	p	
x	1	1	1	0	0	10
*t	0	0	-1	-1	0	10
p	0	0	1	0	1	10

We now find it impossible to find a feasible solution, since there are no positive entries in the starred row. This problem has an empty feasible region, and hence no optimal solution.

4.5 The Simplex Method and Duality

1. Maximize $p = 2x+y$ subject to
$$x + 2y \le 6$$
$$-x + y \le 2$$
$$x \ge 0,\ y \ge 0$$

$$\begin{bmatrix} 1 & 2 & 6 \\ -1 & 1 & 2 \\ 2 & 1 & 0 \end{bmatrix} \to \begin{bmatrix} 1 & -1 & 2 \\ 2 & 1 & 1 \\ 6 & 2 & 0 \end{bmatrix}$$

Minimize $c = 6s+2t$ subject to
$$s - t \ge 2$$
$$2s + t \ge 1$$
$$s \ge 0,\ t \ge 0$$

3. Minimize $c = 2s+t+3u$ subject to
$$s + t + u \ge 100$$
$$2s + t \ge 50$$
$$s \ge 0,\ t \ge 0,\ u \ge 0$$

$$\begin{bmatrix} 1 & 1 & 1 & 100 \\ 2 & 1 & 0 & 50 \\ 2 & 1 & 3 & 0 \end{bmatrix} \to$$
$$\begin{bmatrix} 1 & 2 & 2 \\ 1 & 1 & 1 \\ 1 & 0 & 3 \\ 100 & 50 & 0 \end{bmatrix}$$

Maximize $p = 100x+50y$ subject to
$$x + 2y \le 2$$
$$x + y \le 1$$
$$x \le 3$$
$$x \ge 0,\ y \ge 0$$

5. Maximize $p = x+y+z+w$ subject to
$$x + y + z \le 3$$
$$y + z + w \le 4$$
$$x + z + w \le 5$$
$$x + y + w \le 6$$
$$x \ge 0,\ y \ge 0,\ z \ge 0,\ w \ge 0$$

$$\begin{bmatrix} 1 & 1 & 1 & 0 & 3 \\ 0 & 1 & 1 & 1 & 4 \\ 1 & 0 & 1 & 1 & 5 \\ 1 & 1 & 0 & 1 & 6 \\ 1 & 1 & 1 & 1 & 0 \end{bmatrix}$$
$$\to \begin{bmatrix} 1 & 0 & 1 & 1 & 1 \\ 1 & 1 & 0 & 1 & 1 \\ 1 & 1 & 1 & 0 & 1 \\ 0 & 1 & 1 & 1 & 1 \\ 3 & 4 & 5 & 6 & 0 \end{bmatrix}$$

Minimize $c = 3s+4t+5u+6v$ subject to
$$s + u + v \ge 1$$
$$s + t + v \ge 1$$
$$s + t + u \ge 1$$
$$t + u + v \ge 1$$
$$s \ge 0,\ t \ge 0,\ u \ge 0,\ v \ge 0$$

7. Minimize $c = s+3t+u$ subject to
$$5s - t + v \ge 1000$$
$$u - v \ge 2000$$
$$s + t \ge 500$$
$$s \ge 0,\ t \ge 0,\ u \ge 0,\ v \ge 0$$

$$\begin{bmatrix} 5 & -1 & 0 & 1 & 1000 \\ 0 & 0 & 1 & -1 & 2000 \\ 1 & 1 & 0 & 0 & 500 \\ 1 & 3 & 1 & 0 & 0 \end{bmatrix} \to$$
$$\begin{bmatrix} 5 & 0 & 1 & 1 \\ -1 & 0 & 1 & 3 \\ 0 & 1 & 0 & 1 \\ 1 & -1 & 0 & 0 \\ 1000 & 2000 & 500 & 0 \end{bmatrix}$$

Maximize
$$p = 1000x+2000y+500z$$
subject to
$$5x + z \le 1$$
$$-x + z \le 3$$
$$y \le 1$$
$$x - y \le 0$$
$$x \ge 0,\ y \ge 0,\ z \ge 0$$

9. Minimize $c = s + t$ subject to
$$s + 2t \ge 6$$
$$2s + t \ge 6$$
$$s \ge 0,\ t \ge 0$$

$$\begin{bmatrix} 1 & 2 & 6 \\ 2 & 1 & 6 \\ 1 & 1 & 0 \end{bmatrix} \to \begin{bmatrix} 1 & 2 & 1 \\ 2 & 1 & 1 \\ 6 & 6 & 0 \end{bmatrix}$$

Maximize $p = 6x + 6y$ subject to
$$x + 2y \le 1$$
$$2x + y \le 1$$
$$x \ge 0,\ y \ge 0$$

Solve the dual problem using the standard simplex method:

	x	y	s	t	p	
s	1	2	1	0	0	1

$2R_1 - R_2$

t	$\boxed{2}$	1	0	1	0	1
p	−6	−6	0	0	1	0

$R_3 + 3R_2$

	x	y	s	t	p	
s	0	$\boxed{3}$	2	−1	0	1
x	2	1	0	1	0	1
p	0	−3	0	3	1	3

$3R_2 - R_1$
$R_3 + R_1$

	x	y	s	t	p	
y	0	3	2	−1	0	1
x	6	0	−2	4	0	2
p	0	0	2	2	1	4

The solution to the primal problem is $c = 4/1 = 4$; $s = 2/1 = 2$, $t = 2/1 = 2$.

11. Minimize $c = 6s+6t$
subject to
$$s + 2t \geq 20$$
$$2s + t \geq 20$$
$$s \geq 0, t \geq 0$$

$$\begin{bmatrix} 1 & 2 & 20 \\ 2 & 1 & 20 \\ 6 & 6 & 0 \end{bmatrix} \rightarrow \begin{bmatrix} 1 & 2 & 6 \\ 2 & 1 & 6 \\ 20 & 20 & 0 \end{bmatrix}$$

Maximize $p = 20x+20y$
subject to
$$x + 2y \leq 6$$
$$2x + y \leq 6$$
$$x \geq 0, y \geq 0$$

Solve the dual problem using the standard simplex method:

	x	y	s	t	p	
s	1	2	1	0	0	6
t	$\boxed{2}$	1	0	1	0	6
p	−20	−20	0	0	1	0

$2R_1 - R_2$
$R_3 + 10R_2$

	x	y	s	t	p	
s	0	$\boxed{3}$	2	−1	0	6
x	2	1	0	1	0	6
p	0	−10	0	10	1	60

$3R_2 - R_1$
$3R_3 + 10R_1$

	x	y	s	t	p	
y	0	3	2	−1	0	6
x	6	0	−2	4	0	12
p	0	0	20	20	3	240

The solution to the primal problem is $c = 240/3 = 80$; $s = 20/3$, $t = 20/3$.

13. First, rewrite the given problem with no decimals in the constraints:

Minimize $c = 0.2s + 0.3t$

subject to

$$2s + t \geq 10$$
$$s + 2t \geq 10$$
$$s + t \geq 8$$
$$s \geq 0, t \geq 0$$

$$\begin{bmatrix} 2 & 1 & 10 \\ 1 & 2 & 10 \\ 1 & 1 & 8 \\ 0.2 & 0.3 & 0 \end{bmatrix} \rightarrow$$

$$\begin{bmatrix} 2 & 1 & 1 & 0.2 \\ 1 & 2 & 1 & 0.3 \\ 10 & 10 & 8 & 0 \end{bmatrix}$$

Maximize $p = 10x + 10y + 8z$

subject to

$$2x + y + z \leq 0.2$$
$$x + 2y + z \leq 0.3$$
$$x \geq 0, y \geq 0, z \geq 0$$

Solve the dual problem using the standard simplex method (Note: do not rewrite the constraints without decimals at this stage — clear them in the first step of the simplex method.)

	x	y	z	s	t	p		
s	2	1	1	1	0	0	0.2	$10R_1$
t	1	2	1	0	1	0	0.3	$10R_1$
p	−10	−10	−8	0	0	1	0	

	x	y	z	s	t	p		
s	20	10	10	10	0	0	2	
t	10	20	10	0	10	0	3	$2R_2 - R_1$
p	−10	−10	−8	0	0	1	0	$2R_3 + R_1$

	x	y	z	s	t	p		
x	20	10	10	10	0	0	2	$3R_1 - R_2$
t	0	[30]	10	−10	20	0	4	
p	0	−10	−6	10	0	2	2	$3R_3 + R_2$

	x	y	z	s	t	p		
x	60	0	Erro	40	−20	0	2	
y	0	30	10	−10	20	0	4	$2R_2 - R_1$
p	0	0	−8	20	20	6	10	$5R_3 + 2R_1$

	x	y	z	s	t	p	
z	60	0	20	40	− 20	0	2
y	−60	60	0	−60	60	0	6
p	120	0	0	180	60	30	54

The solution to the primal problem is $c = 54/30 = 1.8$; $s = 180/30 = 6$, $t = 60/30 = 2$.

15. Minimize $c = 2s+t$ subject to

$3s + t \geq 30$
$s + t \geq 20$
$s + 3t \geq 30$
$s \geq 0, t \geq 0$

$$\begin{bmatrix} 3 & 1 & 30 \\ 1 & 1 & 20 \\ 1 & 3 & 30 \\ 2 & 1 & 0 \end{bmatrix} \rightarrow \begin{bmatrix} 3 & 1 & 1 & 2 \\ 1 & 1 & 3 & 1 \\ 30 & 20 & 30 & 0 \end{bmatrix}$$

Maximize $p = 30x+20y+30z$ subject to

$3x + y + z \leq 2$
$x + y + 3z \leq 1$
$x \geq 0, y \geq 0, z \geq 0$

Solve the dual problem using the standard simplex method:

	x	y	z	s	t	p		
s	③	1	1	1	0	0	2	
t	1	1	3	0	1	0	1	$3R_2 - R_1$
p	-30	-20	-30	0	0	1	0	$R_3 + 10R_1$

	x	y	z	s	t	p		
x	3	1	1	1	0	0	2	$8R_1 - R_2$
t	0	2	⑧	-1	3	0	1	
p	0	-10	-20	10	0	1	20	$2R_3 + 5R_2$

	x	y	z	s	t	p		
x	24	6	0	9	-3	0	15	$R_1 - 3R_2$
z	0	②	8	-1	3	0	1	
p	0	-10	0	15	15	2	45	$R_3 + 5R_2$

	x	y	z	s	t	p	
x	24	0	-24	12	-12	0	12
y	0	2	8	-1	3	0	1
p	0	0	40	10	30	2	50

The solution to the primal problem is $c = 50/2 = 25$; $s = 10/2 = 5$, $t = 30/2 = 15$.

17. Minimize $c = s+2t+3u$ subject to

$3s + 2t + u \geq 60$
$2s + t + 3u \geq 60$
$s \geq 0, t \geq 0, u \geq 0$

$$\begin{bmatrix} 3 & 2 & 1 & 60 \\ 2 & 1 & 3 & 60 \\ 1 & 2 & 3 & 0 \end{bmatrix} \rightarrow \begin{bmatrix} 3 & 2 & 1 \\ 2 & 1 & 2 \\ 1 & 3 & 3 \\ 60 & 60 & 0 \end{bmatrix}$$

Maximize $p = 60x+60y$ subject to

$3x + 2y \leq 1$
$2x - y \leq 2$
$x + 3y \leq 3$
$x \geq 0, y \geq 0$

Solve the dual problem using the standard simplex method:

	x	y	s	t	u	p	
s	③	2	1	0	0	0	1
t	2	−1	0	1	0	0	2
u	1	3	0	0	1	0	3
p	−60	−60	0	0	0	1	0

$3R_2 - 2R_1$
$3R_3 - R_1$
$R_4 + 20R_1$

	x	y	s	t	u	p	
x	3	②	1	0	0	0	1
t	0	−7	−2	3	0	0	4
u	0	7	−1	0	3	0	8
p	0	−20	20	0	0	1	20

$2R_2 + 7R_1$
$2R_3 - 7R_1$
$R_4 + 10R_1$

	x	y	s	t	u	p	
y	3	2	1	0	0	0	1
t	21	0	3	6	0	0	15
u	−21	0	−9	0	6	0	9
p	30	0	30	0	0	1	30

The solution to the primal problem is $c = 30/1 = 30$; $s = 30/1 = 30$, $t = 0$, $u = 0$.

19. Minimize $c = 2s+t+3u$
subject to

$s + t + u \geq 100$
$2s + t \geq 50$
$t + u \geq 50$
$s \geq 0, t \geq 0, u \geq 0$

$$\begin{bmatrix} 1 & 1 & 1 & 100 \\ 2 & 1 & 0 & 50 \\ 0 & 1 & 1 & 50 \\ 2 & 1 & 3 & 0 \end{bmatrix}$$

$$\rightarrow \begin{bmatrix} 1 & 2 & 0 & 2 \\ 1 & 1 & 1 & 1 \\ 1 & 0 & 1 & 3 \\ 100 & 50 & 50 & 0 \end{bmatrix}$$

Maximize $p = 100x+50y+50z$
subject to

$x + 2y + z \leq 2$
$x + y + z \leq 1$
$x + z \leq 3$
$x \geq 0, y \geq 0, z \geq 0$

Solve the dual problem using the standard simplex method:

	x	y	z	s	t	u	p	
s	1	2	1	1	0	0	0	2
t	①	1	1	0	1	0	0	1
u	1	0	1	0	0	1	0	3
p	−100	−50	−50	0	0	0	1	0

$R_1 - R_2$

$R_3 - R_2$
$R_4 + 100R_2$

	x	y	z	s	t	u	p	
s	0	1	0	1	−1	0	0	1
x	1	1	1	0	1	0	0	1
u	0	−1	0	0	−1	1	0	2
p	0	50	50	0	100	0	1	100

The solution to the primal problem is $c = 100/1 = 100$; $s = 0$, $t = 100/1 = 100$, $u = 0$.

21. Minimize $c = s+t+u$
subject to
$3s + 2t + u \geq 60$
$2s + t + 3u \geq 60$
$s + 3t + 2u \geq 60$
$s \geq 0, t \geq 0, u \geq 0$

$$\begin{bmatrix} 3 & 2 & 1 & 60 \\ 2 & 1 & 3 & 60 \\ 1 & 3 & 2 & 60 \\ 1 & 1 & 1 & 0 \end{bmatrix} \rightarrow$$

$$\begin{bmatrix} 3 & 2 & 1 & 1 \\ 2 & 1 & 3 & 1 \\ 1 & 3 & 2 & 1 \\ 60 & 60 & 60 & 0 \end{bmatrix}$$

Maximize $p = 60x+60y+60z$
subject to
$3x + 2y + z \leq 1$
$2x + 1y + 3z \leq 1$
$1x + 3y + 2z \leq 1$
$x \geq 0, y \geq 0, z \geq 0$

Solve the dual problem using the standard simplex method:

	x	y	z	s	t	u	p		
s	③	2	1	1	0	0	0	1	
t	2	1	3	0	1	0	0	1	$3R_2 - 2R_1$
u	1	3	2	0	0	1	0	1	$3R_3 - R_1$
p	−60	−60	−60	0	0	0	1	0	$R_4 + 20R_1$

	x	y	z	s	t	u	p		
x	3	2	1	1	0	0	0	1	$7R_1 - R_2$
t	0	−1	⑦	−2	3	0	0	1	
u	0	7	5	−1	0	3	0	2	$7R_3 - 5R_2$
p	0	−20	−40	20	0	0	1	20	$7R_4 + 40R_2$

	x	y	z	s	t	u	p		
x	21	15	0	9	−3	0	0	6	$\frac{1}{3}R_1$
z	0	−1	7	−2	3	0	0	1	
u	0	⑤④	0	3	−15	21	0	9	$\frac{1}{3}R_1$
p	0	−180	0	60	120	0	7	180	

	x	y	z	s	t	u	p		
x	7	5	0	3	−1	0	0	2	$18R_1 - 5R_3$
z	0	−1	7	−2	3	0	0	1	$18R_2 + R_3$

u	0	[18]	0	1	–5	7	0	3
p	0	–180	0	60	120	0	7	180

$R_4 + 10R_3$

	x	y	z	s	t	u	p	
x	126	0	0	49	7	–35	0	21
z	0	0	126	–35	49	7	0	21
y	0	54	0	3	–15	21	0	9
p	0	0	0	70	70	70	7	210

The solution to the primal problem is $c = 210/7 = 30$; $s = 70/7 = 10$, $t = 70/7 = 10$, $u = 70/7 = 10$.

23. Add $k = 2$ to each entry and put 1s to the right and below:

$$\begin{bmatrix} 1 & 3 & 4 & 1 \\ 4 & 1 & 0 & 1 \\ 1 & 1 & 1 & 0 \end{bmatrix}$$

This gives the following LP problem:

Maximize $p = x + y + z$ subject to

$x + 3y + 4z \le 1$

$4x + y \quad\quad \le 1$

$x \ge 0, y \ge 0, z \ge 0$

Here are the tableaux:

	x	y	z	s	t	p	
s	1	3	4	1	0	0	1
t	[4]	1	0	0	1	0	1
p	–1	–1	–1	0	0	1	0

$4R_1 – R_2$

$4R_3 + R_2$

	x	y	z	s	t	p	
s	0	11	[16]	4	–1	0	3
x	4	1	0	0	1	0	1
p	0	–3	–4	0	1	4	1

$4R_3 + R_1$

	x	y	z	s	t	p	
z	0	[11]	16	4	–1	0	3
x	4	1	0	0	1	0	1
p	0	–1	0	4	3	16	7

$11R_2 – R_1$

$11R_3 + R_1$

	x	y	z	s	t	p	
y	0	11	16	4	–1	0	3
x	44	0	–16	–4	12	0	8
p	0	0	16	48	32	176	80

The solution to the primal problem is $p = 80/176 = 5/11$; $x = 8/44 = 2/11$, $y = 3/11$, $z = 0$. So, the column player's optimal strategy is $C = [2/5 \ 3/5 \ 0]^T$ and the value of the game is $e = 11/5 - 2 = 1/5$. The solution to the dual problem is $s = 48/176 = 3/11$, $t = 32/176 = 2/11$, so the row player's optimal strategy is $R = [3/5 \ 2/5]$.

25. Add $k = 2$ to each entry and put 1s to the right and below:

$$\begin{bmatrix} 1 & 3 & 4 & 1 \\ 4 & 1 & 0 & 1 \\ 3 & 4 & 2 & 1 \\ 1 & 1 & 1 & 0 \end{bmatrix}$$

This gives the following LP problem:

Maximize $p = x + y + z$ subject to
$$x + 3y + 4z \le 1$$
$$4x + y \qquad \le 1$$
$$3x + 4y + 2z \le 1$$
$$x \ge 0, \ y \ge 0, \ z \ge 0$$

Here are the tableaux:

	x	y	z	s	t	u	p		
s	1	3	4	1	0	0	0	1	$4R_1 - R_2$
t	[4]	1	0	0	1	0	0	1	
u	3	4	2	0	0	1	0	1	$4R_3 - 3R_2$
p	−1	−1	−1	0	0	0	1	0	$4R_4 + R_2$

	x	y	z	s	t	u	p		
s	0	11	16	4	−1	0	0	3	$R_1 - 2R_3$
x	4	1	0	0	1	0	0	1	
u	0	13	[8]	0	−3	4	0	1	
p	0	−3	−4	0	1	0	4	1	$2R_4 + R_3$

	x	y	z	s	t	u	p		
s	0	−15	0	4	[5]	−8	0	1	
x	4	1	0	0	1	0	0	1	$5R_2 - R_1$
z	0	13	8	0	−3	4	0	1	$5R_3 + 3R_1$
p	0	7	0	0	−1	4	8	3	$5R_4 + R_1$

	x	y	z	s	t	u	p	
t	0	-15	0	4	5	-8	0	1
x	20	20	0	-4	0	8	0	4
z	0	20	40	12	0	-4	0	8
p	0	20	0	4	0	12	40	16

The solution to the primal problem is $p = 16/40 = 2/5$; $x = 4/20 = 1/5$, $y = 0$, $z = 8/40 = 1/5$. So, the column player's optimal strategy is $C = [1/2 \ 0 \ 1/2]^T$ and the value of the game is $e = 5/2 - 2 = 1/2$.

The solution to the dual problem is $s = 4/40 = 1/10$, $t = 0$, $u = 12/40 = 3/10$, so the row player's optimal strategy is $R = [1/4 \ 0 \ 3/4]$.

27. The first row is dominated by the last, so we can remove it. That is as far as we can reduce by dominance. Add $k = 3$ to each entry and put 1s to the right and below:

$$\begin{bmatrix} 5 & 2 & 1 & 0 & 1 \\ 4 & 5 & 3 & 4 & 1 \\ 3 & 5 & 6 & 6 & 1 \\ 1 & 1 & 1 & 1 & 0 \end{bmatrix}$$

This gives the following LP problem:

Maximize $p = x + y + z + w$ subject to
$$5x + 2y + \ \ z \qquad \leq 1$$
$$4x + 5y + 3z + 4w \leq 1$$
$$3x + 5y + 6z + 6w \leq 1$$
$$x \geq 0, \ y \geq 0, \ z \geq 0$$

Here are the tableaux:

	x	y	z	w	s	t	u	p		
s	5	2	1	0	1	0	0	0	1	
t	4	5	3	4	0	1	0	0	1	$5R_2 - 4R_1$
u	3	5	6	6	0	0	1	0	1	$5R_3 - 3R_1$
p	-1	-1	-1	-1	0	0	0	1	0	$5R_4 + R_1$

	x	y	z	w	s	t	u	p		
x	5	2	1	0	1	0	0	0	1	
t	0	17	11	20	-4	5	0	0	1	
u	0	19	27	30	-3	0	5	0	2	$2R_3 - 3R_2$
p	0	-3	-4	-5	1	0	0	5	1	$4R_4 + R_2$

	x	y	z	w	s	t	u	p		
x	5	2	1	0	1	0	0	0	1	$21R_1 - R_3$
w	0	17	11	20	-4	5	0	0	1	$21R_2 - 11R_3$
u	0	-13	$\boxed{21}$	0	6	-15	10	0	1	
p	0	5	-5	0	0	5	0	20	5	$21R_4 + 5R_3$

	x	y	z	w	s	t	u	p	
x	105	55	0	0	15	15	-10	0	20
w	0	500	0	420	-150	270	-110	0	10
z	0	-13	21	0	6	-15	10	0	1
p	0	40	0	0	30	30	50	420	110

The solution to the primal problem is $p = 110/420 = 11/42$; $x = 20/105 = 4/21$, $y = 0$, $z = 1/21$, $w = 10/420 = 1/42$. So, the column player's optimal strategy is $C = [8/11 \ 0 \ 2/11 \ 1/11]^T$ and the value of the game is $e = 42/11 - 3 = 9/11$.

The solution to the dual problem is $s = 30/420 = 1/14$, $t = 30/420 = 1/14$, $u = 50/420 = 5/42$, so the row player's optimal strategy is $R = [0 \ 3/11 \ 3/11 \ 5/11]$. (Note that we put in 0 for the first row that we removed at the beginning.)

29. Unknowns: s = # ounces of fish, t = # ounces of cornmeal

Minimize $c = 5s + 5t$ subject to $8t + 4t \geq 48$, $4s + 8t \geq 48$, $s \geq 0$, $t \geq 0$.

Dualize: $\begin{bmatrix} 8 & 4 & 48 \\ 4 & 8 & 48 \\ 5 & 5 & 0 \end{bmatrix} \rightarrow \begin{bmatrix} 8 & 4 & 5 \\ 4 & 8 & 5 \\ 48 & 48 & 0 \end{bmatrix}$

Dual problem: Maximize $p = 48x + 48y$ subject to $8x + 4y \leq 5$, $4x + 8y \leq 5$, $x \geq 0$, $y \geq 0$.

Solve the dual problem:

	x	y	s	t	p		
s	$\boxed{8}$	4	1	0	0	5	
t	4	8	0	1	0	5	$2R_2 - R_1$
p	-48	-48	0	0	1	0	$R_3 + 6R_1$

	x	y	s	t	p		
x	8	4	1	0	0	5	$3R_1 - R_2$
t	0	$\boxed{12}$	-1	2	0	5	
p	0	-24	6	0	1	30	$R_3 + 2R_2$

	x	y	s	t	p	
x	24	0	4	-2	0	10
y	0	12	-1	2	0	5
p	0	0	4	4	1	40

The solution to the primal problem is $c = 40/1 = 40$, $s = 4/1 = 4$, $t = 4/1 = 4$. Meow should use 4 ounces each of fish and cornmeal, for a total cost of 40¢ per can. The shadow costs are the values of x and y in the final tableau: $x = 10/24 = 5/12$¢ per gram of protein, $y = 5/12$¢ per gram of fat.

31. $s = $ # ounces of chicken, $t = $ # ounces of grain
Minimize $c = 10s + t$ subject to $10s + 2t \geq 200$, $5s + 2t \geq 150$, $s \geq 0$, $t \geq 0$.

Dualize: $\begin{bmatrix} 10 & 2 & 200 \\ 5 & 2 & 150 \\ 10 & 1 & 0 \end{bmatrix} \rightarrow \begin{bmatrix} 10 & 5 & 10 \\ 2 & 2 & 1 \\ 200 & 150 & 0 \end{bmatrix}$

Dual problem: Maximize $p = 200x + 150y$ subject to $10x + 5y \leq 10$, $2x + 2y \leq 1$, $x \geq 0$, $y \geq 0$.
Solve the dual problem:

	x	y	s	t	p		
s	10	5	1	0	0	10	$R_1 - 5R_2$
t	☐2	2	0	1	0	1	
p	−200	−150	0	0	1	0	$R_3 + 100R_2$

	x	y	s	t	p	
s	0	−5	1	−5	0	5
x	2	2	0	1	0	1
p	0	50	0	100	1	100

The solution to the primal problem is $c = 100/1 = 100$; $s = 0$, $t = 100/1 = 100$. Use no chicken and 100 oz of grain for a total cost of \$1. The shadow costs are the values of x and y in the final tableau: $x = 1/2$¢ per g of protein, $y = 0$¢ per gram of fat.

33. Unknowns: $s = $ # servings of cereal, $t = $ # servings of dessert, $u = $ # servings of juice.
Minimize $c = 10s + 53t + 27u$ subject to $60s + 80t + 60u \geq 120$, $45s + 120u \geq 120$, $s \geq 0$, $t \geq 0$, $u \geq 0$.

Dualize: $\begin{bmatrix} 60 & 80 & 60 & 120 \\ 0 & 45 & 120 & 120 \\ 10 & 53 & 27 & 0 \end{bmatrix} \rightarrow \begin{bmatrix} 60 & 0 & 10 \\ 80 & 45 & 53 \\ 60 & 120 & 27 \\ 120 & 120 & 0 \end{bmatrix}$

Dual problem: Maximize $p = 120x + 120y$ subject to $60x \leq 10$, $80x + 45y \leq 53$, $60x + 120y \leq 27$, $x \geq 0$, $y \geq 0$.
Solve the dual problem:

	x	y	s	t	u	p		
s	☐60	0	1	0	0	0	10	
t	80	45	0	1	0	0	53	$12R_2 - 16R_1$
u	60	120	0	0	1	0	27	$R_3 - R_1$
p	−120	−120	0	0	0	1	0	$R_4 + 2R_1$

	x	y	s	t	u	p	
x	60	0	1	0	0	0	10
t	0	540	−16	12	0	0	476
u	0	120	−1	0	1	0	17
p	0	−120	2	0	0	1	20

$2R_2 - 9R_3$

$R_4 + R_3$

	x	y	s	t	u	p	
x	60	0	1	0	0	0	10
t	0	0	−23	24	−9	0	799
y	0	120	−1	0	1	0	17
p	0	0	1	0	1	1	37

The solution to the primal problem is $c = 37/1 = 37$; $s = 1/1 = 1$, $t = 0$, $u = 1$. Prepare one serving of cereal, one serving of juice, and no dessert! for a total cost of 37¢. The shadow costs are the values of x and y in the final tableau: $x = 10/60 = 1/6$¢ per calorie and $y = 17/120$¢ per % U.S. RDA of Vitamin C.

35. Unknowns: s = # mailings to the East Coast, t = # mailings to the Midwest, u = # mailings to the West Coast

Minimize $c = 40s + 60y + 50z$ subject to $100s + 100t + 50u \geq 1500$, $50s + 100t + 100u \geq 1500$, $s \geq 0$, $t \geq 0$, $u \geq 0$.

Dualize: $\begin{bmatrix} 100 & 100 & 50 & 1500 \\ 50 & 100 & 100 & 1500 \\ 40 & 60 & 50 & 0 \end{bmatrix} \rightarrow \begin{bmatrix} 100 & 50 & 40 \\ 100 & 100 & 60 \\ 50 & 100 & 50 \\ 1500 & 1500 & 0 \end{bmatrix}$

Dual problem: Maximize $p = 1500x + 1500y$ subject to $100x + 50y \leq 40$, $100x + 100y \leq 60$, $50x + 100y \leq 50$, $x \geq 0$, $y \geq 0$.

Solve the dual problem:

	x	y	s	t	u	p	
s	100	50	1	0	0	0	40
t	100	100	0	1	0	0	60
u	50	100	0	0	1	0	50
p	−1500	−1500	0	0	0	1	0

$R_2 - R_1$

$2R_3 - R_1$

$R_4 + 15R_1$

	x	y	s	t	u	p	
x	100	50	1	0	0	0	40
t	0	50	−1	1	0	0	20
u	0	150	−1	0	2	0	60
p	0	−750	15	0	0	1	600

$3R_1 - R_3$

$3R_2 - R_3$

$R_4 + 5R_3$

	x	y	s	t	u	p	
x	300	0	4	0	−2	0	60
t	0	0	−2	3	−2	0	0
y	0	150	−1	0	2	0	60
p	0	0	10	0	10	1	900

The solution to the primal problem is $c = 900/1 = 900$; $s = 10/1 = 10$, $t = 0$, $u = 10/1 = 10$. Send 10 mailings to the East coast, none to the Midwest, 10 to the West Coast. Cost: $900. The shadow costs are the values of x and y in the final tableau: $x = 60/300 = 1/5 = 20$¢ per Democrat and $y = 60/150 = 2/5 = 40$¢ per Republican.

Another possible solution: 15 mailings to the Midwest and no mailing to the coasts. Cost: $900. Shadow costs: 20¢ per Democrat and 40¢ per Republican.

37. Unknowns: $s = $ # sleep spells, $t = $ # shock spells
Minimize $c = 500s + 750t$ subject to $2s + 3t \geq 1440$, $3s + 2t \geq 1200$, $-s + 3t \geq 0$, $s - 2t \geq 0$, $s \geq 0$, $t \geq 0$.

Dualize:
$$\begin{bmatrix} 2 & 3 & 1440 \\ 3 & 2 & 1200 \\ -1 & 3 & 0 \\ 1 & -2 & 0 \\ 500 & 750 & 0 \end{bmatrix} \rightarrow \begin{bmatrix} 2 & 3 & -1 & 1 & 500 \\ 3 & 2 & 3 & -2 & 750 \\ 1440 & 1200 & 0 & 0 & 0 \end{bmatrix}$$

Dual problem: Maximize $p = 1440x + 1200y$ subject to $2x + 3y - z + w \leq 500$, $3x + 2y + 3z - 2w \leq 750$, $x \geq 0$, $y \geq 0$, $z \geq 0$, $w \geq 0$.
Solve the dual problem:

	x	y	z	w	s	t	p	
s	[2]	3	−1	1	1	0	0	500
t	3	2	3	−2	0	1	0	750
p	−1440	−1200	0	0	0	0	1	0

$2R_1 - 3R_2$

$R_3 + 480R_2$

	x	y	z	w	s	t	p	
x	2	3	−1	1	1	0	0	500
t	0	−5	[9]	−7	−3	2	0	0
p	0	960	−720	720	720	0	1	360000

$9R_2 + R_1$

$R_3 + 80R_1$

	x	y	z	w	s	t	p	
x	18	22	0	2	6	2	0	4500
z	0	−5	9	−7	−3	2	0	0
p	0	560	0	160	480	160	1	360,000

The solution to the primal problem is $c = 360,000/1 = 360,000$; $s = 480$, $t = 160$. Gillian should use 480 sleep spells and 160 shock spells, costing 360,000 pico-shirleys of energy. Another solution (pivot on the 3 in the first step): 2880/7 sleep spells and 1440/7 shock spells.

39. We observe that the first column dominates the last, so we can eliminate the last. In the matrix that remains, the second row dominates the third, so we eliminate the third. We cannot reduce any further. Add $k = 500$ to each entry and put 1s to the right and below:

$$\begin{bmatrix} 300 & 200 & 1 \\ 0 & 1000 & 1 \\ 1 & 1 & 0 \end{bmatrix}$$

This gives the following LP problem:

Maximize $p = x + y$ subject to
$$300x + 200y \le 1$$
$$1000y \le 1$$
$$x \ge 0, \, y \ge 0$$

Here are the tableaux:

	x	y	s	t	p		
s	300	200	1	0	0	1	
t	0	1000	0	1	0	1	
p	−1	−1	0	0	1	0	$300R_3 + R_1$

	x	y	s	t	p		
x	300	200	1	0	0	1	$5R_1 - R_2$
t	0	**Error!**	0	1	0	1	
p	0	−100	1	0	300	1	$10R_3 + R_2$

	x	y	s	t	p	
x	1500	0	5	−1	0	4
y	0	1000	0	1	0	1
p	0	0	10	1	3000	11

The solution to the primal problem is $p = 11/3000$; $x = 4/1500 = 1/375$, $y = 1/1000$. So, T. N. Spend's optimal strategy is $C = [8/11 \ \ 3/11 \ \ 0]^T$ and the value of the game is $e = 3000/11 - 500 = -2500/11 \approx 227$.

The solution to the dual problem is $s = 10/3000$, $t = 1/3000$, so T. L. Down's optimal strategy is $R = [10/11 \ \ 1/11 \ \ 0]$.

T. N. Spend should spend about 73% of the days in Littleville, 27% in Metropolis, and skip Urbantown.
T. L. Down should spend about 91% of the days in Littleville, 9% in Metropolis, and skip Urbantown.
The expected outcome is that T. L. Down will lose about 227 votes per day of campaigning.

41. With A the row player and B the column player, the payoff matrix is

$$P = \begin{bmatrix} 2 & -1 & 0 \\ -1 & 4 & -1 \\ 0 & -1 & 6 \end{bmatrix}$$

This game does not reduce. Add $k = 1$ to each entry and put 1s to the right and below:

$$\begin{bmatrix} 3 & 0 & 1 & 1 \\ 0 & 5 & 0 & 1 \\ 1 & 0 & 7 & 1 \\ 1 & 1 & 1 & 0 \end{bmatrix}$$

This gives the following LP problem:

Maximize $p = x + y + z$ subject to

$$3x \quad + z \le 1$$
$$5y \quad \le 1$$
$$x \quad +7z \le 1$$
$$x \ge 0, \ y \ge 0, \ z \ge 0$$

Here are the tableaux:

	x	y	z	s	t	u	p	
s	3	0	1	1	0	0	0	1
t	0	5	0	0	1	0	0	1
u	1	0	7	0	0	1	0	1
p	−1	−1	−1	0	0	0	1	0

$3R_3 - R_1$
$3R_4 + R_1$

	x	y	z	s	t	u	p	
x	3	0	1	1	0	0	0	1
t	0	5	0	0	1	0	0	1
u	0	0	20	−1	0	3	0	2
p	0	−3	−2	1	0	0	3	1

$5R_4 + 3R_2$

	x	y	z	s	t	u	p		
x	3	0	1	1	0	0	0	1	$20R_1 - R_3$
y	0	5	0	0	1	0	0	1	
u	0	0	20	−1	0	3	0	2	
p	0	0	−10	5	3	0	15	8	$2R_4 + R_3$

	x	y	z	s	t	u	p	
x	60	0	0	21	0	−3	0	18
y	0	5	0	0	1	0	0	1
z	0	0	20	−1	0	3	0	2
p	0	0	0	9	6	3	30	18

The solution to the primal problem is $p = 18/30 = 3/5$; $x = 18/60 = 3/10$, $y = 1/5$, $z = 2/20 = 1/10$. So, player B's optimal strategy is $C = [1/2 \ 1/3 \ 1/6]^T$ and the value of the game is $e = 5/3 - 1 = 2/3$. The solution to the dual problem is $s = 9/30 = 3/10$, $t = 6/30 = 1/5$, $u = 3/30 = 1/10$, so player A's optimal strategy is $R = [1/2 \ 1/3 \ 1/6]$.

Each player should show one finger with probability 1/2, two fingers with probability 1/3, and three fingers with probability 1/6. The expected outcome is that player A will with 2/3 point per round, on average.

43. Let Colonel Blotto be the row player and Captain Kije the column player. Label possible moves by the number of regiments sent to the first of the two locations; the remaining regiments are sent to the other location. The payoff matrix is then the following:

		Captain Kije			
		0	1	2	3
	0	4	2	1	0
	1	1	3	0	−1
Colonel Blotto	2	−2	2	2	−2
	3	−1	0	3	1
	4	0	1	2	4

We cannot reduce this game. Add $k = 2$ to each entry and put 1s to the right and below:

$$\begin{bmatrix} 6 & 4 & 3 & 2 & 1 \\ 3 & 5 & 2 & 1 & 1 \\ 0 & 4 & 4 & 0 & 1 \\ 1 & 2 & 5 & 3 & 1 \\ 2 & 3 & 4 & 6 & 1 \\ 1 & 1 & 1 & 1 & 0 \end{bmatrix}$$

This gives the following LP problem:

Maximize $p = x + y + z + w$ subject to
$$6x + 4y + 3z + 2w \leq 1$$
$$3x + 5y + 2z + w \leq 1$$
$$4y + 4z \leq 1$$
$$x + 2y + 5z + 3w \leq 1$$
$$2x + 3y + 4z + 6w \leq 1$$
$$x \geq 0,\ y \geq 0,\ z \geq 0,\ w \geq 0$$

Here are the tableaux:

	x	y	z	w	s	t	u	v	r	p		
s	6	4	3	2	1	0	0	0	0	0	1	
t	3	5	2	1	0	1	0	0	0	0	1	$2R_2 - R_1$
u	0	4	4	0	0	0	1	0	0	0	1	
v	1	2	5	3	0	0	0	1	0	0	1	$6R_4 - R_1$
r	2	3	4	6	0	0	0	0	1	0	1	$3R_5 - R_1$
p	−1	−1	−1	−1	0	0	0	0	0	1	0	$6R_6 + R_1$

	x	y	z	w	s	t	u	v	r	p		
x	6	4	3	2	1	0	0	0	0	0	1	$8R_1 - R_5$
t	0	6	1	0	-1	2	0	0	0	0	1	
u	0	4	4	0	0	0	1	0	0	0	1	
v	0	8	27	16	-1	0	0	6	0	0	5	$R_4 - R_5$
r	0	5	9	[16]	-1	0	0	0	3	0	2	
p	0	-2	-3	-4	1	0	0	0	0	6	1	$4R_6 + R_5$

	x	y	z	w	s	t	u	v	r	p		
x	48	27	15	0	9	0	0	0	-3	0	6	$2R_1 - 9R_2$
t	0	[6]	1	0	-1	2	0	0	0	0	1	
u	0	4	4	0	0	0	1	0	0	0	1	$3R_3 - 2R_2$
v	0	3	18	0	0	0	0	6	-3	0	3	$2R_4 - R_2$
w	0	5	9	16	-1	0	0	0	3	0	2	$6R_5 - 5R_2$
p	0	-3	-3	0	3	0	0	0	3	24	6	$2R_6 + R_2$

	x	y	z	w	s	t	u	v	r	p		
x	96	0	21	0	27	-18	0	0	-6	0	3	$10R_1 - 21R_3$
y	0	6	1	0	-1	2	0	0	0	0	1	$10R_2 - R_3$
u	0	0	**Erro**	0	2	-4	3	0	0	0	1	
v	0	0	35	0	1	-2	0	12	-6	0	5	$2R_4 - 7R_3$
w	0	0	49	96	-1	-10	0	0	18	0	7	$10R_5 - 49R_3$
p	0	0	-5	0	5	2	0	0	6	48	13	$2R_6 + R_3$

	x	y	z	w	s	t	u	v	r	p	
x	960	0	0	0	228	-96	-63	0	-60	0	9
y	0	60	0	0	-12	24	-3	0	0	0	9
z	0	0	20	0	4	-8	6	0	0	0	2
v	0	0	0	0	-12	24	-21	24	-12	0	3
w	0	0	0	960	-108	96	-147	0	180	0	21
p	0	0	0	0	12	0	3	0	12	96	27

The solution to the primal problem is $p = 27/96 = 9/32$; $x = 9/960 = 3/320$, $y = 9/60 = 3/20$, $z = 2/20 = 1/10$, $w = 21/960 = 7/320$. So, Captain Kije's optimal strategy is $C = [1/30 \ 8/15 \ 16/45 \ 7/90]^T$ and the value of the game is $e = 32/9 - 2 = 14/9$. In fact, this is only one of several optimal

211

strategies for Captain Kije; which you find depends on the choices of pivots you make in the simplex method. Other optimal strategies are the one given above, in reverse order, and $[1/18 \ 4/9 \ 4/9 \ 1/18]^T$. The solution to the dual problem (which is unique) is $s = 12/96 = 1/8$, $t = 0$, $u = 3/96 = 1/32$, $v = 0$, $r = 12/96 = 1/8$, so Colonel Blotto's optimal strategy is $R = [4/9 \ 0 \ 1/9 \ 0 \ 4/9]$.

Write moves as (x, y) where x represents the number of regiments sent to the first location and y represents the number sent to the second location. Colonel Blotto should play $(0, 4)$ with probability 4/9, $(2, 2)$ with probability 1/9, and $(4, 0)$ with probability 4/9. Captain Kije has several optimal strategies, one of which is to play $(0, 3)$ with probability 1/30, $(1, 2)$ with probability 8/15, $(2, 1)$ with probability 16/45, and $(3, 0)$ with probability 7/90. The expected outcome is that Colonel Blotto will win 14/9 points on average.

45. The dual of a standard minimization problem satisfying the nonnegative objective condition is a standard maximization problem, which can be solved using the standard simplex algorithm, thus avoiding the need to do Phase I.

47. Answers will vary, We use a minimization problem that does not satisfy the nonnegative objective condition. An example is: Minimize $c = x - y$ subject to $x - y \geq 100$, $x + y \geq 200$, $x \geq 0$, $y \geq 0$. this problem can be solved using the techniques in Section 4.4.

49. Unknowns: s = # convention–style hotels, t = # vacation–style hotels, and u = # motels. Minimize $C = 100s + 20t + 4u$ subject to $500s + 200t + 50u \geq 1400$, $s \geq 1$, $u \leq 2$, $s \geq 0$, $t \geq 0$, $u \geq 0$. Rewrite the constraint $z \leq 2$ using "$\geq$":

Minimize $C = 100s + 20t + 4u$ subject to $500x + 200t + 50u \geq 1400$, $s \geq 1$, $-u \leq -2$, $s \geq 0$, $t \geq 0$, $u \geq 0$.

Dualize:
$$\begin{bmatrix} 500 & 200 & 50 & 1400 \\ 1 & 0 & 0 & 1 \\ 0 & 0 & -1 & -2 \\ 100 & 20 & 4 & 0 \end{bmatrix} \rightarrow \begin{bmatrix} 500 & 1 & 0 & 100 \\ 200 & 0 & 0 & 20 \\ 50 & 0 & -1 & 4 \\ 1400 & 1 & -2 & 0 \end{bmatrix}$$

Dual problem: Maximize $P = 1400x + y - 2z$ subject to $500x + y \leq 100$, $200x \leq 20$, $50x - z \leq 4$, $x \geq 0$, $y \geq 0$, $z \geq 0$.

Solve the dual problem:

	x	y	z	s	t	u	p		
s	500	1	0	1	0	0	0	100	$R_1 - 10R_3$
t	200	0	0	0	1	0	0	20	$R_2 - 4R_3$
u	⌐50⌐	0	-1	0	0	1	0	4	
p	-1400	-1	2	0	0	0	1	0	$R_4 + 28R_3$

	x	y	z	s	t	u	p		
s	0	1	10	1	0	-10	0	60	$4R_1 - 10R_2$
t	0	0	⌐4⌐	0	1	-4	0	4	
x	50	0	-1	0	0	1	0	4	$4R_3 + R_2$
p	0	-1	-26	0	0	28	1	112	$4R_4 + 26R_2$

	x	y	z	s	t	u	p	
s	0	$\boxed{4}$	0	4	–10	0	0	200
z	0	0	4	0	1	–4	0	4
x	200	0	0	0	1	0	0	20
p	0	–4	0	0	26	8	4	552

$R_4 + R_1$

	x	y	z	s	t	u	p	
y	0	4	0	4	–10	0	0	200
z	0	0	4	0	1	–4	0	4
x	200	0	0	0	1	0	0	20
p	0	0	0	4	16	8	4	752

The solution to the primal problem is $c = 752/4 = 188$; $s = 4/4 = 1$, $t = 16/4 = 4$, $u = 8/4 = 2$. Build 1 convention-style hotel, 4 vacation style hotels and 2 small motels.

Chapter 4 Review Exercises

1. $2x - 3y \leq 12$

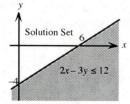

Unbounded

3. $x + 2y \leq 20$
$3x + 2y \leq 30$
$x \geq 0, y \geq 0$

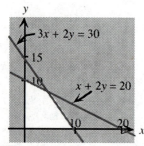

Bounded
Corner points: $(0, 0), (0, 10),$
$(5, 15/2), (10, 0)$

5. Maximize $p = 2x + y$ subject to
$3x + y \leq 30$
$x + y \leq 12$
$x + 3y \leq 30$
$x \geq 0, y \geq 0$

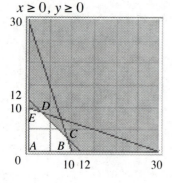

Point	Lines	Coords.	$p = 2x+y$
A	$x = 0$ $y = 0$	$(0, 0)$	0
B	$3x+y = 30$ $y = 0$	$(10, 0)$	20
C	$3x+y = 30$ $x+y = 12$	$(9, 3)$	**21**
D	$x+y = 12$ $x+3y = 30$	$(3, 9)$	15
E	$x+3y = 30$ $x = 0$	$(0, 10)$	10

Maximum value occurs at C:
$p = 21; x = 9, y = 3$

7. Minimize $c = 2x + y$ subject to
$3x + y \geq 30$
$x + 2y \geq 20$
$2x - y \geq 0$
$x \geq 0, y \geq 0$

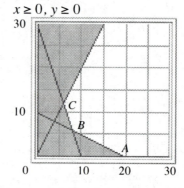

Point	Lines	Coords.	$c = 2x+y$
A	$x+2y = 20$ $y = 0$	$(20, 0)$	40
B	$3x+y = 30$ $x+2y = 20$	$(8, 6)$	**22**
C	$3x+y = 30$ $2x-y = 0$	$(6, 12)$	24

Minimum value occurs at B:
$c = 22; x = 8, y = 6$

9. Introduce slack variables and set up the initial tableau:

$$x + 2y + 2z + s = 60$$
$$2x + y + 3z + t = 60$$
$$-x - y - 2z + p = 0$$

	x	y	z	s	t	p		
s	1	2	2	1	0	0	60	$3R_1 - 2R_2$
t	2	1	③	0	1	0	60	
p	-1	-1	-2	0	0	1	0	$3R_3 + 2R_2$

The most negative entry in the bottom row in the above tableau is the -2 in the z-column. So we use this column as the pivot column. The test ratios are: s: $60/2$, t: $60/3$. The smallest test ratio is t: $60/3$. Thus, we pivot on the 3 in the t-row.

	x	y	z	s	t	p		
s	-1	④	0	3	-2	0	60	
z	2	1	3	0	1	0	60	$4R_2 - R_1$
p	1	-1	0	0	2	3	120	$4R_3 + R_1$

The most negative entry in the bottom row in the above tableau is the -1 in the y-column. So we use this column as the pivot column. The test ratios are: s: $60/4$, z: $60/1$. The smallest test ratio is s: $60/4$. Thus, we pivot on the 4 in the s-row.

	x	y	z	s	t	p	
y	-1	4	0	3	-2	0	60
z	9	0	12	-3	6	0	180
p	3	0	0	3	6	12	540

Optimal Solution: $p = 45$; $x = 0$, $y = 60/4 = 15$, $z = 180/12 = 15$.

11. Introduce slack and surplus variables and set up the initial tableau:

$$x + y + z - s = 100$$
$$y + z + t = 80$$
$$x + z + u = 80$$
$$-x - y - 3z + p = 0$$

	x	y	z	s	t	u	p		
*s	1	1	1	−1	0	0	0	100	$R_1 - R_3$
t	0	1	1	0	1	0	0	80	
u	[1]	0	1	0	0	1	0	80	
p	−1	−1	−3	0	0	0	1	0	$R_4 + R_3$

The first starred row in the above tableau is the s-row, and its (first) largest positive entry is the 1 in the x-column. So we use this column as the pivot column. The test ratios are: s: 100/1, u: 80/1. The smallest test ratio is u: 80/1. Thus, we pivot on the 1 in the u-row.

	x	y	z	s	t	u	p		
*s	0	[1]	0	−1	0	−1	0	20	
t	0	1	1	0	1	0	0	80	$R_2 - R_1$
x	1	0	1	0	0	1	0	80	
p	0	−1	−2	0	0	1	1	80	$R_4 + R_1$

The first starred row in the above tableau remains the s-row, and its largest positive entry is the 1 in the y–column. So we use this column as the pivot column. The test ratios are: s: 20/1, t: 80/1. The smallest test ratio is s: 20/1. Thus, we pivot on the 1 in the s–row.

	x	y	z	s	t	u	p		
y	0	1	0	−1	0	−1	0	20	
t	0	0	[1]	1	1	1	0	60	
x	1	0	1	0	0	1	0	80	$R_3 - R_2$
p	0	0	−2	−1	0	0	1	100	$R_4 + 2R_2$

Since there are no more starred rows in the above tableau, we are in Phase 2, and choose the most negative number in the bottom row to give us the pivot column:

	x	y	z	s	t	u	p	
y	0	1	0	−1	0	−1	0	20
z	0	0	1	1	1	1	0	60
x	1	0	0	−1	−1	0	0	20
p	0	0	0	1	2	2	1	220

Optimal Solution: $p = 220$; $x = 20/1 = 20$, $y = 20/1 = 20$, $z = 60/1 = 60$.

13. Associated maximization problem:
Maximize $p = -x - 2y - 3z$ subject to $3x + 2y + z \geq 60$, $2x + y + 3z \geq 60$, $x \geq 0$, $y \geq 0$, $z \geq 0$.

Phase 1:

	x	y	z	s	t	p	
*s	[3]	2	1	–1	0	0	60
*t	2	1	3	0	–1	0	60
p	1	2	3	0	0	1	0

$3R_2 - 2R_1$
$3R_3 - R_1$

	x	y	z	s	t	p	
x	3	2	1	–1	0	0	60
*t	0	–1	[7]	2	–3	0	60
p	0	4	8	1	0	3	–60

$7R_1 - R_2$
$7R_3 - 8R_2$

Phase 2:

	x	y	z	s	t	p	
x	21	15	0	–9	3	0	360
z	0	–1	7	[2]	–3	0	60
p	0	36	0	–9	24	21	–900

$2R_1 + 9R_2$
$2R_3 + 9R_2$

	x	y	z	s	t	p	
x	42	21	63	0	–21	0	1260
s	0	–1	7	2	–3	0	60
p	0	63	63	0	21	42	–1260

Optimal Solution: $p = -c = 1260/42 = 30$; $x = 1260/42 = 30$, $y = 0$, $z = 0$.

15.

	x	y	z	w	s	t	u	v	$-c$	
*s	1	1	0	0	–1	0	0	0	0	30
*t	1	0	1	0	0	–1	0	0	0	20
u	1	1	0	–1	0	0	1	0	0	10
v	0	1	1	–1	0	0	0	1	0	10
$-c$	1	1	1	1	0	0	0	0	1	0

The first starred row is the s-row, and its (first) largest positive entry is the 1 in the x-column. So we use this column as the pivot column. The test ratios are: s: 30/1, t: 20/1, u: 10/1. The smallest test ratio is u: 10/1. Thus, we pivot on the 1 in the u-row.

	x	y	z	w	s	t	u	v	$-c$		
*s	1	1	0	0	-1	0	0	0	0	30	$R_1 - R_3$
*t	1	0	1	0	0	-1	0	0	0	20	$R_2 - R_3$
u	[1]	1	0	-1	0	0	1	0	0	10	
v	0	1	1	-1	0	0	0	1	0	10	
$-c$	1	1	1	1	0	0	0	0	1	0	$R_5 - R_3$

	x	y	z	w	s	t	u	v	$-c$	
*s	0	0	0	1	-1	0	-1	0	0	20
*t	0	-1	1	1	0	-1	-1	0	0	10
x	1	1	0	-1	0	0	1	0	0	10
v	0	1	1	-1	0	0	0	1	0	10
$-c$	0	0	1	2	0	0	-1	0	1	-10

The first starred row is the s-row, and its largest positive entry is the 1 in the w-column. So we use this column as the pivot column. The test ratios are: s: 20/1, t: 10/1. The smallest test ratio is t: 10/1. Thus, we pivot on the 1 in the t-row.

	x	y	z	w	s	t	u	v	$-c$		
*s	0	0	0	1	-1	0	-1	0	0	20	$R_1 - R_2$
*t	0	-1	1	[1]	0	-1	-1	0	0	10	
x	1	1	0	-1	0	0	1	0	0	10	$R_3 + R_2$
v	0	1	1	-1	0	0	0	1	0	10	$R_4 + R_2$
$-c$	0	0	1	2	0	0	-1	0	1	-10	$R_5 - 2R_2$

	x	y	z	w	s	t	u	v	$-c$	
*s	0	1	-1	0	-1	1	0	0	0	10
w	0	-1	1	1	0	-1	-1	0	0	10
x	1	0	1	0	0	-1	0	0	0	20
v	0	0	2	0	0	-1	-1	1	0	20
$-c$	0	2	-1	0	0	2	1	0	1	-30

The first starred row is the s-row, and its (first) largest positive entry is the 1 in the y-column. So we use this column as the pivot column. There is only one positive ratio, in the s-row. Thus, we pivot on the 1 in the s-row.

	x	y	z	w	s	t	u	v	$-c$		
*s	0	[1]	−1	0	−1	1	0	0	0	10	
w	0	−1	1	1	0	−1	−1	0	0	10	$R_2 + R_1$
x	1	0	1	0	0	−1	0	0	0	20	
v	0	0	2	0	0	−1	−1	1	0	20	
$-c$	0	2	−1	0	0	2	1	0	1	−30	$R_5 - 2R_1$

	x	y	z	w	s	t	u	v	$-c$	
y	0	1	−1	0	−1	1	0	0	0	10
w	0	0	0	1	−1	0	−1	0	0	20
x	1	0	1	0	0	−1	0	0	0	20
v	0	0	2	0	0	−1	−1	1	0	20
$-c$	0	0	1	0	2	0	1	0	1	−50

Since there are no more starred rows, we go to Phase 2, and search for the most negative entry in the bottom row (if any). Since there are no negative entries in the bottom row, we are done.

Optimal Solution: $c = -(-50/1) = 50$; $x = 20/1 = 20$, $y = 10/1 = 10$, $z = 0$, $w = 20/1 = 20$

17. Minimize $c = 2s + t$ subject to $3s + 2t \geq 60$, $2s + t \geq 60$, $x + 3y \geq 60$, $s \geq 0$, $t \geq 0$.

Dualize: $\begin{bmatrix} 3 & 2 & 60 \\ 2 & 1 & 60 \\ 1 & 3 & 60 \\ 2 & 1 & 0 \end{bmatrix} \rightarrow \begin{bmatrix} 3 & 2 & 1 & 2 \\ 2 & 1 & 3 & 1 \\ 60 & 60 & 60 & 0 \end{bmatrix}$

Dual problem: Maximize $p = 60x + 60y + 60z$ subject to $3x + 2y + z \leq 2$, $2x + y + 43z \leq 1$, $x \geq 0$, $y \geq 0$, $z \geq 0$.

Solve the dual problem:

	x	y	z	s	t	p		
s	3	2	1	1	0	0	2	$2R_1 - 3R_2$
t	[2]	1	43	0	1	0	1	
p	−60	−60	−60	0	0	1	0	$R_3 + 30R_2$

	x	y	z	s	t	p		
s	0	1	−127	2	−3	0	1	$R_1 - R_2$
x	2	[1]	43	0	1	0	1	
p	0	−30	1230	0	30	1	30	$R_3 + 30R_2$

	x	y	z	s	t	p	
s	-2	0	-170	2	-4	0	0
y	2	1	43	0	1	0	1
p	60	0	2520	0	60	1	60

Solution to the primal problem: $c = 60$; $s = 0$, $t = 60$. Another possible solution: $c = 60$; $s = 24$, $t = 12$. Since the original unknowns were x and y, we write these optimal solutions as: $c = 60$; $x = 0$, $y = 60$ and $c = 60$; $x = 24$, $y = 12$.

19. First rewrite the given problem as a standard minimization problem with all the constraints using "≥":
Minimize $c = 2s + t$ subject to $3s + 2t \geq 10$, $-2s + t \geq -30$, $s + 3t \geq 60$, $s \geq 0$, $t \geq 0$.

Dualize: $\begin{bmatrix} 3 & 2 & 10 \\ -2 & 1 & -30 \\ 1 & 3 & 60 \\ 2 & 1 & 0 \end{bmatrix} \rightarrow \begin{bmatrix} 3 & -2 & 1 & 2 \\ 2 & 1 & 3 & 1 \\ 10 & -30 & 60 & 0 \end{bmatrix}$

Dual problem: Maximize $p = 10x - 30y + 60z$ subject to $3x - 2y + z \leq 2$, $2x + y + 3z \leq 1$, $x \geq 0$, $y \geq 0$, $z \geq 0$.

Solve the dual problem:

	x	y	z	s	t	p		
s	3	-2	1	1	0	0	2	$3R_1 - R_2$
t	2	1	[3]	0	1	0	1	
p	-10	30	-60	0	0	1	0	$R_3 + 20R_2$

	x	y	z	s	t	p	
s	7	-7	0	3	-1	0	5
z	2	1	3	0	1	0	1
p	30	50	0	0	20	1	20

Solution to the primal problem: $c = 20/1 = 20$; $s = 0$, $t = 20/1 = 20$. Since the original unknowns were x and y, we write the optimal solution as: $c = 20$; $x = 0$, $y = 20$.

21. The first column is dominated by the last, so we can eliminate it. We cannot reduce any further. Add $k = 2$ to each entry and put 1s to the right and below:

$\begin{bmatrix} 4 & 1 & 1 \\ 0 & 3 & 1 \\ 1 & 2 & 1 \\ 1 & 1 & 0 \end{bmatrix}$

This gives the following LP problem:
Maximize $p = x + y$ subject to
$$4x + y \leq 1$$
$$3y \leq 1$$
$$x + 2y \leq 1$$
$$x \geq 0, \, y \geq 0$$

Here are the tableaux:

	x	y	s	t	u	p	
s	[4]	1	1	0	0	0	1
t	0	3	0	1	0	0	1
u	1	2	0	0	1	0	1
p	−1	−1	0	0	0	1	0

$4R_3 - R_1$
$4R_4 + R_1$

	x	y	s	t	u	p	
x	4	1	1	0	0	0	1
t	0	[3]	0	1	0	0	1
u	0	7	−1	0	4	0	3
p	0	−3	1	0	0	4	1

$3R_1 - R_2$

$3R_3 - 7R_2$

$R_4 + R_2$

	x	y	s	t	u	p	
x	12	0	3	−1	0	0	2
y	0	3	0	1	0	0	1
u	0	0	−3	−7	12	0	2
p	0	0	1	1	0	4	2

The solution to the primal problem is $p = 2/4 = 1/2$; $x = 2/12 = 1/6$, $y = 1/3$. So, the column player's optimal strategy is $C = [0 \ 1/3 \ 2/3]^T$ and the value of the game is $e = 2 - 2 = 0$.
The solution to the dual problem is $s = 1/4$, $t = 1/4$, $u = 0$, so the row player's optimal strategy is $R = [1/2 \ 1/2 \ 0]$.

23. This game cannot be reduced. Add $k = 3$ to each entry and put 1s to the right and below:

$$\begin{bmatrix} 0 & 1 & 6 & 1 \\ 4 & 3 & 3 & 1 \\ 1 & 5 & 4 & 1 \\ 1 & 1 & 1 & 0 \end{bmatrix}$$

This gives the following LP problem:

Maximize $p = x + y + z$ subject to
$$y + 6z \le 1$$
$$4x + 3y + 3z \le 1$$
$$x + 5y + 4z \le 1$$
$$x \ge 0, y \ge 0, z \ge 0$$

Here are the tableaux:

	x	y	z	s	t	u	p	
s	0	1	6	1	0	0	0	1
t	[4]	3	3	0	1	0	0	1
u	1	5	4	0	0	1	0	1
p	−1	−1	−1	0	0	0	1	0

$4R_3 - R_2$

$4R_4 + R_2$

	x	y	z	s	t	u	p	
s	0	1	6	1	0	0	0	1
x	4	3	3	0	1	0	0	1
u	0	[17]	13	0	−1	4	0	3
p	0	−1	−1	0	1	0	4	1

$17R_1 - R_3$

$17R_2 - 3R_3$

$17R_4 + R_3$

	x	y	z	s	t	u	p	
s	0	0	[89]	17	1	−4	0	14
x	68	0	12	0	20	−12	0	8
y	0	17	13	0	−1	4	0	3
p	0	0	−4	0	16	4	68	20

$89R_2 - 12R_1$

$89R_3 - 13R_1$

$89R_4 + 4R_1$

	x	y	z	s	t	u	p	
z	0	0	89	17	1	−4	0	14
x	6052	0	0	−204	1768	−1020	0	544
y	0	1513	0	−221	−102	408	0	85
p	0	0	0	68	1428	340	6052	1836

The solution to the primal problem is $p = 1836/6052 = 27/89$; $x = 544/6052 = 8/89$, $y = 85/1513 = 5/89$, $z = 14/89$. So, the column player's optimal strategy is $C = [8/27\ 5/27\ 14/27]^T$ and the value of the game is $e = 89/27 - 3 = 8/27$.
The solution to the dual problem is $s = 68/6052 = 1/89$, $t = 1428/6052 = 21/89$, $u = 340/6052 = 5/89$, so the row player's optimal strategy is $R = [1/27\ 21/27\ 5/27] = [1/27\ 7/9\ 5/27]$.

25. Apply the simplex method to the given problem:

	x	y	s	t	p	
*s	−2	[1]	−1	0	0	1
*t	1	−2	0	−1	0	1
c	1	?	0	0	1	0

$R_2 + 2R_1$

$R_3 - 2R_1$

	x	y	s	t	p	
y	−2	1	−1	0	0	1
*t	−3	0	−2	−1	0	3
c	5	0	2	0	1	−2

Since there are no positive entries in the t-row (other than the right-hand side) we cannot go to the next step. Thus, it is impossible to get into the feasible region by completing Phase 1. In other words, there are no feasible solutions (choice A).

Another way of seeing this would be to graph the constraints. This would show an empty feasible region.

27. First note that the objective function, $Z = x_1 + 4x_2 + 2x_3 - 10$, is not linear because of the "−10". However, maximizing $p = x_1 + 4x_2 + 2x_3$ will also maximize Z, provided we remember to subtract 10 from the optimal value of the objective function when we are done. Thus, we solve the problem using $p = x_1 + 4x_2 + 2x_3$ as our objective function:

	x_1	x_2	x_3	s	t	p		
s	4	1	1	1	0	0	45	$R_1 - R_2$
t	−1	[1]	2	0	1	0	0	
p	−1	−4	−2	0	0	1	0	$R_3 + 4R_2$

	x_1	x_2	x_3	s	t	p		
s	[5]	0	−1	1	−1	0	45	
x_2	−1	1	2	0	1	0	0	$5R_2 + R_1$
p	−5	0	6	0	4	1	0	$R_3 + R_1$

	x_1	x_2	x_3	s	t	p	
x_1	5	0	−1	1	−1	0	45
x_2	0	5	9	1	4	0	45
p	0	0	5	1	3	1	45

Optimal Solution: $p = 45$; $x_1 = 45/5 = 9$, $x_2 = 45/5 = 9$, $x_3 = 0$. Since the maximum value of p is 45, the maximum value of Z is $45 - 10 = 35$.

29. Unknowns: x = # packages from Duffin House, y = # packages from Higgins Press. Minimize $c = 50x + 80y$ subject to $5x + 5y \geq 4000$, $5x + 10y \geq 6000$, $x \geq 0$, $y \geq 0$.

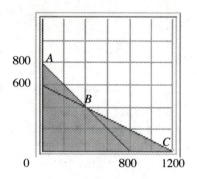

Point	Lines	Coords.	$c = 50x + 80y$
A	$5x+5y = 4000$ $x = 0$	$(0, 800)$	64,000
B	$5x+5y = 4000$ $5x+10y = 6000$	$(400, 400)$	**52,000**
C	$5x+10y = 6000$ $y = 0$	$(1200, 0)$	60,000

Solution: Purchase 400 packages from each for a minimum cost of $52,000.

31. (a) Since the lowest cost is given by ordering the same number of packages from each supplier (point *B* in the feasible region in Exercise 29), changing that by ordering at least 20% more packages from Duffin as from Higgins will result in a higher cost, since the solution will no longer be point *B* but another point in the feasible region (choice B), assuming it is possible to meet all the conditions. On the other hand, it is conceivable that the extra constraint will result in an empty feasible region (choice D). Thus, the correct answers are (B) and (D).

(b) Unknowns: as in Exercise 29. Minimize $c = 50x + 80y$ subject to $5x + 5y \geq 4000$, $5x + 10y \geq 6000$, $x \geq 1.2y$, or $x - 1.2y \geq 0$, $x \geq 0$, $y \geq 0$.

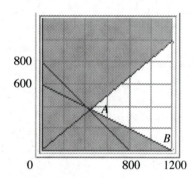

Point	Lines	Coords.	$c = 50x+80y$
A	$5x+10y = 6000$ $x-1.2y = 0$	$(450, 375)$	**52,500**
B	$5x+10y = 6000$ $y = 0$	$(1200, 0)$	60,000

Solution: Purchase 450 packages from Duffin House, 375 from Higgins Press for a minimum cost of $52,500.

33. Unknowns:
$x = $ # shares in EEE
$y = $ # shares in RRR
Minimize $c = 2.0x + 3.0y$ subject to
$\quad 50x + 55y \leq 12{,}100$
$\quad 2.25x + 2.75y \geq 550$
$\quad x \geq 0, y \geq 0$

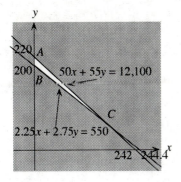

Corner points:

Point	Lines	Coords.	$c = 2.0x + 3.0y$
A	$x = 0$ $50x + 55y = 12,100$	$(0, 220)$	660
B	$x = 0$ $2.25x + 2.75y = 550$	$(0, 200)$	600
C	$50x + 55y = 12,100$ $2.25x + 2.75y = 550$	$(220, 20)$	**500**

Minimum value occurs at the point C: $c = 500$; $x = 220$, $y = 20$.
Solution: Buy 220 shares of EEE and 20 shares of RRR. The minimum total risk index is $c = 500$.

35. Take x = # sprinkles, y = # storms, z = # hurricanes
Maximize $p = x + y + 2z$ subject to $x + 2y + 2z \le 600$, $2x + y + 3z \le 600$, $x + 3y + 6z \le 600$, $x \ge 0$, $y \ge 0$, $z \ge 0$

	x	y	z	s	t	u	p		
s	1	2	2	1	0	0	0	600	$3R_1 - R_3$
t	2	1	3	0	1	0	0	600	$2R_2 - R_3$
u	1	3	[6]	0	0	1	0	600	
p	-1	-1	-2	0	0	0	1	0	$3R_4 + R_3$

	x	y	z	s	t	u	p		
s	2	3	0	3	0	-1	0	1200	$3R_1 - 2R_2$
t	[3]	-1	0	0	2	-1	0	600	
z	1	3	6	0	0	1	0	600	$3R_3 - R_2$
p	-2	0	0	0	0	1	3	600	$3R_4 + 2R_2$

	x	y	z	s	t	u	p	
s	0	11	0	9	-4	-1	0	2400
x	3	-1	0	0	2	-1	0	600
z	0	[10]	18	0	-2	4	0	1200
p	0	-2	0	0	4	1	9	3000

$10R_1 - 11R_3$

$10R_2 + R_3$

$5R_4 + R_3$

	x	y	z	s	t	u	p	
s	0	0	-198	90	-18	-54	0	10800
x	30	0	18	0	18	-6	0	7200
y	0	10	18	0	-2	4	0	1200
p	0	0	18	0	18	9	45	16200

Optimal Solution: $p = 16,200/45 = 360$; $x = 7200/30 = 240$, $y = 1200/10 = 120$, $z = 0$
Make 240 Sprinkles, 120 Storms, and no Hurricanes

37. Unknowns: x = # packages from Duffin House, y = # packages from Higgins Press, z = # packages from Ewing Books. Minimize $c = 50x + 150y + 100z$ subject to $5x + 10y + 5z \geq 4000$, $2x + 10y + 5z \geq 6000$, $y \geq 0.5(x + y + z)$, or $x - y + z \leq 0$, $x \geq 0$, $y \geq 0$, $z \geq 0$
Convert the given problem to a maximization problem:
Maximize $p = -50x - 150y - 100z$ subject to $5x + 10y + 5z \geq 4000$, $2x + 10y + 5z \geq 6000$, $x - y + z \leq 0$, $x \geq 0$, $y \geq 0$, $z \geq 0$.

	x	y	z	s	t	u	p	
*s	5	[10]	5	-1	0	0	0	4000
*t	2	10	5	0	-1	0	0	6000
u	1	-1	1	0	0	1	0	0
p	50	150	100	0	0	0	1	0

$R_2 - R_1$

$10R_3 + R_1$

$R_4 - 15R_1$

	x	y	z	s	t	u	p	
y	5	10	5	-1	0	0	0	4000
*t	-3	0	0	[1]	-1	0	0	2000
u	15	0	15	-1	0	10	0	4000
p	-25	0	25	15	0	0	1	-60000

$R_1 + R_2$

$R_3 + R_2$

$R_4 - 15R_2$

	x	y	z	s	t	u	p	
y	2	10	5	0	-1	0	0	6000
s	-3	0	0	1	-1	0	0	2000
u	12	0	15	0	-1	10	0	6000

p	20	0	25	0	15	0	1	−90000

Optimal Solution: $c = -p = 90,000$; $x = 0$, $y = 6000/10 = 600$, $z = 0$.

39. (a) Take
x = the number of credits of Sciences
y = the number of credits of Fine Arts
z = the number of credits of Liberal Arts
w = the number of credits of Mathematics
Given information:
(1) The total number of credits is at least 120:
$\quad x + y + z + w \geq 120$
(2) At least as many Science credits as Fine Arts credits:
$\quad x \geq y$, or $x - y \geq 0$
(3) At most twice as many Mathematics credits as Science credits. Rephrasing: The number of Mathematics credits is at most twice the sum of the numbers of Science credits:
$\quad w \leq 2x$, or
$\quad -2x + w \leq 0$
(4) Liberal Arts credits exceed Mathematics credits by no more than one third of the number of Fine Arts credits. Rephrasing: the number of Liberal Arts credits minus the number of Mathematics credits is at most one third of the number of Fine Arts credits:
$\quad z - w \leq \frac{1}{3}y$, or
$\quad 3z - 3w \leq y$, or
$\quad -y + 3z - 3w \leq 0$;
Thus, the linear programming problem is: Minimize $C = 300x + 300y + 200z + 200w$ subject to: $x + y + z + w \geq 120$; $x - y \geq 0$; $-2x + w \leq 0$; $-y + 3z - 3w \leq 0$; $x \geq 0$, $y \geq 0$, $z \geq 0$, $w \geq 0$
(b) Using technology (Web Site → Online Urtilities → Simplex Method Tool) We obtain the following solution:
$\quad c = 26400$; $x = 24$, $y = 0$, $z = 48$, $w = 48$.
Billy Sean should take the following combination: Sciences: 24 credits, Fine Arts: no credits, Liberal Arts: 48 credits, Mathematics: 48 credits, for a total cost of $26,400.

41. Unknowns: x = # packages from New York to O'Hagan.com, y = # packages form New York to FantasyBooks.com, z = # packages from Illinois to O'Hagan.com,, w = # packages form Illinois to FantasyBooks.com.
Minimize $c = 20x + 50y + 30z + 40w$ subject to $x + y \leq 600$, $y + z \leq 300$, $x + z \geq 600$, $y + w \geq 200$, $x \geq 0$, $y \geq 0$, $z \geq 0$, $w \geq 0$.
Associated maximization problem: Maximize $p = -20x - 50y - 30z - 40w$ subject to $x + y \leq 600$, $y + z \leq 300$, $x + z \geq 600$, $y + w \geq 200$, $x \geq 0$, $y \geq 0$, $z \geq 0$, $w \geq 0$.

	x	y	z	w	s	t	u	v	p		
s	1	1	0	0	1	0	0	0	0	600	$R_1 - R_3$
t	0	1	1	0	0	1	0	0	0	300	
*u	[1]	0	1	0	0	0	−1	0	0	600	
*v	0	1	0	1	0	0	0	−1	0	200	
p	20	50	30	40	0	0	0	0	1	0	$R_5 - 20R_3$

	x	y	z	w	s	t	u	v	p		
s	0	[1]	−1	0	1	0	1	0	0	0	
t	0	1	1	0	0	1	0	0	0	300	$R_2 - R_1$
x	1	0	1	0	0	0	−1	0	0	600	
*v	0	1	0	1	0	0	0	−1	0	200	$R_4 - R_1$
p	0	50	10	40	0	0	20	0	1	−12,000	$R_5 - 50R_1$

	x	y	z	w	s	t	u	v	p		
y	0	1	−1	0	1	0	1	0	0	0	$2R_1 + R_2$
t	0	0	[2]	0	−1	1	−1	0	0	300	
x	1	0	1	0	0	0	−1	0	0	600	$2R_3 - R_2$
*v	0	0	1	1	−1	0	−1	−1	0	200	$2R_4 - R_2$
p	0	0	60	40	−50	0	−30	0	1	−12,000	$R_5 - 30R_2$

	x	y	z	w	s	t	u	v	p		
y	0	2	0	0	1	1	1	0	0	300	
z	0	0	2	0	−1	1	−1	0	0	300	
x	2	0	0	0	1	−1	−1	0	0	900	
*v	0	0	0	[2]	−1	−1	−1	−2	0	100	
p	0	0	0	40	−20	−30	0	0	1	−21,000	$R_5 - 20R_4$

	x	y	z	w	s	t	u	v	p		
y	0	2	0	0	1	1	1	0	0	300	$R_1 - R_2$
z	0	0	2	0	−1	[1]	−1	0	0	300	
x	2	0	0	0	1	−1	−1	0	0	900	$R_3 + R_2$
w	0	0	0	2	−1	−1	−1	−2	0	100	$R_4 + R_2$
p	0	0	0	0	0	−10	20	40	1	−23,000	$R_5 + 10R_2$

	x	y	z	w	s	t	u	v	p	
y	0	2	−2	0	2	0	2	0	0	0
t	0	0	2	0	−1	1	−1	0	0	300
x	2	0	2	0	0	0	−2	0	0	1200
w	0	0	2	2	−2	0	−2	−2	0	400
p	0	0	20	0	−10	0	10	40	1	−20,000

$2R_2 + R_1$

$R_4 + R_1$

$R_5 + 5R_1$

	x	y	z	w	s	t	u	v	p	
s	0	2	−2	0	2	0	2	0	0	0
t	0	2	2	0	0	2	0	0	0	600
x	2	0	2	0	0	0	−2	0	0	1200
w	0	2	0	2	0	0	0	−2	0	400
p	0	10	10	0	0	0	20	40	1	−20,000

Optimal Solution: $c = -p = 20{,}000$; $x = 1200/2 = 600$, $y = 0$, $z = 0$, $w = 400/2 = 200$. The lowest cost is $20,000; New York to O'Hagan.com: 600 packages, New York to FantasyBooks.com: 0 packages, Illinois to O'Hagan.com: 0 packages, Illinois to FantasyBooks.com: 200 packages.

43. First reduce the game: We can eliminate O'Hagan's option of offering no promotion, and then we can eliminate FantasyBook's no promotion option. The entries in the remaining payoff matrix are nonnegative, so we put 1s to the right and below:

$$\begin{bmatrix} 20 & 10 & 15 & 1 \\ 0 & 15 & 10 & 1 \\ 1 & 1 & 1 & 0 \end{bmatrix}$$

This gives the following LP problem:

Maximize $p = x + y + z$ subject to

$$20x + 10y + 15z \le 1$$
$$15y + 10z \le 1$$
$$x \ge 0,\ y \ge 0,\ z \ge 0$$

Here are the tableaux:

	x	y	z	s	t	p	
s	20	10	15	1	0	0	1
t	0	15	10	0	1	0	1
p	−1	−1	−1	0	0	1	0

$20R_3 + R_1$

	x	y	z	s	t	p	
x	20	10	15	1	0	0	1
t	0	15	10	0	1	0	1
p	0	−10	−5	1	0	20	1

$3R_1 - 2R_2$

$3R_3 + 2R_2$

	x	y	z	s	t	p	
x	60	0	25	3	−2	0	1
y	0	15	10	0	1	0	1
p	0	0	5	3	2	60	5

The solution to the primal problem is $p = 5/60 = 1/12$; $x = 1/60$, $y = 1/15$, $z = 0$. So, FantasyBook's optimal strategy is $C = [0 \ 1/5 \ 4/5 \ 0]^T$ and the value of the game is $e = 12$.

The solution to the dual problem is $s = 3/60 = 1/20$, $t = 2/60 = 1/30$, so O'HaganBook's optimal strategy is $R = [0 \ 3/5 \ 2/5]$

Chapter 5 The Mathematics of Finance

5.1 Simple Interest

1. $PV = 2000$, $r = 0.06$, $t = 1$

$$INT = PVrt$$
$$= 2000(0.06)(1) = \$120$$
$$FV = PV + INT$$
$$= 2000 + 120 = \$2120$$

3. $PV = 20,200$, $r = 0.05$, $t = \frac{1}{2}$

$$INT = PVrt$$
$$= 20,200(0.05)(\tfrac{1}{2}) = \$505$$
$$FV = PV + INT$$
$$= 20,200 + 505 = \$20,705$$

5. $PV = 10,000$, $r = 0.03$, $t = 10/12$

$$INT = PVrt$$
$$= 10,000(0.03)(10/12) = \$250$$
$$FV = PV + INT$$
$$= 10,000 + 250 = \$10,250$$

7. $PV = 10,000/(1 + 0.02 \times 5) = \$9090.91$

9. $PV = 1000/(1 + 0.07 \times 0.5) = \$966.18$

11. $PV = 15,000/(1 + 0.03 \times 15/12) = \$14,457.83$

13. $FV = 5000(1 + 0.08 \times 0.5) = \$5200$

15. $FV = \$1000$, $r = 0.045$, $t = 6$

$$FV = PV(1 + rt)$$
$$1000 = PV(1 + 0.045 \times 6)$$
$$1000 = PV(1.27)$$
$$PV = \frac{1000}{1.27} \approx \$787.40$$

17. We are given the interest and asked to compute r.

$$INT = 250, PV = 1000, t = 5$$
$$INT = PVrt$$
$$250 = 1000 \times r \times 5 = 5000r$$

$$r = \frac{250}{5000} = 0.05, \text{ or } 5\%$$

19. Actual discount $= 0.0025/2 = 0.00125$
Selling price $= 5000 - (0.00125)(5000) = \$4993.75$
$PV = \$4993.75$, $FV = \$5000$, $t = 0.5$

$$FV = PV(1 + rt)$$
$$5000 = 4993.75(1 + 0.5r)$$
$$1 + 0.5r = 5000/4993.75$$
$$0.5r = 5000/4993.75 - 1$$
$$r = 2(5000/4993.75 - 1) \approx 0.002503,$$
or 0.2503%

21. We are given present and future values, and want to compute t.

$$FV = 4640, PV = 4000, r = 0.08$$
$$FV = PV(1 + rt)$$
$$4640 = 4000(1 + 0.08t)$$
$$1 + 0.08t = \frac{4640}{4000} = 1.160$$
$$0.08t = 0.160$$
$$t = \frac{0.160}{0.08} = 2 \text{ years.}$$

23. To say that the discount rate is 3.705% means that its selling price (PV) is 3.705% lower than its maturity value (FV). To simplify the calculation, let us use a T bill with a maturity value of $10,000:

$$PV = 10,000 - 10,000(0.03705/2)$$
$$= 9814.75. \ 10,000$$
$$= 9814.75(1 + 0.5r)$$
$$= 9814.75 + 4907.375r$$
$$r = (10,000 - 9814.75)/4907.375$$
$$= 0.03775, \text{ or } 3.775\%$$

25. $1050 = 1000(1 + 4r/52) = 1000 + 4000r/52$; $r = 50 \times 52/4000 = 0.65$, which is 65% interest.

27. You will pay $5000 \times 0.09 \times 2 = \$900$ in interest on the loan. Adding the $100 fee, you pay the bank a total of $1000 over the two years. To find the effective rate, we solve $6000 = 5000(1 +$

$2r) = 5000 + 10,000r;\ r = 1000/10,000 = 0.1$, so the rate is 10%.

29. $PV = 185.64\ FV = 94.00\ t = 6$ months $= 0.5$ years

$94.00 = 185.64\ (1 + 0.5r) = 185.64 + 92.82r$

$r = (94.00 - 185.64)/\ 92.82 \approx -0.9873 = -98.73\%$

31. The only dates on which would you would have gotten an increase rather than a decrease were May and June. Calculating the annual returns as in the preceding exercises, we get the following figures:

Jan–May: 5.67%

Jan–June: 7.45%

The largest annual return would have been 7.45% if you had sold in June 2008

33. No. Simple interest increase is linear. The graph is visibly not linear in that time period. Further, the slopes of the lines through the successive pairs of marked points are quite different.

35. 1950 population: $PV = 500,000$

2000 population: $FV = 2,800,000$

$INT = FV - PV$

$\quad = 2,800,000 - 500,000$

$\quad = 2,300,000$

$INT = PVrt$

$2,300,000 = 500,000r(50) = 25,000,000r$

$r = \dfrac{2,300,000}{25,000,000} \approx 0.092$ or 9.2%

37. We are given:

$PV = 1950$ population $= 500,000$

$INT = 0.092$ (from Exercise 23)

$t = 60$ (years to 2010)

$FV = PV(1 + rt)$

$\quad = 500,000(1 + 0.092 \times 60)$

$\quad = 3,260,000$

39. $PV = 1950$ population $= 500,000$

$INT = 0.092$ (from Exercise 33)

After t years, the population will be

$FV = PV(1 + rt)$

$\quad = 500,000(1 + 0.092t)$

$\quad = 500,000 + 46t$

(t = time in years since 1950).

Graph:

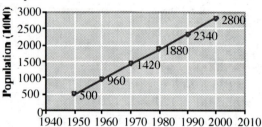

41. (A) is the only possible choice because the equation $FV = PV(1+rt) = PV + PVrt$ gives the future value as a linear function of time.

43. Wrong. In simple interest growth, the change each year is a fixed percentage of the *starting* value, and not the preceding year's value. (Also see the next exercise.)

45. Simple interest is always calculated on a constant amount, PV. If interest is paid into your account, then the amount on which interest is calculated increases, and hence does not remain constant.

5.2 Compound Interest

1. $PV = \$10,000, r = 0.03, m = 1, t = 10$

$FV = PV\left(1 + \dfrac{r}{m}\right)^{mt}$

$= 10,000(1 + 0.03)^{10}$

$= 10,000(1.03)^{10} \approx \$13,439.16$

Technology: `10000*(1+0.03)^10`

3. $PV = \$10,000, r = 0.025, m = 4, t = 5$

$FV = PV\left(1 + \dfrac{r}{m}\right)^{mt}$

$= 10,000\left(1 + \dfrac{0.025}{4}\right)^{(4)(5)}$

$= 10,000(1.00625)^{20} \approx \$11,327.08$

Technology: `10000*(1+0.025/4)^(4*5)`

5. $PV = \$10,000, r = 0.065, m = 365, t = 10$

$FV = PV\left(1 + \dfrac{r}{m}\right)^{mt}$

$= 10,000\left(1 + \dfrac{0.065}{365}\right)^{(365)(10)}$

$= 10,000\left(1 + \dfrac{0.065}{365}\right)^{3650} \approx \$19,154.30$

Technology:
`10000*(1+0.065/365)^(365*10)`

7. Since the monthly interest rate is given, we use the formula $FV = PV(1 + i)^n$.

$PV = \$10,000, i = 0.002, n = 10\times12 = 120$

$FV = PV(1 + i)^n$

$= 10,000(1+0.002)^{120}$

$= 10,000(1.002)^{120} \approx \$12,709.44$

Technology: `10000*(1+0.002)^120`

9. $FV = \$1000, t = 10, r = 0.05, m = 1$

$PV = \dfrac{FV}{\left(1 + \dfrac{r}{m}\right)^{mt}}$

$= \dfrac{1000}{\left(1 + \dfrac{0.05}{1}\right)^{(1)(10)}}$

$= \dfrac{1000}{1.05^{10}} \approx \$613.91$

Technology: `1000/(1+0.05)^10`

11. $FV = \$1000, t = 5, r = 0.042, m = 52$

$PV = \dfrac{FV}{\left(1 + \dfrac{r}{m}\right)^{mt}}$

$= \dfrac{1000}{\left(1 + \dfrac{0.042}{52}\right)^{(52)(5)}}$

$= \dfrac{1000}{\left(1 + \dfrac{0.042}{52}\right)^{260}} \approx \$810.65$

Technology: `1000/(1+0.042/52)^(52*5)`

13. $FV = \$1000, n = 4, i = -0.05$

$PV = \dfrac{FV}{(1 + i)^n}$

$= \dfrac{1000}{(1-0.05)^4} = \dfrac{1000}{0.95^4} \approx \$1,227.74$

Technology: `1000/(1-0.05)^4`

15. $r_{nom} = 0.05, m = 4$

$r_{eff} = \left(1 + \dfrac{r_{nom}}{m}\right)^m - 1$

$= \left(1 + \dfrac{0.05}{4}\right)^4 - 1$

≈ 0.0509, or 5.09%

Technology: `(1+0.05/4)^4-1`

17. $r_{nom} = 0.10, m = 12$

$r_{eff} = \left(1 + \dfrac{r_{nom}}{m}\right)^m - 1$

$= \left(1 + \dfrac{0.10}{12}\right)^{12} - 1$

≈ 0.1047, or 10.47%

Technology: `(1+0.10/12)^12-1`

19. $r_{\text{nom}} = 0.10$, $m = 365 \times 24 = 8760$

$$r_{\text{eff}} = \left(1 + \frac{r_{\text{nom}}}{m}\right)^m - 1$$

$$= \left(1 + \frac{0.10}{8760}\right)^{8760} - 1$$

$$\approx 0.1052, \text{ or } 10.52\%$$

Technology: `(1+0.10/8760)^8760-1`

21. $PV = \$1000$, $r = 0.06$, $m = 4$, $t = 4$

$$FV = PV\left(1 + \frac{r}{m}\right)^{mt}$$

$$= 1000\left(1 + \frac{0.06}{4}\right)^{(4)(4)}$$

$$= 1000(1.015)^{16} \approx \$1268.99$$

The deposit will have grown by
$\$1268.99 - \$1000 = \$268.99$

Technology: `1000*(1+0.06/4)^(4*4)`

23. $PV = \$3000$, $r = -0.376$, $m = 1$, $t = 3$

$$FV = PV\left(1 + \frac{r}{m}\right)^{mt}$$

$$= 3000\left(1 - \frac{0.376}{1}\right)^{(1)(3)}$$

$$= 3000(1 - 0.376)^3 \approx \$728.91$$

Technology: `3000*(1-0.376)^3`

25. $FV = \$5000$, $r = 0.055$, $t = 10$, $m = 1$

$$PV = \frac{FV}{\left(1 + \frac{r}{m}\right)^{mt}}$$

$$= \frac{5000}{\left(1 + \frac{0.055}{1}\right)^{(1)(10)}}$$

$$= \frac{5000}{1.055^{10}} \approx \$2927.15$$

Technology: `5000/(1+0.055)^10`

27. Gold: $PV = \$5000$, $r = 0.10$, $m = 1$, $t = 10$

$$FV = PV\left(1 + \frac{r}{m}\right)^{mt}$$

$$= 5000\left(1 + \frac{0.10}{1}\right)^{(1)(10)}$$

$$= 5000(1.10)^{10} \approx \$12,968.71$$

CDs: $PV = \$5000$, $r = 0.05$, $m = 2$, $t = 10$

$$FV = PV\left(1 + \frac{r}{m}\right)^{mt}$$

$$= 5000\left(1 + \frac{0.05}{2}\right)^{(2)(10)}$$

$$= 5000(1.025)^{20} \approx \$8193.08$$

Combined value after 10 years
$= \$12,968.71 + \$8193.08 = \$21,161.79$

Technology: `5000*(1+0.10)^10+`
`5000*(1+0.05/2)^(2*10)`

29. $PV = \$200,000$, $i = -0.02$, $n = 10$

$$FV = PV(1 + i)^n$$

$$= 200,000(1 - 0.02)^{10}$$

$$= 200,000(0.98)^{10} \approx \$163,414.56$$

Technology: `200000*(1-0.02)^10`

31. $FV = \$100,000$, $r = 0.04$, $m = 1$, $t = 15$

$$PV = \frac{FV}{\left(1 + \frac{r}{m}\right)^{mt}}$$

$$= \frac{100,000}{\left(1 + \frac{0.04}{1}\right)^{(1)(15)}}$$

$$= \frac{100,000}{1.04^{15}} \approx \$55,526.45 \text{ per year}$$

Technology: `100000/(1+0.04)^15`

33. $FV = \$1,000,000$, $r = 0.06$, $m = 1$, $t = 30$

$$PV = \frac{FV}{\left(1 + \frac{r}{m}\right)^{mt}}$$

$$= \frac{1,000,000}{\left(1 + \frac{0.06}{1}\right)^{(1)(30)}}$$

$$= \frac{1,000,000}{1.06^{30}} \approx \$174,110$$

Technology: `1000000/(1+0.06)^30`

35. $FV = \$297.91, n = 6 \times 3 = 18, i = -0.05$

$$PV = \frac{FV}{(1 + i)^n}$$

$$= \frac{297.91}{(1 - 0.05)^{18}} = \frac{297.91}{0.95^{18}} \approx \$750.00$$

Technology: `297.91/(1-0.05)^18`

37. $FV = \$30,000, n = 5, i = 0.02$

$$PV = \frac{FV}{(1 + i)^n}$$

$$= \frac{30,000}{(1 + 0.02)^5} = \frac{30,000}{1.02^5} \approx \$27,171.92$$

Technology: `30000/(1+0.02)^5`

39. $FV = \$200,000, n = 10, i = 0.06$

$$PV = \frac{FV}{(1 + i)^n}$$

$$= \frac{200,000}{(1 + 0.06)^{10}} = \frac{200,000}{1.06^{10}} \approx \$111,678.96$$

Technology: `200000/(1+0.06)^10`

41. Step 1: Calculate the future value of the investment:

$PV = \$1000, i = 0.05, t = 2$

$FV = PV(1 + i)^n$

$\quad = 1000(1 + 0.05)^2$

$\quad = 1000(1.05)^2 \approx \$1102.50$

Technology: `1000*(1+0.05)^2`

Step 2: Discount this value using inflation:

$FV = \$1102.50, i = 0.03, t = 2$

$$PV = \frac{FV}{(1 + i)^n}$$

$$= \frac{1102.50}{(1 + 0.03)^2} = \frac{1102.50}{1.03^2} \approx \$1039.21$$

Technology: `1102.50/(1+0.03)^2`

43. Compare the effective yields of the two investments:

First Investment:

$\quad r_{nom} = 0.12, m = 1$

$\quad r_{eff} = r_{nom} = 0.12,$ or 12%

Second Option:

$r_{nom} = 0.119, m = 12$

$$r_{eff} = \left(1 + \frac{r_{nom}}{m}\right)^m - 1$$

$$= \left(1 + \frac{0.119}{12}\right)^{12} - 1$$

$$\approx 0.1257,$$ or 12.57%

This is the better investment

Technology: `(1+0.119/12)^12-1`

45. $PV = \$24, i = 0.062, n = 2001 - 1626 = 375$

$FV = PV(1 + i)^n$

$\quad = 24(1 + 0.062)^{375}$

$\quad = 24(1.062)^{375} \approx \$150,281$ million

This is more than the 2001 estimated real estate value of \$136,106 million. Thus, the Manhattan Indians could have bought the island back in 2001.

Technology: `24*(1+0.062)^375`

47. $PV = 100$ reals, $i = 0.063, n = 5$

$FV = PV(1 + i)^n$

$\quad = 100(1 + 0.063)^5$

$\quad = 100(1.063)^5 \approx 136$ reals

Technology: `100*(1+0.063)^5`

49. $FV = 1000$ pesos, $i = 0.151$, $n = 10$

$$PV = \frac{FV}{(1 + i)^n}$$

$$= \frac{1000}{(1 + 0.151)^{10}} = \frac{1000}{1.151^{10}}$$

$$\approx 245 \text{ bolivianos}$$

Technology: `1000/(1+0.151)^10`

51. Step 1: Calculate the future value of the investment:

$PV = 1000$ bolivars, $r = 0.08$, $m = 2$, $t = 10$

$$FV = PV\left(1 + \frac{r}{m}\right)^{mt}$$

$$= 1000\left(1 + \frac{0.08}{2}\right)^{(2)(10)}$$

$$= 1000(1.04)^{20} \approx 2191.12 \text{ bolivars}$$

Technology: `1000*(1+0.08/2)^20`

Step 2: Discount this amount using the inflation rate:

$FV = 2191.12$, $i = 0.257$, $n = 10$

$$PV = \frac{FV}{(1 + i)^n}$$

$$= \frac{2191.12}{(1 + 0.257)^{10}} = \frac{2191.12}{1.257^{10}} \approx 223 \text{ bolivars}$$

Technology: `2191.12/(1+0.257)^10`

Note: If we first round the answer of Step 1 to the nearest peso, in Step 2 we get

$$PV = \frac{2191}{1.257^{10}} \approx 222 \text{ bolivars}$$

53. We compare the future values of one unit of the currency for a one-year period:

Mexico: $PV = 1$ peso, $r = 0.053$, $m = 1$, $t = 1$

$$FV = PV\left(1 + \frac{r}{m}\right)^{mt}$$

$$= 1\left(1 + \frac{0.053}{1}\right)^{(1)(1)}$$

$$= 1.053 \text{ pesos}$$

Now discount this using inflation:

$FV = 1.053$, $i = 0.05$, $n = 1$

$$PV = \frac{FV}{(1 + i)^n}$$

$$= \frac{1.053}{(1 + 0.05)^1} = \frac{1.053}{1.05} \approx 1.0029 \text{ pesos}$$

Nicaragua: $PV = 1$ gold cordoba, $r = 0.14$, $m = 2$, $t = 1$

$$FV = PV\left(1 + \frac{r}{m}\right)^{mt}$$

$$= 1\left(1 + \frac{0.14}{2}\right)^{(2)(1)}$$

$$= 1.07^2 \approx 1.1449 \text{ gold cordobas}$$

Now discount this using inflation:

$FV = 1.1449$, $i = 0.138$, $n = 1$

$$PV = \frac{FV}{(1 + i)^n}$$

$$\approx \frac{1.1449}{(1 + 0.138)^1} = \frac{1.1449}{1.138} \approx 1.00606 \text{ gold cordobas}$$

The investment in Nicaragua is better.

55. $PV = 185.64$, $FV = 94.00$, $t = 6$ months $= 1/2$ years

$$94.00 = 185.64 (1 + r)^{1/2}$$

$$r = (94.00/185.64)^2 - 1 \approx -0.7436 \text{ or } -74.36\%$$

57. The only dates on which would you would have gotten an increase rather than a decrease were May and June. Calculating the annual returns as in the preceding exercises, we get the following figures:

Jan–May: 5.77%

Jan–June: 7.61%

The largest annual return would have been 7.61% if you had sold in June 2008

59. No. Compound interest increase is exponential, and exponential curves either increase continually (in the case of appreciation) or decrease continually (in the case of depreciation). The graph of the stock price first decreases and then increases during the given period, so the curve cannot be compound interest change.

61. My investment:
$PV = \$5000$, $r = 0.054$, $m = 2$

$$FV = PV\left(1 + \frac{r}{m}\right)^{mt}$$

$$= 5000\left(1 + \frac{0.054}{2}\right)^{2t} = 5000(1.027)^{2t}$$

Friend's investment:
$PV = \$6000$, $r = 0.048$, $m = 2$

$$FV = PV\left(1 + \frac{r}{m}\right)^{mt}$$

$$= 6000\left(1 + \frac{0.048}{2}\right)^{2t} = 6000(1.024)^{2t}$$

Technology Formulas:
$Y_1 = 5000*1.027^{(2*x)}$
$Y_2 = 6000*1.024^{(2*x)}$
Graph:

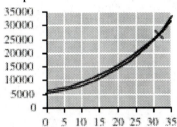

Zoomed-in View:

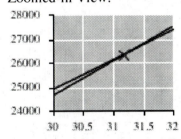

The graphs cross around $t \approx 31$ years. The value of the investment is

$5000(1.027)^{2(31)} \approx \$26100$ (rounded to 3 significant digits)

63. $PV = 40,000$ $i = 1.0$ (a 100% increase per period)

$$FV = PV(1 + i)^n = 40,000(1+1.0)^n$$

$$= 40,000(2)^n$$

$Y_1 = 40000*2^{\wedge}x$
$Y_2 = 1000000$
Graph:

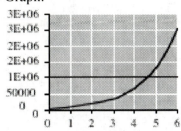

Zoomed-in View:

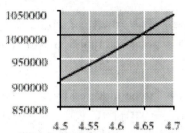

$n \approx 4.65$. Since n measures 6-month periods, this corresponds to $4.65/2 \approx 2.3$ years.

65. (a) The amount you paid for the bond is its present value.
$FV = \$100,000$, $i = 0.15$, $t = 30$

$$PV = \frac{FV}{(1 + i)^n}$$

$$\approx \frac{100,000}{(1 + 0.15)^{30}} = \frac{100,000}{1.15^{30}} \approx \$1510.31$$

(b) The value of the bond at any time is its present value at that time, given the prevailing interest rate. Since it will pay $100,000 in 13 years' time, we have:
$FV = \$100,000$, $i = 0.0475$, $t = 13$

$$PV = \frac{FV}{(1 + i)^n}$$

$$\approx \frac{100,000}{(1 + 0.0475)^{13}} = \frac{100,000}{1.0475^{13}}$$

$$\approx \$54,701.29$$

(c) By part (a) the bond cost you $1510.31 and, by part (b), was worth $54,701.29 after 17 years.
$PV = \$1510.31$, $FV = \$54,701.29$

$$FV = PV(1 + i)^n$$

$$54{,}701.29 = 1510.31(1 + i1)^{17}$$

$$\frac{54{,}701.29}{1510.31} = (1 + i1)^{17}$$

$$1+i = \left(\frac{54{,}701.29}{1510.31}\right)^{1/17}$$

$$i = \left(\frac{54{,}701.29}{1510.31}\right)^{1/17} - 1 \approx 0.2351, \text{ or } 23.51\%$$

Technology:
```
(54701.29/1510.31)^(1/17)-1
```

67. The function $y = P(1 + {}^r/_m)^{mx}$ is not a linear function of x, but an exponential function. Thus, its graph is not a straight line.

69. Wrong. Its growth can be modeled by
$$0.01(1 + 0.10)^t = 0.01(1.10)^t.$$
This is an exponential function of t; not a linear one.

71. A compound-interest investment behaves as though it was being compounded once a year at the effective rate. Thus, if two equal investments have the same effective rates, they grow at the same rate.

73. The effective rate exceeds the nominal rate when the interest is compounded more than once a year, since then interest is being paid on interest accumulated during each year, resulting in a larger effective rate. Conversely, if the interest is compounded less often than once a year, the effective rate is less than the nominal rate.

75. Compare their future values in constant dollars. That is, future compute their future values, and then discount each for inflation. The investment with the larger future value is the better investment.

77. $PV = 100$, $r = 0.10$, $m = 1, 10, 100, \ldots$
```
Y₁ = 100*(1.10)^x
Y₂ = 100*(1+ 0.10/10)^(10*x)
```
$$Y_1 = 100*(1.10)\wedge x$$
$$Y_2 = 100*(1+ 0.10/10)\wedge(10*x)$$

```
Y₃ = 100*(1+ 0.10/100)^(100*x)
```
$$Y_3 = 100*(1+ 0.10/100)\wedge(100*x)$$
...

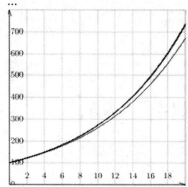

The graphs are approaching a particular curve (shown darker) as m gets larger, approximately the curve given by the largest value of m.

5.3 Annuities, Loans, and Bonds

1. $FV = PMT\dfrac{(1 + i)^n - 1}{i}$

$PMT = \$100$,
i = interest paid each period = $0.05/12$
n = total number of periods = $12 \times 10 = 120$

$FV = 100\dfrac{(1 + 0.05/12)^{120} - 1}{(0.05/12)} \approx \$15{,}528.23$

Technology:
`100*((1+0.05/12)^120-1)/(0.05/12)`

3. $FV = PMT\dfrac{(1 + i)^n - 1}{i}$

$PMT = \$1000$,
i = interest paid each period = $0.07/4$
n = total number of periods = $4 \times 20 = 80$

$FV = 1000\dfrac{(1 + 0.07/4)^{80} - 1}{(0.07/4)} \approx \$171{,}793.82$

Technology:
`1000*((1+0.07/4)^80-1)/(0.07/4)`

5. We need to calculate the sum of
$FV = PV(1 + i)^n$
and
$FV = PMT\dfrac{(1 + i)^n - 1}{i}$
where
$PV = \$5000$, $PMT = \$100$,
i = interest paid each period = $0.05/12$
n = total number of periods = $12 \times 10 = 120$

$FV = 5000(1 + 0.05/12)^{120}$

$\quad + 100\dfrac{(1 + 0.05/12)^{120} - 1}{(0.05/12)} \approx \$23{,}763.28$

Technology:
`5000*(1+0.05/12)^120+`
`100*((1+0.05/12)^120-1)/(0.05/12)`

7. $PMT = FV\dfrac{i}{(1 + i)^n - 1}$

$FV = \$10{,}000$, $i = 0.05/12$, $n = 12 \times 5 = 60$

$PMT = 10{,}000\dfrac{(0.05/12)}{(1 + 0.05/12)^{60} - 1}$

$\quad \approx \$147.05$

Technology:
`10000*0.05/12/((1+0.05/12)^60-1)`

9. $PMT = FV\dfrac{i}{(1 + i)^n - 1}$

$FV = \$75{,}000$, $i = 0.06/4$, $n = 4 \times 20 = 80$

$PMT = 75{,}000\dfrac{(0.06/4)}{(1 + 0.06/4)^{80} - 1}$

$\quad \approx \$491.12$

Technology:
`75000*0.06/4/((1+0.06/4)^80-1)`

11. We first account for the future value of the $10,000 already in the account:
$PV = \$10{,}000$, $i = 0.05/12$, $n = 12 \times 5 = 60$
$FV = PV(1 + i)^n = 10{,}000(1 + 0.05/12)^{60}$
We subtract this from $20,000 to get the future value of the payments, so:

$PMT = FV\dfrac{i}{(1 + i)^n - 1}$ where

$FV = \$20{,}000 - 10{,}000(1 + 0.05/12)^{60}$:
PMT

$\quad = \dfrac{[20{,}000 - 10{,}000(1 + 0.05/12)^{60}]\,(0.05/12)}{(1 + 0.05/12)^{60} - 1}$

$\quad \approx \$105.38$

Technology:
`(20000-10000*(1+0.05/12)^60)*`
`0.05/12/((1+0.05/12)^60-1)`

13. $PMT = \$500$, $i = 0.03/12$, $n = 12 \times 20 = 240$

$PV = PMT\dfrac{1 - (1 + i)^{-n}}{i}$

$\quad = 500\dfrac{1 - (1 + 0.03/12)^{-240}}{0.03/12}$

$\quad \approx \$90{,}155.46$

Technology:
`500*(1-(1+0.03/12)^(-240))/`
`(0.03/12)`

15. $PMT = \$1500, i = 0.06/4, n = 4\times20 = 80$

$$PV = PMT\frac{1 - (1 + i)^{-n}}{i}$$

$$= 1500\frac{1 - (1 + 0.06/4)^{-80}}{0.06/4}$$

$$\approx \$69{,}610.99$$

Technology:
```
1500*(1-(1+0.06/4)^(80))/(0.06/4)
```

17. $FV = \$10{,}000, PMT = \$500, i = 0.03/12,$
$n = 12\times20 = 240$
We need to fund both the future value and the payments, so the present value is the sum

$$PV = FV(1 + i)^{-n} + PMT\frac{1 - (1 + i)^{-n}}{i}$$

$$= 10{,}000(1 + 0.03/12)^{-240}$$

$$+ 500\frac{1 - (1 + 0.03/12)^{-240}}{0.03/12}$$

$$\approx \$95{,}647.68$$

Technology:
```
10000*(1+0.03/12)^(-240)+
500*(1-(1+0.03/12)^(-240))/
(0.03/12)
```

19. $PV = \$100{,}000, i = 0.03/12,$
$n = 12\times20 = 240$

$$PMT = PV\frac{i}{1 - (1 + i)^{-n}}$$

$$= 100{,}000\frac{0.03/12}{1 - (1 + 0.03/12)^{-240}}$$

$$\approx \$554.60$$

Technology:
```
100000*(0.03/12)/
(1-(1+0.03/12)^(-240))
```

21. $PV = \$75{,}000, i = 0.04/4 = 0.01,$
$n = 4\times20 = 80$

$$PMT = PV\frac{i}{1 - (1 + i)^{-n}}$$

$$= 75{,}000\frac{0.01}{1 - (1 + 0.01)^{-80}}$$

$$\approx \$1366.41$$

Technology:
```
75000*0.01/(1-(1+0.01)^(-80))
```

23. Part of the present value has to fund the future value of \$10,000:
$FV = \$10{,}000, i = 0.03/12, n = 12\times20 = 240$
so

$$FV(1 + i)^{-n} = 10{,}000(1 + 0.03/12)^{-240}$$

is required; we subtract this from the present value and use

$$PV = 100{,}000 - 10{,}000(1 + 0.03/12)^{-240}$$

in the payment formula.

$$PMT = PV\frac{i}{1 - (1 + i)^{-n}}$$

$$= \frac{(0.03/12)[100{,}000 - 10{,}000(1 + 0.03/12)^{-240}]}{1 - (1 + 0.03/12)^{-240}}$$

$$\approx \$524.14$$

Technology:
```
(0.03/12)*(100000-
10000*(1+0.03/12)^(-240))/
(1-(1+0.03/12)^(-240))
```

In Exercises 25–28, recall that loans are annuities in the eyes of the lender.

25. $PV = \$10{,}000, i = 0.09/12 = 0.0075,$
$n = 4\times12 = 48$

$$PMT = PV\frac{i}{1 - (1 + i)^{-n}}$$

$$= 10{,}000\frac{0.0075}{1 - (1 + 0.0075)^{-48}}$$

$$\approx \$248.85$$

Technology:
```
10000*0.0075/(1-(1+0.0075)^(-48))
```

27. $PV = \$100{,}000, i = 0.05/4, n = 4\times20 = 80$

$$PMT = PV\frac{i}{1 - (1 + i)^{-n}}$$

$$= 100{,}000\frac{0.05/4}{1 - (1 + 0.05/4)^{-80}}$$

$\approx \$1984.65$

Technology:
```
100000*(0.05/4)/
(1-(1+0.05/4)^(-80))
```

29. The periodic payments are based on a 4.875% annual payment. For payments twice a year, this is
$PMT = 1000(0.04875/2) = \$24.375$
Since the bond yield is 4.880%,
$i = 0.0488/2 = 0.0244$
$n = 2 \times 10 = 20$
The present value comes from the future value of $1000 and the payments, which we treat like an annuity:

$$PV = FV(1 + i)^{-n} + PMT\frac{1 - (1 + i)^{-n}}{i}$$

$$PV = 1000(1 + 0.0244)^{-20}$$
$$+ 24.375\frac{1 - (1 + 0.0244)^{-20}}{0.0244}$$
$$= 1000(1.0244)^{-20} +$$
$$24.375\frac{1 - 1.0244^{-20}}{0.0244}$$
$$\approx \$999.61$$

Technology:
```
1000*1.0244^(-20)+24.375*(1-
1.0244^(-20))/0.0244
```

Online Time Value of Money Utility

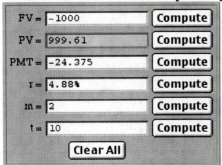

31. The periodic payments are based on a 3.625% annual payment. For payments twice a year, this is

$PMT = 1000(0.03625/2) = \$18.125$
Since the bond yield is 3.705%,
$i = 0.03705/2 = 0.018525$
$n = 2 \times 2 = 4$
The present value comes from the future value of $1000 and the payments, which we treat like an annuity:

$$PV = FV(1 + i)^{-n} + PMT\frac{1 - (1 + i)^{-n}}{i}$$

$$PV = 1000(1 + 0.018525)^{-4}$$
$$+ 18.125\frac{1 - (1 + 0.018525)^{-4}}{0.018525}$$
$$= 1000(1.018525^{-4}$$
$$+ 18.125\frac{1 - 1.018525^{-4}}{0.018525}$$
$$\approx \$998.47$$

Technology:
```
1000*1.018525^(-4)+
18.125*(1-1.018525^(-4))/0.018525
```

Online Time Value of Money Utility

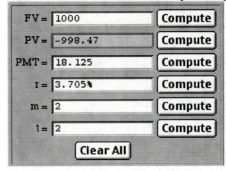

33. FV = maturity value = $1000.
$PMT = 1000(0.035/2) = \$17.50$
PV = Selling price = $994.69
$n = 2 \times 5 = 10$
Using technology, we compute the interest rate to be approximately 3.617%. (The online utility rounds this to two decimal places.)

TI-83/84 Plus

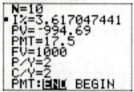

```
N=10
• I%=3.617047441
PV=-994.69
PMT=17.5
FV=1000
P/Y=2
C/Y=2
PMT:END BEGIN
```

Online Time Value of Money Utility

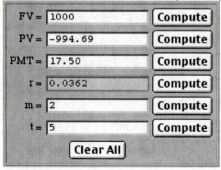

FV = 1000	Compute
PV = -994.69	Compute
PMT = 17.50	Compute
I = 0.0362	Compute
m = 2	Compute
t = 5	Compute
	Clear All

35. FV = maturity value = \$1000.
$PMT = 1000(0.03/2) = \$15.00$
PV = Selling price = \$998.86
$n = 2 \times 2 = 4$

Using technology, we compute the interest rate to be approximately 3.059%. (The online utility rounds this to two decimal places.)

TI-83/84 Plus

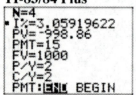

```
N=4
• I%=3.05919622
PV=-998.86
PMT=15
FV=1000
P/Y=2
C/Y=2
PMT:END BEGIN
```

Online Time Value of Money Utility

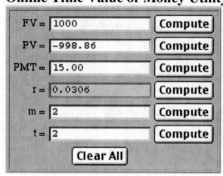

FV = 1000	Compute
PV = -998.86	Compute
PMT = 15.00	Compute
I = 0.0306	Compute
m = 2	Compute
t = 2	Compute
	Clear All

37. This is a two-stage process:

(1) A sinking fund to build the retirement fund.
(2) An annuity depleting the fund during retirement.

For Stage 2, we have
$PMT = \$5000$, $i = 0.03/12 = 0.0025$,
$n = 20 \times 12 = 240$,
and we need to calculate the starting value PV.

$$PV = PMT \frac{1 - (1 + i)^{-n}}{i}$$

$$= 5000 \frac{1 - (1 + 0.0025)^{-240}}{0.0025}$$

$$= 5000 \frac{1 - (1.0025)^{-240}}{0.0025}$$

$$\approx \$901{,}554.57$$

Technology:
```
5000*(1-1.0025^(-240))/0.0025
```

Thus, in Stage 1, you need to accumulate \$901,554.57 in the sinking fund:
$FV = \$901{,}554.57$, $i = 0.03/12 = 0.0025$,
$n = 40 \times 12 = 480$,

$$PMT = FV \frac{i}{(1 + i)^n - 1}$$

$$= \frac{901{,}554.57 \times 0.0025}{1.0025^{480} - 1}$$

$$\approx \$973.54 \text{ per month}$$

Technology:
```
901554.57*0.0025/(1.0025^480-1)
```

39. This is a two-stage process:
(1) A sinking fund to build the retirement fund.
(2) An annuity depleting the fund during retirement.

Stage 1: Building the retirement fund
$PMT = \$1200$, $i = 0.04/4 = 0.01$,
$n = 4 \times 40 = 160$.

$$FV = PMT \frac{(1 + i)^n - 1}{i}$$

$$= 1200 \frac{1.01^{160} - 1}{0.01}$$

$$\approx \$469{,}659.16$$

Technology:
```
1200*(1.01^160-1)/0.01
```

Stage 2: Depleting the fund:
$PV = \$469{,}659.16$, $i = 0.04/4 = 0.01$,
$n = 4 \times 25 = 100$

$$PMT = PV\frac{i}{1 - (1 + i)^{-n}}$$

$$= \frac{469{,}659.16 \times 0.01}{1 - 1.01^{-100}}$$

$$\approx \$7451.49$$

Technology:
```
469659.16*0.01/(1-1.01^(-100))
```

41. Solid Savings & Loan:
$PV = \$10{,}000$, $i = 0.09/12 = 0.0075$,
$n = 12 \times 4 = 48$

$$PMT = PV\frac{i}{1 - (1 + i)^{-n}}$$

$$= \frac{10{,}000 \times 0.0075}{1 - 1.0075^{-48}} \approx \$248.85$$

Technology:
```
10000*0.0075/(1-1.0075^(-48))
```

Fifth Federal Bank & Trust:
$PV = \$10{,}000$, $i = 0.07/12$, $n = 12 \times 3 = 36$

$$PMT = PV\frac{i}{1 - (1 + i)^{-n}}$$

$$= \frac{10{,}000 \times 0.07/12}{1 - (1+0.07/12)^{-36}} \approx \$308.77$$

Technology:
```
10000*(0.07/12)/
(1-(1+0.07/12)^(-36))
```

Answer: You should take the loan from Solid Savings & Loan: it will have payments of $248.85 per month. The payments on the other loan would be more than $300 per month.

43. We can construct an amortization table using the technique outlined in the Technology Guides. For example, using Excel, we might set it up as follows:

	A	B	C	D	E	F
1	Month	Outstanding Principal	Payment on Principle	Interest Payment		
2	0	$50,000			Rate	8%
3	1	=B2-C3	=F$4-D3	=DOLLAR(B2*F$2/12)	Years	15
4	2				Payment	=DOLLAR(-PMT(F2/12,F3*12,B2))
5	3					
6	4					

Adding the payments on principal (Column C) and interest payments (Column D) for each year will give the following table:

Year	Interest	Payment on Principal
1	$3,934.98	$1,798.98
2	$3,785.69	$1,948.27
3	$3,623.97	$2,109.99
4	$3,448.84	$2,285.12
5	$3,259.19	$2,474.77
6	$3,053.77	$2,680.19
7	$2,831.32	$2,902.64
8	$2,590.39	$3,143.57
9	$2,329.48	$3,404.48

10	$2,046.91	$3,687.05
11	$1,740.88	$3,993.08
12	$1,409.47	$4,324.49
13	$1,050.54	$4,683.42
14	$661.81	$5,072.15
15	$240.84	$5,491.80

45. We use the TI-83/84 Plus Finance features. First, calculate the monthly payments for years 1–5:

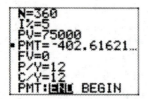

Payments for the first 5 years are therefore $402.62/month.

Now compute the balance at the end of this period (60 months): On the Home Screen, enter

    bal(60,2)

to obtain the remaining balance of $68,871.25.

For years 6–30, compute payments on the remaining balance:

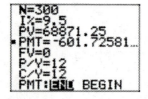

Payments for last 25 years are therefore $601.73/month.

47. We use the TI-83/84 Plus Finance features. First, calculate the monthly payments for the original loan:

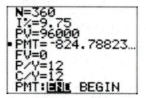

The original monthly payments were therefore $824.79.

Now compute the balance at the end of 4 years (48 months): On the Home Screen, enter

    bal(48,2)

to obtain the remaining balance of $93,383.39.

Had you continued with his loan for the full period of 30 years (360 periods) the remaining interest would have been (use the Home Screen again)

ΣInt(1,360,2)$-\Sigma$Int(1,48,2)

which gives $163,946.49.

You now refinance the outstanding $93,383.39 at 6.875% interest:

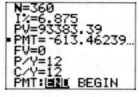

This gives your new monthly payments as $613.46. The total interest you will pay is given by calculating

    ΣInt(1,360,2)

giving $127,464.72. The total savings on interest is therefore

    $163,946.49 − $127,464.72 = $36,481.77.

49. The payments on the loan, ignoring the fee, are

$$PMT = 5000\frac{0.09/12}{1 - (1 + 0.09/12)^{-24}} = \$228.42$$

Add to each payment $100/24 = \$4.17$ to get a new payment of \$232.59. Using technology or trial-and-error, this monthly payment on a 2-year \$5000 loan corresponds to an interest rate of 10.81%.

51.

TI-83/84 Plus

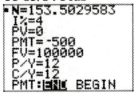

This gives $153.5/12 \approx 13$ years to retirement.

Online Time Value of Money Utility

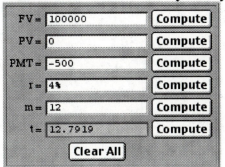

53.

TI-83/84 Plus

This gives $55.798/12 \approx 4.5$ years to repay the debt.

Online Time Value of Money Utility

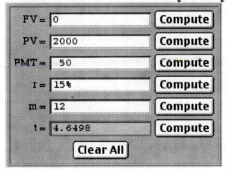

55. Graph the future value of both accounts using the formula for the future value of a sinking fund.

Your account:

$i = 0.045/12 = 0.00375$, $PMT = \$100$

$$FV = PMT\frac{(1 + i)^n - 1}{i}$$

$$= 100\frac{1.00375^n - 1}{0.00375}$$

To graph this, use the technology formula
`100*(1.00375^x-1)/0.00375`

Lucinda's account:

$i = 0.065/12$, $PMT = \$75$.

$$FV = PMT\frac{(1 + i)^n - 1}{i}$$

$$= 75\frac{(1+0.065/12)^n - 1}{(0.065/12)}$$

To graph this, use the technology formula
`75*((1+0.065/12)^x-1)/(0.065/12)`

Graph:

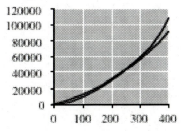

The graphs appear to cross around $n = 300$, so we zoom in there:

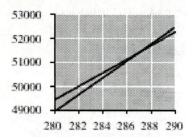

The graphs cross around $n = 287$, so the number of years is approximately $t = 287.5/12 \approx 24$ years.

57. He is wrong because his estimate ignores the interest that will be earned by your annuity—both while it is increasing and while it is decreasing. Your payments will be considerably smaller (depending on the interest earned).

59. He is not correct. For instance, the payments on a $100,000 10-year mortgage 12% are $1434.71, while for a 20-year mortgage at the same rate, they are $1101.09, which is a lot more than half the 10-year mortgage payment.

61. $PV = FV(1 + i)^{-n}$

$\qquad = PMT\dfrac{(1 + i)^n - 1}{i}(1 + i)^{-n}$

$\qquad = PMT\dfrac{1 - (1 + i)^{-n}}{i}$

Chapter 5 Review Exercises

1. $FV = 6000(1 + 0.0475 \times 5) = \$7425.00$

3. $FV = 6000(1 + 0.0475/12)^{60} = \$7604.88$

5. $FV = 100\dfrac{(1 + 0.0475/12)^{60} - 1}{0.0475/12} = \$6757.41$

7. $PV = 6000/(1 + 0.0475 \times 5) = \$4848.48$

9. $PV = 6000(1 + 0.0475/12)^{-60} = \$4733.80$

11. $PV = 100\dfrac{1 - (1 + 0.0475/12)^{-60}}{0.0475/12} =$ $\$5331.37$

13. $PMT = 12{,}000\dfrac{0.0475/12}{(1 + 0.0475/12)^{60} - 1} =$ $\$177.58$

15. $PMT = 6000\dfrac{0.0475/12}{1 - (1 + 0.0475/12)^{-60}} =$ $\$112.54$

17. $PMT = 10{,}000\dfrac{0.0475/12}{1 - (1 + 0.0475/12)^{-60}} =$ $\$187.57$

19. Each interest payment is $10{,}000 \times 0.06/2 = \$300$. For an annuity earning 7% and paying $300 every six months for 5 years, the present value is
$$PV = 300\dfrac{1 - (1 + 0.07/2)^{-10}}{0.07/2} = \$2494.98$$
The present value of the $10,000 maturity value is
$$PV = 10{,}000(1 + 0.07/2)^{-10} = \$7089.19$$
The total price is $\$2494.98 + 7089.19 = \$9584.17$

21. 5.346% (using, for example, the TI-83/84 Plus TVM Solver)

23. $10{,}000 = 6000(1 + 0.0475t) = 6000 + 285t$
$t = (10{,}000 - 6000)/285 = 14.0$ years

25. $10{,}000 = 6000(1 + 0.0475/12)^{12t}$
To solve algebraically requires logarithms:
$$t = \dfrac{\log(10{,}000/6000)}{12\log(1 + 0.0475/12)} \approx 10.8 \text{ years}$$
We could also find this using, for example, the TI-83/84 Plus TVM Solver

27. $10{,}000 = 100\dfrac{(1 + 0.0475/12)^{12t} - 1}{0.0475/12}$
To solve algebraically requires logarithms:
$$t = \dfrac{\log[0.0475 \times 10{,}000/(100 \times 12) + 1]}{12\log(1 + 0.0475/12)}$$
≈ 7.0 years
We could also find this using, for example, the TI-83/84 Plus TVM Solver

29. $PV = 3.28$, $FV = 45.74$, $t = 92$ months $=$ 92/12 years
$45.74 = 3.28(1 + 92r/12) = 3.28 + 25.1467r$
$r = (45.74 - 3.28)/25.1467 = 1.6885 =$ 168.85%

31. The only dates on which would she would have gotten an increase rather than a decrease were November 2004, February 2005, and August 2005. Calculating the annual returns as in the preceding exercises, we get the following figures:
Jan. 2007–Nov. 2009: 60.45%
Jan. 2007–Feb. 2010: 85.28%
Jan. 2007–Aug. 2010: 75.37%
The largest annual return was 85.28% if she sold in February 2010.

33. No. Simple interest increase is linear. We can compare slopes between successive points to see if the slope remained roughly constant: From December 2002 to August 2004 the slope was $(16.31 - 3.28)/(20/12) = 7.818$ while, from August 2004 to March 2005 the slope was $(33.95 - 16.31)/(7/12) = 30.24$. These slopes are quite different.

35. Use the compound interest formula:
$$FV = PV(1 + t)^{n}$$

where $PV = \$150{,}000$, $i = 0.20$, $n = 1, 2, 3, 4,$ 5

2000: $FV = 150{,}000(1.20) = \$180{,}000$

2001: $FV = 150{,}000(1.20)^2 = \$216{,}000$

2002: $FV = 150{,}000(1.20)^3 = \$259{,}200$

2003: $FV = 150{,}000(1.20)^4 = \$311{,}040$

2004: $FV = 150{,}000(1.20)^5 = \$373{,}248$

Revenues first surpass $300,000 in 2003.

37. After the first day of trading, the value of each share will be $6. Thereafter, the shares appreciate in value by 8% per month for 6 months, and you desire a future value of at least $500,000.

$FV = 500{,}000$, $i = 0.08$, $n = 6$

$$PV = FV(1 + i)^{-n}$$
$$= 500{,}000 \times 1.08^{-6} \approx \$315{,}084.81$$

Therefore, since each share will be worth $6 after the first day, the number of shares you must sell is at least

$$\frac{315{,}084.81}{6} \approx 52{,}514.14$$

Since you can offer only a whole number of shares, we must round this up to get the minimum desired future value. Thus, you should offer at least 52,515 shares.

39. $PV = 250{,}000$, $i = 0.095/12$, $n = 12 \times 10 = 120$

$$PMT = PV\frac{i}{1 - (1 + i)^{-n}}$$
$$= \frac{250{,}000 \times 0.095/12}{1 - (1+0.095/12)^{-120}} \approx \$3234.94$$

41. $i = 0.095/12$, $PMT = 3000$, $n = 12 \times 10 = 120$

$$PV = PMT\frac{1 - (1 + i)^{-n}}{i}$$
$$= 3000\frac{1 - (1+0.095/12)^{-120}}{(0.095/12)} \approx \$231{,}844$$

(to the nearest dollar)

43. $PV = 250{,}000$, $FV = 0$, $PMT = 3000$, $n = 12 \times 10 = 120$,
and we are seeking the interest rate

Online Time Value of Money Utility

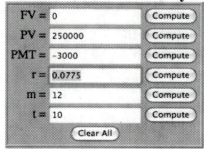

The interest rate would be 7.75%.

45. $PV = 50{,}000$, $PMT = 1000 + 800$ (company contribution) $= 1800$, $i = 0.073/12$, $n = 12 \times 10 = 120$. Considering the contribution of the present $50,000 as well as the payments, we get

$$FV = PV(1 + i)^n + PMT\frac{(1 + i)^n - 1}{i}$$
$$= 50{,}000(1+0.073/12)^{120}$$
$$+ 1800\frac{(1+0.073/12)^{120}-1}{(0.073/12)}$$
$$\approx \$420{,}275$$

(to the nearest dollar)

Technology:
```
50000*(1+0.073/12)^120+
1800*((1+0.073/12)^120-1)/
(0.073/12)
```

47. For the company's contribution, take

$PMT = 800$, $i = 0.073/12$, $n = 12 \times 10 = 120$

$$FV = PMT\frac{(1 + i)^n - 1}{i}$$
$$= 800\frac{(1+0.073/12)^{120}-1}{(0.073/12)} \approx \$140{,}778$$

(to the nearest dollar)

Technology:
```
800*((1+0.073/12)^120-1)/
(0.073/12)
```

49. We first take out the effect of the initial $50,000:

$PV = 50,000$, $i = 0.073/12$, $n = 12 \times 10 = 120$

$FV = PV(1 + i)^n = 50,000(1+0.073/12)^{120}$

Thus, the payments have to result in a future value of

$FV = 500,000 - 50,000(1+0.073/12)^{120}$

$\approx 396,475.163$

51. First compute the amount Callahan needs at the start of retirement:

$PMT = 5000$, $i = 0.087/12$, $n = 12 \times 30 = 360$

$PV = PMT \dfrac{1 - (1 + i)^{-n}}{i}$

$= 5000 \dfrac{1 - (1+0.087/12)^{-360}}{(0.087/12)}$

$\approx \$638,461.93.$

Technology:
```
5000*(1-(1+0.087/12)^(-360))/
(0.087/12)
```

In order to accumulate this amount, using the information from Exercise 45, we first discount the effect of the current $50,000 in the account:

$PV = 50,000$, $FV = 638,461.93$, $i = 0.073/12$, $n = 12 \times 10 = 120$

The initial $50,000 will grow to

$50,000(1+0.073/12)^{120}$

so the payments need to result in a future value of only

$638,461.93 - 50,000(1+0.073/12)^{120}$

$\approx 534,937.093$

Now use the payment formula:

$PMT = FV \dfrac{i}{(1 + i)^n - 1}$

$= \dfrac{534,937.093 \times 0.073/12}{(1+0.073/12)^{120} - 1}$

$\approx \$3039.90$

Since $800 of this is contributed by the company, Callahan's payments should be

$3039.90 - \$800 = \$2239.90$

53. The bond will pay interest every six months amounting to

$50,000(0.072/2) = \$1800$

For someone purchasing this bond after one year, there will be 9 years to maturity, Think of the bond as an investment that will pay the owner $1800 every six months for 9 years, at which time it will pay $50,000. This is exactly the behavior of an annuity paired with an investment with future value $50,000.

$FV = 50,000$, $PMT = 1800$, $i = 0.063/2 = 0.0315$, $n = 2 \times 9 = 18$

The present value has contributions both from the investment and the annuity:

$PV = FV(1 + i)^{-n} + PMT\dfrac{1 - (1 + i)^{-n}}{i}$

$= 50,000(1.0315)^{-18} + 1800\dfrac{1 - 1.0315^{-180}}{0.0315}$

$\approx \$53,055.66$

Technology:
```
50000*1.0315^(-18)+
1800*(1-1.0315^(-18))/0.0315
```

55. Here,

$FV = 50,000$, $PV = 54,000$, $PMT = 1800$, $n = 2 \times 8\frac{1}{2} = 17$,

and we are seeking the interest rate.

TI-83/84 Plus

```
N=17
*I%=5.985470294
PV=-54000
PMT=1800
FV=50000
P/Y=2
C/Y=2
PMT:END BEGIN
```

Online Time Value of Money Utility

FV =	50000	Compute
PV =	-54000	Compute
PMT =	1800	Compute
I =	0.0599	Compute
m =	2	Compute
t =	8.5	Compute
	Clear All	

The interest rate would have to be 5.99%.

Chapter 6 Sets and Counting

6.1 Sets and Set Operations

1. The elements of F are the four seasons: spring, summer, fall, winter. Thus,
$F = \{\text{spring, summer, fall, winter}\}$

3. The elements of I are all the positive integers no greater than 6, namely $1, 2, 3, 4, 5, 6$. Thus,
$I = \{1, 2, 3, 4, 5, 6\}$

5. $A = \{n \mid n$ is a positive integer and $0 \le n \le 3\} = \{1, 2, 3\}$ (Note that 0 is not positive, so we exclude it.)

7. $B = \{n \mid n$ is an even positive integer and $0 \le n \le 8\} = \{2, 4, 6, 8\}$

9. (a) If the coins are distinguishable,
$S = \{(H,H), (H,T), (T,H), (T,T)\}$
(Note that (H, T) and (T, H) are different outcomes, since the first coin is distinguished from the second.)

(b) If the coins are indistinguishable, then (H, T) and (T, H) are the same outcome, and so
$S = \{(H,H), (H,T), (T,T)\}$

11. If the dice are distinguishable, then the outcomes can be thought of as ordered pairs. Thus, since the numbers add to 6,
$S = \{(1,5), (2,4), (3,3), (4,2), (5,1)\}$

13. If the dice are indistinguishable, then the outcomes are characterized only by the numbers that come up, and not by the order. Thus, since the numbers add to 6,
$S = \{(1,5), (2,4), (3,3)\}$

15. Since the numbers facing up can never add to 13 (the largest sum is 12), there are no such outcomes. In other words,
$S = \emptyset$

17.

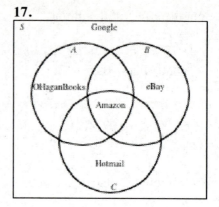

19.

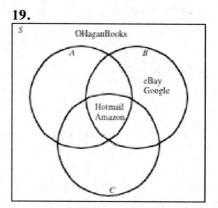

21. $A \cup B$ is the set of all elements that are in A or B (or both): June, Janet, Jill, Justin, Jeffrey, Jello. Thus,
$A \cup B = \{\text{June, Janet, Jill, Justin, Jeffrey, Jello}\}$
$= A$

23. $A \cup \emptyset$ is the set of all elements that are in A or $\emptyset$ (or both). Since $\emptyset$ has no elements,
$A \cup \emptyset = A$
for every set A.

25. $B \cup C = \{\text{Janet, Jello, Justin}\} \cup \{\text{Sally, Solly, Molly, Jolly, Jello}\}$
$= \{\text{Janet, Justin, Sally, Solly, Molly, Jolly, Jello}\}$
Therefore,
$A \cup (B \cup C)$
$= \{\text{June, Janet, Jill, Justin, Jeffrey, Jello}\} \cup \{\text{Janet, Justin, Sally, Solly, Molly, Jolly, Jello}\}$
$= \{\text{June, Janet, Jill, Justin, Jeffrey, Jello, Sally, Solly, Molly, Jolly}\}$

27. $C \cap B$ is the set of all elements that are common to both C and B. Thus,
$C \cap B = \{\text{Jello}\}$

29. $A \cap \emptyset$ is the set of all elements that are common to both A and $\emptyset$. Since $\emptyset$ has no elements,
$A \cap \emptyset = \emptyset$
For every set A.

31. $A \cap B = \{\text{Janet, Jello, Justin}\}$
Therefore,
$(A \cap B) \cap C$
$= \{\text{Janet, Jello, Justin}\} \cap \{\text{Sally, Solly, Molly, Jolly, Jello}\}$
$= \{\text{Jello}\}$

33. $A \cap B = \{\text{Janet, Jello, Justin}\}$
Therefore,
$(A \cap B) \cup C$
$= \{\text{Janet, Jello, Justin}\} \cup \{\text{Sally, Solly, Molly, Jolly, Jello}\}$
$= \{\text{Janet, Justin, Jello, Sally, Solly, Molly, Jolly}\}$

In Exercises 35–42,
$A = \{\text{small, medium, large}\}$
$B = \{\text{blue, green}\}$
$C = \{\text{triangle, square}\}$

35. $A \times C$ is the set of all ordered pairs (a, b) with $a \in A$ and $c \in C$: Therefore,
$A \times C = \{(\text{small, triangle}), (\text{small, square}), (\text{medium, triangle}), (\text{medium, square}), (\text{large, triangle}), (\text{large, square})\}$

37. $A \times B$ is the set of all ordered pairs (a, b) with $a \in A$ and $b \in B$: Therefore,
$A \times B = \{(\text{small, blue}), (\text{small, green}), (\text{medium, blue}), (\text{medium, green}), (\text{large, blue}), (\text{large, green})\}$

39. To represent $B \times C$ we use the elements of $B = \{\text{blue, green}\}$ for the row headings, and the elements of $C = \{\text{triangle, square}\}$ for the column headings:

	A	B	C
1		Triangle	Square
2	Blue	Blue Triangle	Blue Square
3	Green	Green Triangle	Green Square

41. To represent $A \times B$ we use the elements of $A = \{\text{small, medium, large}\}$ for the row headings, and the elements of $B = \{\text{blue, green}\}$ for the column headings:

	A	B	C
1		Blue	Green
2	Small	Small Blue	Small Green
3	Medium	Medium Blue	Medium Green
4	Large	Large Blue	Large Green

43. If a die is rolled and a coin is tossed, each outcome is a pair (b, a) where b is an outcome when a die is rolled and a is an outcome when a coin is tossed. Thus, the set of outcomes is
$B \times A = \{\text{1H, 1T, 2H, 2T, 3H, 3T, 4H, 4T, 5H, 5T, 6H, 6T}\}$

45. If a coin is tossed 3 times, each outcome is a triple (a_1, a_2, a_3) where a_1 and a_2 and a_3 are outcomes when a coin is tossed. Thus, the set of outcomes is
$A \times A \times A = \{\text{HHH, HHT, HTH, HTT, THH, THT, TTH, TTT}\}$

47. E is the set of outcomes in which at least one die shows an even number. Thus, E' is the set of outcomes in which *both* die show an odd number:
$E' = \{(1,1), (1,3), (1,5), (3,1), (3,3), (3,5), (5,1), (5,3), (5,5)\}$

49. $E \cup F$ is the set of outcomes in which either at least one die is even, or at least one is odd. Since this includes *all* the outcomes (in every outcome there is either an odd or even outcome), its complement is empty:
$(E \cup F)' = \emptyset$

51. By Exercise 47, E' is the set of outcomes in which *both* die show an odd number: By Exercise 48, F' is the set of outcomes in which both dice show an even number. Thus, $E' \cup F'$ is the set of outcomes in which either both are even, or both are odd:

$$E'\cup F' = \{(1,1),(1,3),(1,5),(3,1),(3,3),(3,5),$$
$$(5,1),(5,3),(5,5),(2,2),(2,4),(2,6),(4,2),(4,4),$$
$$(4,6),(6,2),(6,4),(6,6)\}$$

53. $(A\cup B)'$ is the region outside $A\cup B$, while $A'\cap B'$ is the overlap of the region outside A with that outside B (the grey region in the diagram below):

$$(A\cup B)' = A'\cap B'$$

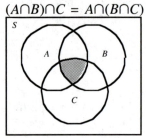

55. $(A\cap B)\cap C$ is the overlap of $A\cap B$ and C, which is the same as the overlap of all three sets: A, B, C. Similarly for $A\cap(B\cap C)$ (the grey region in the diagram below):

$$(A\cap B)\cap C = A\cap(B\cap C)$$

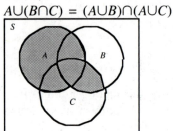

57. $A\cup(B\cap C)$ is the union of A with $B\cap C$, and so consists of all elements in A together with those in both B and C. This is the same as $(A\cup B)\cap(A\cup C)$ (the grey region in the diagram below):

$$A\cup(B\cap C) = (A\cup B)\cap(A\cup C)$$

59. $S' = \varnothing$

61. The set of clients who owe her money is
$$A = \{\text{Acme, Crafts, Effigy, Global}\}$$
The set of clients who have done at least $10,000 worth of business with her is
$$B = \{\text{Acme, Brothers, Crafts, Dion}\}$$
Therefore, the set of clients who owe her money and have done at least $10,000 worth of business with her is
$$A\cap B = \{\text{Acme, Crafts}\}$$

63. The set of clients who have done at least $10,000 worth of business with her is
$$B = \{\text{Acme, Brothers, Crafts, Dion}\}$$
The set of clients who have employed her in the last year is
$$C = \{\text{Acme, Crafts, Dion, Effigy, Global, Hilbert}\}$$
Therefore, the set of clients who have done at least $10,000 worth of business with her or have employed her in the last year is
$$B\cup C = \{\text{Acme, Brothers, Crafts, Dion, Effigy, Global, Hilbert}\}$$

65. The set of clients who do not owe her money is
$$A' = \{\text{Brothers, Dion, Floyd, Global, Hilbert}\}$$
The set of clients who have employed her in the last year is
$$C = \{\text{Acme, Crafts, Dion, Effigy, Global, Hilbert}\}$$
Therefore, the set of clients who do not owe her money and have employed her in the last year is
$$A'\cap C = \{\text{Dion, Hilbert}\}$$

67. The clients who owe her money is
$$A = \{\text{Acme, Crafts, Effigy, Global}\}$$
The set of clients who have not done at least $10,000 worth of business with her is

$B' = \{\text{Effigy, Floyd, Global, Hilbert}\}$
The set of clients who have not employed her in the last year is
$C' = \{\text{Brothers, Floyd}\}$
Therefore, the set of clients who owe her money, have not done at least \$10,000 worth of business with her, and have not employed her in the last year is
$A \cap B' \cap C' = \emptyset$

69. One can organize the spreadsheet as follows: For the row headings, use the years 2003, 2004, 2005, and 2006. For the column headings, use Sail Boats, Motor Boats, and Yachts:

	A	B	C	D
1		**Sail Boats**	**Motor Boats**	**Yachts**
2	**2003**	(2003 Sail Boats)	(2003 Motor Boats)	(2003 Yachts)
3	**2004**	(2004 Sail Boats)	(2004 Motor Boats)	(2004 Yachts)
4	**2005**	(2005 Sail Boats)	(2005 Motor Boats)	(2005 Yachts)
5	**2006**	(2006 Sail Boats)	(2006 Motor Boats)	(2006 Yachts)

This setup gives a tabular representation of the cartesian product
$\{2003, 2004, 2005, 2006\} \times \{\text{Sail Boats, Motor Boats, Yachts}\}$
Alternatively, one could use the years for the column headings and {Sail Boats, Motor Boats, Yachts} for the row headings, and obtain a representation of
$\{\text{Sail Boats, Motor Boats, Yachts}\} \times \{2003, 2004, 2005, 2006\}$

71. The collection of all iPods and jPods combined is the set of all items that are either iPods (that is, in the set I) or jPods (that is, in the set J), and is therefore the union of the two sets:
$I \cup J$.

73. Techno music that is neither European nor Dutch:
Techno AND NOT (European OR Dutch)
Techno $\cap$ (European $\cup$ Dutch)$'$
(Choice B).

75. Let $A = \{1\}$, $B = \{2\}$, and $C = \{1, 2\}$. Then
$(A \cap B) \cup C = \{1, 2\}$
but
$A \cap (B \cup C) = \{1\}$.
In general, $A \cap (B \cup C)$ must be a subset of A, but $(A \cap B) \cup C$ need not be; also, $(A \cap B) \cup C$

must contain C as a subset, but $A \cap (B \cup C)$ need not.

77. A universal set is a set containing all "things" currently under consideration. When discussing sets of positive integers the universe might be the set of all positive integers, or the set of all integers (positive, negative, and 0), or any other set containing the set of all positive integers.

79. $A \cap (B \cup C')$ means A, and either B or not C. Thus, for instance, take A as the set of suppliers who deliver components on time, B as the set of suppliers whose components are known to be of high quality, and C as the set of suppliers who do not promptly replace defective components. Then selecting suppliers in $A \cap (B \cup C')$ means selecting those who deliver components on time and are either companies whose components are of high quality or are suppliers who promptly replace defective components.

81. Let A = movies that are violent, B = movies that are shorter than two hours, C = movies that have a tragic ending, and D = movies that have an unexpected ending. The given sentence can be rewritten as "She prefers movies in $A' \cap B \cap (C \cup D)'$." It can also be rewritten as "She prefers movies in $A' \cap B \cap C' \cap D'$."

253

6.2 Cardinality

1. $n(A)$ = Number of elements in A = 4, $n(B)$ = Number of elements in B = 5. Therefore,
$n(A) + n(B) = 4 + 5 = 9.$

3. $A \cup B$ = {Dirk, Johan, Frans, Sarie, Tina, Klaas, Henrika}
$n(A \cup B) = 7$

5. $B \cap C$ = {Frans}
$A \cup (B \cap C)$ = {Dirk, Johan, Frans, Sarie}
$n(A \cup (B \cap C)) = 4$

7. From Exercise 3, we know that
$n(A \cup B) = 7$
On the other hand,
$n(A) + n(B) - n(A \cap B) = 4 + 5 - 2 = 7$
Therefore,
$n(A \cup B) = n(A) + n(B) - n(A \cap B)$

9. $n(A \times A) = n(A)n(A) = 2 \times 2 = 4$

11. $n(B \times C) = n(B)n(C) = 6 \times 3 = 18$

13. $n(A \times B \times B) = n(A)n(B \times B) =$
$n(A)n(B)n(B) = 2 \times 6 \times 6 = 72$

15. $n(A \cup B) = n(A) + n(B) - n(A \cap B)$
$= 43 + 20 - 3 = 60$

17. $n(A \cup B) = n(A) + n(B) - n(A \cap B)$
$100 = 60 + 60 - n(A \cap B)$
$n(A \cap B) = 60 + 60 - 100 = 20$

Exercises 19–26 use
S = {BA, MU, SO, SU, LI, MS, WA, RT, DU, LY}
A = {SO, LI, MS, RT}
B = {BA, MU, SO}

19. $n(A') = n(S) - n(A) = 10 - 4 = 6$

21. $n((A \cap B)') = n(S) - n(A \cap B) = 10 - 1 = 9$

23. A' = {BA, MU, SU, WA, DU, LY}

B' = {SU, LI, MS, WA, RT, DU, LY}
$A' \cap B'$ = {SU, WA, DU, LY}
$n(A' \cap B') = 4$

25. $n((A \cap B)') = 9$ from Exercise 21
$n(A') + n(B') - n((A \cup B)') = 6 + 7 - 4 = 9$
Therefore, $n((A \cap B)') = n(A') + n(B') - n((A \cup B)').$

27. Assign labels to the regions of the diagram where the quantities are unknown:

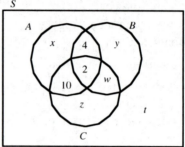

$n(A) = 20 \Rightarrow x + 16 = 20 \Rightarrow \underline{x = 4}$
$n(B \cap C) = 8 \Rightarrow 2 + w = 8 \Rightarrow \underline{w = 6}$
$n(B) = 20 \Rightarrow 6 + y + w = 20$
$\Rightarrow 6 + y + 6 = 20 \Rightarrow \underline{y = 8}$
$n(C) = 28 \Rightarrow 12 + z + w = 28$
$\Rightarrow 12 + z + 6 = 28 \Rightarrow \underline{z = 10}$
$n(S) = 50 \Rightarrow x + y + z + w + t + 16 = 50$
$\Rightarrow 4 + 8 + 10 + 6 + t + 16 = 50 \Rightarrow \underline{t = 6}$
Thus, the completed diagram is

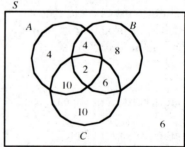

29. Assign labels to the regions of the diagram where the quantities are unknown:

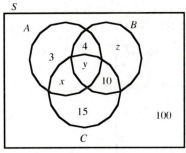

$n(A) = 10 \Rightarrow x + y + 7 = 10 \Rightarrow x + y = 3$
$n(B) = 19 \Rightarrow y + z + 14 = 19 \Rightarrow y + z = 5$
$n(S) = 140 \Rightarrow x + y + z + 132 = 140 \Rightarrow x + y + z = 8$

This is a system of 3 linear equations in 3 unknowns:

$x + y = 3$ (1)
$y + z = 5$ (2)
$x + y + z = 8$ (3)

To solve, we can substitute (1) into (3), giving
$3 + z = 8 \Rightarrow \underline{z = 5}$
(2) now gives $y + 5 = 5 \Rightarrow \underline{y = 0}$
(1) now gives $x + 0 = 3 \Rightarrow \underline{x = 3}$
Thus, the solution is $(x, y, z) = (3, 0, 5)$. The completed diagram is:

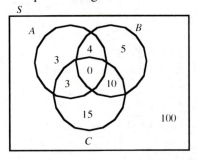

31. Let C be the set of web sites containing "costenoble" and let W be the set of web sites containing the word "waner". We are told that
$n(C) = 93{,}000, n(W) = 428{,}000, n(C \cap W) = 25{,}000$
Therefore,
$n(C \cup W) = n(C) + n(W) - n(C \cap W)$
$= 93{,}000 + 428{,}000 - 25{,}000$
$= 496{,}000$

33. Let B be the set of people who had black hair, and let R be the set of people who had a whole row to themselves. We are told that $n(B \cup R) =$

37, $n(B) = 33$, and $n(R) = 6$. We are asked to find $n(B \cap R)$.
$n(B \cup R) = n(B) + n(R) - n(B \cap R)$
$37 = 33 + 6 - n(B \cap R)$
$n(B \cap R) = 33 + 6 - 37 = 2$

35. $C \cap N$ is the set of authors who are both successful and new.
$C \cup N$ is the set of authors who are either successful or new (or both).
$n(C) = 30, n(N) = 20, n(C \cap N) = 5, n(C \cup N) = 45$
$n(C \cup N) = n(C) + n(N) - n(C \cap N)$
$45 = 30 + 20 - 5$ ✓

37. $C \cap N'$ is the set of authors who are successful but not new—in other words, the set of authors who are successful and established.
$n(C \cap N') = 25$

39. Of the 80 established authors, 25 are successful. Thus, the percentage of established authors who are successful is
$\dfrac{25}{80} = 0.3125$, or 31.25%

Of the 30 successful authors, 25 are established. Thus, the percentage of successful authors who are established is
$\dfrac{25}{30} \approx 0.8333$, or 83.33%

41. The set of sales of existing homes in the Midwest in 2008 is $M \cap C$.
$n(M \cap C) = 1.1$ million

43. The set of sales of existing homes in 2007 excluding sales in the South is $B \cap T'$.
$n(B \cap T') = 5.6 - 2.2 = 3.4$ million

45. The set of sales of existing homes in 2007 in the West and Midwest is $B \cap (W \cup M)$.
$n(B \cap (W \cup M)) = 1.3 + 1.1 = 2.4$ million

47. The set of non-Internet stocks that increased is $V \cap I'$.

255

	P	E	I	Total
V	10	5	15	30
N	30	0	10	40
D	10	5	15	30
Total	50	10	40	100

$n(V \cap I') = 10 + 5 = 15$

49. $n(P' \cup N)$ is the number of stocks that were either not pharmaceutical stocks, or were unchanged in value after a year (or both).

	P	E	I	Total
V	10	5	15	30
N	30	0	10	40
D	10	5	15	30
Total	50	10	40	100

$n(P' \cup N) = 5 + 15 + 30 + 10 + 5 + 15 = 80$

51. $n(V \cap I) = 15; \ n(I) = 40$

	P	E	I	Total
V	10	5	15	30
N	30	0	10	40
D	10	5	15	30
Total	50	10	40	100

$$\frac{n(V \cap I)}{n(I)} = \frac{15}{40} = \frac{3}{8}$$

This is the fraction of Internet stocks that increased in value.

53. Let S be the set of all children in the study, let R be the set of children who had rickets and let U be the set of urban children. We are told that $n(S) = 1556$, $n(R) = 1024$, $n(U) = 243$, $n(U \cap R) = 93$.

Represent the given information in a Venn diagram:

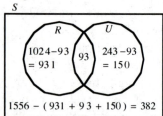

(a) From the diagram, $n(R \cap U') = 931$.

(b) From the diagram, $n(U' \cap R') = 382$.

55. (a) Following is a Venn diagram with most of the unknowns marked with labels:

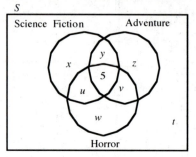

15 had seen a science fiction movie and a horror movie $\Rightarrow 5 + u = 15 \Rightarrow \underline{u = 10}$

5 had seen an adventure movie and a horror movie $\Rightarrow 5 + v = 5 \Rightarrow \underline{v = 0}$

25 had seen a science fiction movie and an adventure movie $\Rightarrow 5 + y = 25 \Rightarrow \underline{y = 20}$

35 had seen a horror movie $\Rightarrow 5 + u + v + w = 35$

$\Rightarrow 5 + 10 + 0 + w = 35 \Rightarrow \underline{w = 20}$

55 had seen an adventure movie $\Rightarrow 5 + y + z + v = 55$

$\Rightarrow 5 + 20 + z + 0 = 55 \Rightarrow \underline{z = 30}$

40 had seen a science fiction movie $\Rightarrow 5 + x + y + u = 40$

$\Rightarrow 5 + x + 20 + 10 = 40 \Rightarrow \underline{x = 5}$

Finally, the total number of people in the survey was 100

$\Rightarrow 5 + x + y + z + u + v + w + t = 100$

$\Rightarrow 5 + 5 + 20 + 30 + 10 + 0 + 20 + t = 100$

$\Rightarrow \underline{t = 10}$

Here is the completed diagram:

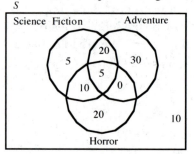

(b) Of the 40 people who had seen science fiction, 15 saw a horror movie. Therefore, the percentage of science fiction movie fans who are also horror movie fans can be estimated as

$$\frac{15}{40} = 0.375, \text{ or } 37.5\%.$$

57. Let R be the set of students who liked rock music, and let C be the set of those who liked classical music.

$n(R) = 22$

$n(R \cap C) = 5$

We are also given other information that we do not need.

Since 5 of the 22 people who liked rock also liked classical, the other 17 did not like classical. Therefore, 17 of those that enjoyed rock did not enjoy classical music

59. Since every element of A is in B, and since B contains at least one more element than A (otherwise they would be the same set) $n(A) < n(B)$.

61. The number of elements in the Cartesian product of two finite sets is the product of the number of elements in the two sets.

63. Answers will vary

65. Since

$n(A \cup B) = n(A) + n(B) - n(A \cap B)$,

$n(A \cup B) \neq n(A) + n(B)$ when $n(A \cap B) \neq 0$; that is, when $A \cap B \neq \varnothing$.

67. Since $A \subseteq A \cup B$, the only way they can have the same number of elements is if $A = A \cup B$; that is, when $B \subseteq A$.

69. $n(A \cup B \cup C) = n(A) + n(B) + n(C) - n(A \cap B) - n(B \cap C) - n(A \cap C) + n(A \cap B \cap C)$

6.3 The Addition and Multiplication Principles

1. Alternative 1: 2 outcomes
Alternative 2: 3 outcomes
Alternative 3: 5 outcomes
Total number of outcomes: 2+3+5 = 10

3. Step 1: 2 outcomes
Step 2: 3 outcomes
Step 3: 5 outcomes
Total number of outcomes: $2 \times 3 \times 5 = 30$

5. Alternative 1: $1 \times 2 = 2$ outcomes
Alternative 2: $2 \times 2 = 4$ outcomes
Total number of outcomes: 2 + 4 = 6

7. Step 1: 1+2 = 3 outcomes
Step 2: 2+2+1 = 5 outcomes
Total number of outcomes: $3 \times 5 = 15$

9. Alternative 1: $(3+1) \times 2 = 8$ outcomes
Alternative 2: 5 outcomes
Total number of outcomes: 8 + 5 = 13

11. Step 1: $(3 \times 1)+2 = 6$ outcomes
Step 2: 5 outcomes
Total number of outcomes: $6 \times 5 = 30$

13. Decision algorithm: start with 4 empty slots, and select slots in which to place the letters.
Step 1: Select a slot to place the b: 4 choices
Step 2: Places the a's in the remaining slots: 1 choice
Total number of outcomes: $4 \times 1 = 4$

15. Decision algorithm:
Step 1: Select a flavor: 31 choices
Step 2: Select cone, cup, or sundae: 3 choices
Total number of outcomes: 31×3 = 93

17. Decision algorithm:
Step 1: Select the first bit: 2 choices
Step 2: Select the second bit: 2 choices
Step 3: Select the third bit: 2 choices
Step 4: Select the fourth bit: 2 choices

Total number of outcomes: $2 \times 2 \times 2 \times 2 = 16$

19. Decision algorithm: start with 6 empty slots, and select slots in which to place the ternary digits.
Step 1: Select a slot to place the 1: 6 choices
Step 2: Select a slot to place the 2: 5 choices
Step 3: Place 0s in the remaining slots: 1 choice
Total number of outcomes: $6 \times 5 \times 1 = 30$

21. Alternative 1 (Gummy candy):
Step 1: Select a size: 3 choices
Step 2: Select a shape: 3 choices
Alternative 2 (Licorice nibs):
Step 1: Select a size: 2 choices
Step 2: Select a color: 2 choices
Total number of outcomes: $3 \times 3 + 2 \times 2 = 13$

23. Alternative 1: Single disc:
Step 1: Select a type: 2 choices
Step 2: Select a color: 5 choices
Alternative 2: Spindle of discs:
Step 1: Select spindle size: 2 choices
Step 2: Select a type: 2 choices
Step 3: Select a color: 2 choices
Total number of choices: $(2 \times 5)+(2 \times 2 \times 2) = 18$

25. Step 1: Answer 1st t/f question: 2 choices
Step 2: Answer 2nd t/f question: 2 choices
. . . .
Step 10: Answer 10th t/f question: 2 choices
Step 11: Answer 1st multiple choice question: 5 choices
Step 12: Answer 2nd multiple choice question: 5 choices
Total number of choices: $(2 \times 2 \times ... \times 2) \times (5 \times 5) = 2^{10} \times 5^2 = 25,600$

27. Alternative 1: Do Part A:
Steps 1–8: Answer the 8 t/f questions; 2 choices each, giving 2^8 possible choices
Alternative 2: Answer Part B:
Steps 1–5: Answer 5 multiple choice questions with 5 choices each, giving 5^5 possible choices
Total number of choices: $2^8 + 5^5 = 3381$

258

29.

(a) Step 1: Select a mutual fund: 4 choices
Step 2: Select a muni. bond fund: 3 choices
Step 3: Select a stock: 8 choices
Step 4: Select a precious metal: 3 choices
Total number of choices: $4 \times 3 \times 8 \times 3 = 288$

(b) Step 1: Select 3 mutual funds: 4 choices
(select which one to leave out)
Step 2: Select 2 muni bond funds: 3 choices
(select which one to leave out)
Step 3: Select one stock: 8 choices
Step 4: Select 2 precious metals 3 choices
(select which one to leave out)
Total number of choices: $4 \times 3 \times 8 \times 3 = 288$

31. The number of possible characters that can be represented equals the number of possible bytes:
Steps 1–8: Select 0 or 1 for each bit:
$2 \times 2 \times ... \times 2 = 2^8$ choices
Thus, the number of possible characters is $2^8 = 256$.

33. Decision algorithm to obtain a symmetry of the five-pointed star:
Alternative 1: Pure rotation: 5 choices
Alternative 2: Rotation followed by a flip: 5 choices
Total number of symmetries: $5 + 5 = 10$

35. Alternative 1: Letter only: 26 choices
Alternative 2: Letter plus digit
Step 1: Choose a letter: 26 choices
Step 2: Choose a digit: 10 choices
Total number of variables $= 26 + 26 \times 10 = 286$

37. Step 1: Select a winner of the North Carolina-Central Connecticut game: 2 choices
Step 2: Select a winner of the Virginia-Syracuse game: 1 choice (We know that Syracuse is the winner)
Step 3: Select the overall winner: 2 choices
Total number of choices: $2 \times 1 \times 2 = 4$

39.

(a) Step 1: Choose the first digit: 8 choices
Steps 2–7 Choose the remaining 6 digits: 10^6 choices
Total number: $8 \times 10^6 = 8,000,000$

(b) To count the numbers beginning with 463, 460, or 400, use the following decision algorithm:
Alternative 1: Start with 463
Steps 1–4: Choose the remaining 4 digits: 10^4 choices
Alternative 2: Start with 460
Steps 1–4: Choose the remaining 4 digits: 10^4 choices
Alternative 3: Start with 400
Steps 1–4: Choose the remaining 4 digits: 10^4 choices
Total number of choices: $10^4 + 10^4 + 10^4 = 30,000$

(c) Use the following decision algorithm:
Step 1: Select the 1st digit (other than 0 or 1): 8 choices
Step 2: Select a 2nd digit different from the 1st digit: 9 choices
Step 2: Select a 3rd digit different from the 2nd digit: 9 choices
. . .
Step 7: Select a 7th digit different from the 6th digit: 9 choices
Total number of choices: $8 \times 9^6 = 4,251,528$

41. (a) Steps 1–3: Choose three bases (4 choices each): $4^3 = 64$ choices

(b) Steps 1–3: Choose n bases (4 choices each): 4^n choices

(c) From part (b) with $n - 2.1 \times 10^{10}$, one obtains a total of
$4^n = 4^{2.1 \times 10^{10}}$ possible DNA chains

43. (a) Steps 1–6: Select 6 hexadecimal digits (16 choices per digit): $16^6 = 16,777,216$ possible colors

(b) Step 1: Choose the 1st repeating pair of digits: 16 choices

Step 2: Choose the 2nd repeating pair of digits: 16 choices

Step 3: Choose the 3rd repeating pair of digits: 16 choices

Total number of colors: $16^3 = 4096$

(c) Step 1: Choose the digit x: 16 choices

Step 2: Choose the digit y: 16 choices

Total number of grayscale shades: $16^2 = 256$

(d) Step 1: Choose the position for the sequence xy: 3 choices

Step 2: Choose the values of x and y: 16^2 choices

Total number of choices: 3×16^2 choices. However, the above decision algorithm leads to the sequence 000000 in three ways. The number of pure colors is therefore $3 \times 16^2 - 2 = 766$.

45. Step 1: Choose a male actor to play Escalus: 10 choices

Step 2: Choose a male actor to play Paris: 9 choices

...

Step 6: Choose a male actor to play Tybalt: 5 choices

Step 7: Choose a male actor to play Friar Lawrence: 4 choices

Step 8: Choose a female actor to play Lady Montague: 8 choices

Step 9: Choose a female actor to play Lady Capulet: 7 choices

Step 10: Choose a female actor to play Juliet: 6 choices

Step 11: Choose a female actor to play Juliet's Nurse: 5 choices

Total number of casts:
$(10 \times 9 \times 8 \times 7 \times 6 \times 5 \times 4) \times (8 \times 7 \times 6 \times 5) = 1,016,064,000$

47.

(a) Steps 1–3 Choose 3 letters: 26^3 choices

Steps 4–6: Choose 3 digits: 10^3 choices

Total number of license plates: $26^3 \times 10^3 = 17,576,000$

(b) Steps 1–2 Choose 2 letters: 26^2 choices

Step 3: Choose a letter other than I, O or Q: 23 choices

Steps 4–6: Choose 3 digits: 10^3 choices

Total number of license plates: $26^2 \times 23 \times 10^3 = 15,548,000$

(c) A decision algorithm for the number of reserved plates:

Step 1: Choose VET, MDZ, or DPZ: 3 choices

Step 2–4: Choose 3 digits: 10^3 choices

Total number of reserved plates: 3×10^3

From part (b), the total number of possible plates is 15,548,000. Therefore, the number of unreserved plates is
$$15,548,000 - 3 \times 10^3 = 15,545,000$$

49.

(a) Start with 4 empty slots in which to place the letters R and D.

Step 1: Select a slot for the single D: 4 choices

Step 2: Place Rs in the remaining slots: 1 choice

Total number of sequences: $4 \times 1 = 4$

(b) Each route in the maze is a sequence of 4 moves: either right (R) or down (D). Since only one down move is possible, the number of such routes equals the number of four-letter sequences that contain only the letters R and D, with D occurring only once: 4 possibilities, by part (a)

(c) If we allowed left and/or up moves, we could get an unlimited number of routes, such as
RRRD, RLRRRD, RLRLRRRD, ...

51.

(a) Step 1: Choose a 1st cylinder: 6 choices

Step 2: Choose a 2nd cylinder on the opposite side: 3 choices

Step 3: Choose a 3rd cylinder on the same side as the first: 2 choices

Step 4: Chose a 4th cylinder on the same side as the second: 2 choices

Step 5: Chose a 5th cylinder on the same side as the first: 1 choice

Step 6: Chose a 6th cylinder on the same side as the second: 1 choice

Total number of firing sequences:

$6 \times 3 \times 2 \times 2 \times 1 \times 1 = 72$

(b) Use the same decision algorithm as for part (a), except that, in Step 1, we only have 3 choices (the cylinders on the left).

Total number of firing sequences:

$3 \times 3 \times 2 \times 2 \times 1 \times 1 = 36$

53. Decision algorithm for producing a painting:
   Steps 1–5: Select blue or grey for each of the odd-numbered lines: 2^5 choices
   Step 6: Select a single color for the remaining lines: 3 choices

Total number of paintings: $2^5 \times 3 = 96$

55. (a) Decision algorithm for incorrect codes that will open the lock:
   Step 1: Decide which digit will be the wrong one: 4 choices
   Step 2: Choose an incorrect digit in that place: 9 choices (one is correct, leaving 9 incorrect ones)

Total number of incorrect codes that will open the lock: $4 \times 9 = 36$

(b) From part (a), there are 36 incorrect codes that will open the lock. Also, there is only 1 correct code. Therefore, the total number of codes that will open the lock is

$36 + 1 = 40$

57. Decision algorithm for creating a calendar:
   Step 1: Choose a day of the week for Jan 1: 7 choices
   Step 2: Decide whether or not it is a leap year: 2 choices

Total number of possible calendars: $7 \times 2 = 14$

59. Decision algorithm to select a particular iteration:
   Step 1: Choose a value for i: 10 choices
   Step 2: Choose a value for j: 19 choices
      (19 integers in the range $2-20$)
   Step 3: Choose a value for k: 10 choices

Total number of iterations: $10 \times 19 \times 10 = 1900$

61. Decision algorithm to place a single 1×1 block:
Step 1: Choose a position in the left-right direction: m choices
Step 2: Choose a position in the front-back direction: n choices
Step 3: Choose a position in the up-down direction: r choices
Total number of possible solids: mnr

63. The number of possible sequences of length 1 is 2 (either a dot: · or a dash: –)
The number of possible sequences of length 2 is $2 \times 2 = 4$ (2 steps, 2 choices per step):

·· ·– –· – –

The number of possible sequences of length 3 is $2^3 = 8$ (3 steps, 2 choices per step)
The number of possible sequences of length 4 is $2^4 = 16$ (3 steps, 2 choices per step)
So far, using different lengths from 1 to 4, we can encode $1 + 2 + 4 + 8 + 16 = 31$ possible letters, which is enough to include the whole alphabet. (Using lengths 1–3 will not work, sine that will give only $1 + 2 + 4 + 8 = 15$ possible letters).
Thus, we need to use sequences of up to 4 dots and dashes.

65. The multiplication principle is based on the cardinality of the <u>Cartesian product</u> of two sets.

67. The decision algorithm produces every pair of shirts twice, first in one order and then in the other. Therefore, it is not valid. The actual number of pairs of shirts is half of the number computed by the algorithm: $90/2 = 45$.

69. Think of placing the five squares in a row of five empty slots. Step1: choose a slot for the blue square 5 choices. Step 2: choose a slot for the green square: 4 choices. Step 3: choose the remaining 3 slots for the yellow squares, 1 choice. Hence there are 20 possible five-square sequences.

6.4 Permutations and Combinations

1. $6! = 6\times5\times4\times3\times2\times1 = 720$

3. $\dfrac{8!}{6!} = \dfrac{8\times7\times\cancel{6}\times\cancel{5}\times\cancel{4}\times\cancel{3}\times\cancel{2}\times\cancel{1}}{\cancel{6}\times\cancel{5}\times\cancel{4}\times\cancel{3}\times\cancel{2}\times\cancel{1}} = 56$

5. $P(6,\,4) = 6\times5\times4\times3 = 360$

7. $\dfrac{P(6,\,4)}{4!} = \dfrac{6\times5\times4\times3}{4\times3\times2\times1} = 15$

9. $C(3,2) = \dfrac{3\times2}{2\times1} = 3$

Alternatively,
$C(3,\,2) = C(3,\,3-2) = C(3,\,1) = 3$

11. $C(10,\,8) = C(10,\,2) \qquad 10-8 = 2$
$= \dfrac{10\times9}{2\times1} = 45$

13. $C(20,\,1) = \dfrac{20}{1} = 20$

15. $C(100,\,98) = C(100,\,2) \quad 100-98 = 2$
$= \dfrac{100\times99}{2\times1} = 4950$

17. The number of ordered lists of 4 items chosen from 6 is
$P(6,\,4) = 6\times5\times4\times3 = 360$

19. The number of unordered sets of 3 objects chosen from 7 is
$C(7,\,3) = \dfrac{7\times6\times5}{3\times2\times1} = 35$

21. Each 5-letter sequence containing the letters b, o, g, e, y is a list of 5 letters chosen from the above 5 (or a permutation of the 5 letters) and the number of these is
$P(5,\,5) = 5! = 5\times4\times3\times2\times1 = 120$

23. Each 3-letter sequence is a list of 3 letters chosen from the given 6, and so the number of 3-letter sequences is
$P(6,\,3) = 6\times5\times4 = 120$

25. Each 3-letter unordered set is a list of 3 letters chosen from the given 6, and so the number of 3-letter sets is
$C(6,\,3) = \dfrac{6\times5\times4}{3\times2\times1} = 20$

27. Since there are repeated letters, we use a decision algorithm to construct such a sequence. Start with 6 empty slots.
  Step 1: Choose a slot for the k: 6 choices
  Step 2: Choose 3 slots from the remaining 5 for the a's: $C(5,\,3)$ choices
  Step 3: Choose 2 slots from the remaining 2 for the u's: $C(2,\,2)$ slots
Total number of sequences:
  $6\times C(5,3)\times C(2,2) = 60$

29. There are a total of 10 marbles in the bag. The number of sets of 4 chosen from 10 marbles is
  $C(10,\,4) = 210$

31. Decision algorithm for assembling a collection of 4 marbles that includes all the red ones:
  Step 1: Choose 3 red marbles: $C(3,\,3) = 1$ choice
  Step 2: Choose 1 non-red marble: $C(7,\,1) = 7$ choices
    (There are 7 non-red marbles to choose from.)
Total number of sets $= C(3,\,3)C(7,\,1) = 7$

33. Decision algorithm for assembling a collection of 4 marbles that includes no red ones:
  Step 1: Choose 4 non-red marble: $C(7,\,4) = 35$ choices
    (There are 7 non-red marbles to choose from.)
Total number of sets $= C(7,\,4) = 35$

35. Decision algorithm for assembling a collection as specified:

   Step 1: Choose 1 red marble: 3 choices

   Step 2: Choose 1 green marble: 2 choices

   Step 3: Choose 1 yellow marble: 2 choices

   Step 4: Choose 1 orange marble: 2 choices

Total number of sets: $3 \times 2 \times 2 \times 2 = 24$

37. A set containing at least 2 red marbles must contain either 2 or 3 red marbles.

   Alternative 1: Decide on 2 red marbles

     Step 1: Choose 2 red marbles: $C(3, 2)$ choices

     Step 2: Choose 3 non-red marbles: $C(7, 3)$ choices

   Alternative 3: Decide on 3 red marbles

     Step 1: Choose 3 red marbles: $C(3, 3)$ choices

   Step 2: Choose 2 non-red marbles: $C(7, 2)$ choices

Total number of choices:

   $C(3, 2)C(7, 3) + C(3, 3)C(7, 2) = 126$

39. A set containing at most 1 yellow marbles must contain either 0 or 1 yellow marbles.

   Alternative 1: Decide on 0 yellow marbles

     Step 1: Choose 5 non-yellow marbles: $C(8, 5)$ choices

   Alternative 2: Decide on 1 yellow marble

     Step 1: Choose 1 yellow marble: $C(2, 1)$ choices

   Step 2: Choose 4 non-yellow marbles: $C(8, 4)$ choices

Total number of choices:

   $C(8,5) + C(2,1)C(8,4) = 196$

41. Decision algorithm for assembling a collection as specified:

   Alternative 1: Use the lavender marble but no yellow ones

     Step 1: Select the lavender marble: $C(1, 1) = 1$ choice

     Step 2: Select 4 non-lavender non-yellow marbles: $C(7, 4)$ choices

   Alternative 2: Use a yellow marble but no lavender ones

     Step 1: Select 1 yellow marble: $C(2, 1)$ choices

     Step 2: Select 4 non-lavender non-yellow marbles: $C(7, 4)$ choices

Total number of sets:

   $C(1, 1)C(7, 4) + C(2, 1)C(7, 4) = 105$

43. Think of the outcome of a sequence of 30 dice throws as a sequence of "words" of length 30 using the "letters" $1, 2, 3, ..., 6$. Using this interpretation, a sequence with 5 ones is a 30-letter word containing 5 ones. For the decision algorithm, start with 30 empty slots and choose numbers to fill the slots:

   Step 1: Choose 5 slots for the ones: $C(30, 5)$ choices

   Steps 2–26: Choose one of $2, 3, 4, 5, 6$ to fill each of the remaining 25 slots: $5 \times 5 \times ... \times 5 = 5^{25}$ choices

Total number of sequences with 5 ones: $C(30, 5) \times 5^{25}$

Since there are 6^{30} different sequences possible, the fraction of sequences with 5 ones is

$$\frac{C(30, 5) \times 5^{25}}{6^{30}} = 0.192$$

45. For the decision algorithm, start with 30 empty slots and choose numbers to fill the slots:

   Step 1: Choose 15 slots for the even numbers: $C(30, 15)$ choices

   Next 15 steps: Choose $2, 4$, or 6 for each of these 15 slots: 3^{15} choices

   Next 15 steps: Choose $1, 3$, or 5 for each of the remaining 15 slots: 3^{15} choices

Since there are 6^{30} different sequences possible, the fraction of sequences with exactly 15 even numbers is

$$\frac{C(30, 15) \times 3^{15} \times 3^{15}}{6^{30}} = 0.144$$

47. Each itinerary is a list of 4 venues chosen from 4. Thus, the number of itineraries is

   $P(4, 4) = 4! = 4 \times 3 \times 2 \times 1 = 24$

49. Decision algorithm for constructing a two pair hand:

   Step 1: Select 2 denominations for the pairs: $C(13, 2)$ choices

   Step 2: Select two cards of the lowest-ranked denomination above: $C(4, 2)$ choices

   Step 3: Select two cards of the other denomination above: $C(4, 2)$ choices

   Step 4: Select a single card that belongs to neither of the two denominations: $C(44, 1) = 44$ choices ($52 - 8 = 44$)

Total number of two pair hands:

   $C(13, 2)C(4, 2)C(4, 2)C(44, 1) = 123{,}552$

51. Decision algorithm for constructing a two of a kind hand:

   Step 1: Select a denomination for the two cards: $C(13, 1) = 13$ choices

   Step 2: Select 2 cards of the above denomination: $C(4, 2)$ choices

   Step 3: Select 3 denominations for the singles: $C(12, 3)$ choices

   Step 4: Select 1 card of the lowest-ranked denomination above: $C(4, 1) = 4$ choices

   Step 5: Select 1 card of the next-ranked denomination: $C(4, 1) = 4$ choices

   Step 6: Select 1 card of the highest-ranked: $C(4, 1) = 4$ choices

Total number of two of a kind hands:

   $13 \times C(4, 2)C(12, 3) \times 4 \times 4 \times 4 = 1{,}098{,}240$

53. A straight is a run of 5 cards of consecutive denominations: $A, 2, 3, 4, 5$ up through $10, J, Q, K, A$, but not all of the same suit. If we ignore the restriction about the suits we get the following decision algorithm:

   Step 1: Select a starting denomination for the straight: $(A, 2, ..., 10)$: 10 choices

   Steps 2–6: Select a card of each of the above denominations: $4 \times 4 \times 4 \times 4 \times 4 = 4^5$ choices

Total number of runs of 5 cards: 10×4^5

However, these runs include hands in which all 5 cards are of the same suit (straight flushes). There are 10×4 of these (Step 1: select a starting card; Step 2: Select a single suit). Excluding the straight flushes gives

Total number of straights:

   $10 \times 4^5 - 10 \times 4 = 10{,}200$

55. (a) Since a portfolio consists of a set of 5 stocks chosen from the 10, the number of possible portfolios is $C(10, 5) = 252$ portfolios.

(b) Decision algorithm for assembling a portfolio as per part (b):

Step 1: Choose BAC and KFT: $C(2, 2) = 1$ choice

Step 2: Choose 3 more stocks from 6 (BAC, KFT, GE, DD excluded): $C(6, 3) = 20$

Total number of choices: $1 \times 20 = 20$ portfolios

(c) Five of the listed stocks have yields above 6%. Decision algorithm for assembling a portfolio as per part (c):

Alternative 1: Exactly 4 stocks have yields above 6%:

   Step 1: Choose 4 stocks with yields above 6%: $C(5, 4) = 5$ choices

   Step 2: Choose 1 stock with a yield not above 6%: $C(5, 1) = 5$ choices

   Total number of choices for Alternative 1: $5 \times 5 = 25$ choices

Alternative 2: All 5 stocks have yields above 6%:

   Step 1: Choose 5 stocks with yields above 6%: $C(5, 5) = 1$ choice

   Total number of choices for Alternative 2: 1 choice

Total number of portfolios = $25 + 1 = 26$

57. (a) Decision algorithm for assembling a collection of stocks:

   Step 1: Choose 3 tech stocks: $C(6, 3) = 20$ choices

   Step 2: Choose 2 non-tech stocks: $C(6, 2) = 15$ choices

Total number of choices: $20 \times 15 = 300$ collections

(b) Decision algorithm for assembling a collection of stocks that declined in value (there were 3 stocks in each category that declined):

Step 1: Choose 3 declining tech stocks: $C(3, 3)$ = 1 choice

Step 2: Choose 2 declining non-tech stocks: $C(3, 2) = 3$ possible choices

Total number of choices: $1×3 = 3$ collections

(c) Since only 3 collections of out a possible 300 consist entirely of stocks that declined in value, the chances of selecting one of those collections is 3 in 300; that is, 1 in 100, or .01.

59. (a) The number of groups of 4 chosen from 10 is $C(10, 4) = 210$

(b) Decision algorithm for assembling a group of 4 movies that satisfies the Lara twins:

Alternative 1: Exactly one of "Death Race" or "Eagle Eye"

Step 1: Choose 1 out of "Death Race" and "Eagle Eye" : $C(2, 1) = 2$ choices

Step 2: Choose 3 from the remaining 7 ("Horton Hears a Who!" is excluded): $C(7, 3)$ = 35 choices

Total number of groups in Alternative 1 is therefore $2×35 = 70$

Alternative 2: Both "Death Race" and "Eagle Eye"

Step 1: Choose 2 out of "Death Race" and "Eagle Eye" : $C(2, 2) = 1$ choice

Step 2: Choose 2 from the remaining 7 ("Death Race", "Eagle Eye", and "Horton Hears a Who!" excluded) $C(7, 2) = 21$ choices

Total number of groups in Alternative 2 is therefore $1×21 = 21$

Total number of groups that make the Lara twins happy is therefore $70 + 21 = 91$

(c) Since 91 of the 210 groups (less than half) make the Lara twins happy, they are less likely than not to be satisfied with a random selection.

61. (a) Each itinerary is a permutation of the 23 listed cities: 23! of them altogether

(b) Once the first five stops are determined, a decision algorithm for completing the itinerary is:

Steps 1–18: Select a different city from the remaining 18 for each remaining stop:

$18×17×...×2×1$ choices

Total number of itineraries: $18×17×...×2×1 = 18!$

(c) Decision algorithm for constructing an itinerary of the required type (think of making an itinerary as filling a sequence of 23 empty slots with different cities):

Step 1: Choose a starting point in the itinerary to insert the sequence of 5 named cities: 1 through 19: 19 choices

Next 18 steps: Select a different city from the remaining 18 for each remaining slot:

$18×17×...×2×1$ choices

Total number of itineraries: $19×18!$

63. Start with 11 empty slots. Decision algorithm:

Step 1: Select a slot for the m: $C(11, 1) = 11$ choices

Step 2: Select 4 slots for the i's: $C(10, 4)$ choices

Step 3: Select 6 slots for the s's: $C(6, 4)$ choices

Step 4: Select 2 slots for the p's: $C(2, 2) = 1$ choice

Total number of sequences:

$C(11, 1)C(10, 4)C(6, 4)C(2, 2)$

65. Start with 11 empty slots. Decision algorithm:

Step 1: Select 2 slots for the m's: $C(11, 2)$ choices

Step 2: Select 1 slot for the e: $C(9, 1) = 9$ choices

Step 3: Select 1 slot for the g: $C(8, 1) = 8$ choices

Step 4: Select 3 slots for the a's: $C(7, 3)$ choices

Step 5: Select 1 slot for the l: $C(4, 1) = 4$ choices

Step 6: Select 1 slot for the o: $C(3, 1) = 3$ choices

Step 7: Select 1 slot for the n: $C(2, 1) = 2$ choices

Step 8: Select 1 slot for the i: $C(1, 1) = 1$ choice

Total number of sequences:

$C(11,2)C(9,1)C(8,1)C(7,3)C(4,1)C(3,1)$

$C(2,1)C(1,1)$

67. Start with 10 empty slots. Decision algorithm:
Step 1: Select 2 slots for the c's: $C(10, 2)$ choices
Step 2: Select 4 slots for the a's: $C(8, 4)$ choices
Step 3: Select 1 slot for the s: $C(4, 1) = 4$ choices
Step 4: Select 1 slot for the b: $C(3, 1) = 3$ choices
Step 5: Select 1 slot for the l: $C(2, 1) = 2$ choices
Step 6: Select 1 slot for the n: $C(1, 1) = 1$ choice

Total number of sequences:
$C(10, 2)C(8, 4)C(4, 1)C(3, 1)C(2, 1)C(1, 1)$

69. We wish to compute the number of groups of 4 chosen from 5 contestants (since we are excluding Ben and Ann): $C(5, 4) = 5$ possible groups (Choice A)

71. Each handshake corresponds to a pair of people, so the question is really asking how many pairs of people there are in a group of 10: $C(10, 2) = 45$ (Choice D).

73. (a) Note: In a set of 3 numbers, all three are different. The number of combinations = the number of sets of 3 numbers chosen from 40:
$C(40, 3) = 9880$

(b) Decision algorithm for constructing a combination in which a number appears twice:
Step 1: Select the number that appears twice: 40 choices
Step 3: Select another number: 39 choices

Total number of new combinations: $40 \times 39 = 1560$

(c) Decision algorithm for constructing a combination in which a number appears 3 times:
Step 1: Select the number that appears 3 times: 40 choices

This gives 40 new combinations.

Total number of combinations:
$9880 + 1560 + 40 = 11,480$

75. (a) The number of combinations of 2 equations chosen from the 20 constraints is $C(20, 2) = 190$ systems of equations to solve
(b) Replace 20 by n, getting $C(n, 2)$.

77. You should choose the multiplication principle because the multiplication principle can be used to solve all problems that call for the formulas for permutations, as well as others.

79. A permutation is an ordered list, or sequence, and only choices (A) and (D) can be represented as ordered lists. (In a presidential cabinet, each portfolio is distinct.)

81. A permutation. Changing the order in a list of driving instructions can result in a different outcome; for instance, "1. Turn left. 2. Drive one mile." and "1. Drive one mile. 2. Turn left." will take you to different locations.

83. Urge your friend not to focus on formulas, but instead learn to formulate decision algorithms and use the principles of counting.

85. It is ambiguous on the following point: are the three students to play different characters, or are they to play a group of three, such as "three guards." This should be made clear in the exercise.

266

Chapter 6 Review Exercises

1. The negative integers greater than or equal to -3 are $-3, -2,$ and -1. Thus,

$N = \{-3, -2, -1\}$

3. If the dice are distinguishable, then the outcomes can be thought of as ordered pairs. Thus, since the numbers are different, S is the set of all ordered pairs of distinct numbers in the range 1–6:

$S = \{(1,2), (1,3), (1,4), (1,5), (1,6), (2,1), (2,3), (2,4), (2,5), (2,6), (3,1), (3,2), (3,4), (3,5), (3,6), (4,1), (4,2), (4,3), (4,5), (4,6), (5,1), (5,2), (5,3), (5,4), (5,6), (6,1), (6,2), (6,3), (6,4), (6,5)\}$

5. $A = \{a, b\}, B = \{b, c\}, S = \{a, b, c, d\}$

$B' = \{a, d\}$

$A \cup B' = \{a, b\} \cup \{a, d\} = \{a, b, d\}$

$A \times B' = \{(a, a), (a, d), (b, a), (b, d)\}$

7. The set of outcomes when a day in August and a time of that day are selected is the set of pairs (day in August, time). That is, it is the set $A \times B$.

9. The set of all integers that are not positive odd perfect squares is the set of integers other than those in P and E' ($E' = $ the odd integers) and Q: $(P \cap E' \cap Q)'$ or $P' \cup E \cup Q'$

11. Let S be the set of all novels in your home; $n(S) = 400$

Let A be the event that you have read a novel: $n(A) = 150$

Let B be the event that Roslyn has read a novel: $n(A) = 200$

We are also told that $n(A \cap B) = 50$.

To compute $n(A \cup B)$ we use

$n(A \cup B) = n(A) + n(B) - n(A \cap B)$
$= 150 + 200 - 50 = 300$

The event that neither of you has read a novel is $(A \cup B)'$, and its cardinality is given by

$n[(A \cup B)'] = n(S) - n(A \cup B)$
$= 400 - 300 = 100$

Note that the formula used for the last calculation is

$n(C') = n(S) - n(C)$

13. The number of outcomes from rolling the dice is $n(A \times B)$, where A is the set of outcomes of the red dice and B is the set of outcomes of the green one.

$n(A \times B) = n(A)n(B) = 6 \cdot 6 = 36$

The set of losing combinations has the form $A \cup B$ where here A is the set of doubles and B is the set of outcomes in which the green die shows an odd number and the red die shows an even number. To compute its cardinality, we use

$n(A \cup B) = n(A) + n(B) - n(A \cap B)$
$= 6 + 3 \times 3 - 0 = 15$

The set of winning combinations is the complement of the set C of losing combinations, and is given by

$n(C') = n(S) - n(C) = 36 - 15 = 21.$

15. Decision algorithm for constructing a two of a kind hand with no Aces:

Step 1: Select a denomination other than Ace for the pair: $C(12, 1)$ choices

Step 2: Select 2 cards of that denomination: $C(4, 2)$ choices

Step 3: Select 3 denominations (other than Ace or the one already selected above) for the remaining singles: $C(11, 3)$ choices

Step 4: Select a suit for the highest-ranked denomination just chosen: $C(4, 1)$ choices

Step 5: Select a suit for the next highest-ranked denomination just chosen: $C(4, 1)$ choices

Step 6: Select a suit for the lowest-ranked denomination just chosen: $C(4, 1)$ choices

Total number of hands:

$C(12, 1)C(4, 2)C(11, 3)C(4, 1)C(4, 1)C(4, 1)$

17. Decision algorithm for constructing a straight flush:

Step 1: Select a suit for the flush: $C(4, 1)$ choices

Step 2. Select a starting denomination for the consecutive run: $C(10, 1)$ choices (A, 2, 3, 4, 5 up through 10, J, Q, K, A)

Total number of hands: $C(4, 1)C(10, 1)$

19. Decision algorithm for constructing a set of 5 marbles that include all the red ones:
Step 1: Select 4 red marbles: $C(4, 4)$ choices
Step 2: Select 1 non-red marble: $C(8, 1)$ choices

Total number of choices: $C(4, 4)C(8, 1) = 8$

21. Decision algorithm for constructing a set of 5 marbles that include at least 2 yellow ones:
Alternative 1: Use 2 yellow ones:
   Step 1: Select 2 yellow marbles: $C(3, 2)$ choices
   Step 2: Select 3 non-yellow marbles: $C(9, 3)$ choices
Alternative 2: Use 3 yellow ones:
   Step 1: Select 3 yellow marbles: $C(3, 3)$ choices
   Step 2: Select 2 non-yellow marbles: $C(9, 2)$ choices

Total number of choices:
   $C(3, 2)C(9, 3) + C(3, 3)C(9, 2) = 288$

23. $S \cup T$ is the set of books that are either sci-fi or stored in Texas (or both)
$$n(S \cup T) = n(S) + n(T) - n(S \cap T)$$
$$= 33{,}000 + 94{,}000 - 15{,}000$$
$$= 112{,}000$$

25. $C \cup S'$ is the set of books that are either stored in California or not sci-fi. To compute its cardinality, add the corresponding entries (shown below in thousands)

	S	H	R	O	Total
W	10	12	12	30	64
C	8	12	6	16	42
T	15	15	20	44	94
Total	33	39	38	90	200

$n(C \cup S') = 175{,}000$

27. $R \cap (T \cup H)$ is the set of romance books that are also horror books or stored in Texas.

	S	H	R	O	Total
W	10	12	12	30	64
C	8	12	6	16	42
T	15	15	20	44	94
Total	33	39	38	90	200

$n(R \cap (T \cup H)) = 20{,}000$

In Exercises 29–33, we let H denote the set of OHaganBooks.com customers, J the set of JungleBooks.com customers, and F the set of FarmerBooks.com customers.

29. We use a Venn diagram:

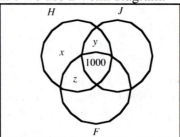

We are told that
OHaganBooks.com had 3500 customers:
   $x + y + z + 1000 = 3500$
2000 customers were shared with JungleBooks.com:
   $y + 1000 = 2000$,
so $y = 1000$
1500 customers were shared with FarmerBooks.com:
   $z + 1000 = 1500$
so $z = 500$
We can now compute x from the first equation:
   $x + 1000 + 500 + 1000 = 3500$
   $x = 1000$
We are asked for the number that are exclusive OHaganBooks.com customers:
   $x = 1000$

31. Let us start with the information from Exercise 29 filled in:

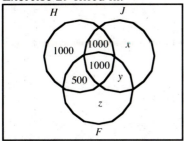

JungleBooks.com has a total of 3600 customers:

$$x + y + 2000 = 3600$$
$$x + y = 1600$$

FarmerBooks.com has 3400 customers:

$$y + z + 1500 = 3400$$
$$y + z = 1900$$

JungleBooks.com and FarmerBooks.com share 1100 customers between them:

$$y + 1000 = 1100$$

so, $y = 100$

giving

$$x + 100 = 1600$$

so, $x = 1500$

$$100 + z = 1900$$

so, $z = 1800$

Thus, OHaganBooks has 1000 exclusive customers, JungleBooks has $x = 1500$ exclusive customers, and FarmerBooks has $z = 1800$ exclusive customers—the most.

33. Here is the complete data from Exercises 29 and 30 above:

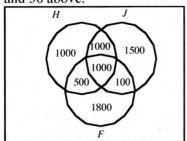

$$n(H \cup J) = 3500 + 1500 + 100 = 5100$$
$$n(H \cup F) = 3500 + 100 + 1800 = 5400$$

Thus, OHaganBooks should merge with FarmerBooks.com for a combined customer base of 5400.

35. Decision algorithm for constructing a 3-letter code:
    Step 1: Choose the 1st letter: 26 choices
    Step 2: Choose the 2nd letter: 26 choices
    Step 3: Choose the 3rd letter: 26 choices

Total number of codes: $26 \times 26 \times 26 = 26^3 = 17{,}576$

37. Decision algorithm for constructing a code:
    Step 1: Choose the 1st letter: 26 choices
    Step 2: Choose the 2nd letter: 25 choices
    Step 3: Choose the 1st digit: 9 choices
    Step 4: Choose the 2nd digit: 10 choices

Total number of codes: $26 \times 25 \times 9 \times 10 = 58{,}500$

39. Decision algorithm for constructing a course schedule meeting the minimum requirements:
    Step 1: Select 3 liberal arts courses: $C(6, 3)$ choices
    Step 2: Select 3 math courses: $C(6, 3)$ choices (Note: At this stage, you have must select a minimum of 4 courses to make up the required 10, and so you must select 2 sciences and 2 fine arts.)
    Step 3: Select 2 science courses: $C(5, 2)$ choices
    Step 4: Select 2 fine arts courses: $C(6, 2)$ choices

Total number of schedules:
    $C(6, 3)C(6, 3)C(5, 2)C(6, 2)$
    $= 20 \times 20 \times 10 \times 15 = 60{,}000$

41. Decision algorithm for constructing a course schedule meeting all of the above requirements, and the physics complication:
    Alternative 1: Take neither Physics I nor Physics II
    Step 1: Select 3 liberal arts courses: $C(6, 3)$ choices
    Step 2: Select Calc I: 1 choice
    Step 3: Select 2 other math courses: $C(5, 2)$ choices
    Step 4: Select 2 science courses other than Physics I and II: $C(3, 2)$ choices

Step 5: Select 2 fine arts courses other than
the single pair that cannot be taken in the first
year: $C(6, 2) - 1$ choices
Total number of schedules for Alternative 1:
$C(6, 3)C(5, 2)C(3, 2)[C(6, 2) - 1]$
$= 20 \times 10 \times 3 \times (15-1) = 8400$

Alternative 2: Take Physics I but not Physics II
Step 1: Select 3 liberal arts courses: $C(6, 3)$
choices
Step 2: Select Calc I: 1 choice
Step 3: Select 2 other math courses: $C(5, 2)$
choices
Step 4: Select Physics I: 1 choice
Step 5: Select 1 science course other than
Physics I and II: $C(3, 1)$ choices
Step 6: Select 2 fine arts courses other than
the single pair that cannot be taken in the first
year: $C(6, 2) - 1$ choices
Total number of schedules for Alternative 1:
$C(6, 3)C(5, 2)C(3, 1)[C(6, 2) - 1]$
$= 20 \times 10 \times 3 \times (15-1) = 8400$

Alternative 3: Take Physics I and Physics II
Step 1: Select 3 liberal arts courses: $C(6, 3)$
choices
Step 2: Select Calc I: 1 choice
Step 3: Select 2 other math courses: $C(5, 2)$
choices
Step 4: Select Physics I & II: 1 choice
Step 5: Select 2 fine arts courses other than
the single pair that cannot be taken in the first
year: $C(6, 2) - 1$ choices
Total number of schedules for Alternative 1:
$C(6, 3)C(5, 2)[C(6, 2) - 1]$
$= 20 \times 10 \times (15-1) = 2800$

Total number of course schedules:
$8400 + 8400 + 2800 = 19{,}600$

Chapter 7 Probability

7.1 Samples Spaces and Events

1. $S = \{HH, HT, TH, TT\}$
   $E = \{HH, HT, TH\}$

3. $S = \{HHH, HHT, HTH, HTT, THH,$
   $\quad THT, TTH, TTT\}$
   $E = \{HTT, THT, TTH, TTT\}$

5. $S = \left\{ \begin{array}{llllll} (1,1) & (1,2) & (1,3) & (1,4) & (1,5) & (1,6) \\ (2,1) & (2,2) & (2,3) & (2,4) & (2,5) & (2,6) \\ (3,1) & (3,2) & (3,3) & (3,4) & (3,5) & (3,6) \\ (4,1) & (4,2) & (4,3) & (4,4) & (4,5) & (4,6) \\ (5,1) & (5,2) & (5,3) & (5,4) & (5,5) & (5,6) \\ (6,1) & (6,2) & (6,3) & (6,4) & (6,5) & (6,6) \end{array} \right\}$
   $E = \{(1, 4), (2, 3), (3, 2), (4, 1)\}$

7. $S = \left\{ \begin{array}{llllll} (1,1) & (1,2) & (1,3) & (1,4) & (1,5) & (1,6) \\ & (2,2) & (2,3) & (2,4) & (2,5) & (2,6) \\ & & (3,3) & (3,4) & (3,5) & (3,6) \\ & & & (4,4) & (4,5) & (4,6) \\ & & & & (5,5) & (5,6) \\ & & & & & (6,6) \end{array} \right\}$

   (We removed all repeating pairs.)
   $E = \{(1, 3), (2, 2)\}$

9. S as in Exercise 7
   $E = \{(2, 2), (2, 3), (2, 5), (3, 3), (3, 5), (5, 5)\}$

11. $S = \{m, o, z, a, r, t\}$
    $E = \{o, a\}$

13. There are $4 \times 3 = 12$ possible sequences of 2 different letters chosen from *sore:*
    $S = \{(s, o), (s, r), (s, e), (o, s), (o, r), (o, e),$
    $\quad (r, s), (r, o), (r, e), (e, s), (e, o), (e, r)\}$
    $E = \{(o, s), (o, r), (o, e), (e, s), (e, o), (e, r)\}$

15. There are $5 \times 4 = 20$ possible sequences of 2 different digits chosen from 0–4:
    $S = \{01, 02, 03, 04, 10, 12, 13, 14, 20, 21,$
    $\quad 23, 24, 30, 31, 32, 34, 40, 41, 42, 43\}$
    $E = \{10, 20, 21, 30, 31, 32, 40, 41, 42, 43\}$

17. $S = \{$domestic car, imported car, van, antique car, antique truck$\}$
    $E = \{$van, antique truck$\}$

19. **(a)** The sample space is the set of all sets of 4 gummy candies chosen from the packet of 12.
    (b) The event that April will get the combination she desires is the set of all sets of 4 gummy candies in which two are strawberry and two are blackcurrant.

21. **(a)** The sample space is the set of all lists of 15 people chosen from 20.
    (b) The event that Hillary Clinton is the Secretary of State is the set of all lists of 15 people chosen from 20, in which Hillary Clinton occupies the eleventh (Secretary of State) position.

23. A: the red die shows 1; B: the numbers add to 4
    "The red die shows 1 *and* the numbers add to 4" is the event $A \cap B$. $\quad$ *And* $= \cap$
    There is only one possible outcome in $A \cap B$:
    $\quad A \cap B = \{(1, 4)\}$
    Therefore, $n(A \cap B) = 1$

25. B: the numbers add to 4
    "The numbers do not add to 4" is the event B'.
    $\quad$ One has $B = \{(1, 3), (2, 2), (3, 1)\}$
    $\quad n(B') = n(S) - n(B) = 36 - 3 = 33$

27. "The numbers do not add to 4" is the event B'.
    "The numbers add to 11" is the event D'.

    Therefore, "The numbers do not add to 4 *but* they do add to 11" is the event $B' \cap D'$.
    $\quad B' \cap D' = \{(6, 5), (5, 6)\}$
    $\quad n(B' \cap D') = 2$

29. "At least one of the numbers is 1" is the event C.
    "The numbers add to 4" is the event B.

    Therefore, "At least one of the numbers is 1 or the numbers add to 4" is the event $C \cup B$
    $\quad C \cup B = \{(1, 1), (1, 2), (1, 3), (1, 4), (1, 5), (1, 6), (6, 1), (5, 1), (4, 1), (3, 1), (2, 1), (2, 2)\}$

271

$n(C \cup B) = 12$

31. *W*: You will use the web site tonight.
   I: Your math grade will improve.
Therefore, "You will use the web site tonight *and* your math grade will improve." is the event $W \cap I$.
$$And = \cap$$

33. *E*: You will use the web site every night.
   I: Your math grade will improve.
Therefore, "Either you will use the web site every night, *or* your math grade will *not* improve." is the event $E \cup I'$.
$$Or = \cup$$

35. *I*: Your math grade will improve.
   W: You will use the web site tonight.
   E: You will use the web site every night.
The given statement is: "Either your math grade will improve, *or* you will use the web site tonight *but* you will not use it every night."
The comma after "improve" breaks the statement into two pieces:
Either
your math grade will improve: (*I*)
Or
you will use the web site tonight *but* you will not use it every night: ($W \cap E'$).
Therefore, the given statement is the event $I \cup (W \cap E')$.

37. *I*: Your math grade will improve.
   W: You will use the web site tonight.
   E: You will use the web site every night.
The given statement is: "Either your math grade will improve *or* you will use the web site tonight, *but* you will not use it not every night."
The comma after "tonight" breaks the statement into two pieces:
Either your math grade will improve *or* you will use the web site tonight, ($I \cup W$)
But
you will not use it every night: E'.
Therefore, the given statement is the event $(I \cup W) \cap E'$

39. Looking at the map, we find the following regions that saw an decrease in housing prices of

4% or more: Pacific, Mountain, South Atlantic. Therefore,
$$E = \{Pacific, Mountain, South Atlantic\}$$

41. *E*: The region you choose saw a decrease in housing prices of 4% or more.
   F: The region you choose is on the east coast.
$E \cup F$ translates to *E or F*: You choose a region that saw a decrease in housing prices of 4% or more *or* is on the east coast.
$E \cup F = \{Pacific, Mountain, New England, Middle Atlantic, South Atlantic\}$
$E \cap F$ translates to *E and F*: You choose a region that saw a decrease in housing prices of 4% or more *and* is on the east coast.
$E \cap F = \{South Atlantic\}$

43. Two events are mutually exclusive if they have no outcomes in common.
(a) *E*: You choose a region from among the two with the highest percentage decrease in housing prices.
$E = \{ Pacific, South Atlantic\}$
F: You choose a region that is not on the east or west coast.
$F = \{ Mountain, West North Central, West South Central, East North Central, East South Central \}$
Since *E* and *F* have no regions in common, $E \cap F = \emptyset$, so they are mutually exclusive.
(b) *E*: You choose a region from among the two with the highest decrease increase in housing prices.
$E = \{ Pacific, South Atlantic\}$
F: You choose a region that is not on the west coast.
$F = \{ Mountain, West North Central, West South Central, East North Central, East South Central, New England, Middle Atlantic, South Atlantic \}$
Since $E \cap F = \{ South Atlantic \}$, $E \cap F \neq \emptyset$, so they are not mutually exclusive.

45. *S*: an author is successful.
   N: an author is new.
$S \cap N$ is the event that an author is successful and new.

272

$S \cup N$ is the event that an author is either successful or new

$n(S \cap N) = 5$ (See table)

	N	E	Total
S	5	25	30
U	15	55	70
Total	20	80	100

$n(S \cup N) = 45$ (See table)

	N	E	Total
S	5	25	30
U	15	55	70
Total	20	80	100

47. Since N and E have no outcomes in common (an author cannot be both new and experienced), they are mutually disjoint.

N and S are not mutually disjoint, since
$n(N \cap S) = 5$ (See table)

S and E are not mutually disjoint, since
$n(S \cap E) = 25$ (See table)

	N	E	Total
S	5	25	30
U	15	55	70
Total	20	80	100

49. S: an author is successful.
N: an author is new.

$S \cap N'$ is the event that an author is successful but not a new author.

$n(S \cap N') = 25$ (See table)

	N	$E = N'$	Total
S	5	25	30
U	15	55	70
Total	20	80	100

51. The total number of established authors is 80. Of these, 25 are successful. Thus, the percentage that are successful is

$$\frac{25}{80} = 31.25\%$$

The total number of successful authors is 30. Of these, 25 are successful. Thus, the percentage that are experienced is

$$\frac{25}{30} \approx 83.33\%$$

53. "A stock's value increased *but* it was not an Internet stock." is the event $V \cap I'$.

$n(V \cap I') = 15$ (See table)

	P (I')	E (I')	I	Total
V	10	5	15	30
N	30	0	10	40
D	10	5	15	30
Total	50	10	40	100

55. $n(P' \cup N) = 80$. (See table)

	P	E	I	Total
V	10	5	15	30
N	30	0	10	40
D	10	5	15	30
Total	50	10	40	100

P': The stock was not a pharmaceutical stock.
N: The stock was unchanged in value.

Therefore, $n(P' \cup N)$ is the number of stocks that were either not pharmaceutical stocks, or were unchanged in value after a year (or both).

57. For a pair of events to be mutually exclusive, they must have no outcomes in common. Three kinds can be identified:

(1) Pairs of events corresponding to separate columns: P and E, P and I, E and I.

(2) Pairs of events corresponding to separate rows: X and N, V and D, N and D.

(3) Pairs corresponding to a row and column whose intersection is empty: N and E.

59. $n(V \cap I) = 15$; $n(I) = 40$

	P	E	I	Total
V	10	5	15	30
N	30	0	10	40
D	10	5	15	30
Total	50	10	40	100

$$\frac{n(V \cap I)}{n(I)} = \frac{15}{40} = \frac{3}{8}$$

This is the fraction of Internet stocks that increased in value.

61. E: The dog's flight drive is strongest.
H: The dog's fight drive is weakest.
G: The dog's fight drive is strongest.
(a) "The dog's flight drive is not strongest *and* its fight drive is weakest" is the event $E' \cap H$.
(b) "The dog's flight drive is strongest *or* its fight drive is weakest" is the event $E \cup H$.
(c) "Neither the dog's flight drive *nor* fight drive are strongest" is the event $E' \cap G' = (E \cup G)'$

63. (a) The dog's fight and flight drives are both strongest.: Intersection of right-most column and bottom row: {9}
(b) The dog's fight drive is strongest, but its flight drive is neither weakest nor strongest:
Intersection of right-most column and middle row: {6}

65. (a) {1, 4, 7} This is the left-most column: The dog's fight drive is weakest.
(b) {1, 9}: The dog's fight and flight drives are either both strongest or both weakest.
(c) {3, 6, 7, 8, 9} This is the union of the right-most column and the bottom row: Either the dog's fight drive is strongest, or its flight drive is strongest.

67. S is the set of sets of 4 gummy bears chosen from 6; $n(S) = C(6, 4) = 15$.
Decision algorithm for constructing a set including the raspberry gummy bear:
   Step 1; Select the raspberry one: $C(1, 1) = 1$ choice
   Step 2: Select 3 non-raspberry ones: $C(5, 3) = 10$ choices

Total number of choices: $n(E) = 1 \times 10 = 10$

69. (a) Decision algorithm for constructing a finish (winner, second place and third place) for the race:
   Step 1: Select the winner: 7 choices
   Step 2: Select second place: 6 choices
   Step 3; Select third place: 5 choices
Total number of finishes: $n(S) = 7 \times 6 \times 5 = 210$
(b) E: Electoral College is in second or third place.
   F: Celera is the winner.
Therefore, $E \cap F$ is the event that Celera wins and Electoral College is in second or third place. In other words, it is the set of all lists of three horses in which Celera is first and Electoral College is second or third.

Decision algorithm for constructing a finish in $E \cap F$:
   Alternative 1: Electoral College is second:
     Step 1: Select Celera as the winner: 1 choice
     Step 2: Select Electoral College for second place: 1 choice
     Step 3: Select third place: 5 choices
Total number of finishes for Alternative 1: $1 \times 1 \times 5 = 5$
   Alternative 2: Electoral College is third:
     Step 1: Select Celera as the winner: 1 choice
     Step 2: Select Electoral College for third place: 1 choice
     Step 3; Select second place: 5 choices
Total number of finishes for Alternative 2: $1 \times 1 \times 5 = 5$
Therefore, $n(E \cap F) = 5 + 5 = 10$.

71. The sample space consists of all sets of 3 marbles chosen from 8.
   $n(S) = C(8, 3) = 56$

73. Using the multiplication principle:
   $n(E) = C(4, 1) \times C(2, 1) \times C(2, 1) = 16$

75. An event is a <u>subset of the sample space</u>.

77. $(E \cap F)'$ is the complement of the event $E \cap F$. Since $E \cap F$ is the event that both E and F occur,

$(E \cap F)'$ is the event that E and F do not both occur.

79. Choice (B): The event should be a subset of the sample space, and only (B) has the property (the collection of tall dark strangers you meet is a subset of the set of people you meet).

81. True; Consider the following experiment: Select an element of the set S at random. Then the sample space is the set of elements of S. In other words, the sample space is S.

83. Answers may vary. Cast a die and record the remainder when the number facing up is divided by 2.

85. Yes. For instance, $E = \{(2, 5), (5, 1)\}$ and $F = \{(4, 3)\}$ are two such events.

7.2 Relative Frequency

1. $P(E) = \dfrac{fr(E)}{N} = \dfrac{40}{100} = .4$

3. $N = 800$, $fr(E) = 640$

$P(E) = \dfrac{fr(E)}{N} = \dfrac{640}{800} = .8$

5. $E = \{1, 2, 3, 4\}$

$P(E) = \dfrac{fr(E)}{N} = \dfrac{8+8+8+12}{60} = \dfrac{36}{60} = .6$

7. We obtain the relative frequency distribution by dividing the frequencies by $N = 4000$:

Outcome	Frequency	Relative Frequency
HH	1100	$\dfrac{1100}{4000} = .275$
HT	950	$\dfrac{950}{4000} = .2375$
TH	1200	$\dfrac{1200}{4000} = .3$
TT	750	$\dfrac{750}{4000} = .1875$

9. Refer to the relative frequency distribution in the solution to Exercise 7. The event the that the second coin lands with heads up is

$E = \{HH, TH\}$

$P(E) = .275 + .3 = .575$

11. From the solution to Exercise 9, the second coin comes up heads approximately 58% of the time. Therefore, the second coin *seems* slightly biased in favor of heads, especially in view of the fact that the coin was tossed 4000 times. On the other hand, it is conceivable that the coin is fair and that heads came up 58% of the time purely by chance. Deciding which conclusion is more reasonable requires some knowledge of inferential statistics.

13. Yes: Since the relative frequencies are between 0 and 1 (inclusive) and add up to 1, the

given distribution can be a relative frequency distribution.

15. No: Relative frequencies cannot be negative.

17. Yes: Since the relative frequencies are between 0 and 1 (inclusive) and add up to 1, the given distribution can be a relative frequency distribution.

19.

Outcome	1	2	3	4	5
Rel. Frequency	.2	.3	.1	.1	x

Since the probabilities of all the outcomes add to 1,

$.2 + .3 + .1 + .1 + x = 1$

$.7 + x = 1$

$x = .3$

(a) From the properties of relative frequency distributions,

$P(\{1, 3, 5\}) = P(1) + P(3) + P(5)$

$= .2 + .1 + .3 = .6$

(b) Since $E = \{1, 2, 3\}$, $E' = \{4, 5\}$ and so

$P(E') = P(\{4, 5\}) = P(4) + P(5)$

$= .1 + .3 = .4$

21. Use the following formulas to generate binary digits (0 represents heads, say, and 1 represents tails):

TI-83/84 Plus: `randInt(0,1)`

To obtain `randInt`, follow MATH →PRB.

Excel: =RANDBETWEEN(0,1)

Answers will vary. The estimated probability that heads comes up should be around .5.

23. Simulate tossing two coins by generating two random binary digits (0 represents heads, say, and 1 represents tails). If the outcome is one head and one tail, the digits should add to 1. Otherwise, they will add to 0 or 2. The sum of these digits can be obtained as follows:

TI-83/84 Plus:

```
randInt(0,1)+randInt(0,1)
```

To obtain `randInt`, follow [MATH] ⟶PRB.

Excel:

```
=RANDBETWEEN(0,1)
      +RANDBETWEEN(0,1)
```

Then count how many 1s you get.
Answers will vary. The estimated probability that the outcome is one head and one tail should be around .5.

25. (a) $P(E) = \dfrac{fr(E)}{N} = \dfrac{160}{500} = .32$

(b) $P(E) = \dfrac{fr(E)}{N} = \dfrac{20+55}{500} = .15$

(c) $P(E) = \dfrac{fr(E)}{N} = \dfrac{500-160}{500} = \dfrac{340}{500} = .68$

27. (a) We obtain the relative frequency distribution by dividing the frequencies by $N = 134+52+9+5 = 200$:

Outcome	Frequency	Rel. Frequency
Current	134	$\dfrac{134}{200} = .67$
Past Due	52	$\dfrac{52}{200} = .26$
In Foreclosure	9	$\dfrac{9}{200} = .045$
Repossessed	5	$\dfrac{5}{200} = .025$

(b) Take E to be the event that a randomly selected subprime mortgage in Texas was not current.

$E = \{$Past Due, In Foreclosure, Repossessed$\}$
$P(E) = .26 + .045 + .025 = .33$

29. (a) Using the result of Quick Example #3, the relative frequency distribution is obtained by converting the percentages into decimals:

Age	18 25	26–35	36–45	46–55	> 55
Rel. Frequency	.29	.47	.20	.03	.01

(b) The event E that an employed adult resident of Mexico is *not* between 26 and 45 is shown by the shaded parts of the distribution:

Age	18–25	26–35	36–45	46–55	> 55
Rel. Frequency	.29	.47	.20	.03	.01

Thus the relative frequency is
$P(E) = .29 + .03 + .01 = .33$

31.(a) We obtain the relative frequency distribution by dividing the frequencies by $N = 1+4+4+1 = 10$:

Outcome	Frequency	Rel. Frequency
3	1	$\dfrac{1}{10} = .1$
2	4	$\dfrac{4}{10} = .4$
1	4	$\dfrac{4}{10} = .4$
0	1	$\dfrac{1}{10} = .1$

(b) Take E to be the event that a randomly selected small SUV will have a crash test rating of "Acceptable" (2) or better.

$E = \{2, 3\}$
$P(E) = .4 + .1 = .5$

33. The number households is $N = 2000$. To calculate $fr(E)$ in each case, we use

$P(E) = \dfrac{fr(E)}{N}$, giving

$fr(E) = P(E) \times N$

(a) Here is a chart showing the frequencies and their calculation:

Outcome	Rel. Frequency	Frequency
Dial-up	.628	.628×2000 = 1256
Cable Modem	.206	.206×2000 = 412
DSL	.151	.151×2000 = 302
Other	.015	.015×2000 = 30

35. Using the given classification of the data, we get the following:

Outcome	Surge	Plunge	Steady
Frequency	4	6	10
Rel. Frequency	$\frac{4}{20} = .2$	$\frac{6}{20} = .3$	$\frac{10}{20} = .5$

37. $P(E \cap S) = \frac{fr(E \cap S)}{N} = \frac{25}{100} = .25$

	N	E	Total
S	5	25	30
U	15	55	70
Total	20	80	100

39. $P(N) = \frac{fr(N)}{N} = \frac{20}{100} = .2$

	N	E	Total
S	5	25	30
U	15	55	70
Total	20	80	100

41. $P(U) = \frac{fr(U)}{N} = \frac{70}{100} = .7$

	N	E	Total
S	5	25	30
U	15	55	70
Total	20	80	100

43. Restrict attention to the successful authors only:

$P(\text{Successful author is established}) = \frac{25}{30} = \frac{5}{6}$

		N	E	Total
	S	5	25	30

45. Restrict attention to the established authors only:

$P(\text{Established author is successful})$
$= \frac{25}{80} = \frac{5}{16}$

		E
	S	25
	U	55
	Total	80

47. $P(\text{Contaminated}) = .8$
Therefore,
$$P(U) = 1 - .8 = .2$$

Of the contaminated chicken, 20% had the strain resistant to antibiotics. Therefore,
$$P(R) = .2 \times .8 = .16 \qquad (20\% \text{ of } 80\%)$$

The other 80% of contaminated chicken is not resistant. Therefore,
$$P(C) = .8 \times .8 = .64 \qquad (80\% \text{ of } 80\%)$$

This gives the following relative frequency distribution:

Outcome	U	C	R
Rel. Frequency	.2	.64	.16

49. Conventionally grown produce:
The events are:
   No pesticide (NP)
   Single pesticide (SP)
   Multiple pesticides (MP)
We are told that 73% of conventionally grown foods had residues from at least one pesticide. Therefore,
$$P(NP) = 1 - .73 = .27$$

278

We are also told that conventionally grown foods were 6 times as likely to contain multiple pesticides as organic foods, and that 10% of organic foods had multiple pesticides. Therefore,
$$P(MP) = 6 \times .10 = .60$$
This leaves
$$P(SP) = 1 - (.27 + .60) = .13$$
Probability distribution for conventionally grown produce:

Outcome	NP	SP	MP
Probability	.27	.13	.60

Organic produce:
We are told that
$$P(MP) = .10$$
and that 23% had either single or multiple pesticide residues. Therefore,
$$P(SP) = .23 - .10 = .13$$
This leaves
$$P(NP) = 1 - (.10 + .13) = .77$$
Probability distribution for organic produce:

Outcome	NP	SP	MP
Probability	.77	.13	.10

51. $P(\text{False negative}) = \dfrac{10}{400} = .025$

$P(\text{False positive}) = \dfrac{10}{200} = .05$

53. Answers will vary.

55. The estimated probability of an event E is defined to be the fraction of times E occurs.

57. 101; $fr(E)$ can be any number between 0 and 100 inclusive, so the possible answers are $0/100 = 0$, $1/100 = .01$, $2/100 = .02$, ..., $99/100 = .99$, $100/100 = 1$

59. Wrong. For a pair of fair dice, the theoretical probability of a pair of matching numbers is 1/6, as Ruth says. However, it is quite possible, although not very likely, that if you cast a pair of fair dice 20 times, you will never obtain a matching pair (in fact, there is approximately a 2.6% chance that this will happen). In general, a nontrivial claim about theoretical probability can never be absolutely validated or refuted experimentally. All we can say is that the evidence suggests that the dice are not fair.

61. For a (large) number of days, record the temperature prediction for the next day and then check the actual high temperature the next day. Record whether the prediction was accurate (within, say, $2°F$ of the actual temperature). The fraction of times the prediction was accurate is the relative frequency.

7.3 Probability and Probability Models

1.

Outcome	a	b	c	d	e
Probability	.1	.05	.6	.05	x

Since the probabilities of all the outcomes add to 1,

$.1 + .05 + .6 + .05 + x = 1$

$.8 + x = 1$

$x = .2$

(a) $P(\{a, c, e\}) = P(a) + P(c) + P(e)$

$= .1 + .6 + .2 = .9$

(b) $P(E \cup F) = P(\{a, b, c, e\})$

$= P(a) + P(b) + P(c) + P(e)$

$= .1 + .05 + .6 + .2 = .95$

(c) $P(E') = P(\{b, d\}) = P(b) + P(d)$

$= .05 + .05 = .1$

(d) $P(E \cap F) = P\{(c, e)\} = P(c) + P(e)$

$= .6 + .2 = .8$

3. $P(E) = \dfrac{n(E)}{n(S)} = \dfrac{5}{20} = \dfrac{1}{4}$

5. $P(E) = \dfrac{n(E)}{n(S)} = \dfrac{10}{10} = 1$

7. $n(S) = 4$, $n(E) = 3$

$P(E) = \dfrac{n(E)}{n(S)} = \dfrac{3}{4}$

9. $S = \{HH, HT, TH, TT\}$; $n(S) = 4$

$E = \{HH, HT, TH\}$; $n(E) = 3$

$P(E) = \dfrac{n(E)}{n(S)} = \dfrac{3}{4}$

11. $S = \{HHH, HHT, HTH, HTT, THH,$ $THT, TTH, TTT\}$; $n(S) = 8$

$E = \{HTT, THT, TTH, TTT\}$; $n(E) = 4$

$P(E) = \dfrac{n(E)}{n(S)} = \dfrac{4}{8} = \dfrac{1}{2}$

13. $n(S) = 36$, $n(E) = 4$

$P(E) = \dfrac{n(E)}{n(S)} = \dfrac{4}{36} = \dfrac{1}{9}$

15. $n(S) = 36$, $n(E) = 0$

$P(E) = \dfrac{n(E)}{n(S)} = \dfrac{0}{36} = 0$

17. $n(S) = 36$,

$E = \{(2, 2), (2, 3), (2, 5), (3, 2), (3, 3),$ $(3, 5), (5, 2), (5, 3), (5, 5)\}$; $n(E) = 9$

$P(E) = \dfrac{n(E)}{n(S)} = \dfrac{9}{36} = \dfrac{1}{4}$

19. $E = \{(4, 4), (2, 3)\}$

The outcomes for a pair of indistinguishable dice are not equally likely, so we cannot use $P(E) = n(E)/n(S)$. Using Example 7 in the textbook,

$P(4, 4) = \dfrac{1}{36}$

$P(2, 3) = \dfrac{1}{18}$

Therefore, $P(E) = \dfrac{1}{36} + \dfrac{1}{18} = \dfrac{1}{12}$

The corresponding event for distinguishable dice is

$\{(4, 4), (2, 3), (3, 2)\}$

(We can also use this to compute $P(E)$:

$P(E) = \dfrac{n(E)}{n(S)} = \dfrac{3}{36} = \dfrac{1}{12}$)

21. Start with

Outcome	1	2	3	4	5	6
Probability	x	2x	x	2x	x	2x

Since the sum of the probabilities of the outcomes must be 1, we get

$9x = 1$,

so $x = \dfrac{1}{9}$

This gives:

Outcome	1	2	3	4	5	6
Probability	$\dfrac{1}{9}$	$\dfrac{2}{9}$	$\dfrac{1}{9}$	$\dfrac{2}{9}$	$\dfrac{1}{9}$	$\dfrac{2}{9}$

$P(\{1, 2, 3\}) = \dfrac{1}{9} + \dfrac{2}{9} + \dfrac{1}{9} = \dfrac{4}{9}$

23. Start with

Outcome	1	2	3	4
Probability	$8x$	$4x$	$2x$	x

Since the sum of the probabilities of the outcomes must be 1, we get

$$15x = 1,$$

so $x = \dfrac{1}{15}$

This gives:

Outcome	1	2	3	4
Probability	$\dfrac{8}{15}$	$\dfrac{4}{15}$	$\dfrac{2}{15}$	$\dfrac{1}{15}$

25. $P(A) = .1,\ P(B) = .6,\ P(A\cap B) = .05$
$P(A\cup B) = P(A) + P(B) - P(A\cap B)$
$\qquad = .1 + .6 - .05 = .65$

27. $A\cap B = \varnothing,\ P(A) = .3,\ P(A\cup B) = .4$
$P(A\cup B) = P(A) + P(B)$ Mutually exclusive
$4 = .3 + P(B)$
$P(B) = .4 - .3 = .1$

29. $A\cap B = \varnothing,\ P(A) = .3,\ P(B) = .4$
$P(A\cup B) = P(A) + P(B)$ Mutually exclusive
$\qquad = .3 + .4 = .7$

31. $P(A\cup B) = .9,\ P(B) = .6,\ P(A\cap B) = .1$
$P(A\cup B) = P(A) + P(B) - P(A\cap B)$
$.9 = P(A) + .6 - .1$
$P(A) = .9 - .6 + .1 = .4$

33. $P(A) = .75$
$P(A') = 1 - P(A) = 1 - .75 = .25$

35. $P(A) = .3,\ P(B) = .4,\ P(C) = .3$
Since $A,\ B$ and C are mutually exclusive,
$P(A\cup B\cup C) = P(A) + P(B) + P(C)$
$\qquad = .3 + .4 + .3 = 1$

37. $P(A) = .3,\ P(B) = .4$
Since $A,$ and B are mutually exclusive,
$P(A\cup B) = P(A) + P(B)$
$\qquad = .3 + .4 = .7$
$P((A\cup B)') = 1 - P(A\cup B)$

$\qquad = 1 - .7 = .3$

39. Since $A\cup B = S,$
$\quad P(A\cup B) = 1$
Since $A\cap B = \varnothing,\ A$ and B are mutually exclusive, so
$\quad P(A\cup B) = P(A) + P(B)$
Substituting $P(A\cup B) = 1$ gives
$\quad 1 = P(A) + P(B)$

41. $P(A) = .2,\ P(B) = .1;\ P(A\cup B) = .4$
$P(A\cup B)$ is more than $P(A) + P(B).$ But, since $P(A\cup B) = P(A) + P(B) - P(A\cap B),\ P(A\cup B)$ cannot be more than $P(A) + P(B).$ Therefore, the given information does not describe a probability distribution.

43. $P(A) = .2,\ P(B) = .4;\ P(A\cap B) = .2$
Since $P(A\cap B)$ is $\le$ both $P(A)$ and $P(B),$ the given information is consistent with a probability distribution.

45. $P(A) = .1,\ P(B) = 0;\ P(A\cup B) = 0$
No; $P(A\cup B)$ should be $\ge P(A),$ since $A\cup B \supseteq A$

47. (a) $E = \{\text{Current, Past Due}\};$
$\quad n(E) = 136{,}330 + 53{,}310 = 189{,}640$
$$P(E) = \frac{n(E)}{n(S)} = \frac{189{,}640}{203{,}480} \approx .93$$
(b) $E = \{\text{Past Due, In Foreclosure, Repossessed}\};$
$\quad n(E) = 203{,}480 - 136{,}330 = 67{,}150$
$$P(E) = \frac{n(E)}{n(S)} = \frac{67{,}150}{203{,}480} \approx .33$$

49. The probability distribution is obtained by converting the percentages to decimals:

Outcome	Probability
Hispanic or Latino	.48
White (not Hispanic)	.29
African American	.08
Asian	.08
Other	.07

$P(\text{Neither White nor Asian}) = 1 - (.29 + .08) = .63$

51. (a) S = {stock market success, sold to other concern, fail}

(b) P(Stock market success) = $\frac{2}{10}$ = .2

P(Sold to other concerns) = $\frac{3}{10}$ = .3

P(Fail) = 1 − (.2 + .3) = .5

(c) To realize profits for early investors, a start-up venture must be either a stock market success or sold to another concern, so
P(Profit) = .3 + .2 = .5

53. The outcomes of the experiment are: SUV, pickup, passenger car, and minivan. The given information tells us that P(SUV) = .25 and P(pickup) = .15. We are also told that P(passenger car) is five times P(minivan). take P(minivan) = x. Then we have:

Outcome	SUV	Pickup	Passenger Car	Minivan
Probability	.25	.15	5x	x

Since the sum of the probabilities must be 1, we get

$$.25 + .15 + 5x + x = 1$$
$$.40 + 6x = 1$$
$$6x = 1 - .40 = .60$$
$$x = .10$$

So, P(minivan) = .10, and P(passenger car) = $5x$ = .50. This gives the required distribution:

Outcome	SUV	Pickup	Passenger Car	Minivan
Probability	.25	.15	.50	.10

55. Start with

Outcome	1	2	3	4	5	6
Probability	0	x	x	x	x	0

$4x = 1$, so $x = 1/4$ = .25 giving

Outcome	1	2	3	4	5	6
Probability	0	.25	.25	.25	.25	0

57. Start with

Outcome	1	2	3	4	5	6
Probability	x	2x	2x	2x	2x	x

$10x = 1$, so $x = .1$, giving

Outcome	1	2	3	4	5	6
Probability	.1	.2	.2	.2	.2	.1

P(Odd) = .1 + .2 + .2 = .5

59. Let x = P(matching numbers). Then
P(Mismatching numbers) = $2x$
There are 6 matching pairs and 30 mismatching ones. Since the probabilities add to 1,

$$6x + 30(2x) = 1$$
$$66x = 1, \text{ so } x = \frac{1}{66}.$$

This gives:

$$P(1,1) = P(2,2) = \ldots = P(6,6) = \frac{1}{66}$$

$$P(1,2) = \ldots = P(6,5) = \frac{2}{66} = \frac{1}{33}$$

To obtain an odd sum, the pair must consist of an even and odd number, and there are nine such (mismatching) pairs:

$$P(\text{odd sum}) = 9 \times \frac{1}{33} = \frac{6}{11}$$

61. Take $P(2)$ = $15x$. This gives

Outcome	1	2	3	4	5	6
Probability	5x	15x	5x	3x	5x	5x

$38x = 1$, so $x = \frac{1}{38}$, giving

Outcome	1	2	3	4	5	6
Probability	$\frac{5}{38}$	$\frac{15}{38}$	$\frac{5}{38}$	$\frac{3}{38}$	$\frac{5}{38}$	$\frac{5}{38}$

$$P(\text{Odd}) = \frac{5}{38} + \frac{5}{38} + \frac{5}{38} = \frac{15}{38}$$

63. Take

D: I will meet a tall dark stranger; $P(D) = \frac{1}{3}$

T: I will travel; $P(T) = \frac{2}{3}$

$P(D \cap T) = \frac{1}{6}$

$P(D \cup T) = P(D) + P(T) - P(D \cap T)$

$$= \frac{1}{3} + \frac{2}{3} - \frac{1}{6} = \frac{5}{6}$$

65. Take A: A randomly selected person polled ranked jobs or health care as the top domestic priority.
We are told that $P(A) = .61$, and we are asked to find $P(A')$.

$$P(A') = 1 - P(A)$$
$$= 1 - .61 = .39$$

67. Take
G: A randomly chosen new automobile was manufactured by GM; $P(G) = .21$
T: It was manufactured by Toyota; $P(T) = .18$
The events G and T are mutually exclusive, so

$$P(G \cup T) = P(G) + P(T)$$
$$= .21 + .18 = .39$$

We are asked to find
$$P[(G \cup T)'] = 1 - P(G \cup T) = 1 - .39 = .61$$

69. S = set of all applicants; $n(S) = 49{,}060$
E = set of admitted applicants; $n(E) = 11{,}685$

$$P(E) = \frac{n(E)}{n(S)} = \frac{11 \cdot 685}{49 \cdot 060} \approx .24$$

71. S = set of all applicants; $n(S) = 49{,}060$
E = set of admitted applicants who had a Math SAT below 400; $n(E) = 49$

$$P(E) = \frac{n(E)}{n(S)} = \frac{49}{49 \cdot 060} \approx .0010 \approx .00$$

73. A: An applicant was admitted.

$$P(A) = \frac{n(A)}{n(S)} = \frac{11 \cdot 685}{49 \cdot 060} \approx .24$$

$$P(A') = 1 - P(A) \approx 1 - .24 = .76$$

75. R: An applicant had a Math SAT in the range 500–599.
A: An applicant was admitted.

$$P(R) = \frac{12 \cdot 267}{49 \cdot 060} \approx .250$$

$$P(A) = \frac{11 \cdot 685}{49 \cdot 060} \approx .238$$

$$P(R \cap A) = \frac{1470}{49 \cdot 060} \approx .030$$
$$P(R \cup A) = P(R) + P(A) - P(R \cap A)$$
$$\approx .250 + .238 - .030 \approx .46$$

77. R: An applicant had a Math SAT in the range 500–599.
A: An applicant was admitted.
We are asked to find $P[(R \cup A)']$. In Exercise **75** we computed

$$P(R \cup A) \approx .46$$

So
$$P[(R \cup A)'] = 1 - P(R \cup A) \approx 1 - .46 = .54$$

79. S = set of all applicants who had a Math SAT below 400; $n(S) = 1322$
E = set of admitted applicants who had a Math SAT below 400; $n(E) = 49$

$$P(E) = \frac{n(E)}{n(S)} = \frac{49}{1322} \approx .04$$

81. S = set of all admitted applicants; $n(S) = 11{,}685$
E = set of admitted applicants who had a math SAT of 700 or above; $n(E) = 5919$

$$P(E) = \frac{n(E)}{n(S)} = \frac{5919}{11 \cdot 685} \approx .51$$

83. S = set of all rejected applicants; $n(S) = 37{,}375$
E = set of rejected applicants who had a math SAT below 600;
$$n(E) = 10{,}797 + 4755 + 1273 = 16{,}825$$

$$P(E) = \frac{n(E)}{n(S)} = \frac{16 \cdot 825}{37 \cdot 375} \approx .45$$

85. Use the following events:
E: Agree that it should be the government's responsibility to provide a decent standard of living for the elderly.
F: Agree that it would be a good idea to invest part of their Social Security taxes on their own.

$$P(E \cup F) = P(E) + P(F) - P(E \cap F)$$
$$P(E \cup F) = .79 + .43 - P(E \cap F)$$
$$P(E \cup F) = 1.22 - P(E \cap F) \qquad \dots \qquad (1)$$

Since $P(E \cup F)$ must be less than or equal to 1, $P(E \cap F)$ must be at least .22, so the smallest percentage of people that could have agreed with both statements is 22%.

For the second part of the question, we ask how large $P(E \cap F)$ can be for the formula (1) to make sense. The largest conceivable value for $P(E \cap F)$ is .43 (because $E \cap F$ is a subset of E and F and hence its probability cannot exceed either .79 or .43). Thus, the largest percentage of people that could have agreed with both statements is 43%.

87. Use the following events:

B: Failed it backwards; $P(B) = \dfrac{2}{3}$

D: Failed it sideways; $P(D) = \dfrac{3}{4}$

We are also told that $P(B \cap D) = \dfrac{5}{12}$

$$P(B \cup D) = P(B) + P(D) - P(B \cap D)$$
$$= \dfrac{2}{3} + \dfrac{3}{4} - \dfrac{5}{12} = 1$$

Therefore, all of them failed it either backwards or sideways, so all were disqualified.

89. Use the following events:

V: Vaccinated; $P(V) = ,80$

F: Gets flu; $P(F) = .10$

We are also told that 2% of the vaccinated population gets the flu. Therefore, the percentage that are vaccinated and also get the flu is

$P(V \cap F) = .02P(V) = .02 \times .80 = .016$

Now

$$P(V \cup F) = P(V) + P(F) - P(V \cap F)$$
$$= .80 + .10 - .016 = .884$$

Thus, 88.4% of the population either gets vaccinated or gets he disease.

91. Here is one possible experiment: roll a die, and observe which of the following outcomes occurs: Outcome A: 1 or 2 facing up; $P(A) = 1/3$, Outcome B: 3 or 4 facing up; $P(B) = 1/3$, Outcome C: 5 or 6 facing up; $P(C) = 1/3$.

93. He is wrong. It is possible to have a run of losses of any length. Each time he plays the game,. the chances of losing are the same,

regardless of his history of wins or losses. Tony may have grounds to *suspect* that the game is rigged, but no proof.

95. The probability of the union of two events is the sum of the probabilities of the two events if <u>they are mutually exclusive</u>.

97. Wrong. For example, the modeled probability of winning a state lotto is small but nonzero. However, the vast majority of people who play lotto every day of their lives never win, no mater how frequently they play, so the relative frequency is zero for these people.

99. When $A \cap B = \emptyset$ we have
$P(A \cap B) = P(\emptyset) = 0,$
so $P(A \cup B) = P(A) + P(B) - P(A \cap B)$
$= P(A) + P(B) - 0$
$= P(A) + P(B).$

101. Zero. According to the assumption, no matter how many thunderstorms occur, lightning cannot strike a given spot more than once, and so, after n trials the relative frequency will never exceed $1/n$, and so will approach zero as the number of trials gets large. Since the modeled probability models the limit of relative frequencies as the number of trials gets large, it must therefore be zero.

103. $P(A \cup B \cup C) = P(A) + P(B) + P(C) - P(A \cap B) - P(A \cap C) - P(B \cap C) + P(A \cap B \cap C)$

7.4 Probability and Counting Techniques

1. $n(S) = C(10, 5) = 252$
$n(E) = C(4, 4)C(6, 1) = 6$
4 red, 1 non-red
$$P(E) = \frac{n(E)}{n(S)} = \frac{6}{252} = \frac{1}{42}$$

3. $n(S) = C(10, 5) = 252$
E: At least 1 white marble
The complementary event is
F: No white ones
$n(F) = C(8, 5) = 56$
   6 non-white
$$P(F) = \frac{n(F)}{n(S)} = \frac{56}{252} = \frac{2}{9}$$
Therefore,
$$P(E) = 1 - P(F) = 1 - \frac{2}{9} = \frac{7}{9}$$

5. $n(S) = C(10, 5) = 252$
$n(E) = C(4, 2)C(3, 1)C(2, 1)C(1, 1) = 36$
   2 red, 1 green, 1 white, 1 purple
$$P(E) = \frac{n(E)}{n(S)} = \frac{36}{252} = \frac{1}{7}$$

7. $n(S) = C(10, 5) = 252$
$n(E) = C(7, 5) + C(3, 1)C(7, 4)$
   5 non-green or 1 green, 4 non-green
$= 21 + 105 = 126$
$$P(E) = \frac{n(E)}{n(S)} = \frac{126}{252} = \frac{1}{2}$$

9. $n(S) = C(10, 5) = 252$
E: She does not have all the red ones
The complementary event is
F: She has all the red ones.
In Exercise 1, we computed this probability:
$$P(F) = \frac{1}{42}$$
Therefore,
$$P(E) = 1 - \frac{1}{42} = \frac{41}{42}$$

11. $n(S) = C(10, 2) = 45$

There are only 3 stocks listed with yields of 7%
or more. Therefore,
$n(E) = C(3, 2) = 3$
$$P(E) = \frac{n(E)}{n(S)} = \frac{3}{45} = \frac{1}{15}$$

13. $n(S) = C(10, 4) = 210$
$n(E) = C(1, 1)\ C(8, 3) = 56$
(Choose BAC and then 3 out of the 8 that remain
when you exclude BAC and KFT.)
$$P(E) = \frac{n(E)}{n(S)} = \frac{56}{210} = \frac{4}{15}$$

15. $n(S) = C(10, 2) = 45$
Ending up 200 shares of PFE means that PFE was
one of the stocks you chose:
$n(E) = C(1, 1)\ C(9, 1) = 9$
(Choose PFE and then 1 out of the 9 that remain.)
$$P(E) = \frac{n(E)}{n(S)} = \frac{9}{45} = \frac{1}{5}$$

17. Number of possible completed tests:
$n(S) = 2^8 \times 5^5 \times 5!$
   8 true/false, 5 multiple choice, 5 matching
Number of correct answers: $n(E) = 1$
$$P(E) = \frac{n(E)}{n(S)} = \frac{1}{2^8 \times 5^5 \times 5!}$$

19. $n(S) = C(52, 5) = 2,598,960$
$n(E) = 1,098,240$ (6.4 Exercise 47)
$$P(E) = \frac{n(E)}{n(S)} = \frac{1,098,240}{2,598,960} \approx .4226$$

21. $n(S) = C(52, 5) = 2,598,960$
$n(E) = 123,552$ (6.4 Exercise 49)
$$P(E) = \frac{123,552}{2,598,960} \approx .0475$$

23. $n(S) = C(52, 5) = 2,598,960$
Decision algorithm for computing the number of
flushes:
First compute the number of hands in which all 5
cards are of the same suit:
   Step 1: Select a suit: 4 choices
   Step 2: Select 5 cards of that suit: $C(13, 5)$
   choices
Total number of choices: $4 \times C(13, 5)$

285

Among these are hands the 5 cards are consecutive: A, 2, 3, 4, 5 up through 10, J, Q, K, A (straight flushes). The number of straight flushes is 4×10 (Step 1: Select a suit; Step 2: Select a starting card for the run). Excluding these gives
Total number of flushes:

$$n(E) = 4 \times C(13, 5) - 4 \times 10 = 5108$$
$$P(E) = \frac{n(E)}{n(S)} = \frac{5108}{2,598,960} \approx .0020$$

25. $n(S) = 27^{39}$
$n(E) = 1$ (The correct sequence)
$$P(E) = \frac{n(E)}{n(S)} = \frac{1}{27^{39}}$$

27. $n(S) = C(7, 4) = 35$
$n(E) = C(5, 4) = 5$
$$P(E) = \frac{n(E)}{n(S)} = \frac{5}{35} = \frac{1}{7}$$

29. $n(S) = C(50, 5) = 2,118,760$
Big Winner: $n(E) = 1$
$$P(E) = \frac{n(E)}{n(S)} = \frac{1}{2,118,760} \approx .000\,000\,472$$
Small-Fry Winner:
$n(F) = C(5, 4)C(45, 1) = 225$
　　4 winning numbers & 1 losing number
$$P(F) = \frac{n(F)}{n(S)} = \frac{225}{2,118,760} \approx .000\,106\,194$$

$P(E \cup F) = P(E) + P(F)$ Mutually exclusive
$$= \frac{226}{2,118,760} \approx .000\,106\,666$$

31. $n(S) = C(700, 400)$
(a) $n(E) = C(100, 100)C(600, 300) = C(600, 300)$
　　　100 managers, 300 non-managers
$$P(E) = \frac{n(E)}{n(S)} = \frac{C(600, 300)}{C(700, 400)}$$

(b) $n(E) = C(1, 1)C(699, 399) = C(699, 399)$
　　You, 399 others
$$P(E) = \frac{n(E)}{n(S)} = \frac{C(699, 399)}{C(700, 400)}$$

33. $n(S) = 10 \times 10 \times 10 = 10^3$
$n(E) = 10 \times 9 \times 8 = P(10, 3) = 720$
$$P(E) = \frac{n(E)}{n(S)} = \frac{720}{1000} = .72$$

35. A perfect progression is a sequence of 8 "Won" scores in the range 0–7 that are all different. S is the set of all sequences of 8 digits in the range 0–7.
$$n(S) = 8^8; \ n(E) = 8!$$
$$P(E) = \frac{n(E)}{n(S)} = \frac{8!}{8^8}$$

37. Each of the two random moves from the starting position be go up, down, left, or right. Since this gives 4 choices per move, there are
$$n(S) = 4 \times 4 = 16$$
possible sequences of two moves. Only two of these sequences will get you to the Finish node: right+down and down+right. Therefore,
$$n(E) = 2$$
$$P(E) = \frac{n(E)}{n(S)} = \frac{2}{16} = \frac{1}{8}$$

39. The number of possible outcomes is
$$n(S) = 2 \times 2 \times 2 = 8 \quad \text{Select a winner 3 times}$$
The only way North Carolina (NC) will beat Central Connecticut (CC) but lose to Virginia is if NC beats CC, Virginia beats Syracuse, and CC loses to Virginia. Therefore,
$$n(E) = 1$$
$$P(E) = \frac{n(E)}{n(S)} = \frac{1}{8}$$

41. The number of possible sequences of 4 digits in the range 0–9 is
$$n(S) = 10 \times 10 \times 10 \times 10 = 10,000$$
Number of correct codes: 1
Number of codes that are correct except for a single digit:
　Select a slot for the incorrect digit: 4 choices
　Select the incorrect digit: 9 choices
　Fill in the remaining slots with the correct digits: 1 choice
Total number of incorrect codes: 4×9×1 = 36

Therefore, $n(E) = 1 + 36 = 37$

$$P(E) = \frac{n(E)}{n(S)} = \frac{37}{10,000} = .0037$$

43. **(a)** Decision algorithm for forming a committee:

Select a chief investigator: $C(6, 1)$ choices
Select an assistant investigator: $C(6, 1)$ choices
Select 2 at-large investigators: $C(10, 2)$ choices
Select 5 ordinary members: $C(8, 5)$ choices

This gives

$$n(S) = C(6, 1)C(6, 1)C(10, 2)C(8, 5) = 90,720$$

(b) Decision algorithm to make Larry happy:

Alternative 1: Larry is chief and Otis is assistant:

Select Larry for chief and Otis for assistant: 1 choice
Select 2 at-large investigators: $C(10, 2)$ choices
Select 5 ordinary members: $C(8, 5)$ choices

This gives $C(10, 2)C(8, 5)$ choices for this alternative

Alternative 2: Larry is not on the committee:

Select a chief investigator: $C(5, 1)$ choices
Select an assistant investigator: $C(6, 1)$ choices
Select 2 at-large investigators: $C(9, 2)$ choices
Select 5 ordinary members: $C(7, 5)$ choices

This gives $C(5, 1)C(6, 1)C(9, 2)C(7, 5)$ choices for this alternative

Adding gives a total of

$n(E)$
$= C(10,2)C(8,5) + C(5,1)C(6,1)C(9,2)C(7,5)$
$= 25,200$

(c) $P(E) = \dfrac{n(E)}{n(S)} = \dfrac{25,200}{90,720} \approx .28$

The difference between permutations and combinations cancels when we compute the probability.

45. The four outcomes listed are not equally likely; for example, (red, blue) can occur in four ways. The methods of this section yield a probability for (red, blue) of $C(2,2)/C(4,2) = 1/6$

47. No. If we do not pay attention to order, the probability is $C(5,2)/C(9,2) = 10/36 = 5/18$. If we do pay attention to order, the probability is $P(5,2)/P(9,2) = 20/72 = 5/18$ again.

7.5 Conditional Probability and Independence

1. $P(A \mid B) = \dfrac{P(A \cap B)}{P(B)} = \dfrac{.2}{.5} = .4$

3. $P(A \cap B) = P(A \mid B)P(B) = (.2)(.4) = .08$

5. $P(A \cap B) = P(A \mid B)P(B)$
$.3 = .4P(sB)$
$P(B) = \dfrac{.3}{.4} = .75$

7. If A and B are independent, then
$P(A \cap B) = P(A)P(B) = (.5)(.4) = .2$

9. If A and B are independent, then
$P(A \mid B) = P(A) = .5$

11.

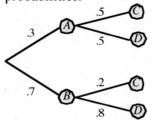

Since the probabilities leaving each branching point must add to 1, we can fill in the missing probabilities:

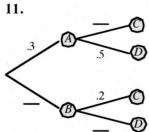

The probability of each outcome is obtained by multiplying the probabilities along the branches leading to the corresponding node:

$P(A \cap C) = .3 \times .5 = .15$
$P(A \cap D) = .3 \times .5 = .15$
$P(B \cap C) = .7 \times .2 = .14$
$P(B \cap D) = .7 \times .8 = .56$

13.

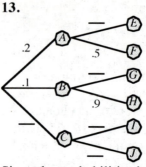

Since the probabilities leaving each branching point must add to 1, we can fill in some of the missing probabilities:

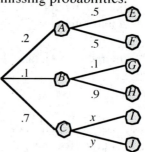

To get the remaining two, use the given information about the probabilities of the outcomes:

$P(C \cap I) = .7 \times x = .14 \Rightarrow x = \dfrac{.14}{.7} = .2$

And so $y = 1 - .2 = .8$

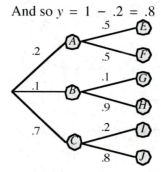

The probability of each outcome is obtained by multiplying the probabilities along the branches leading to the corresponding node:

$P(A \cap E) = .2 \times .5 = .10$
$P(A \cap F) = .2 \times .5 = .10$
$P(B \cap G) = .1 \times .1 = .01$
$P(B \cap H) = .1 \times .9 = .09$
$P(C \cap I) = .7 \times .2 = .14$
$P(C \cap J) = .7 \times .8 = .56$

15. A: The sum is 5.
   B: The green one is not a 1.

$P(A \cap B) = P\{(3, 2), (2, 3), (1, 4)\} = \dfrac{3}{36} = \dfrac{1}{12}$

$P(B) = \dfrac{30}{36} = \dfrac{5}{6}$

$P(A|B) = \dfrac{P(A \cap B)}{P(B)} = \dfrac{1/12}{5/6} = \dfrac{1}{10}$

17. A: The red one is 5.
   B: The sum is 6.

$P(A \cap B) = P\{(5, 1)\} = \dfrac{1}{36}$

$P(B) = P(\{1, 5), (2, 4), (3, 3), (4, 2), (5, 1)\}$

$= \dfrac{5}{36}$

$P(A|B) = \dfrac{P(A \cap B)}{P(B)} = \dfrac{1/36}{5/36} = \dfrac{1}{5}$

19. A: The sum is 5.
   B: The dice have opposite parity.

$P(A \cap B) = P\{(4, 1), (3, 2), (2, 3), (1, 4)\}$

$= \dfrac{4}{36} = \dfrac{1}{9}$

$P(B) = \dfrac{18}{36} = \dfrac{1}{2}$

$P(A|B) = \dfrac{P(A \cap B)}{P(B)} = \dfrac{1/9}{1/2} = \dfrac{2}{9}$

21. A: She gets all 3 red ones.
   B: She gets the fluorescent pink one.

$n(S) = C(10, 4) = 210$

$P(A \cap B) = \dfrac{C(1, 1)C(3, 3)}{210} = \dfrac{1}{210}$

$P(B) = \dfrac{C(1, 1)C(9, 3)}{210} = \dfrac{84}{210}$

$P(A|B) = \dfrac{P(A \cap B)}{P(B)} = \dfrac{1/210}{84/210} = \dfrac{1}{84}$

23. A: She gets no red ones.
   B: She gets the fluorescent pink one.

$n(S) = C(10, 4) = 210$

$P(A \cap B) = \dfrac{C(1, 1)C(6, 3)}{210} = \dfrac{20}{210}$

$P(B) = \dfrac{C(1, 1)C(9, 3)}{210} = \dfrac{84}{210}$

$P(A|B) = \dfrac{P(A \cap B)}{P(B)} = \dfrac{20/210}{84/210} = \dfrac{5}{21}$

25. A: She gets one of each color other than
   fluorescent pink.
   B: She gets at least one red one.

Notice that $A \cap B = A$

$n(S) = C(10, 4) = 210$

$P(A \cap B) = P(A)$

$= \dfrac{C(3, 1)C(2, 1)C(2, 1)C(2, 1)}{210} = \dfrac{24}{210}$

$P(B) = 1 - \dfrac{C(7, 4)}{210} = 1 - \dfrac{35}{210} = \dfrac{175}{210}$

$P(A|B) = \dfrac{P(A \cap B)}{P(B)} = \dfrac{24/210}{175/210} = \dfrac{24}{175}$

27. A: Your new skateboard design is a success.
   B: Your new skateboard design is a failure.
 If your new skateboard design is a success, it
cannot be a failure, and *vice-versa*. Therefore, A
and B are mutually exclusive.

29. A: Your new skateboard design is a success.
   B: Your competitor's new skateboard design
   is a failure.
 Since The likelihood of A does effect the
likelihood of B, the events are not independent.
Since it is possible for A and B to occur together,
they are not mutually exclusive. Therefore, the
correct choice is "neither".

31. A: The red die is 1, 2, or 3; $P(A) = \dfrac{1}{2}$

   B: The green die is even; $P(B) = \dfrac{1}{2}$

   $A \cap B$: The red die is 1, 2, or 3 and the green
   one is even; $P(A \cap B) = \dfrac{9}{36} = \dfrac{1}{4}$

$P(A)P(B) = \dfrac{1}{2} \times \dfrac{1}{2} = \dfrac{1}{4}$

$P(A \cap B) = \dfrac{1}{4}$

Since $P(A \cap B) = P(A)P(B)$, A and B are
independent.

33. A: Exactly one die is 1; $P(A) = \dfrac{10}{36} = \dfrac{5}{18}$

B: The sum is even; $P(B) = \dfrac{1}{2}$

$A \cap B$: Exactly one die is 1 and the sum is even; $P(A \cap B) = \dfrac{4}{36} = \dfrac{1}{9}$

$P(A)P(B) = \dfrac{5}{18} \times \dfrac{1}{2} = \dfrac{5}{36}$

$P(A \cap B) = \dfrac{1}{9}$

Since $P(A \cap B) \neq P(A)P(B)$, A and B are dependent.

35. A: Neither die is 1; $P(A) = \dfrac{25}{36}$

B: Exactly one die is 2; $P(B) = \dfrac{10}{36}$

$A \cap B$: Neither die is 1 and exactly one is 2; $P(A \cap B) = \dfrac{8}{36}$

$P(A)P(B) = \dfrac{25}{36} \times \dfrac{10}{36} \approx .1929$

$P(A \cap B) = \dfrac{8}{36} = \dfrac{2}{9} \approx .2222$

Since $P(A \cap B) \neq P(A)P(B)$, A and B are dependent.

37. The outcome of each coin toss is independent of the others. Therefore,

$P(HTTHHHTHHTT) = P(H)P(T)P(H) \ .. \ P(T)$

$$= \dfrac{1}{2} \cdot \dfrac{1}{2} \ \cdots \ \cdot \dfrac{1}{2} = \dfrac{1}{2^{11}} = \dfrac{1}{2048}$$

39. 10% of all Anchovians detest anchovies (D).
Rewording: The probability that an Anchovian detests anchovies is .10
   $P(D) = .10$
30% of all married Anchovians (M) detest anchovies.
Rewording this as in Example 2, we get:
The probability that an Anchovian detests anchovies, given that he or she is married, is .30;
   $P(D|M) = .30$

41. 30% of all lawyers who lost clients (L) were antitrust lawyers (A).
Rewording this as in Example 2, we get:
The probability that a lawyer was an antitrust lawyer, given that she lost clients, is .30
   $P(A|L) = .30$
10% of all antitrust lawyers lost clients.
Rewording this as in Example 2, we get:
The probability that a lawyer lost clients, given that she was an antitrust lawyer, was .10;
   $P(L|A) = .10$

43. 55% of those who go out in the midday sun (M) are Englishmen (E).
Rewording this as in Example 2, we get:
The probability that someone is an Englishman, given that he goes out in the midday sun, is .55
   $P(E|M) = .55$
5% of those who do not go out in the midday sun are Englishmen.
Rewording this as in Example 2, we get:
The probability that someone is an Englishman, given that he does not go out in the midday sun, is .05
   $P(E|M') = .05$

45. The two events we are interested in are:
B: A person in the US declared personal bankruptcy.
E: A person in the US recently experienced a "big three" event.
We are told that:
The probability that a person in the US would declare personal bankruptcy was .006. That is,
   $P(B) = .006.$
The probability that a person in the US would declare personal bankruptcy *and* had recently experienced a "big three" event was .005. That is,
   $P(E \cap B) = .005$
We are asked to find the probability that a person had recently experienced one of the "big three" events given that she had declared personal bankruptcy. That is, we are asked to find $P(E \mid B)$.

$$P(E \mid B) = \dfrac{P(E \cap B)}{P(B)} = \dfrac{.005}{.006} \approx .8$$

47. Take

A: A new listing on eBay was in the U.S.

B: A new listing on eBay was posted during the fourth quarter.

(a) We are asked for $P(A|B)$.

$A \cap B$ is the event that a new listing on eBay was in the U.S. and posted in the fourth quarter.

$$P(A \cap B) = \frac{275}{2750}; \; P(B) = \frac{730}{2750}$$

$$P(A|B) = \frac{P(A \cap B)}{P(B)} = \frac{275/2750}{730/2750} = \frac{275}{730} \approx .38$$

(b) We are asked for $P(B|A)$.

$$P(A \cap B) = \frac{275}{2750}; \; P(A) = \frac{1130}{2750}$$

$$P(B|A) = \frac{P(A \cap B)}{P(A)} = \frac{275/2750}{1130/2750} = \frac{275}{1130} \approx .24$$

49. The two events are

A: Agreed that it should be the government's responsibility to provide a decent standard of living for the elderly.

$P(A) = .79$

B: Agreed that it would be a good idea to invest part of their Social Security taxes on their own. $P(B) = .43$

For independence,

$$P(A \mid B) = P(A)P(B) = (.79)(.43) \approx .34$$

51. Take

X: Used Brand X; $P(X) = .40$

G: Gave up doing laundry; $P(G) = .05$

$P(X \cap G) = .04$

$P(X)P(G) = .40 \times .05 = .02 \neq P(X \cap G)$

Therefore X and G are not independent.

$$P(G|X) = \frac{P(G \cap X)}{P(X)} = \frac{.04}{.10} = .10,$$

which is larger than $P(G)$. Therefore, a user of Brand X is more likely to give up doing laundry than a randomly chosen person.

53. Use the following events:

D: Involved in deadly accident.

T: Involved in tire-related accident.

We are told that

$P(D \cap T) = .000003$

We are also told that the probability that a tire-related accident would prove deadly was .02. Rephrasing this,

The probability that an accident is deadly, *given that* it is tire related, is .02.

$P(D|T) = .02$

We are asked to find $P(T)$. Start with the formula:

$$P(D|T) = \frac{P(D \cap T)}{P(T)}$$

$$.02 = \frac{.000\,003}{P(T)}$$

$$.02P(T) = .000003$$

$$P(T) = \frac{.000\,003}{.02} = .00015$$

55. $P(E|S) = \dfrac{P(E \cap S)}{P(S)} = \dfrac{.25}{.30} = \dfrac{5}{6}$

or $P(E|S) = \dfrac{fr(E \cap S)}{fr(S)} = \dfrac{25}{30} = \dfrac{5}{6}$

	N	E	Total
S	5	25	30
U	15	55	70
Total	20	80	100

57. $P(U|N) = \dfrac{P(U \cap N)}{P(N)} = \dfrac{.15}{.20} = .75$

or $P(U|N) = \dfrac{fr(U \cap N)}{fr(N)} = \dfrac{15}{20} = .75$

	N	E	Total
S	5	25	30
U	15	55	70
Total	20	80	100

59. $P(U|E) = \dfrac{P(U \cap E)}{P(E)} = \dfrac{.55}{.80} = \dfrac{11}{16}$

or $P(U|E) = \dfrac{fr(U \cap E)}{fr(E)} = \dfrac{55}{80} = \dfrac{11}{16}$

	N	E	Total
S	5	25	30
U	15	55	70
Total	20	80	100

61. P(An unsuccessful author is established)

$= P(E|U)$

$$P(E|U) = \frac{P(E \cap U)}{P(U)} = \frac{.55}{.70} = \frac{11}{14}$$

$$or \quad P(E|U) = \frac{fr(E \cap U)}{fr(U)} = \frac{55}{70} = \frac{11}{14}$$

	N	E	Total
S	5	25	30
U	15	55	70
Total	20	80	100

63.

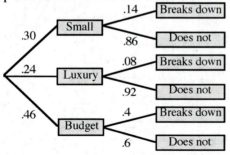

$P(\text{Sell 2-door}) = .12$

$P(\text{Sell 4-door}) = .28$

$P(\text{No sale}) = .6$

Note that the probability of each outcome was obtained by multiplying the probabilities of the corresponding edges.

65. Notice that the total number of vehicles add up to 100, so that the numbers give the probabilities for the fist branches of the tree:

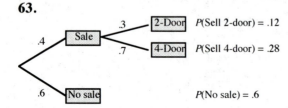

The probability of each outcome was obtained by multiplying the probabilities of the corresponding edges:

$P(\text{Small and breaks down})$
  $= .30 \times .14 = .042$
$P(\text{Small and does not break down})$
  $= .30 \times .86 = .258$
$P(\text{Luxury and breaks down})$
  $= .24 \times .08 = .0192$
$P(\text{Luxury and does not break down})$
  $= .24 \times .92 = .2208$
$P(\text{Budget and breaks down})$
  $= .46 \times .4 = .184$
$P(\text{Budget and does not break down})$

$= .46 \times .6 = .276$

67.

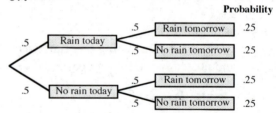

(The probability of each outcome was obtained by multiplying the probabilities of the corresponding edges.)
From the tree,
  $P(\text{No rain today or tomorrow}) = .25$

69. Use the following events:
  A: Employed
  B: Bachelor's degree or higher
$$P(A|B) = \frac{fr(A \cap B)}{fr(B)} = \frac{44.1}{57.1} \approx .77$$

71. Use the following events:
  A: Bachelor's degree or higher
  B: Employed
$$P(A|B) = \frac{fr(A \cap B)}{fr(B)} = \frac{44.1}{127.1} \approx .35$$

73. Rephrasing the question, we want to find the probability that someone was not in the labor force, given that they had not completed a Bachelor's degree or higher.
  A: Not in labor force
  B: Not completed a Bachelor's degree or higher
$$P(A|B) = \frac{fr(A \cap B)}{fr(B)} = \frac{62.7 - 12.1}{195.3 - 57.1} \approx .37$$

75. Rephrasing the question, we want to find the probability that someone was employed, given that he or she had completed a Bachelor's degree or higher and was in the labor force.
A: Employed
B: Bachelor's degree or higher and in the labor force

$$P(A|B) = \frac{fr(A \cap B)}{fr(B)} = \frac{44.1}{57.1\text{-}12.1} \approx .98$$

77. We compare

$P(\text{High school diploma only}|\text{Unemployed})$

with

$P(\text{High school diploma only}|\text{Employed})$

$P(\text{High school diploma only}|\text{Unemployed})$

$$= \frac{1.7}{5.5} \approx .31$$

$P(\text{High school diploma only}|\text{Employed})$

$$= \frac{36.9}{127.1} \approx .29$$

Your friend is right: The probability that an unemployed person has a high school diploma only is .31, while the corresponding figure for an employed person is .29.

79. $P(K|D) = 1.31P(K|D')$

81. (a) If the course has a positive effect on productivity, then the probability of a productivity increase is greater for a person who took the course. That is,

$P(I|T) > P(I)$

(b) If T and I are independent, than $P(I|T) = P(I)$, so that taking the course had no effect on productivity: The course was ineffective.

83. (a) Use the following events:

A: The person was an Internet user
B: The person's family income was at least $35,000

$$P(A|B) = \frac{P(A \cap B)}{P(B)} = \frac{.35}{.35+.24} \approx .59$$

(b) We compare

$P(\text{Internet User} \mid \text{Income} \geq \$35\text{K})$

with

$P(\text{Internet User} \mid \text{Income} < \$35\text{K})$

By part (a),

$P(\text{Internet User} \mid \text{Income} < \$35\text{K}) = .59$

By a similar calculation

$P(\text{Internet User} \mid \text{Income} < \$35\text{K})$

$$= \frac{.11}{.11+.30} \approx .27$$

Therefore, a person was more likely to be an Internet user if his or her family income was $35,000 or more.

85. From the table, the probability that Jeep Wrangler was reported stolen was .0170. In other words,

The probability that a vehicle was reported stolen given that it was a Jeep Wrangler was .0170.

$P(R|J) = .0170$

87. From the table, the probability that Toyota Land Cruiser was reported stolen was .0143. In other words,

The probability that a vehicle was reported stolen given that it was a Toyota Land Cruiser was .0143.

This is exactly what Choice (D) says.

89. Consider the following events:

B: My BMW 3-series will be stolen;
$P(B) = .0077$
L: My Lexus GS300 will be stolen;
$P(L) = .0074$

(a) Since the events are independent,

$P(B \cap L) = P(B)P(L) = .0077 \times .0074 \approx .000057$

(b) $P(B \cup L) = P(B) + P(L) - P(B \cap L)$

$$= .0077 + .0074 - .000057$$

$$= .015043$$

91. Use the following events:

V: Tests positive
U: Use steroids
$P(U|V) = .90, \; P(U \cap V) = .10$

$$P(U|V) = \frac{P(U \cap V)}{P(V)}$$

$$.90 = \frac{.10}{P(V)} \Rightarrow .90P(V) = .10$$

$$P(V) = \frac{.10}{.90} \approx .11, \text{ or } 11\%$$

93. Use the following events:

A: Contaminated by salmonella
B: Contaminated by a strain of salmonella resistant to at least three antibiotics

We are told that

293

$P(A) = .20,$

and also that the probability that a salmonella-contaminated sample was contaminated by a strain resistant to at least three antibiotics was .53. Rewriting this gives:

The probability that a sample was contaminated by a strain resistant to at least three antibiotics *given that* it was contaminated by salmonella is .53.

$P(B|A) = .53$

We are asked to find $P(B \cap A)$ (which is actually the same as $P(B)$), so we use the formula:

$$P(B|A) = \frac{P(B \cap A)}{P(A)}$$

$$.53 = \frac{P(B \cap A)}{.20}$$

$$P(B \cap A) = .53 \times .20 = .106$$

95. Take

A: Contaminated by salmonella
B1: Contaminated by a strain of salmonella resistant to at least one antibiotic
B3: Contaminated by a strain of salmonella resistant to at least three antibiotics

We are told:

$P(A) = .20$, $P(B1|A) = .84$, $P(B3|A) = .53$

We are asked to find $P(B3|B1)$.

$$P(B3|B1) = \frac{P(B3 \cap B1)}{P(B1)} = \frac{P(B3)}{P(B1)} = \frac{P(B3 \cap A)}{P(B1 \cap A)}$$

because $B3 \cap B1 = B3 = B3 \cap A$, and $B1 = B1 \cap A$. But we have

$$P(B1|A) = \frac{P(B1 \cap A)}{P(A)} = .84$$

$$P(B3|A) = \frac{P(B3 \cap A)}{P(A)} = .53$$

Taking their ratio gives

$$\frac{P(B3 \cap A)}{P(B1 \cap A)} = \frac{.53}{.84} \approx .631$$

Therefore,

$$P(B3|B1) = \frac{P(B3 \cap A)}{P(B1 \cap A)} \approx .631$$

97. Answers will vary. Here is a simple one: E: the first toss is a head, F: the second toss is a head, G: the third toss is a head.

99. The probability you seek is $P(E|F)$, or should be. If, for example, you were going to place a wager on whether E occurs or not, it is crucial to know that the sample space has been reduced to F (you know that F did occur). If you base your wager on $P(E)$ rather than $P(E|F)$ you will misjudge your likelihood of winning.

101. You might explain that the conditional probability is not the *a priori* probability of E, but it is the probability of E in a hypothetical world in which the outcomes are restricted to be what is given. In the example she is citing, yes, the probability of thowing a double six is 1/36, in the absence of any other knowledge. However by the "conditional probability" of throwing a double six given that the sum is larger than 7, we mean the probability of a double six in a situation in which the dice have already been thrown and we know that the sum is greater than 7. Since there are only 15 ways in which that can happen, the conditional probability is 1/15. For a more extreme case, consider the conditional probability of throwing a double six given that the sum is 12.

103. If $A \subseteq B$ then $A \cap B = A$, so

$$P(A \cap B) = P(A)$$

and

$$P(A|B) = \frac{P(A \cap B)}{P(B)} = \frac{P(A)}{P(B)}$$

105. Your friend is correct. If A and B are mutually exclusive then

$$P(A \cap B) = 0.$$

On the other hand, if A and B are independent then

$$P(A \cap B) = P(A)P(B).$$

Thus,

$$P(A)P(B) = 0.$$

If a product is 0 then one of the factors must be 0, so either $P(A) = 0$ or $P(B) = 0$. Thus, it cannot be true that A and B are mutually exclusive, have nonzero probabilities, and are independent all at the same time.

107. Suppose that A and B are independent, so that

$P(A \cap B) = P(A)P(B)$

Then,

$P(A' \cap B') = 1 - P(A \cup B)$

$= 1 - [P(A) + P(B) - P(A \cap B)]$

$= 1 - [P(A) + P(B) - P(A)P(B)]$

[Since $P(A \cap B) = P(A)P(B)$]

$= (1 - P(A))(1 - P(B))$

$= P(A')P(B')$

Therefore,

$P(A' \cap B') = P(A')P(B')$

and so A' and B' are independent.

7.6 Bayes' Theorem and Applications

1. $P(A|B) = .8$, $P(B) = .2$, $P(A|B') = .3$
$P(B') = 1 - P(B) = 1 - .2 = .8$

$$P(B|A) = \frac{P(A|B)P(B)}{P(A|B)P(B) + P(A|B')P(B')}$$

$$= \frac{(.8)(.2)}{(.8)(.2) + (.3)(.8)} = \frac{.16}{.16 + .24} = .4$$

3. $P(X|Y) = .8$, $P(Y') = .3$, $P(X|Y') = .5$
$P(Y) = 1 - P(Y') = 1 - .3 = .7$,

$$P(Y|X) = \frac{P(X|Y)P(Y)}{P(X|Y)P(Y) + P(X|Y')P(Y')}$$

$$= \frac{(.8)(.7)}{(.8)(.7) + (.5)(.3)} = \frac{.56}{.56 + .15}$$

$$\approx 7887$$

5. $P(X|Y_1) = .4$, $P(X|Y_2) = .5$, $P(X|Y_3) = .6$,
$P(Y_1) = .8$, $P(Y_2) = .1$
$P(Y_3) = 1 - (.8+.1) = .1$
$P(Y_1|X)$

$$= \frac{P(X|Y_1)P(Y_1)}{P(X|Y_1)P(Y_1) + P(X|Y_2)P(Y_2) + P(X|Y_3)P(Y_3)}$$

$$= \frac{(.4)(.8)}{(.4)(.8) + (.5)(.1) + (.6)(.1)}$$

$$= \frac{.32}{.32 + .05 + .06} \approx .7442$$

7. $P(X|Y_1) = .4$, $P(X|Y_2) = .5$, $P(X|Y_3) = .6$,
$P(Y_1) = .8$, $P(Y_2) = .1$
$P(Y_3) = 1 - (.8+.1) = .1$
$P(Y_2|X)$

$$= \frac{P(X|Y_2)P(Y_2)}{P(X|Y_1)P(Y_1) + P(X|Y_2)P(Y_2) + P(X|Y_3)P(Y_3)}$$

$$= \frac{(.5)(.1)}{(.4)(.8) + (.5)(.1) + (.6)(.1)}$$

$$= \frac{.05}{.32 + .05 + .06} \approx .1163$$

9. Use the following events:
D: Decreased spending on music
I: Internet user

$P(D|I) = .11$, $P(I) = .40$, $P(D|I') = .2$
We are asked to compute $P(I|D)$

$$P(I|D) = \frac{P(D|I)P(I)}{P(D|I)P(I) + P(D|I')P(I')}$$

$$= \frac{(.11)(.4)}{(.11)(.4) + (.2)(.6)} = \frac{.044}{.044 + .12}$$

$$\approx .268, \text{ or } 26.8\%$$

11. Use the following events:
A: Snows in Greenland;
B: Glaciers grow;

$$P(A) = \frac{1}{25} = .04$$

$P(B|A) = .20$, $P(B|A') = .04$
We are asked to compute $P(A|B)$.

$$P(A|B) = \frac{P(B|A)P(A)}{P(B|A)P(A) + P(B|A')P(A')}$$

$$= \frac{(.2)(.04)}{(.2)(.04) + (.04)(.96)} = \frac{.008}{.008 + .0384}$$

$$\approx .1724$$

13. Use the following events
C: Driving a car
D: Driver dies (in a severe side-impact)
We are given $P(C) = .454$. Also, the information in the table tells us that $P(D|C) = 1$, $P(D|C') = .3$. We are asked to find $P(C|D)$. Bayes' theorem states

$$P(C|D) = \frac{P(D|C)P(C)}{P(D|C)P(C) + P(D|C')P(C')}$$

$$= \frac{.454}{.454+(.3)(1-.454)} \approx .73$$

15. Use the following events:
A: Fit enough to play
B: Failed the fitness test and therefore dropped from the team
$P(B|A) = 5$, $P(B|A') = 1$
$P(A) = .45$, so $P(A') = .55$
We are asked to compute the probability that Mona was justifiably dropped.

$$P(A|B) = \frac{P(B|A)P(A)}{P(B|A)P(A) + P(B|A')P(A')}$$

$$= \frac{(.5)(.45)}{(.5)(.45) + (1)(.55)} = \frac{.225}{.225 + .55}$$

$$\approx .2903$$

This is the probability that she was unjustifiably dropped. Therefore, the probability that Mona was justifiably dropped is

$P(A'|B) = 1 - P(A|B) \approx 1 - .2309 = .7097$

17. Use the following events
  L: Driving a light truck
  V: Driving an SUV
  C: Driving a car
  D: Driver dies (in a severe side-impact)
We are given the following information:
  $P(L) = .273, P(V) = .273, P(C) = .454$. Also, the information in the table tells us that
  $P(D|T) = .210, P(D|V) = .271, P(D|C) = 1$.
We are asked to find the probability that a driver was driving an SUV, given that the driver died (in a severe side-impact), that is, $P(V|D)$. Bayes' theorem states

$P(V|D)$

$= \dfrac{P(D|V)P(V)}{P(D|L)P(L) + P(D|V)P(V) + P(D|C)P(C)}$

$= \dfrac{(.371)(.273)}{(.210)(.273) + (.371)(.273) + (1)(.454)}$

$\approx .165$

19. Use the following events:
  A: Admitted into UCLA
  C: California applicant
  U: Applicant from another U.S. state
  I: International applicant
  $P(A|C) = .22, P(A|U) = .28, P(A|I) = .22$
  $P(C) = .84, P(U) = .10, P(I) = .06$
We are asked to compute $P(C|A)$

$P(C|A)$

$= \dfrac{P(A|C)P(C)}{P(A|C)P(C) + P(A|U)P(U) + P(A|I)P(I)}$

$= \dfrac{(.22)(.84)}{(.22)(.84) + (.28)(.10) + (.22)(.06)}$

$\approx .82$, or approximately 82%

21. Use the following events:
  I: Used the Internet for e-mail
  C: Caucasian
  A: African American
  H: Hispanic
  R: Other

$P(I|C) = .86, P(I|A) = .77, P(I|H) = .77,$
$P(I|R) = .85$
$P(C) = .69, P(A) = .12, P(H) = .13,$
$P(R) = 1 - (.69+.12+.13) = .06$
We are asked to compute $P(H|I)$
  $P(H|I)$
$=$

$\dfrac{P(I|H)P(H)}{P(I|C)P(C)+P(I|A)P(A)+P(I|H)P(H)+P(I|R)P(R)}$

$= \dfrac{(.77)(.13)}{(.86)(.69)+(.77)(.12)+(.77)(.13)+(.85)(.06)}$

$\approx .1196$, or approximately 12%

23. Use the following events:
  M: Married
  L: Have pool
  $P(M|L) = .86, P(L) = .15$
(a) Given that $P(M|L') = .90$, we are asked to find $P(L|M)$

$P(L|M) = \dfrac{P(M|L)P(L)}{P(M|L)P(L) + P(M|L')P(L')}$

$= \dfrac{(.86)(.15)}{(.86)(.15) + (.9)(.85)} = \dfrac{.129}{.129 + .765}$

$\approx .1443$, or 14.43%

(b) From part (a), 14.43% of married couples have pools. The percentage of single people with pools is obtained as follows: Take
  S: Single
  L: Have pool
  $P(L) = .15,$
  $P(S|L) = 1 - .86 = .14$
  $P(S|L') = 1 - .90 = .10$ (From part (b))

$P(L|S) = \dfrac{P(S|L)P(L)}{P(S|L)P(L) + P(S|L')P(L')}$

$= \dfrac{(.14)(.15)}{(.14)(.15) + (.1)(.85)} = \dfrac{.021}{.021 + .085}$

$\approx .1981$, or 19.81%

Therefore, 19.81% of single homeowners have pools. Thus pool manufacturers should go after the single homeowners.

25. Use the following events:
  A: Former student of Prof. A.
  C: Earned C– or lower
All of Professor A's former students wound up with a C– or lower. In other words,
  $P(C|A) = 1$

297

Two thirds of students not from Prof A's class got better than a C–. In other words,

$$P(C|A') = \frac{1}{3}$$

Three quarters of Professor F's class consisted of former students of Professor A, so

$$P(A) = .75$$

We are asked to find what percentage of the students who got C– or worse were former students of Prof. A: $P(A|C)$.

$$P(A|C) = \frac{P(C|A)P(A)}{P(C|A)P(A) + P(C|A')P(A')}$$

$$= \frac{(1)(.75)}{(1)(.75) + (1/3)(.25)} = \frac{.75}{.75 + 1/3}$$

$$= .9$$

Therefore, we estimate that 9 of the 10 students in the delegation were former students of Prof. A.

27. Use the following events:
  H: Husband employed
  W: wife employed
  $P(H) = .95$, $P(W|H) = .71$
Since either the husband or wife in a couple with earnings had to be employed,
  $P(W|H') = 1$
We are asked to find $P(H|W)$.

$$P(H|W) = \frac{P(W|H)P(H)}{P(W|H)P(H) + P(W|H')P(H')}$$

$$= \frac{(.71)(.95)}{(.71)(.95) + (1)(.05)} = \frac{.6745}{.6745 + .05}$$

$$\approx .9310$$

29. Use the following events:
  A: Arrested by age 14
  C: Become a chronic offender
  $P(C|A) = 17.9P(C|A')$
  $P(A) = .001$

$$P(A|C) = \frac{P(C|A)P(A)}{P(C|A)P(A) + P(C|A')P(A')}$$

$$= \frac{17.9P(C|A')(.001)}{17.9P(C|A')(.001) + P(C|A')(.999)}$$

Now cancel the $P(C|A')$ to get

$$\frac{(17.9)(.001)}{(17.9)(.001) + .999} = \frac{.0179}{.0179 + .999}$$

$$\approx .0176, \text{ or } 17.6\%$$

31. Use the following events:

D: has diabetes
A: very active
$P(D|A) = .5P(D|A')$

$$P(A) = \frac{1}{3}$$

$$P(A|D) = \frac{P(D|A)P(A)}{P(D|A)P(A) + P(D|A')P(A')}$$

$$= \frac{.5P(D|A')P(A)}{.5P(D|A')P(A) + P(D|A')P(A')}$$

Cancel the $P(D|A')$ to obtain

$$\frac{.5P(A)}{.5P(A) + P(A')}$$

$$= \frac{(.5)(1/3)}{(.5)(1/3) + 2/3} = \frac{1/6}{1/6 + 2/3} = .2$$

33. Use the following events:
  K: The child was killed
  D: The airbag deployed
  $P(K|D) = 1.31P(K|D')$
  $P(D) = .25$
We are asked to compute $P(D|K)$.

$$P(D|K) = \frac{P(K|D)P(D)}{P(K|D)P(D) + P(K|D')P(D')}$$

$$= \frac{1.31P(K|D')P(D)}{1.31P(K|D')P(D) + P(K|D')P(D')}$$

Cancel the terms $P(K|D')$ to obtain

$$\frac{1.31P(D)}{1.31P(D) + P(D')}$$

$$= \frac{(1.31)(.25)}{(1.31)(.25) + .75} = \frac{.3275}{.3275 + .75}$$

$$\approx .30$$

35. Show him an example like Example 1 of this section, where $P(T|A) = .95$ but $P(A|T) \approx .64$.

37. Suppose that the steroid test gives 10% false negatives and that only 0.1% of the tested population uses steroids. Then the probability that an athlete uses steroids, given that he or she has tested positive, is $\approx .083$.

39. Draw a tree in which the first branching shows which of R_1, R_2, or R_3 occurred, and the second branching shows which of T or T' then occurred. There are three final outcomes in which T occurs:

$$P(R_1 \cap T) = P(T \mid R_1)P(R_1)$$

$$P(R_2 \cap T) = P(T \mid R_2)P(R_2)$$
$$P(R_3 \cap T) = P(T \mid R_3)P(R_3)$$

In only one of these, the first, does R_1 occur. Thus,

$$P(R_1 \mid T) = \frac{P(R_1 \cap T)}{P(T)}$$

$$= \frac{P(T|R_1)P(R_1)}{P(T|R_1)P(R_1) + P(T|R_2)P(R_2) + P(T|R_3)P(R_3)}$$

41. The reasoning is flawed. Let A be the event that a Democrat agrees with Safire's column, and let F and M be the events that a Democrat reader is female and male respectively. Then A. D. makes the following argument:

$$P(M|A) = .9, \ P(F|A') = .9.$$

Therefore,

$$P(A|M) = .9.$$

According to Bayes' Theorem we cannot conclude anything about $P(A|M)$ unless we know $P(A)$, the percentage of all Democrats who agreed with Safire's column. This was not given.

7.7 Markov Systems

1. The transition matrix P is organized as follows:

$$P = \begin{bmatrix} 1\to1 & 1\to2 \\ 2\to1 & 2\to2 \end{bmatrix}$$

Therefore, the diagram

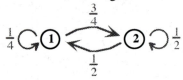

has $P = \begin{bmatrix} 1/4 & 3/4 \\ 1/2 & 1/2 \end{bmatrix}$

3. The transition matrix P is organized as follows:

$$P = \begin{bmatrix} 1\to1 & 1\to2 \\ 2\to1 & 2\to2 \end{bmatrix}$$

Therefore, the diagram

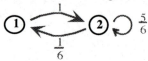

has $P = \begin{bmatrix} 0 & 1 \\ 1/6 & 5/6 \end{bmatrix}$

5. The transition matrix P is organized as follows:

$$P = \begin{bmatrix} 1\to1 & 1\to2 & 1\to3 \\ 2\to1 & 2\to2 & 2\to3 \\ 3\to1 & 3\to2 & 3\to3 \end{bmatrix}$$

Therefore, the diagram

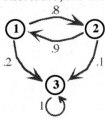

has $P = \begin{bmatrix} 0 & .8 & .2 \\ .9 & 0 & .1 \\ 0 & 0 & 1 \end{bmatrix}$

7. The transition matrix P is organized as follows:

$$P = \begin{bmatrix} 1\to1 & 1\to2 & 1\to3 \\ 2\to1 & 2\to2 & 2\to3 \\ 3\to1 & 3\to2 & 3\to3 \end{bmatrix}$$

Therefore, the diagram

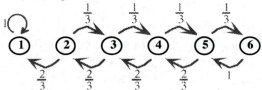

has $P = \begin{bmatrix} 1 & 0 & 0 \\ 0 & 1 & 0 \\ 0 & 0 & 1 \end{bmatrix}$

9. The diagram

has $P = \begin{bmatrix} 1 & 0 & 0 & 0 & 0 & 0 \\ 2/3 & 0 & 1/3 & 0 & 0 & 0 \\ 0 & 2/3 & 0 & 1/3 & 0 & 0 \\ 0 & 0 & 2/3 & 0 & 1/3 & 0 \\ 0 & 0 & 0 & 2/3 & 0 & 1/3 \\ 0 & 0 & 0 & 0 & 1 & 0 \end{bmatrix}$

11. (a) $P^2 = \begin{bmatrix} .5 & .5 \\ 0 & 1 \end{bmatrix} \begin{bmatrix} .5 & .5 \\ 0 & 1 \end{bmatrix}$

$= \begin{bmatrix} .25 & .75 \\ 0 & 1 \end{bmatrix}$

(b) distribution after one step:

$$vP = \begin{bmatrix} 1 & 0 \end{bmatrix} \begin{bmatrix} .5 & .5 \\ 0 & 1 \end{bmatrix} = \begin{bmatrix} .5 & .5 \end{bmatrix}$$

after two steps:

$$vP^2 = \begin{bmatrix} .5 & .5 \end{bmatrix} \begin{bmatrix} .5 & .5 \\ 0 & 1 \end{bmatrix} = \begin{bmatrix} .25 & .75 \end{bmatrix}$$

after three steps:

$$vP^3 = \begin{bmatrix} .25 & .75 \end{bmatrix} \begin{bmatrix} .5 & .5 \\ 0 & 1 \end{bmatrix} = \begin{bmatrix} .125 & .875 \end{bmatrix}$$

13. (a) $P^2 = \begin{bmatrix} .2 & .8 \\ .4 & .6 \end{bmatrix} \begin{bmatrix} .2 & .8 \\ .4 & .6 \end{bmatrix} =$
$\begin{bmatrix} .36 & .64 \\ .32 & .68 \end{bmatrix}$

(b) distribution after one step:

$vP = [.5 \ \ .5] \begin{bmatrix} .2 & .8 \\ .4 & .6 \end{bmatrix} = [.3 \ \ .7]$

after two steps:

$vP^2 = [.3 \ \ .7] \begin{bmatrix} .2 & .8 \\ .4 & .6 \end{bmatrix} = [.34 \ \ .66]$

after three steps:

$vP^3 = [.34 \ \ .66] \begin{bmatrix} .2 & .8 \\ .4 & .6 \end{bmatrix} = [.332 \ \ .668]$

15. (a) $P^2 = \begin{bmatrix} 1/2 & 1/2 \\ 1 & 0 \end{bmatrix} \begin{bmatrix} 1/2 & 1/2 \\ 1 & 0 \end{bmatrix}$
$= \begin{bmatrix} 3/4 & 1/4 \\ 1/2 & 1/2 \end{bmatrix}$

(b) distribution after one step:

$vP = [2/3 \ \ 1/3] \begin{bmatrix} 1/2 & 1/2 \\ 1 & 0 \end{bmatrix} = [2/3 \ \ 1/3]$

after two and three steps, the result will remain the same: $[2/3 \ \ 1/3]$

17. (a) $P^2 = \begin{bmatrix} 3/4 & 1/4 \\ 3/4 & 1/4 \end{bmatrix} \begin{bmatrix} 3/4 & 1/4 \\ 3/4 & 1/4 \end{bmatrix}$
$= \begin{bmatrix} 3/4 & 1/4 \\ 3/4 & 1/4 \end{bmatrix}$

(b) distribution after one step:

$vP = [1/2 \ \ 1/2] \begin{bmatrix} 3/4 & 1/4 \\ 3/4 & 1/4 \end{bmatrix} = [3/4 \ \ 1/4]$

after two steps:

$vP^2 = [3/4 \ \ 1/4] \begin{bmatrix} 3/4 & 1/4 \\ 3/4 & 1/4 \end{bmatrix} = [3/4 \ \ 1/4]$

after three steps, the result will remain the same:
$[3/4 \ \ 1/4]$

19. (a) $P^2 = \begin{bmatrix} .5 & .5 & 0 \\ 0 & 1 & 0 \\ 0 & .5 & .5 \end{bmatrix} \begin{bmatrix} .5 & .5 & 0 \\ 0 & 1 & 0 \\ 0 & .5 & .5 \end{bmatrix}$
$= \begin{bmatrix} .25 & .75 & 0 \\ 0 & 1 & 0 \\ 0 & .75 & .25 \end{bmatrix}$

(b) distribution after one step:

$vP = [1 \ \ 0 \ \ 0] \begin{bmatrix} .5 & .5 & 0 \\ 0 & 1 & 0 \\ 0 & .5 & .5 \end{bmatrix} = [.5 \ \ .5 \ \ 0]$

after two steps:

$vP^2 = [.5 \ \ .5 \ \ 0] \begin{bmatrix} .5 & .5 & 0 \\ 0 & 1 & 0 \\ 0 & .5 & .5 \end{bmatrix} = [.25 \ \ .75$
$0]$

after three steps:

$vP^3 = [.25 \ \ .75 \ \ 0] \begin{bmatrix} .5 & .5 & 0 \\ 0 & 1 & 0 \\ 0 & .5 & .5 \end{bmatrix}$
$= [.125 \ \ .875 \ \ 0]$

21. (a) $P^2 = \begin{bmatrix} 0 & 1 & 0 \\ 1/3 & 1/3 & 1/3 \\ 1 & 0 & 0 \end{bmatrix}$
$\begin{bmatrix} 0 & 1 & 0 \\ 1/3 & 1/3 & 1/3 \\ 1 & 0 & 0 \end{bmatrix}$
$= \begin{bmatrix} 1/3 & 1/3 & 1/3 \\ 4/9 & 4/9 & 1/9 \\ 0 & 1 & 0 \end{bmatrix}$

(b) distribution after one step:

$vP = [1/2 \ \ 0 \ \ 1/2] \begin{bmatrix} 0 & 1 & 0 \\ 1/3 & 1/3 & 1/3 \\ 1 & 0 & 0 \end{bmatrix}$
$= [1/2 \ \ 1/2 \ \ 0]$

after two steps:

$vP^2 = [1/2 \ \ 1/2 \ \ 0] \begin{bmatrix} 0 & 1 & 0 \\ 1/3 & 1/3 & 1/3 \\ 1 & 0 & 0 \end{bmatrix}$
$= [1/6 \ \ 2/3 \ \ 1/6]$

after three steps:

$vP^3 = [1/6 \ \ 2/3 \ \ 1/6] \begin{bmatrix} 0 & 1 & 0 \\ 1/3 & 1/3 & 1/3 \\ 1 & 0 & 0 \end{bmatrix}$
$= [7/18 \ \ 7/18 \ \ 2/9]$

23. (a) $P^2 = \begin{bmatrix} .1 & .9 & 0 \\ 0 & 1 & 0 \\ 0 & .2 & .8 \end{bmatrix} \begin{bmatrix} .1 & .9 & 0 \\ 0 & 1 & 0 \\ 0 & .2 & .8 \end{bmatrix}$
$= \begin{bmatrix} .01 & .99 & 0 \\ 0 & 1 & 0 \\ 0 & .36 & .64 \end{bmatrix}$

(b) distribution after one step:

$vP = [.5 \ \ 0 \ \ .5] \begin{bmatrix} .1 & .9 & 0 \\ 0 & 1 & 0 \\ 0 & .2 & .8 \end{bmatrix} = [.05 \ \ .55$
$.4]$

after two steps:

$$vP^2 = [.05 \quad .55 \quad .4] \begin{bmatrix} .1 & .9 & 0 \\ 0 & 1 & 0 \\ 0 & .2 & .8 \end{bmatrix}$$

$$= [.005 \quad .675 \quad .32]$$

after three steps:

$$vP^3 = [.005 \quad .675 \quad .32] \begin{bmatrix} .1 & .9 & 0 \\ 0 & 1 & 0 \\ 0 & .2 & .8 \end{bmatrix}$$

$$= [.0005 \quad .7435 \quad .256]$$

25. To find the steady-state vector $v_\infty = [x \quad y]$ we solve

$$x + y = 1$$

$$[x \quad y] \begin{bmatrix} 1/2 & 1/2 \\ 1 & 0 \end{bmatrix} = [x \quad y]$$

The above equations give:

$$x + y = 1$$
$$x/2 + y = x$$
$$x/2 = y$$

Rewriting in standard form and dropping the last equation (which is the same as the next-to-last) we get

$$x + y = 1$$
$$-x/2 + y = 0$$

Solving this system gives

$$x = 2/3, \ y = 1/3$$

So, $v_\infty = [2/3 \quad 1/3]$

27. To find the steady-state vector $v_\infty = [x \quad y]$ we solve

$$x + y = 1$$

$$[x \quad y] \begin{bmatrix} 1/3 & 2/3 \\ 1/2 & 1/2 \end{bmatrix} = [x \quad y]$$

The above equations give:

$$x + y = 1$$
$$x/3 + y/2 = x$$
$$2x/3 + y/2 = y$$

Rewriting in standard form and dropping the last equation (which is the same as the next-to-last) we get

$$x + y = 1$$
$$-2x/3 + y/2 = 0$$

Solving this system gives

$$x = 3/7, \ y = 4/7$$

So, $v_\infty = [3/7 \quad 4/7]$

29. To find the steady-state vector $v_\infty = [x \quad y]$ we solve

$$x + y = 1$$

$$[x \quad y] \begin{bmatrix} .1 & .9 \\ .6 & .4 \end{bmatrix} = [x \quad y]$$

The above equations give:

$$x + y = 1$$
$$.1x + .6y = x$$
$$.9x + .4y = y$$

Rewriting in standard form and dropping the last equation (which is the same as the next-to-last) we get

$$x + y = 1$$
$$-.9x + .6y = 0$$

Solving this system gives

$$x = 2/5, \ y = 3/5$$

So, $v_\infty = [2/5 \quad 3/5]$

31. To find the steady-state vector $v_\infty = [x \quad y \quad z]$ we solve

$$x + y + z = 1$$

$$[x \quad y \quad z] \begin{bmatrix} .5 & 0 & .5 \\ 1 & 0 & 0 \\ 0 & .5 & .5 \end{bmatrix} = [x \quad y \quad z]$$

The above equations give:

$$x + y + z = 1$$
$$.5x + y = x$$
$$.5z = y$$
$$.5x + .5z = z$$

Rewriting in standard form we get

$$x + y \quad + z = 1$$
$$-.5x + y \quad = 0$$
$$-y + .5z = 0$$
$$.5x \quad - .5z = 0$$

Solving this system gives

$$x = 2/5, \ y = 1/5, \ z = 2/5$$

So, $v_\infty = [2/5 \quad 1/5 \quad 2/5]$

33. To find the steady-state vector $v_\infty = [x \quad y \quad z]$ we solve

$$x + y + z = 1$$

$$[x \quad y \quad z] \begin{bmatrix} 0 & 1 & 0 \\ 1/3 & 1/3 & 1/3 \\ 1 & 0 & 0 \end{bmatrix} = [x \quad y \quad z]$$

The above equations give:

$$x + y + z = 1$$
$$y/3 + z = x$$

$x + y/3 = y$

$y/3 = z$

Rewriting in standard form we get

$x + y + z = 1$

$-x + y/3 + z = 0$

$x - 2y/3 = 0$

$y/3 - z = 0$

Solving this system gives

$x = 1/3,\ y = 1/2,\ z = 1/6$

So, $v_\infty = [1/3\ \ 1/2\ \ 1/6]$

35. To find the steady-state vector $v_\infty = [x\ \ y\ \ z]$ we solve

$x + y + z = 1$

$[x\ \ y\ \ z] \begin{bmatrix} .1 & .9 & 0 \\ 0 & 1 & 0 \\ 0 & .2 & .8 \end{bmatrix} = [x\ \ y\ \ z]$

The above equations give:

$x + y + z = 1$

$.1x = x$

$.9x + y + .2z = y$

$.8z = z$

Rewriting in standard form we get

$x + y + z = 1$

$-.9x = 0$

$.9x + .2z = 0$

$-.2z = 0$

Solving this system gives

$x = 0,\ y = 1,\ z = 0$

So, $v_\infty = [0\ \ 1\ \ 0]$

37. Take 1 = Sorey State, 2 = C&T

$P = \begin{bmatrix} 1/2 & 1/2 \\ 1/4 & 3/4 \end{bmatrix}$

We are asked to find the (1, 1) entry of the 2-step transition probability matrix.

$P^2 = \begin{bmatrix} 1/2 & 1/2 \\ 1/4 & 3/4 \end{bmatrix} \begin{bmatrix} 1/2 & 1/2 \\ 1/4 & 3/4 \end{bmatrix}$

$= \begin{bmatrix} 3/8 & 5/8 \\ 5/16 & 11/16 \end{bmatrix}$

The (1, 1) entry is $3/8 = .375$

39. (a) Take 1 = Not checked in, 2 = Checked in

$P = \begin{bmatrix} .4 & .6 \\ 0 & 1 \end{bmatrix}$

$P^2 = \begin{bmatrix} .4 & .6 \\ 0 & 1 \end{bmatrix} \begin{bmatrix} .4 & .6 \\ 0 & 1 \end{bmatrix} = \begin{bmatrix} .16 & .84 \\ 0 & 1 \end{bmatrix}$

$P^3 = \begin{bmatrix} .4 & .6 \\ 0 & 1 \end{bmatrix} \begin{bmatrix} .16 & .84 \\ 0 & 1 \end{bmatrix}$

$= \begin{bmatrix} .064 & .936 \\ 0 & 1 \end{bmatrix}$

(b) 1 hour: $P_{12} = .6$

2 hours: $(P^2)_{12} = .84$

3 hours: $(P^3)_{12} = .936$

(c) Eventually, all the roaches will have checked in.

41. Take 1 = High risk, 2 = Low risk

$P = \begin{bmatrix} .50 & .50 \\ .10 & .90 \end{bmatrix}$

To find the steady-state vector $[x\ \ y]$ we solve

$x + y = 1$

$[x\ \ y] \begin{bmatrix} .50 & .50 \\ .10 & .90 \end{bmatrix} = [x\ \ y]$

The above equations give:

$x + y = 1$

$.5x + .1y = x$

$.5x + .9y = y$

Rewriting in standard form and dropping the last equation, we get

$x + y = 1$

$-.5x + .1y = 0$

Solving this system gives

$x = 1/6,\ y = 5/6$

So, $[x\ \ y] = [1/6\ \ 5/6]$

In the long-term, $1/6 \approx 16.67\%$ fall into the high-risk category and $5/65 \approx 83.33\%$ into the low risk category.

43. (a) Take 1 = User, 2 = Non-User

From Exercise 25 in Section 9.1,

$P = \begin{bmatrix} 2/3 & 1/3 \\ 1/10 & 9/10 \end{bmatrix}$

The 2-year transition matrix is

$P^2 = \begin{bmatrix} 2/3 & 1/3 \\ 1/10 & 9/10 \end{bmatrix} \begin{bmatrix} 2/3 & 1/3 \\ 1/10 & 9/10 \end{bmatrix}$

$= \begin{bmatrix} 43/90 & 47/90 \\ 47/300 & 253/300 \end{bmatrix}$

Non-user$\rightarrow$User in 2 steps:

$(P^2)_{21} = 47/300 \approx .156667$

(b) To find the steady-state vector $[x \ y]$ we solve

$$x + y = 1$$

$$[x \ y] \begin{bmatrix} 2/3 & 1/3 \\ 1/10 & 9/10 \end{bmatrix} = [x \ y]$$

The above equations give:

$$x + y = 1$$
$$(2/3)x + (1/10)y = x$$
$$(1/3)x + (9/10)y = y$$

Rewriting in standard form and dropping the last equation, we get

$$x + y = 1$$
$$-(1/3)x + (1/10)y = 0$$

Solving this system gives

$$x = 3/13, \ y = 10/13$$

So, $[x \ y] = [3/13 \ \ 10/13]$

In the long-term, 3/13 of the college instructors will be users of this book.

45. Take 1 = Paid up, 2 = 0–90 Days, 3 = Bad debt

$$P = \begin{bmatrix} .5 & .5 & 0 \\ .5 & .3 & .2 \\ 0 & .5 & .5 \end{bmatrix}$$

Steady-state vector:

$$x + y + z = 1$$

$$[x \ y \ z] \begin{bmatrix} .5 & .5 & 0 \\ .5 & .3 & .2 \\ 0 & .5 & .5 \end{bmatrix} = [x \ y \ z]$$

The above equations give:

$$x + y + z = 1$$
$$.5x + .5y = x$$
$$.5x + .3y + .5z = y$$
$$.2y + .5z = z$$

Rewriting in standard form and dropping the last equation, we get

$$x + y + z = 1$$
$$-.5x + .5y = 0$$
$$.5x - .7y + .5z = 0$$

Solving gives

$$x = 5/12, \ y = 5/12, \ z = 1/6$$

So, $[x \ y \ z] = [5/12 \ \ 5/12 \ \ 1/6]$

Answer:
5/12 or approximately 41.67% of the customers will be in the Paid up category, 5/12 or approximately 41.67% in the 0-90 days category,

and 1/6 or approximately 16.67% in the bad debt category.

47. (a) Take 1 = Affluent, 2 = Middle class, 3 = Poor

$$p_{11} = 1-.729 = .271, \ p_{12} = .729, \ p_{13} = 0$$
$$p_{21} = .075, \ p_{22} = 1-(.075+.085) = .84, \ p_{23} = .085$$
$$p_{31} = 0, \ p_{32} = .304, \ p_{33} = 1-.304 = .696$$

$$P = \begin{bmatrix} .729 & .271 & 0 \\ .075 & .84 & .085 \\ 0 & .304 & .696 \end{bmatrix}$$

(b) The period 1980–2002 is a 22-year period, corresponding to two 11-year transition steps.

$$P^2 = \begin{bmatrix} .729 & .271 & 0 \\ .075 & .84 & .085 \\ 0 & .304 & .696 \end{bmatrix} \begin{bmatrix} .729 & .271 & 0 \\ .075 & .84 & .085 \\ 0 & .304 & .696 \end{bmatrix}$$

$$\approx \begin{bmatrix} .552 & .425 & .023 \\ .118 & .752 & .131 \\ .023 & .467 & .51 \end{bmatrix}$$

Affluent→Poor in 2 steps:

$$(P^2)_{12} = .023, \text{ or } 2.3\%$$

(c) Let us use technology to compute a high enough power of P directly so that the row are approximately the same:

$$P^\infty \approx P^{64} \approx \begin{bmatrix} .178 & .643 & .18 \\ .178 & .643 & .18 \\ .178 & .643 & .18 \end{bmatrix}$$

(We used the 64th power because powers of 2 are computed most efficiently in the online Matrix Algebra Tool.)

Answer:
Affluent: 17.8%; Middle class: 64.3%; Poor: 18.0%

49. Take 1 = Bottom 10%, 2 = 10–50%, 3 = 50–99%, 4 = Top 10%.

The transition matrix is given by the table:

$$P = \begin{bmatrix} .3 & .52 & .17 & .01 \\ .1 & .48 & .38 & .04 \\ .04 & .38 & .48 & .1 \\ .01 & .17 & .52 & .3 \end{bmatrix}$$

Let us use technology to compute a high enough power of P directly so that the row are approximately the same:

$$P^{\infty} \approx P^{64} \approx \begin{bmatrix} .0843 & .4157 & .4157 & .0843 \\ .0843 & .4157 & .4157 & .0843 \\ .0843 & .4157 & .4157 & .0843 \\ .0843 & .4157 & .4157 & .0843 \end{bmatrix}$$

(We used the 64th power because powers of 2 are computed most efficiently in the online Matrix Algebra Tool.)

Thus, the percentages in each category are given by [8.43 41.57 41.57 8.43].

51. (a) Number the states as follows: 1: Verizon, 2: Cingular, 3: AT&T, 4: Other. The figures next to the arrows show all the transition probabilities (obtained by dividing by 100) except the ones from each state to itself. To compute the missing probabilities, use the fact that the entries in each row add to 1. This leads to:

$$P = \begin{bmatrix} .981 & .005 & .005 & .009 \\ .01 & .972 & .006 & .012 \\ .01 & .006 & .973 & .011 \\ .008 & .006 & .005 & .981 \end{bmatrix}$$

(b) Each time-step is one quarter. At the *end* of the third quarter, the distribution is given as
$$w = [0.297 \quad 0.193 \quad 0.181 \quad 0.329]$$
We are asked to find the distribution v at the beginning of that quarter: one time-step earlier. We know that

Distribution after one quarter
$$= vP = [0.297 \quad 0.193 \quad 0.181 \quad 0.329]$$
To obtain v, multiply both sides on the right by P^{-1}.
$$v = [0.297 \quad 0.193 \quad 0.181 \quad 0.329]P^{-1}$$
On the Matrix Algebra Tool, use the format `w*P^-1`: We obtain (rounding to 3 decimal places:
$$v = [0.296 \quad 0.194 \quad 0.182 \quad 0.328].$$
Thus, the market shares at the beginning of the quarter were:
Verizon: 29.6%, Cingular: 19.3%, AT&T: 18.1%, Other: 32.8%.

(c) The end of 2005 is 9 quarters from the end of the third quarter in 2003. Therefore, the distribution is predicted as
$$wP^9 = [0.303 \quad 0.186 \quad 0.176 \quad 0.335]$$
Verizon: 30.3%, Cingular: 18.6%, AT&T: 17.6%, Other: 33.5%. (Technology format: `w*P^9`) The

biggest gainers are Verizon and Other, each gaining 0.6%

53. From the diagram, the transition matrix is

$$P = \begin{bmatrix} 1/2 & 1/2 & 0 & 0 & 0 \\ 1/2 & 0 & 1/2 & 0 & 0 \\ 0 & 1/2 & 0 & 1/2 & 0 \\ 0 & 0 & 1/2 & 0 & 1/2 \\ 0 & 0 & 0 & 1/2 & 1/2 \end{bmatrix}$$

For the steady-state vector, we solve:
$$x + y + z + u + v = 1$$
$$(1/2)x + (1/2)y = x$$
$$(1/2)x + (1/2)z = y$$
$$(1/2)y + (1/2)u = z$$
$$(1/2)z + (1/2)v = u$$
$$(1/2)u + (1/2)v = v$$
Rewriting in standard form and dropping the last equation, we get the system
$$x + y + z + u + v = 1$$
$$-(1/2)x + (1/2)y = 0$$
$$(1/2)x - y + (1/2)z = 0$$
$$(1/2)y - z + (1/2)u = 0$$
$$(1/2)z - u + (1/2)v = 0$$
Solving gives
$$x = 1/5, \ y = 1/5, \ z = 1/5, \ u = 1/5, \ v = 1/5$$
Hence the steady-state vector is
$$[1/5 \quad 1/5 \quad 1/5 \quad 1.5].$$
The system spends an average of 1/5 of the time in each state.

55. Answers will vary.

57. There are two assumptions made by Markov systems that may not be true about the stock market: the assumption that the transition probabilities do not change over time, and the assumption that the transition probability depends only on the current state.

59. If q is a row of Q, then by assumption $qP = q$. Thus, when we multiply the rows of Q by P, nothing changes, and $QP = Q$.

61. At each step only 0.4 of the population in state 1 remains there, and nothing enters from any other state. Thus, when the first entry in the steady-state distribution vector is multiplied by

0.4 it must remain unchanged. The only number
for which this true is 0.

63. An example is

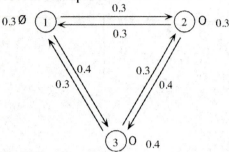

Its transition matrix is

$$P = \begin{bmatrix} .3 & .3 & .4 \\ .3 & .3 & .4 \\ .3 & .3 & .4 \end{bmatrix}$$

and is therefore already in a steady state!

65. If $vP = v$ and $wP = w$, then

$$\tfrac{1}{2}(v + w) \, P = \tfrac{1}{2} \, vP + \tfrac{1}{2} \, wP$$

$$= \tfrac{1}{2} \, v + \tfrac{1}{2} \, w = \tfrac{1}{2}(v + w) \ .$$

Further, if the entries of v and w add up to 1, then
so do the entries of $(v + w)/2$.

Chapter 7 Review Exercises

1. $n(S) = 2 \times 2 \times 2 = 8$

$E = \{HHT, HTH, HTT, THH, THT, TTH, TTT\}$

$P(E) = \dfrac{n(E)}{n(S)} = \dfrac{7}{8}$

3. $n(S) = 6 \times 6 = 36$

$E = \{(1,6), (2,5), (3,4), (4,3), (5,2), (6,1)\}$

$P(E) = \dfrac{n(E)}{n(S)} = \dfrac{6}{36} = \dfrac{1}{6}$

5. $n(S) = 6$ and $E = \{2\}$. However, the outcomes are not equally likely, so we need the probability distribution. Start with

Outcome	1	2	3	4	5	6
Probability	$2x$	x	x	x	x	$2x$

Since the sum of the probabilities of the outcomes must be 1, we get

$8x = 1$, so $x = \dfrac{1}{8} = .125$

This gives:

Outcome	1	2	3	4	5	6
Probability	.25	.125	.125	.125	.125	.25

$P(E) = P(\{2\}) = .125$ (or 1/8)

7. $P(2 \text{ heads}) = \dfrac{fr(2 \text{ heads})}{N} = \dfrac{12}{50}$

$P(\text{At least 1 tail}) = 1 - P(2 \text{ heads})$

$= 1 - \dfrac{12}{50} = \dfrac{38}{50} = .76$

9. We use the following events for a randomly selected novel:

U: You have read the novel.

R: Roslyn has read the novel.

$P(U \cup R) = P(U) + P(R) - P(U \cap R)$

$= \dfrac{150}{400} + \dfrac{200}{400} - \dfrac{50}{400} = \dfrac{300}{400} = .75$

Therefore, the probability that a novel has been read by neither you nor your sister is

$P[(U \cup R)'] = 1 - P(U \cup R) = 1 - .75 = .25$

11. Let us use the following events:

A: A student is in category A

B: A student is in category B

$P(A \cup B) = 1$, $P(A) = P(B) = \dfrac{24}{32} = .75$

We are asked to find $P(A \cap B)$

$P(A \cup B) = P(A) + P(B) - P(A \cap B)$

$1 = .75 + .75 - P(A \cap B)$

$P(A \cap B) = .75 + .75 - 1 = .5$

13. Use the following events:

A: Is a Model A

R: Is orange

$P(A \cup R) = P(A) + P(R) - P(A \cap R)$

$= \dfrac{1}{3} + \dfrac{1}{5} - \dfrac{1}{15} = \dfrac{7}{15}$

15. $n(S) = C(12, 5) = 792$

$P(E) = \dfrac{n(E)}{n(S)} = \dfrac{C(4,4)C(8,1)}{792} = \dfrac{8}{792}$

17. $n(S) = C(12, 5) = 792$

$P(E) = \dfrac{n(E)}{n(S)}$

$= \dfrac{C(4,1)C(2,1)C(1,1)C(3,1)C(2,1)}{792}$

$= \dfrac{48}{792}$

19. $n(S) = C(12, 5) = 792$

Decision algorithm for constructing s set with at least 2 yellow

Alternative 1: 2 yellow: $C(3, 2)C(9, 3) = 252$ choices

Alternative 2: 3 yellow: $C(3, 3)C(9, 2) = 36$ choices

This gives a total of $252 + 36 = 288$ choices.

$P(E) = \dfrac{n(E)}{n(S)} = \dfrac{288}{792}$

21. $n(S) = C(52, 5)$

$P(E) = \dfrac{n(E)}{n(S)} = \dfrac{C(8, 5)}{C(52, 5)}$

23. $n(S) = C(52, 5)$

307

$$P(E) = \frac{n(E)}{n(S)} = \frac{C(4, 3)C(1, 1)C(3, 1)}{C(52, 5)}$$

25. $n(S) = C(52, 5)$

Decision algorithm for constructing a full house of commons:

   Step 1: Select the denomination for the triple (2, 3, 4, 5, 6, 7, 8, 9, 10): $C(9, 1)$ choices

   Step 2: Select one of the remaining 8 denominations for the double: $C(8, 1)$ choices

   Step 3: Select 3 cards of the chosen denomination for the triple: $C(4, 3)$ choices

   Step 4: Select 2 cards of the chosen denomination for the double: $C(4, 2)$ choices

$$P(E) = \frac{n(E)}{n(S)}$$
$$= \frac{C(9, 1)C(8, 1)C(4, 3)C(4, 2)}{C(52, 5)}$$

27. A: The sum is 5.

  B: The green one is not a 1 and the yellow one is 1.

  $A \cap B$: The sum is 5, the green one is not 1 and the yellow one is 1.

  $B = \{(2, 1), (3, 1), (4, 1), (5, 1), (6, 1)\}$

  $A \cap B = \{((4, 1)\}$

$$P(A \cap B) = \frac{1}{36}, \; P(B) = \frac{5}{36}$$

$$P(A|B) = \frac{P(A \cap B)}{P(B)} = \frac{1/36}{5/36} = \frac{1}{5}$$

$$P(A) = \frac{4}{36} = \frac{1}{9}$$

Since $P(A|B) \neq P(A)$, the events A and B are dependent.

29. A: The yellow one is 4.

  B: The green one is 4.

  $A \cap B$: Both the yellow and green dice are 4.

$$P(A \cap B) = \frac{1}{36}, \; P(B) = \frac{1}{6}$$

$$P(A|B) = \frac{P(A \cap B)}{P(B)} = \frac{1/36}{1/6} = \frac{1}{6}$$

$$P(A) = \frac{1}{6} = P(A|B). \text{ Therefore, the events } A$$

and B are independent.

31. A: The dice have the same parity.

  B: Both dice are odd.

  $A \cap B$: The dice have the same parity and are both odd.

  Note that $A \cap B = B$

$$P(A \cap B) = \frac{9}{36}, P(B) = \frac{9}{36}$$

$$P(A|B) = \frac{P(A \cap B)}{P(B)} = \frac{9/36}{9/36} = 1$$

$$P(A) = \frac{18}{36} = \frac{1}{2}$$

Since $P(A|B) \neq P(A)$, the events A and B are dependent.

33. Take 1 = Brand A, 2 = Brand B.

$$P = \begin{bmatrix} 1/2 & 1/2 \\ 1/4 & 3/4 \end{bmatrix}$$

35. From Exercise 33,

$$P = \begin{bmatrix} 1/2 & 1/2 \\ 1/4 & 3/4 \end{bmatrix}$$

$$P^2 = \begin{bmatrix} 1/2 & 1/2 \\ 1/4 & 3/4 \end{bmatrix} \begin{bmatrix} 1/2 & 1/2 \\ 1/4 & 3/4 \end{bmatrix}$$

$$= \begin{bmatrix} 3/8 & 5/8 \\ 5/16 & 11/16 \end{bmatrix}$$

$$P^3 = \begin{bmatrix} 1/2 & 1/2 \\ 1/4 & 3/4 \end{bmatrix} \begin{bmatrix} 3/8 & 5/8 \\ 5/16 & 11/16 \end{bmatrix}$$

$$= \begin{bmatrix} 11/32 & 21/32 \\ 21/64 & 43/64 \end{bmatrix}$$

Take $v = [2/3 \; 1/3]$

Distribution after 3 years is

$$vP^3 = [2/3 \; 1/3] \begin{bmatrix} 11/32 & 21/32 \\ 21/64 & 43/64 \end{bmatrix}$$

$$= [65/192 \; 127/192]$$

Brand A: $65/192 \approx .339$, Brand B: $127/192 \approx .661$

Exercises 37–42

In Exercises 37–42 we take the first letter of each category shown in the table to stand for the corresponding event:

  S: The book is Sci-Fi

  W: The book is stored in Washington,

and so on. Here is the table, with all figures shown in thousands:

	S	H	R	O	Total
W	10	12	12	30	64
C	8	12	6	16	42
T	15	15	20	44	94
Total	33	39	38	90	200

37. $P(S \cup T) = \dfrac{94 + 33 - 15}{200} = \dfrac{112}{200} = \dfrac{14}{25}$

	S	H	R	O	Total
W	10	12	12	30	64
C	8	12	6	16	42
T	15	15	20	44	94
Total	33	39	38	90	200

39. $P(S|T) = \dfrac{n(S \cap T)}{n(T)} = \dfrac{15}{94}$

	S	H	R	O	Total
W	10	12	12	30	64
C	8	12	6	16	42
T	15	15	20	44	94
Total	33	39	38	90	200

41. $P(T|S') = \dfrac{n(T \cap S')}{n(S')} = \dfrac{79}{167}$

	S	H	R	O	Total
W	10	12	12	30	64
C	8	12	6	16	42
T	15	15	20	44	94
Total	33	39	38	90	200

Exercises 43–50

In Exercises 43–50 we use the following events:
  H: Visited OHaganBooks.com; $P(H) = .02$
  C: Visited a competitor; $P(C) = .05$

43. $P(H') = 1 - P(H) = 1 - .02 = .98$, or 98%

45. Since *H* and *C* are independent,
  $P(H \cap C) = P(H)P(C) = .02 \times .05 = .001$
Therefore,
  $P(H \cup C) = P(H) + P(C) - P(H \cap C)$
  $\qquad = .02 + .05 - .001 = .069$, or 6.9%

47. $P[(H \cup C)'] = 1 - P(H \cup C) = 1 - .069 = .931$

49. We are told that an online shopper visiting a competitor was more likely to visit OHaganBooks.com that a randomly selected online shopper. That is,
  $P(H|C) > P(H)$
Multiplying both sides by $P(C)$ gives
  $P(H|C)P(C) > P(H)P(C)$
That is,
  $P(H \cap C) > P(H)P(C)$
Therefore, $P(H \cap C)$ is greater.

Exercises 51–54

In Exercises 51–54 we use the following events:
  H: visited OHaganBooks.com
  B: purchased books
  $P(B|H) = .08$
  $P(H) = .02$
  $P(B|H') = .005$

51. We compare $P(B|H)$ and $P(B|H')$. From the figures above,
  $P(B|H) = 16P(B|H')$
Therefore, online shoppers who visit the OHaganBooks.com web site are <u>16</u> times as likely to purchase books than shoppers who do not.

53. We are asked to compute $P(H|B)$. Using Bayes theorem,
$$P(H|B) = \frac{P(B|H)P(H)}{P(B|H)P(H) + P(B|H')P(H')}$$
$$= \frac{(.08)(.02)}{(.08)(.02) + (.005)(.98)} = \frac{.0016}{.0016 + .0049}$$
$$\approx .2462$$

Exercises 55–58

In Exercises 55–58 we number the states in the order shown in the table:
$$P = \begin{bmatrix} .8 & .1 & .1 \\ .4 & .6 & 0 \\ .2 & 0 & .8 \end{bmatrix}$$

55. Starting distribution (July 1):
  $v = [2000 \quad 4000 \quad 4000]$
Distribution after 1 month:

$$vP = [2000 \quad 4000 \quad 4000] \begin{bmatrix} .8 & .1 & .1 \\ .4 & .6 & 0 \\ .2 & 0 & .8 \end{bmatrix}$$

$$= [4000 \quad 2600 \quad 3400]$$

57. Here are three factors that the Markov model does not take into account:
(1) It is possible for someone to be a customer at two different enterprises.
(2) Some customers may stop using all three of the companies.
(3) New customers can enter the field.

Chapter 8 Random Variables and Statistics

8.1 Random Variables and Distributions

1. X = the sum of the numbers facing up when you roll two dice. The possible sums are 2, 3, ..., 12. Thus, X is finite and has the set $\{2, 3, ..., 12\}$ of possible values.

3. X is the profit, to the nearest dollar, earned in a year if you purchase one share of a stock selected review at random. The possible values of X are 0, $\pm 1, \pm 2, ...$ (negative numbers indicate a loss). This gives an infinite number of discrete possible values for X. Therefore, X is a discrete infinite random variable with the set of possible values
$\{0, 1, -1, 2, -2, ...\}$

5. X is the time the second hand of your watch reads in seconds. This can be any real number between 0 and 60. Therefore, X is a continuous random variable that can assume any value between 0 and 60.

7. The total number of goals that can be scored, up to a maximum of 10, is 1, 2, 3, ..., 9, or 10. Therefore, X is a finite random variable with values in the set
$\{0, 1, 2, ..., 10\}$.

9. The possible energies of an electron in a hydrogen atom are $k/1$, $k/4$, $k/9$, $k/16$, ... Thus, X is a discrete infinite random variable with set of possible values
$\{k/1, k/4, k/9, k/16, ...\}$

11. (a) $S = \{HH, HT, TH, TT\}$
(b) X is the rule that assigns to each outcome the number of tails.
(c) Counting the number of tails in each outcome gives us the following values of X:

Outcome	HH	HT	TH	TT
Value of X	0	1	1	2

13. (a) $S = \{(1,1), (1,2), ..., (1,6), (2,1), (2,2), ..., (6,6)\}$
(b) X is the rule that assigns to each outcome the sum of the two numbers.
(c) Computing the sum of the numbers in each outcome gives us the following values of X:

Outcome	$(1, 1)$	$(1, 2)$	$(1, 3)$	$(1, 4)$
Value of X	2	3	4	5
Outcome	...	$(6, 4)$	$(6, 5)$	$(6, 6)$
Value of X	...	10	11	12

15. Take each outcome to be a pair of numbers (# of red marbles, # of green ones). Thus,
$S = \{(4,0), (3,1), (2,2)\}$
(The pairs $(1, 3)$ and $(0, 4)$ are impossible since there are only 2 green marbles.)
(b) X is the rule that assigns to each outcome the number of red marbles.
(c) Writing down the number of red marbles (the first coordinate) in each outcome gives the following values of X.

Outcome	$(4, 0)$	$(3, 1)$	$(2, 2)$
Value of X	4	3	2

17. (a) S = the set of students in the study group.
(b) X is the rule that assigns to each student his or her final exam score.
(c) The values of X, in the order given, are 89%, 85%, 95%, 63%, 92%, 80%.

19. (a) Assign letters to the missing values (we use the same letter for both since they are given to be equal):

x	2	4	6	8	10
$P(X = x)$	.1	.2	x	x	.1

Since the probabilities add to 1, we get
$.1 + .2 + 2x + .1 = 1$

311

$.4 + 2x = 1$

$2x = .6$

$x = .3$

The completed table is:

x	2	4	6	8	10
P(X = x)	.1	.2	.3	.3	.1

(b) $P(X \geq 6) = P(X = 6) + P(X = 8) + P(X = 10)$

$$= .3 + .3 + .1 = .7$$

21. Since the probability that any specific number faces up is 1/6, the probability distribution for X is given by the following table and histogram:

x	1	2	3	4	5	6
P(X = x)	$\frac{1}{6}$	$\frac{1}{6}$	$\frac{1}{6}$	$\frac{1}{6}$	$\frac{1}{6}$	$\frac{1}{6}$

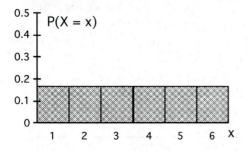

23. The number of heads showing when you toss 3 fair coins is 0, 1, 2, or 3. The corresponding probabilities are 1/8, 3/8, 3/8, and 1/8 (see Example 3). The corresponding values of X are the squares of 0, 1, 2, and 3; that is, 0, 1, 4, and 9. This gives us the following distribution and histogram:

x	0	1	4	9
P(X = x)	$\frac{1}{8}$	$\frac{3}{8}$	$\frac{3}{8}$	$\frac{1}{8}$

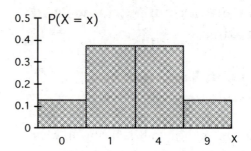

25. When two distinguishable dice are thrown, there are 36 possible outcomes. The values of X are the possible sums of the numbers facing up: 2, 3, ..., 11, 12. To calculate their probabilities, we can use

$$P(X = 2) = P\{(1, 1)\} = \tfrac{1}{36}$$

$$P(X = 3) = P\{(1, 2), (2, 1)\} = \tfrac{2}{36}$$

$$P(X = 4) = P\{(1, 3), (2, 2), (3, 1)\} = \tfrac{3}{36}$$

and so on. The completed probability distribution is as follows:

x	2	3	4	5	6	7
P(X = x)	$\frac{1}{36}$	$\frac{2}{36}$	$\frac{3}{36}$	$\frac{4}{36}$	$\frac{5}{36}$	$\frac{6}{36}$
x	8	9	10	11	12	
P(X = x)	$\frac{5}{36}$	$\frac{4}{36}$	$\frac{3}{36}$	$\frac{2}{36}$	$\frac{1}{36}$	

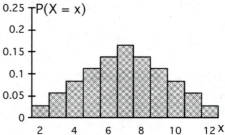

27. The possible values of X are 1, 2, ..., 6 (the larger of the two numbers facing up). We calculate their probabilities as follows:

$$P(X = 1) = P\{(1, 1)\} = \tfrac{1}{36}$$

$$P(X = 2) = P\{(1, 2), (2, 2), (2, 1)\} = \tfrac{3}{36}$$

$P(X = 3) = P\{(1,3),(2,3),(3,3),(3,2),(3,1)\}$

$$= \tfrac{5}{36}$$

and so on. The completed probability distribution is as follows:

x	1	2	3	4	5	6
$P(X = x)$	$\tfrac{1}{36}$	$\tfrac{3}{36}$	$\tfrac{5}{36}$	$\tfrac{7}{36}$	$\tfrac{9}{36}$	$\tfrac{11}{36}$

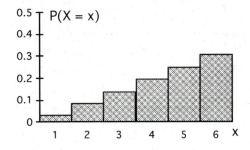

29. (a) For the values of X, use the rounded midpoints of the measurement classes given:
$(19,999-0)/2 \approx 10,000$
$(39,999-20,000)/2 \approx 30,000$
$(59,999-40,000)/2 \approx 50,000$
$(79,999-60,000)/2 \approx 70,000$
$(99,999-80,000)/2 \approx 90,000$
To obtain the probabilities, divide each frequency by the sum of all the frequencies, 1000:

x	10,000	30,000	50,000	70,000	90,000
Freq	240	270	220	160	110
$P(X=x)$	.24	.27	.22	.16	.11

(b) Here is the histogram with the area corresponding to $X > 50,000$ shaded:

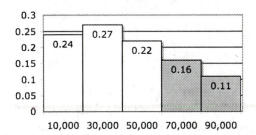

$P(X > 50,000) = .16 + .11 = .27$

31. The associated random variable is shown on the x axis of the histogram: X = age of a working adult in Mexico

To determine the probability distribution of X, use a frequency table. For the values of X, use the rounded midpoints of the measurement classes given:
$(16+25)/2 \approx 20$
$(26+35)/2 \approx 30$
$(36+45)/2 \approx 40$
$(46+55)/2 \approx 50$
$(56+65)/2 \approx 60$
To obtain the relative frequencies, divide each frequency by the sum of all the frequencies, 250:

x	20	30	40	50	60
Freq	72	118	50	8	2
$P(X=x)$	.288	.472	.200	.032	.008

33. (a) The values of X are the possible tow ratings: 2000, 3000, 4000, 5000, 6000, 7000, 8000 (7000 is optional)
(b) The following table shows the frequency (number of models with each tow rating) and the resulting probabilities (divide each frequency by the sum of all the frequencies):

x	2000	3000	4000	5000
Freq	2	1	1	1
$P(X = x)$	.2	.1	.1	.1
x	6000	7000	8000	
Freq	2	0	3	
$P(X = x)$	.2	.0	.3	.2

(c) $P(X \le 5000) = .2+.1+.1+.1 = .5$
(Add the probabilities in the first row of the above table.)

35. (a) The values of X are the (rounded) changes of the Dow: -700, -600, -500, -400, -300, -200, -100, 0, 100, 200, 300, 400, 500, 600, 700, 800, 900 (-600, 0, 100, 300, 500, 600, 700, and 800 are optional)

(b) The following table shows the frequency (number of days with each specified change in the Dow) and the resulting probabilities (divide each frequency by the sum of all the frequencies, 20):

x	-700	-600	-500	-400	-300	-200	-100	0
Freq	2	0	2	1	1	4	4	0
P(X = x)	.1	0	.1	.05	.05	.2	.2	0

x	100	200	300	400	500	600	700	800	900
Freq	0	2	0	2	0	0	0	0	2
P(X = x)	0	.1	0	.1	0	0	0	0	.1

(c) $P(X < -200) = .05 + .05 + .1 + .1 = .3$

37. To obtain the frequency table, count how many of the scores fall in each measurement class:

Class	1.1–2.0	2.1–3.0	3.1–4.0
Freq	4	7	9

To obtain the probability distribution, divide each frequency by the sum of all the frequencies, 20:

x	1.5	2.5	3.5
P(X = x)	.20	.35	.45

39. First, construct the probability distribution:

x	0	1	2	3
Freq	140	350	450	650
P(X = x)	.07	.175	.225	.325
x	4	5	6	7
Freq	200	140	50	10
P(X = x)	.1	.06	.025	.005
x	8	9	10	
Freq	5	15	10	
P(X = x)	.0025	.0075	.005	

$P(X < 6) = .07 + .175 + .225 + .325 + .1 + .06 = .955$

Therefore, 95.5% of cars are newer than 6 years old.

41. The frequency distribution for X is obtained from the first row (small cars) of the given table, and we can use that to compute the probability distribution (divide each frequency by the sum of the frequencies, 16):

x	3	2	1	0
Freq	1	11	2	2
P(X = x)	.0625	.6875	.125	.125

43. $P(X \geq 2) = P(X = 2) + P(X = 3)$
$= .6875 + .0625 = .75$

The probability that a randomly selected small car is rated Good or Acceptable is .75.

45. The following tables give the frequency distributions for Y and Z:

y	3	2	1	0
Freq	1	4	4	1
P(Y = y)	.1	.4	.4	.1

z	3	2	1	0
Freq	3	5	3	4
P(Z = z)	.2	.3333	.2	.3333

$P(Y \geq 2) = P(Y = 2) + P(Y = 3)$
$= .4 + .1 = .5$
$P(Z \geq 2) = P(Z = 2) + P(Z = 3)$
$\approx .333 + .2 = .533$

Since a crash rating of at least 2 indicates "acceptable" or "good," the data suggest that medium SUVs are safer than small SUVs in frontal crashes.

47. $P(X = 3) = \dfrac{1}{16} = .0625$

$P(Y = 3) = \dfrac{1}{10} = .1$

$P(Z = 3) = \dfrac{3}{15} = .2$

$P(U = 3) = \dfrac{3}{13} \approx .2308$

$P(V = 3) = \dfrac{3}{15} = .2$

$P(W = 3) = \dfrac{9}{19} \approx .4737$

The lowest is $P(Y = 3)$, for small cars.

49. From Exercise 41, the probability that a randomly selected small car will be rated at least 2 is

$P(X \geq 2) = .75$

From Exercise 43, the probability that a randomly selected small SUV will be rated at least 2 is

$P(Y \geq 2) = .5$

Since these two events are independent (both vehicles are selected at random), the probability that both will be rated at least 2 is

$P(X \geq 2) \times P(Y \geq 2) = .75 \times .5 = .375$

51. The sample space is the set of all sets of 4 tents selected from 7; $n(S) = C(7, 4) = 35$.

The possible values of X are 1, 2, 3, and 4. (X cannot equal 0, since that would require 4 green tents, and there are only 3.)

$P(X = 1) = \dfrac{C(4, 1)C(3, 3)}{35} = \dfrac{4}{35}$

$P(X = 2) = \dfrac{C(4, 2)C(3, 2)}{35} = \dfrac{18}{35}$

$P(X = 3) = \dfrac{C(4, 3)C(3, 1)}{35} = \dfrac{12}{35}$

$P(X = 4) = \dfrac{C(4, 4)C(3, 0)}{35} = \dfrac{1}{35}$

Probability distribution:

x	1	2	3	4
$P(X = x)$	$\dfrac{4}{35}$	$\dfrac{18}{35}$	$\dfrac{12}{35}$	$\dfrac{1}{35}$

$P(X \geq 2) = 1 - P(X = 1)$

$= 1 - \dfrac{4}{35} = \dfrac{31}{35} \approx 0.886$

53. Various answers possible

55. No; examples are infinite discrete random variables. For example, if X is the number of times you must toss a coin until heads comes up, then X is infinite but not continuous.

57. By measuring the values of X for a large number of outcomes, and then using the estimated probability (relative frequency).

59. Here are two examples:
(1) Let X be the number of times you have read a randomly selected book.
(2) Let X be the number of days a diligent student waits before beginning to study for an exam scheduled in 10 days' time.

61. The bars should be 1 unit wide, so that their height is numerically equal to their area.

63. Answers may vary. If we are interested in exact page-counts, then the number of possible values is very large and the values are (relatively speaking) close together, so using a continuous random variable might be advantageous. In general, the finer and more numerous the measurement classes, the more likely it becomes that a continuous random variable could be advantageous..

8.2 Bernoulli Trials and Binomial Random Variables

1. $n = 5$, $p = .1$, $q = .9$
$P(X = 2) = C(5, 2)(.1)^2(.9)^3 = .0729$

3. $n = 5$, $p = .1$, $q = .9$
$P(X = 0) = C(5, 0)(.1)^0(.9)^5 = .59049$

5. $n = 5$, $p = .1$, $q = .9$
$P(X = 5) = C(5, 5)(.1)^5(.9)^0 = .00001$

7. $n = 5$, $p = .1$, $q = .9$
$P(X \le 2) = P(X = 0) + P(X = 1) + P(X = 2)$
$P(X = 0) = C(5, 0)(.1)^0(.9)^5 = .59049$
$P(X = 1) = C(5, 1)(.1)^1(.9)^4 = .32805$
$P(X = 2) = C(5, 2)(.1)^2(.9)^3 = .0729$
Therefore,
$P(X \le 2) = .59049 + .32805 + .0729 = .99144$

9. $n = 5$, $p = .1$, $q = .9$
$P(X \ge 3) = P(X = 3) + P(X = 4) + P(X = 5)$
$P(X = 3) = C(5, 3)(.1)^3(.9)^2 = .0081$
$P(X = 4) = C(5, 4)(.1)^4(.9)^1 = .00045$
$P(X = 5) = C(5, 5)(.1)^5(.9)^0 = .00001$
Therefore,
$P(X \ge 3) = .0081 + .00045 + .00001 = .00856$

11. $n = 6$, $p = .4$, $q = 1-p = .6$
$P(X = 3) = C(6, 3)(.4)^3(.6)^3 = .27648$

13. $n = 6$, $p = .4$, $q = 1-p = .6$
$P(X \le 2) = P(X = 0) + P(X = 1) + P(X = 2)$
$P(X = 0) = C(6, 0)(.4)^0(.6)^6 = .046656$
$P(X = 1) = C(6, 1)(.4)^1(.6)^5 = .18662$
$P(X = 2) = C(6, 2)(.4)^2(.6)^4 = .31104$
Therefore,
$P(X \le 2) = .046656 + .18662 + .31104$

$= .54432$

15. $n = 6$, $p = .4$, $q = 1-p = .6$
$P(X \ge 5) = P(X = 5) + P(X = 6)$
$P(X = 5) = C(6, 5)(.4)^5(.6)^1 = .036864$
$P(X = 6) = C(6, 6)(.4)^6(.6)^0 = .004096$
Therefore,
$P(X \ge 5) = .036864 + .004096 = .04096$

17. $n = 6$, $p = .4$, $q = 1-p = .6$
$P(1 \le X \le 3) = P(X = 1) + P(X = 2) + P(X = 3)$
$P(X = 1) = C(6, 1)(.4)^1(.6)^5 = .18662$
$P(X = 2) = C(6, 2)(.4)^2(.6)^4 = .31104$
$P(X = 3) = C(6, 3)(.4)^3(.6)^3 = .27648$
Therefore,
$P(1 \le X \le 3) = .18662 + .31104 + .27648$
$= .77414$

19. $n = 5$, $p = \frac{1}{4}$, $q = \frac{3}{4}$
We used Excel to generate the distribution.
Format: BINOMDIST(x,n,p,0)

	H	I
1	x	P(X=x)
2	0	BINOMDIST(H2,5,0.25,0)
3	1	
4	2	
5	3	
6	4	
7	5	

TI-83/84 Plus: `binompdf(5,0.25,x)`

The resulting values are shown on the histogram:

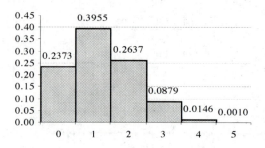

21. $n = 4$, $p = \frac{1}{3}$, $q = \frac{2}{3}$

We used Excel to generate the distribution.

Format: BINOMDIST(x,n,p,0)

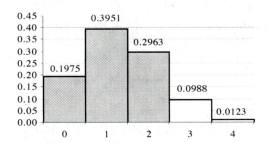

	A	B
1	x	P(X=x)
2		0 BINOMDIST(A2,4,1/3,0)
3	1	
4	2	
5	3	
6	4	

TI-83/84 Plus: binompdf(4,1/3,x)

The resulting values are shown on the histogram, with the portion corresponding to $P(X \le 2)$ shaded.

$P(X \le 2) \approx .1975 + .3951 + .2963 = 0.8889$

23. Take "success" = connect to the Internet immediately upon waking.

$n = 5$, $p = .25$, $q = 1 - .25 = .75$

$P(X = 2) = C(5, 2)(.25)^2(.75)^3 \approx .2637$

25. Take "success" = a stock market success.

$n = 10$, $p = .2$, $q = 1 - .2 = .8$

$P(X \ge 1) = 1 - P(X = 0)$

$P(X = 0) = C(10, 0)(.2)^0(.8)^{10} \approx .10737$

Therefore,

$P(X \ge 1) \approx 1 - 10737 = .8926$

27. Take "success" = selecting a male.

$n = 3$, $p = .5$, $q = 1 - .5 = .5$

$P(X \ge 1) = 1 - P(X = 0)$

$P(X = 0) = C(3, 0)(.5)^0(.5)^3 = .125$

Therefore,

$P(X \ge 1) = 1 - 125 = .875$

29. Take "success" = selecting a defective bag.

$n = 5$, $p = .1$, $q = 1 - .1 = .9$

(a) $P(X = 3) = C(5, 3)(.1)^3(.9)^2 = .0081$

(b) $P(X \ge 2) = 1 - P(X \le 1)$

$P(X = 0) = C(5, 0)(.1)^0(.9)^5 = .59049$

$P(X = 1) = C(5, 1)(.1)^1(.9)^4 = .32805$

Therefore,

$P(X \ge 2) = 1 - (.59049 + .32805) = .08146$

31. Take "success" = watching a rented video at least once.

$n = 10$, $p = .71$, $q = 1 - .71 = .29$

$P(X \ge 8) = P(X = 8) + P(X = 9) + P(X = 10)$

$P(X = 8) = C(10, 8)(.71)^8(.29)^2 \approx .244$

$P(X = 9) = C(10, 9)(.71)^9(.29)^1 \approx .133$

$P(X = 10) = C(10, 10)(.71)^{10}(.29)^0 \approx .033$

Therefore,

$P(X \ge 8) \approx .244 + .133 + .033$

$\approx .41$

33.(a) Take "success" = in foreclosure.

$n = 10$, $p = .24$, $q = 1 - .24 = .76$

$P(X = 5) = C(10, 5)(.24)^5(.76)^5 \approx .0509$

(b) Following is the distribution generated by the Web Site utility at

Web site $\rightarrow$ On-line Utilities $\rightarrow$ Binomial Distribution Utility:

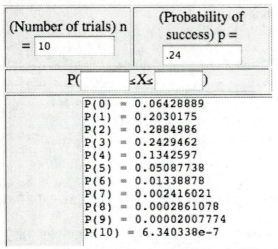

(Number of trials) n = 10	(Probability of success) p = .24

$$P(\qquad \leq X \leq \qquad)$$

```
P(0) = 0.06428889
P(1) = 0.2030175
P(2) = 0.2884986
P(3) = 0.2429462
P(4) = 0.1342597
P(5) = 0.05087738
P(6) = 0.01338878
P(7) = 0.002416021
P(8) = 0.0002861078
P(9) = 0.00002007774
P(10) = 6.340338e-7
```

(c) The value of X with the largest probability is $X = 2$. So, the most likely number of homes to have been in foreclosure was $\underline{2}$.

35. Take "success" = computer malfunction.
$n = 3$, $p = .01$, $q = 1 - .01 = .99$
$P(X \geq 2) = P(X = 2) + P(X = 3)$
$P(X = 2) = C(3, 2)(.01)^2(.99)^1 \approx .000297$
$P(X = 3) = C(3, 3)(.01)^3(.99)^0 \approx .000001$
Therefore,
$P(X \geq 2) \approx .000297 + .000001 = .000298$

37. Take "success" = answering a question correctly.
$n = 100$, $p = .80$
We use technology to compute
$P(75 \leq X \leq 85) = P(X \leq 85) - P(X \leq 74)$
TI-83/84 Plus:
```
binomcdf(100,0.8,85)-
binomcdf(100,0.8,74)
```
Excel:
```
BINOMDIST(85,100,0.8,1)-
BINOMDIST(74,100,0.8,1)
```
Answer: $P(75 \leq X \leq 85) \approx .8321$

39. Take "success" = containing more than 10 grams of fat.
$n = 50$, $p = .43$
We use technology to generate the probability distribution:

Excel: `BINOMDIST(x,50,0.43,0)`
TI-83/84 Plus: `binompdf(50,0.43,x)`

	A	B
1	x	P(X=x)
2	0	6.21932E-13
3	1	2.34588E-11
4	2	4.33577E-10
5		
49		
50	48	1.01478E-15
51	49	3.12463E-17
52	50	4.71436E-19

(a) Since 43% of all the burgers contain more than 10 grams of fat, we can expect about
$$0.43 \times 50 = 21$$
of them to contain more than 10 grams of fat.
(b) The probability that k or more patties contain more than 10 grams of fat is
$$P(X \geq k) = 1 - P(X \leq k-1)$$
We want this to equal approximately .71:
$$1 - P(X \leq k-1) \approx .71$$
$$P(X \leq k-1) \approx .29$$
To answer this question, use the cumulative probability distribution:

Excel: `BINOMDIST(x,50,0.43,1)`
TI-83/84 Plus: `binomcdf(50,0.43,x)`

	A	B
1	k	P(X ≤ k)
2	0	6.21932E-13
3	1	2.40808E-11
4	2	4.57658E-10
5		
19		
20	18	0.19634337
21	19	0.285669736
22	20	0.390118898
23	21	0.502683159
24	22	0.61461907

From the table, $k - 1 = 19$, so $k = 20$.
There is approximately a 71% chance that a batch of 50 ZeroFat patties contains $\underline{20}$ or more patties with at least 10 grams of fat.

(c) Graphs

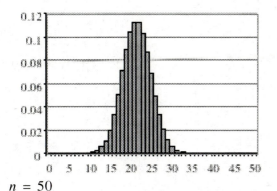

$n = 50$

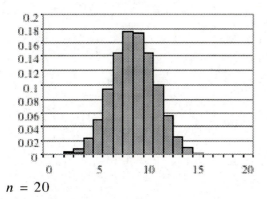

$n = 20$

The graph for $n = 50$ trials is more widely distributed than the graph for $n = 20$

41. Take "success" = bad bulb.

$p = .01, n = ?$

We want $P(X \geq 1) \geq .5$

$1 - P(X = 0) \geq .5$

$P(X = 0) \leq .5$

$C(n, 0) (.01)^0 (.99)^n \leq .5$

$1 \times 1 \times .9^n \leq .5$

$.99^n \leq .5$

Using technology, we compute the values of $.9^n$ for $n = 0, 1, 2, \ldots$:

n	$.9^n$
0	1
1	0.99
. . .	. . .
68	0.50488589
69	0.49983703

. . .	. . .

The probability first dips below .5 when $n = 69$ trials.

43. The estimated probability of an accident in a given mile is

$$p = \frac{\text{Number of accidents}}{\text{Number of miles}} = \frac{562}{100{,}000{,}000}$$

$$= 0.562 \times 10^{-5}$$

Since this is the estimated probability of a success, it is the probability that a male driver will have an accident in a one-mile trip.

45. Let X be the number of cases of mad cow disease found among the 243,000 tested. Then X is a binomial random variable with

$$p = \frac{5}{45{,}000{,}000} \approx 1.1111 \times 10^{-7}$$

and $n = 243{,}000$

We are asked to compute the probability of at least one "success": $P(X \geq 1)$.

$P(X \geq 1) \qquad = 1 - P(X = 0)$

$\qquad = 1 - C(n, 0)p^0(1-p)^{n-0}$

$\qquad = 1 - (1)(1)(1 - 1.1111 \times 10^{-7})^{243{,}000}$

$\qquad = 1 - (.999\,999\,889)^{243{,}000}$

$\qquad \approx 1 - .9734 = .0266$

Since there is only a 2.66% chance of detecting the disease in a given year, the government's claim seems dubious.

47. No; in the given scenario, the probability of success depends on the outcome of the previous shot. However, in a sequence of Bernoulli trials, the occurrence of one success does not effect the probability of success on the next attempt.

49. No; if life is a sequence of Bernoulli trials, then the occurrence of one misfortune ("success") does not affect the probability of a misfortune on the next trial. Hence, misfortunes may very well not "occur in threes."

51. Think of performing the experiment as a Bernoulli trial with "success" being the occurrence of E. Performing the experiment n times independently in succession would then be a sequence of n Bernoulli trials.

53. The probability of selecting a red marble changes after each selection, as the number of marbles left in the bag decreases. This violates the requirement that, in a sequence of Bernoulli trials, the probability of "success" does not change.

8.3 Measures of Central Tendency

1. $\bar{x} = \dfrac{-1 + 5 + 5 + 7 + 14}{5} = 6$

To compute the median, arrange the scores in order and take the (average of the) middle score(s).

$-1, 5, 5, 7, 14$

Median = middle score = 5

Mode = most frequent score(s) = 5

3. $\bar{x} = \dfrac{2 + 5 + 6 + 7 - 1 - 1}{6} = 3$

To compute the median, arrange the scores in order and take the (average of the) middle score(s).

$-1. -1, 2, 5, 6, 7$

Median = average of middle scores = $\dfrac{2+5}{2} = 3.5$

Mode = most frequent score(s) = -1

5. In decimal notation, the given scores are:

$0.5, 1.5, -4, 1.25$

$\bar{x} = \dfrac{0.5 + 1.5 - 4 + 1.25}{4} = -0.1875$

To compute the median, arrange the scores in order and take the (average of the) middle score(s).

$-4, 0.5, 1.25, 1.5$

Median = average of middle scores =

$\dfrac{0.5 + 1.25}{2} = 0.875$

Mode = most frequent score(s) = Every value

7. $\bar{x} = \dfrac{2.5 - 5.4 + 4.1 - 0.1 - 0.1}{5} = 0.2$

To compute the median, arrange the scores in order and take the (average of the) middle score(s).

$-5.4, -0.1, -0.1, 2.5, 4.1$

Median = middle score = -0.1

Mode = most frequent score(s) = -0.1

9. Answers may vary. One example is:

$0, 0, 0, 0, 0, 6$

$\bar{x} = 1$, Median = 0

Another example is:

$0, 0, 0, 1, 2, 3$

$\bar{x} = 1$, Median = 0

11. We use the tabular method described in Example 3 in the textbook:

x	0	1	2	3
$P(X = x)$	.5	.2	.2	.1
$xP(X = x)$	0	0.2	0.4	0.3

$E(X)$ = Sum of entries in the bottom row = 0.9

13. We first convert the fractions into decimals, and then use the tabular method described in Example 3 in the textbook:

x	10	20	30	40
$P(X = x)$	.3	.4	.2	.1
$xP(X = x)$	3	8	6	4

$E(X)$ = Sum of entries in the bottom row = 21

15. We use the tabular method described in Example 3 in the textbook:

x	-5	-1	0	2	5	10
$P(X=x)$	.2	.3	.2	.1	.2	.0
$xP(X=x)$	-1	-0.3	0	0.2	1	0

$E(X)$ = Sum of entries in the bottom row = -0.1

17. The probability distribution of X is

x	1	2	3	4	5	6
$P(X=x)$	1/6	1/6	1/6	1/6	1/6	1/6

Using the tabular method described in Example 3, we get

x	1	2	3	4	5	6
$P(X=x)$	1/6	1/6	1/6	1/6	1/6	1/6
$x \cdot P(X=x)$	1/6	2/6	3/6	4/6	5/6	6/6

$E(X)$ = sum of numbers in bottom row = 3.5

19. X is a binomial random variable with $n = 2$ and $p = .5$. Therefore,
$$E(X) = np = 2(.5) = 1$$

21. We compute the probability distribution of X (see the solution to Exercise 27 in Section 8.1) and then use the tabular method of Example 3 to compute $E(X)$:

x	1	2	3	4	5	6
Freq	1	3	5	7	9	11
$P(X=x)$	$\frac{1}{36}$	$\frac{3}{36}$	$\frac{5}{36}$	$\frac{7}{36}$	$\frac{9}{36}$	$\frac{11}{36}$
$xP(X=x)$	$\frac{1}{36}$	$\frac{6}{36}$	$\frac{15}{36}$	$\frac{28}{36}$	$\frac{45}{36}$	$\frac{66}{36}$

$E(X)$ = Sum of entries in the bottom row = $\frac{161}{36}$
≈ 4.4722

23. Number of sets of 4 marbles = $C(6, 4) = 15$. If X is the number of red marbles, then the possible values of X are 2, 3, 4, and
$$P(X = x) = \frac{C(4, x)C(2, 4-x)}{15}$$

x	2	3	4
$P(X = x)$	$\frac{6}{15}$	$\frac{8}{15}$	$\frac{1}{15}$
$xP(X = x)$	$\frac{12}{15}$	$\frac{24}{15}$	$\frac{4}{15}$

$E(X)$ = Sum of entries in the bottom row = $\frac{40}{15}$
≈ 2.6667

25. X is a binomial random variable with $n = 20$ and $p = .1$. Therefore,
$$E(X) = np = 20(.1) = 2$$

27. The number of possible hands of 5 cards is $C(52, 5) = 2,598,960$. The possible values of X are 0, 1, 2, 3, 4, and
$$P(X = x) = \frac{C(4, x)C(48, 5-x)}{2,598,960}$$

x	0	1	2
$P(X=x)$	.6588	.2995	.0399
$xP(X=x)$	0.0000	0.2995	0.0799
x	3	4	
$P(X=x)$	.0017	.0000	
$xP(X=x)$	0.0052	0.0001	

$E(X)$ = Sum of entries $xP(X=x) \approx 0.3846$

29. The sum of the given Dow changes is -1500. Therefore,
$$\bar{x} = \frac{-1500}{10} = -150$$
If we arrange the Dow changes in order, we get
$$-700, -700, -500, -400, -200,$$
$$-100, -100, -100, 400, 900$$
The two middle scores are -200 and -100. Therefore,
$$\text{Median} = \frac{-200 + (-100)}{2} = -150$$
There were as many days with a change in the Dow above -150 points as there were with changes below that. (See the definition of the median.)

31. The sum of the scores is 9183. Therefore,
$$\bar{x} = \frac{9183}{10} = \$918.30$$
Arranging the scores in order gives:
$$895, 905, 905, 910, 913, 918, 920, 936,$$
$$938, 943$$
The two middle scores are 913 and 918. Therefore,
$$\text{Median} = \frac{913+918}{2} = \$915.50$$
The mode is the most frequently occurring score: 905.

Over the 10-business day period sampled, the price of gold averaged $915.30 per ounce. It was above $915.50 as many times as it was below that, and stood at $905 per ounce more often than any other price.

33. (a) We use the tabular method described in Example 3 in the textbook:

x	1	2	3	4	5
$P(X=x)$	.01	.04	.04	.08	.10
$xP(X=x)$	0.01	0.08	0.12	0.32	0.5
x	6	7	8	9	10
$P(X=x)$	.15	.25	.20	.08	.05
$xP(X=x)$	0.9	1.75	1.6	0.72	0.5

$\mu = E(X)$ = sum of values $xP(X=x) = 6.5$

There were an average of 6.5 checkout lanes in a supermarket that was surveyed.

(b) $P(X < \mu) = P(X < 6.5)$

$\qquad = .01 + .04 + .04 + .08 + .10 + .15$

$\qquad = .42$

$P(X > \mu) = P(X > 6.5)$

$\qquad = .25 + .20 + .08 + .05$

$\qquad = .58,$

and is thus larger. Most supermarkets have more than the average number of checkout lanes.

35. Using the rounded midpoints of the given measurement classes, we get the following table. (The probabilities are obtained by dividing the given frequencies by their sum, 72 and then rounding to 2 decimal places.)

x	5	10	15
$P(X=x)$	.17	.33	.21
$xP(X=x)$	0.85	3.3	3.15
x	20	25	35
$P(X=x)$	.19	.03	.07
$xP(X=x)$	3.8	0.75	2.45

$E(X)$ = sum of values $xP(X=x) = 14.3$

Interpretation: The average age of a student in 1998 was 14.3.

37. The associated random variable is shown on the x axis of the histogram: X = age of a working adult in Mexico. We use the tabular method described in Example 3 in the textbook:

x	20	30	40	50	60
Freq	72	118	50	8	2
$P(X=x)$	0.288	0.472	0.2	0.032	0.008
$xP(X=x)$	5.76	14.16	8	1.6	0.48

Expected working age = sum of values $xP(X=x)$
= 30.

39. Using the rounded midpoints of the measurement classes, we get the following table:

x	10,000	30,000	50,000	70,000	90,000
$P(X = x)$	.27	.28	.20	.15	.10
$x \cdot P(X = x)$	2700	8400	10000	10500	9000

The sum of the entries in the bottom row is 40,600, representing an average income of $40,600 \approx $41,000$

41. The probabilities in the tables below are obtained by dividing the given frequencies by the sum, 16 for X and 10 for Y:

x	3	2	1	0
$P(X = x)$	.0625	.6875	.125	.125
$xP(X = x)$	0.1875	1.375	0.125	0

$E(X)$ = sum of values in the bottom row = 1.6875

y	3	2	1	0
$P(Y = y)$	.1	.4	.4	.1
$P(Y = y)$	0.3	0.8	0.4	0

$E(Y)$ = sum of values in the bottom row = 1.5

Since small cars (X) have a higher average rating, small cars performed better in frontal crashes.

43. Small cars:

x	3	2	1	0
$P(X = x)$	.0625	.6875	.125	.125
$xP(X = x)$	0.1875	1.375	0.125	0

$E(X) = 1.6875$

Midsize cars:

v	3	2	1	0
$P(V = v)$	.2	.333	0	.467
$vP(V = v)$	0.6	0.667	0	0

$E(V) \approx 1.267$

Large cars:

w	3	2	1	0
$P(W = w)$	.474	.263	.158	.105
$wP(W = w)$	1.421	0.526	0.158	0

$E(W) \approx 2.105$

Of the three large cars (W) performed best.

45. Either you lose \$1 ($X = -1$) or win \$1 ($X = 1$).
There are 20 losing numbers out of 38, so
$$P(X = -1) = \frac{20}{38}$$
There are 18 winning numbers out of 38, so
$$P(X = 1) = \frac{18}{38}$$

x	-1	1
$P(X = x)$	$\frac{20}{38}$	$\frac{18}{38}$
$xP(X = x)$	$-\frac{20}{38}$	$\frac{18}{38}$

$$E(X) = -\frac{20}{38} + \frac{18}{38} = -\frac{2}{38} \approx -0.53$$

Expect to lose 53¢.

47. The given experiment consists of $n = 40$ Bernoulli trials with $p = .63$. Thus, the expected number of students that will shop at a mall during the next week is
$$\mu = np = 40 \times .63 = 25.2 \text{ students}$$

49. (a) The given experiment consists of $n = 20$ Bernoulli trials with $p = .10$. Therefore, the expected number of defective air bags is
$$\mu = np = 20 \times .10 = 2$$
(b) $p = .10$, $\mu = 12$ and n is unknown
$$\mu = np$$
$$12 = .10n$$
$$n = \frac{12}{.10} = 120 \text{ airbags}$$

51. The probability distribution for X = number of red tents is derived in the solution to Set 8.1, Exercise 47. Here, we add a new row to compute the expected value:

x	1	2	3	4
$P(X = x)$	$\frac{4}{35}$	$\frac{18}{35}$	$\frac{12}{35}$	$\frac{1}{35}$
$xP(X = x)$	$\frac{4}{35}$	$\frac{36}{35}$	$\frac{36}{35}$	$\frac{4}{35}$

$E(X)$ = Sum of bottom row entries
$$= \frac{80}{35} \approx 2.2857 \text{ tents}$$

53. Let X = rate of return of Fastforward Funds, and let Y = rate of return of SolidState Securities. The following worksheets show the computation of the expected values of X and Y:

	A	B	C
1	x	P(X = x)	xP(X = x)
2	-0.4	0.015	-0.006
3	-0.3	0.025	-0.0075
4	-0.2	0.043	-0.0086
5	-0.1	0.132	-0.0132
6	0	0.289	0
7	0.1	0.323	0.0323
8	0.2	0.111	0.0222
9	0.3	0.043	0.0129
10	0.4	0.019	0.0076
11		Total:	0.0397

	E	F	G
1	y	P(Y = y)	yP(Y = y)
2	-0.4	0.012	-0.0048
3	-0.3	0.023	-0.0069
4	-0.2	0.05	-0.01
5	-0.1	0.131	-0.0131
6	0	0.207	0
7	0.1	0.33	0.033
8	0.2	0.188	0.0376
9	0.3	0.043	0.0129
10	0.4	0.016	0.0064
11		Total:	0.0551

From the worksheets, we read off the following expected rates of return: FastForward: 3.97%; SolidState: 5.51%; SolidState gives the higher expected return.

55. If a driver wrecks a car, the net cost to the insurance company is
$100,000 – $5000 = $95,000
so $X = -95,000$
If a driver does not wreck a car, the net profit for the insurance company is the premium: $5000, so
$X = 5000$
To compute the probability distribution, use the following tree:

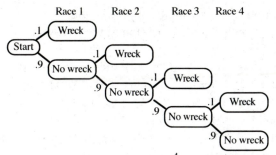

Probability of no wreck is $.9^4 = .6561$
Probability of a wreck is $1 - .9^4 = .3439$

x	-95,000	5000
$P(X = x)$	.3439	.6561
$xP(X = x)$	-32670.5	3280.5

$E(X)$ = sum of entries in bottom row = -29390

The company can expect to lose $29,390 per driver.

57. Since there are as many scores above the median as below the median, the correct choice is (A): The median and mean are equal.

59. He is wrong; for example, the collection 0, 0, 300 has mean 100 and median 0.

61. No. The expected number of times you will hit the dart-board is the average number of times you will hit the bull's eye per 50 shots; the average of a set of whole numbers need not be a whole number.

63. Wrong. It might be the case that only a small fraction of people in the class scored better than you but received a exceptionally high scores that raised the class average. Suppose, for instance, that there are 10 people in the class. Four received 100%, you received 80%, and the rest received 70%. Then the class average is 83%, 5 people have lower scores that you, but only four have higher scores.

65. No; the mean of a very large sample is only an *estimate* of the population mean. The means of

larger and larger samples *approach* the
population mean as the sample size increases.

67. Wrong. the statement attributed to President
Bush asserts that the mean tax refund would be
$1000, whereas the statements referred to as "The
Truth" suggest that the *median* tax refund would
be close to $100 [and that the 31st percentile
would be zero].

69. Select a U.S. household at random, and let X
be the income of that household. The expected
value of X is then the population mean of all U.S.
household incomes.

8.4 Measures of Dispersion

1. For ease of computation, we arrange the data in a table. The first column lists the values of X, giving us the mean $\bar{x}$. The second column lists the numbers $(x_i - \bar{x})$, and the third column lists their squares.

x	$(x - \bar{x})$	$(x - \bar{x})^2$
-1	-7	49
5	-1	1
5	-1	1
7	1	1
14	8	64
30	0	116

The bottom row shows the sums.

$$\bar{x} = \frac{30}{5} = 6 \qquad \sum_{i=1}^{n} (x_i - \bar{x})^2 = 116$$

$$s^2 = \frac{\sum_{i=1}^{n} (x_i - \bar{x})^2}{n-1} = \frac{116}{5-1} = 29$$

$$s = \sqrt{s^2} \approx 5.39$$

3.

x	$(x - \bar{x})$	$(x - \bar{x})^2$
2	-1	1
5	2	4
6	3	9
7	4	16
-1	-4	16
-1	-4	16
18	0	62

The bottom row shows the sums.

$$\bar{x} = \frac{18}{6} = 3 \qquad \sum_{i=1}^{n} (x_i - \bar{x})^2 = 62$$

$$s^2 = \frac{\sum_{i=1}^{n} (x_i - \bar{x})^2}{n-1} = \frac{62}{6-1} = 12.4$$

$$s = \sqrt{s^2} \approx 3.52$$

5. In the following table, we first converted all the fractions to decimals.

x	$(x - \bar{x})$	$(x - \bar{x})^2$
0.5	0.6875	0.4727
1.5	1.6875	2.8477
-4	-3.8125	14.5352
1.25	1.4375	2.0664
-0.75	0	19.9219

The bottom row shows the sums.

$$\bar{x} = \frac{-0.75}{4} = -0.1875$$

$$\sum_{i=1}^{n} (x_i - \bar{x})^2 \approx 19.9219$$

$$s^2 = \frac{\sum_{i=1}^{n} (x_i - \bar{x})^2}{n-1} \approx \frac{19.9219}{4-1} \approx 6.64$$

$$s = \sqrt{s^2} \approx 2.58$$

7.

x	$(x - \bar{x})$	$(x - \bar{x})^2$
2.5	2.3	5.29
-5.4	-5.6	31.36
4.1	3.9	15.21
-0.1	-0.3	0.09
-0.1	-0.3	0.09
1	0	52.04

The bottom row shows the sums.

$$\bar{x} = \frac{1}{5} = 0.2 \qquad \sum_{i=1}^{n} (x_i - \bar{x})^2 = 52.04$$

$$s^2 = \frac{\sum_{i=1}^{n} (x_i - \bar{x})^2}{n-1} = \frac{52.04}{5-1} = 13.01$$

$$s = \sqrt{s^2} \approx 3.61$$

9. We use the tabular method of arranging the data described in Example 3:

x	0	1	2	3
$P(X = x)$	.5	.2	.2	.1
$xP(X = x)$	0	0.2	0.4	0.3

μ = Sum of entries in bottom row = 0.9

$x - \mu$	-0.9	0.1	1.1	2.1
$(x-\mu)^2$	0.81	0.01	1.21	4.41
$P(X=x)$ $\times(x-\mu)^2$	0.405	0.002	0.242	0.441

σ^2 = Sum of entries in bottom row = 1.09

$\sigma = \sqrt{\sigma^2} \approx 1.04$

11. We first convert all fractions into decimals, and then use the tabular method described in Example 3:

x	10	20	30	40
$P(X = x)$	.3	.4	.2	.1
$xP(X = x)$	3	8	6	4

μ = Sum of entries in bottom row = 2.8

$x - \mu$	-11	-1	9	19
$(x-\mu)^2$	121	1	81	361
$P(X=x)$ $\times(x-\mu)^2$	36.3	0.4	16.2	36.1

σ^2 = Sum of entries in bottom row = 89

$\sigma = \sqrt{\sigma^2} \approx 9.43$

13. We use the tabular method of arranging the data described in Example 3. (Note that, for ease of layout, we are arranging the data in columns.)

x	$P(X = x)$	$xP(X = x)$
-5	.2	-1
-1	.3	-0.3
0	.2	0
2	.1	0.2
5	.2	1
10	0	0

μ = Sum of entries in right-hand column = $-$ 0.1

$x - \mu$	$(x - \mu)^2$	$P(X = x) \times (x - \mu)^2$
-4.9	24.01	4.802
-0.9	0.81	0.243
0.1	0.01	0.002
2.1	4.41	0.441
5.1	26.01	5.202
10.1	102.01	0

σ^2 = Sum of entries in right-hand column = 10.69

$\sigma = \sqrt{\sigma^2} \approx 3.27$

15. The probability distribution and expected value calculated in Exercise 17 of Section 8.3:

x	$P(X = x)$	$xP(X = x)$
1	1/6	1/6
2	1/6	1/3
3	1/6	1/2
4	1/6	2/3
5	1/6	5/6
6	1/6	1

μ = Sum of entries in right-hand column = 3.5

$x - \mu$	$(x - \mu)^2$	$P(X = x) \times (x - \mu)^2$
-2.5	6.25	1.042
-1.5	2.25	0.375
-0.5	0.25	0.042
0.5	0.25	0.042
1.5	2.25	0.375
2.5	6.25	1.042

σ^2 = Sum of entries in right-hand column $\approx$ 2.918

$\sigma = \sqrt{\sigma^2} \approx 1.71$

17. X is a binomial random variable with $n = 2$ and $p = .5$.

$\mu = E(X) = np = 2(.5) = 1$

$\sigma^2 = npq = 2(.5)(.5) = 0.5$

$\sigma = \sqrt{\sigma^2} \approx 0.71$

19. The probability distribution was calculated in Exercise 27 of Section 8.1. To continue the calculation, we use decimal approximations of the fractions:

x	$P(X = x)$	$xP(X = x)$
1	.0278	0.0278
2	.0833	0.1667
3	.1389	0.4167
4	.1944	0.7778
5	.2500	1.2500
6	.3056	1.8333

μ = Sum of entries in right-hand column ≈ 4.47

$x - \mu$	$(x - \mu)^2$	$P(X = x) \times (x - \mu)^2$
-3.4722	12.0563	0.3349
-2.4722	6.1119	0.5093
-1.4722	2.1674	0.3010
-0.4722	0.2230	0.0434
0.5278	0.2785	0.0696
1.5278	2.3341	0.7132

σ^2 = Sum of entries in right-hand column $\approx$ 1.9715

$\sigma = \sqrt{\sigma^2} \approx 1.40$

21. The probability distribution was calculated in Exercise 21 of Section 8.3. To continue the calculation, we use decimal approximations of the fractions:

x	$P(X = x)$	$xP(X = x)$
2	.40	0.8000
3	.5333	1.6000

4	.0667	0.2667

μ = Sum of entries in right-hand column $\approx$ 2.67

$x - \mu$	$(x - \mu)^2$	$P(X = x) \times (x - \mu)^2$
-0.6667	0.4444	0.1778
0.3333	0.1111	0.0593
1.3333	1.7778	0.1185

σ^2 = Sum of entries in right-hand column $\approx$ 0.36

$\sigma = \sqrt{\sigma^2} \approx 0.60$

23. X is a binomial random variable with $n = 20$ and $p = .1$

$\mu = np = 20(.1) = 2$

$\sigma^2 = npq = 20(.1)(.9) = 1.8$

$\sigma = \sqrt{\sigma^2} \approx 1.34$

25. (a)

x	$(x - \bar{x})$	$(x - \bar{x})^2$
3	0	0
2	-1	1
0	-3	9
9	6	36
1	-2	4
15	0	50

(The bottom row shows the sums.)

$\bar{x} = \dfrac{15}{5} = 3$

$\displaystyle\sum_{i=1}^{n} (x_i - \bar{x})^2 = 50$

$s^2 = \dfrac{\displaystyle\sum_{i=1}^{n} (x_i - \bar{x})^2}{n-1} \approx \dfrac{50}{5-1} \approx 12.5$

$s = \sqrt{s^2} \approx 3.54$

(b) The empirical rule states that approximately 68% of the class will rank you between

$\bar{x} - s = 3 - 3.54 = -0.54$

and

$\bar{x} + s = 3 + 3.54 = 6.54$

That is, in the interval $[0, 6.54]$ (We replaced the negative score by 0, since no rankings can be negative.

We must assume that the population distribution is bell-shaped and symmetric.

27. (a)

x	$(x-\bar{x})$	$(x-\bar{x})^2$
4.2	–0.8	0.6
4.7	–0.3	0.09
5.4	0.4	0.16
5.8	0.8	0.64
4.9	–0.1	0.01
25	0	1.54

(The bottom row shows the sums.)

$\bar{x} = \dfrac{25}{5} = 5.0$

$\displaystyle\sum_{i=1}^{n} (x_i - \bar{x})^2 = 1.54$

$s^2 = \dfrac{\displaystyle\sum_{i=1}^{n} (x_i - \bar{x})^2}{n-1} = \dfrac{1.54}{5-1} = 0.385$

$s = \sqrt{s^2} \approx 0.6$

(b) The empirical rule states that approximately 95% of the data will fall between

$\bar{x} - 2s \approx 5.0 - 1.2 = 3.8$

and

$\bar{x} + 2s \approx 5.0 + 1.2 = 6.2$

29. (a)

x	$(x-\bar{x})$	$(x-\bar{x})^2$
–400	–250	62,500
–500	–350	122,500
–200	–50	2500
–700	–550	302,500
–100	50	2,500
900	1,050	1,102,500
–100	50	2,500
–700	–550	302,500
400	550	302,500
–100	50	2,500
–1,500	0	2,205,000

(The bottom row shows the sums.)

$\bar{x} = \dfrac{-1500}{10} = -150$

$\displaystyle\sum_{i=1}^{n} (x_i - \bar{x})^2 = 2,205,000$

$s^2 = \dfrac{\displaystyle\sum_{i=1}^{n} (x_i - \bar{x})^2}{n-1} = \dfrac{2,205,000}{10-1} = 245,000$

$s = \sqrt{s^2} \approx 495$

(b) The empirical rule states that approximately 68% of the data will fall between

$\bar{x} - s \approx -150 - 495 = -645$

and

$\bar{x} + s \approx -150 + 495 = 345$

Thus, $100 - 68 = 32\%$ of the time the data fall outside that range, and so, by symmetry, 16% of the data will be below –645. That is, 16% of the time, the Dow will drop by more than 645 points. To obtain the actual percentage of times the Dow fell by more than 645 points, count how many times this actually happened: twice.

Since 2 scores is 20% of the original 10, we conclude that the Dow actually fell by more than 645 points 20% of the time.

31. (a) Following is a worksheet computation of the variance:

	A	B
1	x	(x - mu)^2
2	17	0.5625
3	18	0.0625
4	17	0.5625
5	18	0.0625
6	21	10.5625
7	16	3.0625
8	21	10.5625
9	18	0.0625
10	16	3.0625
11	14	14.0625
12	15	7.5625
13	22	18.0625
14	17	0.5625
15	19	1.5625
16	17	0.5625
17	18	0.0625
18	**mean (mu)**	**variance**
19	**17.75**	**4.733333**

From the worksheet,
$s^2 \approx 4.7333$
$s = \sqrt{s^2} \approx 2.1756 \approx 2.18$

(b) Chebyshev's inequality predicts that at least 8/9 of the scores will fall between
$\bar{x} - 3s \approx 17.75 - 3(2.1756) \approx 11.22$
and
$\bar{x} + 3s \approx 17.75 + 3(2.1756) \approx 24.28$
That is, they will fall in the interval
$[11.22, 24.28]$.

(c) Every single score, or <u>100%</u> of the scores fall in the range $[11.22, 24.28]$. Since the empirical rule predicts that 99.7% of the scores will fall in the given range, making it a more accurate predictor.

33. We use the tabular method of arranging the data described in Example 3. (Note that, for ease of layout, we are arranging the data in columns.)

x	$P(X = x)$	$xP(X = x)$
0	0.4	0
1	0.1	0.1
2	0.2	0.4
3	0.2	0.6
4	0.1	0.4

μ = Sum of entries in right-hand column = 1.5

$x - \mu$	$(x - \mu)^2$	$P(X = x) \times (x - \mu)^2$
-1.5	2.25	0.9
-0.5	0.25	0.025
0.5	0.25	0.05
1.5	2.25	0.45
2.5	6.25	0.625

σ^2 = Sum of entries in right-hand column = 2.05
$\sigma = \sqrt{\sigma^2} \approx 1.43$

The range of X within 2 standard deviations of the mean is given by
Lower limit $= \mu - 2\sigma \approx 2.05 - 2(1.43) = -0.81$
Upper limit $= \mu + 2\sigma \approx 2.05 + 2(1.43) = 4.91$

All (100%) of the values of X are within the interval $[-0.81, 4.91]$. Therefore, 100% of malls have a number of movie theater screens within two standard deviations of μ.

35. The probabilities for the distribution are computed by dividing the frequencies by the total: 100.

x	$P(X = x)$	$xP(X = x)$
10	.27	2.7
30	.28	8.4
50	.20	10.0
70	.15	10.5

90	.10	9.0

μ = Sum of entries in right-hand column = 40.6, representing \$40,600

$x - \mu$	$(x - \mu)^2$	$P(X = x) \times (x - \mu)^2$
−30.6	936.36	252.8172
−10.6	112.36	31.4608
9.4	88.36	17.672
29.4	864.36	129.654
49.4	2440.36	244.036

σ^2 = Sum of entries in right-hand column = 675.64

$\sigma = \sqrt{\sigma^2} \approx 26$, representing \$26,000

The range of X within 1 standard deviation of the mean is given by

Lower limit = $\mu - \sigma \approx 40.6 - 26 \approx 14.6$

Upper limit = $\mu + \sigma \approx 40.6 + 26 \approx 66.6$

The difference is $66.6 - 14.6 = 52$ (2 standard deviations), representing an income gap of \$52,000.

37. (a) The rounded midpoints of the measurement classes (ages) are:

$(15+24.9)/2 \approx 20$

$(25+54.9)/2 \approx 40$

$(55 + 64.9)/2 \approx 60$

In the following table, all intermediate calculations have been rounded to two decimal places.

X	20	40	60	**Total**
N	16,000	13,000	1600	30600
$P(X=x)$	.52	.42	.05	
$x \cdot P(X=x)$	10.4	16.8	3	30.2
$x - \mu$	-10.2	9.8	29.8	
$(x - \mu)^2$	104.04	96.04	888.04	
$P(X=x) \cdot (x - \mu)^2$	54.1	40.34	44.4	138.84

Expected value = μ = Sum of entries in 4th row ≈ 30.2 yrs old

Variance = σ^2 = Sum of entries in bottom row ≈ 138.84

St. Deviation = $\sigma = \sqrt{138.84} \approx 11.78$ years

(b) According to the empirical rule approximately 68% of all male Hispanic workers fall in the interval

$[\mu - \sigma, \mu + \sigma] = [30.2 - 11.78, 30.2 + 11.78]$

$= [18.42, 41.98] \approx [18, 42]$

So, approximately 68% of all male Hispanic workers are 18–42 years old

39. Since the probability distribution is highly skewed, we need to use Chebyshev's rule.

$\mu = 2, \sigma = 0.15$

Following is a representation of the mean ± several standard deviations (each division is one standard deviation in width):

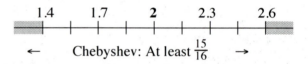

Chebyshev: At least $\frac{15}{16}$

The range $[1.4, 2.6]$ represents the interval $\mu \pm 4\sigma$, so by Chebyshev's rule, at least 15/16 of companies have a lifespan in this range. Therefore, *at most* 1/16 have a lifespan outside this range (the gray regions above). Since the distribution is not symmetric, we cannot conclude that half of the 1/16 is in the range on the right (companies at least as old as yours).

Therefore, all we can say is that at most 1/16, or 6.25% of all companies are in the range on the right (at least as old as yours).

41. Since the probability distribution is not known to be bell-shaped, we need to use Chebyshev's rule.

$\mu = 9, \sigma = 2$

Following is a representation of the mean ± several standard deviations (each division is one standard deviation in width):

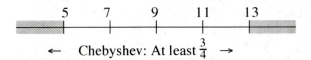

← Chebyshev: At least $\frac{3}{4}$ →

The range [5, 13] represents the interval $\mu \pm 2\sigma$, so by Chebyshev's rule, at least 3/4 of all Batmobiles have a lifespan in this range. Therefore, *at most* 1/4 have a lifespan outside this range (the gray regions above). Since the distribution is symmetric, at most half of these: 1/8 of all Batmobiles, have life spans more than 13 years (the gray region on the right).

Therefore, there is a*t most* (Choice B) a 12.5% chance that your new Batmobile will last 13 years or more.

43. (a) Take "success" to mean shopping at a mall. The given distribution is a binomial distribution with

$$n = 40, \ p = .63, \ q = 1 - .63 = .37$$
$$\mu = np = 40 \times .63 = 25.2 \text{ students}$$
$$\sigma = \sqrt{npq} = \sqrt{40 \times .63 \times .37} \approx 3.05$$

(b) Since the binomial distribution is symmetric and bell-shaped, we can use the Empirical rule, which says there is a 95% chance that that between

$$\mu - 2\sigma = 25.2 - 2(3.05) \approx 19$$

and

$$\mu + 2\sigma = 25.2 + 2(3.05) \approx 31$$

teenagers in the sample will shop at a mall during the next week.

Therefore, there is a 5% chance of this not happening—that is, a 5% chance that either 19 or fewer students will shop at a mall, or that 31 or more will.

Since the distribution is symmetric, there is a 2.5% chance that <u>31</u> or more students in the group will shop at a mall during the next week.

45. (a) Take "success" to mean *not* having a checking account. The given distribution is a binomial distribution with

$$n = 1000, \ p = 1 - .22 = .78, \ q = .22$$

$$\mu = np = 1000 \times .78 = 780$$
$$\sigma = \sqrt{npq} = \sqrt{1000 \times .78 \times .22} \approx 13.1$$

(b) Since the binomial distribution is symmetric and bell-shaped, we can use the Empirical rule, which says there is a 95% chance that that between $\mu - 2\sigma$ and $\mu + 2\sigma$ teenagers in the sample will not have checking accounts. These numbers, are respectively,

$$\mu - 2\sigma \approx 780 - 2(13.1) \approx 754$$
$$\mu + 2\sigma \approx 780 + 2(13.1) \approx 806$$

47. (a) Following is a worksheet computation of the variance (the bottom row gives the column sums):

	A	B	C	D
1	x	P(x)	x*P(x)	(x-mu)^2*P(x)
2	1	0.01	0.01	0.3025
3	2	0.04	0.08	0.81
4	3	0.04	0.12	0.49
5	4	0.08	0.32	0.5
6	5	0.1	0.5	0.225
7	6	0.15	0.9	0.0375
8	7	0.25	1.75	0.0625
9	8	0.2	1.6	0.45
10	9	0.08	0.72	0.5
11	10	0.05	0.5	0.6125
12	55	1	6.5	3.99

From the worksheet,

$$\mu = 6.5$$
$$\sigma^2 = 3.99 \approx 4.0$$
$$\sigma = \sqrt{\sigma^2} \approx 2.0$$

(b) According to Chebyshev's inequality, at least 3/4 or 75% of all supermarkets will have between

$$\mu - 2\sigma \approx 6.5 - 2(2.0) = 2.5$$

and

$$\mu + 2\sigma \approx 6.5 + 2(2.0) = 10.5$$

checkout lanes.

The smallest (whole) number of checkout lanes in this range is 3.

49. The household income of a poor family in the U.S. is

$$38,000 - 1.3(21,000) = \$10,700 \text{ or less.}$$

51. The household income of a rich family in the U.S. is

38,000 + 1.3(21,000) = $65,300 or more.

53. Cutoffs for poor families (1.3 standard deviations):

U.S.: 38,000 – 1.3(21,000) = $10,700
Canada: 35,000 – 1.3(17,000) = $12,900
Switzerland: 39,000 – 1.3(16,000) = $18,200
Germany: 34,000 – 1.3(14,000) = $15,800
Sweden: 32,000 – 1.3(11,000) = $17,700

The U.S. has the poorest households.

55. The gap between rich and poor is measured by 2×1.3 = 2.6 standard deviations. Since the U.S. has the largest standard deviation listed, it has the largest gap between rich and poor.

57. An income of $17,000 is 1 standard deviation below the U.S. mean income. By the empirical rule, approximately 68% earned within 1 standard deviation, so that approximately 32% earn outside the 1-standard deviation interval. Half of that, approximately 16%, earn less.

59. By the empirical rule, approximately 99.7% earned within 3 standard deviations of the mean. This is the range

34,000 – 3(14,000) = –8000

to

34,000 + 3(14,000) = 76,000

Since income can't be negative, the answer is 0–$76,000.

61. Using technology, μ = 12.56%, $\sigma \approx$ 1.8885%.

63. The empirical rule predicts approximately 68%. The 1-standard deviation interval based on the calculations in Exercise 61 is

$\mu - \sigma$ = 12.56 – 1.888 = 10.672
$\mu + \sigma$ = 12.56 + 1.888 = 14.448

The scores in that range are shown in bold:
6, 9, 10, 10, 10, **11, 11, 11, 11, 11, 11, 11,** 12, 12, **12, 12, 12, 12, 12, 12, 12,** 13, 13, 13, 13,

13, 13, 13, 13, 13, 13, 13, 13, 13, 13, **14, 14, 14, 14, 14, 14, 14,** 15, 15, 15, 15, 16, 18
These are 39 states, representing 39/50 = 78% of all states. This differs substantially from the empirical rule prediction. One reason for the discrepancy is that the associated probability distribution is roughly bell-shaped but not symmetric:

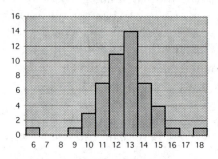

65. The 2-standard deviation interval based on the calculations in Exercise 61 is

$\mu - \sigma$ = 12.56 – 2(1.888) $\approx$ 8.78
$\mu + \sigma$ = 12.56 + 2(1.888) $\approx$ 16.34

The scores in that range are shown in bold:
6, **9, 10, 10, 10, 11, 11, 11, 11, 11, 11, 11, 12, 12, 12, 12, 12, 12, 12, 12, 12, 12, 13, 13, 13, 13, 13, 13, 13, 13, 13, 13, 13, 13, 13, 14, 14, 14, 14, 14, 14, 14, 15, 15, 15, 15, 16,** 18
These are 48 states, representing 48/50 = 96% of all states. Chebyshev's rule is valid, since it predicts that *at least* 75% of the scores are in this range.

67. (A) The graph shows standard deviations, not the actual power grid frequency, so we cannot conclude (A).
(B) The standard deviation indicates the variability in the power supply frequency. Since it was lower in mid-1999 than in 1995, this indicates greater stability in mid-1999 than in 1995, so the assertion is true.
(C) The standard deviation indicates the variability in the power supply frequency. Since it was higher in mid-2002 than in mid-1995, this indicates *less* stability in mid-2002, so we cannot conclude (C).

(D) The standard deviation was greatest in 2001–2002, indicating that the greatest fluctuations in the power grid frequency occurred during that period, so the assertion is true.

(E) Around January 1995, the standard deviation was closest to its average of 0.9, but was lower around January 1999, so the power grid was more stable around January 1999, than around January 1995. Thus, (E) is false.

69. The sample standard deviation is bigger; the formula for sample standard deviation involves division by the smaller term $n-1$ instead of n, which makes the resulting number larger.

71. The grades in the first class were clustered fairly close to 75. By Chebyshev's inequality, at least 88% of the class had grades in the range 60–90. On the other hand, the grades in the second class were widely dispersed. The second class had a much wider spread of ability than did the first class.

73. Since the standard deviation is 0, there is no variability at all; that is, the variable must be constant. Therefore, the variable must take on only the value 10, with probability 1.

75. Since there are 2 data points, the mean is midway between them, at a distance of $(y-x)/2$ from each. Summing the square of this distance twice gives

$$\Sigma(x_i - \mu)^2 = \frac{(y-x)^2}{4} + \frac{(y-x)^2}{4} = \frac{(y-x)^2}{2}$$

Dividing by 2 gives the variance:

$$\sigma^2 = \frac{(y-x)^2}{4}$$

Therefore, $\sigma = \frac{y-x}{2}$.

8.5 Normal Distributions

Note: Answers for Section 8.5 were computed using 4-digit tables, and may differ slightly from (more accurate) answers generated using technology.

1. We use the table.
$P(0 \le Z \le 0.5) = .1915$

0.4	0.1554	0.1591
0.5	0.1915	0.1950
0.6	0.2257	0.2291

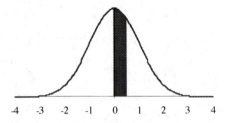

3. The table gives us
$P(0 \le Z \le 0.71) = .2611$

0.6	0.2257	0.2291
0.7	0.2580	0.2611
0.8	0.2881	0.2910

$P(-0.71 \le Z \le 0.71)$ is twice this area, as shown:

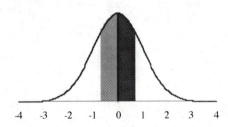

$P(-0.71 \le Z \le 0.71) = 2(.2611) = .5222$

5. The table gives us
$P(0 \le Z \le 1.34) = .4099$

1.2	0.3849	0.3869	0.3888	0.3907	0.3925
1.3	0.4032	0.4049	0.4066	0.4082	0.4099
1.4	0.4192	0.4207	0.4222	0.4236	0.4251

$P(0 \le Z \le 0.71) = .2611$

0.6	0.2257	0.2291	0.2324	0.2357	0.2389
0.7	0.2580	0.2611	0.2642	0.2673	0.2704
0.8	0.2881	0.2910	0.2939	0.2967	0.2995

$P(-0.71 \le Z \le 1.34)$ is obtained by adding these areas (see figure).

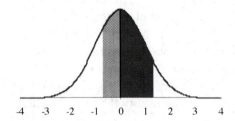

$P(-0.71 \le Z \le 1.34) = .4099 + .2611 = .6710$

7. From the table,
$P(0 \le Z \le 1.5) = .4332$
$P(0 \le Z \le 0.5) = .1915$
To obtain $P(0.5 \le Z \le 1.5)$, subtract the smaller from the larger (see figure).

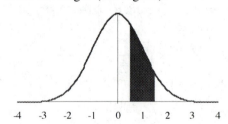

$P(0.5 \le Z \le 1.5) = .4332 - .1915 = .2417$

9. $\mu = 50$, $\sigma = 10$
Standardize the given problem:
$$P(a \le X \le b) = P\left(\frac{a-\mu}{\sigma} \le Z \le \frac{b-\mu}{\sigma}\right)$$

336

$P(35 \leq X \leq 65) =$
$$P\left(\frac{35-50}{10} \leq Z \leq \frac{65-50}{10}\right)$$
$$= P(-1.5 \leq Z \leq 1.5)$$
$$= 2(.4332) = .8664$$

11. $\mu = 50$, $\sigma = 10$
Standardize the given problem:
$$P(a \leq X \leq b) = P\left(\frac{a-\mu}{\sigma} \leq Z \leq \frac{b-\mu}{\sigma}\right)$$
$$P(30 \leq X \leq 62) = P\left(\frac{30-50}{10} \leq Z \leq \frac{62-50}{10}\right)$$
$$= P(-2 \leq Z \leq 1.2)$$
$$= .3849 + .4772 = .8621$$

13. $\mu = 100$, $\sigma = 15$
Standardize the given problem:
$$P(a \leq X \leq b) = P\left(\frac{a-\mu}{\sigma} \leq Z \leq \frac{b-\mu}{\sigma}\right)$$
$$P(110 \leq X \leq 130)$$
$$=$$
$$P\left(\frac{110-100}{15} \leq Z \leq \frac{130-100}{15}\right)$$
$$\approx P(0.67 \leq Z \leq 2)$$
$$= .4772 - .2486 = .2286$$

15. The Z-value measures the number of standard deviations form the mean, Therefore, the given problem translates to
$$P(-0.5 \leq Z \leq 0.5) = 2(.1915) = .3830$$

17. This is the probability that Z is either $> \frac{2}{3}$ or $<$ $-\frac{2}{3}$. The complement of this event is the event that
$$-\frac{2}{3} \leq Z \leq \frac{2}{3}$$
$$P(-0.67 \leq Z \leq 0.67) = 2(.2486) = .4972$$
Therefore, the desired probability is
$$1 - .4972 = .5028$$

19. $P(10 \leq X \leq 15) = P(9.5 \leq Y \leq 15.5)$
where Y has a mean of
$$\mu = np = 100 \times \frac{1}{6} \approx 16.6667$$
$$\sigma = \sqrt{npq} = \sqrt{100 \times \frac{1}{6} \times \frac{5}{6}} \approx 3.7268$$

We now standardize Y:
$$P(9.5 \leq Y \leq 15.5)$$
$$= P\left(\frac{9.5-16.6667}{3.7268} \leq Z \leq \frac{15.5-16.6667}{3.7268}\right)$$
$$\approx P(-1.92 \leq Z \leq -0.31)$$
$$= .4726 - .1217 = .3509 \approx .35$$

21. $P(X < 25) = P(0 \leq X \leq 24)$
$$= P(-0.5 \leq Y \leq 24.5)$$
where Y has a mean of
$$\mu = np = 200 \times \frac{1}{6} \approx 33.3333$$
$$\sigma = \sqrt{npq} = \sqrt{200 \times \frac{1}{6} \times \frac{5}{6}} \approx 5.2705$$
We now standardize Y:
$$P(-0.5 \leq Y \leq 24.5)$$
$$= P\left(\frac{-0.5-33.3333}{5.2705} \leq Z \leq \frac{24.5-33.3333}{5.2705}\right)$$
$$\approx P(-6.42 \leq Z \leq -1.68)$$
$$= .5000 - .4535 = .0465 \approx .05$$

23. $\mu = 500$, $\sigma = 100$
$$P(450 \leq X \leq 550)$$
$$= P\left(\frac{450-500}{100} \leq Z \leq \frac{550-500}{100}\right)$$
$$= P(-0.5 \leq Z \leq 0.5)$$
$$= .1915 + .1915 = .3830$$

25. $\mu = 500$, $\sigma = 100$
$$P(300 \leq X \leq 550)$$
$$= P\left(\frac{300-500}{100} \leq Z \leq \frac{550-500}{100}\right)$$
$$= P(-2 \leq Z \leq 0.5)$$
$$= .4772 + .1915 = .6687$$

27. $\mu = 100$, $\sigma = 16$
$$P(110 \leq X \leq 140)$$
$$= P\left(\frac{110-100}{16} \leq Z \leq \frac{140-100}{16}\right)$$
$$\approx P(0.63 \leq Z \leq 2.5)$$
$$= .4938 - .2357 = .2581,$$
approximately 26%

29. $\mu = 100$, $\sigma = 16$
$$P(X \geq 120) = P\left(Z \geq \frac{120-100}{16}\right)$$

$= P(Z \geq 1.25) = .5 - .3944 = .1056$
The total number of such people in the U.S. is
.1056×280,000,000 ≈ 29,600,000.

31. $\mu = 0.25$, $\sigma = 0.03$

$$P(X \geq 0.4) = P\left(Z \geq \frac{0.4-0.25}{0.03}\right)$$

$= P(Z \geq 5) = .5 - .500 = .0000$
The total number of such batters is therefore 0.

33. $\mu = 7.5$, $\sigma = 1$

$$P(X \geq 9) = P\left(Z \geq \frac{9-7.5}{1}\right)$$

$= P(Z \geq 1.5) = .5 - .4332 = .0668$
The total number of jars is therefore
.0668×100,000 ≈ 6680

35. $\mu = 38$, $\sigma = 21$ (in thousands of dollars)

$$P(X \geq 50) = P\left(Z \geq \frac{50-38}{21}\right)$$

$\approx P(Z \geq 0.57) = .5 - .2157 = .2843$;
approximately 28%

37. $\mu = 39$, $\sigma = 16$ (in thousands of dollars)

Very rich: $P(X \geq 100) = P\left(Z \geq \frac{100-39}{16}\right) \approx$

$P(Z \geq 3.81) = .5 - .5000 = 0$

Very poor: $P(X \leq 12) = P\left(Z \leq \frac{12-39}{16}\right)$

$\approx P(Z \leq -1.69) = .5 - .4545 = .0455$;
approximately 5%

39. U.S:
$\mu = 38$, $\sigma = 21$ (in thousands of dollars)

$$P(X \leq 12) = P\left(Z \leq \frac{12-38}{21}\right)$$

$\approx P(Z \leq -1.24) = .5 - .3925 = .1075$

Canada:
$\mu = 35$, $\sigma = 17$ (in thousands of dollars)

$$P(X \leq 12) = P\left(Z \leq \frac{12-35}{17}\right)$$

$\approx P(Z \leq -1.35) = .5 - .4115 = .0885$

The U.S. has a higher proportion of very poor
people.

41. Wechsler. Since this test has a smaller
standard deviation, a greater percentage of scores
fall within 20 points of the mean.

43. $\mu = 6$, $\sigma = 1$
The Z-value corresponding to $X = 1$ month is

$$Z = \frac{1-6}{1} = -5$$

This is surprising, because the time between
failures was more than 5 standard deviations
away from the mean, which happens with an
extremely small probability.

45. Task 1: $\mu = 11.4$, $\sigma = 5.0$

$$P(X \geq 10) = P\left(Z \geq \frac{10-11.4}{5}\right)$$

$\approx P(Z \geq -0.28) = .5 + .1103 = .6103$

47. Task 1: By Exercise 41,
$P(X_1 \leq 10) = .6103$

Task 2: $\mu = 11.9$, $\sigma = 9$

$$P(X_2 \geq 10) = P\left(Z \geq \frac{10-11.9}{9}\right)$$

$\approx P(Z \geq -0.21) = .5 + .0832 = .5832$

Since the times taken to complete the tasks are
independent,
$P(X_1 \leq 10 \text{ and } X_2 \leq 10)$
$= P(X_1 \leq 10) \times P(X_2 \leq 10)$
$= .6103 \times .5832 \approx .3559$

49. Tasks 1&2:
$\mu = 11.4 + 11.9 = 23.3$
$\sigma = \sqrt{5^2 + 9^2} \approx 10.2956$

$$P(X \geq 20) = P\left(Z \geq \frac{20-23.3}{10.2956}\right)$$

$\approx P(Z \geq -0.32) = .5 + .1255 = .6255$

51. $\mu = np = 800 \times .51 = 408$
$\sigma = \sqrt{npq} = \sqrt{800 \times .51 \times .49} \approx 14.139$
$\mu - 3\sigma \approx 366$
$\mu + 3\sigma \approx 450$

Since these are between 0 and $n = 800$, the normal approximation of the binomial distribution is valid.

$P(400 \le X \le 800)$

$= P(399.5 \le Y \le 800.5)$

$= P\left(\dfrac{399.5-408}{14.139} \le Z \le \dfrac{800.5-408}{14.139}\right)$

$\approx P(-0.6 \le Z \le 27.76) = .5 + .2257 = 0.7257$

53. $n = 100,000,000$, $p = .00000165$

$\mu = np = 165$

$\sigma = \sqrt{npq} = \sqrt{165 \times .99999835} \approx 12.845$

$\mu - 3\sigma \approx 126$

$\mu + 3\sigma \approx 204$

Since these are between 0 and $n = 100,000,000$, the normal approximation of the binomial distribution is valid.

$P(X < 180) = P(X \le 179)$

$= P(Y \le 179.5)$

$= P\left(Z \le \dfrac{179.5-165}{12.845}\right)$

$\approx P(Z \le 1.13)$

$= .3708 + .5 = .8708$

55. $n = 100,000,000$, $p = .00000165$

$\mu = np = 165$

$\sigma = \sqrt{npq} = \sqrt{165 \times .99999835} \approx 12.845$

$\mu - 3\sigma \approx 126$

$\mu + 3\sigma \approx 204$

Since these are between 0 and $n = 100,000,000$, the normal approximation of the binomial distribution is valid.

Suppose there are X crashes. Since 10 people buy insurance,

Payout $= 10 \times 1,000,000$ per crash; that is,

10,000,000X

Premium $= 10 \times 2 \times 100,000,000$

$= 2,000,000,000$

For break-even

$10,000,000X = 2,000,000,000$

$X = 200$ flights

To lose money, $X > 200$

$P(X > 200) = P(X \ge 201)$

$= P(Y \ge 200.5)$

$= P\left(Z \ge \dfrac{200.5-165}{12.845}\right)$

$\approx P(Z \ge 2.76)$

$= .5 - .4971 = .0029$

57. Let "success" = preferring Goode.

$p = .9 \times .55 + .1 \times .45 = .54$

$n = 1000$

$\mu = np = 540$

$\sigma = \sqrt{npq} = \sqrt{1000 \times .54 \times .46} \approx 15.761$

$\mu - 3\sigma \approx 493$

$\mu + 3\sigma \approx 587$

Since these are between 0 and $n = 1000$, the normal approximation of the binomial distribution is valid.

$P(X > 520) = P(X \ge 521) = P(Y \ge 520.5)$

$= P\left(Z \ge \dfrac{520.5-540}{15.761}\right) \approx P\ (Z \ge -1.25)$

$= .5 + .3925 = .8925$

59. $\mu = 100$, σ unknown.

Let the Z-score of someone who just barely qualifies be k.

To be in the top 2%, $P(Z \ge k) = .02$

Therefore, $P(0 \le z \le k) = .5 - .02 = .48$

Looking in the table for the closest score to .48 (which is .4798) gives

$k \approx 2.05$

Since $Z = \dfrac{148-100}{\sigma}$, we have

$k = 2.05 = \dfrac{148-100}{\sigma} = \dfrac{48}{\sigma}$

$2.05\sigma = 48$, so $\sigma = \dfrac{48}{2.05} \approx 23.4$

Note: A more accurate guess at k would be midway between 2.05 and 2..06. Using $k = 2.055$ gives the same answer to one decimal place.

61. Since the empirical rule is based on the normal distribution (see the remarks before Example 3 in the textbook), the empirical rule gives the exact results when the distribution is exactly normal.

63. Neither. they are equal, since they differ by $P(X = a)$, which is zero for a continuous random variable.

65. The total area under the curve must be equal to 1, and the area is a rectangle with width $(b-a)$. Therefore, its height must be $\frac{1}{b-a}$.

67. A normal distribution with standard deviation 0.5, since it is narrower near the mean, but must enclose the same amount of area as the standard curve, and so it must be higher.

Chapter 8 Review Exercises

1. X = the number of boys. X is a binomial random variable with $p = .5$, $n = 2$

$$P(X = 0) = C(2, 0)(.5)^0(.5)^2 = \tfrac{1}{4}$$

$$P(X = 1) = C(2, 1)(.5)^1(.5)^1 = \tfrac{1}{2}$$

$$P(X = 2) = C(2, 2)(.5)^2(.5)^0 = \tfrac{1}{4}$$

Probability distribution:

x	0	1	2
$P(X = x)$	$\frac{1}{4}$	$\frac{1}{2}$	$\frac{1}{4}$

Histogram:

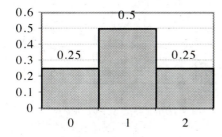

3. X = the sum of the two numbers when a four-sided dice is rolled twice

There are $4 \times 4 = 16$ possible outcomes

$X = 2$: Outcomes $\{(1, 1)\}$

$$P(X = 2) = \tfrac{1}{16}$$

$X = 3$: Outcomes $\{(1, 2), (2, 1)\}$

$$P(X = 3) = \tfrac{2}{16}$$

$X = 4$: Outcomes $\{(1, 3), (2, 2), (3, 1)\}$

$$P(X = 4) = \tfrac{3}{16}$$

$X = 5$: Outcomes $\{(1, 4), (2, 3), (3, 2), (4, 1)\}$

$$P(X = 5) = \tfrac{4}{16}$$

$X = 6$: Outcomes $\{(2, 4), (3, 3), (4, 2)\}$

$$P(X = 6) = \tfrac{3}{16}$$

$X = 7$: Outcomes $\{3, 4), (4, 3)\}$

$$P(X = 7) = \tfrac{2}{16}$$

$X = 8$: Outcomes $\{(4, 4)\}$

$$P(X = 2) = \tfrac{1}{16}$$

Probability distribution:

x	2	3	4	5	6	7	8
$P(X = x)$	$\frac{1}{16}$	$\frac{2}{16}$	$\frac{3}{16}$	$\frac{4}{16}$	$\frac{3}{16}$	$\frac{2}{16}$	$\frac{1}{16}$

Histogram:

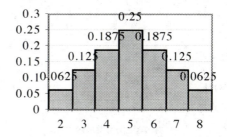

5. X = number of defective joysticks chosen when 3 are selected from a bin that contains 20 defective joysticks and 30 good ones

$$P(X = 0) = \frac{C(20,0)C(30,3)}{C(50,3)} \approx .2071$$

$$P(X = 1) = \frac{C(20,1)C(30,2)}{C(50,3)} \approx .4439$$

$$P(X = 2) = \frac{C(20,2)C(30,1)}{C(50,3)} \approx .2908$$

$$P(X = 3) = \frac{C(20,3)C(30,1)}{C(50,3)} \approx .0582$$

Histogram:

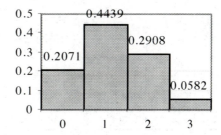

7. The first column lists the values of X, giving us the mean $\bar{x}$. The second column lists the numbers $(x_i - \bar{x})$, and the third column lists their squares.

x	$(x-\bar{x})$	$(x-\bar{x})^2$
-1	-3	9
2	0	0
0	-2	4
3	1	1
6	4	16
10	0	30

The bottom row shows the sums.

$$\bar{x} = \frac{10}{5} = 2 \qquad \sum_{i=1}^{n} (x_i-\bar{x})^2 = 30$$

$$s^2 = \frac{\sum_{i=1}^{n} (x_i-\bar{x})^2}{n-1} = \frac{30}{5-1} = 7.5$$

$$s = \sqrt{s^2} \approx 2.7386$$

For the median m, arrange the scores in order and select the middle score:

$-1, 0, 2, 3, 6$

$m = 2$

9. Two examples are:

0, 0, 0, 4

and

$-1, -1, 1, 5$

11. An example is $-1, -1, -1, 1, 1, 1$.
(Also see Exercise 68 in Section 8.4.)
Here is the calculation of the population standard deviation (the bottom row shows the sums):

x	$(x-\bar{x})$	$(x-\bar{x})^2$
-1	-1	1
-1	-1	1
-1	-1	1
1	1	1
1	1	1
1	1	1
0	0	6

$$\bar{x} = \frac{0}{6} = 0 \qquad \sum_{i=1}^{n} (x_i-\bar{x})^2 = 6$$

$$\sigma^2 = \frac{\sum_{i=1}^{n} (x_i-\bar{x})^2}{n} = \frac{6}{6} = 1$$

$$\sigma = \sqrt{\sigma^2} = 1$$

13–20. *To construct the probability distribution for the weighted die, take p to be the probability of a 1. The given information implies that the probability distribution for the die is*

x	1	2	3	4	5	6
$P(X = x)$	p	p	p	p	p	$2p$

Since $7p = 1$, we have $p = 1/7$, so the probability distribution for a single die is

x	1	2	3	4	5	6
$P(X = x)$	$\frac{1}{7}$	$\frac{1}{7}$	$\frac{1}{7}$	$\frac{1}{7}$	$\frac{1}{7}$	$\frac{2}{7}$

Throwing the die 4 times is a sequence of 4 Bernoulli trials with $p = 2/7$ and $q = 1-2/7 = 5/7$.

13. $P(X = 1) = C(4, 1) \left(\frac{2}{7}\right)^1 \left(\frac{5}{7}\right)^3 \approx .4165$

15. The probability that 6 comes up at most twice is

$P(X \le 2) = P(X = 0) + P(X = 1) + P(X = 2)$

$P(X = 0) = C(4, 0) \left(\frac{2}{7}\right)^0 \left(\frac{5}{7}\right)^4 \approx .2603$

$P(X = 1) = C(4, 1) \left(\frac{2}{7}\right)^1 \left(\frac{5}{7}\right)^3 \approx .4165$

$P(X = 2) = C(4, 2) \left(\frac{2}{7}\right)^2 \left(\frac{5}{7}\right)^2 \approx .2499$

Therefore,

$P(X \le 2) \approx .2603 + .4165 + .2499 = .9267$

17. The probability that X is more than 3 is

$P(X > 3) = P(X = 4)$

$= C(4, 4) \left(\frac{2}{7}\right)^4 \left(\frac{5}{7}\right)^0 \approx .0067$

19. $P(1 \le X \le 3) = P(X = 1) + P(X = 2) + P(X = 3)$

We computed $P(X = 1)$ and $P(X = 2)$ in Exercise 15.

$$P(X = 3) = C(4, 3) \left(\frac{2}{7}\right)^3 \left(\frac{5}{7}\right)^1 \approx .0666$$

Therefore,

$$P(1 \le X \le 3) = P(X = 1) + P(X = 2) + P(X = 3)$$
$$= .4165 + .2499 + .0666 = .7330$$

21. Think of the experiment as a sequence of 3 Bernoulli trials with "success" = girl, $n = 3$ and $p = .5$.

$$\mu = np \ 3(.5) = 1.5$$
$$\sigma = \sqrt{npq} = \sqrt{3 \times .5 \times .5} \approx 0.8660$$

To answer the last part,

$$[\mu - \sigma, \mu + \sigma] = [0.634, 2.366]$$

This interval does not include all 3.

$$[\mu - 2\sigma, \mu + 2\sigma] = [-0.232, 3.232]$$

This interval includes all the scores, so all values of X lie within 2 standard deviations of the expected value.

23. The frequencies add up to 16, so we obtain the probabilities by dividing the frequencies by 16:

x	$P(X = x)$	$xP(X = x)$
−3	.0625	−0.1875
−2	.125	−0.25
−1	.1875	−0.1875
0	.25	0
1	.1875	0.1875
2	.125	0.25
3	.0625	0.1875

$\mu =$ Sum of entries in right-hand column $= 0$.

$x - \mu$	$(x - \mu)^2$	$P(X = x) \times (x - \mu)^2$
−3	9	0.5625
−2	4	0.5
−1	1	0.1875
0	0	0
1	1	0.1875
2	4	0.5
3	9	0.5625

$\sigma^2 =$ Sum of entries in right-hand column $= 2.5$

$$\sigma = \sqrt{\sigma^2} \approx 1.5811$$

For the last part, note that 14 of the 16 scores in the frequency table (exclude the first and last value) are in the interval $[-2, 2]$; that is, in the interval

$$[\mu - 2, \mu + 2]$$

because $\mu = 0$. The number of standard deviations that 2 represents is approximately

$$\frac{2}{1.5811} \approx 1.3 \text{ standard deviations}$$

Therefore, 87.5% (or 14/16) of the time, X is within 1.3 standard deviations of the expected value.

25. By Chebyshev's rule, X is guaranteed to lie within k standard deviations with of μ with a probability of at least $1 - \dfrac{1}{k^2}$. Thus,

$$1 - \frac{1}{k^2} = .90$$

$$\frac{1}{k^2} = 1 - .90 = .10$$

$$k^2 = \frac{1}{.10} = 10$$

$$k = \sqrt{10} \approx 3.162$$

The associated interval is therefore

$$[\mu - k\sigma, \mu + k\sigma]$$
$$\approx [100 - 3.162(16), 100 + 3.162(16)]$$
$$\approx [49.4, 150.6]$$

27. Since X is bell-shaped and symmetric, the empirical rule applies so approximately 99.7% of samples of X are within the interval
  $[\mu - 3\sigma, \mu + 3\sigma]$.
This means that approximately $0.3/2 = 0.15\%$ of the samples of X are greater than $\mu + 3\sigma$
  $= 200 + 3(20) = 260$

29. From the table,
  $P(0 \le Z \le 1.5) = .4332$

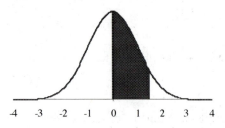

31. $P(|Z| \ge 2.1) = 2(.5 - .4821) = .0358$

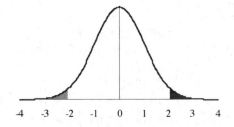

33. Standardize the given problem:
  $$P(X \le b) = P\!\left(Z \le \frac{b - \mu}{\sigma}\right)$$
  $$P(X \le -1) = P\!\left(Z \le \frac{-1 - 0}{2}\right)$$
  $$= P(Z \le -0.5) = .5 - .1915 = .3085$$

35. The frequency distribution for the price is

x	5.50	10	12	15
$fr(X = x)$	1	2	3	4

Dividing the frequencies by the sum, 10 gives the probability distribution. We also add a row for the computation of $E(X)$:

x	5.50	10	12	15
$P(X = x)$	.1	.2	.3	.4
$xP(X = x)$	0	0.2	0.4	0.3

$E(X) = $ Sum of entries in bottom row $= \$12.15$

37. Revenue = Price per copy sold $\times$ Number of copies sold. The values of the weekly revenue are obtained by multiplying the prices by the weekly sales:

x **(Revenue)**	34,100	35,000	36,000	15,000
$P(X = x)$	.1	.2	.3	.4
$xP(X = x)$	3410	7000	10,800	6000

$E(X) = $ Sum of entries in bottom row $=$
$\$27,210$

39. False; let $X =$ price and $Y =$ weekly sales. Then weekly revenue $= XY$. However, $27,210 \ne 12.15 \times 2620$. In other words, $E(XY) \ne E(X)E(Y)$.

41.(a) Take X to be the number of orders per million residents. For the values of X, use the rounded midpoints of the measurement classes given. To obtain the probabilities, divide each frequency by the sum of all the frequencies, 100:

x	2	4	6	8	10
Freq	25	35	15	15	10
$P(X = x)$	.25	.35	.15	.15	.10

To compute the expected value, add the $xP(X = x)$ row as usual:

x	2	4	6	8	10
$P(X = x)$	.25	.35	.15	.15	.10
$xP(X = x)$	0.5	1.4	0.9	1.2	1

$\mu = $ Sum of entries in bottom row $= 5$

$x-\mu$	−3	−1	1	3	5
$(x-\mu)^2$	9	1	1	9	25
$P(X=x)$					
$\times(x-\mu)^2$	2.25	0.35	0.15	1.35	2.5

σ^2 = Sum of entries in bottom row = 6.6

$\sigma = \sqrt{\sigma^2} \approx 2.5690$

(b) The empirical rule predicts that 68% of all orders will lie in the interval

$$[\mu-\sigma, \mu+\sigma] \approx [5-2.569, 5+2.569]$$
$$= [2.431, 7.569]$$

The empirical rule does not apply because the distribution is not symmetric. (However, it does give a fairly accurate prediction in this case.)

(c) The original frequency distribution is

x	1–2.9	3–4.9	5–6.9	7–8.9	9–10.9
Freq	25	35	15	15	10

The shaded cells correspond to the cities from which you obtain between 3 and 8 orders per million residents. These cells account for a total of 35% + 15% = 50% of the cities. Therefore, *at least* 50% of the cities gave orders between 3 and 8 per million residents.

However, we know more: The additional 15 cities in the 7–8.9 category could conceivably all be ≤ 8, so we can conclude that *up to* 50% + 15% = 65% of the cities gave orders between 3 and 8 per million residents.

Conclusion: Between 50% and 65% of the cities gave orders between 3 and 8 per million residents [choice (A)].

43–48. Let X be the number of Mac OS users who will order books in the next hour, and let Y be the number of Windows users who will order books in the next hour. X and Y are both binomial random variables with the following properties:
  X: n = 10, p = .05
  Y: n = 20, p = .10

43. $P(Y = 3) = C(20, 3)(.10)^3(.90)^{17} \approx .190$

45. $P(X = 1) = C(10, 1)(.05)^1(.95)^9 \approx .3151$
$P(Y = 3) = C(20, 3)(.10)^3(.90)^{17} \approx .1901$
By independence,
$P(X = 1 \text{ and } Y = 3) = P(X = 1)P(Y = 3)$
$= .3151 \times .1901 \approx .060$

47. $E(X) = np = (10)(.05) = 0.5$

49. Skin cream: $\mu = 38$, $\sigma = 21$ (in thousands of dollars)
$$P(X \geq 50) = P\left(Z \geq \frac{50-38}{21}\right) \approx P(Z \geq 0.571)$$
$$= .5 - .2157 \approx .284$$

51. Skin cream: $\mu = 38$, $\sigma = 21$ (in thousands of dollars)
$$P(X \leq 12) = P\left(Z \leq \frac{12-38}{21}\right) \approx P(Z \leq -1.238) = .5 - .3925 \approx .108$$

53. People in the 3 Sigma Club have IQs with Z-values of at least 3.
$P(Z \geq 3) \approx .5 - .4987 = .0013$
Given a population of 280 million, this gives
$.0013 \times 280,000,000 = 364,000$
people. (A more accurate answer using technology is 378,000 people.)

55. Given $\mu = 100$, $\sigma = 16$, to get into the 3 Sigma club, your IQ must be at least 3 standard deviations above the mean; that is
$\mu + 3\sigma = 100 + 3(16) = 148$

Chapter 9 Nonlinear Functions and Models

9.1 Quadratic Functions and Models

1. $f(x) = x^2 + 3x + 2$
$a = 1$, $b = 3$, $c = 2$; $-b/(2a) = -3/2$;
$f(-3/2) = -1/4$, so: vertex: $(-3/2, -1/4)$
y intercept $= c = 2$
$x^2 + 3x + 2 = (x + 1)(x + 2)$, so:
x intercepts: $-2, -1$
$a > 0$ so the parabola opens upward

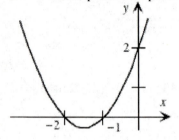

3. $f(x) = -x^2 + 4x - 4$
$a = -1$, $b = 4$, $c = -4$; $-b/(2a) = 2$;
$f(2) = 0$, so: vertex: $(2,0)$
y intercept $= c = -4$
$-x^2 + 4x - 4 = -(x - 2)^2$, so:
x intercept: 2
$a < 0$ so the parabola opens downward

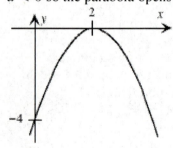

5. $f(x) = -x^2 - 40x + 500$
$a = -1$, $b = -40$, $c = 500$; $-b/(2a) = -20$;
$f(-20) = 900$, so: vertex: $(-20, 900)$
y intercept $= c = 500$
$-x^2 - 40x + 500 = -(x + 50)(x - 10)$, so:
x intercepts: $-50, 10$
$a < 0$ so the parabola opens downward

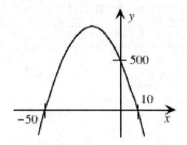

7. $f(x) = x^2 + x - 1$
$a = 1$, $b = 1$, $c = -1$; $-b/(2a) = -1/2$;
$f(-1/2) = -5/4$, so: vertex: $(-1/2, -5/4)$
y intercept $= c = -1$
from the quadratic formula:
x intercepts: $-1/2 \pm \sqrt{5}/2$
$a > 0$ so the parabola opens upward

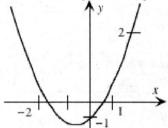

9. $f(x) = x^2 + 1$
$a = 1$, $b = 0$, $c = 1$; $-b/(2a) = 0$;
$f(0) = 1$, so: vertex: $(0, 1)$
y intercept $= c = 1$
$b^2 - 4ac = -4 < 0$, so no x intercept
$a > 0$ so the parabola opens upward

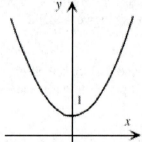

346

11. $q = -4p + 100$
$R = pq = p(-4p + 100) = -4p^2 + 100p$;
maximum revenue when $p = -b/(2a) = \$12.50$

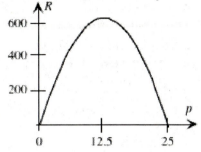

13. $q = -2p + 400$
$R = pq = p(-2p + 400) = -2p^2 + 400p$
maximum revenue when $p = -b/(2a) = \$100$

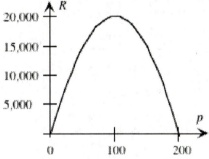

15. $y = -0.7955x^2 + 4.4591x - 1.6000$

17. $y = -1.1667x^2 - 6.1667x - 3$

19. (a) Positive because the data suggest a curve that is concave up.
(b) The data suggest a parabola that is concave up (a positive) and with y-intercept around 1150. Only choice **(C)** has both properties.
(c) $-b/(2a) = 80/10 = 8$, which is in 1998. Extrapolating in the positive direction leads one to predict more and more steeply rising military expenditure, which may or may not occur; extrapolating in the negative direction predicts more and more steeply increasing military expenditure as we go back in time, contradicting history.

21. The given function $I(t) = -0.015t^2 + 0.1t + 1.4$ is quadratic with graph a concave down

parabola (as $a = -0.015$ is negative). Thus its vertex is the highest point on the graph, and occurs when

$$t = \frac{-b}{2a} = \frac{0.1}{0.03} \approx 3.3,$$

partway through 2003. At that time, imports were about

$$I(3.3) = -0.015(3.3)^2 + 0.1(3.3) + 1.4$$
$$\approx 1.6 \text{ million barrels/day}$$

23. $-b/(2a) = 5000$ pounds. The model is not trustworthy for vehicle weights larger than 5000 pounds because it predicts increasing fuel economy with increasing weight. Also, 5000 is close to the upper limit of the domain of the function.

25. $q = -0.5p + 140$.
Revenue is $R = pq = -0.5p^2 + 140p$.
Maximum revenue occurs when
$p = -b/(2a) = \$140$; the corresponding revenue is $R = \$9800$.

27. The given data points are $(p, q) = (40, 200{,}000)$ and $(60, 160{,}000)$. The line passing through these points is $q = -2000p + 280{,}000$. Revenue is $R = pq = -2000p^2 + 280{,}000p$. Maximum revenue occurs when $p = -b/(2a) = 70$ houses; the corresponding revenue is $R = \$9{,}800{,}000$.

29. (a) A linear demand function has the form
$\quad q = mp + b$.
(x is the price p, and y is the demand q). We are given two points on its graph: $(3, 28{,}000)$ and $(5, 19{,}000)$.
Slope:

$$m = \frac{q_2 - q_1}{p_2 - p_1} = \frac{19{,}000 - 28{,}000}{5 - 3} = \frac{-9000}{2}$$
$$= -4500$$

Intercept:

$$b = q_1 - mp_1 = 28000 - (-4500)(3)$$
$$= 28{,}000 + 13{,}500 = 41{,}500$$

Thus, the demand equation is
$q = mp + b$
$q = -4500p + 41{,}500$

347

(b) Revenue
$$R = pq = p(-4500p + 41,500)$$
$$= -4500p^2 + 41,500p$$

For maximum revenue,
$$p = \frac{-b}{2a} = \frac{-41,500}{-9000} \approx \$4.61$$

The corresponding daily revenue is
$$R = -4500(4.61)^2 + 41,500(4.61)$$
$$= \$95,680.55$$

(c) The maximum annual revenue the company could have earned was
$$95,680.55 \times 365 \approx \$34,923,400,$$
which is about $10 million short of what it needed to break even. Therefore, it would not have been possible to break even.

31. (a) The data points are
$(x, q) = (2, 280)$ and $(1.5, 560)$.
Thus, $q = -560x + 1400$ and
$R = xq = -560x^2 + 1400x$.
(b) $P = R - C = -560x^2 + 1400x - 30$. The largest monthly profit occurs when
$x = -b/(2a) = \$1.25$ and then
$P = \$845$ per month.

33. As a function of q,
$C = 0.5q + 20$.
Substituting $q = -400x + 1200$, we get
$C = 0.5(-400x + 1200) + 20$
$$= -200x + 620.$$
The profit is
$P = R - C = xq - C$
$$= -400x^2 + 1400x - 620.$$
The profit is largest when
$x = -b/(2a) = \$1.75$ per log on; the corresponding profit is
$P = \$605$ per month.

35. (a) The data points are
$(p, q) = (10, 300)$ and $(15, 250)$,
so $q = -10p + 400$.
(b) $R = pq = -10p^2 + 400p$.
(c) $C = 3q + 3000 = 3(-10p + 400) + 3000$
$= -30p + 4200$

(d) $P = R - C = -10p^2 + 430p - 4200$. The maximum profit occurs when
$$p = -b/(2a) = \$21.50.$$

37. Here is the Excel tabulation of the data, together with the scatter plot and the quadratic (polynomial order 2) trendline (with the option "Display equation on chart" checked):

◇	A	B	C
1	t	C	
2	4	900	
3	8	800	
4	16	1200	
5			
6			
7			
8			
9			
10			
11			
12			
13			
14			
15			

From the trendline, the quadratic model is
$$f(t) = 6.25t^2 - 100t + 1200.$$
To estimate world military expenditure in 2008, substitute the corresponding value $t = 18$, to obtain
$$f(18) = 6.25(18)^2 - 100(18) + 1200 = 1425,$$
representing $1425 billion. The actual figure (from Exercise 19) is $1450 billion, so the predicted value is $25 billion lower than the actual value.

39. (a) Here is the Excel tabulation of the data, together with the scatter plot and the quadratic (polynomial order 2) trendline (with the option "Display equation on chart" checked):

	A	B
1	t	S
2	2	270
3	3	1119
4	4	2315
5	5	1703
6	6	717
7		
8		
9		
10	$y = -391.29x^2 +$	
11	$3278.1x - 4844.4$	
12		
13		
14		
15		
16		
17		
18		
19		
20		
21		

We round the coefficients to the nearest whole number

$$S(t) = -391t^2 + 3278t - 4844$$

(b) To predict third quarter of 2008 ($t = 7$) compute

$$S(7) = -391(7)^2 + 3278(7) - 4844$$
$$= -1057 \text{ thousand units}$$

Clearly this makes no sense, and shows the danger of extrapolating mathematical models beyond their domains. Actual sales in the third quarter of 2008 were 6.9 million.

41. If $a = 0$, then $f(x) = bx + c$, a linear function, so its graph is a straight line.

43. Since the curve is concave up, a is positive. Since the y-intercept is negative, c is negative. Hence the correct choice is (C).

45. Positive; The x-coordinate of the vertex is negative, so $-b/2a$ must be negative. Since a is positive (the parabola is concave up), this means that b must also be positive to make $-b/2a$ negative.

47. The x coordinate of the vertex represents the unit price that leads to the maximum revenue, the y coordinate of the vertex gives the maximum possible revenue, the x intercepts give the unit prices that result in zero revenue, and the y intercept gives the revenue resulting from zero unit price (which is obviously zero).

49. Graph the data to see whether the points suggest a curve rather than a straight line. If the curve suggested by the graph is concave up or concave down, then a quadratic model would be a likely candidate.

51. No; The graph of a quadratic function is a parabola. In the case of a concave-up parabola, the curve would unrealistically predict sales increasing extremely fast and becoming unrealistically large in the future. In the case of a concave-down parabola, the curve would predict "negative" sales from some point on.

53. If $q = mp + b$ (with $m < 0$), then the revenue is given by $R = pq = mp^2 + bp$. This is the equation of a parabola with $a = m < 0$, and so is concave down. Thus, the vertex is the highest point on the parabola, showing that there is a single highest value for R, namely the y coordinate of the vertex.

55. Because $R = pq$, the demand must be given by

$$q = \frac{R}{p} = \frac{-50p^2 + 60p}{p} = -50p + 60.$$

349

9.2 Exponential Functions and Models

1. `4^x`

x	-3	-2	-1	0	1	2	3
$f(x)$	$\frac{1}{64}$	$\frac{1}{16}$	$\frac{1}{4}$	1	4	16	64

3. `3^(-x)`

x	-3	-2	-1	0	1	2	3
$f(x)$	27	9	3	1	$\frac{1}{3}$	$\frac{1}{9}$	$\frac{1}{27}$

5. `2*2^x or 2*(2^x)`

x	-3	-2	-1	0	1	2	3
$f(x)$	$\frac{1}{4}$	$\frac{1}{2}$	1	2	4	8	16

7. `-3*2^(-x)`

x	-3	-2	-1	0	1	2	3
$f(x)$	-24	-12	-6	-3	$-\frac{3}{2}$	$-\frac{3}{4}$	$-\frac{3}{8}$

9. `2^x-1`

x	-3	-2	-1	0	1	2	3
$f(x)$	$-\frac{7}{8}$	$-\frac{3}{4}$	$-\frac{1}{2}$	0	1	3	7

11. `2^(x-1)`

x	-3	-2	-1	0	1	2	3
$f(x)$	$\frac{1}{16}$	$\frac{1}{8}$	$\frac{1}{4}$	$\frac{1}{2}$	1	2	4

13.

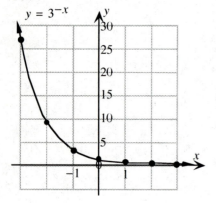

$y = 3^{-x}$

15.

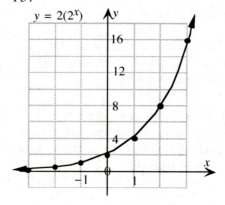

$y = 2(2^x)$

17.

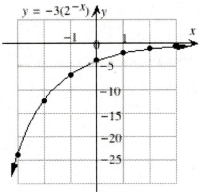

$$y = -3(2^{-x})$$

19.

x	−2	−1	0	1	2
f(x)	0.5	1.5	4.5	13.5	40.5
g(x)	8	4	2	1	$\frac{1}{2}$

For every increase in x by one unit, the value of f is multiplied by 3, so f is exponential.
Since f(0) = 4.5, the exponential model is f(x) = 4.5(3^x).
For every increase in x by one unit, the value of g is multiplied by 1/2, so g is exponential.
Since g(0) = 2, the exponential model is g(x) = 2(1/2)^x, or 2(2^{-x})

21.

x	−2	−1	0	1	2
f(x)	22.5	7.5	2.5	7.5	22.5
g(x)	0.3	0.9	2.7	8.1	16.2

When x increases from −1 to 0, the value of f is multiplied by 1/3, but when x is increased from 0 to 1, the value of f is multiplied by 3. So f is not exponential.
When x increases from −1 to −0, the value of g is multiplied by 3, but when x is increased from 1 to 2, the value of g is multiplied by 2. So g is not exponential.

23.

x	−2	−1	0	1	2
f(x)	100	200	400	600	800
g(x)	100	20	4	0.8	0.16

The values of f(x) double for every one-unit increase in x except when x increases from 1 to 2, when the value of f is multiplied by 4/3. Hence f is not linear.
When g is multiplied by 0.2 for every increase by one unit in x, so g is exponential.
Since g(0) = 4, the exponential model is g(x) = 4(0.2)^x.

25. e^(-2*x) or EXP(-2*x)

x	−3	−2	−1	0	1	2	3
f(x)	403.4	54.60	7.389	1	0.1353	0.01832	0.002479

27. 1.01*2.02^(-4*x)

x	−3	−2	−1	0	1	2	3
f(x)	4662	280.0	16.82	1.01	0.06066	0.003643	0.0002188

29. 50*(1+1/3.2)^(2*x)

x	−3	−2	−1	0	1	2	3
f(x)	9.781	16.85	29.02	50	86.13	148.4	255.6

The following solutions also show some common errors you should avoid.

31. `2^(x-1)` *not* `2^x-1`

33. `2/(1-2^(-4*x))` *not* `2/1-2^-4*x` *and not* `2/1-2^(-4*x)`

35. `(3+x)^(3*x)/(x+1)` *or* `((3+x)^(3*x))/(x+1)` *not* `(3+x)^(3*x)/x+1` *and not* `(3+x^(3*x))/(x+1)`

37. `2*e^((1+x)/x)` *or* `2*EXP((1+x)/x)` *not* `2*e^1+x/x` *and not* `2*e^(1+x)/x` *and not* `2*EXP(1+x)/x`

In the following solutions, f_1 is black and f_2 is gray.

39.

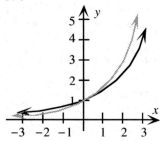

41. (Note that the *x*-axis shown here crosses at 200, not 0.)

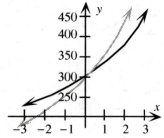

43.

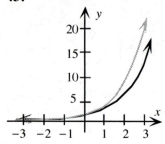

45. (Note that the *x*-axis shown here crosses at 900.)

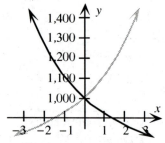

47. Each time *x* increases by 1, $f(x)$ is multiplied by 0.5. Also, $f(0) = 500$. So, $f(x) = 500(0.5)^x$.

49. Each time *x* increases by 1, $f(x)$ is multiplied by 3. Also, $f(0) = 10$. So, $f(x) = 10(3)^x$.

51. Each time *x* increases by 1, $f(x)$ is multiplied by $225/500 = 0.45$. Also, $f(0) = 500$. So, $f(x) = 500(0.45)^x$.

53. Write $f(x) = Ab^x$. We have $Ab^1 = -110$ and $Ab^2 = -121$. Dividing, $b = -121/(-110) = 1.1$. Substituting, $A(1.1) = -110$, so $A = -100$. Thus, $f(x) = -100(1.1)^x$.

55. We want an equation of the form $y = Ab^x$. Substituting the coordinates of the given points gives

$$36 = Ab^2$$
$$324 = Ab^4$$

Dividing the second equation by the first gives

$$324/36 = 9 = b^2$$

so $b = 3$.
Substituting into the first equation now gives

$36 = A(3)^2 = 9A$

so $A = 36/9 = 4$.

Hence the model is

$$y = Ab^x = 4(3^x)$$

57. We want an equation of the form $y = Ab^x$. Substituting the coordinates of the given points gives

$$-25 = Ab^{-2}$$
$$-0.2 = Ab$$

Dividing the second equation by the first gives

$$-0.2/(-25) = 0.008 = b^3$$

so $b = 0.2$.

Substituting into the second equation now gives

$$-0.2 = 0.2A$$

so $A = -1$.

Hence the model is

$$y = Ab^x = -1(0.2^x)$$

59. Write $f(x) = Ab^x$. We have $Ab^1 = 3$ and $Ab^3 = 6$. Dividing, $b^2 = 6/3 = 2$, so $b = \sqrt{2} \approx 1.4142$. Substituting, $A\sqrt{2} = 3$, so $A = 3/\sqrt{2} \approx 2.1213$. Thus, $y = 2.1213(1.4142^x)$.

61. Write $f(x) = Ab^x$. We have $Ab^2 = 3$ and $Ab^6 = 2$. Dividing, $b^4 = 2/3$, so $b = \sqrt[4]{2/3} \approx 0.9036$. Substituting, $A\left(\sqrt[4]{2/3}\right)^2 = 3$, so $A = 3/\left(\sqrt[4]{2/3}\right)^2 \approx 3.6742$. Thus, $y = 3.6742(0.9036^x)$.

63. Use the formula $f(t) = Pe^{rt}$ with $P = 5000$ and $r = 0.10$, giving $f(t) = 5000e^{0.10t}$.

65. Use the formula $f(t) = Pe^{rt}$ with $P = 1000$ and $r = -0.063$, giving $f(t) = 1000e^{-0.063t}$.

67. $y = 1.0442(1.7564)^x$

69. $y = 15.1735(1.4822)^x$

71. We want a model of the form $f(t) = Ab^x$ We are given two points on the graph: $(0, 300)$ and $(2, 75)$. Substituting the coordinates, we get

$$300 = Ab^0 = A \quad \text{Substitute } (0, 300)$$
$$75 = Ab^2 \quad \text{Substitute } (2, 75)$$

This gives

$$A = 300$$
$$75 = 300b^2$$

so $b^2 = 75/300 = 0.25$

giving $b = 0.25^{1/2} = 0.5$,

and the model is

$$f(t) = 300(0.5)^t.$$

After 5 hours,

$$f(5) = 300(0.5)^5 = 9.375 \text{ mg}$$

73. (a) Linear Model: $F = mt + n$

Points: $(4, 200)$ and $(10, 590)$

Slope: $m = \dfrac{F_2 - F_1}{t_2 - t_1}$

$\qquad = \dfrac{590 - 200}{10 - 4} = 65$

Intercept: $b = F_1 - mt_1$

$\qquad = 200 - (65)(4) = -60$

Model: $F = mt + b = 65t - 60$

Exponential Model: $F = Ab^t$

Substitute $(4, 200)$: $200 = Ab^4$

Substitute $(10, 590)$: $590 = Ab^{10}$

Now divide the second equation by the first:

$$\frac{590}{200} = \frac{b^{10}}{b^4} = b^6$$
$$2.95 = b^6$$
$$b = (2.95)^{1/6} \approx 1.19758 \approx 1.20$$

Substituting the value of b in the first equation now gives

$$200 \approx A(1.19758)^4 \approx 2.0569A$$

$$\text{So } A \approx \frac{200}{2.0569} \approx 97.2$$

Model: $F = Ab^t = 97.2(1.20)^t$

The exponential model is applicable: The successive ratios of the values of C are not too far from $b^2 \approx 1.44$. The successive *differences* climb steadily, making a linear model not appropriate.

(b) The year 2008 corresponds to $t = 8$, and so
$F = Ab^t = 97.2(1.20)^8 \approx 418$ tons,
not too far from the projected figure.

75. (a) Let $P = Ab^t$ million people. We are given the data points $(0, 180)$ and $(48, 303)$.

Substituting them into the equation $P = Ab^t$ gives

$A = 180$

$303 = 180\, b^{48}$

so $b^{48} = 303/180$, which gives

$b = (303/180)^{1/48} \approx 1.01091$.

Thus, the model is

$P(t) = 180(1.01091)^t$ million people.

(b) Rounding to 4 significant digits gives
$180(1.011)^t$. Putting $t = 48$ *gives*

$P(48) = 180(1.011)^{48} \approx 304 \neq 303$.

Rounding to 5 significant digits gives
$180(1.0109)^t$. Putting $t = 48$ *gives*

$P(48) = 180(1.0109)^{48} \approx 303$,

which is accurate to 3 significant digits. Therefore, we should round to at least 5 significant digits.

(c) When $t = 60$, $P = 180(1.01091)^{60} \approx 345$ million people.

77. (a) Let y be the number of frogs in year t, with $t = 0$ representing two years ago; we seek a model of the form $y = Ab^t$. We are given the initial value of $A = 50{,}000$ and are told that $50{,}000b^2 = 75{,}000$. This gives $b = (75{,}000/50{,}000)^{1/2} = 1.5^{1/2}$. Thus, $y = 50{,}000(1.5^{1/2})^t = 50{,}000(1.5^{t/2})$. **(b)** When $t = 3$, $y = 50{,}000(1.5)^{3/2} = 91{,}856$ tags.

79. If y represents the size of the culture at time t, then $y = Ab^t$. We are told that the initial size is 1000, so $A = 1000$. We are told that the size doubles every 3 hours, so $b^3 = 2$, or $b = 2^{1/3}$. Thus, $y = 1000(2^{1/3})^t = 1000(2^{t/3})$. There will be $1000(2^{48/3}) = 65{,}536{,}000$ bacteria after 2 days.

81. The desired model is $A(t) = Ab^t$.
At time $t = 0$ (March 17, 2003) the number of cases was 167, so $A = 167$.
Since the number was increasing by 18% each day, the number of cases is multiplied by 1.18 each day, so $b = 1.18$. Hence, the model is

$A(t) = 167(1.18)^t$.

Since March 31, 2003 corresponds to $t = 14$, the number of cases was about

$A(14) = 167(1.18)^{14} \approx 1695$

83. Apply the formula

$A(t) = P\left(1 + \dfrac{r}{m}\right)^{mt}$

with $P = 5000$, $r = 0.0494$, and $m = 12$. We get the model

$A(t) = 5000(1 + 0.0494/12)^{12t}$

At the end of 2014 ($t = 7$), the deposit would be worth $5000(1 + 0.0494/12)^{12(7)} \approx \$7061$

85. From the answer to Exercise 83, the value of the investment after t years is

$A(t) = 5000(1 + 0.0494/12)^{12t}$.

TI-83/84 Plus: Enter

`Y1 = 5000*(1+0.0494/12)^(12*X),`

Press [2nd] [TBLSET], and set `Indpnt` to Ask. (You do this once and for all; it will permit you to specify values for x in the table screen.) Then, press [2nd] [TABLE], and you will be able to evaluate the function at several values of x. Here are some values of x and the resulting values of Y_1.

x	Y₁
1	5252.67
2	5518.11
3	5796.96
4	6089.9
5	6397.65
6	6720.95
7	7060.59
8	7417.39
9	7792.22
10	8185.99

354

Notice that Y_1 first exceeds 7500 when $x = 9$. Since $x = 0$ represents the beginning of 2007, $x = 9$ represents the start of 2016, so the investment will first exceed \$7500 at the beginning of 2016.

87. From the continuous compounding formula, $1000e^{0.04 \times 10} = \$1491.82$, of which \$491.82 is interest.

89. We use the continuous compounding formula $A = Pe^{rt}$. Since sales were depreciating, we use
$$r = -0.3$$
At time $t = 0$ (2005) sales were 1.3 million, so
$$A = 1.3$$
Hence the model is
$$A(t) = 1.3e^{-0.3t}$$
2008: $t = 3$
$$A(3) = 1.3e^{-0.3(3)} \approx 0.53 \text{ million}$$
2009: $t = 4$
$$A(4) = 1.3e^{-0.3(4)} \approx 0.39 \text{ million}$$

91. (a)

year	1950	2000	2050	2100
$C(t)$ **parts per million**	561	669	799	953

(b) Testing the decades between 2000 ($t = 250$) and 2050, we find that $C(260) \approx 694$ and $C(270) \approx 718$. Testing individual years between $t = 260$ and $t = 270$, we find that the level surpasses 700 parts per million for the first time when $t = 263$. Thus, to the nearest decade, the level passes 700 in 2010 ($t = 260$).

93. (a) Here is the Excel tabulation of the data, together with the scatter plot and the quadratic (polynomial order 2) trendline (with the option "Display equation on chart" checked):

◇	A	B	C
1	t	c	
2	0	0.38	
3	2	0.4	
4	4	0.6	
5	6	0.95	
6	8	1.2	
7	10	1.6	
8			
9			
10			
11			
12			
13			
14			
15			
16			
17			
18			
19			

Note that the displayed model has the form $P(t) = 0.3394e^{0.1563t}$. To express this in the form $A(b)^t$ we write
$$0.339e^{0.1563t} = 0.339(e^{0.1563})^t$$
$$\approx 0.339(1.169)^t.$$
so the model is
$$P(t) = 0.339(1.169)^t.$$
(b) In 2005, $t = 11$, so the predicted cost is
$$P(15) = 0.339(1.169)^{11}$$
$$\approx \$1.9 \text{ million}$$

95. (a) Here is the Excel tabulation of the data, together with the scatter plot and the quadratic (polynomial order 2) trendline (with the option "Display equation on chart" checked):

	A	B
1	t	n
2	0	1
3	0.5	2
4	1	5.5
5	1.5	7
6	2	12
7	2.5	30
8	3	58
9	3.5	80
10		

Note that the displayed model has the form $n(t) = 1.1274e^{1.2652t}$. To express this in the form $A(b)^t$ we write

$$1.1274e^{1.2652t} = 1.1274(e^{1.2652})^t.$$

$$\approx 1.1274(3.5438)^t$$

which rounds to $1.127(3.544)^t$

and so the model is

$$n(t) = 1.127(3.544)^t$$

(b) At the start of 2009, $t = 4$, so the predicted membership is

$$n(4) = 1.127(3.544)^4 \approx 178 \text{ million}$$

members.

97. (B) An exponential function eventually becomes larger than any polynomial.

99. Exponential functions of the form $f(x) = Ab^x$ ($b > 0$) increase rapidly for large values of x. In real-life situations, such as population growth, this model is reliable only for relatively short periods of growth. Eventually, population growth

tapers off because of pressures such as limited resources and overcrowding.

101. The article was published about a year before the "housing bubble" burst in 2006, whereupon house prices started to fall, contrary to the prediction of the graph, as documented in Exercise 90. This shows the danger of using any mathematical model to extrapolate. The blogger was, however, cautious in the choice of words, claiming only to be estimating what the future U.S. median house price "might be."

103. Linear functions are better for cost models where there is a fixed cost and a variable cost and for simple interest, where interest is paid only on the original amount invested. Exponential models are better for compound interest and population growth. In both of these latter examples, the rate of growth depends on the current number of items, rather than on a fixed initial quantity.

105. Take the ratios y_2/y_1 and y_3/y_2. If they are the same, the points fit on an exponential curve.

107. This reasoning is suspect—the bank need not use its computer resources to update all the accounts every minute, but can instead use the continuous compounding formula to calculate the balance in any account at any time as needed.

9.3 Logarithmic Functions and Models

1.

Exponential form	$10^4 = 10,000$	$4^2 = 16$	$3^3 = 27$	$5^1 = 5$	$7^0 = 1$	$4^{-2} = \frac{1}{16}$
Logarithmic form	$\log_{10} 10,000 = 4$	$\log_4 16 = 2$	$\log_3 27 = 3$	$\log_5 5 = 1$	$\log_7 1 = 0$	$\log_4 \frac{1}{16} = -2$

3.

Exponential form	$(0.5)^2 = 0.25$	$5^0 = 1$	$10^{-1} = 0.1$	$4^3 = 64$	$2^8 = 256$	$2^{-2} = \frac{1}{4}$
Logarithmic form	$\log_{0.5} 0.25 = 2$	$\log_5 1 = 0$	$\log_{10} 0.1 = -1$	$\log_4 64 = 3$	$\log_2 256 = 8$	$\log_2 \frac{1}{4} = -2$

5. $x = \log_3 5 = \log 5 / \log 3 = 1.4650$

7. $-2x = \log_5 40 = \log 40 / \log 5$, so $x = -(\log 40 / \log 5)/2 = -1.1460$

9. $e^x = 2/4.16$, so $x = \ln(2/4.16) = -0.7324$

11. $1.06^{2x+1} = 11/5$, so $2x + 1 = \log_{1.06}(11/5)$, $x = (\log(11/5)/\log(1.06) - 1)/2 = 6.2657$

13. $f(x) = \log_4 x$

We plot some points and draw the graph:

x	$y = \log_4(x)$
1/16	-2
1/4	-1
1/2	$-1/2$
1	0
2	1/2
4	1

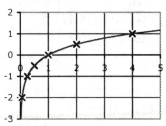

15. $f(x) = \log_4(x-1)$

We plot some points and draw the graph:

x	$y = \log_4(x-1)$
17/16	-2
5/4	-1
3/2	$-1/2$
2	0
3	1/2
5	1

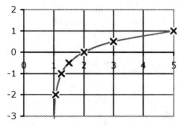

17. $f(x) = \log_{1/4} x$

We plot some points and draw the graph:

x	$y = \log_{1/4}(x)$
1/16	1
1/4	1
1/2	1/2
1	0
2	$-1/2$
4	-1

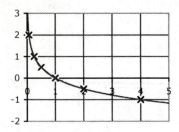

19. $Q = 1000$ when $t = 0$; Half-life = 1

We want a model of the form $Q = Q_0 e^{-kt}$ for suitable Q_0 and k. We are given $Q_0 = 1000$. For k, we use the formula $t_h k = \ln 2$ with $t_h = $ half-life = 1, so $k = \ln 2$, and the model is
$$Q = Q_0 e^{-kt} = 1000 e^{-t \ln 2}$$

21. $Q = 1000$ when $t = 0$; Doubling time = 2

We want a model of the form $Q = Q_0 e^{kt}$ for suitable Q_0 and k. We are given $Q_0 = 1000$. For k, we use the formula $t_d k = \ln 2$ with $t_d = $ doubling time = 2, so $2k = \ln 2$, giving $k = (\ln 2)/2$ and the model is
$$Q = Q_0 e^{kt} = 1000 e^{t(\ln 2)/2}$$

23. $Q = 1000 e^{0.5t}$

Since the exponent is positive, the model represents exponential growth. We use the formula $t_d k = \ln 2$, giving $t_d(0.5) = \ln 2$, giving doubling time = $t_d = (\ln 2)/0.5 = 2 \ln 2$

25. $Q = Q_0 e^{-4t}$

Since the exponent is negative, the model represents exponential decay. We use the formula $t_h k = \ln 2$, giving $t_h(4) = \ln 2$, giving half-life = $t_h = (\ln 2)/4$

27. $f(x) = 4e^{2x} = 4(e^2)^x \approx 4(7.389)^x$

29. $f(t) = 2.1(1.001)^t = 2.1 e^{kt}$
$$1.001^t = e^{kt}$$
$$t \ln(1.001) = kt$$
$$k = \ln(1.001) \approx 0.0009995$$
so $f(t) = 2.1 e^{0.000\,9995t}$

31. $f(t) = 10(0.987)^t = 10 e^{-kt}$
$$0.987^t = e^{-kt}$$
$$t \ln(0.987) = -kt$$
$$k = -\ln(0.987) \approx 0.013\,09$$
so $f(t) = 10 e^{-0.013\,09t}$

33. Substitute $A = 700, P = 500$, and $r = 0.10$ in the continuous compounding formula: $700 = 500 e^{0.10t}$. Solve for t: $e^{0.10t} = 700/500 = 7/5$, $0.10t = \ln(7/5)$, $t = 10\ln(7/5) \approx 3.36$ years.

35. Substitute $A = 3P$ and $r = 0.10$ in the continuous compounding formula: $3P = P e^{0.10t}$. Solve for t: $e^{0.10t} = (3P)/P = 3, 0.10t = \ln 3, t = 10\ln 3 \approx 11$ years

37. We want a model of the form $A = P e^{rt}$. For the rate of interest r, we use the formula $t_d r = \ln 2$ with $t_d = $ doubling time = 3, so $3r = \ln 2$, giving $r = (\ln 2)/3 \approx 0.231$, or 23.1%

39. If 99.95% has decayed, then 0.05% remains, so $C(t) = 0.0005A$. Therefore, $0.0005A = A(0.999879)^t, (0.999879)^t = 0.0005$, so $t = \log_{0.999879} 0.0005 = \log(0.0005)/\log(0.999879) \approx 63{,}000$ years old.

41. Substitute $A = 15{,}000, P = 10{,}000, r = 0.039$, and $m = 1$ into the compound interest formula: $15{,}000 = 10{,}000(1 + 0.039)^t, (1.039)^t = 1.5, t = \log_{1.039} 1.5 = \log(1.5)/\log(1.039) \approx 11$ years.

43. Substitute $A = 20{,}000, P = 10{,}400, r = 0.038$, and $m = 12$ into the compound interest formula: $20{,}000 = 10{,}400(1 + 0.038/12)^{12t}, (1 + 0.038/12)^{12t} = 200/104, 12t = \log(200/104)/\log(1 + 0.038/12) \approx 207$ months.

45. Substitute $A = 2P, r = 0.059$, and $m = 2$ into the compound interest formula: $2P = P(1 + 0.059/2)^{2t}, (1.0295)^{2t} = 2, 2t = \log 2/\log 1.0295$ $t = (\log 2/\log 1.0295)/2 \approx 12$ years.

47. Substitute $A = 2P$ and $r = 0.039$ in the continuous compounding formula: $2P = Pe^{0.039t}$, $e^{0.039t} = 2, 0.039\ t = \ln 2, t = (\ln 2)/0.039 \approx 17.77$ years.

49. We want to find the value of t for which $C(t)$ = the weight of undecayed radium-226 left = half the original weight = $0.5A$. Substituting, we get

$$0.5A = A(0.999\ 567)^t$$

so $\qquad 0.5 = 0.999\ 567^t$

$$t = \log_{0.999567} 0.5 \approx 1600 \text{ years}$$

51. (a) If $y = Ab^t$, substitute $y = 3A$ and $t = 6$ to get $3A = Ab^6$, so $b = 3^{1/6} \approx 1.20$.
(b) Now substitute $y = 2A$ and solve for t: $2A = A(1.20)^t$, so $t = \log 2/\log 1.20 \approx 3.8$ months.

53. (a) $t_h k = \ln 2$
$\quad 5k = \ln 2$
$\quad k = (\ln 2)/5 \approx 0.139$
$\quad Q(t) = Q_0 e^{-kt} \approx Q_0 e^{-0.139t}$
(b) One third has decayed when 2/3 is left:

$$\frac{2}{3} Q_0 = Q_0 e^{-0.139t}$$

$$\frac{2}{3} = e^{-0.139t}$$

$$\ln(2/3) = -0.139t$$

$$t = \ln(2/3)/(-0.139) \approx 3 \text{ years}$$

55. Let $Q(t)$ be the amount left after t million years. We first find a model of the form $Q(t) = Q_0 e^{-kt}$.
To find k use
$t_h k = \ln 2$
$\quad 710k = \ln 2$
$\quad k = (\ln 2)/710 \approx 0.000\ 976\ 3$
$\quad Q(t) = Q_0 e^{-kt} \approx Q_0 e^{-0.000\ 976\ 3t}$
We now answer the question: $Q_0 = 10$g, $Q(t) =$ 1g. Substituting in the model gives

$$1 = 10e^{-0.000\ 976\ 3t}$$

$$e^{-0.000\ 976\ 3t} = 0.1$$

$$-0.000\ 976\ 3t = \ln(0.1)$$

$$t = -\ln 0.1/0.000\ 976\ 3$$

$$\approx 2360 \text{ million years}$$

(rounded to 3 significant digits).

57. Let $Q(t)$ be the amount left after t hours. We first find a model of the form $Q(t) = Q_0 e^{-kt}$.
To find k use
$t_h k = \ln 2$
$\quad 2k = \ln 2$
$\quad k = (\ln 2)/2 \approx 0.3466$
$\quad Q(t) = Q_0 e^{-kt} \approx Q_0 e^{-0.3466t}$
We now answer the question: $Q_0 = 300$mg, $Q(t) =$ 100mg. Substituting in the model gives

$$100 = 300e^{-0.3466t}$$

$$e^{-0.3466t} = 1/3$$

$$-0.3466t = \ln(1/3)$$

$$t = -\ln(1/3)/0.3466 \approx 3.2 \text{ hours}$$

59. We first find a model of the form $A(t) = Pb^t$. We are given the data points $(0, 1)$ and $(2, 0.7)$, so $P = 1$ and $b^2 = 0.7$, so $b = (0.7)^{1/2}$; the model is $A(t) = (0.7)^{t/2}$. We wish to find when $A(t) = 0.5$:

$$0.5 = (0.7)^{t/2}$$

$$t/2 = \log_{0.7}(0.5)$$

$$t = 2\log(0.5)/\log(0.7) \approx 3.89 \text{ days}.$$

61. (a) To obtain the logarithmic regression equation for the data in Excel, do a scatter plot and add a Logarithmic trendline with the option "Display equation on chart" checked. On he TI-83/84 Plus, use $\boxed{\text{STAT}}$, select CALC, and choose the option **LnReg**. The resulting regression equation with coefficients rounded to 4 digits is

$$P(t) = 6.591 \ln(t) - 17.69$$

b. The year 1940 is represented by $t = 40$, so

$$P(40) = 6.591 \ln(40) - 17.69 \approx 6.6,$$

which is accurate only to one digit.
c. The logarithm increases without bound (choice (A)).

63. Following is the Excel tabulation of the data, together with the scatter plot and the logarithmic

359

trendline (with the option "Display equation on chart" checked):

◇	A	B	C
1	t	y	
2	5	118	
3	6	129	
4	7	140	
5	8	150	
6	9	165	
7	10	183	
8	11	181	
9	12	170	
10	13	172	
11	14	172	
12	15	182	
13	16	191	
14	17	194	
15	18	197	
16	19	200	
17			
18			
19			

y = 57.508Ln(x) + 30.96

From the regression output, the model is

$$S(t) = 57.51\ln t + 30.96$$

(coefficients rounded to 2 decimal places). Extrapolating in the positive direction results in a prediction of ever-increasing R&D expenditures by industry. This is reasonable to a point, as expenditures cannot reasonably be expected to increase without bound. Extrapolating in the negative direction eventually leads to negative values, which does not model reality.

65. (a) Substitute:

$$8.2 = (2/3)(\log E - 11.8).$$

Solve for log E:

$$\log E = (3/2) \times 8.2 + 11.8 = 24.1,$$
$$E = 10^{24.1} \approx 1.259 \times 10^{24} \text{ ergs}$$

(b) Compute the energy released as in (a):

$$7.1 = (2/3)(\log E - 11.8)$$
$$\log E = (3/2) \times 7.1 + 11.8 = 22.45$$
$$E = 10^{22.4} \approx 2.818 \times 10^{22} \text{ ergs.}$$

Comparing, the energy released in 1989 was

$$\frac{2.818 \times 10^{22}}{1.259 \times 10^{24}} \approx 2.24\% \text{ of the energy released in}$$

1906.

(c) Start with the given equation

$$R = \tfrac{2}{3}\left(\log E - 11.8\right)$$
$$\tfrac{3}{2}R = \log E - 11.8$$
$$\log E = 1.5R + 11.8$$

In exponential form, this is

$$E = 10^{1.5R + 11.8}.$$

(d) Applying part 9c) to each of the two earthquakes gives

$$E_2 = 10^{1.5R_2 + 11.8}$$
$$E_1 = 10^{1.5R_1 + 11.8}$$

Dividing gives

$$\frac{E_2}{E_1} = \frac{10^{1.5R_2 + 11.8}}{10^{1.5R_1 + 11.8}} = 10^{1.5(R_2 - R_1)}$$

(e) If $R_2 - R_1 = 2$, then $E_2/E_1 = 10^{1.5(2)} = 10^3 = 1000$.

67. (a) By substitution: 75 dB, 69 dB, 61 dB (b) $D = 10 \log(320 \times 10^7) - 10 \log r^2 \approx 95 - 20 \log r$ (c) Solve $0 = 95 - 20 \log r$: $\log r = 95/20$, $r = 10^{95/20} \approx 57,000$ feet (rounding up so that we're beyond the point where the decibel level is 0).

69. Graph:

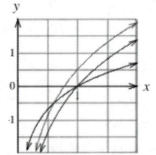

The lowest graph on the right is $y = \ln x$. The middle graph is $y = 2\ln x$, and the upmost graph is $y = 2\ln x + 0.5$. Multiplying by A stretches the graph in the y-direction by a factor of A. Adding C moves the graph C units vertically up.

71. The logarithm of a negative number, were it defined, would be the power to which a base must be raised to give that negative number. But raising a base to a power never results in a negative number, so there can be no such number as the logarithm of a negative number.

73. Any logarithmic curve $y = \log_b t + C$ will eventually surpass 100%, and hence not be suitable as a long-term predictor of market share.

75. $\log_4 y$

77. 8. Note that $b^{\log_b a} = a$ for any a and b, since $\log_b a$ is the power to which you raise b to get a.

79. x. To what power to you raise e to get $e^{x?}$ x!

81. Time is increasing logarithmically with population: Solving $P = Ab^t$ for t gives $t = \log_b(P/A) = \log_b P - \log_b A$, which is of the form $t = \log_b P + C$.

83. For the two function, write

$$Q_1 = A_1 \ln t + B_1$$
$$Q_2 = A_2 \ln t + B_2$$

Adding gives

$$Q_1 + Q_2 = A_1 \ln t + B_1 + A_2 \ln t + B_2$$
$$= (A_1 + A_2)\ln t + (B_1 + B_2),$$

which is of the form

Constant$\times\ln t$ + Constant

and is thus a logarithmic function.

9.4 Logistic Functions and Models

1. $N = 7$, $A = 6$, $b = 2$; `7/(1+6*2^-x)`

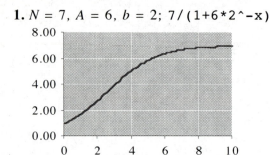

3. $N = 10$, $A = 4$, $b = 0.3$;
`10/(1+4*0.3^-x)`

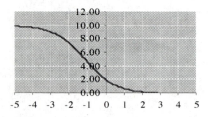

5. $N = 4$, $A = 7$, $b = 1.5$; `4/(1+7*1.5^-x)`

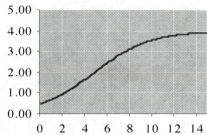

7. $N = 200$, the limiting value. $10 = f(0) = N/(1 + A) = 200/(1 + A)$, so $A = 19$. For small values of x, $f(x) \approx 10b^x$; to double with every increase of 1 in x we must therefore have $b = 2$. This gives $f(x) = \dfrac{200}{1 + 19 \cdot (2^{-x})}$.

9. $N = 6$, the limiting value. $3 = f(0) = N/(1 + A) = 6/(1 + A)$, so $A = 1.4 = f(1) = 6/(1 + b^{-1})$, $1 + b^{-1} = 6/4 = 3/2$, $b^{-1} = 1/2$, $b = 2$. So, $f(x) = \dfrac{6}{1 + 2^{-x}}$.

11. B: The limiting value is 9, eliminating (A). The initial value is $2 = 9/(1 + 3.5)$, not $9/(1 + 0.5)$, so the answer is (B), not (C).

13. B: The graph decreases, so $b < 1$, eliminating (C). The y intercept is $2 = 8/(1 + 3)$ as in (B).

15. C: The initial value is 2, as in (A) or (C). If b were 5, an increase in 1 in x would multiply the value of $f(x)$ by approximately 5 when x is small. However, increasing x from 0 to 10 does not quite double the value. Hence b must be smaller, as in (C).

17. $y = \dfrac{7.2}{1 + 2.4\,(1.04)^{-x}}$

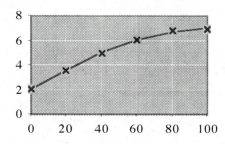

19. $y = \dfrac{97}{1 + 2.2(0.942)^{-x}}$

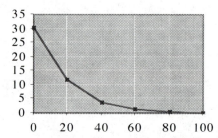

21. (a) We can eliminate (B) and (D) because $b = 0.8$ is less than 1, giving a decreasing function of t. The value $N = 2.0$ in (C) predicts a leveling-off value of around 2%, which is clearly too small. Thus, we are left with choice (A).

(b) The logistic curve is steepest when $t = \frac{\ln A}{\ln b}$.

This occurs when $t = \frac{\ln 8.6}{\ln 1.8} \approx 3.7$ years since 2000; during 2003.

23. (a) We can eliminate (C) because $b = 0.8$ is less than 1, giving a decreasing function of t. We can eliminate (D) because $b = 6.2$, indicating a rate of growth of 520% per year—far faster than suggested by the data. The value $N = 4.0$ in (B) predicts a leveling-off value of around 4000 articles, which is clearly too small. Thus, we are left with choice (A).

(b) 1985 is in the initial period of the model, when the rate of growth is governed by exponential growth. Since $b = 1.2$, the rate of growth is 20% per year.

25. (a) For the extremely wealthy we look at large values of x, where $P(x)$ is close to $N = 91\%$. **(b)** $P(x) \approx [91/(1 + 5.35)](1.05)^x \approx 14.33(1.05)^x$. **(c)** Set $P(x) = 50$ and solve for x: $50 = 91/[1 + 5.35(1.05)^{-x}]$, $1 + 5.35(1.05)^{-x} = 91/50$, $x = -\log[41/(50\times5.35)]/\log(1.05) \approx \$38,000$.

27. $N = 10,000$, the susceptible population. $1000 = 10,000/(1 + A)$, so $A = 9$. In the initial stages, the rate of increase was 25% per day, so $b = 1.25$. Thus, $N(t) = \frac{10,000}{1 + 9(1.25)^{-t}}$. $N(7) \approx$ 3463 cases.

29. $N = 3000$, the total available market. $100 = 3000/(1 + A)$, so $A = 29$. Sales are initially doubling every 5 days, so $b^5 = 2$, or $b = 2^{1/5}$. Thus, $N(t) = \frac{3000}{1 + 29(2^{1/5})^{-t}}$. Set $N(t) = 700$ and solve for t: $700 = 3000/[1 + 29(2^{1/5})^{-t}]$, $1 + 29(2^{1/5})^{-t} = 30/7$, $(2^{1/5})^{-t} = 23/(29\times7)$, $t = -\log[23/(29\times7)]/\log(2^{1/5}) \approx 16$ days.

31. (a) Following Example 2, if using Excel, start with initial rough estimates of N, A, and b. (Their exact value is not important.)

N = leveling off value ≈ 7
$b = 1.1$ (slightly larger than 1, since N is increasing with t)
A: Use y-intercept $= N/(1+A)$

$$1.2 = \frac{7}{1+A}$$

$$1+A = 7/1.2 \approx 6$$

So, $A \approx 5$.
Solver then gives the following solution:
$N \approx 6.3$, $A \approx 4.8$, $b \approx 1.2$.
Thus, the regression model is
$$A(t) = \frac{6.3}{1 + 4.8(1.2)^{-t}}.$$
The model predicts that the number will level off around $N = 6.3$ thousand, or 6300 articles.

(b) $A(17) = \frac{6.3}{1 + 4.8(1.2)^{-17}} \approx 5.2$,

or 5200 articles.

33. (a) Following Example 2, if using Excel, start with initial rough estimates of N, A, and b. (Their exact value is not important.)

N = leveling off value ≈ 1000
$b = 1.1$ (slightly greater than 1, since N is increasing with t)
A: Use y-intercept $= N/(1+A)$

$$767 = \frac{1000}{1+A}$$

$$1+A = 1000/767 \approx 1.3$$

So, $A \approx 0.3$
Solver then gives the following solution:
$N \approx 1070$, $A \approx 0.391$, $b \approx 1.10$.
Thus, the regression model is
$$B(t) = \frac{1070}{1 + 0.391 \, (1.10)^{-t}}.$$
The model predicts that the number will level off around $N = 1070$ teams

(b) The logistic curve is steepest when $t = \frac{\ln A}{\ln b}$.

This occurs when $t = \frac{\ln 0.391}{\ln 1.1} \approx -9.85$. This means that, according to the model, the number of teams was rising fastest about 9.85 years *prior* to 1990; that is, some time during 1980.

(c) Since the coefficient of b is 1.1, the number of men's basketball teams was growing by about 10% per year around 1990.

35. For a limiting value of 4.5 million, we take $N = 4500$.

Here is the set of values $(x, N/y - 1)$ with $N = 4500$.

x	N/y-1
0	1.02703
10	0.90678
20	0.65441
30	0.40187
40	0.27119
50	0.17801
55	0.1509
60	0.12782
65	0.11111
70	0.10294
75	0.09756

The exponential regression equation is $y \approx 1.1466(0.96551)^x$

so that

$$A = 1.1466$$

$b^{-1} = 0.96551$, giving

$$b = 1/0.96551 \approx 1.0357$$

Thus the logistic model is

$$y = \frac{4500}{1 + 1.1466\,(1.0357)^{-t}}$$

To predict when $y = 4000$, we set

$$4000 = \frac{4500}{1 + 1.1466\,(1.0357)^{-t}}$$

and solve for t:

$$1 + 1.1466\,(1.0357)^{-t} = \frac{4500}{4000} = 1.125$$

so $(1.0357)^{-t} = (1.125 - 1)/1.1466 \approx 0.1090$

giving $t = -\ln(0.1090)/\ln(1.0357) \approx 63$, or 2013

37. Just as diseases are communicated via the spread of a pathogen (such as a virus), new technology is communicated via the spread of information (such as advertising and publicity). Further, just as the spread of a disease is ultimately limited by the number of susceptible individuals, so the spread of a new technology is ultimately limited by the size of the potential market.

39. It can be used to predict where the sales of a new commodity might level off.

41. The curve is still a logistic curve, but decreases when $b > 1$ and increases when $b < 1$.

43. Substituting $t = \dfrac{\ln A}{\ln b}$ in the formula for y gives

$$y = \frac{N}{1 + Ab^{-\ln A/\ln b}} = \frac{N}{1 + Ab^{-\ln_b A}}$$

But $\quad b^{-\ln_b A} = \dfrac{1}{b^{\ln_b A}} = \dfrac{1}{A}$

So y becomes

$$\frac{N}{1 + A(1/A)} = \frac{N}{1+1} = N/2.$$

364

Chapter 9 Review Exercises

1. $a = 1$, $b = 2$, $c = -3$, so $-b/(2a) = -1$; $f(-1) = -4$, so the vertex is $(-1, -4)$. y intercept: $c = -3$. $x^2 + 2x - 3 = (x + 3)(x - 1)$, so the x intercepts are -3 and 1. $a > 0$, so the parabola opens upward.

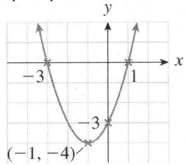

3. For every increase in x by one unit, the value of f is multiplied by $1/2$, so f is exponential. Since $f(0) = 5$, the exponential model is $f(x) = 5(1/2)^x$, or $5(2^{-x})$.

$g(2) = 0$, whereas exponential functions are never zero, so g is not exponential.

5. We use the following table of values:

x	$f(x)$	$g(x)$
-3	$1/54$	$27/2$
-2	$1/18$	$9/2$
-1	$1/6$	$3/2$
0	$1/2$	$1/2$
1	$3/2$	$1/6$
2	$9/2$	$1/18$
3	$27/2$	$1/54$

Graphing these gives the following curves:

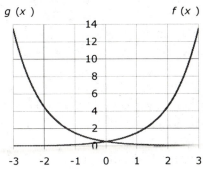

7. The technology formulas for the two functions are
TI-83/84 Plus: `e^x; e^(0.8*x)`
Excel: `=EXP(x); =EXP(0.8*x)`
Technology gives us the following graphs:

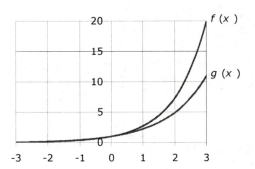

9. $A = P(1 + \frac{r}{m})^{mt}$;
$P = 3000$, $r = 0.03$; $m = 12$; $t = 5$
$A = 3000(1+0.03/12)^{60} \approx \$3484.85$

11. $A = P(1 + \frac{r}{m})^{mt}$;
$A = 5000$, $r = 0.03$; $m = 12$; $t = 10$
$5000 = P(1+0.03/12)^{120}$
$P = 5000/(1+0.03/12)^{120} \approx \$3705.48$

13. $A = Pe^{rt}$;
$P = 3000$, $r = 0.03$; $t = 5$
$A = 3000e^{0.15} \approx \$3485.50$

15. Increasing x by $1/2$-unit triples the value of f. Therefore, increasing x by 1-unit multiples the value of f by 9, giving $b = 9$. $A = f(0) = 4.5$. Therefore,
$$f(x) = Ab^x = 4.5(9^x)$$

17. $2 = Ab^1$ and $18 = Ab^3$. Dividing, $b^2 = 9$, so $b = 3$. Then $2 = 3A$, so $A = 2/3$: $f(x) = \frac{2}{3} 3^x$

19. $-2x = \log_3 4$, so $x = -\frac{1}{2}\log_3 4$

21. $10^{3x} = 315/300 = 1.05$, $3x = \log 1.05$, $x = \frac{1}{3}\log 1.05$

23. We use the following table of values:

x	f(x)	g(x)
1/27	–3	3
1/9	–2	2
1/3	–1	1
1	0	0
3	1	–1
9	2	–2
27	3	–3

Graphing these gives the following curves:

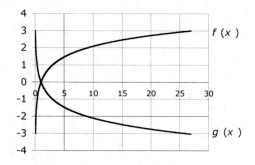

25. $Q_0 = 5$ (given)

$t_h k = \ln 2$

$100k = \ln 2$

$k = (\ln 2)/100 \approx 0.00693$

$Q = Q_0 e^{-kt}$

$Q = 5e^{-0.00693t}$

27. $Q_0 = 2.5$ (given)

$t_d k = \ln 2$

$2k = \ln 2$

$k = (\ln 2)/2 \approx 0.347$

$Q = Q_0 e^{kt}$

$Q = 2.5e^{0.347t}$

29. $3000 = 2000(1 + 0.04/12)^{12t}$

$t = [\log(3/2)/\log(1 + 0.04/12)]/12$

≈ 10.2 years

31. $3000 = 2000e^{0.0375t}$

$t = \ln(3/2)/0.0375$

≈ 10.8 years

33. $N = 900$ is given, $100 = 900/(1 + A)$ gives $A = 8$, initially increasing 50% per unit increase in x gives $b = 1.5$: $f(x) = \dfrac{900}{1 + 8(1.5)^{-x}}$

35. $N = 20$ is given, $20/(1 + A) = 5$, so $A = 3$, decreasing at a rate of 20% per unit of x near 0 means $b = 1 - 0.2 = 0.8$: $f(x) = \dfrac{20}{1 + 3(0.8)^{-x}}$

37. (a) The largest volume will occur at the vertex: $-b/(2a) = 0.085/(2 \times 0.000005) \approx$ \$8500 per month; substituting $c = 8500$ gives h = an average of approximately 2100 hits per day. **(b)** Solve $h = 0$ using the quadratic formula: $c \approx$ \$29,049 per month (the other solution given by the quadratic formula is negative). **(c)** The fact that -0.000005, the coefficient of c^2, is negative.

39. (a) $R = pq = -60p^2 + 950p$. The maximum revenue occurs at the vertex: $p = -b/(2a) = 950/(2 \times 60) \approx$ \$7.92 per novel. At that price the monthly revenue is $R =$ \$3760.42.
(b) $C = 900 + 4q = 900 + 4(-60p + 950) = -240p + 4700$, so $P = R - C = -60p^2 + 950p - (-240p + 4700) = -60p^2 + 1190p - 4700$. The maximum monthly profit occurs at the vertex:
$p = -b/(2a) = 1190/(2 \times 60) \approx$ \$9.92 per novel. At that price, the monthly profit is $P =$ \$1200.42.

41. (a) 1997 corresponds to $t = 0$, and so the harvest was $n = 10(0.66^0) = 10$ million pounds. The value of the base $b = 0.66$ of the exponential model tells us that the value each year 0.66 times, or 66% of the value the previous year. In other words, the value decreases by $100 - 66 = 34\%$ each year.
(b) 2005 corresponds to $t = 8$, and
$$n(8) = 10(0.66^8) \approx 0.36,$$
or about 360,000 pounds.

43. The model for the lobster harvest from Exercise 39 is $n(t) = 10(0.66^t)$ million pounds.

We need to find the value of t when it first dips below 100,000, which is 0.1 million pounds:

$$0.1 = 10(0.66^t)$$

$$0.01 = 0.66^t$$

$$\log(0.01) = t \log(0,66)$$

$$t = \log(0.01)/\log(0.66) \approx 11.08$$

Since $t = 0$ corresponds to June, 1997, $t = 11$ corresponds to June, 2008.

45. In 2007 ($t = 10$), the harvest was

$$10(0.66^{10}) \approx 0.1568$$

From that point on, the model is

$$0.1568(1.24)^t$$

where t is now time in years since 2007. In 2010, the size of the harvest is

$$0.1568(1.24)^3 \approx 0.2990 \text{ million pounds, or}$$

299,000 pounds.

47. We use the data shown in the following table:

t	n
0	8.2
1	7.5
2	6.4
3	2.6
4	1.8
5	1.4
6	0.8

The regression model is

Excel: $n(t) = 10.318e^{-0.4145t} = 10.318(e^{-0.4145})^t \approx$ $10.318(0.6607)^t$

TI-83/84 Plus and Web Site: $10.3182(0.660642)^t$

Rounding to 2 digits gives $n(t) = 10(0.66^t)$ million pounds of lobster.

49. C: (A) is true because $L_1 = L_0 e^{-1/t} < L_0$. (B) is true because $L_{3t} = L_0 e^{-3} > 0$. (D) is true because increasing t decreases x/t, which increases $e^{-x/t}$ (makes it closer to 1), hence increases L_x for every x. (E) is true because $e^{-x/t}$ is never 0. On the other hand, (C) is false because $L_t = L_0 e^{-1} \approx 0.37 L_0 < L_0/2$.

Chapter 10 Introduction to the Derivative

10.1 Limits: Numerical and Graphical Approaches

1. 0:

x	f(x)
−0.1	0.01111111
−0.01	0.00010101
−0.001	1.001×10^{-6}
−0.0001	1.0001×10^{-6}
0	
0.0001	9.999×10^{-9}
0.001	9.99×10^{-7}
0.01	9.901×10^{-5}
0.1	0.00909091

3. 4:

x	f(x)
1.9	3.9
1.99	3.99
1.999	3.999
1.9999	3.9999
2	
2.0001	4.0001
2.001	4.001
2.01	4.01
2.1	4.1

5. Does not exist:

x	f(x)
−1.1	−22.1
−1.01	−202.01
−1.001	−2002.001
−1.0001	−20002
−1	
−0.9999	19998.0001
−0.999	1998.001
−0.99	198.01
−0.9	18.1

7. 1.5:

x	f(x)
10	2.66
100	1.58969231
1000	1.50877143
10,000	1.50087521
100,000	1.5000875

9. 0.5:

x	f(x)
−10	53.1578947
−1000	1
−100,000	0.505
−10,000,000	0.50005
−1,000,000,000	0.5000005

11. Diverges to $+\infty$:

x	f(x)
10	76.9423077
100	258.966135
1000	2059.17653
10000	20059.1977
100000	200059.2

13. 0:

x	f(x)
10	0.79988005
100	0.02600019
1000	0.00206
10000	0.0002006
100000	2.0006×10^{-5}

15. 1:

x	f(x)
1.9	0.90483742
1.99	0.99004983
1.999	0.9990005
1.9999	0.9999
2	
2.0001	1.00010001
2.001	1.0010005
2.01	1.01005017
2.1	1.10517092

17. 0:

x	f(x)
10	0.000453999
100	3.72008×10^{-42}
1000	0
10000	0
100000	0

(The last three values of f(x), while not mathematically 0, are too small to be represented in Excel, which just gives the values as 0.)

19. (a) −2 **(b)** −1

21. (a) 2 **(b)** 1 **(c)** 0 **(d)** +∞

23. (a) 0 **(b)** 2: As x approaches 0 from the right, $f(x)$ approaches the solid dot at height 2. **(c)** −1: As x approaches 0 from the left, $f(x)$ approaches the open dot at height −1. The fact that $f(0) = 2$ is irrelevant. **(d)** Does not exist: Parts (b) and (c) show that the one-sided limits, though they both exist, do not agree. **(e)** 2: The solid dot indicates the actual value of $f(0)$. **(f)** +∞

25. (a) 1 **(b)** 1: Similar to Exercise 23. **(c)** 2 **(d)** Does not exist **(e)** 1 **(f)** 2

27. (a) 1 **(b)** +∞ **(c)** +∞ **(d)** +∞ **(e)** Not defined **(f)** −1

29. (a) −1: Approaching from the left or the right, the value of $f(x)$ approaches the height of the open dot, −1. **(b)** +∞ **(c)** −∞ **(d)** Does not exist

(e) 2 **(f)** 1: The value of the function is given by the closed dot on the graph.

31. Here is a table of values for $v(t)$
Technology formulas:
TI-83/84 Plus: `210-62*e^(-0.05*x)`
Excel: `210-62*EXP(-0.05*x)`

t	v(t)
0	148
20	187.191475
40	201.609212
60	206.913202
80	208.86443
100	209.582247
120	209.846317
140	209.943463
160	209.979201
180	209.992349
200	209.997185

As t gets larger and larger, the values of $v(t)$ are getting closer and closer to 210. Thus, we estimate
$$\lim_{t \to +\infty} v(t) = 210$$
Since $v(t)$ represents the value, in trillions of pesos, of sold goods in Mexico each month, we interpret the result as follows: In the long term, the value of sold goods in Mexico will approach 210 trillion pesos per month.

33. Here is a table of values for $A(t)$
Technology formula:
`7.0/(1+5.4*1.2^(-x))`

t	A(t)
0	1.09375
10	3.7390563
20	6.135755
30	6.8443011
40	6.9743759
50	6.9958488
60	6.9993292
70	6.9998917
80	6.9999825
90	6.9999972
100	6.9999995

At t gets larger and larger, the values of $A(t)$ are getting closer and closer to 7.0. Thus, we estimate
$$\lim_{t \to +\infty} A(t) = 7.0$$
Since $A(t)$ represents the number of thousands of research articles per year in *Physics Review*

written by researchers in Europe, we interpret the result as follows: In the long term, the number of research articles in *Physics Review* written by researchers in Europe approaches 7000 per year.

35. Here is a table of values for $S(x)$
Technology formula:
```
470-136*0.974^x
```

x	S(x)
0	334
50	433.567686
100	460.240342
150	467.385537
200	469.299626
250	469.81238
300	469.94974
350	469.986536
400	469.996393
450	469.999034
500	469.999741

At t gets larger and larger, the values of $S(x)$ are getting closer and closer to 470. Thus, we estimate

$$\lim_{x \to +\infty} S(x) = 470$$

Since $S(x)$ represents the average SAT verbal score of a student whose income is x thousand dollars per year, we interpret the result as follows: Students whose parents earn an exceptionally large income score an average of 470 on the SAT verbal test.

37. If you move your pencil point along the graph to the extreme right, the y-coordinate approaches 15. Therefore, $\lim_{t \to +\infty} p(t) = 15$. Since $p(t)$ is the percentage change (from 2003) of the home price index, we conclude that, assuming the trend shown in the graph continues indefinitely, the home price index will approach a value 15% above the 2003 level in the long term.

39. $\lim_{t \to 1-} C(t) = 0.06$, $\lim_{t \to 1+} C(t) = 0.08$, so that $\lim_{t \to 1} C(t)$ does not exist.

41. $\lim_{t \to +\infty} I(t) = +\infty$, $\lim_{t \to +\infty} (I(t)/E(t)) \approx 2.5$. In the long term, U.S. imports from China will rise without bound and be 2.5 times U.S.

exports to China. In the real world, imports and exports cannot rise without bound. Thus, the given models should not be extrapolated far into the future.

43. To approximate $\lim_{x \to a} f(x)$ numerically, choose values of x closer and closer to and on either side of $x = a$, and evaluate $f(x)$ for each of them. The limit (if it exists) is then the number that these values of $f(x)$ approach. A disadvantage of this method is that it may never give the exact value of the limit, but only an approximation. (However, we can make the approximation as accurate as we like.)

45. Any situation in which there is a sudden change can be modeled by a function in which $\lim_{t \to a+} f(t)$ is not the same as $\lim_{t \to a-} f(t)$ One example is the value of a stock market index before and after a crash: $\lim_{t \to a-} f(t)$ is the value immediately before the crash at time $t = a$, while $\lim_{t \to a+} f(t)$ is the value immediately after the crash. Another example might be the price of a commodity that is suddenly increased from one level to another.

47. It is possible for $\lim_{x \to a} f(x)$ to exist even though $f(a)$ is not defined. An example is
$$\lim_{x \to 1} \frac{x^2 - 3x + 2}{x - 1}.$$

49. The limit may not be defined and, even if it is, may not equal $f(a)$. See, for example, Exercises 23 and 24.

51. An example is $f(x) = (x-1)(x-2)$.

10.2 Limits and Continuity

1. Continuous on its domain

3. Continuous on its domain

5. Discontinuous at $x = 0$:
$$\lim_{x \to 0^+} f(x) \neq \lim_{x \to 0^-} f(x), \text{ so } \lim_{x \to 0} f(x) \text{ does not}$$
exist.

7. Discontinuous at $x = -1$:
$$\lim_{x \to -1^+} f(x) \neq \lim_{x \to -1^-} f(x), \text{ so } \lim_{x \to -1} f(x) \text{ does}$$
not exist.

9. Continuous on its domain:
Note that $f(0)$ is not defined, so 0 is not in the domain of f.

11. Discontinuous at $x = -1$ and 0:
$$\lim_{x \to -1} f(x) = -1 \neq f(-1)$$
and $\lim_{x \to 0} f(x)$ does not exist.
Note that $f(0)$ is defined [$f(0) = 2$] so 0 is in the domain of f.

13. (A), (B), (D), (E) are continuous on their domains:
Note that 1 is not in the domain in (B) and (D) and that the domain in (E) is
$(-\infty, -1] \cup (1, +\infty)$; the "horizontal break" in the graph in (E) does not make the function discontinuous. (C) is discontinuous at 1 because the limit there does not equal the function's value.

In Exercises 15–22, either a graph of the function or a table of values can be used to compute the indicated limit. We show a graph drawn with technology in each case. Note: The vertical lines near discontinuities in some of the graphs below is typical behavior of graphing technology.

15. Technology Formula:
`(x^2-2*x+1)/(x-1)`
Graph drawn with technology:

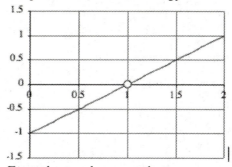

From the graph we see that
$$\lim_{x \to 1} \frac{x^2 - 2x + 1}{x - 1} = 0,$$
so setting $f(1) = 0$ makes it continuous at 1.

17. Technology Formula:
`x/(3*x^2-x)`
Graph drawn with technology:

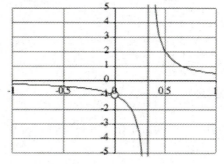

From the graph we see that
$$\lim_{x \to 0} \frac{x}{3x^2 - x} = -1,$$
so setting $f(0) = -1$ makes it continuous at 0.

371

19. Technology Formula:
`3/(3*x^2-x)`
Graph drawn with technology:

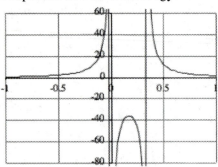

From the graph we see that

$$\lim_{x \to 0} \frac{3}{3x^2 - x} \text{ is undefined,}$$

so no value of $f(0)$ will make it continuous at 0.

21. Technology Formula:
TI-83/84 Plus: `(1-e^x)/x`
Excel: `(1-exp(x))/x`
Graph drawn with technology:

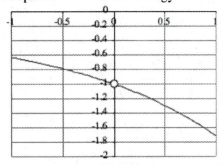

From the graph we see that

$$\lim_{x \to 0} \frac{1 - e^x}{x} = -1,$$

so setting $f(0) = -1$ will make it continuous at 0.

Note: The vertical lines near discontinuities in some of the graphs below is typical behavior of graphing technology.

23. Continuous on its domain:

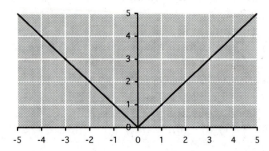

25. Continuous on its domain: Note that 1 and -1 are not in the domain of g.

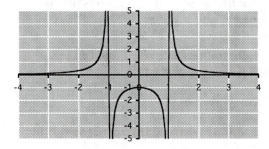

27. Discontinuity at $x = 0$:

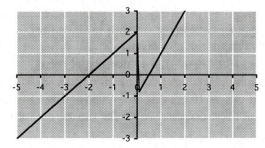

29. Discontinuity at $x = 0$: The limit of $h(x)$ as $x \longrightarrow 0$ does not exist, but 0 is now in the domain of h.

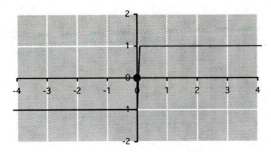

31. Continuous on its domain:

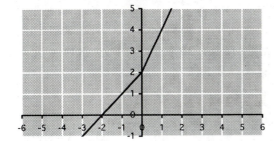

33. Not unless the domain of the function consists of all real numbers. (It is impossible for a function to be continuous at points not in its domain.) For example, $f(x) = 1/x$ is continuous on its domain—the set of nonzero real numbers—but not at $x = 0$.

35. True. If the graph of a function has a break in its graph at any point a, then it cannot be continuous at the point a.

37. Answers may vary. $f(x) = 1/[(x-1)(x-2)(x-3)]$ is such a function; it is undefined at $x = 1, 2, 3$ and so its graph consists of three distinct curves.

39. Answers may vary.

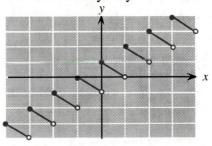

41. Answers may vary. The price of OHaganBooks.com stock suddenly drops by $10 as news spreads of a government investigation. Let $f(x)$ = the price of OHaganBooks.com stock.

10.3 Limits and Continuity: Algebraic Approach

1. The natural domain of $f(x) = \dfrac{1}{x-1}$ consists of all real numbers except $x = 1$. Therefore, f is continuous for all x except <u>$x = 1$.</u>

3. Since $x = 2$ is in the domain of $f(x) = \sqrt{x+1}$, we can obtain the limit by substituting:
$$\lim_{x \to 3} \sqrt{x+1} = \sqrt{3+1} = \sqrt{4} = 2.$$

5. Substituting $x = 0$ leads to $60/0$, which is the determinate form $k/0$. Since x^4 is non-negative, the quantity $60/x^4$ is non-negative, so the limit diverges to $+\infty$.

7. Substituting $x = 0$ leads to $-1/0$, which is the determinate form $k/0$. Since the denominator x^3, and hence the entire expression $(x^3-1)/x^3$, could be either positive or negative depending on whether $x \to 0^+$ or $x \to 0^-$, the limit does not exist.

9. Substituting $x = -\infty$ leads to $-\infty + 5$, which is the determinate form $k - \infty = -\infty$, so the limit diverges to $-\infty$.

11. Substituting $x = +\infty$ leads to $4^{-\infty}$, which is the determinate form $k^{-\infty} = 0$, so the limit is 0.

13. Substituting $x = 0$ leads to the indeterminate form $0/0$. Canceling the x^3 gives is the limit of $-1/3$.

15. Substituting $x = -\infty$ leads to the indeterminate form ∞/∞. Canceling the x^3 gives is $-1/3x^3$, resulting in the determinate form $(-1)/(-\infty) = 0$.

17. Substituting $x = -\infty$ leads to the determinate form $4/(\infty+2) = 4/\infty = 0$. Thus, the limit is 0.

19. Substituting $x = -\infty$ leads to the determinate form $60/(e^{-\infty} - 1) = 60/(0-1) = -60$. Thus, the limit is -60.

21. $\displaystyle\lim_{x \to 0} (x + 1) = 0 + 1 = 1$

23. $\displaystyle\lim_{x \to 2} \dfrac{2 + x}{x} = \dfrac{2 + 2}{2} = 2$

25. $\displaystyle\lim_{x \to -1} \dfrac{x + 1}{x} = \dfrac{-1 + 1}{-1} = 0$

27. $\displaystyle\lim_{x \to 8} (x - \sqrt[3]{x}) = 8 - \sqrt[3]{8} = 6$

29. $\displaystyle\lim_{h \to 1} (h^2 + 2h + 1) = 1^2 + 2(1) + 1 = 4$

31. $\displaystyle\lim_{h \to 3} 2 = 2$

33. $\displaystyle\lim_{h \to 0} \dfrac{h^2}{h + h^2} = \lim_{h \to 0} \dfrac{h}{1 + h} = \dfrac{0}{1 + 0} = 0$

35. $\displaystyle\lim_{x \to 1} \dfrac{x^2 - 2x + 1}{x^2 - x} = \lim_{x \to 1} \dfrac{(x-1)^2}{x(x-1)} = $
$\displaystyle\lim_{x \to 1} \dfrac{x - 1}{x} = \dfrac{1 - 1}{1} = 0$

37. $\displaystyle\lim_{x \to 2} \dfrac{x^3 - 8}{x - 2} = $
$\displaystyle\lim_{x \to 2} \dfrac{(x - 2)(x^2 + 2x + 4)}{x - 2} = $
$\displaystyle\lim_{x \to 2} (x^2 + 2x + 4) = 2^2 + 2(2) + 4 = 12$

39. $\displaystyle\lim_{x \to 0^+} \dfrac{1}{x^2}$ diverges to $+\infty$ because it has the form "$k/0$" and $1/x^2$ is positive.

41. $\displaystyle\lim_{x \to -1} \dfrac{x^2 + 1}{x + 1}$ does not exist: The limit from the left is $-\infty$ whereas the limit from the right is $+\infty$.

43. $\lim\limits_{x \to -2^+} \dfrac{x^2 + 8}{x^2 + 3x + 2}$

$= \lim\limits_{x \to -2^+} \dfrac{x^2 + 8}{(x + 2)(x + 1)}$.

As $x \to -2^+$, the numerator approaches 12 and the factor in the denominator $(x+2)$ is positive and approaches zero. Thus, the limit has the

determinate form $\dfrac{12}{(0^+)(-1)}$ which is negative, and

so diverges to $-\infty$.

45. $\lim\limits_{x \to -2} \dfrac{x^2 + 8}{x^2 + 3x + 2} = \lim\limits_{x \to -2}$

$\dfrac{x^2 + 8}{(x + 2)(x + 1)}$. As $x \to -2^+$, the numerator

approaches 12 and the factor in the denominator $(x+2)$ is positive and approaches zero. Thus, the

limit has the determinate form $\dfrac{12}{(0^+)(-1)}$ which is

negative, and so the limit diverges to $-\infty$. As $x \to$

-2^-, the factor $(x+2)$ is negative and approaches zero. Thus, the limit has the determinate form

$\dfrac{12}{(0^-)(-1)}$ which is positive, and so the limit

diverges to $+\infty$. Since the left and right limits

disagree, $\lim\limits_{x \to -2} \dfrac{x^2 + 8}{x^2 + 3x + 2}$ does not exist.

47. $\lim\limits_{x \to 2} \dfrac{x^2 + 8}{x^2 - 4x + 4} = \lim\limits_{x \to 2} \dfrac{x^2 + 8}{(x-2)^2}$.

As $x \to 2$ from either side, the numerator approaches 12 and the denominator is positive and approaches zero, so the limit has the determinate form $12/0^+$, telling us that the limit diverges to $+\infty$.

49. $\lim\limits_{x \to +\infty} \dfrac{3x^2 + 10x - 1}{2x^2 - 5x}$

$= \lim\limits_{x \to +\infty} \dfrac{3x^2}{2x^2} = \lim\limits_{x \to +\infty} \dfrac{3}{2} = 3/2$

51. $\lim\limits_{x \to +\infty} \dfrac{x^5 - 1000x^4}{2x^5 + 10{,}000}$

$= \lim\limits_{x \to +\infty} \dfrac{x^5}{2x^5} = \lim\limits_{x \to +\infty} \dfrac{1}{2} = 1/2$

53. $\lim\limits_{x \to +\infty} \dfrac{10x^2 + 300x + 1}{5x + 2}$

$= \lim\limits_{x \to +\infty} \dfrac{10x^2}{5x} = \lim\limits_{x \to +\infty} (2x)$

diverges to $+\infty$

55. $\lim\limits_{x \to +\infty} \dfrac{10x^2 + 300x + 1}{5x^3 + 2} = \lim\limits_{x \to +\infty} \dfrac{10x^2}{5x^3}$

$= \lim\limits_{x \to +\infty} \dfrac{2}{x} = 0$

57. $\lim\limits_{x \to -\infty} \dfrac{3x^2 + 10x - 1}{2x^2 - 5x} = \lim\limits_{x \to -\infty} \dfrac{3x^2}{2x^2} =$

$\lim\limits_{x \to -\infty} \dfrac{3}{2} = 3/2$

59. $\lim\limits_{x \to -\infty} \dfrac{x^5 - 1000x^4}{2x^5 + 10{,}000}$

$= \lim\limits_{x \to -\infty} \dfrac{x^5}{2x^5} = \lim\limits_{x \to -\infty} \dfrac{1}{2} = 1/2$

61. $\lim\limits_{x \to -\infty} \dfrac{10x^2 + 300x + 1}{5x + 2}$

$= \lim\limits_{x \to -\infty} \dfrac{10x^2}{5x} = \lim\limits_{x \to -\infty} \dfrac{10x}{5}$

diverges to $-\infty$

63. $\lim\limits_{x \to -\infty} \dfrac{10x^2 + 300x + 1}{5x^3 + 2} = \lim\limits_{x \to -\infty} \dfrac{10x^2}{5x^3}$

$= \lim\limits_{x \to -\infty} \dfrac{10}{5x} = 0$

65. As $x \to +\infty, 4e^{-3x} = \dfrac{4}{e^{3x}} \to 0 \left[\dfrac{4}{+\infty} = 0 \right]$

so $\lim\limits_{x \to +\infty} (4e^{-3x} + 12) = 0 + 12 = 12$

67. As $t \to +\infty$, $3^{3t} \to +\infty$ as well, so the denominator $5 - 5.3(3^{3t}) \to -\infty$. Thus

$$\lim_{t \to +\infty} \frac{2}{5 - 5.3(3^{3t})} = 0 \quad \left[\frac{2}{-\infty} = 0\right]$$

69. As $t \to +\infty$. the numerator becomes large without bound, and so diverges to $+\infty$. The term $5.3e^{-t}$ in the denominator is $5.3/e^t$ and so approaches 0, so the denominator approaches $1 + 0 = 1$. Since the numerator is diverging to $+\infty$ and the denominator is approaching 1, the ratio diverges to $+\infty$. $\left[\frac{+\infty}{1} = +\infty\right]$

71. As $x \to -\infty$. the term $-3^{2x} \to 0$, and the term e^x in the denominator approaches 0. So,

$$\lim_{x \to -\infty} \frac{-3^{2x}}{2 + e^x} = \frac{0}{2 + 0} = 0$$

73. The only possible discontinuity is at $x = 0$. There, $\lim_{x \to 0^-} f(x) = 0 + 2 = 2$ whereas $\lim_{x \to 0^+} f(x) = 2(0) - 1 = -1$. Since these disagree, there is a discontinuity at $x = 0$.

75. The only possible discontinuities are at $x = 0$ and $x = 2$. We have $\lim_{x \to 0^-} g(x) = 0 + 2 = 2$ and $\lim_{x \to 0^+} g(x) = 2(0) + 2 = 2 = g(0)$, hence g is continuous at 0. We have $\lim_{x \to 2^-} g(x) = 2(2) + 2 = 8$ and $\lim_{x \to 2^+} g(x) = 2^2 + 2 = 8 = g(2)$.

Hence, g is continuous everywhere.

77. The only possible discontinuity is at $x = 0$. We have $\lim_{x \to 0^-} h(x) = 0 + 2 = 2 \neq h(0) = 0$, so there is a discontinuity at $x = 0$. (Note that we don't have to bother computing the limit from the right, which also equals 2.)

79. The only possible discontinuities are at $x = 0$ and $x = 2$. We have $\lim_{x \to 0^-} f(x) = \lim_{x \to 0^-} (1/x) =$

$-\infty$, so there is a discontinuity at $x = 0$. On the other hand, $\lim_{x \to 2^-} f(x) = 2 = f(2)$ and $\lim_{x \to 2^+} f(x) = 2^{2-1} = 2$ also, so f is continuous at 2.

81. (a) Since $N(t) = 0.22t + 3$ when $t < 5$ and is a closed-form function in this range, we compute the limit $\lim_{t \to 5^-} N(t)$ by substituting $t = 5$ to get

$$\lim_{t \to 5^-} N(t) = \lim_{t \to 5^-} 0.22t + 3 = 0.22(5) + 3$$
$$= 4.1$$

Thus, shortly before 2000 ($t = 5$) the number of workers employed in manufacturing jobs in Mexico was close to 4.1 million.

Since $N(t) = -0.15t + 4.85$ when $t > 5$ and is a closed-form function in this range, we compute the limit $\lim_{t \to 5^+} N(t)$ by substituting $t = 5$ to get

$$\lim_{t \to 5^+} N(t) = \lim_{t \to 5^+} -0.15t + 4.85$$
$$= -0.15(5) + 4.85 = 4.1$$

Thus, shortly after 2000 ($t = 5$) the number of workers employed in manufacturing jobs in Mexico was close to 4.1 million.

(b) Since $\lim_{t \to 5^-} N(t) = \lim_{t \to 5^+} N(t)$, N is continuous at $t = 5$. Further, N is continuous at every other point of its domain because it is closed-form everywhere else. Thus, the number $N(t)$ of workers employed in manufacturing jobs in Mexico is continuous on its domain $[0, 9]$, and hence does not change abruptly at any point.

83. (a) Since $f(t) = 0.04t + 0.33$ when $t < 4$ and is a closed-form function in this range, we compute the limit $\lim_{t \to 4^-} f(t)$ by substituting $t = 4$ to get

$$\lim_{t \to 4^-} f(t) = \lim_{t \to 4^-} 0.04t + 0.33$$
$$= 0.04(4) + 0.33 = 0.49$$

Thus, shortly before 1999 ($t = 4$), annual advertising expenditures were close to $0.49 billion.

Since $f(t) = -0.01t + 1.2$ when $t > 4$ and is a closed-form function in this range, we compute the limit $\lim_{t \to 4^+} f(t)$ by substituting $t = 4$ to get

$$\lim_{t \to 4^+} f(t) = \lim_{t \to 4^+} -0.01t + 1.2$$
$$= -0.01(4) + 1.2 = 1.16$$

Thus, shortly after 1999 ($t = 4$), annual advertising expenditures were close to $1.16 billion.

(b) By part (a), $\lim_{t \to 4^-} f(t) \neq \lim_{t \to 4^+} f(t)$, and so f is not continuous at $t = 4$. Interpretation: Movie advertising expenditures jumped suddenly in 1999.

85. $\lim_{t \to +\infty} \dfrac{P(t)}{C(t)} = \lim_{t \to +\infty} \dfrac{1.745t + 29.84}{1.097t + 10.65}$

$= \lim_{t \to +\infty} \dfrac{1.745t}{1.097t} = \lim_{t \to +\infty} \dfrac{1.745}{1.097} \approx 1.59$

If the trend continued indefinitely, the annual spending on police would be 1.59 times the annual spending on courts in the long run.

87. As $t \to +\infty$, the term $240e^{-0.4t} = \dfrac{240}{e^{0.4t}}$ being 240 divided by a very large number, approaches zero (in terms of determinate forms, $240/\infty = 0$). Thus

$$\lim_{t \to +\infty} R(t) = \lim_{t \to +\infty} (825 - 240e^{-0.4t})$$
$$= 825 - 0 = 825$$

Interpretation: In the long term, annual revenues will approach $825 million.

89. $\lim_{t \to +\infty} I(t) = \lim_{t \to +\infty} (t^2 + 3.5t + 50) = +\infty$

$\lim_{t \to +\infty} \dfrac{I(t)}{E(t)} = \lim_{t \to +\infty} \dfrac{t^2 + 3.5t + 50}{0.4t^2 - 1.6t + 14} =$

$\lim_{t \to +\infty} \dfrac{t^2}{0.4t^2}$

$= \lim_{t \to +\infty} \dfrac{1}{0.4} = 2.5.$

In the long term, U.S. imports from China will rise without bound and be 2.5 times U.S. exports to China. In the real world, imports and exports cannot rise without bound. Thus, the given

models should not be extrapolated far into the future.

91. $\lim_{t \to +\infty} p(t) = \lim_{t \to +\infty} 100\left(1 - \dfrac{12{,}200}{t^{4.48}}\right)$
$= 100(1 - 0) = 100.$

The percentage of children who learn to speak approaches 100% as their age increases.

93. To evaluate $\lim_{x \to a} f(x)$ algebraically, first check whether $f(x)$ is a closed-form function. Then check whether $x = a$ is in its domain. If so, the limit is just $f(a)$; that is, it is obtained by substituting $x = a$. If not, then try to first simplify $f(x)$ in such a way as to transform it into a new function such that $x = a$ is in its domain, and then substitute. A disadvantage of this method is that it is sometimes extremely difficult to evaluate limits algebraically, and rather sophisticated methods are often needed.

95. She is wrong. Closed-form functions are continuous only at points in their domains, and $x = 2$ is not in the domain of the closed-form function $f(x) = 1/(x-2)^2$.

97. Answers may vary. (1) See Example 2: $\lim_{x \to 2} \dfrac{x^3 - 8}{x - 2}$ which leads to the indeterminate form 0/0 but the limit is 12. (2) $\lim_{x \to +\infty} \dfrac{60x}{2x}$ which leads to the indeterminate form ∞/∞, but where the limit exists and equals 30.

99. The statement may not be true. For example, if $f(x) = \begin{cases} x+2 & \text{if } x < 0 \\ 2x-1 & \text{if } x \geq 0, \end{cases}$
then $f(0)$ is defined and equals -1, and yet $\lim_{x \to 0} f(x)$ does not exist. The statement can be corrected by requiring that f be a closed-form function: "If f is a closed form function, and $f(a)$ is defined, then $\lim_{x \to a} f(x)$ exists and equals $f(a)$."

101. Answers may vary. For example, $f(x) =$
$\begin{cases} 0 & \text{if } x \text{ is any number other than } 1 \text{ or } 2 \\ 1 & \text{if } x = 1 \text{ or } 2 \end{cases}$

103. Answers may vary.

(1) $\lim\limits_{x \to +\infty} [(x+5) - x] = \lim\limits_{x \to +\infty} 5 = 5$

(2) $\lim\limits_{x \to +\infty} [x^2 - x] = \lim\limits_{x \to +\infty} x(x-1) = +\infty$

(3) $\lim\limits_{x \to +\infty} [(x-5) - x] = \lim\limits_{x \to +\infty} -5 = -5$

10.4 Average Rate of Change

1. $[f(3) - f(1)]/(3 - 1) = (-1 - 5)/2 = -3$

3. $[f(-1) - f(-1)]/[-1 - (-3)] = [-1.5 - (-2.1)]/2 = 0.3$

5. $[R(6) - R(2)]/(6 - 2) = (20.1 - 20.2)/4 = -\$25,000$ per month

7. $[q(5.5) - q(5)]/(5.5 - 5) = (300 - 400)/0.5 = -200$ items per dollar

9. $[S(5) - S(2)]/(5 - 2) = (27 - 23)/3 \approx \$1.33$ per month

11. $[U(4) - U(0)]/(4 - 0) = (8 - 5)/4 = 0.75$ percentage point increase in unemployment per 1 percentage point increase in the deficit

13. $[f(3) - f(1)]/(3 - 1) = [6 - (-2)]/2 = 4$

15. $[f(0) - f(-2)]/[0 - (-2)] = (4 - 0)/2 = 2$

17. $[f(3) - f(2)]/(3 - 2) = [9/2 + 1/3 - (2 + 1/2)]/1 = 7/3$

19. $f(x) = 2x^2$
Average rate of change
$$= \frac{f(a+h) - f(a)}{h} = \frac{f(h) - f(0)}{h}$$
because $a = 0$.

$h = 1$: $\dfrac{f(h) - f(0)}{h} = \dfrac{f(1) - f(0)}{1} = \dfrac{2 - 0}{1} = 2$

$h = 0.1$: $\dfrac{f(h) - f(0)}{h} = \dfrac{f(0.1) - f(0)}{0.1} = \dfrac{0.02 - 0}{0.1} = 0.2$

Technology can be used to compute the remaining cases. All the values are shown in the following table:

h	Ave. Rate of Change
1	2
0.1	0.2
0.01	0.02
0.001	0.002
0.0001	0.0002

21. $f(x) = 1/x$
Average rate of change
$$= \frac{f(a+h) - f(a)}{h} = \frac{f(2+h) - f(2)}{h}$$
because $a = 2$.

$h = 1$: $\dfrac{f(2+h) - f(2)}{h} = \dfrac{f(3) - f(2)}{1} = \dfrac{1/3 - 1/2}{1} = -1/6$
≈ -0.1667

$h = 0.1$: $\dfrac{f(2+h) - f(2)}{h} = \dfrac{f(2.1) - f(2)}{0.1} = \dfrac{1/2.1 - 1/2}{0.1}$
≈ -0.2381

Technology can be used to compute the remaining cases. All the values are shown in the following table:

h	Ave. Rate of Change
1	−0.1667
0.1	−0.2381
0.01	−0.2488
0.001	−0.2499
0.0001	−0.24999

23. $f(x) = x^2 + 2x$
Average rate of change
$$= \frac{f(a+h) - f(a)}{h} = \frac{f(3+h) - f(3)}{h}$$
because $a = 3$.

$h = 1$: $\dfrac{f(3+h) - f(3)}{h} = \dfrac{f(4) - f(3)}{1} = \dfrac{24 - 15}{1} = 9$

$h = 0.1$: $\dfrac{f(3+h) - f(3)}{h} = \dfrac{f(3.1) - f(3)}{0.1} = \dfrac{15.81 - 15}{0.1}$
$= 8.1$

Technology can be used to compute the remaining cases. All the values are shown in the following table:

h	Ave. Rate of Change
1	9
0.1	8.1
0.01	8.01
0.001	8.001
0.0001	8.0001

25. (a) Average rate of change over $[-6, 6]$

$$= \frac{C(6) - C(-6)}{6 - (-6)} = \frac{1200 - 900}{6 - (-6)}$$

$$= \frac{300}{12} = \$25 \text{ billion per year;}$$

World military expenditure increased at an average rate of $25 billion per year during 1994–2006.

(b) Average rate of change over $[-2, 6]$

$$= \frac{C(6) - C(-2)}{6 - (-2)} = \frac{1200 - 800}{6 - (-2)}$$

$$= \frac{400}{8} = \$50 \text{ billion per year;}$$

World military expenditure increased at an average rate of $50 billion per year during 1998–2006.

27. (a) The period 2002–2007 is represented by $[2, 7]$. The average rate of change of $P(t)$ is

$$\frac{P(7) - P(2)}{7 - 2} = \frac{3.2 - 3.3}{5} = -\frac{0.1}{5}$$

$$= -0.02 \text{ million barrels/year,}$$

or –20,000 barrels/year.

Interpretation: During 2002–2007, daily oil production by Pemex was decreasing at an average rate of 20,000 barrels of oil per year.

(b) The one-year average rates of change from 2001 on are the successive differences:

$$P(2) - P(1), P(3) - P(2), ..., P(9) - P(8)$$

From the table, these differences are:

0.2, 0.1, 0, 0, –0.1, –0.1, –0.1, –0.1

and thus never increased in value (Choice (C))

29. (a) From the graph:

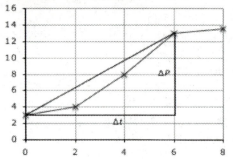

Rate of change over $[0, 6]$

$$= \frac{\Delta P}{\Delta t} \approx \frac{10}{6} \approx 1.7$$

Interpretation: The percentage of mortgages classified as subprime was increasing at an average rate of around 1.7 percentage points per year between 2000 and 2006.

(b) The rates of change of $P(t)$ over successive 2-year periods are given by the slopes of the individual line segments that make up the graph in the figure. Thus, the greatest average rate of change over a single year corresponds to the segment with the largest slope: the segment over $[4, 6]$, corresponding to 2004–2006.

31. The rates of change over possible two-quarter periods are represented by the slopes of the lines joining points separated by two quarters as shown:

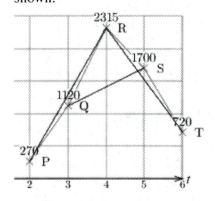

(a) The greatest average rate of change corresponds to the line segment with the greatest slope: PR, with a slope of $(2315 - 270)/2 = 1022.5$ thousand iPhones per quarter. Interpretation: The segment PR corresponds to the second and third quarter of 2007. During the second and third

quarter of 2007 iPhone sales were increasing at an average rate of 1,022,500 phones per quarter.
(b) The least average rate of change corresponds to the line segment with the least slope: RT, with a slope of (720–2315)/2 = –797.5 thousand iPhones per quarter. Interpretation: The segment RT corresponds to the fourth quarter of 2007 and first quarter of 2008. During the fourth quarter of 2007 and first quarter of 2008 iPhone sales were decreasing at an average rate of 797,500 phones per quarter.

33. (a) Check each interval ([3, 5], [3, 7], and so on). You will find the most negative average drop, in [3, 5], is (4.6–5.1)/(5–3) = –0.25 thousand articles per year. So, during the period 1993–1995, the number of articles authored by U.S. researchers decreased at an average rate of 250 articles per year.
(b) The percentage rate of change is
  $[N(13) - N(3)]/N(3)$
  = (4.2–5.1)/5.1
  ≈ –0.1765;
the average rate of change is
  $[N(13) - N(3)]/(13 - 3)$
  = (4.2–5.1)/10
  = –0.09 thousand articles per year.
Over the period 1993–2003, the number of articles authored by U.S. researchers decreased at an average rate of 90 per year, representing an 17.65% decrease over that period.

35. (a) (984 – 936)/4 = 17 teams per year
(b) It decreased: The slope of the graph from $t = 2$ to $t = 6$ is steeper than that from $t = 3$ to $t = 7$ and in turn is steeper than the slope from $t = 4$ to $t = 8$.

37. We can estimate the slope of the regression line using two grid points it passes through: (2, 1.25) and (6, 1.3).
  Slope of regression line
  ≈ (1.3–1.25)/(6–2)
  = 0.0125 billion dollars per year
(a) (C): On the interval [0, 4], the average rate of change of government funding was approximately

  (1.3–1.24)/(4–0) = 0.0125,
the same as the estimated slope of the regression line.
(b) (A): On the interval [4, 8], the average rate of change of government funding was approximately
  (1.2–1.3)/(8–4) = –0.025,
less than the estimated slope of the regression line.
(c) (B): On the interval [3, 6], the average rate of change of government funding was approximately
  (1.4–1.22)/(6–3) = 0.06,
greater than the estimated slope of the regression line.
(d) Over the interval [0, 8] the average rate of change of government funding was approximately
  (1.2 – 1.25)/(8 – 0)
  = –0.00625 ≈ –0.0063
(to two significant digits) billion dollars per year, (–$6,300,000 per year). This is much less than the (positive) slope of the regression line, 0.0125 ≈ 0.013 billion dollars per year, ($13,000,000 per year).

39. From 1991 to 1995 the volatility decreased at an average rate of 0.2 points per year, so decreased a total of 4×0.2 = 0.8 points. Since its value in 1995 was 1.1, its value in 1991 must have been 1.1 + 0.8 = 1.9. Similarly, from 1995 to 1999 the volatility increased a total of 4×0.3 = 1.2 to end at 1.1 + 1.2 = 2.3. In between these points we've found almost anything could happen, but the graph might look something like the following:

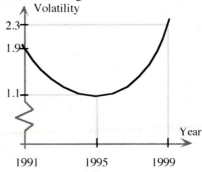

41. $[I(2) - I(0)]/(2 - 0)$
$= (1600 - 1000)/2 = 300.$

The index was increasing at an average rate of 300 points per day.

43. (a) Average rate of change of P is

$$\frac{P(26) - P(1)}{26 - 1}$$

$$= \frac{97.2-93.45}{25} = \$0.15 \text{ per year.}$$

The price per barrel of crude oil in constant 2008 dollars was growing at an average rate of about 15¢ per year over the period 1981–2006.
(b) No; According to the model, during that 25-year period the price of oil went down from around \$93 to a low of around \$25 in 1993 before climbing back up. In general, the average rate of change of a function over $[a, b]$ is not effected by the values of the function between a and b.

45. (a) The period March 17 to March 23 corresponds to $[0, 6]$. Average rate of change of A is

$$\frac{A(6) - A(0)}{6 - 0}$$

$$\approx \frac{450.8-167}{6} = 47.3 \text{ new cases per day.}$$

Interpretation: The number of SARS cases was growing at an average rate of 47.3 new cases per day over the period March 17 to March 23.
(b) The graph of $A(t) = 167(1.18)^t$ is increasingly steep as t increases. Therefore, the average rates of change increase with t, so the number of reported cases was increasing at a faster and faster rate (choice (A)).

47. (a) $[f(6) - f(5)]/(6 - 5)$
$= (27.6 - 18.75)/1$
$= 8.85$ manatee deaths per 100,000 boats;
$[f(8) - f(7)]/(8 - 7)$
$= (66.6 - 43.55)/1$
$= 23.05$ manatee deaths per 100,000 boats
(b) More boats result in more manatee deaths per additional boat.

49. (a) Here is an Excel worksheet that computes the successive rates of change:

◇	A	B	C
1	x	p(x)	Rate of Change
2	39	=0.092*A2^2-8.1*A2+190	
3	39.5		=(B3-B2)/0.5
4	40		
5	40.5		
6	41		
7	41.5		
8	42		

This worksheet leads to the following values, showing the desired rates of change in the rightmost column.

◇	A	B	C
1	x	p(x)	Rate of Change
2	39	14.032	
3	39.5	13.593	-0.88
4	40	13.2	-0.79
5	40.5	12.853	-0.69
6	41	12.552	-0.60
7	41.5	12.297	-0.51
8	42	12.088	-0.42

(b) From the table, the average rate of change of p over $[40, 40.5]$ is -0.69. The unites of measurement are units of p per unit ot x: percentage points per \$1000 of household income. Thus, for household incomes between \$40,000 and \$40,500, the poverty rate decreases at an average rate of 0.69 percentage points per \$1000 increase in the median household income.
(c) All the rates of change in obtained in part (a) are negative, showing that the poverty rate decreases as the median houshold income increases (Choice B).
(d) Although all the rates of change in the table are negative, they become less so as the household income increases. Thus, the effect is decreasing in magnitude (Choice B).

51. The average rate of change of f over an interval $[a, b]$ can be determined numerically; using a table of values; graphically, by measuring the slope of the corresponding line segment through two points on the graph; or algebraically, using an algebraic formula for the function. Of these, the least precise is the graphical method, because it relies on reading coordinates of points on a graph.

53. No, the formula for the average rate of a function f over $[a, b]$ depends only on $f(a)$ and $f(b)$, and not on any values of f between a and b.

55. Answers will vary. Here is one possibility:

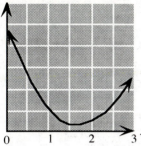

57. For every change of 1 in C, B changes by 3, so A changes by $2 \times 3 = 6$ units of quantity A per unit of quantity C

59. (A): The secant line given by $x = 1$ and $x = 1 + h$ is steeper for smaller values of h.

61. Yes. Here is an example, in which the average rate of growth for 2000–2003 is negative, but the average rates of growth for 2000–2001 and 2001–2002 are positive:

Year	2000	2001	2002	2003
Revenue (billion)	$10	$20	$30	$5

63. (A): This can be checked by algebra:
$\{[f(2) - f(1)]/(2 - 1) + [f(3) - f(2)]/(3 - 2)\}/2$
$= [f(3) - f(1)]/(3 - 1)$.

10.5 Derivatives: Numerical and Graphical Viewpoints

1. 6: The average rates of change are approaching 6 for both positive and negative values of h approaching 0.

3. −5.5: The average rates of change are approaching −5.5 for both positive and negative values of h approaching 0.

5. The average rate of change is $\dfrac{R(a+h) - R(a)}{h}$.

Here, $a = 5$.

$h = 1$:
$$\frac{R(5+1) - R(5)}{1} = \frac{R(6) - R(5)}{1}$$
$$= \frac{39}{1} = 39$$

$h = 0.1$:
$$\frac{R(5+0.1) - R(5)}{0.1} = \frac{R(5.1) - R(5)}{0.1}$$
$$= \frac{3.99}{0.1} = 39.9$$

$h = 0.01$:
$$\frac{R(5+0.01) - R(5)}{0.01} = \frac{R(5.01) - R(5)}{0.01}$$
$$= \frac{0.3999}{0.01} = 39.99$$

Table:

h	1	0.1	0.01
Ave. rate	39	39.9	39.99

The average rates are approaching an instantaneous rate of $40 per day.

7. The average rate of change is $\dfrac{R(a+h) - R(a)}{h}$.

Here, $a = 1$.

$h = 1$:
$$\frac{R(1+1) - R(1)}{1} = \frac{R(2) - R(1)}{1}$$
$$= \frac{140}{1} = 140$$

$h = 0.1$:
$$\frac{R(1+0.1) - R(1)}{0.1} = \frac{R(1.1) - R(1)}{0.1}$$

$$= \frac{6.62}{0.1} = 66.2$$

$h = 0.01$:
$$\frac{R(1+0.01) - R(1)}{0.01} = \frac{R(1.01) - R(1)}{0.01}$$
$$= \frac{0.60602}{0.01} = 60.602$$

Table:

h	1	0.1	0.01
Ave. rate	140	66.2	60.602

The average rates are approaching an instantaneous rate of $60 per day.

9. The average cost to manufacture h more items is the average rate of change: $\dfrac{C(a+h) - C(a)}{h}$.

Here, $a = 1000$.

$h = 10$:
$$\frac{C(1000+10) - C(1000)}{10}$$
$$= \frac{C(1010) - C(1000)}{10} = \frac{47.99}{10} = 4.799$$

$h = 1$:
$$\frac{C(1000+1) - C(1000)}{1}$$
$$= \frac{C(1001) - C(1000)}{1} = \frac{4.7999}{1} = 4.7999$$

Table:

h	10	1
C_{ave}	4.799	4.7999

$C'(1,000) = \$4.8$ per item

11. The average cost to manufacture h more items is the average rate of change: $\dfrac{C(a+h) - C(a)}{h}$. Here, $a = 100$.

$h = 10$:
$$\frac{C(100+10) - C(100)}{10} = \frac{C(110) - C(100)}{10}$$
$$= \frac{999.0909091}{10} \approx 99.91$$

$h = 1$:
$$\frac{C(100+1) - C(100)}{1} = \frac{C(101) - C(100)}{1}$$
$$= \frac{99.9009901}{1} \approx 99.90$$

Table:

384

h	10	1
C_{ave}	99.91	99.90

$C'(100) = \$99.9$ per item

In each of 13–16, the answer is obtained by estimating the slope of the tangent line shown.
13. In the graph, the tangent line passes through $(0, 2)$ and $(6, 5)$. Therefore its slope is $(5-2)/(6-0) = 1/2$.

15. In the graph, the tangent is horizontal. Therefore its slope is 0.

In each of 17–22, the answer to (a) is the point at which the graph is rising with the steepest slope or falling with the shallowest slope; the answer to (b) is the point at which the graph is rising with the shallowest slope or falling with the steepest slope.
17. (a) R **(b)** P

19. (a) P **(b)** R

21. (a) Q **(b)** P

In each of 23–26, the answer may be obtained by estimating the slopes of the tangent lines to the given points and comparing these slopes to the given numbers.
23. (a) Q **(b)** R **(c)** P

25. (a) R **(b)** Q **(c)** P

27. (a) The only point where the tangent line has slope 0 is $(1, 0)$.
(b) None; the graph never rises.
(c) The only point where the tangent line has slope -1 is $(-2, 1)$.

29. (a) The points where the tangent line has slope 0 are $(-2, 0.3)$, $(0, 0)$, and $(2, -0.3)$.
(b) None; the graph never rises that steeply.
(c) None; the graph never falls that steeply.

31. $(a, f(a)); f'(a)$.

33. (B): The derivative is the slope of the tangent line. It is not any particular average rate of change or difference quotient; these only *approximate* the derivative.

35. (a) (A): The graph rises above the tangent line at $x = 2$.
(b) (C): The secant line is roughly parallel to the tangent line at $x = 0$.
(c) (B): The slopes of the tangent lines are decreasing.
(d) (B): The slopes of the tangent lines decrease to 0 then increase again.
(e) (C): The height of the graph is approximately 0.7 while the slope of the tangent line at $x = 4$ is approximately 1.

37. $[f(2 + 0.0001) - f(2 - 0.0001)]/0.0002 = -2$

39. $[f(-1 + 0.0001) - f(-1 - 0.0001)]/0.0002 \approx -1.5$

In each of 41–48 we use the "quick approximation" method of estimating the derivative using the balanced difference quotient. You could also use the ordinary difference quotient or a table of average rates of change with values of h approaching 0.
41. $[g(t + 0.0001) - g(t - 0.0001)]/0.0002 \approx -5$

43. $[y(2 + 0.0001) - y(2 - 0.0001)]/0.0002 = 16$

45. $[s(-2 + 0.0001) - s(-2 - 0.0001)]/0.0002 = 0$

47. $[R(20 + 0.0001) - R(20 - 0.0001)]/0.0002 \approx -0.0025$

49. (a) $[f(-1 + 0.0001) - f(-1 - 0.0001)]/0.0002 \approx 3$

(b) The equation of the line through $(-1, -1)$ with slope 3 is $y = 3x + 2$.

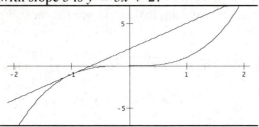

51. **(a)** $[f(2 + 0.0001) - f(2 - 0.0001)]/0.0002 \approx \frac{3}{4}$

(b) The equation of the line through $(2, 2.5)$ with slope $\frac{3}{4}$ is $y = \frac{3}{4}x + 1$.

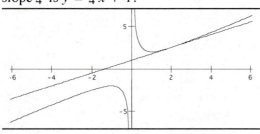

53. **(a)** $[f(4 + 0.0001) - f(4 - 0.0001)]/0.0002 \approx \frac{1}{4}$

(b) The equation of the line through $(4, 2)$ with slope $\frac{1}{4}$ is $y = \frac{1}{4}x + 1$.

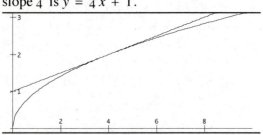

55. $[e^{0.0001} - e^{-0.0001}]/0.0002 \approx 1.000$

57. $[\ln(1 + 0.0001) - \ln(1 - 0.0001)]/0.0002 \approx 1.000$

59. (C): The graph of f is a falling straight line, so must have the same negative slope (derivative) at every point; f' must be a negative constant.

61. (A): The function f decreases until $x = 0$, where it turns around and starts to increase. Its derivative must be negative until $x = 0$, where the derivative is 0; past that the derivative becomes positive. This is exactly what (A) illustrates.

63. (F): The function f increases slowly at first, it becomes steeper around $x = 0$, then it returns to slowly rising. Its derivative starts as a small positive number, increases to become largest around $x = 0$, then decreases back toward 0. This is the behavior seen in (F).

65. Increasing (sloping up) for $x < 0$; Decreasing (sloping down) for $x > 0$.

67. Increasing (sloping up) for $x < -1$ and $x > 1$ Decreasing (sloping down) for $-1 < x < 1$

69. Increasing (derivative positive) for $x > 1$; Decreasing (derivative negative) for $x < 1$.

71. Increasing (derivative positive) for $x < 0$; Decreasing (derivative negative) for $x > 0$.

73. The derivative:

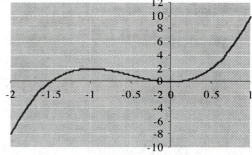

$x = -1.5$, $x = 0$: These are the points where the derivative is 0 (crosses the x axis).

75. Set up the spreadsheet as in Example 4:

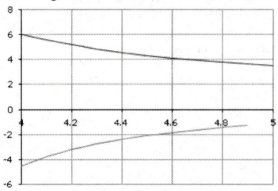

To graph $f(x)$ and $f'(x)$, highlight columns A through C and use the Chart Wizard to create the following Scatter Plot:

The top curve is $y = f(x)$; the bottom curve is $y = f'(x)$.

77. $q(100) = 50{,}000$, $q'(100) = -500$ (use one of the quick approximations). A total of 50,000 pairs of sneakers can be sold at a price of $100, but the demand is decreasing at a rate of 500 pairs per $1 increase in the price.

79. (a) The tangent line passes through the two points $(1, 1.8)$ and $(5, 1.6)$ as shown on the graph:

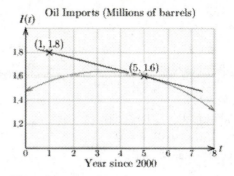

Oil Imports (Millions of barrels)

Therefore its slope is

$$\frac{\Delta y}{\Delta x} = \frac{1.6 - 1.8}{5 - 1} = -0.05.$$

Since $I(5) = 1.6$ (value on graph) and $I'(5) \approx -0.05$ (slope of tangent), we conclude that, in 2005, daily oil imports from Mexico in 2005 were 1.6 million barrels and declining at a rate of 0.05 million barrels (or 50,000 barrels) per year.
(b) Decreasing throughout, since the slope is decreasing (Even through the function first increases then decreases, the *slope* is decreasing.)

81. (a) (B): The graph is getting less steep.
(b) (B): The graph goes from above the tangent line to below it, so the slope of the tangent line is greater than the average rate of change.
(c) (A): From 0 to about 12 the graph is getting steeper, so the instantaneous rate of change is increasing; from that point on the graph is getting less steep, so the instantaneous rate of change is decreasing.
(d) 1992: This is the point ($t = 12$) where the graph is steepest.
(e) Reading values from the graph, we get the approximation $(1.2 - 0.8)/8 = 0.05$: In 1996, the total number of state prisoners was increasing at a rate of approximately 50,000 prisoners per year.

83. (a) $[s(4) - s(2)]/(4 - 2) = -96$ ft/sec
(b) $[s(4 + 0.0001) - s(4 - 0.0001)]/0.0002 = -128$ ft/sec

85. (a) Average rate of change of P is
$$\frac{P(28) - P(0)}{28 - 0}$$

387

$$= \frac{121.8-105}{28} = \$0.60 \text{ per year.}$$

The price per barrel of crude oil in constant 2008 dollars was growing at an average rate of about 60¢ per year over the 28-year period beginning at the start of 1980.

(b) Instantaneous rate of change of $P(t)$ at $t = 0$ is (using the quick approximation)

$$\frac{P(0 + 0.0001) - P(0 - 0.0001)}{0.0002}$$

$$\approx \frac{104.9988 - 105.0012}{0.0002}$$

$$= -12.0$$

The price per barrel of crude oil in constant 2008 dollars was dropping at an instantaneous rate of about \$12 per year at the start of 1980.

(c) The price of oil was decreasing in January 1980, but eventually began to increase (making the average rate of change in part (a) positive).

87. (a) Instantaneous rate of change of $A(t)$ at $t = 10$ is (using the quick approximation)

$$\frac{A(10 + 0.0001) - A(10 - 0.0001)}{0.0002}$$

$$\approx \frac{874.0650 - 874.03607}{0.0002}$$

$$\approx 144.7$$

Interpretation: The number of SARS cases was growing at a rate of about 144.7 new cases per day on March 27.

(b) The slope of the tangent to the graph of $A(t) = 167(1.18)^t$ is increasingly steep as t increases. Therefore, the instantaneous rates of change increase with t (choice (A)).

89. $S(5) \approx 109,$

$$\frac{dS}{dt}\Big|_{t=5} \approx [S(5 + 0.0001) -$$

$$S(5 - 0.0001)]/0.0002 \approx 9.1.$$

After 5 weeks, sales are 109 pairs of sneakers per week, and sales are increasing at a rate of 9.1 pairs per week each week.

91. $A(0) = 4.5$ million because A gives the number of subscribers; $A'(0) = 60,000$ subscribers per week because A' gives the rate at which the number of subscribers is changing.

93. (a) 60% of children can speak at the age of 10 months. At the age of 10 months, this percentage is increasing by 18.2 percentage points per month.

(b) As t increases, p approaches 100 percentage points (almost all children eventually learn to speak), and dp/dt approaches zero because the percentage stops increasing.

95. (a) $A(6) = \dfrac{15}{1 + 8.6(1.8)^{-6}} \approx 12.0$

$$A'(6) \approx \frac{A(6 + 0.0001) - A(6 - 0.0001)}{0.0002}$$

$$\approx \frac{11.97284169 - 11.97255763}{0.0002}$$

$$\approx 1.4$$

Interpretation: At the start of 2006, about 12% of U.S. mortgages were subprime, and this percentage was increasing at a rate of about 1.4 percentage points per year.

(b) Graphs:

Graph of A:

Graph of A':

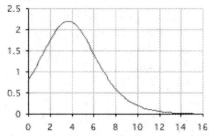

From the graphs, $A(t)$ approaches 15 as t becomes large (in terms of limits, $\lim\limits_{x \to +\infty} A(t) = 15$) and $A'(t)$ approaches 0 as t becomes large (in terms of limits, $\lim\limits_{x \to +\infty} A'(t) = 0$). Interpretation: If the trend modeled by the function A had continued

388

indefinitely, in the long term 15% of U.S. mortgages would have been subprime, and this percentage would not be changing.

97.

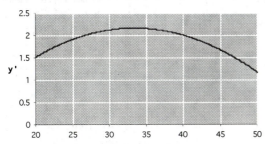

(a) (D): The graph of the derivative is rising.
(b) 33 days after the egg was laid: That is where the graph of the derivative is highest.
(c) 50 days after the egg was laid: That is where the graph of the derivative is lowest in the range $20 \leq t \leq 50$.

99. $L(0.95) \approx 31.2$ meters and $L'(0.95) \approx$ $[L(0.95 + 0.0001) - L(0.95 - 0.0001)]/0.0002 \approx -304.2$ meters/warp. Thus, at a speed of warp 0.95, the spaceship has an observed length of 31.2 meters and its length is decreasing at a rate of 304.2 meters per unit warp, or 3.042 meters per increase in speed of 0.01 warp.

101. The difference quotient is not defined when $h = 0$ because there is no such number as 0/0.

103. $H(10) = 50$ tells you that the membership in 2030 ($t = 10$) was 50 million.
$H'(10) = -6$ tells you that the rate of change of membership in 2030 ($t = 10$) was -6 million per year; that is, that the membership was decreasing at a rate of 6 million per year. These statements give choice (D) as the correct answer.

105. The derivative is positive (sales are still increasing) and decreasing toward zero (sales are leveling off).

107. Company B. Although the company is currently losing money, the derivative is positive,

showing that the profit is increasing. Company A, on the other hand, has profits that are declining.

109. (C) is the only graph in which the instantaneous rate of change on January 1 is greater than the one-month average rate of change.

111. The tangent to the graph is horizontal at that point, and so the graph is almost horizontal near that point.

113. Various graphs are possible.

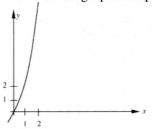

115. If $f(x) = mx + b$, then its average rate of change over any interval $[x, x+h]$ is
$$\frac{m(x+h) + b - (mx + b)}{h} = m.$$
Because this does not depend on h, the instantaneous rate is also equal to m.

117. Increasing, because the average rate of change appears to be rising as we get closer to 5 from the left (see the bottom row).

119. Answers may vary.

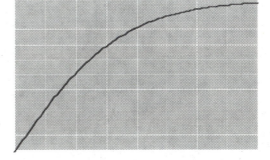

121.

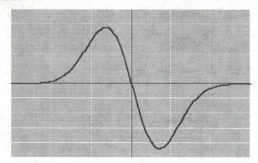

123. (B): His average speed was 60 miles per hour. If his instantaneous speed was always 55 mph or less, he could not have averaged more than 55 mph.

125. Answers will vary. Graph:

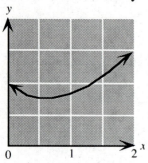

10.6 The Derivative: Algebraic Viewpoint

1. $f'(2) = \lim_{h \to 0} \frac{f(2+h) - f(2)}{h} = \lim_{h \to 0} \frac{(2+h)^2 + 1 - (2^2 + 1)}{h} = \lim_{h \to 0} \frac{4 + 4h + h^2 + 1 - 5}{h} =$

$\lim_{h \to 0} \frac{4h + h^2}{h} = \lim_{h \to 0} (4 + h) = 4$

3. $f'(-1) = \lim_{h \to 0} \frac{f(-1+h) - f(-1)}{h} = \lim_{h \to 0} \frac{3(-1 + h) - 4 - (-3 - 4)}{h} =$

$\lim_{h \to 0} \frac{-3 + 3h - 4 + 7}{h} = \lim_{h \to 0} \frac{3h}{h} = \lim_{h \to 0} 3 = 3$

5. $f'(1) = \lim_{h \to 0} \frac{f(1+h) - f(1)}{h} = \lim_{h \to 0} \frac{3(1+h)^2 + (1+h) - (3 + 1)}{h} =$

$\lim_{h \to 0} \frac{3 + 6h + 3h^2 + 1 + h - 4}{h} = \lim_{h \to 0} \frac{7h + 3h^2}{h} = \lim_{h \to 0} (7 + 3h) = 7$

7. $f'(-1) = \lim_{h \to 0} \frac{f(-1+h) - f(-1)}{h} = \lim_{h \to 0} \frac{2(-1+h) - (-1+h)^2 - (-2 - 1)}{h} =$

$\lim_{h \to 0} \frac{-2 + 2h - 1 + 2h - h^2 + 3}{h} = \lim_{h \to 0} \frac{4h - h^2}{h} = \lim_{h \to 0} (4 - h) = 4$

9. $f'(2) = \lim_{h \to 0} \frac{f(2+h) - f(2)}{h} = \lim_{h \to 0} \frac{(2+h)^3 + 2(2+h) - (8 + 4)}{h} =$

$\lim_{h \to 0} \frac{8 + 12h + 6h^2 + h^3 + 4 + 2h - 12}{h} = \lim_{h \to 0} \frac{14h + 6h^2 + h^3}{h} = \lim_{h \to 0} (14 + 6h + h^2) = 14$

11. $f'(1) = \lim_{h \to 0} \frac{f(1+h) - f(1)}{h} = \lim_{h \to 0} \frac{-1/(1+h) - (-1)}{h} = \lim_{h \to 0} \frac{-1 + (1 + h)}{h(1 + h)} =$

$\lim_{h \to 0} \frac{h}{h(1 + h)} = \lim_{h \to 0} \frac{1}{1 + h} = 1$

13. $f'(43) = \lim_{h \to 0} \frac{f(43+h) - f(43)}{h} = \lim_{h \to 0} \frac{m(43+h) + b - (43m + b)}{h} =$

$\lim_{h \to 0} \frac{43m + mh + b - 43m - b}{h} = \lim_{h \to 0} \frac{mh}{h} = \lim_{h \to 0} m = m$

15. $f'(x) = \lim_{h \to 0} \frac{f(x+h) - f(x)}{h} = \lim_{h \to 0} \frac{(x+h)^2 + 1 - (x^2 + 1)}{h} =$

$\lim_{h \to 0} \frac{x^2 + 2xh + h^2 + 1 - x^2 - 1}{h} = \lim_{h \to 0} \frac{2xh + h^2}{h} = \lim_{h \to 0} (2x + h) = 2x$

17. $f'(x) = \lim_{h \to 0} \dfrac{f(x+h) - f(x)}{h} = \lim_{h \to 0} \dfrac{3(x + h) - 4 - (3x - 4)}{h} = \lim_{h \to 0} \dfrac{3x + 3h - 4 - 3x + 4}{h}$

$= \lim_{h \to 0} \dfrac{3h}{h} = \lim_{h \to 0} 3 = 3$

19. $f'(x) = \lim_{h \to 0} \dfrac{f(x+h) - f(x)}{h} = \lim_{h \to 0} \dfrac{3(x+h)^2 + (x+h) - (3x^2 + x)}{h}$

$\lim_{h \to 0} \dfrac{3x^2 + 6xh + 3h^2 + x + h - 3x^2 - x}{h} = \lim_{h \to 0} \dfrac{6xh + 3h^2 + h}{h} = \lim_{h \to 0} (6x + 3h + 1) = 6x + 1$

21. $f'(x) = \lim_{h \to 0} \dfrac{f(x+h) - f(x)}{h} = \lim_{h \to 0} \dfrac{2(x+h) - (x+h)^2 - (2x - x^2)}{h}$

$\lim_{h \to 0} \dfrac{2x + 2h - x^2 - 2xh - h^2 - 2x + x^2}{h} = \lim_{h \to 0} \dfrac{2h - 2xh - h^2}{h} = \lim_{h \to 0} (2 - 2x - h) = 2 - 2x$

23. $f'(x) = \lim_{h \to 0} \dfrac{f(x+h) - f(x)}{h} = \lim_{h \to 0} \dfrac{(x+h)^3 + 2(x+h) - (x^3 + 2x)}{h} =$

$\lim_{h \to 0} \dfrac{x^3 + 3x^2h + 3xh^2 + h^3 + 2x + 2h - x^3 - 2x}{h} = \lim_{h \to 0} \dfrac{3x^2h + 3xh^2 + h^3 + 2h}{h} =$

$\lim_{h \to 0} (3x^2 + 3xh + h^2 + 2) = 3x^2 + 2$

25. $f'(x) = \lim_{h \to 0} \dfrac{f(x+h) - f(x)}{h} = \lim_{h \to 0} \dfrac{-1/(x+h) - (-1/x)}{h} = \lim_{h \to 0} \dfrac{-x + (x + h)}{hx(x + h)} =$

$\lim_{h \to 0} \dfrac{h}{hx(x + h)} = \lim_{h \to 0} \dfrac{1}{x(x + h)} = \dfrac{1}{x^2}$

27. $f'(x) = \lim_{h \to 0} \dfrac{f(x+h) - f(x)}{h} = \lim_{h \to 0} \dfrac{m(x+h) + b - (mx + b)}{h} =$

$\lim_{h \to 0} \dfrac{mx + mh + b - mx - b}{h} = \lim_{h \to 0} \dfrac{mh}{h} = \lim_{h \to 0} m = m$

29. $R'(2) = \lim_{h \to 0} \dfrac{R(2+h) - R(2)}{h} = \lim_{h \to 0} \dfrac{-0.3(2+h)^2 - (-0.3 \times 2^2)}{h} =$

$\lim_{h \to 0} \dfrac{-1.2 - 1.2h - 0.3h^2 + 1.2}{h} = \lim_{h \to 0} \dfrac{-1.2h - 0.3h^2}{h} = \lim_{h \to 0} (-1.2 - 0.3h) = -1.2$

31. $U'(3) = \lim\limits_{h \to 0} \dfrac{U(3+h) - U(3)}{h} = \lim\limits_{h \to 0} \dfrac{5.1(3+h)^2 + 5.1 - (5.1 \times 9 + 5.1)}{h} =$

$\lim\limits_{h \to 0} \dfrac{45.9 + 30.6h + 5.1h^2 + 5.1 - 51}{h} = \lim\limits_{h \to 0} \dfrac{30.6h + 5.1h^2}{h} = \lim\limits_{h \to 0} (30.6 + 5.1h) = 30.6$

33. $U'(1) = \lim\limits_{h \to 0} \dfrac{U(1+h) - U(1)}{h} = \lim\limits_{h \to 0} \dfrac{-1.3(1+h)^2 - 4.5(1+h) - (-1.3 - 4.5)}{h} =$

$\lim\limits_{h \to 0} \dfrac{-1.3 - 2.6h - 1.3h^2 - 4.5 - 4.5h + 5.8}{h} = \lim\limits_{h \to 0} \dfrac{-7.1h - 1.3h^2}{h} = \lim\limits_{h \to 0} (-7.1 - 1.3h) = -7.1$

35. $L'(1.2) = \lim\limits_{h \to 0} \dfrac{L(1.2+h) - L(1.2)}{h} = \lim\limits_{h \to 0} \dfrac{4.25(1.2+h) - 5.01 - (4.25 \times 1.2 - 5.01)}{h} =$

$\lim\limits_{h \to 0} \dfrac{5.1 + 4.25h - 5.01 - 5.1 + 5.01}{h} = \lim\limits_{h \to 0} \dfrac{4.25h}{h} = \lim\limits_{h \to 0} 4.25 = 4.25$

37. $q'(2) = \lim\limits_{h \to 0} \dfrac{q(2+h) - q(2)}{h} = \lim\limits_{h \to 0} \dfrac{2.4/(2+h) - 2.4/2}{h} = \lim\limits_{h \to 0} \dfrac{4.8 - 2.4(2 + h)}{2h(2 + h)} =$

$\lim\limits_{h \to 0} \dfrac{-2.4h}{2h(2 + h)} = \lim\limits_{h \to 0} \dfrac{-1.2}{h(2 + h)} = -0.6$

39. Find the slope by finding the derivative:

$m = f'(2) = \lim\limits_{h \to 0} \dfrac{f(2+h) - f(2)}{h} = \lim\limits_{h \to 0} \dfrac{(2+h)^2 - 3 - (2^2 - 3)}{h} = \lim\limits_{h \to 0} \dfrac{4 + 4h + h^2 - 3 - 1}{h} =$

$\lim\limits_{h \to 0} \dfrac{4h + h^2}{h} = \lim\limits_{h \to 0} (4 + h) = 4.$ The tangent line has slope 4 and goes through $(2, f(2)) = (2, 1)$, so has equation $y = 4x - 7$.

41. Find the slope by finding the derivative:

$m = f'(3) = \lim\limits_{h \to 0} \dfrac{f(3+h) - f(3)}{h} = \lim\limits_{h \to 0} \dfrac{-2(3+h) - 4 - (-2 \times 3 - 4)}{h} =$

$\lim\limits_{h \to 0} \dfrac{-6 - 2h - 4 + 10}{h} = \lim\limits_{h \to 0} \dfrac{-2h}{h} = \lim\limits_{h \to 0} (-2) = -2.$ The tangent line has slope -2 and goes through $(3, f(3)) = (3, -10)$, so has equation $y = -2x - 4$.

43. Find the slope by finding the derivative:

$m = f'(-1) = \lim\limits_{h \to 0} \dfrac{f(-1+h) - f(-1)}{h} = \lim\limits_{h \to 0} \dfrac{(-1+h)^2 - (-1 + h) - [(-1)^2 - (-1)]}{h} =$

$\lim\limits_{h \to 0} \dfrac{1 - 2h + h^2 + 1 - h - 2}{h} = \lim\limits_{h \to 0} \dfrac{-3h + h^2}{h} = \lim\limits_{h \to 0} (-3 + h) = -3.$ The tangent line has slope -3 and goes through $(-1, f(-1)) = (-1, 2)$, so has equation $y = -3x - 1$.

45. $s'(4) = \lim\limits_{h \to 0} \dfrac{s(4+h) - s(4)}{h} = \lim\limits_{h \to 0} \dfrac{400 - 16(4+h)^2 - (400 - 16(4)^2)}{h} =$

$\lim\limits_{h \to 0} \dfrac{-128h - 16h^2}{h} = -128$ ft/sec

47. $\dfrac{dI}{dt} = \lim\limits_{h \to 0} \dfrac{I(t+h) - I(t)}{h}$

$= \lim\limits_{h \to 0} \dfrac{-0.015(t+h)^2 + 0.1(t+h) + 1.4 - [-0.015t^2 + 0.1t + 1.4]}{h}$

$= \lim\limits_{h \to 0} \dfrac{-0.015(t^2 + 2th + h^2) + 0.1(t+h) + 1.4 - [-0.015t^2 + 0.1t + 1.4]}{h}$

$= \lim\limits_{h \to 0} \dfrac{-0.030th - 0.015h^2 + 0.1h}{h}$ Rest of the terms cancel

$= \lim\limits_{h \to 0} \dfrac{h(-0.030t - 0.015h + 0.1)}{h}$

$= -0.030t + 0.1$ Cancel the h and then let $h \to 0$

At time $t = 7$, this becomes

$\dfrac{dI}{dt}\Big|_{t=7} = -0.030(7) + 0.1 = -0.11$ million barrels per year

Daily oil imports were decreasing at a rate of 0.11 million barrels per year.

49. $R'(t) = \lim\limits_{h \to 0} \dfrac{R(t+h) - R(t)}{h}$

$= \lim\limits_{h \to 0} \dfrac{12(t + h)^2 + 500(t + h) + 4700 - (12t^2 + 500t + 4700)}{h}$

$= \lim\limits_{h \to 0} \dfrac{24th + 12h^2 + 500h}{h} = 24t + 500;\ R'(5) = 620;$ annual U.S. sales of bottled water were

increasing by 620 million gallons per year in 2005.

51. $f'(8) = \lim\limits_{h \to 0} \dfrac{f(8+h) - f(8)}{h} =$

$\lim\limits_{h \to 0} \dfrac{3.55(8+h)^2 - 30.2(8+h) + 81 - (3.55 \times 8^2 - 30.2 \times 8 + 81)}{h} = \lim\limits_{h \to 0} \dfrac{26.6h + 3.55h^2}{h} =$

26.6 manatee deaths per 100,000 additional boats. At a level of 800,000 boats, the number of manatee deaths is increasing at a rate of 26.6 manatees per 100,000 additional boats.

53. (a) The only possible point of discontinuity could be at $t = 8$. However,

$C(8) = 1.24$ and

$\lim\limits_{t \to 8^-} C(t) = 0.08(8) + 0.6 = 1.24$

$$\lim_{t \to 8^+} C(t) = 0.13(8) + 0.2 = 1.24$$

So $\lim_{t \to 8} C(t) = 1.24 = C(8)$,

showing that the function is continuous at $t = 8$.

(b) The graph of C comes to a point at $t = 8$ (the slope immediately before that point is 0.08, and the slope immediately after is 0.13). Thus, C is not differentiable at $t = 8$. $C'(t) = 0.08$ if $t < 8$, so

$$\lim_{t \to 8^-} C'(t) = 0.08$$

$C'(t) = 0.13$ if $t > 8$, so

$$\lim_{t \to 8^+} C'(t) = 0.13$$

Until 1998, the cost of a Super Bowl ad was increasing at a rate of $80,000 per year. Immediately thereafter, it was increasing at a rate of $130,000 per year.

55. The algebraic method, because it gives the exact value of the derivative. The other two approaches give only approximate values (except in some special cases).

57. The error is in the second line: $f(x+h)$ is *not* equal to $f(x) + h$. For instance, if $f(x) = x^2$, then $f(x+h) = (x+h)^2$, whereas $f(x) + h = x^2 + h$.

59. The error is in the second line: One could only cancel the h if it were a *factor* of both the numerator and denominator; it is not a factor of the numerator.

61. Because the algebraic computation of $f'(a)$ is exact, and not an approximation, it makes no difference whether one uses the balanced difference quotient or the ordinary difference quotient in the algebraic computation.

63. The computation results in a limit that cannot be evaluated.

Chapter 10 Review Exercises

1. 5:

x	$f(x)$
2.9	4.9
2.99	4.99
2.999	4.999
2.9999	4.9999
3	
3.0001	5.0001
3.001	5.001
3.01	5.01
3.1	5.1

3. Does Not Exist:

x	$f(x)$
−1.1	0.3226
−1.01	0.3322
−1.001	0.3332
−1.0001	0.3333
−1	
−0.9999	−0.3333
−0.999	−0.3334
−0.99	−0.3344
−0.9	−0.3448

5. (a) −1: As x approaches 0 from the left or right, $f(x)$ approaches the open dot at height −1. The fact that $f(0) = 3$ is irrelevant.
(b) 3: As x approaches 1 from the left or right, $f(x)$ approaches the point on the graph corresponding to $x = 1$, whose y-coordinate is 3.
(c) Does not exist: As x approaches 2 from the left, $f(x)$ approaches the open dot at height 2. As x approaches 2 from the right, $f(x)$ approaches the solid dot at height 1. Thus, the one-sided limits, though they both exist, do not agree.

7. $f(x) = \dfrac{x^2}{x-3}$ is a closed-form function whose domain includes $x = -2$. Therefore

$$\lim_{x\to-2} \frac{x^2}{x-3} = \frac{(-2)^2}{-2-3} = -4/5$$

9. $f(x) = \dfrac{x}{2x^2 - x}$ is a closed-form function but its domain does not include $x = 0$, but we can simplify:

$$\lim_{x\to 0} \frac{x}{2x^2 - x} = \lim_{x\to 0} \frac{x}{x(2x-1)}$$

$$\lim_{x\to 0} \frac{1}{(2x-1)} = \frac{1}{(0-1)} = -1$$

11. $\displaystyle\lim_{x\to -1} \frac{x^2 + 3x}{x^2 - x - 2} =$

$\displaystyle\lim_{x\to -1} \frac{x(x+3)}{(x-2)(x+1)}$. As $x \to -1^-$, the ratio has the determinate form $-2/[(-3)(0^-)]$, so

$\displaystyle\lim_{x\to -1^-} \frac{x(x+3)}{(x-2)(x+1)} = -\infty$. As $x \to -1^+$, the ratio has the determinate form $-2/[(-3)(0^+)]$, so

$\displaystyle\lim_{x\to -1^-} \frac{x(x+3)}{(x-2)(x+1)} = +\infty$. Since the left and right limits disagree, $\displaystyle\lim_{x\to -1} \frac{x^2 + 3x}{x^2 - x - 2}$ does not exist

13. $\displaystyle\lim_{x\to 4} \frac{x^2 + 8}{x^2 - 2x - 8} = \lim_{x\to 4} \frac{x^2 + 8}{(x+2)(x-4)}$.
As $x \to 4^+$, the numerator approaches 24 and the factor in the denominator $(x-4)$ is positive and approaches zero. Thus, the limit has the determinate form $\dfrac{24}{(6)(0^+)}$ which is positive, and so the limit diverges to $+\infty$. As $x \to 4^-$, the factor $(x-4)$ is negative and approaches zero. Thus, the limit has the determinate form $\dfrac{24}{(6)(0^-)}$ which is negative, and so the limit diverges to $-\infty$. Since the left and right limits disagree, $\displaystyle\lim_{x\to 4} \frac{x^2 + 8}{x^2 - 2x - 8}$ does not exist.

15. $\displaystyle\lim_{x\to 1/2} \frac{x^2 + 8}{4x^2 - 4x + 1} = \lim_{x\to 1/2} \frac{x^2 + 8}{(2x-1)^2}$.

As $x \to 1/2$ from either side, the numerator approaches 8.25 and the denominator is positive and approaches zero, so the limit has the determinate form $8.25/0^+$, telling us that the limit diverges to $+\infty$.

17. Ignoring all the highest terms in numerator and denominator, we get:

$$\lim_{x \to -\infty} \frac{x^2 - x - 6}{x - 3} = \lim_{x \to -\infty} \frac{x^2}{x}$$

$$= \lim_{x \to -\infty} x = -\infty,$$

So the limit diverges to $-\infty$.

19. As $t \to +\infty$, $3^{2t} \to +\infty$ as well, so the denominator $5 + 5.3(3^{2t}) \to +\infty$. Thus

$$\lim_{t \to +\infty} \frac{-5}{5 + 5.3(3^{2t})} = 0 \quad \left[\frac{-5}{+\infty} = 0\right]$$

21. As $x \to +\infty$, $4e^{-3x} = \frac{4}{e^{3x}} \to 0 \quad \left[\frac{4}{+\infty} = 0\right]$

so $\lim_{x \to +\infty} \frac{2}{5 + 4e^{-3x}} = \frac{2}{5 + 0} = 2/5$

23. As $t \to +\infty$. the term 2^{-3t} is $1/2^{3t}$ and so approaches zero. Thus the entire numerator approaches $1 + 0 = 1$. The term $5.3e^{-t}$ in the denominator is $5.3/e^t$ and so approaches 0, so the denominator approaches $1 + 0 = 1$ as well. Hence, the entire expression converges to $1/1 = 1$.

25. $f(x) = \dfrac{1}{x + 1}$; $a = 0$

The average rate of change is $\dfrac{f(a+h) - f(a)}{h}$.

$h = 1$:

$$\frac{f(0+1) - f(0)}{1} = \frac{f(1) - f(0)}{1}$$

$$= \frac{-0.5}{1} = -0.5$$

$h = 0.01$:

$$\frac{f(0+0.01) - f(0)}{0.01} = \frac{f(0.01) - f(0)}{0.01}$$

$$= \frac{-0.009901}{0.01} = -0.99001$$

$h = 0.001$:

$$\frac{f(0+0.001) - f(0)}{0.001} = \frac{f(0.001) - f(0)}{0.001}$$

$$= \frac{-0.000999}{0.001} = -0.9990$$

Table:

	Ave. Rate of Change
h	
1	-0.5
0.01	-0.9901
0.001	-0.9990

The slope of the tangent is the limit as $h \to 0$, which appears to be -1.
Slope ≈ -1

27. $f(x) = e^{2x}$; $a = 0$
Technology formula for $f(x)$:

$\quad$ `Y₁= e^(2x)` TI-83/84 Plus

$\quad$ `EXP(2*x)` $\quad$ Excel

To compute the average rate of change on the TI-83/84 Plus, use the following formulas:

$\quad h = 1$: `(Y₁(0+1)-Y₁(0))/1`

$\quad h = 0.01$: `(Y₁(0+.01)-Y₁(0))/.01`

$\quad h = 0.001$: `(Y₁(0+.001)-Y₁(0))/.001`

Table:

	Ave. Rate of Change
h	
1	6.3891
0.01	2.0201
0.001	2.0020

The slope of the tangent is the limit as $h \to 0$, which appears to be 2.
Slope ≈ 2

29. (i) P **(ii)** Q **(iii)** R **(iv)** S

31. (i) Q **(ii)** None **(iii)** None **(iv)** None

33. (a) (B): The graph starts on the tangent line and falls below it.
(b) (B): The graph starts above the tangent line and ends below it.
(c) (B): The graph is getting less steep.
(d) (A): The graph gets steeper until $x = 0$ and then gets less steep.

(e) (C): The value of $f(2)$ is the height of the graph at $x = 2$, which is about 2.5; the rate of change is the slope of the tangent line at that point, which is approximately 1.5.

35. $f(x) = x^2 + x$

$$f'(x) = \lim_{h \to 0} \frac{f(x+h) - f(x)}{h}$$

$$= \lim_{h \to 0} \frac{(x+h)^2 + (x+h) - (x^2 + x)}{h}$$

$$= \lim_{h \to 0} \frac{x^2 + 2xh + h^2 + x + h - x^2 - x}{h}$$

$$= \lim_{h \to 0} \frac{2xh + h^2 + h}{h} =$$

$$\lim_{h \to 0} (2x + h + 1) = 2x + 1$$

37. $f(x) = 1 - \dfrac{2}{x}$

$$f'(x) = \lim_{h \to 0} \frac{f(x+h) - f(x)}{h}$$

$$= \lim_{h \to 0} \frac{1 - 2/(x+h) - (1 - 2/x)}{h}$$

$$= \lim_{h \to 0} \frac{-2/(x+h) + 2/x}{h}$$

$$= \lim_{h \to 0} \frac{-2/(x+h) + 2/x}{h}$$

$$= \lim_{h \to 0} \frac{-2x + 2(x+h)}{h(x+h)x}$$

$$= \lim_{h \to 0} \frac{2h}{h(x+h)x}$$

$$= \lim_{h \to 0} \frac{2}{(x+h)x} = 2/x^2$$

39. `y1 = 10x^5+(1/2)x^4-x+2`
`y2 = nDeriv(Y1,X,X)` (TI-83/84 Plus)
`y2 = deriv(y1)` (Web site)

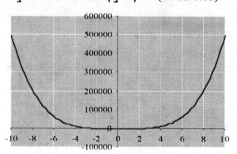

41. `y1 = 3x^3+3x^(1/3)`
`y2 = nDeriv(Y1,X,X)` (TI-83/84 Plus)
`y2 = deriv(y1)` (Web site)

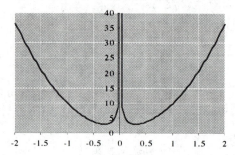

43. (a) $P(3)$ = value of $P(t)$ at $t = 3$
$= 25$.

As t approaches 3 from the left. the y-coordinate of the corresponding point on the graph approaches 25. Therefore,

$$\lim_{t \to 3^-} P(t) = 25$$

As t approaches 3 from the right. the y-coordinate of the corresponding point on the graph approaches 10. Therefore,

$$\lim_{t \to 3^+} P(t) = 10$$

Since the left and right-limits do not agree, $\lim_{t \to 3} P(t)$ does not exist.

Interpretation: $P(3) = 3$: O'Hagan purchased the stock at \$25. $\lim_{t \to 3^-} P(t) = 25$: The value of the stock had been approaching \$25 up the time he bought it. $\lim_{t \to 3^+} P(t) = 10$: The value of the stock dropped to \$10 immediately after he bought it.

398

(b) As t approaches 6 from either side, $P(t)$ approaches 5, which is also the value of $P(6)$. In other words,

$$\lim_{t \to 6} P(t) = 5 = P(6)$$

showing that P is continuous at $t = 6$. On the other hand, the graph comes to a sharp point at $t = 6$ so P is not differentiable at $t = 6$. Interpretation: the stock changed continuously but suddenly reversed direction (and started to go up) the instant O'Hagan sold it.

45. (a) $\lim_{t \to 3} p(t) \approx 40 \quad \lim_{t \to \infty} p(t) = +\infty$. Close to 2007 ($t = 3$), the home price index was about 40. In the long term, the home price index will rise without bound.
(b) 10; (The slope of the linear portion of the curve is 10.) In the long term, the home price index will rise about 10 points per year.

47. (a) $(9000 - 6500)/5 = 500$ books per week
(b) [3, 4] (600 books per week), [4, 5] (700 books per week)
(c) [3, 5], when the average rate of increase was 650 books per week.

49. (a) The ten year period beginning 2004 corresponds to the interval [0, 10] for t.

Average rate of change $= \dfrac{p(10) - p(0)}{10 - 0}$

$\approx \dfrac{40 - 10}{10}$

$= 3$ percentage points per year

(b) $\dfrac{p(10) - p(3)}{10 - 3} \approx \dfrac{40 - 40}{7}$

$= 0$ percentage points per year
(c) Choice (D): The slope of the tangent decreases from 4 to around 5, and then it starts to increase (the graph curves less steeply downward). This means that the rate of change of the index first decreased, and then increased.

51. (a) $w'(t) = \lim_{h \to 0} \dfrac{w(t+h) - w(t)}{h} =$

$\lim_{h \to 0} \dfrac{36(t + h)^2 + 250(t + h) + 6240 - (36t^2 + 250t + 6240)}{h}$

$= \lim_{h \to 0} \dfrac{72th + 36h^2 + 250h}{h} = 72t + 250;$

(b) $w'(1) = 72(1) + 250 = 322$ books per week
(c) $w'(7) = 72(7) + 250 = 754$ books per week

Chapter 11 Techniques of Differentiation with Applications

11.1 Derivatives of Powers, Sums, and Constant Multiples

1. $5x^4$

3. $2(-2x^{-3}) = -4x^{-3}$

5. $-0.25x^{-0.75}$

7. $4(2x^3) + 3(x^2) - 0 = 8x^3 + 9x^2$

9. $-1 - 1/x^2$

11. $\dfrac{dy}{dx} = 10(0) = 0$ (constant multiple and power rule)

13. $\dfrac{dy}{dx} = \dfrac{d}{dx}(x^2) + \dfrac{d}{dx}(x)$ (sum rule) $= 2x + 1$ (power rule)

15. $\dfrac{dy}{dx} = \dfrac{d}{dx}(4x^3) + \dfrac{d}{dx}(2x) - \dfrac{d}{dx}(1)$ (sum and difference) $= 4\dfrac{d}{dx}(x^3) + 2\dfrac{d}{dx}(x) - \dfrac{d}{dx}(1)$ (constant multiples) $= 12x^2 + 2$ (power rule)

17. $f'(x) = 2x - 3$

19. $f'(x) = 1 + 0.5x^{-0.5}$

21. $g'(x) = -2x^{-3} + 3x^{-2}$

23. $g'(x) = \dfrac{d}{dx}(x^{-1} - x^{-2}) = -x^{-2} + 2x^{-3} = -\dfrac{1}{x^2} + \dfrac{2}{x^3}$

25. $h'(x) = \dfrac{d}{dx}(2x^{-0.4}) = -0.8x^{-1.4} = -\dfrac{0.8}{x^{1.4}}$

27. $h'(x) = \dfrac{d}{dx}(x^{-2} + 2x^{-3}) = -2x^{-3} - 6x^{-4} = -\dfrac{2}{x^3} - \dfrac{6}{x^4}$

29. $r'(x) = \dfrac{d}{dx}\left(\dfrac{2}{3}x^{-1} - \dfrac{1}{2}x^{-0.1}\right) = -\dfrac{2}{3}x^{-2} + \dfrac{0.1}{2}x^{-1.1} = -\dfrac{2}{3x^2} + \dfrac{0.1}{2x^{1.1}}$

31. $r'(x) = \dfrac{d}{dx}\left(\dfrac{2}{3}x - \dfrac{1}{2}x^{0.1} + \dfrac{4}{3}x^{-1.1} - 2\right) = \dfrac{2}{3} - \dfrac{0.1}{2}x^{-0.9} - \dfrac{4.4}{3}x^{-2.1} = \dfrac{2}{3} - \dfrac{0.1}{2x^{0.9}} - \dfrac{4.4}{3x^{2.1}}$

33. $t'(x) = \dfrac{d}{dx}(|x| + x^{-1}) = |x|/x - x^{-2} = |x|/x - 1/x^2$

35. $s'(x) = \dfrac{d}{dx}(x^{1/2} + x^{-1/2}) = \dfrac{1}{2}x^{-1/2} - \dfrac{1}{2}x^{-3/2} = \dfrac{1}{2\sqrt{x}} - \dfrac{1}{2x\sqrt{x}}$

37. $s'(x) = \dfrac{d}{dx}(x^3 - 1) = 3x^2$

39. $t'(x) = \dfrac{d}{dx}(x - 2x^2) = 1 - 4x$

41. $2.6x^{0.3} + 1.2x^{-2.2}$

43. $1.2(1 - |x|/x)$

45. $3at^2 - 4a$ (Remember to treat a as a constant, i.e., a number.)

47. $5.15x^{9.3} - 99x^{-2}$

49. $\dfrac{ds}{dt} = \dfrac{d}{dt}(2.3 + 2.1t^{-1.1} - \dfrac{1}{2}t^{0.6}) = -2.31t^{-2.1} + 0.3t^{-0.4} = -\dfrac{2.31}{t^{2.1}} - \dfrac{0.3}{t^{0.4}}$

51. $4\pi r^2$

In 53–58, we need to find the derivative at the indicated value of x or t.

53. $f'(x) = 3x^2$, so $f'(-1) = 3$

55. $f'(x) = -2$, so $f'(2) = -2$

57. $g'(t) = \dfrac{d}{dt} t^{-5} = -5t^{-6} = -\dfrac{5}{t^6}$, so $g'(1) = -5$

59. $f'(x) = 3x^2$, so $f'(-1) = 3$. The line with slope 3 passing through $(-1, f(-1)) = (-1, -1)$ is $y = 3x + 2$.

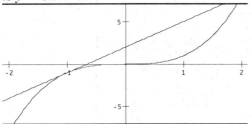

61. $f(x) = x + x^{-1}$, so $f'(x) = 1 - x^{-2} = 1 - \dfrac{1}{x^2}$; $f'(2) = 1 - 1/4 = 3/4$. The line with slope 3/4 passing through $(2, f(2)) = (2, 5/2)$ is $y = \dfrac{3}{4} x + 1$

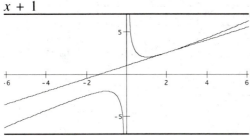

63. $f(x) = x^{1/2}$, so $f'(x) = \dfrac{1}{2} x^{-1/2} = \dfrac{1}{2\sqrt{x}}$; $f'(4) = 1/4$. The line with slope 1/4 passing through $(4, f(4)) = (4, 2)$ is $y = \dfrac{1}{4} x + 1$.

In 65–70 we need to find all values of x (if any) where the derivative is 0.

65. $y' = 4x + 3 = 0$ when $x = -3/4$

67. $y' = 2$ is never 0, so there are no such values of x

69. $y' = 1 - x^{-2} = 1 - 1/x^2 = 0$ when $x^2 = 1$, so $x = 1$ or -1

71. $\dfrac{d}{dx} x^4 = \lim_{h \to 0} \dfrac{(x+h)^4 - x^4}{h}$

$= \lim_{h \to 0} \dfrac{x^4 + 4x^3h + 6x^2h^2 + 4xh^3 + h^4 - x^4}{h}$

$= \lim_{h \to 0} \dfrac{4x^3h + 6x^2h^2 + 4xh^3 + h^4}{h} =$

$\lim_{h \to 0} (4x^3 + 6x^2h + 4xh^2 + h^3) = 4x^3$

73. The derivative:

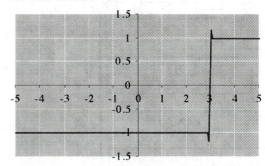

(a) $x = 3$: The sudden jump in value is a discontinuity and the derivative is not defined at $x = 3$.
(b) None: The derivative is never 0.

75. The derivative:

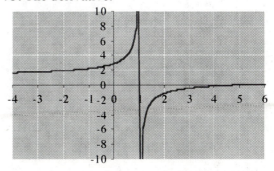

(a) $x = 1$: The sudden jump in value is a discontinuity and the derivative is not defined at $x = 1$.

(b) $x = 4.2$: The derivative is 0 at approximately 4.2. The derivative is not 0 at $x = 1$; that is just a defect of the graphing technology.

77. (a) $f'(1) = 1/3$:

h	$\dfrac{f(a+h) - f(a)}{h}$
-1	1
-0.1	0.34510615
-0.01	0.33445066
-0.001	0.33344451
-0.0001	0.33334445
1	0.25992105
0.1	0.32280115
0.01	0.33222835
0.001	0.33322228
0.0001	0.33332222

(b) f is not differentiable at 0:

h	$\dfrac{f(a+h) - f(a)}{h}$
-1	1
-0.1	4.64158883
-0.01	21.5443469
-0.001	100
-0.0001	464.158883
1	1
0.1	4.64158883
0.01	21.5443469
0.001	100
0.0001	464.158883

79. (a) Not differentiable at 1:

h	$\dfrac{f(a+h) - f(a)}{h}$
-1	0
-0.1	-4.4814047
-0.01	-21.472292
-0.001	-99.966656
-0.0001	-464.14341
1	-1.259921
0.1	-4.7914199
0.01	-21.615923
0.001	-100.03332
0.0001	-464.17435

(b) Not differentiable at 0:

h	$\dfrac{f(a+h) - f(a)}{h}$
-1	1.25992105
-0.1	4.79141986
-0.01	21.6159233
-0.001	100.033322
-0.0001	464.174355
1	0
0.1	4.48140475
0.01	21.4722917
0.001	99.9666555
0.0001	464.143411

81. Since putting $x = 1$ yields 0/0, L'Hospital's rule applies.

$$\lim_{x \to 1} \frac{x^2 - 2x + 1}{x^2 - x} = \lim_{x \to 1} \frac{2x - 2}{2x - 1} = \frac{0}{1} = 0$$

83. Since putting $x = 1$ yields 0/0, L'Hospital's rule applies.

$$\lim_{x \to 2} \frac{x^3 - 8}{x - 2} = \lim_{x \to 2} \frac{3x^2}{1} = \frac{12}{1} = 12$$

85. Since putting $x = 1$ yields 6/2 = 3. Since this is not an indeterminate form, L'Hospital's rule does not apply, and the limits is 3 (closed-form function).

87. Since putting $x = -\infty$ yields ∞/∞, L'Hospital's rule applies.

$$\lim_{x \to -\infty} \frac{3x^2 + 10x - 1}{2x^2 - 5x} = \lim_{x \to -\infty} \frac{6x + 10}{4x - 5} =$$

$$\lim_{x \to -\infty} \frac{6}{4} = 3/2$$

89. Since putting $x = -\infty$ yields ∞/∞, L'Hospital's rule applies.

$$\lim_{x \to -\infty} \frac{10x^2 + 300x + 1}{5x + 2} =$$

$$\lim_{x \to -\infty} \frac{20x + 300}{5} \text{ diverges to } -\infty$$

91. Since putting $x = -\infty$ yields ∞/∞, L'Hospital's rule applies.

$$\lim_{x \to -\infty} \frac{x^3 - 100}{2x^2 + 500} = \lim_{x \to -\infty} \frac{3x^2}{4x}$$

Since this still has the form ∞/∞, we can use L'Hospital's rule again to get

$$\lim_{x \to -\infty} \frac{3x^2}{4x} = \lim_{x \to -\infty} \frac{6x}{4} = -\infty/4 = -\infty$$

93. $P(t) = 0.45t^2 - 12t + 105$
So $P'(t) = 0.9t - 12$
$P'(20) = 0.9(20) - 12 = \$6$ per year
The price of a barrel of crude oil was increasing at a rate of $6 per year in 2000 ($t = 20$).

95. (a) $n(t) = -0.56t^2 + 14t + 930$
so $n'(t) = -1.12t + 14$
(b) 2006 is represented by $t = 6$.
$\quad n'(6) = -1.12(6) + 14 = 7.28 \approx 7$
teams/year
The number of teams was increasing at a rate of about 7 teams per year.
(c) The rate of increase of the number of teams is the derivative,
$\quad n'(t) = -1.12t + 14$
this is a linear function with negative slope; as t increases, $n'(t)$ decreases; The rate of increase of the number of teams decreases with time.

97. To find the rate of change of spending on food (y) with rrespect to x we take the derivative using the power rule:

$$y = \frac{35}{x^{0.35}} = 35x^{-0.35}$$

$$\frac{dy}{dx} = (35)(-0.35)x^{-0.35-1} = 12.25x^{-1.35}$$

We evaluate this at $x = 10\%$:

$$\left.\frac{dy}{dx}\right|_{x=10} = 12.25(10)^{-1.35} \approx 0.5472$$

≈ 0.55 percentage points per one percentage point increase in spending on education.

99. (a) $s'(t) = -32t$; $s'(0) = 0$, $s'(1) = -32$, $s'(2) = -64$, $s'(3) = -96$, $s'(4) = -128$ ft/sec
(b) $s(t) = 0$ when $400 - 16t^2 = 0$, so $t^2 = 400/16 = 25$, so at $t = 5$ seconds; the stone is traveling at the velocity $s'(5) = -160$, so downward at 160 ft/sec.

101. (a) $S(t) = -390t^2 + 3300t - 4800$
$S'(t) = -780t + 3300$
$S'(5) = -780(5) + 3300 = -600$
Sales of iPhones were dropping at a rate of 600,000 units per quarter.
(b) Geometrically: Notice that the graph is rising over the interval $[2, 4]$ (iPhone sales were increasing) and that the slope of the tangent is getting smaller (the rate of increase was slowing). Therefore, the value of S increased at a slower and slower rate (Choice (B)).
Algebraically: The slope of the sales curve is the derivative: $S'(t) = -780t + 3300$. Its value on $[2, 4]$ is positive, so the rate of change of sales is positive (sales were increasing). However, as t increases from 2 to 4, $S'(t)$ decreases in value, so the rate of increase of change is decreasing (that is, sales were increasing more slowly). Therefore, the value of S increased at a slower and slower rate (Choice (B)).

103. (a) $f'(x) = 7.1x - 30.2$ manatees per 100,000 boats.
(b) $f'(x)$ is increasing; the number of manatees killed per additional 100,000 boats increases as the number of boats increases.

(c) $f'(8) = 26.6$ manatees per 100,000 additional boats. At a level of 800,000 boats, the number of manatee deaths is increasing at a rate of 26.6 manatees per 100,000 additional boats.

105. (a) $c(t)$ measures the combined market share (including MSN), while $m(t)$ measures the share due to MSN. Therefore, $c(t)-m(t)$ measures the combined market share of the other three providers (Comcast, earthlink, and AOL). Similarly, $c'(t)-m'(t)$ measures the rate of change of the combined market share of the other three providers.

(b) We can visualize $c(t) - m(t)$ on the graph as the vertical distance from the lower curve to the higher one:

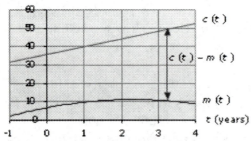

As we move from $t = 3$ to $t = 4$, that distance is increasing (choice A).

(c) From part (b), $c(t) - m(t)$ measures the vertical distance from the lower curve to the higher one, since this distance is increasing on [3, 4], its rate of change (derivative) $c'(t) - m'(t)$ is positive (choice A).

(d) $\quad m(t) = -0.83t^2 + 3.8t + 6.8$

so $\quad m'(t) = -1.66t + 3.8$

and $\quad m'(2) = -1.66(2) + 3.8 = 0.48\%$ per year.

$c(t) = 4.2t + 36,$

so $\quad c'(t) = c'(2) = 4.2\%$ per year.

Therefore, $c'(2) - m'(2) = 4.2 - 0.48 = 3.72\%$ per year. In 1992, the combined market share of the other three providers was increasing at a rate of about 3.72 percentage points per year.

107. After graphing the curve $y = 3x^2$, draw the line passing through $(-1, 3)$ with slope -6.

109. The slope of the tangent line of g is twice the slope of the tangent line of f because $g(x) = 2f(x)$, so $g'(x) = 2f'(x)$.

111. $g'(x) = -f'(x)$

113. The left-hand side is not equal to the right-hand side. The *derivative* of the left-hand side is equal to the right-hand side, so your friend should have written

$$\frac{d}{dx}(3x^4 + 11x^5) = 12x^3 + 55x^4.$$

115. The derivative of a constant times a function is the constant times the derivative of the function, so $f'(x) = (2)(2x) = 4x$. Your enemy mistakenly computed the *derivative* of the constant times the derivative of the function. (The derivative of a product of two functions is not the product of the derivative of the two functions. The rule for taking the derivative of a product is discussed later in the chapter.).

117. For a general function f, the derivative of f is defined to be $f'(x) = \lim\limits_{h \to 0} \dfrac{f(x+h) - f(x)}{h}$. One then finds by calculation that the derivative of the specific function x^n is nx^{n-1}. In short, nx^{n-1} is the derivative of a specific function: $f(x) = x^n$; it is not the *definition* of the derivative of a general function or even the definition of the derivative of the function $f(x) = x^n$.

119. Answers may vary; here is one possibility. At one point its derivative is not defined but it has a tangent line: The tangent line at that point is vertical so has undefined slope.

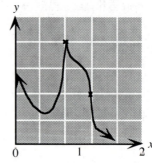

11.2 A First Application: Marginal Analysis

1. $C'(x) = 5 - 0.0002x$; $C'(1000) = \$4.80$ per item

3. $C'(x) = 100 - 1000/x^2$; $C'(100) = \$99.90$ per item

5. $C(x) = 4x$, so $C'(x) = 4$;
$R(x) = 8x - 0.001x^2$, so $R'(x) = 8 - x/500$;
$P(x) = R(x) - C(x) = R(x) = 8x - 0.001x^2 - 4x$
$= R(x) = 4x - 0.001x^2$, so $P'(x) = 4 - 0.002x$;
$P'(x) = 0$ when $x = 2000$. Thus, at a production level of 2000, the profit is stationary (neither increasing nor decreasing) with respect to the production level. This may indicate a maximum profit at a production level of 2000.

7. (a) (B): The slope of the graph decreases and then increases.
(b) (C): This is where the slope of the graph is least.
(c) (C): At $x = 50$, the height of the graph is about 3000, so the cost is \$3000. The tangent line at that point passes roughly through (100, 4000), so has a slope of roughly 20; hence the cost is increasing at a rate of about \$20 per item.

9. (a) $C(x) = 150 + 2250x - 0.02x^2$, so
$\quad C'(x) = 2250 - 0.04x.$
$\quad C'(4) = 2250 - 0.04(4)$
$\qquad = 2249.84$ thousand dollars;
that is, \$2,249,840. The cost is going up at a rate of \$2,249,840 per television commercial. The exact cost of airing the fifth television commercial is
$\quad C(5) - C(4) = 11399.5 - 9149.68$
$\qquad = 2249.82$ thousand dollars, or \$2,249,820.
(b) $\bar{C}(x) = 150/x + 2250 - 0.02x$; $\bar{C}(4) = $ \$2,287,420 per television commercial. The average cost of airing the first four television commercials is \$2,287,420.

11. (a) $R(x) = 0.90x$
Marginal revenue $= R'(x) = 0.90$
$\quad P(x) = R(x) - C(x)$
$\qquad = 0.90x - (70 + 0.10x + 0.001x^2)$
$\qquad = -70 + 0.80x - 0.001x^2$
Marginal Profit $= P'(x) = 0.80 - 0.002x$
(b) $R(x) = 0.90x$
$\quad R(500) = 0.90(500) = \$450$
The total revenue from the sale of 500 copies is \$450.
$\quad P(x) = -70 + 0.80x - 0.001x^2$
$\quad P(500) = -70+0.80(500)-0.001(500)^2$
$\qquad = \$80$
The profit from the production and sale of 500 copies is \$80.
$\quad R'(x) = 0.90,$
so $R'(500) = 0.90$
Approximate revenue from the sale of the 501st copy is 90¢.
$\quad P'(x) = 0.80 - 0.002x$
$\quad P'(500) = 0.80-0.002(500) = -0.2$
Approximate loss from the sale of the 501st copy is 20¢. (Negative marginal profit indicates a loss.)
(c) The marginal profit $P'(x)$ is zero when
$\quad 0.80 - 0.002x = 0$
$\quad x = 0.80/0.002 = 400$ copies.
The graph of the profit function is a parabola with a vertex at $x = 400$, so the profit is a maximum when you produce and sell 400 copies.

13. $P(1000)$ represents the profit on the sale of 1000 DVDs. $P(1000) = 3000$, so the profit on the sale of 1000 DVDs is \$3000. $P'(1000)$ represents the rate of increase of the profit as a function of x. $P'(1000) = -3$, so the profit is decreasing at a rate of \$3 per additional DVD sold.

15. $P(x) = 5x + \sqrt{x}$. Your current profit is
$\quad P(50) = 5(50) + \sqrt{50} \approx \$257.07.$
The marginal profit is
$\quad P'(x) = dP/dx = 5 + 1/(2\sqrt{x})$
$\quad P'(50) = 5 + 1/(2\sqrt{50}) \approx 5.07$
The derivative is measured in dollars per additional magazine sold. Thus, your current

405

profit is $257.07 per month and this would increase at a rate of $5.07 per additional magazine in sales.

17. (a) $q = 20,000/q^{1.5}$. When $q = 400, p = 20,000/(400)^{1.5} \approx \$2.50$ per pound.
(b) $R(q) = pq = (20,000/q^{1.5})q$
$$= 20,000/q^{0.5}$$
(c) $R(400) = \$1000$. This is the monthly revenue that will result from setting the price at $2.50 per pound.
$$R'(q) = -10,000/q^{1.5},$$
so $R'(400) = -10,000/(400)^{1.5}$
$$= -\$1.25 \text{ per pound of tuna.}$$
Thus, at a demand level of 400 pounds per month, the revenue is decreasing at a rate of $1.25 per pound.
(d) Since the revenue goes down with increasing demand, the fishery should raise the price to reduce the demand and hence increase revenue.

19. $P'(n) = 400 - n$, so $P'(50) = \$350$. This means that, at an employment level of 50 workers, the firm's daily profit will increase at a rate of $350 per additional worker it hires.

21. (a) (B): $C'(x) = -0.002x + 0.3$ decreases as x increases.
(b) (B): $\bar{C}(x) = -0.001x + 0.3 + 500/x$ decreases as x increases.
(c) (C): $C'(100) = 0.1, \bar{C}(100) = 5.2$.

23. (a) $C(x) = 500,000 + 1,600,000x - 100,000\sqrt{x}$

$$C'(x) = 1,600,000 - \frac{50,000}{\sqrt{x}}$$

$$\bar{C}(x) = C(x)/x = \frac{500,000}{x} + 1,600,000 - \frac{100,000}{\sqrt{x}}$$

(b) $C'(3) \approx \$1,570,000$ per spot, $\bar{C}(3) \approx \$1,710,000$ per spot. Since the marginal cost is less than the average cost, the cost of the fourth ad is lower than the average cost of the first three, so the average cost will decrease as x increases.

25. (a) $C'(q) = 200q$ so $C'(10) = \$2000$ per one-pound reduction in emissions.
(b) $S'(q) = 500$. Thus $S'(q) = C'(q)$ when $500 = 200q$, or $q = 2.5$ pounds per day reduction.
(c) $N(q) = C(q) - S(q) = 100q^2 - 500q + 4000$. This is a parabola with lowest point (vertex) given by $q = 2.5$. The net cost at this production level is $N(2.5) = \$3375$ per day. The value of q is the same as that for part (b). The net cost to the firm is minimized at the reduction level for which the cost of controlling emissions begins to increase faster than the subsidy. This is why we get the answer by setting these two rates of increase equal to each other.

27. $M'(x) = \dfrac{3600x^{-2} - 1}{(3600x^{-1} + x)^2}$. So,

$$M'(10) = \frac{3600(10)^{-2} - 1}{(3600(10)^{-1} + 10)^2}$$
$$\approx 0.0002557 \text{ mpg/mph.}$$

This means that, at a speed of 10 mph, the fuel economy is increasing at a rate of 0.0002557 miles per gallon per 1-mph increase in speed.

$$M'(60) = \frac{3600(60)^{-2} - 1}{(3600(60)^{-1} + 60)^2} = 0$$

mpg/mph.
This means that, at a speed of 60 mph, the fuel economy is neither increasing nor decreasing with increasing speed.

$$M'(70) = \frac{3600(70)^{-2} - 1}{(3600(70)^{-1} + 70)^2}$$
$$\approx -0.00001799.$$

This means that, at 70 mph, the fuel economy is decreasing at a rate of 0.00001799 miles per gallon per 1-mph increase in speed. Thus 60 mph is the most fuel-efficient speed for the car.

29. (C): If the marginal cost were lower in one plant than another, moving some production from the higher cost plant to the lower would result in a lower cost for the same production level.

31. (D): The marginal product per dollar of salary is $2/1.5 \sim 1.33$ times as high for a junior professor as compared to a senior professor. Therefore, discharging senior professors and

406

hiring more junior professors will result in a higher quantity of output for the same amount of money.

33. (B): (In most cases) This is why we use the marginal cost as an estimate of the actual cost of the item.

35. Cost is often measured as a function of the number of items x. Thus, $C(x)$ is the cost of producing (or purchasing, as the case may be) x items.
(a) The average cost function $\bar{C}(x)$ is given by $\bar{C}(x) = C(x)/x$. The marginal cost function is the derivative, $C'(x)$, of the cost function.
(b) The average cost $\bar{C}(r)$ is the slope of the line through the origin and the point on the graph where $x = r$. The marginal cost of the rth unit is the slope of the tangent to the graph of the cost function at the point where $x = r$.
(c) The average cost function $\bar{C}(x)$ gives the average cost of producing the first x items. The marginal cost function $C'(x)$ is the rate at which cost is changing with respect to the number of items x, or the incremental cost per item, and approximates the cost of producing the $(x+1)$st item.

37. The marginal cost: If the average cost is rising, then the cost of the next piano must be larger than the average cost of the pianos already built.

39. Not necessarily. For example, it may be the case that the marginal cost of the 101st item is larger than the average cost of the first 100 items (even though the marginal cost is decreasing). Thus, adding this additional item will *raise* the average cost.

41. The circumstances described suggest that the average cost function is at a relatively low point at the current production level, and so it would be appropriate to advise the company to maintain current production levels; raising or lowering the production level will result in increasing average costs.

11.3 The Product and Quotient Rules

The solutions to Exercises 1–11 show the calculation of the derivative using the product or quotient rule as appropriate.

1. Product rule: $f'(x) = (0)x + 3(1) = 3$

3. Product rule: $g'(x) = (1)x^2 + x(2x) = 3x^2$

5. Product rule: $h'(x) = (1)(x + 3) + x(1) = 2x + 3$

7. Product rule: $r'(x) = (0)x^{2.1} + 100(2.1x^{1.1}) = 210x^{1.1}$

9. Quotient rule: $s'(x) = \dfrac{(0)x - 2(1)}{x^2} = -\dfrac{2}{x^2}$

11. Quotient rule: $u'(x) = \dfrac{(2x)3 - x^2(0)}{3^2} = \dfrac{2x}{3}$

13. $\dfrac{dy}{dx} = 3(4x^2 - 1) + 3x(8x) = 36x^2 - 3$

15. $\dfrac{dy}{dx} = 3x^2(1 - x^2) + x^3(-2x) = 3x^2 - 5x^4$

17. $\dfrac{dy}{dx} = 2(2x + 3) + (2x + 3)(2) = 8x + 12$

19. $\dfrac{dy}{dx} = \sqrt{x} + \dfrac{x}{2\sqrt{x}} = \sqrt{x} + \dfrac{\sqrt{x}}{2} = \dfrac{3\sqrt{x}}{2}$

21. $\dfrac{dy}{dx} = (x^2 - 1) + (x + 1)(2x) = (x + 1)(3x - 1)$

23. $\dfrac{dy}{dx} = (x^{-0.5} + 4)(x - x^{-1}) + (2x^{0.5} + 4x - 5) \cdot (1 + x^{-2})$

25. $\dfrac{dy}{dx} = (4x - 4)(2x^2 - 4x + 1) + (2x^2 - 4x + 1)(4x - 4) = 8(2x^2 - 4x + 1)(x - 1)$

27. $\dfrac{dy}{dx} = \left(\dfrac{1}{3.2} - \dfrac{3.2}{x^2}\right)(x^2 + 1) + \left(\dfrac{x}{3.2} + \dfrac{3.2}{x}\right)(2x)$

29. $\dfrac{dy}{dx} = 2x(2x + 3)(7x + 2) + 2x^2(7x + 2) + 7x^2(2x + 3)$

31. $\dfrac{dy}{dx} = 5.3(1 - x^{2.1})(x^{-2.3} - 3.4) - 2.1x^{1.1}(5.3x - 1)(x^{-2.3} - 3.4) - 2.3x^{-3.3}(5.3x - 1)(1 - x^{2.1})$

33. $\dfrac{dy}{dx} = \dfrac{1}{2\sqrt{x}}\left(\sqrt{x} + \dfrac{1}{x^2}\right) + (\sqrt{x} + 1)\left(\dfrac{1}{2\sqrt{x}} - \dfrac{2}{x^3}\right)$

35. $\dfrac{dy}{dx} = \dfrac{2(3x - 1) - 3(2x + 4)}{(3x - 1)^2} = \dfrac{-14}{(3x - 1)^2}$

37. $\dfrac{dy}{dx} = \dfrac{(4x + 4)(3x - 1) - 3(2x^2 + 4x + 1)}{(3x - 1)^2} = \dfrac{6x^2 - 4x - 7}{(3x - 1)^2}$

39. $\dfrac{dy}{dx} = \dfrac{(2x - 4)(x^2 + x + 1) - (x^2 - 4x + 1)(2x + 1)}{(x^2 + x + 1)^2} = \dfrac{5x^2 - 5}{(x^2 + x + 1)^2}$

41. $\dfrac{dy}{dx} = \dfrac{(0.23x^{-0.77} - 5.7)(1 - x^{-2.9}) - 2.9x^{-3.9}(x^{0.23} - 5.7x)}{(1 - x^{-2.9})^2}$

43. $\dfrac{dy}{dx} = \dfrac{\frac{1}{2}x^{-1/2}(x^{1/2}-1) - \frac{1}{2}x^{-1/2}(x^{1/2}+1)}{(x^{1/2}-1)^2} = \dfrac{-1}{\sqrt{x}\,(\sqrt{x}-1)^2}$

45. $\dfrac{dy}{dx} = \dfrac{d}{dx}\left[\dfrac{\frac{1}{2}(x+1)}{x(1+x)}\right] = \dfrac{d}{dx}\left(\dfrac{1}{x^3}\right) = -\dfrac{3}{x^4}$ (sometimes it pays to simplify first)

47. $\dfrac{dy}{dx} = \dfrac{[(x+1)+(x+3)](3x-1) - 3(x+3)(x+1)}{(3x-1)^2} = \dfrac{3x^2 - 2x - 13}{(3x-1)^2}$

49. $\dfrac{dy}{dx} = \dfrac{[(x+1)(x+2)+(x+3)(x+2)+(x+3)(x+1)](3x-1) - 3(x+3)(x+1)(x+2)}{(3x-1)^2}$

$= \dfrac{6x^3 + 15x^2 - 12x - 29}{(3x-1)^2}$

51. $\dfrac{d}{dx}\,[(x^2+x)(x^2-x)]$

$= (2x+1)(x^2-x) + (x^2+x)(2x-1)$

$= 4x^3 - 2x$

53. $\dfrac{d}{dx}\,[(x^3+2x)(x^2-x)]\Big|_{x=2}$

$= [(3x^2+2)(x^2-x) + (x^3+2x)(2x-1)]\Big|_{x=2}$

$= (5x^4 - 4x^3 + 6x^2 - 4x)\Big|_{x=2} = 64$

55. $\dfrac{d}{dt}\,[(t^2 - t^{0.5})(t^{0.5} + t^{-0.5})]\Big|_{t=1}$

$= [(2t - 0.5t^{-0.5})(t^{0.5} + t^{-0.5})$

$\qquad + (t^2 - t^{0.5})(0.5t^{-0.5} - 0.5t^{-1.5})]\Big|_{t=1}$

$= (2.5t^{1.5} + 1.5t^{0.5} - 1)\Big|_{t=1} = 3$

57. The calculation thought experiment tells us that the expression for y is a difference:

$\dfrac{d}{dx}[x^4 - (x^2+120)(4x-1)]$

$= \dfrac{d}{dx}[x^4] - \dfrac{d}{dx}[(x^2+120)(4x-1)]$

Now use the power rule for the first expression and the product rule for the second:

$= 4x^3 - [(2x)(4x-1) + (x^2+120)(4)]$

$= 4x^3 - 12x^2 + 2x - 480$

59. The calculation thought experiment tells us that the expression for y is a sum:

$\dfrac{d}{dx}\left[x+1 + 2\left(\dfrac{x}{x+1}\right)\right]$

$= \dfrac{d}{dx}[x] + \dfrac{d}{dx}[1] + 2\dfrac{d}{dx}\left(\dfrac{x}{x+1}\right)$

Now use the power rule for the expressions on the left and the quotient rule for the one on the right:

$= 1 + 0 + 2\dfrac{(1)(x+1) - x(1)}{(x+1)^2}$

$= 1 + \dfrac{2}{(x+1)^2}$

61. Since the last operation one would do is multiply, the CTE tells us that the given expression is a product. By the product rule:

$\dfrac{dy}{dx} = (1)\left[\dfrac{x}{x+1}\right] + (x+2)\dfrac{d}{dx}\left[\dfrac{x}{x+1}\right]$

$= \left[\dfrac{x}{x+1}\right] + (x+2)\,\dfrac{1}{(x+1)^2}$

63. The calculation thought experiment tells us that the expression for y is a difference:

$\dfrac{d}{dx}\left[(x+1)(x-2) - 2\left(\dfrac{x}{x+1}\right)\right]$

$= \dfrac{d}{dx}[(x+1)(x-2)] - 2\dfrac{d}{dx}\left(\dfrac{x}{x+1}\right)$

Now use the product rule for the expressions on the left and the quotient rule for the one on the right:

$$= (1)(x-2) + (x+1)(1) - 2\frac{(1)(x+1) - x(1)}{(x+1)^2}$$

$$= 2x - 1 - \frac{2}{(x+1)^2}$$

65. $f'(x) = 2x(x^3 + x) + (x^2 + 1)(3x^2 + 1)$
$= 5x^4 + 6x^2 + 1$,
so $f'(1) = 12$ is the slope. The tangent line passes through $(1, f(1)) = (1, 4)$, so its equation is $y = 12x - 8$.

67. $f'(x) = \frac{(x + 2) - (x + 1)}{(x + 2)^2} = \frac{1}{(x + 2)^2}$, so
$f'(0) = 1/4$. The tangent line passes through $(0, f(0)) = (0, 1/2)$, so its equation is $y = x/4 + 1/2$.

69. $f'(x) = \frac{2x(x) - (x^2 + 1)}{x^2} = \frac{x^2 - 1}{x^2}$, so
$f'(-1) = 0$. The tangent line passes through $(-1, f(-1)) = (-1, -2)$, so its equation is $y = -2$.

71. Rate of change of monthly sales = $q'(t) = 2000 - 200t$.
When $t = 5$: $q'(5) = 2000 - 200(5) = 1000$ units/month
Therefore, sales are increasing at a rate of 1000 units per month).
Rate of change of price = $p'(t) = -2t$
When $t = 5$: $p'(5) = -2(5) = -\$10/$month.
Therefore, The price of a sound system is dropping at a rate of $10 per month.
Revenue:
$$R(t) = p(t)q(t)$$
$$= (1000 - t^2)(2000t - 100t^2)$$
Rate of change of revenue:
$$R'(t) = p'(t)q(t) + p(t)q'(t)$$
$$= (-2t)(2000t - 100t^2)$$
$$+ (1000 - t^2)(2000 - 200t)$$
$$R'(5) = [-2(5)][2000(5) - 100(5)^2]$$
$$+ [1000 - (5)^2][2000 - 200(5)]$$
$$= \$900,000/$month

Therefore, revenue is increasing at a rate of $900,000 per month.

73. Revenue $\quad R(t) = P(t)Q(t)$ million dollars.
In 2001, $t = 1$, so
$$R(1) = P(1)Q(1)$$
$$= [5(1) + 25][0.082(1)^2 - 0.22(1) + 8.2]$$
$$= 241.86 \approx \$242 \text{ million}$$
To obtain the rate of change of daily revenue, we compute $R'(t)$.
By the product rule,
$$R'(t) = P'(t)Q(t) + P(t)Q'(t)$$
$$= (5)(0.082t^2 - 0.22t + 8.2)$$
$$+ (5t + 25)(0.164t - 0.22)$$
Therefore,
$$R'(1) = (5)(0.082(1)^2 - 0.22(1) + 8.2)$$
$$+ (5(1) + 25)(0.164(1) - 0.22)$$
$$= 38.63 \approx \$39 \text{ million per year.}$$

75. Let $S(t)$ be the number of T-shirts sold per day. If $t = 0$ is now, we are told that $S(0) = 20$ and $S'(0) = -3$. Let $p(t)$ be the price of T-shirts. We are told that $p(0) = 7$ and $p'(0) = 1$. The revenue is then $R(t) = S(t)p(t)$, so $R'(0) = S'(0)p(0) + S(0)p'(0) = -3(7) + 20(1) = -1$. So, revenue is decreasing at a rate of $1 per day.

77. The cost per passenger is $Q(t) = C(t)/P(t) = (10,000 + t^2)/(1000 + t^2)$. So, $Q'(t) = \frac{2t(1000 + t^2) - 2t(10,000 + t^2)}{(1000 + t^2)^2} = \frac{-18,000t}{(1000 + t^2)^2}$ and $Q'(6) \approx -0.10$. The cost per passenger is decreasing at a rate of $0.10 per month.

79. $M'(x) = \frac{3000(3600x^{-2} - 1)}{(x + 3600x^{-1})^2}$, so $M'(10) \approx$ 0.7670 mpg/mph. This means that, at a speed of 10 mph, the fuel economy is increasing at a rate of 0.7670 miles per gallon per one mph increase in speed. $M'(60) = 0$ mpg/mph. This means that, at a speed of 60 mph, the fuel economy is neither increasing nor decreasing with increasing speed. $M'(70) \approx -0.0540$. This means that, at 70 mph, the fuel economy is decreasing at a rate of 0.0540 miles per gallon per one mph increase in speed. 60 mph is the most fuel-efficient speed for the car. (In the next chapter we shall discuss how to locate largest values in general.)

81. (a) $P(t) - I(t)$ = Production – Imports by the U.S., and therefore represents the daily production of oil in Mexico that was not exported to the U.S.

$I(t)/P(t)$ = (Imports by U.S.)/(Production), U.S; imports of oil from Mexico as a fraction of the total produced there.

(b) By the quotient rule

$$\frac{d}{dt}\left(\frac{I(t)}{P(t)}\right) = \frac{I'(t)P(t) - I(t)P'(t)}{[P(t)]^2}$$

$$= \frac{(-0.11)(3.9 - 0.10t) - (2.1 - 0.11t)(-0.10)}{(3.9 - 0.10t)^2}$$

Instead of simplifying this, we can just evaluate directly at $t = 8$:

$$\frac{d}{dt}\left(\frac{I(t)}{P(t)}\right)\Big|_{t=8} =$$

$$\frac{(-0.11)(3.9 - 0.10(8)) - (2.1 - 0.11(8))(-0.10)}{(3.9 - 0.10(8))^2}$$

≈ -0.023 per year, or -2.3 percentage points per year.

At the start of 2008, $(t = 8)$ the fraction of oil produced in Mexico that was imported by the U.S. was decreasing at a rate of 0.023 (or 2.3 percentage points) per year.

83. Cost: $10,000t/3 + 80,000$
Personnel: $-12,500t + 1,500,000$ (t since 1995).
Rate of change at $t = 7$ is
$(10,000/3)(-12,500(7) + 1,500,000) + (10,000(7)/3 + 80,000)(-12,500)$
$= 3416,666,667 \approx 3,420,000,000.$
Total military personnel costs were increasing at a rate of about \$3420 million per year in 2002.

85. $R'(p) = -\dfrac{5.625}{(1 + 0.125p)^2}$ so $R'(4) = -2.5$
thousand organisms per hour, per 1000

organisms. This means that the reproduction rate of organisms in a culture containing 4000 organisms is declining at a rate of 2500 organisms per hour, per 1000 additional organisms.

87. Let $P(t)$ be the number of eggs; then
$P(t) = 30 - t.$
The total oxygen consumption is $P(t)C(t)$ and its rate of change is
$P'(t)C(t) + P(t)C'(t)$
$= (-1)(-0.016t^4 + 1.1t^3 - 11t^2 + 3.6t)$
$+ (30-t)(-0.064t^3 + 3.3t^2 - 22t + 3.6)$
At $t = 25$ this is approximately $-1572 \approx -1600.$
Thus, oxygen consumption is decreasing at a rate of about 1600 milliliters per day. This must be due to the fact that the number of eggs is decreasing, because $C'(25)$ is positive.

89. We are given that $f(3) = 5$, $f'(3) = 2$, $g(3) = 4$, $g'(3) = 5$.
Thus, fg has the value $f(3)g(3) = (5)(4) = 20$
By the product rule,
$(fg)'(3) = f'(3)g(3) + f(3)g'(3)$
$= (2)(4) + (5)(5) = 33$ units per second
fg equals $\underline{20}$ and is rising at a rate of $\underline{33}$ units per second.

91. We are given that $f(3) = 5$, $f'(3) = 2$, $g(3) = 4$, $g'(3) = 5$.
Thus, f/g has the value $f(3)/g(3) = 5/4$
By the quotient rule,
$$(f/g)'(3) = \frac{f'(3)g(3) - f(3)g'(3)}{g(3)^2}$$
$$= \frac{(2)(4) - (5)(5)}{4^2} = -17/16 \text{ units per second}$$
f/g equals $\underline{5/4}$ and is changing at a rate of $\underline{-17/16}$ units per second

93. The analysis is suspect, because it seems to be asserting that the annual increase in revenue, which we can think of as dR/dt, is the product of the annual increases, dp/dt in price, and dq/dt in sales. However, because $R = pq$, the product rule implies that dR/dt is not the product of dp/dt and dq/dt, but is instead

411

$$\frac{dR}{dt} = \frac{dp}{dt} \cdot q + p \cdot \frac{dq}{dt} \ .$$

95. Answers will vary; $q = -p + 1000$ is one example: $R(p) = pq = -p^2 + 1000p$, so $R'(p) = -2p + 1000$ and $R'(100) = 800 > 0$.

97. Mine; it is increasing twice as fast as yours. The rate of change of revenue is given by $R'(t) = p'(t)q(t)$ because $q'(t) = 0$ for both of us. Thus, in this case, $R'(t)$ does not depend on the selling price $p(t)$.

99. (A): If the marginal product was greater than the average product it would force the average product to increase. (See the formula for the rate of change of the average given in the solution to Exercise 82.)

11.4 The Chain Rule

1. $2(2x + 1)(2) = 4(2x + 1)$

3. $-(x - 1)^{-2}(1) = -(x - 1)^{-2}$

5. $-2(2 - x)^{-3}(-1) = 2(2 - x)^{-3}$

7. $0.5(2x + 1)^{-0.5}(2) = (2x + 1)^{-0.5}$

9. $-(4x - 1)^{-2}(4) = -4(4x - 1)^{-2}$

11. $-(3x - 1)^{-2}(3) = -3/(3x - 1)^2$

13. $4(x^2 + 2x)^3 \dfrac{d}{dx}(x^2 + 2x) =$
$4(x^2 + 2x)^3(2x + 2)$

15. $-(2x^2 - 2)^{-2}\dfrac{d}{dx}(2x^2 - 2) = -4x(2x^2 - 2)^{-2}$

17. $-5(x^2 - 3x - 1)^{-6}\dfrac{d}{dx}(x^2 - 3x - 1) =$
$-5(2x - 3)(x^2 - 3x - 1)^{-6}$

19. $-3(x^2 + 1)^{-4}\dfrac{d}{dx}(x^2 + 1) = -6x/(x^2 + 1)^4$

21. $1.5(0.1x^2 - 4.2x + 9.5)^{0.5}\dfrac{d}{dx}(0.1x^2 - 4.2x + 9.5)$
$= 1.5(0.2x - 4.2)(0.1x^2 - 4.2x + 9.5)^{0.5}$

23. $4(s^2 - s^{0.5})^3\dfrac{d}{dx}(s^2 - s^{0.5}) =$
$4(2s - 0.5s^{-0.5})(s^2 - s^{0.5})^3$

25. $\dfrac{d}{dx}\sqrt{1 - x^2} = \dfrac{d}{dx}(1 - x^2)^{1/2}$
$= \dfrac{1}{2}(1 - x^2)^{-1/2}(-2x) = -\dfrac{x}{\sqrt{1 - x^2}}$

27. $\left(-\dfrac{1}{2}\right)2[(x + 1)(x^2 - 1)]^{-3/2}\dfrac{d}{dx}[(x + 1)(x^2 - 1)]$
$= -[(x + 1)(x^2 - 1)]^{-3/2}[(x^2 - 1) + (x + 1)(2x)]$
$= -[(x + 1)(x^2 - 1)]^{-3/2}(3x - 1)(x + 1)$

29. $2(3.1x - 2)^1(3.1) - (-2)(3.1x - 2)^{-3}(3.1)$
$= 6.2(3.1x - 2) + 6.2/(3.1x - 2)^3$

31. $2[(6.4x - 1)^2 + (5.4x - 2)^3]\dfrac{d}{dx}[(6.4x - 1)^2 + (5.4x - 2)^3]$
$= 2[(6.4x - 1)^2 + (5.4x - 2)^3][12.8(6.4x - 1) + 16.2(5.4x - 2)^2]$

33. $\dfrac{d}{dx}[(x^2 - 3x)^{-2}](1 - x^2)^{0.5} +$
$(x^2 - 3x)^{-2}\dfrac{d}{dx}[(1 - x^2)^{0.5}]$
$= -2(x^2 - 3x)^{-3}(2x - 3)(1 - x^2)^{0.5} -$
$x(x^2 - 3x)^{-2}(1 - x^2)^{-0.5}$

35. $2\left(\dfrac{2x + 4}{3x - 1}\right)\dfrac{d}{dx}\left(\dfrac{2x + 4}{3x - 1}\right)$
$= \left(\dfrac{4(x + 2)}{3x - 1}\right) \cdot \dfrac{2(3x - 1) - (2x + 4)(3)}{(3x - 1)^2} =$
$\dfrac{-56(x + 2)}{(3x - 1)^3}$

37. $3\left(\dfrac{z}{1 + z^2}\right)^2\dfrac{d}{dz}\left(\dfrac{z}{1 + z^2}\right)$
$= \left(\dfrac{3z^2}{(1 + z^2)^2}\right) \cdot \dfrac{(1 + z^2) - z(2z)}{(1 + z^2)^2}$
$= \dfrac{3z^2(1 - z^2)}{(1 + z^2)^4}$

39. $3[(1 + 2x)^4 -$
$(1 - x)^2]^2\dfrac{d}{dx}[(1 + 2x)^4 - (1 - x)^2]$
$= 3[(1 + 2x)^4 -$
$(1 - x)^2]^2[8(1 + 2x)^3 + 2(1 - x)]$

41. $4.3[2 + (x + 1)^{-0.1}]^{3.3}\dfrac{d}{dx}[2 + (x + 1)^{-0.1}]$
$= -0.43(x + 1)^{-1.1}[2 + (x + 1)^{-0.1}]^{3.3}$

413

43. $-(\sqrt{2x+1} - x^2)^{-2}\dfrac{d}{dx}[(2x + 1)^{1/2} - x^2]$

$= -(\sqrt{2x+1} - x^2)^{-2}\left(\dfrac{1}{2}(2x + 1)^{-1/2}(2) - 2x\right)$

$= -\dfrac{\dfrac{1}{\sqrt{2x+1}} - 2x}{(\sqrt{2x+1} - x^2)^2}$

45. $3\{1 + [1 + (1 + 2x)^3]^3\}^2$

$\dfrac{d}{dx}\{1 + [1 + (1 + 2x)^3]^3\}$

$= 9\{1 + [1 + (1 + 2x)^3]^3\}^2[1 + (1 + 2x)^3]^2$

$\dfrac{d}{dx}[1 + (1 + 2x)^3]$

$= 27\{1 + [1 + (1 + 2x)^3]^3\}^2[1 + (1 + 2x)^3]^2(1 + 2x)^2(2)$

$= 54(1 + 2x)^2[1 + (1 + 2x)^3]^2\{1 + [1 + (1 + 2x)^3]^3\}^2$

47. $\dfrac{dy}{dt} = 100x^{99}\dfrac{dx}{dt} - 99x^{-2}\dfrac{dx}{dt}$

$= (100x^{99} - 99x^{-2})\dfrac{dx}{dt}$

49. $\dfrac{ds}{dt} = \dfrac{d}{dt}(r^{-3} + r^{0.5})$

$= -3r^{-4}\dfrac{dr}{dt} + 0.5r^{-0.5}\dfrac{dr}{dt}$

$= (-3r^{-4} + 0.5r^{-0.5})\dfrac{dr}{dt}$

51. $\dfrac{dV}{dt} = 4\pi r^2\dfrac{dr}{dt}$

53. $\dfrac{dy}{dt} = 3x^2\dfrac{dx}{dt} - x^{-2}\dfrac{dx}{dt}$,

so $\dfrac{dy}{dt}\Big|_{t=1} = 3(2)^2(-1) - 2^{-2}(-1) = -47/4$

55. $\dfrac{dx}{dy} = \dfrac{1}{dy/dx} = \dfrac{1}{3}$

57. $\dfrac{dy}{dx} = \dfrac{dy/dt}{dx/dt} = \dfrac{-5}{3}$

59. $\dfrac{dx}{dy} = \dfrac{1}{dy/dx}$

$= \dfrac{1}{6x-2}\dfrac{dx}{dy}\Big|_{x=1}$

$= \dfrac{1}{6(1)-2} = \dfrac{1}{4}$

61. $\dfrac{dP}{dq} = 5000 - 0.5q$ and $\dfrac{dq}{dn} = 30 + 0.02n$.

When $n = 10$,

$q = 30(10) + 0.01(10)^2 = 301$.

Hence, $\dfrac{dP}{dn}\Big|_{n=10} = \dfrac{dP}{dq}\dfrac{dq}{dn}$

$= (5000 - 0.5(301))(30 + 0.02(10))$

$= 146,454.9$.

At an employment level of 10 engineers, Paramount will increase its profit at a rate of $146,454.90 per additional engineer hired.

63. To express y as a function of t, take the given equation for y:

$y = 35x^{-0.25}$

and substitute for x using $x = 7 + 0.2t$:

$y = 35(7 + 0.2t)^{-0.25}$

The rate of change of spending on food is given by dy/dt. By the chain rule,.

$\dfrac{dy}{dt} = 35(-0.25)(7 + 0.2t)^{-1.25}\dfrac{d}{dt}(7 + 0.2t)$

$= -8.75(7 + 0.2t)^{-1.25}(0.2)$

$= -1.75(7 + 0.2t)^{-1.25}$ percentage points per month.

(The units of dy/dt are units of y are unit of t; that is, percentage points per month.)

January 1 is represented by $t = 0$, and so November 1 is given by $t = 10$:

$\dfrac{dy}{dt}\Big|_{t=10} = -1.75(7 + 0.2\times10)^{-1.25}$

≈ -0.11 percentage points per month.

65. We are told that

$\dfrac{dR}{dp} = \$40$ per $1 increase in the price

$\dfrac{dp}{dq} = -\$0.75$ per additional ruby sold

Marginal revenue $= \dfrac{dR}{dq} = \dfrac{dR}{dp}\dfrac{dp}{dq} = (40)(-0.75) = -\$30$ per additional ruby sold.

Interpretation: The revenue is decreasing at a rate of $30 per additional ruby sold.

67. We are given $dx/dt = -3$ and we need to find dy/dt. From the linear relation we have

$$\frac{dy}{dt} = \frac{dy}{dx}\frac{dx}{dt}$$

$$= (1.5)(-2) = -3$$

Hence, $\frac{dy}{dt} = -3$ murders per 100,000 residents/yr each year.

69. Using the Quick Example in the text:

$$\frac{dS}{dP} = \frac{dS/dt}{dP/dt}$$

$$= \frac{3t - 11}{2t - 10}.$$

Evaluating at $t = 2$ gives

$$\frac{dS}{dP}\bigg|_{t=2} = \frac{3(2) - 11}{2(2) - 10} = \frac{5}{6} \approx 0.833$$

Interpretation: Note that units of dS/dP are percentage points of home sales per percentage points of price, relative to the 2003 levels. So we can say that, relative to the 2003 levels, home sales in 2008 ($t = 2$) were changing at a rate of 0.833 percentage points per percentage point change in price. Equivalently, home sales were dropping at a rate of 0.833 percentage points per percentage point drop in price.

71. $\dfrac{dA}{dt} = 2\pi r\dfrac{dr}{dt} = 2\pi(3)(2) = 12\pi$ mi^2/h

73. $\dfrac{dV}{dt} = 4\pi r^2\dfrac{dr}{dt} = 4\pi(10)^2(0.5) = 200\pi$

ft^2/week;

$C = 1000V$, so

$$\frac{dC}{dt} = 1000\frac{dV}{dt} = \$200,000\pi/\text{week}$$

$$\approx \$628,000/\text{week}$$

75. (a) $q'(4) \approx (q(4 + 0.0001) - q(4 - 0.0001))/0.0002 \approx 333$ units per month
(b) $R = 800q$, so $dR/dq = \$800/\text{unit}$: The marginal revenue is the selling price.

(c) $\dfrac{dR}{dt} = \dfrac{dR}{dq}\dfrac{dq}{dt} \approx 800(333) \approx \$267,000$ per month.

77. Keeping r and p fixed,

$$\frac{dM}{dt} = 1.2y^{-0.4}r^{-0.3}p\frac{dy}{dt},$$

so $\dfrac{dM/dt}{M} = \dfrac{1.2y^{-0.4}r^{-0.3}p}{2y^{0.6}r^{-0.3}p}\dfrac{dy}{dt} = 0.6\dfrac{dy/dt}{y}$. We are

given $\dfrac{dy/dt}{y} = 5\%$ per year, so

$$\frac{dM/dt}{M} = 0.6(5) = 3\% \text{ per year}$$

79. Keeping r fixed,

$$\frac{dM}{dt} = 1.2y^{-0.4}r^{-0.3}p\frac{dy}{dt} + 2y^{0.6}r^{-0.3}\frac{dp}{dt},$$

so $\dfrac{dM/dt}{M} = 0.6\dfrac{dy/dt}{y} + \dfrac{dp/dt}{p}$

$$= 0.6(5) + 5 = 8\% \text{ per year.}$$

81. The glob squared, times the derivative of the glob.

83. The derivative of a quantity cubed is three times the *original quantity* squared, times the derivative of the quantity, not three times the derivative of the quantity squared. Thus, the correct answer is $3(3x^3 - x)^2(9x^2 - 1)$.

85. First, the derivative of a quantity cubed is three times the *original quantity* squared times the derivative of the quantity; not three times the derivative of the quantity squared. Second, the derivative of a quotient is not the quotient of the derivatives; the quotient rule needs to be used in

calculating the derivative of $\dfrac{3x^2 - 1}{2x - 2}$. Thus, the

correct result (before simplifying) is

$$3\left(\frac{3x^2 - 1}{2x - 2}\right)^2\left(\frac{6x(2x - 2) - (3x^2 - 1)(2)}{(2x - 2)^2}\right).$$

87. Following the calculation thought experiment, pretend that you are evaluating the function at a specific value of x. If the last operation you would perform is addition or subtraction, look at each

415

summand separately. If the last operation is multiplication, use the product rule first; if it is division, use the quotient rule first; if it is any other operation (such as raising a quantity to a power or taking a radical of a quantity), then use the chain rule first.

89. An example is

$$f(x) = \sqrt{x + \sqrt{x + \sqrt{x + \sqrt{x + \sqrt{x + 1}}}}} \, .$$

11.5 Derivative of Logarithmic and Exponential Functions

1. $\dfrac{d}{dx}\,[\ln(x-1)] = \dfrac{1}{x-1}\dfrac{d}{dx}(x-1)$

$\qquad = \dfrac{1}{x-1}\cdot 1 = \dfrac{1}{x-1}$

3. $1/(x\ln 2)$ (from the formula in the textbook)

5. $\dfrac{d}{dx}\,[\ln(x^2+3)] = \dfrac{1}{x^2+3}\dfrac{d}{dx}(x^2+3)$

$\qquad = \dfrac{1}{x^2+3}\cdot 2x = \dfrac{2x}{x^2+3}$

7. $\dfrac{d}{dx}\,e^{x+3} = e^{x+3}\dfrac{d}{dx}(x+3)$

$\qquad = e^{x+3}(1) = e^{x+3}$

9. $\dfrac{d}{dx}\,e^{-x} = e^{-x}\dfrac{d}{dx}(-x)$

$\qquad = e^{-x}(-1) = -e^{-x}$

11. $4^x\ln 4$ (from the formula in the textbook)

13. $\dfrac{d}{dx}\,2^u = 2^u\ln 2\,\dfrac{du}{dx}$

$\qquad \dfrac{d}{dx}\,2^{x^2-1} = 2^{x^2-1}\ln 2\,\dfrac{d}{dx}(x^2-1)$

$\qquad = 2^{x^2-1}\,2x\ln 2$

15. $(1)\ \ln x + x\dfrac{d}{dx}\ln x$

$\qquad = \ln x + x\!\left(\dfrac{1}{x}\right) = 1+\ln x$

17. $2x\ln x + (x^2+1)\!\left(\dfrac{1}{x}\right)$

$\qquad = 2x\ln x + \dfrac{x^2+1}{x}$

19. $5(x^2+1)^4(2x)\ln x + (x^2+1)^5\!\left(\dfrac{1}{x}\right)$

$\qquad = 10x(x^2+1)^4\ln x + \dfrac{(x^2+1)^5}{x}$

21. $\dfrac{d}{dx}\,[\ln|3x-1|] = \dfrac{1}{3x-1}\dfrac{d}{dx}(3x-1)$

$\qquad = \dfrac{1}{3x-1}\cdot 3 = \dfrac{3}{3x-1}$

23. $\dfrac{d}{dx}\,[\ln|2x^2+1|] = \dfrac{1}{2x^2+1}\dfrac{d}{dx}(2x^2+1)$

$\qquad = \dfrac{1}{2x^2+1}\cdot 4x = \dfrac{4x}{2x^2+1}$

25. $\dfrac{d}{dx}\,[\ln|x^2-2.1x^{0.3}|]$

$\qquad = \dfrac{1}{x^2-2.1x^{0.3}}\dfrac{d}{dx}(x^2-2.1x^{0.3})$

$\qquad = \dfrac{1}{x^2-2.1x^{0.3}}\cdot(2x-0.63x^{-0.7})$

$\qquad = \dfrac{2x-0.63x^{-0.7}}{x^2-2.1x^{0.3}}$

27. $\dfrac{d}{dx}\,[\ln(-2x+1)+\ln(x+1)]$

$\qquad = \dfrac{-2}{-2x+1} + \dfrac{1}{x+1}$

29. $\dfrac{d}{dx}\,[\ln(3x+1)-\ln(4x-2)]$

$\qquad = \dfrac{3}{3x+1} - \dfrac{4}{4x-2}$

31. $\dfrac{d}{dx}\,[\ln|x+1|+\ln|x-3|-$

$\qquad\qquad\qquad \ln|-2x-9|]$

$\qquad = \dfrac{1}{x+1} + \dfrac{1}{x-3} - \dfrac{-2}{-2x-9}$

$\qquad = \dfrac{1}{x+1} + \dfrac{1}{x-3} - \dfrac{2}{2x+9}$

33. $\dfrac{d}{dx}\,[1.3\ln(4x-2)] = \dfrac{5.2}{4x-2}$

35. $\dfrac{d}{dx}\,[\ln|x+1|^2 - \ln|3x-4|^3 - \ln|x-9|]$

$\qquad = \dfrac{d}{dx}\,[2\ln|x+1| - 3\ln|3x-4| - \ln|x-9|]$

$\qquad = \dfrac{2}{x+1} - \dfrac{9}{3x-4} - \dfrac{1}{x-9}$

37. $\dfrac{d}{dx}\log_2(x+1) = \dfrac{1}{(x+1)\ln 2}\dfrac{d}{dx}(x+1)$

$= \dfrac{1}{(x+1)\ln 2}$

39. $\dfrac{d}{dt}\log_3(t+1/t) = \dfrac{1}{(t+1/t)\ln 3}\dfrac{d}{dt}((t+1/t))$

$= \dfrac{1}{(t+1/t)\ln 3}(1-1/t^2)$

$= \dfrac{1-1/t^2}{(t+1/t)\ln 3}$

41. $2(\ln|x|)\left(\dfrac{1}{x}\right) = \dfrac{2\ln|x|}{x}$

43. $\dfrac{d}{dx}\{2\ln x - [\ln(x-1)]^2\}$

$= \dfrac{2}{x} - 2[\ln(x-1)]\dfrac{1}{x-1}$

$= \dfrac{2}{x} - \dfrac{2\ln(x-1)}{x-1}$

45. $e^x + xe^x = e^x(1+x)$

47. $\dfrac{1}{x+1} + 9x^2e^x + 3x^3e^x$

$= \dfrac{1}{x+1} + 3e^x(x^3 + 3x^2)$

49. $e^x\ln|x| + e^x\left(\dfrac{1}{x}\right) = e^x(\ln|x| + 1/x)$

51. $2e^{2x+1}$

53. $(2x-1)e^{x^2-x+1}$

55. $2xe^{2x-1} + x^2(2e^{2x-1}) = 2xe^{2x-1}(1+x)$

57. $2e^{2x-1}\dfrac{d}{dx}e^{2x-1} = 2e^{2x-1}(2e^{2x-1}) = 4(e^{2x-1})^2$

OR

$\dfrac{d}{dx}[(e^{2x-1})^2] = \dfrac{d}{dx}(e^{4x-2}) = 4e^{4x-2} = 4(e^{2x-1})^2$

59. $t(x) - 3^{2x-4}$

$t'(x) = 3^{2x-4}\ln 3\dfrac{d}{dx}[2x-4]$

$= 2\cdot 3^{2x-4}\ln 3$

61. $v(x) = 3^{2x+1} + e^{3x+1}$

$v'(x) = 3^{2x+1}\ln 3(2) + e^{3x+1}(3)$

$= 2\cdot 3^{2x+1}\ln 3 + 3e^{3x+1}$

63. $u(x) = \dfrac{3^{x^2}}{x^2+1}$

$u'(x) = \dfrac{3^{x^2}\ln 3\,(2x)(x^2+1) - 3^{x^2}(2x)}{(x^2+1)^2}$

$= \dfrac{2x3^{x^2}[(x^2+1)\ln 3 - 1]}{(x^2+1)^2}$

65. $\dfrac{(e^x - e^{-x})(e^x - e^{-x}) - (e^x + e^{-x})(e^x + e^{-x})}{(e^x - e^{-x})^2}$

$= \dfrac{e^{2x} - 2 + e^{-2x} - (e^{2x} + 2 + e^{-2x})}{(e^x - e^{-x})^2}$

$= \dfrac{-4}{(e^x-e^{-x})^2}$

67. $\dfrac{d}{dx}[e^{(3x-1)+(x-2)+x}] = \dfrac{d}{dx}e^{5x-3} = 5e^{5x-3}$

69. $\dfrac{d}{dx}(x\ln x)^{-1} = -(x\ln x)^{-2}[\ln x + x(1/x)]$

$= -\dfrac{\ln x + 1}{(x\ln x)^2}$

71. Note that $\ln(e^x) = x$, so $f(x) = x^2 - 2\ln(e^x)$
$= x^2 - 2x$, so $f'(x) = 2x - 2 = 2(x-1)$

73. $\dfrac{1}{\ln x}\dfrac{d}{dx}\ln x = \dfrac{1}{x\ln x}$

75. $\dfrac{d}{dx}\ln(\ln x)^{1/2} = \dfrac{d}{dx}\left(\dfrac{1}{2}\ln(\ln x)\right) = \dfrac{1}{2}\left(\dfrac{1}{\ln x}\right)\left(\dfrac{1}{x}\right)$

$= \dfrac{1}{2x\ln x}$

418

77. $\frac{dy}{dx} = e^x \log_2 x + \frac{e^x}{x \ln 2} = \frac{e}{\ln 2}$ when $x = 1$.

Therefore the slope of the desired line is $e/\ln 2$, and a point on the line is $(1, 0)$.

So, the equation of the line is $y = (e/\ln 2)(x - 1)$ $\approx 3.92(x - 1)$.

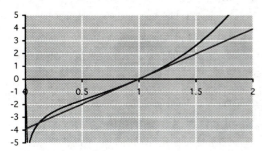

79. $\frac{dy}{dx} = \frac{d}{dx}\left(\frac{1}{2}\ln(2x + 1)\right) = \frac{1}{2x + 1} = 1$ when

$x = 0$. The tangent line has slope 1 and passes through $(0, 0)$, so its equation is $y = x$.

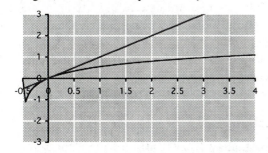

81. $\frac{dy}{dx} = 2xe^{x^2} = 2e$ when $x = 1$. The line at

right angles to the graph has slope $-1/(2e)$ and passes through $(1, e)$, so its equation is

$y = -[1/(2e)](x - 1) + e \approx -0.1839x + 2.9022$.

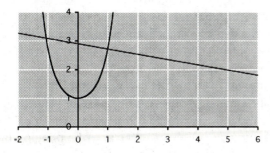

83. Total spent in 2000 ($t = 10$) is given by

$S(10) = 57.5 \ln 10 + 31 \approx \$163$ billion

The rate of change of spending on research and development in 2000 is given by the derivative:

$$S'(t) = 57.5 \cdot \frac{1}{t}$$

$$S'(10) = \frac{57.5}{10} = \$5.75 \text{ billion per year}$$

In 2000, spending was increasing at a rate of $5.75 billion per year.

85. Total spent in 2000 ($t = 0$) is given by

$S(0) = 57.5 \ln(17.1) \approx \$163$ billion

The rate of change of spending on research and development in 2000 is given by the derivative:

$$S'(t) = 57.5 \cdot \frac{1.71}{1.71t + 17.1}$$

$$S'(0) = \frac{(57.5)(1.71)}{17.1} = \$5.75 \text{ billion per year}$$

In 2000, spending was increasing at a rate of $5.75 billion per year.

87. $\frac{dy}{dx} = \frac{d}{dx}[\log_{0.999879}(0.1x)]$

$= \frac{1}{0.1x \ln 0.999879} \frac{d}{dx}(0.1x)$

$= \frac{1}{0.1x \ln 0.999879} 0.1 = \frac{1}{x \ln 0.999879}$

Thus

$$\frac{dy}{dx}\bigg|_{x=5} = \frac{1}{5 \ln 0.999879} \approx -1653 \text{ years per}$$

gram

The age of the specimen is decreasing at a rate of about 1653 years per additional one gram of carbon-14 present in the sample. (Equivalently, the age of the specimen is increasing at a rate of about 1653 years per additional one gram less of carbon-14 in the sample.)

89. In 2003, $t = 9$, and so the price of an apartment was

$p(3) = 0.33e^{0.16(9)} = 0.33e^{1.44} \approx \$1.4$ million dollars

For the rate of change, we compute the derivative:

$$p'(t) = \frac{d}{dt}[0.33e^{0.16t}]$$

$= (0.33)(0.16)e^{0.16t} = 0.0528e^{0.16t}$
In 2003,

$p'(9) = 0.0528e^{0.16(9)} = 0.0528e^{1.44}$

≈ 0.22 million dollars/year,

or \$220,000 per year. Thus, in 2003 the average price of a two-bedroom apartment in downtown New York City was increasing at a rate of about \$220,000 per year.

91. (a) $N(t) = 820e^{0.051t}$

$N(15) = 820e^{0.051(15)} \approx 1762 \approx 1800$

(rounded to 2 significant digits)

$N'(t) = 820e^{0.051t}(0.051) = 41.82e^{0.051t}$

$N'(15) = 41.82e^{0.051(15)} \approx 89.87 \approx 90$

wiretap orders per year (rounded to 2 significant digits). The constants in the model are specified to 2 significant digits, so we cannot expect the answer to be accurate to more than 2 digits. In other words, all digits from the third on are probably meaningless.
(b) From the answer in part (a), the number of wiretap orders was about 1800 in 2005 ($t = 15$) and increasing at a rate of about 90 per year. Thus, in 2005, the number of people whose communications were intercepted was about $100 \times 1800 = 180,000$ and increasing at a rate of about $100 \times 90 = 9000$ people per year.
(c) The rate of change of N is the derivative $N'(t)$ $= 41.82e^{0.051t}$, which is also an exponential function . Hence the number of wiretap orders increased at an exponential rate.

93. From the continuous compounding formula, the value of the balance at time t years is $A(t) = 10,000e^{0.04t}$. Its derivative is $A'(t) = 400e^{0.04t}$, so, after 3 years, the balance is growing at the rate of $A'(3) = \$451.00$ per year.

95. From the compound interest formula, the value of the balance at time t years is $A(t) = 10,000(1 + 0.04/2)^{2t} = 10,000(1.02)^{2t}$. Its derivative is $A'(t) = [20,000 \ln(1.02)] (1.02)^{2t}$, so, after 3 years, the balance is growing at the rate of $A'(3) = \$446.02$ per year.

97. The desired model is $A(t) = Ab^{t}$.
At time $t = 0$ (March 17, 2003) the number of cases was 167, so $A = 167$.
Since the number was increasing by 18% each day, the number of cases is multiplied by 1.18 each day, so $b = 1.18$. Hence, the model is

$A(t) = 167(1.18)^{t}$.

The rate of spread of the epidemic is the rate of change, dA/dt:

$\dfrac{dA}{dt} = 167(1.18)^{t} \ln 1.18$ cases per year

Since March 31, 2003 corresponds to $t = 14$, the rate of spread of the epidemic on April 30 was

$\left. \dfrac{dA}{dt} \right|_{t=14} = 167(1.18)^{14} \ln 1.18 \approx 280$

new cases per day

99. (a) (A): The data would best be modeled by a function that approaches a value in the high 400s exponentially. The function in (A) is the only one that does that: the one in (B) decays to 0, the one in (C) increases without bound, and the one in (D) decreases without bound.
(b) $S'(x) = 136(0.0000264)e^{-0.0000264x}$, so $S'(45,000) \approx 0.001$. At an income level of \$45,000, the average verbal SAT increases by approximately $1000(0.001) = 1$ point for each \$1000 increase in income.
(c) $S'(x)$ decreases with increasing x, so that as parental income increases, the effect on SAT scores decreases.

101. (a) $a'(x) = 120(0.172)^{x} \ln(0.172)$
$a'(2) = 120(0.172)^{2} \ln(0.172)$
≈ -6.25 years/child
(We rounded to 3 significant digits because the given coefficients are only given to 3 digits.) The answer tells us that, when the fertility rate is 2 children per woman, the average age of a population is dropping at a rate of 6.26 years per one-child increase in the fertility rate.
(b) From part (a), the average age of a population is dropping at a rate of 6.26 years per one-child increase in the fertility rate. In other words:

420

Given: 1 child per woman increase → 6.26 year drop in average age
Want: x child per woman increase → 1 year drop in average age
Solution: $x = 1/6.25 = 0.160$ children per woman.

103. $P(t) = \dfrac{150}{1 + 15{,}000e^{-0.35t}}$

$= 150(1 + 15{,}000e^{-0.35t})^{-1}$.

$P'(t) = -150(1 + 15{,}000e^{-0.35t})^{-2}$
$[15{,}000(-0.35)e^{-0.35t}]$

$= 787{,}500e^{-0.35t}/(1 + 15{,}000e^{-0.35t})^{2}$.

$P'(20) \approx 3.3$ million cases/week, or 330,000 cases/week

$P'(30) \approx 11.0$ million cases/week, or 11,000,000 cases/week

$P'(40) \approx 0.64$ million cases/week, or 640,000 cases/week

105. $A(t) = \dfrac{15}{1 + 8.6e^{-0.59t}}$

$= 15(1 + 8.6e^{-0.59t})^{-1}$

The rate of change of the percentage is the derivative,

$A'(t) = -15(1 + 8.6e^{-0.59t})^{-2}(8.6e^{-0.59t})$
(-0.59)

$= \dfrac{(15)(8.6)(0.59)\, e^{-0.59t}}{(1 + 8.6e^{-0.59t})^{2}}$

$= \dfrac{76.11e^{-0.59t}}{(1 + 8.6e^{-0.59t})^{2}}$

The start of 2003 corresponds to $t = 3$ so

$A'(3) = \dfrac{76.11\, e^{-0.59(3)}}{(1 + 8.6e^{-0.59(3)})^{2}} \approx 2.1$

percentage points per year
Referring to the graph, the rate of change is the slope of the tangent at $t = 3$. This is also approximately the average rate of change over [2, 4] which is about 4/2 = 2, in approximate agreement with the answer.

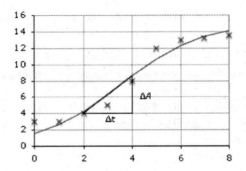

107. $A(t) = \dfrac{15.0}{1 + 8.6(1.8)^{-t}}$

$= 15(1 + 8.6(1.8)^{-t})^{-1}$

The rate of change of the percentage is the derivative,

$A'(t) = -15(1 + 8.6(1.8)^{-t})^{-2}$
$(-8.6(1.8)^{-t})\ln(1.8)$

$= \dfrac{(15)(8.6)\ln(1.8)\,(1.8)^{-t}}{(1 + 8.6(1.8)^{-t})^{2}}$

$\approx \dfrac{75.82(1.8)^{-t}}{(1 + 8.6(1.8)^{-t})^{2}}$

The start of 2003 corresponds to $t = 3$ so

$A'(3) = \dfrac{75.82(1.8)^{-3}}{(1 + 8.6(1.8)^{-3})^{2}} \approx 2.1$

percentage points per year

109. For an exponential model $P(t) = Ab^{t}$ of the population where t is time in years since 2010, use the two points on its graph: $(0, 4{,}000{,}000)$ and $(10, 8{,}000{,}000)$. Substituting gives

$4{,}000{,}000 = A$

$8{,}000{,}000 = Ab^{10} = 4{,}000{,}000\, b^{10}$

so $b^{10} = 2$

$b = 2^{1/10}$

and the model is

$P(t) = 4{,}000{,}000(2^{1/10})^{t}$

$= 4{,}000{,}000(2^{t/10})$

Its derivative is

$P'(t) = [400{,}000 \ln 2]\, 2^{t/10}$,

so, at the start of 2010, the population was growing at the rate of

421

$P'(0) \approx 277{,}000$ people per year.

111. The amount of Plutonium-239 left after t years is $P(t) = 10(0.5)^{t/24{,}400}$. Its derivative is $P'(t) = [(10/24{,}400) \ln 0.5] (0.5)^{t/24{,}400}$, so, after 100 years, the rate of change is $P'(100) \approx -0.000283$ g/year. That is, the Plutonium is decaying at the rate of 0.000283 g/year.

113. (a)

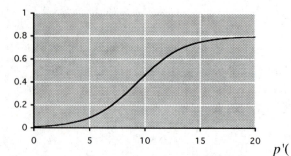

$p'($
$10) \approx 0.09$, so the percentage of firms using numeric control is increasing at a rate of 9 percentage points per year after 10 years.
(b) $\lim_{t \to +\infty} p(t) = 0.80$. Thus, in the long run, 80% of all firms will be using numeric control.
(c) $p'(t) = 0.3816e^{4.46 \, -0.477t}/(1 + e^{4.46 \, -0.477t})^2$.
$p'(10) = 0.0931$. Graph:

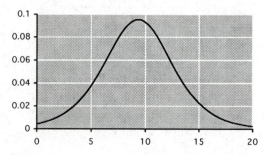

(d) $\lim_{t \to +\infty} p'(t) = 0$. Thus, in the long run, the percentage of firms using numeric control will stop increasing.

115. $R(t) = 350e^{-0.1t}(39t + 68)$ million dollars, so $R(2) \approx \$42{,}000$ million $= \$42$ billion.
$R'(t) = -35e^{-0.1t}(39t + 68) + 350(39)e^{-0.1t}$
$\quad = (11{,}270 - 1365t)e^{-0.1t}$,
so $R'(2) \approx \$7000$ million per year $= \$7$ billion per year.

117. e raised to the glob, times the derivative of the glob.

119. 2 raised to the glob, times the derivative of the glob, times the natural logarithm of 2.

121. The derivative of $\ln |u|$ is not $\dfrac{1}{|u|} \dfrac{du}{dx}$; it is $\dfrac{1}{u}$ $\dfrac{du}{dx}$. Thus, the correct result is $\dfrac{3}{3x + 1}$.

123. The power rule does not apply when the exponent is not constant. The derivative of 3 raised to a quantity is 3 raised to the quantity, times the derivative of the quantity, times $\ln 3$. Thus, the correct answer is $3^{2x} \, 2 \ln 3$

125. No. If $N(t)$ is exponential, so is its derivative.

127. If $f(x) = e^{kx}$, then the fractional rate of change is $\dfrac{f'(x)}{f(x)} = \dfrac{ke^{kx}}{e^{kx}} = k$, the fractional growth rate.

129. If $A(t)$ is growing exponentially, then $A(t) = A_0 e^{kt}$ for constants A_0 and k. Its percentage rate of change is then
$$\frac{A'(t)}{A(t)} = \frac{kA_0 e^{kt}}{A_0 e^{kt}} = k,$$ a constant.

11.6 Implicit Differentiation

1. Implicit differentiation:

$$2 + 3\frac{dy}{dx} = 0$$

$$\frac{dy}{dx} = -2/3$$

Solving for y first and then taking the derivative:

$$y = (7 - 2x)/3$$

$$\frac{dy}{dx} = -2/3$$

3. Implicit differentiation:

$$2x - 2\frac{dy}{dx} = 0$$

$$\frac{dy}{dx} = x$$

Solving for y first and then taking the derivative:

$$y = (x^2 - 6)/2$$

$$\frac{dy}{dx} = x$$

5. Implicit differentiation:

$$2 + 3\frac{dy}{dx} = y + x\frac{dy}{dx}$$

$$(3 - x)\frac{dy}{dx} = y - 2$$

$$\frac{dy}{dx} = (y - 2)/(3 - x)$$

Solving for y first and then taking the derivative:

$$y = -2x/(3 - x)$$

$$\frac{dy}{dx} = \frac{-2(3 - x) - 2x}{(3 - x)^2}$$

$$= \frac{-2}{3 - x} - \frac{2x}{(3 - x)^2}$$

$$= \frac{-2}{3 - x} + \frac{y}{3 - x} = \frac{y - 2}{3 - x}$$

7. Implicit differentiation:

$$e^x y + e^x \frac{dy}{dx} = 0$$

$$\frac{dy}{dx} = -y$$

Solving for y first and then taking the derivative:

$$y = 1/e^x = e^{-x},$$

$$\frac{dy}{dx} = -e^{-x} = -y$$

9. Implicit differentiation:

$$\frac{dy}{dx} \ln x + \frac{y}{x} + \frac{dy}{dx} = 0$$

$$(1 + \ln x)\frac{dy}{dx} = -\frac{y}{x}$$

$$\frac{dy}{dx} = -\frac{y}{x(1 + \ln x)}$$

Solving for y first and then taking the derivative:

$$y = 2/(1 + \ln x),$$

$$\frac{dy}{dx} = -2(1 + \ln x)^{-2}\left(\frac{1}{x}\right)$$

$$= -\frac{2}{1 + \ln x} \cdot \frac{1}{x(1 + \ln x)} = -\frac{y}{x(1 + \ln x)}$$

11. $2x + 2y\frac{dy}{dx} = 0$, $\frac{dy}{dx} = -x/y$

13. $2xy + x^2\frac{dy}{dx} - 2y\frac{dy}{dx} = 0$

$$(x^2 - 2y)\frac{dy}{dx} = -2xy$$

$$\frac{dy}{dx} = -2xy/(x^2 - 2y)$$

15. $3y + 3x\frac{dy}{dx} - \frac{1}{3}\frac{dy}{dx} = -\frac{2}{x^2}$

$$(x^3 - x^2)\frac{dy}{dx} = -(6 + 9x^2 y)$$

$$\frac{dy}{dx} = -(6 + 9x^2 y)/(9x^3 - x^2)$$

17. $2x\frac{dx}{dy} - 6y = 0$

$$\frac{dx}{dy} = 3y/x$$

19. $2p\frac{dp}{dq} - q\frac{dp}{dq} - p = 10pq^2\frac{dp}{dq} + 10p^2 q$

$$(2p - q - 10pq^2)\frac{dp}{dq} = p + 10p^2 q$$

$$\frac{dp}{dq} = (p + 10p^2 q)/(2p - q - 10pq^2)$$

21. $e^y + xe^y\frac{dy}{dx} - e^x\frac{dy}{dx} - ye^x = 0$

$$(xe^y - e^x)\frac{dy}{dx} = ye^x - e^y$$

423

$$\frac{dy}{dx} = (ye^x - e^y)/(xe^y - e^x)$$

23. $e^{st}\left(t\dfrac{ds}{dt} + s\right) = 2s\dfrac{ds}{dt}$

$(2s - te^{st})\dfrac{ds}{dt} = se^{st}$

$\dfrac{ds}{dt} = se^{st}/(2s - te^{st})$

25. $\dfrac{e^x y^2 - 2ye^x(dy/dx)}{y^4} = e^y \dfrac{dy}{dx}$

$(2e^x + y^3 e^y)\dfrac{dy}{dx} = ye^x$

$\dfrac{dy}{dx} = ye^x/(2e^x + y^3 e^y)$

27. $\dfrac{2y - 1}{y^2 - y}\dfrac{dy}{dx} + 1 = \dfrac{dy}{dx}$

$[(2y - 1) - (y^2 - y)]\dfrac{dy}{dx} = -(y^2 - y)$

$\dfrac{dy}{dx} = (y - y^2)/(-1 + 3y - y^2)$

29. $\dfrac{y + x\dfrac{dy}{dx} + 2y\dfrac{dy}{dx}}{xy + y^2} = e^y\dfrac{dy}{dx}$

$[x + 2y - e^y(xy + y^2)]\dfrac{dy}{dx} = -y$

$\dfrac{dy}{dx} = -y/(x + 2y - xye^y - y^2 e^y)$

31. (a) $4x^2 + 2y^2 = 12$

$8x + 4y\dfrac{dy}{dx} = 0$

$\dfrac{dy}{dx} = -2x/y$

$\left.\dfrac{dy}{dx}\right|_{(1,-2)} = -2(1)/(-2) = 1$

(b) Point: $(1, -2)$ Slope: 1

Intercept: $b = y_1 - mx_1 = -2 - (1)(1) = -3$

Equation of tangent line:

$y = x - 3$

33. (a) $2x^2 - y^2 = xy$

$4x - 2y\dfrac{dy}{dx} = y + x\dfrac{dy}{dx}$

$\dfrac{dy}{dx} = \dfrac{4x - y}{x + 2y}$

$\left.\dfrac{dy}{dx}\right|_{(-1,2)} = \dfrac{4(-1) - 2}{-1 + 2(2)} = -2$

(b) Point: $(-1, 2)$ Slope: -2

Intercept: $b = y_1 - mx_1 = 2 - (-2)(-1) = 0$

Equation of tangent line:

$y = -2x$

35. (a) $x^2 y - y^2 + x = 1$

$2xy + x^2\dfrac{dy}{dx} - 2y\dfrac{dy}{dx} + 1 = 0$

$\dfrac{dy}{dx}(x^2 - 2y) = -(1 + 2xy)$

$\dfrac{dy}{dx} = -\dfrac{1 + 2xy}{x^2 - 2y}$

$\left.\dfrac{dy}{dx}\right|_{(1,0)} = -\dfrac{1 + 2(1)(0)}{1^2 - 2(0)} = -1$

(b) Point: $(1, 0)$ Slope: -1

Intercept: $b = y_1 - mx_1 = 0 - (-1)(1) = 1$

Equation of tangent line:

$y = -x + 1$

37. (a) $xy - 2000 = y$

$y + x\dfrac{dy}{dx} = \dfrac{dy}{dx}$

$y = \dfrac{dy}{dx}(1 - x)$

$\dfrac{dy}{dx} = \dfrac{y}{1 - x}$

When $x = 2$, the corresponding value of y is obtained from the original equation $xy - 2000 = y$:

$xy - 2000 = y$

$2y - 2000 = y$

$y = 2000$

We now evaluate the derivative:

$\dfrac{dy}{dx} = \dfrac{y}{1 - x} = \dfrac{2000}{1 - 2} = -2000$

(b) Point: $(2, 2000)$ Slope: -2000

Intercept: $b = y_1 - mx_1$

$= 2000 - (-2000)(2) = 6000$

Equation of tangent line:

$y = -2000x + 6000$

39. (a) $\ln(x+y) - x = 3x^2$

424

$$\frac{1 + dy/dx}{x + y} - 1 = 6x$$

$$\frac{dy}{dx} = (6x + 1)(x + y) - 1$$

When $x = 0$, the corresponding value of y is obtained from the original equation $\ln(x+y) - x = 3x^2$ by substituting:

$\ln y = 0$, $y = e^0 = 1$.

We can now evaluate the derivative:

$$\frac{dy}{dx} = (0 + 1)(0 + 1) - 1 = 0$$

(b) Point: $(0, 1)$ Slope: 0

Intercept: $b = y_1 - mx_1 = 1$

Equation of tangent line:

$y = 0x + 1$

$y = 1$

41. (a) $e^{xy} - x = 4x$

$$e^{xy}(y + x\frac{dy}{dx}) - 1 = 4$$

$$\frac{dy}{dx} = \frac{5 - ye^{xy}}{xe^{xy}}$$

When $x = 3$, the corresponding value of y is obtained from the original equation $e^{xy} - x = 4x$ by substituting:

$e^{3y} - 3 = 12$,

$e^{3y} = 15$

$3y = \ln 15$

$y = \frac{1}{3}\ln 15 \approx 0.902683$,

So $\dfrac{dy}{dx} \approx \dfrac{5 - (0.902683)(15)}{3(15)} \approx -0.1898$

(using the fact that $e^{3y} = 15$)

(b) Point: $(3, 0.9027)$ Slope: -0.1898

Intercept: $b \approx 0.9027 - 3(-0.1898) = 1,4721$

Equation of tangent line:

$y = -0.1898x + 1,4721$

43. $y = \dfrac{2x+1}{4x-2}$

$\ln y = \ln\left(\dfrac{2x+1}{4x-2}\right) = \ln(2x+1) - \ln(4x-2)$

Take d/dx of both sides:

$$\frac{1}{y}\frac{dy}{dx} = \frac{2}{2x+1} - \frac{4}{4x-2}$$

$$\frac{dy}{dx} = y\left[\frac{2}{2x+1} - \frac{4}{4x-2}\right]$$

$$= \frac{2x+1}{4x-2}\left[\frac{2}{2x+1} - \frac{4}{4x-2}\right]$$

45. $y = \dfrac{(3x+1)^2}{4x(2x-1)^3}$

$\ln y = \ln\left[\dfrac{(3x+1)^2}{4x(2x-1)^3}\right]$

$= \ln((3x+1)^2 - \ln(4x) - \ln(2x-1)^3$

$= 2\ln(3x+1)) - \ln(4x) - 3\ln(2x-1)$

Take d/dx of both sides:

$$\frac{1}{y}\frac{dy}{dx} = \frac{6}{3x+1} - \frac{1}{x} - \frac{6}{2x-1}$$

$$\frac{dy}{dx} = y\left[\frac{6}{3x+1} - \frac{1}{x} - \frac{6}{2x-1}\right]$$

$$= \frac{(3x+1)^2}{4x(2x-1)^3}\left[\frac{6}{3x+1} - \frac{1}{x} - \frac{6}{2x-1}\right]$$

47. $y = (8x-1)^{1/3}(x-1)$

$\ln y = \ln[(8x-1)^{1/3}(x-1)]$

$= \ln(8x-1)^{1/3} + \ln(x-1)$

$= \dfrac{1}{3}\ln(8x-1) + \ln(x-1)$

$$\frac{1}{y}\frac{dy}{dx} = \frac{8}{3(8x-1)} + \frac{1}{x-1}$$

$$\frac{dy}{dx} = y\left[\frac{8}{3(8x-1)} + \frac{1}{x-1}\right]$$

$$= (8x-1)^{1/3}(x-1)\left[\frac{8}{3(8x-1)} + \frac{1}{x-1}\right]$$

49. $\ln y = \ln\left[(x^3 + x)\sqrt{x^3 + 2}\right]$

$= \ln(x^3 + x) + \dfrac{1}{2}\ln(x^3 + 2)$

$$\frac{1}{y}\frac{dy}{dx} = \frac{3x^2 + 1}{x^3 + x} + \frac{1}{2}\frac{3x^2}{x^3 + 2}$$

$$\frac{dy}{dx} = (x^3 + x)\sqrt{x^3 + 2}\left(\frac{3x^2 + 1}{x^3 + x} + \frac{1}{2}\frac{3x^2}{x^3 + 2}\right)$$

51. $\ln y = \ln(x^x) = x\ln x$

$$\frac{1}{y}\frac{dy}{dx} = \ln x + x\left(\frac{1}{x}\right) = 1 + \ln x$$

$$\frac{dy}{dx} = x^x(1 + \ln x)$$

53. $P = x^{0.6}y^{0.4}$
Taking d/dx of both sides gives

$$0 = 0.6x^{-0.4}y^{0.4} + 0.4x^{0.6}y^{-0.6}\frac{dy}{dx}$$

$$-0.6x^{-0.4}y^{0.4} = 0.4x^{0.6}y^{-0.6}\frac{dy}{dx}$$

$$\frac{dy}{dx} = -\frac{0.6x^{-0.4}y^{0.4}}{0.4x^{0.6}y^{-0.6}} = -\frac{3y}{2x}$$

$$\frac{dy}{dx}\Big|_{x=100,\,y=200,000}$$
$$= -\frac{3(200,000)}{2(100)}$$
$$= -\$3000 \text{ per worker}$$

The monthly budget to maintain production at the fixed level P is decreasing by approximately $3000 per additional worker at an employment level of 100 workers and a monthly operating budget of $200,000. In other words, increasing the workforce by one worker will result in a saving of approximately $3000 per month.

55. $xy - 2000 = y$
Taking d/dx of both sides gives

$$y + x\frac{dy}{dx} = \frac{dy}{dx}$$

$$y = \frac{dy}{dx}(1-x)$$

$$\frac{dy}{dx} = \frac{y}{1-x}$$

When $x = 5$, the corresponding value of y is obtained from the original equation $xy - 2000 = y$:

$$xy - 2000 = y$$
$$5y - 2000 = y$$
$$4y = 2000$$
$$y = 500$$

We now evaluate the derivative:

$$\frac{dy}{dx} = \frac{y}{1-x} = \frac{500}{1-5} = -125 \text{ T-shirts per}$$

dollar. Thus, when the price is set at $5, the demand is dropping by 125 T-shirts per $1 increase in price.

57. Set $C = 200,000$ and differentiate the equation with respect to e:

$$0 = 100k\frac{dk}{de} + 120e$$

$$\frac{dk}{de} = -\frac{120e}{100k} = -\frac{6e}{5k}.$$

When $e = 15$
$$200,000 = 15,000 + 50k^2 + 60(15)^2$$
$$= 50k^2 + 28,500,$$
$$50k^2 = 171,500,$$

$$k \approx 58.57, \text{ so } \frac{dk}{de}\Big|_{e=15} = -\frac{6(15)}{5(58.57)}$$

$$\approx -0.307 \text{ carpenters per electrician.}$$

This means that, for a $200,000 house whose construction employs 15 electricians, adding one more electrician would cost as much as approximately 0.307 additional carpenters. In other words, one electrician is worth approximately 0.307 carpenters.

59. (a) Set $x = 3.0$ and $g = 80$ and solve for t:
$80 = 12t - 0.2t^2 - 90$, $0.2t^2 - 12t + 170 = 0$, $t \approx 22.93$ hours by the quadratic formula. (The other root is rejected because it is larger than 30.)
(b) Set $g = 80$ and differentiate the equation with respect to x:

$$0 = 4x\frac{dt}{dx} + 4t - 0.4t\frac{dt}{dx} - 20x,$$

$$\frac{dt}{dx} = \frac{4t - 20x}{0.4t - 4x} = \frac{t - 5x}{0.1t - x},$$

so $\frac{dt}{dx}\Big|_{x=3.0} \approx \frac{22.93 - 5(3.0)}{0.1(22.93) - 3.0} \approx -11.2$ hours

per grade point. This means that, for a 3.0 student who scores 80 on the examination, 1 grade point is worth approximately 11.2 hours.

61. $0 = 1.2y^{-0.4}r^{-0.3}p - 0.6y^{0.6}r^{-1.3}p\frac{dr}{dy}$

$$\frac{dr}{dy} = 2\frac{r}{y}, \text{ so } \frac{dr}{dt} = \frac{dr}{dy}\frac{dy}{dt} = 2\frac{r}{y}\frac{dy}{dt}$$

by the chain rule.

63. x, y, y, x

65. Let $y = f(x)g(x)$. Then
$$\ln y = \ln f(x) + \ln g(x),$$
and $\frac{1}{y}\frac{dy}{dx} = \frac{f'(x)}{f(x)} + \frac{g'(x)}{g(x)},$

so $\frac{dy}{dx} = y\left(\frac{f'(x)}{f(x)} + \frac{g'(x)}{g(x)}\right)$

$$= f(x)g(x)\left(\frac{f'(x)}{f(x)} + \frac{g'(x)}{g(x)} \right)$$
$$= f'(x)g(x) + f(x)g'(x)$$

67. Writing $y = f(x)$ specifies y as an explicit function of x. This can be regarded as an equation giving y as an *implicit* function of x. The procedure of finding dy/dx by implicit differentiation is then the same as finding the derivative of y as an explicit function of x: we take d/dx of both sides.

69. Differentiate both sides of the equation $y = f(x)$ with respect to y to get
$$1 = f'(x) \cdot \frac{dx}{dy} \ ,$$
giving
$$\frac{dx}{dy} = \frac{1}{f'(x)} = \frac{1}{dy/dx}$$

Chapter 11 Review Exercises

1. $f'(x) = 50x^4 + 2x^3 - 1$

3. $f'(x) = (3x^3 + 3x^{1/3})' = 9x^2 + x^{-2/3}$

5. $f'(x) = \dfrac{d}{dx}\left(x + \dfrac{1}{x^2}\right) = \dfrac{d}{dx}(x + x^{-2})$

$= 1 - 2x^{-3} = 1 - \dfrac{2}{x^3}$

7. $f'(x) = \dfrac{d}{dx}\left(\dfrac{4}{3x} - \dfrac{2}{x^{0.1}} + \dfrac{x^{1.1}}{3.2} - 4\right)$

$= \dfrac{d}{dx}\left(\dfrac{4}{3}x^{-1} - 2x^{-0.1} + \dfrac{1}{3.2}x^{1.1} - 4\right)$

$= -\dfrac{4}{3}x^{-2} + 0.2x^{-1.1} + \dfrac{1.1}{3.2}x^{0.1}$

$= -\dfrac{4}{3x^2} + \dfrac{0.2}{x^{1.1}} + \dfrac{1.1x^{0.1}}{3.2}$

9. $f(x) = e^x(x^2 - 1)$
Product rule:
$f'(x) = e^x(x^2 - 1) + e^x(2x)$
$\quad = e^x(x^2 + 2x - 1)$

11. $f(x) = (x^2 - 1)^{10}$
Chain rule:
$f'(x) = 10(x^2 - 1)^9(2x)$
$\quad = 20x(x^2 - 1)^9$

13. $f(x) = e^x(x^2 + 1)^{10}$
Product rule:
$f'(x) = e^x(x^2 + 1)^{10} + e^x \cdot 10(x^2 + 1)^9(2x)$
$\quad = e^x(x^2 + 1)^9(x^2 + 20x + 1)$

15. $f(x) = \dfrac{3^x}{x-1}$
Quotient rule:
$f'(x) = \dfrac{3^x \ln 3\ (x-1) - 3^x(1)}{(x-1)^2}$

$\quad = \dfrac{3^x[(x-1)\ln 3 - 1]}{(x-1)^2}$

17. $f(x) = e^{x^2-1}$
Chain rule:

$f'(x) = 2xe^{x^2-1}$

19. $\ln(x^2 - 1)$
Chain rule:
$f'(x) = \dfrac{2x}{x^2 - 1}$

21. Since the slope of the tangent line is the derivative, the tangent line is horizontal when its slope is 0; that is, $dy/dx = 0$.
$$y = -3x^2 + 7x - 1$$
$$\frac{dy}{dx} = -6x + 7 = 0$$
$$x = 7/6$$

23. Since the slope of the tangent line is the derivative, the tangent line is horizontal when its slope is 0; that is, $dy/dx = 0$.
$$y = \frac{x}{2} + \frac{2}{x}$$
$$\frac{dy}{dx} = \frac{1}{2} - \frac{2}{x^2} = 0$$
$$\frac{1}{2} = \frac{2}{x^2}$$
$$x^2 = 4$$
$$x = \pm 2$$

25. $y = x - e^{2x-1}$
$y' = 1 - 2e^{2x-1}$
Set $y' = 0$:
$1 - 2e^{2x-1} = 0$
$2x - 1 = \ln(1/2) = -\ln 2$
$x = (1 - \ln 2)/2$

27. $y = \dfrac{x}{x+1}$
$y' = \dfrac{(x+1) - x}{(x+1)^2} = \dfrac{1}{(x+1)^2}$
This is never 0, so there are no points at which the tangent line is horizontal.

29. $x^2 - y^2 = x$
Take d/dx of both sides:
$$2x - 2y\frac{dy}{dx} = 1$$

Solve for $\dfrac{dy}{dx}$:

$$\frac{dy}{dx} = \frac{2x-1}{2y}$$

31. $e^{xy} + xy = 1$

Take d/dx of both sides:

$$\left(y + x\frac{dy}{dx}\right)e^{xy} + y + x\frac{dy}{dx} = 0,$$

Solve for $\dfrac{dy}{dx}$:

$$\frac{dy}{dx} = \frac{-y(e^{xy}+1)}{x(e^{xy}+1)} = -\frac{y}{x}$$

33. $y = \dfrac{(2x-1)^4(3x+4)}{(x+1)(3x-1)^3}$

We use logarithmic differentiation:

$$\ln y = \ln\left[\frac{(2x-1)^4(3x+4)}{(x+1)(3x-1)^3}\right]$$

$= 4\ln(2x-1) + \ln(3x+4) - \ln(x+1) - 3\ln(3x-1)$

$$\frac{1}{y}\frac{dy}{dx} = \frac{8}{2x-1} + \frac{3}{3x+4} - \frac{1}{x+1} - \frac{9}{3x-1}$$

$$\frac{dy}{dx} = y\left[\frac{8}{2x-1} + \frac{3}{3x+4} - \frac{1}{x+1} - \frac{9}{3x-1}\right]$$

$$= \frac{(2x-1)^4(3x+4)}{(x+1)(3x-1)^3}$$

$$\left[\frac{8}{2x-1} + \frac{3}{3x+4} - \frac{1}{x+1} - \frac{9}{3x-1}\right]$$

35. $y = (x^2-3x)^{-2}$

When $x = 1$, $y = (1-3)^{-2} = 1/4$

$$\frac{dy}{dx} = -2(x^2-3x)^{-3}(2x-3)$$

$$\frac{dy}{dx}\bigg|_{x=1} = -2(1-3)^{-3}(2-3) = -1/4$$

Tangent line:

Slope: $m = -1/4$

Intercept: $b = y_1 - mx_1 = 1/4 - (-1/4)(1) = 1/2$

$y = -x/4 + 1/2$

37. $y = x^2e^{-x}$

When $x = -1$, $y = e^1 = e$

$$\frac{dy}{dx} = 2xe^{-x} + x^2e^{-x}(-1) = e^{-x}(2x - x^2)$$

$$\frac{dy}{dx}\bigg|_{x=-1} = e(-2-1) = -3e$$

Tangent line:

Slope: $m = -3e$

Intercept: $b = y_1 - mx_1 = e - (-3e)(-1) = -2e$

$y = -3ex - 2e$

39. $xy - y^2 = x^2 - 3$

$$y + x\frac{dy}{dx} - 2y\frac{dy}{dx} = 2x$$

$$\frac{dy}{dx}[x-2y] = 2x-y$$

$$\frac{dy}{dx} = \frac{2x-y}{x-2y}$$

$$\frac{dy}{dx}\bigg|_{(-1,1)} = \frac{2(-1)-1}{-1-2(1)} = 1$$

Tangent line:

Point: $(-1, 1)$ Slope: 1

Intercept: $b = y_1 - mx_1 = 1 - (1)(-1) = 2$

Equation of tangent line:

$y = x + 2$

41. (a) $w'(t) = -11.1t^2 + 149.2t + 135.5$, so $w'(1) \approx 274$ books per week

(b) $w'(7) = 636$ books per week

(c) It would not be realistic to use the function w through week 20: It begins to decrease after $t = 14$. Graph:

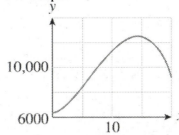

(d) Since the data suggest an upward curving parabola, the long-term prediction of sales for a quadratic model would be that sales will increase without bound, in sharp contrast to (c)

43. (a) $C'(x) = -0.00004x + 3.2$, so $C'(8000) = \$2.88$ per book

(b) $\bar{C}(x) = -0.00002x + 3.2 + 5400/x$, $\bar{C}(8000) = \$3.715$ per book

(c) $\bar{C}'(x) = -0.00002 - 5400/x^2$, so $\bar{C}'(8000) \approx -\$0.000104$ per book, per additional book sold.

(d) At a sales level of 8000 books per week, the cost is increasing at a rate of $2.88 per book (so that the 8001st book costs approximately $2.88 to sell), and it costs an average of $3.715 per book to sell the first 8000 books. Moreover, the average cost is decreasing at a rate of $0.000104 per book, per additional book sold.

45. (a) Let $q(t)$ be the number of books sold per week and let $p(t)$ be the price per book. We are given that $q(0) = 1000$ and $q'(0) = 200$ (taking $t = 0$ as now), and that $p(0) = 20$ and $p'(0) = -1$. Since revenue is $R(t) = p(t)q(t)$, we have

$$R'(0) = p'(0)q(0) + p(0)q'(0)$$
$$= (-1)(1000) + 20(200)$$
$$= \$3000 \text{ per week (rising).}$$

(b) Let $q(t)$ be the number of books sold per week and let $p(t)$ be the price per book. We are given that $q(0) = 1000$, $p(0) = 20$ and $p'(0) = -1$ (taking $t = 0$ as now). We desire $R'(0) = 5000$ and need to compute $q'(0)$. Since $R(t) = p(t)q(t)$, we have

$$(-1)(1000) + 20q'(0) = 5000$$
$$q'(0) = 300 \text{ books per week}$$

47. $R = pq$ gives $R' = p'q + pq'$. Thus, $R'/R = R'/(pq) = (p'q + pq')/pq = p'/p + q'/q$.

49. Let P be the stock price and E the earnings. We are given $P = 100$, $E = 1$, $E' = 0.10$, and

$$\frac{d}{dt}\left(\frac{P}{E}\right) = 100.$$

$$\frac{d}{dt}\left(\frac{P}{E}\right) = \frac{P'E - PE'}{E^2}$$

$$100 = \frac{P' - 100(0.10)}{1^2}$$

Solve for p' to get
$$P' = \$110 \text{ per year.}$$

51. (a) The rate of increase of weekly sales is

$$s'(t) = -\frac{4500(-0.55)\,e^{-0.55(t-4.8)}}{(1 + e^{-0.55(t-4.8)})^2}$$

$$= \frac{2475\,e^{-0.55(t-4.8)}}{(1 + e^{-0.55(t-4.8)})^2}$$

$t = 6$:
$$s'(6) \approx 556 \text{ books per week}$$

(b) As $t \to +\infty$, the expression $e^{-0.55(t-4.8)} \to 0$, so

$$\lim_{t \to +\infty} s'(t) = \lim_{t \to +\infty} \frac{2475\,e^{-0.55(t-4.8)}}{(1 + e^{-0.55(t-4.8)})^2}$$

$$= \frac{0}{(1+0)^2} = 0.$$

In the long term, the rate of increase of weekly sales slows to zero.

53. If $H(t)$ is the daily number of hits t weeks into the year, then
$$H(t) = 1000(1.05)^t.$$
So,
$$H'(t) = 1000\,\ln(1.05)\,(1.05)^t$$
and
$$H'(52) = 1000 \cdot \ln(1.05)\,(1.05)^{52}$$
$$\approx 616.8 \text{ hits per day per week.}$$

55. (a) $250q + 250p\dfrac{dq}{dp} + 2q\dfrac{dq}{dp} = 0$

$$\frac{dq}{dp} = \frac{-250q}{250p + 2q}$$

When $p = 50$ and $q = 1000$
$$\frac{dq}{dp} = \frac{-250(1000)}{250(50) + 2(1000)}$$
$$\approx -17.24 \text{ copies per \$1.}$$

The demand for the gift edition of *The Complete Harry Potter* is dropping at a rate of about 17.24 copies per $1 increase in the price.

(b) Since $R = pq$,
$$\frac{dR}{dp} = q + p\frac{dq}{dp} \approx 1000 + 50(-17.24)$$
$$\approx \$138 \text{ per dollar increase in price.}$$

The derivative is positive, so the price should be raised.

Chapter 12 Further Applications of the Derivative

12.1 Maxima and Minima

1. Absolute min.: $(-3, -1)$, relative max: $(-1, 1)$, relative min: $(1, 0)$, absolute max: $(3, 2)$

3. Absolute min: $(3, -1)$ and $(-3, -1)$, absolute max: $(1, 2)$

5. Absolute min: $(-3, 0)$ and $(1, 0)$, absolute max: $(-1, 2)$ and $(3, 2)$

7. Relative min: $(-1, 1)$

9. Absolute min: $(-3, -1)$, relative max: $(-2, 2)$, relative min: $(1, 0)$, absolute max: $(3, 3)$

11. Relative max: $(-3, 0)$, absolute min: $(-2, -1)$, stationary non-extreme point: $(1, 1)$.

13. $f(x) = x^2 - 4x + 1$ with domain $[0, 3]$
$$f'(x) = 2x - 4$$
$f'(x) = 0$ when
$$2x - 4 = 0$$
$$x = 2$$
$f'(x)$ is defined for all x in $[0, 3]$. Thus, we have the stationary point at $x = 2$ and the endpoints $x = 0$ and $x = 3$:

x	0	2	3
$f(x)$	1	-3	-2

The graph must decrease from $x = 0$ until $x = 2$, then increase again until $x = 3$.

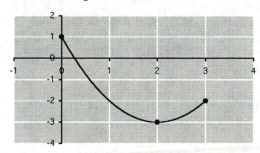

This gives an absolute max at $(0, 1)$, an absolute min at $(2, -3)$, and a relative max at $(3, -2)$.

15. $g(x) = x^3 - 12x$ with domain $[-4, 4]$
$$g'(x) = 3x^2 - 12$$
$g'(x) = 0$ when
$$3x^2 - 12 = 0$$
$$x = \pm 2$$
$g'(x)$ is defined for all x in $[-4, 4]$. Thus, we have stationary points at $x = \pm 2$ and the endpoints $x = \pm 4$:

x	-4	-2	2	4
$g(x)$	-16	16	-16	16

The graph increases from $x = -4$ until $x = -2$, then decreases until $x = 2$, then increases until $x = 4$.

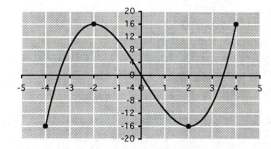

This gives an absolute min at $(-4, -16)$, an absolute max at $(-2, 16)$, an absolute min at $(2, -16)$, and an absolute max at $(4, 16)$.

17. $f(t) = t^3 + t$ with domain $[-2, 2]$
$$f'(t) = 3t^2 + 1$$
$f'(t) = 0$ when
$$3t^2 + 1 = 0,$$
which has no solution; $f'(t)$ is defined for all t in $[-2, 2]$. Thus, there are no critical points in the domain, just the endpoints ± 2:

t	-2	2
$f(t)$	-10	10

The graph increases from $x = -2$ until $x = 2$.

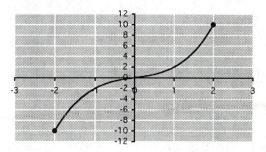

431

This gives an absolute min at $(-2, -10)$ and an absolute max at $(2, 10)$.

19. $h(t) = 2t^3 + 3t^2$ with domain $[-2, +\infty)$
$\quad h'(t) = 6t^2 + 6t$
$h'(t) = 0$ when
$\quad 6t^2 + 6t = 0$
$\quad t = 0$ or $t = -1$
$h'(t)$ is defined for all t in $[-2, \infty)$. Thus, we have stationary points at $t = 0$ and $t = -1$ and the endpoint at $t = -2$. In addition to these we test one point to the right of $t = 0$:

t	-2	-1	0	1
$h(t)$	-4	1	0	5

The graph increases from $x = -2$ until $x = -1$, then decreases until $x = 0$, then increases from that point on. (Remember that $x = 1$ is just a test point to see whether the graph is increasing or decreasing to the right of $x = 0$.)

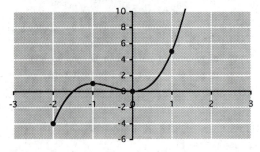

This gives an absolute min at $(-2, -4)$, a relative max at $(-1, 1)$, and relative min at $(0, 0)$.

21. $f(x) = x^4 - 4x^3$ with domain $[-1, +\infty)$
$\quad f'(x) = 4x^3 - 12x^2$
$f'(x) = 0$ when
$\quad 4x^3 - 12x^2 = 0$
$\quad 4x^2(x - 3) = 0$
$\quad x = 0$ or $x = 3$
$f'(x)$ is defined for all x in $[-1, \infty)$. Thus, we have stationary points at $x = 0$ and $x = 3$ and the endpoint at $x = -1$. In addition to these we test one point to the right of $x = 3$:

x	-1	0	3	4
$f(x)$	5	0	-27	0

The graph decreases from $x = -1$ until $x = 0$, continues to decrease until $x = 3$, then increases from that point on.

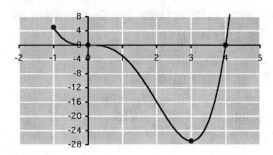

This gives a relative max at $(-1, 5)$ and an absolute min at $(3, -27)$.

23. $g(t) = \frac{1}{4}t^4 - \frac{2}{3}t^3 + \frac{1}{2}t^2$ with domain $(-\infty, +\infty)$
$\quad g'(t) = t^3 - 2t^2 + t$
$g'(t) = 0$ when
$\quad t^3 - 2t^2 + t = 0$
$\quad t(t - 1)^2 = 0$
$\quad t = 0$ or $t = 1$
$g'(t)$ is defined for all t. Thus, we have stationary points at $t = 0$ and $t = 1$. Since we have no endpoints, we test a point to the left of $t = 0$ and a point to the right of $t = 1$:

t	-1	0	1	2
$g(t)$	$17/12$	0	$1/12$	$2/3$

The graph decreases until $t = 0$, increases until $t = 1$, then continues to increase to the right of $t = 1$.

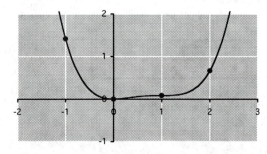

This gives an absolute min at $(0, 0)$ and no other extrema.

25. $h(x) = (x-1)^{2/3}$ with domain $[0, 2]$
Endpoints: $0, 2$
$h'(x) = \frac{2}{3}(x-1)^{-1/3} = \frac{2}{3(x-1)^{1/3}}$

Stationary points: $h'(x) = 0$ when

$$\frac{2}{3(x-1)^{1/3}} = 0,$$

which is impossible, so there are no stationary points.

Singular points: $h'(x)$ is undefined when $x = 1$, so we have a singular point at $x = 1$.

x	0	1	2
$h(x)$	1	0	1

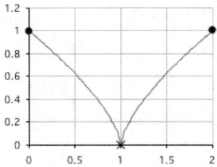

This gives absolute maxima at $(0, 1)$ and $(2, 1)$ and an absolute min at $(1, 0)$.

27. $k(x) = \dfrac{2x}{3} + (x+1)^{2/3}$ with domain $(-\infty, 0]$

Endpoint: 0

$$k'(x) = \frac{2}{3} + \frac{2}{3}(x+1)^{-1/3}$$

$$= \frac{2}{3} + \frac{2}{3(x+1)^{1/3}}$$

Stationary points: $k'(x) = 0$ when

$$\frac{2}{3} + \frac{2}{3(x+1)^{1/3}} = 0$$

$$\frac{2}{3} = -\frac{2}{3(x+1)^{1/3}}$$

cross-multiply:

$$6(x+1)^{1/3} = -6$$

$$(x+1)^{1/3} = -1$$

$$(x+1) = (-1)^3 = -1$$

$$x = -2,$$

so we have a stationary point at $x = -2$.

Singular points: $k'(x)$ is undefined when $x = -1$, so we have a singular point at $x = -1$.

Thus we have an endpoint at $x = 0$, a stationary point at $x = -2$ and a singular point at $x = -1$.

In addition we use a test point to the left of $x = -2$

x	-3	-2	-1	0
$f(x)$	-0.41	$-1/3$	$-2/3$	1

Notice that the singular point $(-1, -2/3)$ is not an absolute minimum because the graph eventually gets lower on the left.

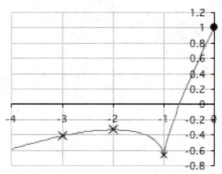

From the graph we see that we have a relative maximum at $(-2, -1/3)$, a relative minimum at $(-1, -2/3)$ and an absolute maximum at $(0, 1)$.

29. $f(t) = \dfrac{t^2 + 1}{t^2 - 1}; \ -2 \le t \le 2, \ t \ne \pm1$

$$f'(t) = \frac{2t(t^2 - 1) - 2t(t^2 + 1)}{(t^2 - 1)^2}$$

$$= \frac{-4t}{(t^2 - 1)^2}$$

$f'(t) = 0$ when $t = 0$; $f'(t)$ is not defined when $t = \pm1$, but these points are not in the domain of f. Thus, we have a stationary point at $t = 0$ and the endpoints at $t = \pm2$. We also test points on either side of $t = \pm1$ to see how f behaves near these points where it goes to $\pm\infty$ (that is, it has vertical asymptotes there, because the denominator goes to 0):

t	-2	$-3/2$	$-1/2$	0
$f(t)$	5/3	13/5	$-5/3$	-1
t	1/2	3/2	2	
$f(t)$	$-5/3$	13/5	5/3	

The graph increases from $t = -2$, approaching the vertical asymptote at $t = -1$; on the other side of the asymptote it increases until $t = 0$ then decreases as it approaches another vertical asymptote at $t = 1$; on the other side of the second asymptote it decreases until $t = 2$.

433

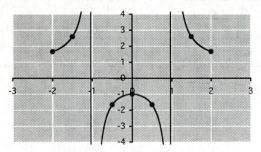

This gives a relative min at $(-2, 5/3)$, a relative max at $(0, -1)$, and a relative min at $(2, 5/3)$.

31. $f(x) = \sqrt{x}(x-1)$; $x \geq 0$

$$f'(x) = \frac{3}{2} x^{1/2} - \frac{1}{2} x^{-1/2}$$

$f'(x) = 0$ when

$$\frac{3}{2} x^{1/2} - \frac{1}{2} x^{-1/2} = 0$$

$$3x - 1 = 0$$

$$x = 1/3$$

$f'(x)$ is not defined at $x = 0$. Thus, we have a stationary point at $x = 1/3$ and a singular point at $x = 0$, which is also an endpoint. In addition, we test a point to the right of $x = 1/3$:

x	0	1/3	1
$f(x)$	0	$-2\sqrt{3}/9$	0

The graph decreases from $t = 0$ until $t = 1/3$, then increases from that point on.

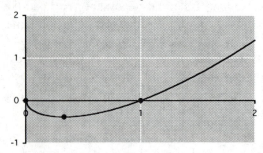

This gives a relative max at $(0, 0)$ and an absolute min at $(1/3, -2\sqrt{3}/9)$.

33. $g(x) = x^2 - 4\sqrt{x}$

Since the domain of g is not specified, we take it to be the set of all x for which g is defined, which is $[0, \infty)$.

$$g'(x) = 2x - 2x^{-1/2}$$

$g'(x) = 0$ when

$$2x - 2x^{-1/2} = 0$$

$$x = x^{-1/2}$$

$$x^{3/2} = 1$$

$$x = 1$$

$g'(x)$ is not defined when $x = 0$, which is also an endpoint. Thus, we have a singular endpoint at $x = 0$ and a stationary point at $x = 1$. We test one point to the right of $x = 1$:

x	0	1	2
$g(x)$	0	-3	-1.7

The graph decreases from $x = 0$ to $x = 1$, then increases from that point on.

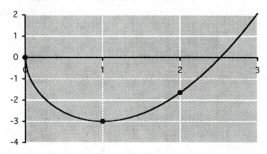

Thus, there is a relative max at $(0, 0)$ and an absolute min at $(1, -3)$.

35. $g(x) = \dfrac{x^3}{x^2 + 3}$

The domain of g is $(-\infty, \infty)$.

$$g'(x) = \frac{3x^2(x^2 + 3) - x^3(2x)}{(x^2 + 3)^2}$$

$$= \frac{x^4 + 9x^2}{(x^2 + 3)^2}$$

$g'(x) = 0$ when

$$x^4 + 9x^2 = 0$$

$$x^2(x^2 + 9) = 0$$

$$x = 0$$

$g'(x)$ is always defined. Thus, we have only one stationary point, at $x = 0$. In addition, we test a point on either side:

x	-1	0	1
$g(x)$	$-1/4$	0	$1/4$

The graph is always increasing.

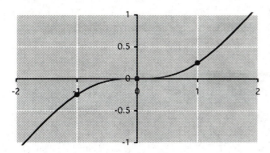

Thus, there are no relative extrema.

37. $f(x) = x - \ln x$ with domain $(0, +\infty)$

$f'(x) = 1 - 1/x$

$f'(x) = 0$ when

$1 - 1/x = 0$

$x = 1$

$f'(x)$ is defined for all x in the domain of f. We test the one stationary point at $x = 1$ and a point on either side:

x	1/2	1	2
$g(x)$	1.19	1	1.31

The graph decreases until $x = 1$ and then increases from that point on.

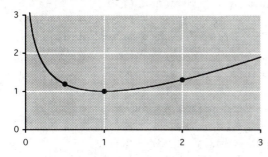

Thus, there is an absolute min at $(1, 1)$.

39. $g(t) = e^t - t$ with domain $[-1, 1]$

$g'(t) = e^t - 1$

$g'(t) = 0$ when

$e^t - 1 = 0$

$e^t = 1$

$t = \ln 1 = 0$

$g'(t)$ is always defined. We need to test only the stationary point and the endpoints:

t	-1	0	1
$g(t)$	1.37	1	1.72

The graph decreases from $t = -1$ to $t = 0$ and then increases to $t = 1$.

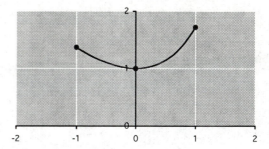

Thus, there is a relative max at $(-1, e^{-1} + 1)$, an absolute min at $(0, 1)$, and an absolute max at $(1, e - 1)$.

41. $f(x) = \dfrac{2x^2 - 24}{x + 4}$

The domain of f is all $x \neq 4$.

$f'(x) = \dfrac{4x(x + 4) - (2x^2 - 24)}{(x + 4)^2}$

$\quad = \dfrac{2x^2 + 16x + 24}{(x + 4)^2}$

$f'(x) = 0$ when

$2x^2 + 16x + 24 = 0$

$2(x + 2) \cdot (x + 6) = 0$

$x = -2$ or -6

$f'(x)$ is defined for all x in the domain of f. Thus, we have stationary points at $x = -2$ and $x = -6$. We test points on either side of the stationary points and between them and the asymptote at $x = -4$:

x	-7	-6	-5
$f(x)$	-24.7	-24	-26

x	-3	-2	-1
$f(x)$	-6	-8	-7.3

The graph increases to $x = -6$ then decreases approaching the asymptote at $x = -4$. On the other side of the asymptote the graph decreases to $x = -2$ and then increases again.

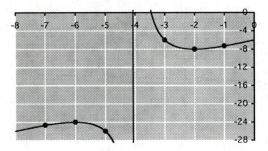

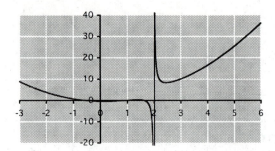

Thus, there is a relative max at $(-6, -24)$ and a relative min at $(-2, -8)$.

43. $f(x) = xe^{1-x^2}$

$f'(x) = e^{1-x^2} + x(-2x)e^{1-x^2}$
$\quad = (1 - 2x^2)e^{1-x^2}$

$f'(x) = 0$ when
$\quad 1 - 2x^2 = 0$
$\quad x = \pm 1/\sqrt{2}$

$f'(x)$ is always defined. We test the two stationary points and a point on either side:

x	-1	$-\dfrac{1}{\sqrt{2}}$	$1/\sqrt{2}$	1
$f(x)$	-1	-1.17	1.17	1

The graph decreases until $x = -1/\sqrt{2}$, increases until $x = 1/\sqrt{2}$, then decreases again.

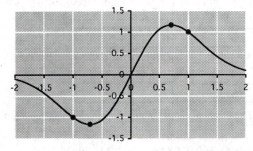

Thus, there is an absolute max at $(1/\sqrt{2}, \sqrt{e/2}\,)$ and an absolute min at $(-1/\sqrt{2}, -\sqrt{e/2}\,)$.

45. $y = x^2 + \dfrac{1}{x-2}$

$y' = 2x - \dfrac{1}{(x - 2)^2}$.

The graphs of f and of f' look like this:

Looking closely at the graph of f' we can see three places where it crosses the x axis. Zooming in we can locate these points approximately as $x \approx 0.15$, $x \approx 1.40$, and $x \approx 2.45$. Thus, we have relative minima at $(0.15, -0.52)$ and $(2.45, 8.22)$ and a relative maximum at $(1.40, 0.29)$.

47. $f(x) = (x-5)^2(x+4)(x-2)$ with domain $[-5, 6]$

$f'(x) = 2(x - 5)(x + 4)(x - 2)$
$\qquad + (x - 5)^2 \cdot (x - 2) + (x - 5)^2(x + 4)$
$\quad = (x - 5)[2(x + 4) \cdot (x - 2)$
$\qquad + (x - 5)(x - 2) + (x - 5)(x + 4)]$
$\quad = (x - 5)(4x^2 - 4x - 26).$

The graphs of f and f' look like this:

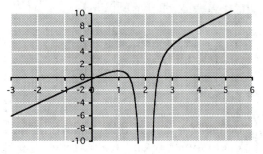

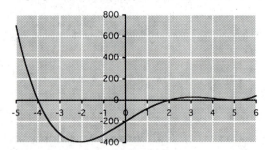

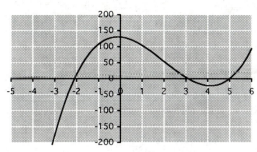

The graph of f' crosses the x axis in three places, hence f has three stationary points. Zooming in we can approximate these as $x \approx -2.10$, $x \approx 3.10$, and $x = 5$. Substituting these and the endpoints -5 and 6 into f, we find that we have an absolute maximum at $(-5, 700)$, relative maxima at $(3.10, 28.19)$ and $(6, 40)$, an absolute minimum at $(-2.10, -392.69)$, and a relative minimum at $(5, 0)$.

49. The derivative is zero at $x = -1$. To the left of $x = -1$ the derivative is negative, indicating that the graph of f is decreasing. To the right of $x = -1$ the derivative is positive, indicating that the graph of f is increasing. Thus, f has a stationary minimum at $x = -1$.

51. Stationary minima at $x = -2$ and $x = 2$, stationary maximum at $x = 0$ (see the solution to #49).

53. Singular minimum at $x = 0$. At $x = 1$, the derivative of also zero, but to both the left and right of $x = 0$ the derivative is positive, indicating that the graph of f is increasing on both sides of $x = 0$. Thus, f has a stationary non-extreme point at $x = 1$

55. Stationary minimum at $x = -2$, singular non-extreme point at $x = -1$, singular non-extreme point at $x = 1$, stationary maximum at $x = 2$

57.

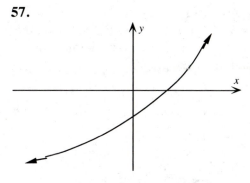

59.

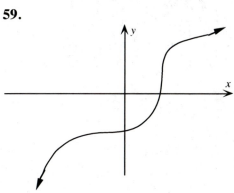

61. Not necessarily; it could be neither a relative maximum nor a relative minimum, as in the graph of $y = x^3$ at the origin.

63. Answers will vary.

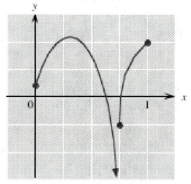

65. The graph oscillates faster and faster above and below zero as it approaches the end-point at 0, so 0 cannot be either a relative minimum or maximum. Here is the graph:

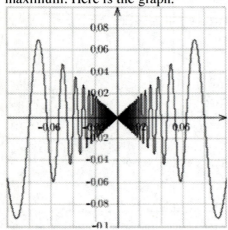

12.2 Applications of Maxima and Minima

1. Solve for $y = 10 - x$ and substitute:
$$P = x(10 - x) = 10x - x^2.$$
Stationary points: Set $P'(x) = 0$ and solve for x:
$$P'(x) = 10 - 2x$$
$P'(x) = 0$ when $x = 5$. Since the graph of $P(x)$ is a parabola opening downward, $x = 5$ must be its vertex, hence gives the maximum. The corresponding value of y is $y = 10 - x = 5$. Hence, $x = y = 5$ and $P = 25$.

3. Solve for $y = 9/x$ and substitute:
$$S = x + 9/x$$
with domain $x > 0$.
Stationary points: Set $S'(x) = 0$ and solve for x:
$$S'(x) = 1 - 9/x^2$$
$S'(x) = 0$ when $x = 3$ ($x = -3$ is not in the domain). Testing points on either side of $x = 3$ we see that we have the minimum. Hence, $x = y = 3$ and $S = 6$.

5. Solve for $x = 10 - 2y$ and substitute:
$$F(y) = (10 - 2y)^2 + y^2$$
$$= 100 - 40y + 5y^2.$$
Stationary points: Set $F'(y) = 0$ and solve for y:
$$F'(y) = -40 + 10y$$
$F'(y) = 0$ when $y = 4$. Since the graph of F is a parabola opening upward, $y = 4$ must be its vertex, hence gives the minimum. Hence, $x = 2$, $y = 4$, and $F = 20$.

7. Since y appears in both constraints, we can solve for the other two variables in terms of y: $x = 30 - y$ and $z = 30 - y$. Substitute:
$$P = (30 - y) \cdot y \cdot (30 - y)$$
$$= y(30 - y)^2.$$
Since all the variables must be nonnegative, we have $0 \le y \le 30$ as the domain.
Stationary points: Set $P'(y) = 0$ and solve for y:
$$P'(y) = (30 - y)^2 - 2y(30 - y)$$
$$= (30 - y)(30 - y - 2y)$$
$$= (30 - y)(30 - 3y)$$
$P'(y) = 0$ when $y = 10$ or $y = 30$. Substituting these values and the other endpoint $y = 0$ into P,

we find that the maximum occurs when $y = 10$. Thus, $x = 20$, $y = 10$, $z = 20$, and $P = 4000$.

9. Let x and y be the dimensions. Then we want to maximize $A = xy$ subject to $2x + 2y = 20$. Solve for $y = 10 - x$ and substitute:
$$A = x(10 - x) = 10x - x^2.$$
Stationary points: Set $A'(x) = 0$ and solve for x:
$$A'(x) = 10 - 2x;$$
$A'(x) = 0$ when $x = 5$. Since the graph of $A(x)$ is a parabola opening downward, $x = 5$ must be its vertex, hence gives the maximum. The corresponding value of y is $y = 10 - x = 5$. Hence, the rectangle should have dimensions 5×5.

11. $C(x) = 22{,}500 + 100x + 0.01x^2$
$$\bar{C}(x) = \frac{22{,}500}{x} + 100 + 0.01x$$
$$\bar{C}'(x) = -\frac{22{,}500}{x^2} + 0.01$$
Stationary points:
$$-\frac{22{,}500}{x^2} + 0.01 = 0$$
$$\frac{22{,}500}{x^2} = 0.01$$
$$0.01x^2 = 22{,}500$$
$$x^2 = \frac{22{,}500}{0.01} = 2{,}250{,}000$$
$$x = \sqrt{2{,}250{,}000} = 1500 \text{ iPods per day}$$

The resulting average cost is
$$\bar{C}(1500) = \frac{22{,}500}{1500} + 100 + 0.01(1500) = \$130.$$

13. Unknown: q, the number of pounds of pollutant removed per day
Objective function: average cost
$$\bar{C}(q) = \frac{C(q)}{q} = \frac{4000 + 100q^2}{q}$$
$$= \frac{4000}{q} + 100q$$
with domain $q > 0$.

Stationary points: Set $\bar{C}'(q) = 0$ and solve for q:

$$\bar{C}'(q) = -\frac{4000}{q^2} + 100 = 0$$

$$q^2 = 4000/100 = 40$$

$$q = \sqrt{40} \approx 6.32 \text{ pounds of pollutant per day.}$$

Testing points on either side shows that this gives a minimum and a consideration of the graph shows that this minimum is absolute. Thus, average cost is minimized when we remove about 6.32 pounds per day, giving an average cost of

$$\bar{C}(\sqrt{40}) = \frac{4000 + 100(40)}{\sqrt{40}}$$

$$\approx \$1265 \text{ per pound.}$$

15. Net cost is
$$N = C(q) - 500q = 4000 + 100q^2 - 500q,$$
with $q \geq 0$.
Stationary points: Set $N'(q) = 0$ and solve for q:
$$N'(q) = 200q - 500$$
$N'(q) = 0$ when $q = 2.5$; $N'(q)$ is defined for all q. Testing the endpoint $q = 0$, the stationary point $q = 2.5$, and one more point to the right of 2.5, we see that the net cost is minimized when $q = 2.5$ pounds of pollutant per day.

17. Let x be the length of the east and west sides and let y be the length of the north and south sides. The area is $A = xy$ and the cost of the fence is $2 \cdot 4x + 2 \cdot 2y = 8x + 4y$. (We multiply by 2 because there are two sides of length x and two sides of length y.) So, our problem is to maximize $A = xy$ subject to $8x + 4y = 80$. Solve for y:
$$y = 20 - 2x$$
and substitute:
$$A(x) = x(20 - 2x) = 20x - 2x^2.$$
Since x and y must both be nonnegative, we have $0 \leq x \leq 10$.
Stationary points: Set $A'(x) = 0$ and solve for x:
$$A'(x) = 20 - 4x$$
$A'(x) = 0$ when $20 - 4x = 0$,
$x = 5$; $A'(x)$ is always defined. Testing the endpoints 0 and 10 as well as the stationary point 5, we see that the maximum area occurs when $x = 5$. The corresponding value of y is $y = 10$, so the largest area possible is $5 \times 10 - 50$ square feet.

19. Let x be the length of the bottom fence and let y be the length of the fence on the left. The area is $A = xy/2$ and the cost of the fence is $x + 5y$. So our problem is:
Maximize $A = xy/2$ subject to $x + 5y = 100$.
Solve the constraint equation for x and substitute into the objective:
$$x = 100 - 5y$$
Since x and y must be nonnegative, we must have $0 \leq y \leq 20$.
giving $A = (100 - 5y)y/2$
$$= 50y - 2.5y^2 \text{ with } 0 \leq y \leq 20$$
Stationary points:
$$A' = 50 - 5y = 0$$
when $y = 10$.
$A(10) = 50(10) - 2.5(10)^2 = 250$ sq. ft.
Endpoints: Each of the endpoints $y = 0$ and 20 gives an area of zero, so the stationary point gives the maximum. When $y = 10$, $x = 100 - 5(10) = 50$.
Thus the dimensions (E-W)×(N-S) are 50 ft×10 ft for an area of 250 sq. ft.

21. Let x be the length of the east and west sides and let y be the length of the north and south sides. The area is
$$A = xy = 242$$
and the cost of the fence is
$$C = 2(4x) + 2(2y) = 8x + 4y$$
(We multiply by 2 because there are two sides of length x and two sides of length y.) Since we want to minimize cost, this time the cost is the objective, and the constraint is $xy = 242$. Solving the constraint equation for y gives
$$y = 242/x$$
Substituting in the objective:
$$C = 8x + 4(242/x) = 8x + 968/x$$
Since x cannot be negative nor zero, the domain of C is $x > 0$.
Stationary points:
$$C' = 8 - 968/x^2$$
This is zero when
$$8 = \frac{968}{x^2}$$
$$8x^2 = 968$$

$x^2 = 121$

$x = 11$.

From the graph of C we see that $x = 11$ is the absolute minimum of C:

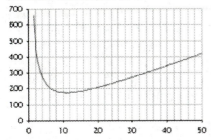

Thus, for the cheapest fencing, $x = 11$, $y = 242/11 = 22$.

23. $R = pq = p(200,000 - 10,000p)$
$= 200,000p - 10,000p^2$.

For p and q to both be nonnegative, we must have $0 \le p \le 20$.

Stationary points: Set $R'(p) = 0$ and solve for p:
$R'(p) = 200,000 - 20,000p$

$R'(p) = 0$ when $p = 10$; $R'(p)$ is always defined. Testing the endpoints 0 and 20 as well as the stationary point 10, we see that the maximum revenue occurs when the price is $p = \$10$.

25. The optimization problem is:
Maximize
$R = pq$
subject to the constraint
$q = -p + 156$
For p and q to both be nonnegative, we must have
$0 \le p \le 156$
To solve, substitute the constraint equation in the objective function:
$R(p) = p(-p + 156)$
$= -p^2 + 156p$ million dollars
$R'(p) = -2p + 156$
$R'(p) = 0$ when $p = 156/2 = \$78$
$R'(p)$ is always defined, so there are no singular points.
Testing the endpoints 0 and 156 as well as the stationary point 78, we see that the maximum revenue occurs when the price is $p = \$78$. At this price, the quarterly revenue would be

$R(78) = -(78)^2 + 156(78) = \$6084$ million , or \$6.084 billion

27. The optimization problem is:
Maximize
$R = pq$
subject to the constraint
$q = -4500p + 41,500$
For p and q to both be nonnegative, we must have
$0 \le p \le 41,500/4500 = 83/9 \approx 9.22$
To solve, substitute the constraint equation in the objective function:
$R(p) = p(-4500p + 41,500)$
$= -4500p^2 + 41,500p$
$R'(p) = -9000p + 41,500$
$R'(p) = 0$ when $p = 41,500/9000 \approx \$4.61$
$R'(p)$ is always defined., so there are no singular points.
Testing the endpoints 0 and 9.22 as well as the stationary point 4.61, we see that the maximum revenue occurs when the price is $p = \$4.61$. At this price, the annual revenue would be
$R(80) = -4500(4.61)^2 + 41,500(4.61)$
$= \$95,680.55$

29. (a) $R = pq = \dfrac{500,000}{q^{1.5}} \cdot q$

$= \dfrac{500,000}{q^{0.5}} = 500,000q^{-0.5}$.

We are given that $q \ge 5000$.
Stationary points: Set $R'(q) = 0$ and solve for q:
$R'(q) = -250,000q^{-1.5}$
$R'(q)$ is never 0; $R'(q)$ is defined for all $q > 0$.
We test the endpoint $q = 5000$ and a point to the right: $R(5000) = 7071.07$ and $R(10,000) = 5000$. Thus, the revenue is decreasing and its maximum value occurs at the endpoint $q = 5000$. The corresponding price is $p = \$1.41$ per pound.
(b) As found in part (a), the maximum occurs at $q = 5000$ pounds.
(c) The maximum revenue is $R(5000) = \$7071.07$ per month.

31. We are given two points: $(p, q) = (25, 22)$ and $(14, 27.5)$. The equation of the line through these two points is

441

$q = -p/2 + 34.5$.

The revenue is

$R = pq = p(-p/2 + 34.5)$

$= -p^2/2 + 34.5p$.

For p and q to both be nonnegative we must have $0 \le p \le 69$.

Stationary points: Set $R'(p) = 0$ and solve for p:

$R'(p) = -p + 34.5$

$R'(p) = 0$ when $p = 34.5$; $R'(p)$ is defined for all p. Testing the endpoints 0 and 69 as well as the stationary point 34.5, we find that the revenue is maximized when the price is $p = 34.5$¢ per pound, for an annual (per capita) revenue of $5.95.

33. The profit function is

$P = R - C$ (Revenue – Cost)

$= pq - 40q$

The optimization problem is:

Maximize

$P = pq - 40q$

subject to the constraint

$q = -p + 156$

For p and q to both be nonnegative, we must have $0 \le p \le 156$

To solve, substitute the constraint equation in the objective function:

$P(p) = p(-p + 156) - 40(-p + 156)$

$= -p^2 + 196p - 6240$ million dollars

$P'(p) = -2p + 196$

$P'(p) = 0$ when $p = 196/2 = \$98$

$P'(p)$ is always defined., so there are no singular points.

Testing the endpoints 0 and 156 as well as the stationary point 98, we see that the maximum revenue occurs when the price is $p = \$98$. At this price, the quarterly profit would be

$P(98) = -(98)^2 + 196(98) - 6240 = \$3364$

million, or $3.364 billion

35. (a) $R = pq$ and $C = 100q$, so

$P = R - C$

$= pq - 100q$

$= \dfrac{1000}{q^{0.3}} \cdot q - 100q$

$= 1000q^{0.7} - 100q$.

For p and q to be defined and nonnegative we need $q > 0$.

Stationary points: Set $P'(q) = 0$ and solve for q:

$P'(q) = 700q^{-0.3} - 100$

$P'(q) = 0$ when $q = 7^{1/0.3} \approx 656$; $P'(q)$ is defined for all $q > 0$. Testing the stationary point at approximately 656 and points on either side, we see that the profit is maximized when you sell 656 headsets, for a profit of $P(656) \approx \$28{,}120$.

(b) The corresponding price is $p \approx \$143$ per headset.

37. Take the height of the can as h and the radius of the base as r. The objective function is the total surface area S of the can:

$S = \pi r^2 + 2\pi rh$ Bottom disc plus side

Constraint:

$V = \pi r^2 h = 27{,}000$

Solve the constraint for h:

$h = \dfrac{27{,}000}{\pi r^2}$

Substituting in the objective function, we get

$S = \pi r^2 + 2\pi r \dfrac{27{,}000}{\pi r^2} = \pi r^2 + \dfrac{54{,}000}{r}$

where $0 < r < +\infty$

Stationary points:

$S'(r) = 2\pi r - \dfrac{54{,}000}{r^2} = 0$

$2\pi r = \dfrac{54{,}000}{r^2}$

$2\pi r^3 = 54{,}000$

$r^3 = \dfrac{27{,}000}{\pi}$

so $r = \sqrt[3]{\dfrac{27{,}000}{\pi}} = \dfrac{30}{\sqrt[3]{\pi}}$

$\approx 20.4835 \approx 20.48$ cm

$h = \dfrac{27{,}000}{\pi r^2} \approx \dfrac{27{,}000}{\pi (20.4835)^2}$

≈ 20.48 cm

(There are no singular points or endpoints in the domain.)

39. Take the height of the can as h and the radius of the base as r. The objective function is the cost C of the metal in the can:

$$C = 0.02 \times 2\pi r^2 + 0.01 \times 2\pi rh \quad \text{Top \&}$$

bottom plus side

$$= 0.04\pi r^2 + 0.02\pi rh$$

Constraint:

$$V = \pi r^2 h = 250$$

Solve the constraint for h:

$$h = \frac{250}{\pi r^2}$$

Substituting in the objective function, we get

$$C = 0.04\pi r^2 + 0.02\pi r \frac{250}{\pi r^2} = 0.04\pi r^2 + \frac{5}{r}$$

where $0 < r < +\infty$

Stationary points:

$$S'(r) = 0.08\pi r - \frac{5}{r^2} = 0$$

$$0.08\pi r = \frac{5}{r^2}$$

$$0.08\pi r^3 = 5$$

$$r^3 = \frac{5}{0.08\pi}$$

so $\quad r = \sqrt[3]{\dfrac{5}{0.08\pi}}$

$$\approx 2.7096 \approx 2.71 \text{ cm}$$

$$h = \frac{250}{\pi r^2} \approx \frac{250}{\pi(2.7096)^2}$$

$$\approx 10.84 \text{ cm}$$

(There are no singular points or endpoints in the domain.)

The ratio Height/Radius is about $10.84/2.71 = 4$

41. Let x be the length of one side of the square cut out of each corner, as in the figure:

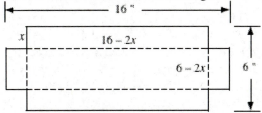

When the sides are folded up, the resulting box will have volume

$$V = x(16 - 2x)(6 - 2x)$$
$$= 4x^3 - 44x^2 + 96x.$$

For the sides to have nonnegative lengths, we must have $0 \le x \le 3$.

Stationary points: Set $V'(x) = 0$ and solve for x:

$$V'(x) = 12x^2 - 88x + 96$$
$$= 2(x - 6)(6x - 8)$$

$V'(x) = 0$ when $x = 4/3$ or $x = 6$, but $x = 6$ is outside of the domain. Testing the endpoints 0 and 3 and the stationary point $4/3$, we find that the largest volume occurs when $x = 4/3$". Thus, thee box with the largest volume has dimensions 13 1/3" $\times$ 3 1/3" $\times$ 1 1/3" and it has volume

$$V(4/3) = 1600/27 \approx 59 \text{ cubic inches.}$$

43. Let x be the width and depth of the box and let y be the height. The amount of material used will be

$$S = 2x^2 + 4xy,$$

counting the top, bottom (each of which has area x^2) and four sides (each of which has area xy). We are told that the volume is 125 cm^3, so we must have

$$x^2y = 125.$$

So, our problem is to maximize

$$S = 2x^2 + 4xy \text{ subject to}$$
$$x^2y = 125$$

Also, $x > 0$ for x and y to be nonnegative. Solve for $y = 125/x^2$ and substitute:

$$S = 2x^2 + 500/x.$$

Stationary points: Set $S'(x) = 0$ and solve for x:

$$S'(x) = 4x - 500/x^2$$

$S'(x) = 0$ when $x^3 = 500/4 = 125$, so $x = 5$. (The corresponding value of y is $y = 125/x^2 = 5$ also.) Testing this stationary point and a point on either side, we see that the least material is used building a box of dimensions $5 \times 5 \times 5$ cm.

45. Let l = length, w = width, and h = height. We want to maximize the volume

$$V = lwh$$

but we are restricted by

$$l + w + h \le 62.$$

Since we're looking for the largest volume, we can assume that

$$l + w + h = 62.$$

We are also told that $h = w$. Thus, our problem is to maximize

$V = lwh$ subject to
$l + w + h = 62$ and
$h = w$.
Substitute $h = w$ in the other constraint to get
$l + 2w = 62$,
then solve for l:
$l = 62 - 2w$.
Substitute to get
$V = (62 - 2w)w^2 = 62w^2 - 2w^3$.
For all dimensions to be nonnegative we need $0 \le w \le 31$.
Stationary points:
$V'(w) = 124w - 6w^2$
$V'(w) = 0$ when
$124w - 6w^2 = 0$
$2w(62 - 3w) = 0$,
so $w = 0$ or $w = 62/3$.
Singular points: None; $V'(w)$ is defined for all w. Testing the endpoints 0 and 31 as well as the (other) stationary point 62/3, we see that the volume is maximized when $w = 62/3$. The corresponding values of the other dimensions are $l = h = 62/3$. Thus, the largest volume bag has dimensions $l = w = h \approx 20.67$ in, and volume $V \approx 8827$ in^3.

47. Let l = length, w = width, and h = height. We want to maximize the volume $V = lwh$ but we have the constraints $l + w = 45$ and $w + h = 45$. Solve for $l = 45 - w$ and $h = 45 - w$. Substitute to get $V = w(45 - w)^2$. For all dimensions to be nonnegative we need $0 \le w \le 45$. Stationary points: $V'(w) = (45 - w)^2 - 2w(45 - w) = (45 - w)(45 - w - 2w) = (45 - w)(45 - 3w)$; $V'(w) = 0$ when $w = 15$ or $w = 45$; $V'(w)$ is defined for all w. Testing the endpoints 0 and 45 as well as the stationary point 15, we see that the volume is maximized when $w = 15$. The corresponding values of the other dimensions are $l = h = 30$. Thus, the largest volume bag has dimensions $l = 30$ in, $w = 15$ in, and $h = 30$ in.

49. Let l = length, w = width, and h = height. We want to maximize the volume $V = lwh$ but we are restricted by $l + 2(w + h) \le 108$. Since we're looking for the largest volume, we can assume that $l + 2(w + h) = 108$. We are also told that $w = h$. Substitute to get $l + 2(2w) = 108$, or $l + 4w = 108$. Solve for $l = 108 - 4w$ and substitute to get $V = w^2(108 - 4w) = 108w^2 - 4w^3$. For all the dimensions to be nonnegative we need $0 \le w \le 27$.
Stationary points: $V'(w) = 216w - 12w^2 = 12w(18 - w)$; $V'(w) = 0$ when $w = 0$ or $w = 18$; $V'(w)$ is defined for all w. Testing the endpoints 0 and 27 as well as the stationary point 18, we see that the volume is maximized when $w = 18$. The corresponding values of the other dimensions are $h = 18$ and $l = 36$. Thus, the largest volume package has dimensions $l = 36$ in and $w = h = 18$ in, and has volume $V = 11,664$ in^3.

51. (a) Stationary points:
$R'(t) = 350(39)e^{-0.3t} - 0.3(350)(39t + 68)e^{-0.3t}$
$\quad = 350(39 - 11.7t - 20.4)e^{-0.3t}$
$\quad = 350(18.6 - 11.7t)e^{-0.3t}$.
$R'(t) = 0$ when $t \approx 1.6$; $R'(t)$ is defined for all t. Testing points on either side of the stationary point, we see that the maximum revenue occurs at $t \approx 1.6$ years, or year 2001.6.
(b) The maximum revenue is $R(1.6) \approx \$28{,}241$ million.

53. Let $C(t) = \dfrac{D(t)}{S(t)} = \dfrac{10 + t}{2.5e^{0.08t}}$
$\quad = 0.4(10 + t)e^{-0.08t}$.
Stationary points:
$C'(t) = 0.4e^{-0.08t} - 0.032(10 + t)e^{-0.08t}$
$\quad = (0.4 - 0.32 - 0.032t)e^{-0.08t}$
$\quad = (0.08 - 0.032t)e^{-0.08t}$
$C'(t) = 0$ when $t = 2.5$; $C'(t)$ is defined for all t. Testing points on either side of 2.5 we see that the maximum revenue occurs at $t = 2.5$ or midway through 1972, when $D(2.5)/S(2.5) \approx 4.09$. Midway through 1972 the number of new (approved) drugs per \$1 billion dollars of spending on research and development reached a high of around 4 approved drugs per \$1 billion.

55. $p = ve^{-0.05t} = (300{,}000 + 1000t^2)e^{-0.05t}$.
  $(t \geq 5)$
Endpoint: $t = 5$
Stationary points:
$p' - 2000te^{-0.05t} - 0.05(300{,}000 + 1000t^2)e^{-0.05t}$

  $= (-15{,}000 + 2000t - 50t^2)e^{-0.05t}$
$p' = 0$ when
  $-15{,}000 + 2000t - 50t^2 = 0$
  $-50(t - 30)(t - 10) = 0$
so $t = 10$ or $t = 30$; p' is always defined.
Testing $t = 5$, 10, 30, and 40, we see that the maximum discounted value occurs $t = 30$ years from now.

57. $R = (100 + 2t)(400{,}000 - 2500t)$
  $= 40{,}000{,}000 + 550{,}000t - 5000t^2$
Stationary points:
  $R' = 550{,}000 - 10{,}000t$
$R' = 0$ when $t = 55$; R' is defined for all t.
Testing $t = 55$ and points on either side of it (or recognizing that the graph of R is a parabola), we see that the release should be delayed for 55 days.

59. $R = 500\sqrt{x}$, $C = 10{,}000 + 2x$, so the total profit is
  $P = R - C = 500\sqrt{x} - (10{,}000 + 2x)$
and the average profit per copy is
  $\overline{P} = 500x^{-1/2} - 10{,}000x^{-1} + 2$.
Stationary points:
  $\overline{P}' = -250x^{-3/2} + 10{,}000x^{-2}$
$\overline{P}' = 0$ when
  $x = (10{,}000/250)^2 = 1600$
$\overline{P}'$ is defined for all $x > 0$. Testing the stationary point 1600 and a point on either side, we see that the average profit is maximized at $x = 1600$ copies. For this many copies, the average profit is
  $\overline{P}(1600) = \$8.25/\text{copy}$.
Since $P' = 250x^{-1/2} + 2$, the marginal profit is
  $P'(1600) = \$8.25/\text{copy}$ also.
At this value of x, average profit equals marginal profit; beyond this the marginal profit is smaller than the average and so the average declines.

61. Minimize cost
  $C = 100x + 16y$

subject to
  $xy = 10{,}000$
Solve the constraint for y:
  $y = 10{,}000/x$
Substitute in the objective:
  $C = 100x + 160{,}000/x \quad (x > 0)$
Stationary points:
  $C'(x) = 100 - \dfrac{160{,}000}{x^2} = 0$
when
  $100 = \dfrac{160{,}000}{x^2}$
  $100x^2 = 160{,}000$
  $x^2 = 1600$
  $x = 40$ laborers
This represents the absolute minimum because the objective function increases without bound as x approaches either 0 or $+\infty$ (graph it to see). The corresponding value of y is given by the constraint equation:
  $y = 10{,}000/40 = 250$ robots

63. We want to minimize $C = 20{,}000x + 365y$
subject to $x^{0.4}y^{0.6} = 1000$. Solve for y:
  $y = (1000x^{-0.4})^{1/0.6} = 1000^{5/3}x^{-2/3}$
and substitute:
  $C = 20{,}000x + 365(1000^{5/3})x^{-2/3}$.
The domain is $x > 0$.
Stationary points:
  $C'(x) = 20{,}000 - \dfrac{730}{3}1000^{5/3}x^{-5/3}$
$C'(x) = 0$ when
  $x = \left(\dfrac{60{,}000}{730 \cdot 1000^{5/3}}\right)^{-3/5} \approx 71$.
Testing the stationary point at 71 and a point on either side, we see that C has its minimum at $x \approx 71$. So, you should hire 71 employees.

65. We are being asked to find the extreme values of the derivative, $N'(t)$. Call this function $M(t)$.
  $M(t) = N'(t) = 0.12t^2 - 4t + 40$
  $M'(t) = 0.24t - 4$
$M'(t) = 0$ when
  $0.24t = 4$
  $t \approx 4/0.24 \approx 16.667 \approx 17$
(representing 2007) to the nearest year.

Testing the endpoints 0 and 18 as well as the stationary point 16.667, we see that M has an absolute minimum of $M(16.667) \approx 6.667$ and an absolute maximum of $M(0) = 40$. Hence, N was increasing most rapidly in 1990 and increasing least rapidly in 2007.

67. Let $r(t) = c'(t) = -0.195t^3 + 6.8t - 22$.
$r'(t) = -0.39t + 6.8$; $r'(t) = 0$ when $t \approx 17$.
Testing the endpoints 8 and 30 as well as the stationary point 17, we see that $c'(t)$ has its maximum when $t = 17$ days. This means that the embryo's oxygen consumption is increasing most rapidly 17 days after the egg is laid.

69. Graph of derivative:

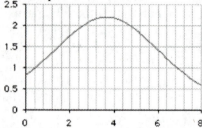

If we zoom in, we can determine that the maximum value of around 2.2 occurs around $t = 3.7$; during the year 2003. Since $A'(t)$ represents the rate of change of the percentage of mortgages that were subprime, we can say that the percentage of mortgages that were subprime was increasing most rapidly during 2003, when it increased at a rate of around 2.2 percentage points per year.

71. $p = v(1.05)^{-t} = \dfrac{10,000(1.05)^{-t}}{1 + 500e^{-0.5t}}$.
Technology formula:
Excel:
`10000*(1.05)^(-x)/(1+500*exp(-0.5*x))`
TI-83/84 Plus:
`10000*(1.05)^(-x)/(1+500*e^(-0.5*x))`
Here are the graphs of $p(t)$ and $p'(t)$:

Graph of $p(t)$;

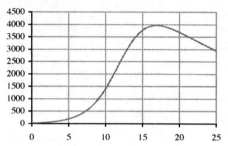

Graph of $p'(t)$:

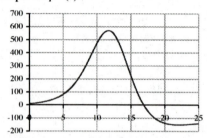

From the graph, we see that the maximum occurs between $t = 15$ and $t = 20$. The maximum is more accurately seem in the graph of $p'(t)$ where it crosses the t-axis. Zooming in on the graph of $p'(t)$, we see the following:

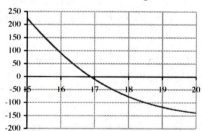

and so the maximum is very close to $t = 17$ years. To obtain the value, substitute $t = 17$ in the formula for $p(t)$ or else use the trace feature to see the y-coordinate of the highest point in the graph of $p(t)$: approximately \$3960.

73. Let $y(x)$ be the annual yield per tree when there are x trees;
$$y(x) = 100 - (x - 50) = 150 - x.$$
If $Y(x)$ is the total annual yield from x trees, then
$$Y(x) = xy(x) = x(150 - x) = 150x - x^2$$
$$Y'(x) = 150 - 2x$$
$Y'(x) = 0$ when $x = 75$. Since the graph of Y is a parabola opening downward, we know that this gives the maximum value of Y. Hence, the total

annual yield is largest when there are 75 trees, or 25 additional trees beyond the 50 we already have.

75. $\dfrac{d(TR)}{dQ} = b - 2cQ;\ \dfrac{d(TR)}{dQ} = 0$ when $Q = \dfrac{b}{2c}$, which is (D).

77. The objective is to maximize the height, so the objective function is H (choice (A)). One of the constraints is that the height H should not exceed eight times the distance D from the road, so for maximum height, we can say $8D = H$ (choice (B)).

79. The problem is uninteresting because the company can accomplish the objective by cutting away the entire sheet of cardboard, resulting in a box with surface area zero. Put another way, if it doesn't cut away everything, it can make the surface area be as close to zero as you like.

81. Not all absolute extrema occur at stationary points; some may occur at an end-point or singular point of the domain, as in Exercises 29, 30, 65, and 66..

83. The minimum of dq/dp is the fastest that the demand is dropping in response to increasing price.

12.3 Higher Order Derivatives: Acceleration and Concavity

1. $\dfrac{dy}{dx} = 6x;\ \dfrac{d^2y}{dx^2} = 6$

3. $\dfrac{dy}{dx} = -\dfrac{2}{x^2};\ \dfrac{d^2y}{dx^2} = \dfrac{4}{x^3}$

5. $\dfrac{dy}{dx} = 1.6x^{-0.6} - 1;\ \dfrac{d^2y}{dx^2} = -0.96x^{-1.6}$

7. $\dfrac{dy}{dx} = -e^{-(x-1)} - 1;\ \dfrac{d^2y}{dx^2} = e^{-(x-1)}$

9. $\dfrac{dy}{dx} = -\dfrac{1}{x^2} - \dfrac{1}{x};\ \dfrac{d^2y}{dx^2} = \dfrac{2}{x^3} + \dfrac{1}{x^2}$

11. (a) $s = 12 + 3t - 16t^2$

$v = \dfrac{ds}{dt} = 3 - 32t$

$a = \dfrac{dv}{dt} = -32 \text{ ft/sec}^2$

(b) $a(2) = -32 \text{ ft/sec}^2$

13. (a) $s = \dfrac{1}{t} + \dfrac{1}{t^2}$

$v = \dfrac{ds}{dt} = -\dfrac{1}{t^2} - \dfrac{2}{t^3}$

$a = \dfrac{dv}{dt} = \dfrac{2}{t^3} + \dfrac{6}{t^4} \text{ ft/sec}^2$

(b) $a(1) = 8 \text{ ft/sec}^2$

15. (a) $s = \sqrt{t} + t^2$

$v = \dfrac{ds}{dt} = \dfrac{1}{2} t^{-1/2} + 2t$

$a = \dfrac{dv}{dt} = -\dfrac{1}{4} t^{-3/2} + 2 \text{ ft/sec}^2$

(b) $a(4) = \dfrac{63}{32} \text{ ft/sec}^2$

17. (1, 0): changes from concave down to concave up

19. (1, 0): changes from concave down to concave up

21. None: always concave down

23. (−1, 0): changes from concave up to concave down; (1, 1): changes from concave down to concave up

For exercises 25–28, remember that a point of inflection of f corresponds to a relative extreme point of f′ that is internal, not an endpoint.

25. Points of inflection at $x = -1$ (relative max of f') and $x = 1$ (relative min of f')

27. One point of inflection, at $x = -2$ (relative min of f'). Note that f' has a stationary point at $x = 1$ but not a relative extreme point there.

For exercises 29–34, remember that a point of inflection of f corresponds to a point at which f″ changes sign, from positive to negative or vice versa. This could be a point where its graph crosses the x-axis, or a point where its graph is broken: positive on one side of the break and negative on the other.

29. Points of inflection where the graph of f'' crosses the x-axis: at $x = -2$, $x = 0$, and $x = 2$.

31. Points of inflection where the graph of f'' crosses the x-axis at $x = -2$ and $x = 2$.

33. $f(x) = x^2 - 4x + 1$
$f'(x) = 2x - 4$
$f''(x) = 2$
Stationary points:
$\quad 2x - 4 = 0$
$\quad x = 2$
$f''(2) = 2$, which is positive, so f has a relative minimum at $x = 2$.

35. $g(x) = x^3 - 12x$
$g'(x) = 3x^2 - 12$
$g''(x) = 6x$

Stationary points:

$3x^2 - 12 = 0$

$3(x-2)(x+2) = 0$

$x = -2, 2$

$g''(-2) = 6(-2) = -12$, which is negative, so g has a relative maximum at $x = -2$.

$g''(2) = 6(2) = 12$, which is positive, so g has a relative minimum at $x = 2$.

37. $f(t) = t^3 - t$

$f'(t) = 3t^2 - 1$

$f''(t) = 6t$

Stationary points:

$3t^2 - 1 = 0$

$3t^2 = 1$

$t = -1/\sqrt{3}, 1/\sqrt{3}$

$f''(-1/\sqrt{3}) = 6(-1/\sqrt{3}) = -6/\sqrt{3}$ which is negative, so f has a relative maximum at $t = -1/\sqrt{3}$.

$f''(1/\sqrt{3}) = 6(1/\sqrt{3}) = 6/\sqrt{3}$ which is positive, so f has a relative minimum at $t = 1/\sqrt{3}$.

39. $f(x) = x^4 - 4x^3$

$f'(x) = 4x^3 - 12x^2$

$f''(x) = 12x^2 - 24x$

Stationary points:

$4x^3 - 12x^2 = 0$

$4x^2(x-3) = 0$

$x = 0, 3$

$f''(0) = 0$; the second derivative test is inconclusive. However the first derivative is $4x^2(x-3)$ and is negative on both sides of $x = 0$, so the First derivative test tells us that f has neither a relative maximum nor minimum at 0.

$f''(3) = 12(3)^2 - 24(3) = 36$, which is positive, so f has a relative minimum at $x = 3$.

41. $f(x) = e^{-x^2}$

$f'(x) = -2xe^{-x^2}$

$f''(x) = e^{-x^2}[-2 + 4x^2]$

Stationary points:

$-2xe^{-x^2} = 0$

$x = 0$

since the exponential term is never zero.

$f''(0) = e^0[-2 + 4(0)] = -2$, which is negative, so f has a relative maximum at $x = 0$.

43. $f(x) = xe^{1-x^2}$

$f'(x) = e^{1-x^2}[1 - 2x^2]$

$f''(x) = e^{1-x^2}[-4x - 2x(1 - 2x^2)]$

$\quad = e^{1-x^2}[4x^3 - 6x]$

$\quad = e^{1-x^2}[2x(2x^2 - 3)]$

Stationary points:

$e^{1-x^2}[1 - 2x^2] = 0$

$x = -1/\sqrt{2}, 1/\sqrt{2},$

since the exponential term is never zero.

$f''(-1/\sqrt{2}) = e^{1-1/2}[4/\sqrt{2}]$, which is positive, so f has a relative minimum at $x = -1/\sqrt{2}$.

$f''(1/\sqrt{2}) = e^{1-1/2}[-4/\sqrt{2}]$, which is negative, so f has a relative maximum at $x = 1/\sqrt{2}$.

45. Take the derivative repeatedly:

$f(x) = 4x^2 - x + 1$

$f'(x) = 8x - 1$

$f''(x) = 8$

$f'''(x) = f^{(4)}(x) = \ldots = f^{(n)}(x) = 0$

47. Take the derivative repeatedly:

$f(x) = -x^4 + 3x^2$

$f'(x) = -4x^3 + 6x$

$f''(x) = -12x^2 + 6$

$f'''(x) = -24x$

$f^{(4)}(x) = -24$

$f^{(5)}(x) = f^{(6)}(x) = \ldots = f^{(n)}(x) = 0$

49. Take the derivative repeatedly:

$f(x) = (2x + 1)^4$

$f'(x) = 4(2x + 1)^3(2) = 8(2x + 1)^3$

$f''(x) = 24(2x + 1)^2(2) = 48(2x + 1)^2$

$f'''(x) = 96(2x + 1)(2) = 192(2x + 1)$

$f^{(4)}(x) = 384$

$f^{(5)}(x) = f^{(6)}(x) = \ldots = f^{(n)}(x) = 0$

449

51. Take the derivative repeatedly:

$f(x) = e^{-x}$

$f'(x) = e^{-x}(-1) = -e^{-x}$

$f''(x) = -e^{-x}(-1) = e^{-x}$

$f'''(x) = e^{-x}(-1) = -e^{-x}$

$f^{(4)}(x) = -e^{-x}(-1) = e^{-x}$

...

$f^{(n)}(x) = (-1)^n e^{-x}$

53. Take the derivative repeatedly:

$f(x) = e^{3x-1}$

$f'(x) = e^{3x-1}(3) = 3e^{3x-1}$

$f''(x) = 3e^{3x-1}(3) = 9e^{3x-1}$

$f'''(x) = 9e^{3x-1}(3) = 27e^{3x-1}$

$f^{(4)}(x) = 27e^{3x-1}(3) = 81e^{3x-1}$

...

$f^{(n)}(x) = 3^n e^{3x-1}$

55. $s(t) = 40 - 1.9t^2$

$v(t) = s'(t) = -3.8t$

$a(t) = s''(t) = -3.8 \text{ m/s}^2$

57. $s(t) = t^3 - t^2$

$v(t) = s'(t) = 3t^2 - 2t$

$a(t) = s''(t) = 6t - 2 \text{ ft/s}^2$

$a(1) = 6(1) - 2 = 4 \text{ ft/s}^2$

Since this is positive, the velocity is increasing.

59. $R(t) = 12t^2 + 500t + 4700$

$R'(t) = 24t + 500$

$R''(t) = 24 \text{ million gals/yr}^2$

Accelerating by 24 million gals/yr^2

61. (a) $c(t) = -0.065t^3 + 3.4t^2 - 22t + 3.6$

$c(20) = -0.065(20)^3 + 3.4(20)^2 - 22(20) + 3.6$

$\approx 400 \text{ ml}$

(b) $c'(t) = -0.195t^2 + 6.8t - 22$

$c'(20) = -0.195(20)^2 + 6.8(20) - 22$

$= 36 \text{ ml/day}$

(c) $c''(t) = -0.39t + 6.8$

$c''(20) = -0.39(20) + 6.8 = -1 \text{ ml/day}^2$

63. (a) $I(t) = -0.04t^3 + 0.4t^2 + 0.1t + 202$

$I'(t) = -0.12t^2 + 0.8t + 0.1$

Inflation Rate:

$$\frac{I'(t)}{I(t)} = \frac{-0.12t^2 + 0.8t + 0.1}{-0.04t^3 + 0.4t^2 + 0.1t + 202}$$

Februrary 2007 ($t = 2$):

$$\frac{I'(2)}{I(2)} =$$

$$\frac{-0.12(2)^2 + 0.8(2) + 0.1}{-0.04(2)^3 + 0.4(2)^2 + 0.1(2) + 202}$$

$$= \frac{1.22}{203.48} \approx 0.006, \text{ or } 0.6\%$$

(b) $I'(t) = -0.12t^2 + 0.8t + 0.1$

$I''(t) = -0.24t + 0.8$

$I''(2) = -0.24(2) + 0.8 = 0.32$

Since $I''(2)$ is positive, inflation was speeding up in Februrary 2007.

(c) Points of inflection occur when $I''(t) = 0$

$-0.24t + 0.8 = 0$

$$t = \frac{0.8}{0.24} \approx 3.33$$

Looking at the graph we see that it is concave up to the left of $t = 3.33$ and concave down to the right. Therefore, inflation was speeding up for $t < 3.33$ (prior to 1/3 of the way through March) and slowing for $t > 3.33$ (after that time).

65. (a) $I(t) = 0.06t^3 - 0.8t^2 + 3.1t + 195$

$I'(t) = 0.18t^2 - 1.6t + 3.1$

Inflation Rate:

$$\frac{I'(t)}{I(t)} = \frac{0.18t^2 - 1.6t + 3.1}{0.06t^3 - 0.8t^2 + 3.1t + 195}$$

December 2005 ($t = 5$):

$$\frac{I'(5)}{I(5)} = \frac{0.18(5)^2 - 1.6(5) + 3.1}{0.06(5)^3 - 0.8(5)^2 + 3.1(5) + 195}$$

$$= \frac{-0.4}{198} \approx -0.00202, \text{ or } -0.202\%$$

(deflation rate of 0.202%)

February 2006 ($t = 7$):

$$\frac{I'(7)}{I(7)} = \frac{0.18(7)^2 - 1.6(7) + 3.1}{0.06(7)^3 - 0.8(7)^2 + 3.1(7) + 195}$$

$$= \frac{0.72}{198.08} \approx 0.00363, \text{ or } 0.363\%$$

(b) $I'(t) = 0.18t^2 - 1.6t + 3.1$

$I''(t) = 0.36t - 1.6$

$I''(7) = 0.36(7) - 1.6 = 0.92$

Since $I''(7)$ is positive, inflation was speeding up in February 2006.

(c) Points of inflection occur when $I''(t) = 0$

$0.36t - 1.6 = 0$

$t = \frac{1.6}{0.36} \approx 4.44444$

Looking at the graph we see that it is concave down to the left of $t = 4.44$ and concave up to the right. Therefore, inflation was decreasing for $t < 4.44$ (prior to mid-November) and increasing for $t > 4.44$ (after that time).

67. The graph of P is concave up when P'' is positive, and concave down when P'' is negative. From the graph of P'', we see that it is negative until about $t = 8$, at which time it turns positive. Therefore: The graph of P is concave up for $8 < t < 20$, concave down for $0 < t < 8$, and there is a point of inflection around $t = 8$ (when $P'' = 0$). Interpretation: From the graph of P' we see that P' has a minimum at around $t = 8$, meaning that the percentage of articles written by researchers in the U.S. was decreasing most rapidly at around $t = 8$ (1991).

69. (a) The graph of c' is concave down throughout the range $[8, 30]$, and therefore has no points of inflection: Choice (B).

(b) At $t = 18$ (the point of inflection) the graph of c' has a maximum. Since c' measures the rate of change of daily oxygen consumption, it means that oxygen consumption is increasing at a maximum rate at around $t = 18$: Choice (B).

(c) For $t > 18$, the graph of c is increasing but concave down, so that oxygen consumption is increasing at a decreasing rate: Choice (A).

71. Graphs: $A(t)$:

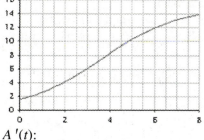

$A'(t)$:

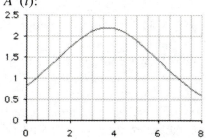

$A''(t)$:

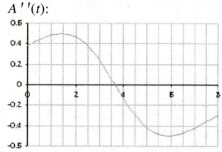

The point of inflection occurs when the second derivative is zero: Around $t \approx 4$. It is concave up when $t < 4$ (where $A''(t)$ is positive) and concave down when $t > 4$ (where $A''(t)$ is negative). The rate of change of $A(t)$ (that is, the derivative $A'(t)$) is a maximum at that value of t, meaning that the percentage of U.S. mortgages that were subprime was increasing fastest at the beginning of 2004.

73. (a) Where the graph is steepest: 2 years into the epidemic.

(b) At the point at inflection 2 years into the epidemic: There the steepness stops increasing and starts to decline, so the rate of new infections starts to drop.

75. (a) 2024 ($t = 4$): The point where the graph is increasing and steepest.

(b) 2026 ($t = 6$): The point where the graph is decreasing and steepest.

(c) 2022 ($t = 2$): Since the graph is steepest downward at $t = 2$ compared with nearby points, the rate of change of industrial output reached a minimum in 2022 compared with nearby years (choice (A)).

77. (a) $S'(n) \approx -1757(n - 180)^{-2.325}$; $S''(n) \approx 4085(n - 180)^{-3.325}$; $S''(n)$ is never zero and is always defined for $n > 180$. So, there are no points of inflection in the graph of S.
(b) Since the graph is concave up ($S''(n) > 0$ for $n > 180$), the derivative of S is increasing, and so the rate of *decrease* of SAT scores with increasing numbers of prisoners is diminishing. In other words, the apparent effect on SAT scores of increasing numbers of prisoners is diminishing.

79. (a) $\dfrac{dn}{ds} = 144.42 - 47.72s + 4.371s^2$; $\dfrac{d^2n}{ds^2}$

$= -47.72 + 8.742s$; $\dfrac{d^2n}{ds^2}\Big|_{s=3} = -21.494$.

Thus, for a firm with annual sales of $3 million, the rate at which new patents are produced decreases with increasing firm size. This means that the returns (as measured in the number of new patents per increase of $1 million in sales) are diminishing as the firm size increases.
(b) $\dfrac{d^2n}{ds^2}\Big|_{s=7} = 13.474$. Thus, for a firm with annual sales of $7 million, the rate at which new patents are produced increases with increasing firm size by 13.474 new patents per $1 million increase in annual sales.
(c) There is a point of inflection when $s \approx 5.4587$, so that in a firm with sales of $5,458,700 per year, the number of new patents produced per additional $1 million in sales is a minimum.

81. $I(t)/P(t)$ represents the fraction, or percentage, of Mexico-produced oil exported to the U.S.
Graphs: $I(t)/P(t)$:

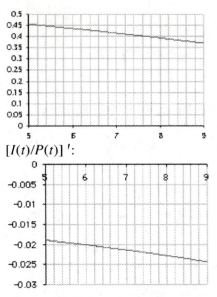

$[I(t)/P(t)]'$:

The graph of $I(t)/P(t)$ is decreasing, so the percentage of oil produced in Mexico that was exported to the U.S. was decreasing. Since the *derivative* of $I(t)/P(t)$ is also decreasing, the graph of $I(t)/P(t)$ is concave down. The concavity tells us that $I(t)/P(t)$ was decreasing at a faster rate. Thus, the percentage of oil produced in Mexico that was exported to the U.S. was decreasing at a faster rate. (Choice (D).)

83. To locate candidates for points of inflection we set $f''(x) = 0$ and solve for x.

$$f(x) = \frac{N}{1 + Ab^{-x}}$$

$$f'(x) = -\frac{N}{(1 + Ab^{-x})^2}\left(Ab^{-x}\ln b\right)(-1)$$

$$= NA\ln b\frac{b^{-x}}{(1 + Ab^{-x})^2}$$

$$f''(x) = NA\ln b \cdot$$
$$\frac{-b^{-x}\ln b(1 + Ab^{-x})^2 - b^{-x}2(1 + Ab^{-x})(Ab^{-x}\ln b)(-1)}{(1 + Ab^{-x})^4}$$

$$= NA(\ln b)^2(1 + Ab^{-x})b^{-x} \cdot$$
$$\frac{-(1 + Ab^{-x}) + 2Ab^{-x}}{(1 + Ab^{-x})^4}$$

$$= NA(\ln b)^2 b^{-x}\frac{-1 + Ab^{-x}}{(1 + Ab^{-x})^3}$$

For this to be zero, the numerator of the fraction on the right is zero:
$$-1 + Ab^{-x} = 0$$

or $Ab^{-x} = 1$

$b^{-x} = 1/A$

Taking natural logarithms and solving for x:

$-x\ln b = \ln(1/A) = -\ln A$

or $x = \ln A/\ln b$

which is the only possible candidate for a point of inflection. We already know from the shape of the logistic curve that there must be a point of inflection somewhere, and so it must occur at $x = \ln A/\ln b$. (Alternatively, notice that $f''(x)$ changes sign when the numerator $(-1 + Ab^{-x})$ changes sign, meaning that the graph is concave up on one side of $x = \ln A/\ln b$ and concave down on the other, so, again, there must be a point of inflection at $x = \ln A/\ln b$.)

85. Exercise 83 tells us that the point of inflection occurs when

$t = \ln A/\ln b$

where here $A = 1.1466$ and $b = 1.0357$. Thus,

$t = \ln 1.1466/\ln 1.0357 \approx 3.900 \approx 4$

Since the rate of change of P reaches a maximum at the point of inflection, was can say that the population of Puerto Rico was increasing fastest in 1954.

87. $p = v(1.05)^{-t} = \dfrac{10,000(1.05)^{-t}}{1 + 500e^{-0.5t}}$.

Technology formula:

Excel:

```
10000*(1.05)^(-x)/(1+500*exp(-
0.5*x))
```

TI-83/84 Plus:

```
10000*(1.05)^(-x)/(1+500*e^(-0.5*x))
```

Here are the graphs of $p(t)$ and $p'(t)$:

Graph of $p(t)$;

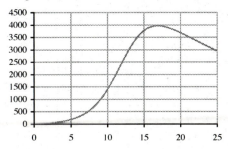

Graph of $p'(t)$:

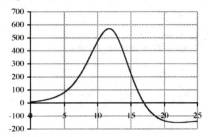

Graph of $p''(t)$:

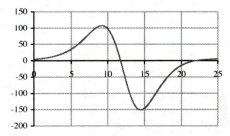

The greatest rate of increase of p occurs when the derivative is greatest. This high point on the graph of $p'(t)$ is located accurately by determining where the graph of $p''(t)$ crosses the t-axis: at approximately $t = 12$. The value of the greatest rate of increase at this point is the y-coordinate of $p'(t)$ at $t = 12$, which we can determine from the graph of $p'(t)$ as approximately \$570 per year.

89. $p(t) = ve^{-0.05t} = (300,000 + 1000t^2)\, e^{-0.05t}$

Technology formula:

Excel:

```
(300000+1000*x^2)*exp(-0.05*x)
```

TI-83/84 Plus:

```
(300000+1000*x^2)*e^(-0.05*x)
```

Graph of $p(t)$:

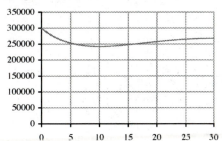

Graph of $p'(t)$:

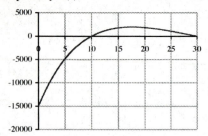

Graph of $p''(t)$:

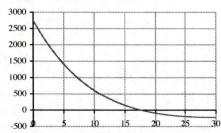

The greatest rate of increase of p occurs when the the derivative is greatest. This high point on the graph of $p'(t)$ is located accurately by determining where the graph of $p''(t)$ crosses the t-axis: at approximately $t = 17.7$.

   $p(t)$ is decreasing most rapidly at the point where the derivative $p'(t)$ is a minimum, which occurs at $t = 0$ (see the graph of $p'(t)$).

91. Nonnegative

93. Daily sales were decreasing most rapidly in June 2002.

95.

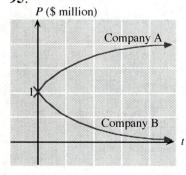

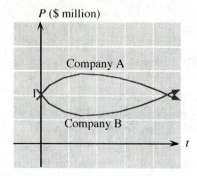

97. At a point of inflection, the graph of a function changes either from concave up to concave down, or vice-versa. If it changes from concave up to concave down, then the derivative changes from increasing to decreasing, and hence has a relative maximum. Similarly, if it changes from concave down to concave up, the derivative has a relative minimum.

12.4 Analyzing Graphs

1. (a) x-intercepts:
$x^2 + 2x + 1 = 0$ when
$(x+1)^2 = 0$
so $x = -1$
y-intercept: $f(0) = 1$.
(b) Extrema:
The only extrema are stationary points (no endpoints of singular points)
$f'(x) = 2x + 2$
$f'(x) = 0$ when $x = -1$
Absolute minimum at $(0, -1)$
(c) Points of inflection:
$f''(x) = 2 \neq 0$
No points of inflection
(d) No points where the function is not defined.
(e) $y \to +\infty$ as $x \to \pm\infty$
Graph:

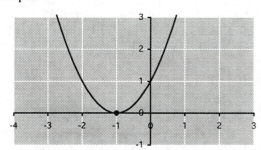

3. (a) x-intercepts:
$x^3 - 12x = 0$ when
$x(x^2 - 12) = 0$
so $x = 0, \pm\sqrt{12}$
y-intercept: $g(0) = 0$.
(b) Extrema:
Stationary points:
$g'(x) = 3x^2 - 12 = 3(x^2 - 4)$
$g'(x) = 0$ when $x = \pm 2$
Absolute max at $(-2, 16)$, absolute min at $(2, -16)$
Endpoints:
$x = -4, 4$
Absolute min at $(-4, -16)$ and absolute max at $(4, 16)$
No singular points (the derivative is defined for all x in the domain).

(c) Points of inflection:
$g''(x) = 6x$; $g''(x) = 0$ when $x = 0$
Point of inflection at $(0, 0)$
(d) None; the domain is $[-4, 4]$.
Graph:

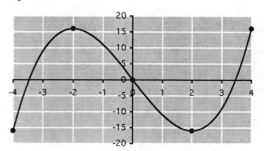

5. (a) x-intercepts (from the graph) $-3.6, 0, 5.1$
y-intercept: $h(0) = 0$
(b) The only extrema are the stationary points:
$h'(x) = 6x^2 - 6x - 36$
$= 6(x-3)(x+2)$
$6(x-3)(x+2) = 0$ when $x = -2$ and 3.
Relative max at $(-2, 44)$, relative min at $(3, -81)$
(c) $h''(x) = 12x - 6$
$12x - 6 = 0$ when $x = 0.5$.
Point of inflection at $(0.5, -18.5)$
(d) None
(e) $y \to -\infty$ as $x \to -\infty$; $y \to +\infty$ as $x \to +\infty$
Graph:

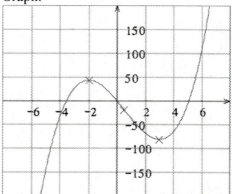

7. (a) x-intercepts (from the graph) $0.1, 1.8$
y-intercept: $f(0) = 1$
(b) The only extrema are the stationary points:
$f'(x) = 6x^2 + 6x - 12 = 6(x + 2)(x - 1)$;
$6(x + 2)(x - 1) = 0$ when $x = -2$ or $x = 1$
Relative max at $(-2, 21)$, relative min at $(1, -6)$

(c) $f''(x) = 12x + 6$; $f''(x) = 0$ when $x = -1/2$
Point of inflection at $(-1/2, 15/2)$
(d) None
(e) $y \to -\infty$ as $x \to -\infty$; $y \to +\infty$ as $x \to +\infty$
Graph:

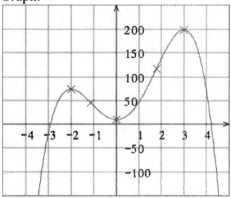

9. (a) x-intercepts (from the graph) -2.9, 4.2
y-intercept: $k(0) = 10$
(b) The only extrema are the stationary points:
$k'(x) = -12x^3 + 12x^2 + 72x$
$= -12x(x^2 - x - 6)$
$= -12x(x-3)(x+2)$
$-12x(x-3)(x+2) = 0$ when $x = 0, -2,$ or 3
Relative max at $(-2, 74)$, relative min at $(0, 10)$,
absolute max at $(3, 199)$
(c) $k''(x) = -12(3x^2 - 2x - 6)$
$k''(x) = 0$ when $3x^2 - 2x - 6 = 0$
The quadratic formula gives the solutions as
$\quad x \approx -1.12, 1.79$
Point of inflection at $(-1.12, 44.8)$, $(1.79, 117.3)$
(d) None
(e) $y \to -\infty$ as $x \to -\infty$; $y \to -\infty$ as $x \to +\infty$
Graph:

11. (a) t-intercepts
$\frac{1}{4}t^4 - \frac{2}{3}t^3 + \frac{1}{2}t^2 = 0$ when
$t^2(\frac{1}{4}t^2 - \frac{2}{3}t + \frac{1}{2}) = 0$
The quadratic in parentheses has no solution (its discriminant $b^2 - 4ac$) is negative. Thus, the only solution, and hence t-intercept is $t = 0$..
y-intercept: $g(0) = 0$
(b) The only extrema are the stationary points:
$g'(t) = t^3 - 2t^2 + t$
$= t(t^2 - 2t + 1) = t(t-1)^2$
$t(t-1)^2 = 0$ when $t = 0, 1$
Note that g increases from the stationary point at $t = 0$ to the one at $t = 1$, then continues to increase to the right (as is clear from the graph or from a test point to the right of $t = 1$). So, g has an absolute min at $(0, 0)$.
(c) $g''(t) = 3t^2 - 4t + 1$
$3t^2 - 4t + 1 = 0$ when
$(3t - 1)(t - 1) = 0$, or $t = 1/3$ or $t = 1$
Point of inflection at $(1/3, 11/324)$ and $(1, 1/12)$.
(d) None
(e) $y \to +\infty$ as $t \to -\infty$; $y \to +\infty$ as $t \to +\infty$
Graph:

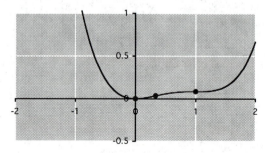

13. (a) x-intercepts:
$0 = x + \frac{1}{x}$
Multiply both sides by x:
$x^2 + 1 = 0$
This equation has no real solutions, so there are no x-intercepts
y-intercept: $f(0)$ is not defined; notice that the domain of f includes all numbers except 0. Thus, there is no y-intercept.
(b) The only extrema are the stationary points:

$f'(x) = 1 - \dfrac{1}{x^2}$

$f'(x) = 0$ when $x^2 = 1$, $x = \pm 1$

f has relative min at $(1, 2)$, relative max at $(-1, -2)$

(c) $f''(x) = \dfrac{2}{x^3}$ and thus is never zero. So there are no points of inflection.

(d) f is not defined at $x = 0$. As $x \to 0^-$ $y = x + \dfrac{1}{x} \to -\infty$. As $x \to 0^+$ $y = x + \dfrac{1}{x} \to +\infty$, so there is a vertical asymptote at $x = 0$.

(e) $y \to -\infty$ as $x \to -\infty$; $y \to +\infty$ as $x \to +\infty$

Graph:

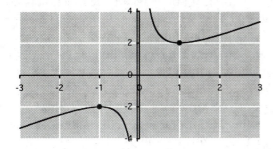

15. (a) x-intercepts: $y = 0$ when $x^3/(x^2+3) = 0$ so the numerator must be zero: $x = 0$

y-intercept: $g(0) = 0$

(b) The only extrema are the stationary points:

$g'(x) = \dfrac{3x^2(x^2 + 3) - x^3(2x)}{(x^2 + 3)^2} = \dfrac{x^4 + 9x^2}{(x^2 + 3)^2}$

$g'(x) = 0$ when

$x^4 + 9x^2 = 0$

$x^2(x^2 + 9) = 0$

$x = 0$

From the graph (or by using test points or the first derivative test) we see that $x = 0$ is a non-extreme stationary point. Thus, there are no local extrema.

(c) $g''(x) =$

Error!

$= \dfrac{(x^2 + 3)(-6x^3 + 54x)}{(x^2 + 3)^4}$

$= \dfrac{-6x(x^2 - 9)}{(x^2 + 3)^3}$

$g''(x) = 0$ when $x = 0$ or $x = \pm 3$

It is difficult to tell from the graph, but the points at $x = \pm 3$ are points of inflection. We can tell by computing, for example, $g''(2) = 60/7^3 > 0$ and $g''(4) = -168/19^3 < 0$, which shows that the concavity changes at $x = 3$. So, g has points of inflection at $(0, 0)$, $(-3, -9/4)$, and $(3, 9/4)$.

(d) No points where the function is undefined.

(e) $y \to -\infty$ as $x \to -\infty$; $y \to +\infty$ as $x \to +\infty$

Graph:

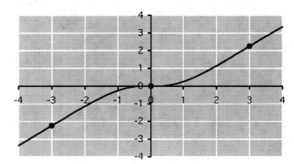

17. (a) t-intercepts:

$0 = \dfrac{t^2+1}{t^2-1}$

Thus the numerator must be zero:

$t^2 + 1 = 0$

This equation has no real solutions, so there are no t-intercepts

y-intercept: $f(0) = -1$.

(b) The only extrema are the stationary points:

$f'(t) = \dfrac{2t(t^2 - 1) - (t^2 + 1)(2t)}{(t^2 - 1)^2}$

$= \dfrac{-4t}{(t^2 - 1)^2}$

and is zero when the numerator, $t = 0$. This gives a relative max at $(0, -1)$.

Endpoints: $t = -2, 2$. These give a relative min at $(-2, 5/3)$ and at $(2, 5/3)$.

(c) $f''(t) = \dfrac{-4(t^2 - 1)^2 + 4t(2)(t^2 - 1)(2t)}{(t^2 - 1)^4}$

$= \dfrac{(t^2 - 1)[-4(t^2 - 1) + 16t^2]}{(t^2 - 1)^4}$

$= \dfrac{12t^2 + 4}{(t^2 - 1)^3}$

and thus is never zero (because the numerator cannot be zero). So there are no points of inflection.

457

(d) f is not defined at $t = \pm 1$.

As $t \to -1^{-}$ $y \to +\infty$. As $t \to -1^{+}$ $y \to -\infty$, so there is a vertical asymptote at $t = -1$.

As $t \to 1^{-}$ $y \to -\infty$. As $t \to 1^{+}$ $y \to +\infty$, so there is a vertical asymptote at $t = 1$.

(e) The domain of f is $[-2, 2]$ so there is no limiting behavior at infinity.

Graph:

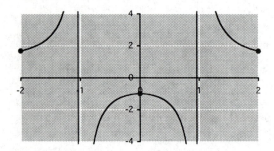

19. (a) x−intercepts: From the graph: $x \approx -0.6$
y-intercept: $k(0) = 1$

(b) Relative Extrema:

$$k(x) = \frac{2x}{3} + (x+1)^{2/3}$$

$$k'(x) = \frac{2}{3} + \frac{2}{3}(x+1)^{-1/3} = \frac{2}{3} + \frac{2}{3(x+1)^{1/3}}$$

Stationary points: $k'(x) = 0$ when

$$\frac{2}{3} + \frac{2}{3(x+1)^{1/3}} = 0$$

$$\frac{2}{3} = -\frac{2}{3(x+1)^{1/3}}$$

cross-multiply:

$$6(x+1)^{1/3} = -6$$

$$(x+1)^{1/3} = -1$$

$$(x+1) = (-1)^{3} = -1$$

$$x = -2,$$

so we have a stationary point at $x = -2$.

Singular points: $k'(x)$ is undefined when $x = -1$, so we have a singular point at $x = -1$.

In addition we use a test point to the left of $x = -2$ and to the right of $x = -1$.

x	−3	−2	−1	0
$f(x)$	− 0.4	− 1/3	− 2/3	1

	1			

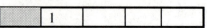

From the graph we see that we have a relative maximum at $(-2, -1/3)$ and a relative minimum at $(-1, -2/3)$.

(c) Points of Inflection:

$$k'(x) = \frac{2}{3} + \frac{2}{3(x+1)^{1/3}}$$

$$k''(x) = -\frac{2}{9(x+1)^{4/3}}$$

$k''(x)$ is never zero and not defined when $x = -1$. So, the only candidate for a point of inflection is $x = -1$ which we see from the graph is not one. No points of inflection.

(d) Behavior near points where the function is not defined:
The domain of this function consists of all real numbers, so there are no such points.

(e) Behavior at infinity:

$$\lim_{x \to -\infty} k(x) = \lim_{x \to -\infty} [2x/3 + (x+1)^{2/3}] = -\infty$$

(computing $k(x)$ for large negative values of x gives large negative numbers).

$$\lim_{x \to +\infty} k(x) = \lim_{x \to +\infty} [2x/3 + (x+1)^{2/3}] = +\infty$$

Graph:

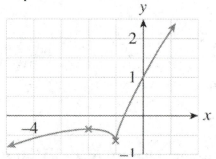

Technology format:
`2*x/3+((x+1)^2)^(1/3)`

21. (a) x-intercepts: $y = 0$ when $x - \ln x = 0$, so $x = \ln x$. In exponential form this equation is $e^x = x$. However, $e^x > x$ for every x, and so the equation has no solution. Hence there are no x-intercepts.

y-intercept: $f(0)$ is not defined, so there is no y-intercept either.

(b) Extrema:

458

Stationary:

$$f'(x) = 1 - \frac{1}{x}$$

$f'(x) = 0$ when $x = 1$,

so there is a stationary point at $x = 1$: an absolute minimum at $(1, 1)$.

Since $f'(x)$ is defined for all x in the domain of f, there are no singular points.

(c) $f''(x) = \frac{1}{x^2}$ so $f''(x)$ is never 0, and hence there are no points of inflection.

(d) As $x \to 0^+$ $y \to +\infty$ so there is a vertical asymptote at $x = 0$

(e) As $x \to +\infty$ $y \to +\infty$

Graph:

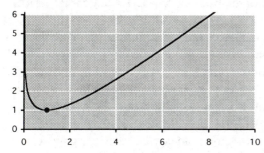

23. (a) x-intercepts: From the graph, $x \approx \pm 0.8$.

y-intercept: $f(0)$ is not defined, so there is no y-intercept.

(b) Extrema:

Stationary points:

$$f'(x) = 2x + \frac{2x}{x^2} = 2x + \frac{2}{x}$$

$2x + \frac{2}{x} = 0$ gives

$2x = -\frac{2}{x}$ and so $2x^2 = -2$,

which has no real solutions, so

$f'(x)$ is never 0 and there are no stationary points.

Since $f'(x)$ is defined for all x in the domain of f, there are no singular points.

There are no endpoints either, and hence no extrema.

(c) $f''(x) = 2 - \frac{2}{x^2}$

$f''(x) = 0$ when

$2 - \frac{2}{x^2} = 0$

$$2 = \frac{2}{x^2}$$

$2 = 2x^2$

$x^2 = 1$, $x = \pm 1$

Points of inflection at $(1,1)$ and $(-1, 1)$.

(d) As $x \to 0$ $y \to -\infty$ so there is a vertical asymptote at $x = 0$

(e) As $x \to \pm\infty$ $y \to +\infty$

Graph:

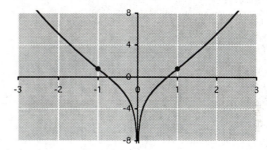

25. (a) t-intercepts: $e^t - t = 0$

when $e^t = t$, which has no real solutions as $e^t > t$ for every t.

y-intercept: $g(0) = e^0 = 1$

(b) Extrema:

Stationary points:

$g'(t) = e^t - 1$

$g'(t) = 0$ when

$e^t = 1$

$t = 0$

Absolute minimum at $(0, 1)$.

Since $g'(t)$ is defined for all x in the domain of g there are no singular points.

Endpoints: $t = -1$, $t = 1$.

absolute max at $(1, e - 1)$, and a relative max at $(-1, e^{-1} + 1)$.

(c) $g''(t) = e^t$, which can never be zero.

(d) No points where the function is not defined.

(e) None (domain is $[-1, 1]$).
Graph:

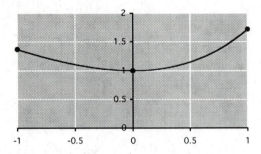

27. $f(x) = x^4 - 2x^3 + x^2 - 2x + 1$
$f'(x) = 4x^3 - 6x^2 + 2x - 2$; by graphing $f'(x)$ we see that $f'(x) = 0$ for $x \approx 1.40$; $f'(x)$ is always defined. $f''(x) = 12x^2 - 12x + 2$; $f''(x) = 0$ for $x = \dfrac{1}{2} \pm \dfrac{\sqrt{3}}{6}$ (by the quadratic formula), $x \approx 0.21$ or $x \approx 0.79$; $f''(x)$ is always defined.

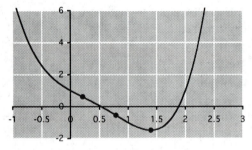

f has an absolute min at $(1.40, -1.49)$ and points of inflection at $(0.21, 0.61)$ and $(0.79, -0.55)$.

29. $f(x) = e^x - x^3$
$f'(x) = e^x - 3x^2$; by graphing $f'(x)$ we see that $f'(x) = 0$ for $x \approx -0.46$, $x \approx 0.91$, and $x \approx 3.73$; $f'(x)$ is always defined. $f''(x) = e^x - 6x$; by graphing $f''(x)$ we see that $f''(x) = 0$ for $x \approx 0.20$ and $x \approx 2.83$; $f''(x)$ is always defined.

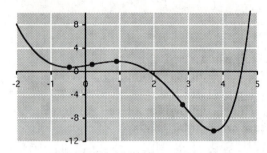

f has a relative min at $(-0.46, 0.73)$, a relative max at $(0.91, 1.73)$, an absolute min at $(3.73, -10.22)$, and points of inflection at $(0.20, 1.22)$ and $(2.83, -5.74)$.

31. y-intercept: 10; t-intercepts: None
The home price index was 10 percentage points at the start of 2004 ($t = 0$)
Extrema: Absolute minimum at $(0, 10)$; absolute maximum at $(2, 40)$
The home price index was lowest in 2004 ($t = 0$) when it stood at 10 percentage points. The index was at its highest in 2006 ($t = 2$) at 40 points.
Point of inflection at $(3.5, 30)$
The home price index was decreasing most rapidly midway through 2007 ($t \approx 3.5$) when it stood at 30 percentage points.
Points where the function is not defined: None
Behavior at infinity: As $t \to +\infty$ $y \to 15$;
Assuming the trend shown in the graph continues indefinitely, the home price index will approach a value of 15 percentage points in the long term.

33. (a) Intercepts: No t-intercept (seen from the graph); y-intercept at $I(0) = 195$.
The CPI was never zero during the period under consideration; in July 2005 ($t = 0$) the CPI was 195.
Extrema:
Stationary points:
$$I'(t) = 0.18t^2 - 1.6t + 3.1$$
$I'(t) = 0$ when
$$t = \frac{1.6 \pm \sqrt{1.6^2 - 4(0.18)(3)}}{2(0.18)}$$
$$\approx 2.9, 6.0$$
The corresponding values of $I(t)$ are 198.7 and 197.8

Endpoints: 0, 8
The corresponding values of $I(t)$ are 195 and 199.3
Absolute min at (0, 195), absolute max at (8, 199.3), relative max at (2.9, 198.7), relative min at (6.0, 197.8).
The CPI was at a low of 195 in July 2005, rose to 198.7 around October 2005, dipped to 197.8 around January 2006, and then rose to a high of 199.3 in March 2006.
Points of inflection occur when $I''(t) = 0$

$$0.36t - 1.6 = 0$$

$$t = \frac{1.6}{0.36} \approx 4.44444$$

$$I(4.44444) \approx 198.2$$

There is a point of inflection at (4.4, 198.2) The rate of change of the CPI (inflation) reached a minimum around $t = 4.4$ (Mid-November, 2005) when the CPI was 198.2.
(b) At the stationary points, $I'(t) = 0$, and the inflation rate, $I'(t)/I(t)$ is zero at precisely these points as well.

35. Extrema:
Endpoint: $t = 0$
Stationary points:
$s'(t) = 6t^2 - 6t = 6t(t - 1)$
$s'(t) = 0$ when $t = 0$ and $t = 1$
No singular points.
The corresponding values of s are:
$s(0) = 100$
$s(1) = 2(1)^3 - 3(1)^2 + 100 = 99$
Relative max at (0, 100), absolute min at (1, 99).
Points of inflection:
$s''(t) = 12t - 6$
$s''(t) = 0$ when $t = 1/2$
The corresponding value of s is:
$s(1/2) = 2(1/2)^3 - 3(1/2)^2 + 100 = 199/2 = 99.5$
Point of inflection at (0.5, 99.5)
Behavior at infinity: As $t \to +\infty$, $s(t) = t^2(2t - 3) + 100 \to +\infty$
At time $t = 0$ seconds, the UFO is stationary, 100 ft away from the observer, and begins to move closer. At time $t = 0.5$ seconds, when the UFO is 99.5 feet away, its distance is decreasing most

rapidly (it is moving toward the observer most rapidly). It then slows down to a stop at $t = 1$ second when it is at its closest point (99 ft away) and then begins to move further and further away.
Graph:

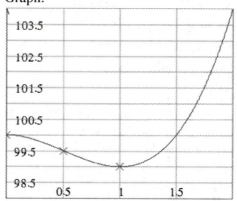

37. $C(x) = 22{,}500 + 100x + 0.01x^2$
$\bar{C}(x) = \dfrac{22{,}500}{x} + 100 + 0.01x \ (x > 0)$
Intercepts:
x-intercepts occur when $\bar{C}(x) = 0$
However, since $x > 0$, $\bar{C}(x)$ is positive, and hence never zero. So, there are no x-intercepts.
y-intercept: $\bar{C}(0)$ is not defined; no y-intercept.
Extrema:
Stationary points:
$\bar{C}'(x) = -\dfrac{22{,}500}{x^2} + 0.01$

$$-\frac{22{,}500}{x^2} + 0.01 = 0$$

$$\frac{22{,}500}{x^2} = 0.01$$

$$0.01x^2 = 22{,}500$$

$$x^2 = \frac{22{,}500}{0.01} = 2{,}250{,}000$$

$$x = \sqrt{2{,}250{,}000} = 1500 \text{ iPods per day}$$

The resulting average cost is

$$\bar{C}(1500) = \frac{22{,}500}{1500} + 100 + 0.01(1500) =$$

$130
Endpoints: None; Singular points: None
Absolute minimum at (1500, 130).

Points of inflection:
$$\overline{C}''(x) = \frac{45,000}{x^3},$$
which can never be zero. No points of inflection. Behavior near points where $\overline{C}(x)$ is not defined: $\overline{C}(x)$ is not defined when $x = 0$. As $x \to 0^+$, $\overline{C}(x) \to +\infty$, so there is a vertical asymptote at $x = 0$.
Behavior at infinity:
As $x \to +\infty$, $\overline{C}(x) \to +\infty$.
Interpretation: The average cost is never zero, nor is it defined for zero iPods. The average cost is a minimum ($130) when 1500 iPods are manufactured per day. The average cost becomes extremely large for very small or very large numbers of iPods. Graph:

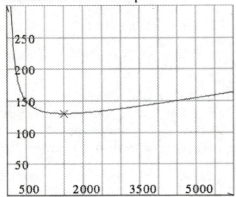

As $t \to +\infty$, $A'(t) \to 0$; In the long term, assuming the trend shown in the model continues, the rate of change of the percentage of mortgages that were subprime approaches zero; that is, the percentage of mortgages that were subprime approaches a constant value.

41. No; Yes. Near a vertical asymptote the value of y increases without bound, and so the graph could not be included between two horizontal lines; hence no vertical asymptotes are possible. Horizontal asymptotes are possible, as for instance in the graph in Exercise 31

43. It too has a vertical asymptote at $x = a$; the magnitude of the derivative increases without bound as $x \to a$.

45. No. If the leftmost critical point is a relative maximum, the function will decrease from there until it reaches the rightmost critical point, so can't have a relative maximum there.

47. Between every pair of zeros of $f(x)$ there must be a local extremum, which must be a stationary point of $f(x)$, hence a zero of $f'(x)$.

39. Graph of derivative:

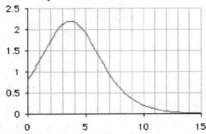

If we zoom in, we can determine that the maximum occurs around $t = 3.7$; during the year 2003. Since $A'(t)$ represents the rate of change of the percentage of mortgages that were subprime, we can say that the percentage of mortgages that were subprime was increasing most rapidly during 2003, when it increased at a rate of around 2.2 percentage points per year.

12.5 Related Rates

1. The population P is currently 10,000 and its rate of change is 1000 per year:

$$P = 10,000 \text{ and } \frac{dP}{dt} = 1000.$$

3. Let R be the annual revenue of my company and let q be annual sales. R is currently $7000 but and its rate of change is $-\$700$ each year. Find how fast q is changing:

$$R = 7000 \text{ and } \frac{dR}{dt} = -700.$$

Find $\frac{dq}{dt}$.

5. Let p be the price of a pair of shoes and let q be the demand for shoes.

$$\frac{dp}{dt} = 5. \text{ Find } \frac{dq}{dt}.$$

7. Let T be the average global temperature and let q be the number of Bermuda shorts sold per year.

$$T = 60 \text{ and } \frac{dT}{dt} = 0.1. \text{ Find } \frac{dq}{dt}.$$

9. (a) Changing quantities: the radius r and the area A.
The problem:

$$\frac{dA}{dt} = 1200. \text{ Find } \frac{dr}{dt} \text{ when } r = 10,000.$$

The relationship:

$$A = \pi r^2$$

$\frac{d}{dt}$ of both sides:

$$\frac{dA}{dt} = 2\pi r \frac{dr}{dt}.$$

Substitute:

$$1200 = 2\pi(10,000)\frac{dr}{dt}$$

so $\frac{dr}{dt} = 6/(100\pi) \approx 0.019$ km/sec.

(b) This time the problem is to find $\frac{dr}{dt}$ when $A = 640,000$. From part (a) we have the derived equation

$$\frac{dA}{dt} = 2\pi r \frac{dr}{dt}.$$

Since r appears in the derived equation but not A, we need to find r from $A = \pi r^2$:

$$640,000 = \pi r^2$$
$$r = \sqrt{640,000/\pi} = 800\sqrt{\pi}.$$

Substituting these values in the derived equation:

$$\frac{dA}{dt} = 2\pi r \frac{dr}{dt}.$$

$$1200 = 2\pi(800\sqrt{\pi})\frac{dr}{dt}$$

$$\frac{dr}{dt} = 6/(8\sqrt{\pi}) \approx 0.4231 \text{ km/sec.}$$

11. Changing quantities: the volume V and the radius r.
The problem:

$$\frac{dV}{dt} = 3. \text{ Find } \frac{dr}{dt} \text{ when } r = 1.$$

The relationship:

$$V = \frac{4}{3}\pi r^3$$

$\frac{d}{dt}$ of both sides:

$$\frac{dV}{dt} = 4\pi r^2 \frac{dr}{dt}$$

Substitute:

$$3 = 4\pi(1)^2 \frac{dr}{dt}$$

$$\frac{dr}{dt} = \frac{3}{4\pi} \approx 0.24 \text{ ft/min}$$

13. Changing quantities: $b =$ the distance of the base of the ladder from the wall and $h =$ the height of the top of the ladder.
The problem:

$$\frac{db}{dt} = 10. \text{ Find } \frac{dh}{dt} \text{ when } b = 30$$

The relationship:

$$b^2 + h^2 = 50^2$$

$\frac{d}{dt}$ of both sides:

$$2b\frac{db}{dt} + 2h\frac{dh}{dt} = 0.$$

We need the value of h:

$$30^2 + h^2 = 50^2$$
$$h = \sqrt{2500 - 900} = 40.$$

Substitute:

$$2(30)(10) + 2(40)\frac{dh}{dt} = 0$$

$$\frac{dh}{dt} = -600/80 = -7.5,$$

so the top of the ladder is sliding down at 7.5 ft/sec.

15. Changing quantities: the average cost $\bar{C}$ and the number of CD players x.

The problem:

$$x = 3000 \text{ and } \frac{dx}{dt} = 100. \text{ Find } \frac{d\bar{C}}{dt}.$$

The relationship:

$$\bar{C}(x) = 150{,}000x^{-1} + 20 + 0.0001x \quad \frac{d}{dt} \text{ of}$$

both sides:

$$\frac{d\bar{C}}{dt} = -150{,}000x^{-2}\frac{dx}{dt} + 0.0001\frac{dx}{dt}$$

Substitute:

$$\frac{d\bar{C}}{dt} = -150{,}000(3000)^{-2}(100) +$$

$$0.0001(100)$$

$$\approx -1.66.$$

The average cost is decreasing at a rate of $1.66 per player per week.

17. Changing quantities: the number q of T shirts sold per month and the price p per T-shirt

The problem:

$$p = 15 \text{ and } \frac{dp}{dt} = 2. \text{ Find } \frac{dq}{dt}$$

The relationship: This is the given demand equation:

$$q = 500 - 100p^{0.5}$$

$\frac{d}{dt}$ of both sides:

$$\frac{dq}{dt} = -50p^{-0.5}\frac{dp}{dt}.$$

Substitute:

$$\frac{dq}{dt} = -50(15)^{-0.5}(2) \approx -26.$$

Monthly sales will drop at a rate of 26 T-shirts per month.

19. Changing quantities: The price p, the weekly demand q, and the weekly revenue R.

The problem:

$$q = 50, \, p = 30¢, \text{ and } \frac{dq}{dt} = -5.$$

Find $\frac{dp}{dt}$ if $\frac{dR}{dt} = 0$.

The relationship:

$$R = pq \qquad \text{Revenue = price} \times \text{quantity}$$

$\frac{d}{dt}$ of both sides:

$$\frac{dR}{dt} = \frac{dp}{dt}q + p\frac{dq}{dt}.$$

Substitute:

$$0 = \frac{dp}{dt}(50) + (30)(-5)$$

$$\frac{dp}{dt} = 150/50 = 3.$$

You must raise the price by 3¢ per week.

21. Changing quantities: The price p, the weekly demand q, and the weekly revenue R.

The problem:

$$p = 90, \, q = -0.022t^2 + 0.2t + 2.9,$$

and $\frac{dp}{dt} = 80$.

Find $\frac{dR}{dt}$

The relationship:

$$R = pq \qquad \text{Revenue = price} \times \text{quantity}$$

$\frac{d}{dt}$ of both sides:

$$\frac{dR}{dt} = \frac{dp}{dt}q + p\frac{dq}{dt}.$$

We need to know q and $\frac{dq}{dt}$, which we can get from the formula for q:

$$q(t) = -0.022t^2 + 0.2t + 2.9$$

$$q'(t) = -0.044t + 0.2$$

In 2008:

$$q(8) = -0.022(8)^2 + 0.2(8) + 2.9 = 3.092$$

$$q'(8) = -0.044(8) + 0.2 = -0.152$$

Substitute:

$$\frac{dR}{dt} = (80)(3.092) + (90)(-0.152)$$

$$= \$233.68 \text{ million per year}$$

23. Changing quantities: The number x of laborers, and the number y of robots.
The problem:

$$y = 400 \text{ and } \frac{dy}{dt} = 16.$$

Find $\frac{dx}{dt}$.

The relationship:

$$xy = 10{,}000$$

$\frac{d}{dt}$ of both sides:

$$\frac{dx}{dt} y + x \frac{dy}{dt} = 0$$

We need the value of x when $y = 400$:

$$x(400) = 10{,}000$$
$$x = 25.$$

Substitute:

$$\frac{dx}{dt} 400 + 25(16) = 0$$

$$\frac{dx}{dt} = -\frac{400}{400} = -1$$

You are laying off 1 laborer per month.

25. Changing quantities: the number P of automobiles produced per year, the number x of employees, and the daily operating budget y.
The problem:

P is constant at 1000. $x = 150$ and $\frac{dx}{dt} = 10$.

Find $\frac{dy}{dt}$.

The relationship:

$$P = 10x^{0.3}y^{0.7}$$

$\frac{d}{dt}$ of both sides:

$$0 = 3x^{-0.7}y^{0.7} \frac{dx}{dt} + 7x^{0.3}y^{-0.3} \frac{dy}{dt}$$

(We are told that P is constant so its derivative is zero.)
The solution to the problem is a bit simpler if we first solve for $\frac{dy}{dt}$ before substituting values:

$$\frac{dy}{dt} = -\frac{3x^{-0.7}y^{0.7}}{7x^{0.3}y^{-0.3}} = -\frac{3y}{7x} \frac{dx}{dt}.$$

We need the value of y:

$$1000 = 10(150)^{0.3}y^{0.7}$$
$$y = (100/150^{0.3})^{1/0.7} = 100^{10/7}/150^{3/7}$$

Substitute:

$$\frac{dy}{dt} = -\frac{3(100^{1/0.7})/150^{3/7}}{7(150)} \quad (10)$$

$$= -\frac{30(100^{10/7})}{7(150)^{10/7}} \approx -2.40.$$

The daily operating budget is dropping at a rate of $2.40 per year.

27. Changing quantities: the number q of pounds of tuna that can be sold in one month, the price p in dollars per pound.
The problem:

$q = 900$ and $\frac{dq}{dt} = 100$. Find $\frac{dp}{dt}$.

The relationship: This is the demand equation:

$$pq^{1.5} = 50{,}000$$

$\frac{d}{dt}$ of both sides:

$$\frac{dp}{dt} q^{1.5} + 1.5pq^{0.5} \frac{dq}{dt} = 0$$

We will also need the value of p:

$$p(900)^{1.5} = 50{,}000$$
$$p = 50{,}000/(900)^{1.5}$$

Substitute:

$$\frac{dp}{dt}(900)^{1.5} + 1.5[50{,}000/(900)^{1.5}](900)^{0.5}(100) = 0$$

$$\frac{dp}{dt} = -(75{,}000/9)/(900)^{1.5} \approx -0.31$$

The price is decreasing at a rate of approximately 31¢ per pound per month.

29. Changing quantities: Let x be the distance of the Mona Lisa from Montauk and let y be the distance of the Dreadnaught from Montauk. Let z be the distance between the two ships.
The problem:

$x = 50$, $\frac{dx}{dt} = 30$, $y = 40$, and $\frac{dy}{dt} = 20$.

Find $\frac{dz}{dt}$.

The relationship:

$$z^2 = x^2 + y^2$$

$\frac{d}{dt}$ of both sides:

$$2z \frac{dz}{dt} = 2x \frac{dx}{dt} + 2y \frac{dy}{dt}$$

We need the value of z:
$$z^2 = 50^2 + 40^2 = 4100$$
$$z = \sqrt{4100}$$
Substitute:
$$2\sqrt{4100}\,\frac{dz}{dt} = 2(50)(30) + 2(40)(20)$$
$$= 4600$$
$$\frac{dz}{dt} = \frac{2300}{\sqrt{4100}} \approx 36 \text{ mph.}$$

31. Changing quantities: Let x be the distance of the batter from home base, and let h be the distance from third base as shown here:

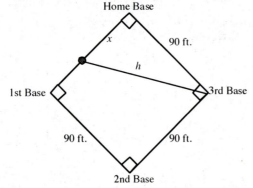

The problem:

Given that $\frac{dx}{dt} = 24$, find $\frac{dh}{dt}$ when $x = 45$.

Equation relating the changing quantities:
$$x^2 + 90^2 = h^2$$
$$x^2 + 8100 = h^2$$
Derived equation:
$$2x\frac{dx}{dt} = 2h\frac{dh}{dt}$$
$$\frac{dh}{dt} = \frac{x}{h}\frac{dx}{dt}$$
$$h^2 = 8100 + 45^2$$
$$= 8100 + 2025 = 10{,}125$$
$$h = \sqrt{10{,}125} \approx 100.62$$
$$\frac{dh}{dt} = \frac{x}{h}\frac{dx}{dt} = \frac{45}{\sqrt{10{,}125}}(24) \approx 10.7 \text{ ft/sec}$$

33. Changing quantities: the x-coordinate x and the y-coordinate y of a point on the graph.
The problem:

$\frac{dx}{dt} = 4$ and $y = 2$. Find $\frac{dy}{dt}$.
The relationship:
$$y = x^{-1}$$
$\frac{d}{dt}$ of both sides:
$$\frac{dy}{dt} = -x^{-2}\frac{dx}{dt}$$
We need the value of x:
$$2 = 1/x$$
$$x = 1/2.$$
Substitute:
$$\frac{dy}{dt} = -(1/2)^{-2}(4) = -16$$
The y coordinate is decreasing at a rate of 16 units per second.

35. Changing quantities: I and n.
The problem:

$n = 13$ and $\frac{dn}{dt} = \frac{1}{3}$. Find $\frac{dI}{dn}$.
The relationship:
$$I = 2928.8n^3 - 115{,}860n^2 + 1{,}532{,}900n - 6{,}760{,}800$$
$\frac{d}{dt}$ of both sides:
$$\frac{dI}{dt} = 8786.4n^2\frac{dn}{dt} - 231{,}720n\frac{dn}{dt} +$$
$$1{,}532{,}900\frac{dn}{dt}$$
Substitute:
$$\frac{dI}{dt} = 8786.4(13)^2(1/3)$$
$$- 231{,}720(13)(1/3) + 1{,}532{,}900(1/3)$$
$$\approx \$1814 \text{ per year.}$$

37. Changing quantities: V, e, g.
The problem:

V is constant at 200. $g = 3.0$ and $\frac{dg}{dt} = -0.2$.

Find $\frac{de}{dt}$.
The relationship:
$$V = 3e^2 + 5g^3$$
$\frac{d}{dt}$ of both sides:
$$0 = 6e\frac{de}{dt} + 15g^2\frac{dg}{dt}$$

We need the value of e:
$$200 = 3e^2 + 5(3.0)^3$$
$$e = \sqrt{65/3} \approx 4.655$$

Substitute:
$$0 = 6(4.655)\frac{de}{dt} + 15(3.0)^2(-0.2)$$
$$\frac{de}{dt} = \frac{27}{27.93} \approx 0.97.$$

Their prior experience must increase at a rate of approximately 0.97 years every year.

39. Changing quantities: V, h
The problem:
$$\frac{dV}{dt} = 100 \text{ and } V = 200\pi. \text{ Find } \frac{dh}{dt}.$$

The relationship:
The given formula expresses V in terms of both h and r. To get the relationship between V and h we need to know how r is related to h. Looking at the vessel from the side, we can see that, for any given value of h, the corresponding radius r must satisfy
$$r/h = 30/50$$
the ratio at the brim of the vessel. So,
$$r = \frac{3}{5}h.$$

Substituting into $V = \frac{1}{3}\pi r^2 h$ we get the relationship we want:
$$V = \frac{3}{25}\pi h^3$$
$\frac{d}{dt}$ of both sides:
$$\frac{dV}{dt} = \frac{9}{25}\pi h^2 \frac{dh}{dt}$$
We need the value of h:
$$200\pi = \frac{3}{25}\pi h^3$$
$$h = (5000/3)^{1/3}$$
Substitute:
$$100 = \frac{9\pi}{25}\left(\frac{5000}{3}\right)^{2/3}\frac{dh}{dt}$$
$$\frac{dh}{dt} = \frac{2500}{9\pi}\left(\frac{3}{5000}\right)^{2/3} \approx 0.63 \text{ m/sec.}$$

41. Changing quantities: V and h.
The problem:
r is constant at 2. $V = 4t^2 - t$ and $h = 2$.

Find $\frac{dh}{dt}$.

The relationship:
From $V = \pi r^2 h = 4\pi h$ we get
$$h = \frac{V}{4\pi}$$
$\frac{d}{dt}$ of both sides:
$$\frac{dh}{dt} = \frac{1}{4\pi}\frac{dV}{dt} = \frac{1}{4\pi}(8t - 1).$$
We need the value of t when $h = 2$. We first find
$$V = 4\pi h = 8\pi$$
so $8\pi = 4t^2 - t$
$$4t^2 - t - 8\pi = 0$$
$$t = \frac{1 \pm \sqrt{1 + 128\pi}}{8}$$
by the quadratic formula. We take the positive solution
$$t = \frac{1 + \sqrt{1 + 128\pi}}{8},$$
where the volume is rising. Substituting:
$$\frac{dh}{dt} = \frac{\sqrt{1 + 128\pi}}{4\pi} \approx 1.6 \text{ cm/sec.}$$

43. Changing quantities: q and x.
The problem:
$$x = 30,000 \text{ and } \frac{dx}{dt} = 2000. \text{ Find } q \text{ and } \frac{dq}{dt}.$$
The relationship:
$$q = 0.3454 \ln x - 3.047$$
$\frac{d}{dt}$ of both sides:
$$\frac{dq}{dt} = \frac{0.3454}{x}\frac{dx}{dt}$$
Substitute:
$$\frac{dq}{dt} = \frac{0.3454}{30,000}(2000) \approx 0.0230.$$
We are also asked for the value of q (the number of computers per household) so we substitute $x = 30,000$ in the original equation:
$$q = 0.3454 \ln(30,000) - 3.047 \approx 0.5137$$
So, there are approximately 0.5137 computers per household, increasing at a rate of 0.0230 computers per household per year.

45. Changing quantities: S and n.
The problem:

$n = 475$ and $\dfrac{dn}{dt} = 35$. Find S and $\dfrac{dS}{dt}$.

The relationship:

$$S = 904 + \dfrac{1326}{(n-180)^{1.325}}$$

$S(475) \approx 904.71$ from the formula.

Derived relationship:

$$\dfrac{dS}{dt} = 1326(-1.325)(n - 180)^{-2.325} \dfrac{dn}{dt} \ .$$

Substitute:

$$\dfrac{dS}{dt} = -1756.95(475 - 180)^{-2.325}(35)$$

$$\approx -0.11$$

The average SAT score was 904.71, decreasing at a rate of 0.11 per year.

47. $r = 1.1$ and $\dfrac{dr}{dt} = 0.05$. Find $\dfrac{d}{dt} d(r)$. Note that $r = 1.1$ is in the range where $d(r) = -40r + 74$ and that we stay in that range because we are interested in the slope at that point. Thus,

$$\dfrac{d}{dt} d(r) = -40 \dfrac{dr}{dt} = -40(0.05) = -2. \text{ The}$$

divorce rate is decreasing by 2 percentage points per year.

49. The section is called "related rates" because the goal is to compute the rate of change of a quantity based on a knowledge of the rate of change of a related quantity. The relationship between the quantities gives a relationship between their rates of change.

51. Answers may vary. For example: A rectangular solid has dimensions 2 cm $\times$5 cm $\times$10 cm and each side is expanding at a rate of 3 cm/second. How fast is the volume increasing?

53. $R = pq$, so $R' = p'q + pq'$, where the derivatives are with respect to time t. Divide by $R = pq$ to get $\dfrac{R'}{R} = \dfrac{p'q}{pq} + \dfrac{pq'}{pq} = \dfrac{p'}{p} + \dfrac{q'}{q}$ as claimed.

55. The derived equation is linear in the unknown rate X. This follows from the chain rule, since if Q is a quantity and $f(Q)$ is any expression in Q,

we have $\dfrac{d}{dt} f(Q) = f'(Q) \dfrac{dQ}{dt}$, which is linear in the derivative $\dfrac{dQ}{dt}$. The presence of other variables may add terms not containing $\dfrac{dQ}{dt}$ but those maintain the linearity.

57. Let x = my grades and y = your grades. If $dx/dt = 2 \ dy/dt$, then $dy/dt = (1/2) \ dx/dt$

12.6 Elasticity

1. $E = -\dfrac{dq}{dp} \cdot \dfrac{p}{q}$

$= -(-20)\dfrac{p}{1000 - 20p} = \dfrac{20p}{1000 - 20p}$

When $p = 30$, $E = 1.5$: The demand is going down 1.5% per 1% increase in price at that price level. $E = 1$ when

$\dfrac{20p}{1000 - 20p} = 1$

$20p = 1000 - 20p$

$p = 25$

Revenue is maximized when $p = \$25$; weekly revenue at that price is $R = pq = 25(1000 - 20 \cdot 25) = \$12{,}500$.

3. (a) $E = -\dfrac{dq}{dp} \cdot \dfrac{p}{q}$

$= -(-2)(100 - p)\dfrac{p}{(100 - p)^2}$

$= \dfrac{2p}{100 - p}.$

When $p = 30$, $E = 6/7$. The demand is going down 6% per 7% increase in price at that price level. Thus, a price increase is in order.

(b) $E = 1$ when

$\dfrac{2p}{100 - p} = 1$

$2p = 100 - p$

$p = 100/3$

Revenue is maximized when $p = 100/3 \approx \$33.33$.

(c) Demand would be $(100 - 100/3)^2 = (200/3)^2 \approx 4444$ cases per week.

5. (a) $E = -\dfrac{dq}{dp} \cdot \dfrac{p}{q}$

$= -(-4p + 33) \cdot \dfrac{p}{-2p^2 + 33p}$

$= \dfrac{4p - 33}{-2p + 33}$

(b) $E(10) = \dfrac{4(10) - 33}{-2(10) + 33} = \dfrac{7}{13} \approx 0.54$

Interpretation: The demand for $E = mc^2$ T-shirts is going down by about 0.54% per 1% increase in the price.

(c) Revenue is maximized when $E = 1$:

$\dfrac{4p - 33}{-2p + 33} = 1$

$4p - 33 = -2p + 33$

$6p = 66$

$p = \$11$

Therefore, the club should charge \$11 per T shirt to maximize revenue.

At this price, the total revenue is given by

$R = pq = p(-2p^2 + 33p)$

$= (11)(-2(11)^2 + 33(11))$

$= \$1331$

7. (a) $E = -\dfrac{dq}{dp} \cdot \dfrac{p}{q}$

$= -(-2.2) \cdot \dfrac{p}{9900 - 2.2\,p}$

$= \dfrac{2.2p}{9900 - 2.2p}$

$E(2900) = \dfrac{2.2(2900)}{9900 - 2.2(2900)}$

≈ 1.81

Thus, the demand is elastic at the given tuition level, showing that a decrease in tuition will result in an increase in revenue.

(c) Revenue is maximized when $E = 1$:

$\dfrac{2.2p}{9900 - 2.2p} = 1$

$2.2p = 9900 - 2.2p$

$4.4p = 9900$

$p = 9900/4.4 = \$2250$ per student, and this will result in an enrollment of about

$q = 9900 - 2.2(2250) = 4950$ students, giving a revenue of about

$pq = 2250 \times 4950 = \$11{,}137{,}500$.

9. (a) $E = -\dfrac{dq}{dp} \cdot \dfrac{p}{q}$

$= -(-6p + 1)100e^{-3p^2 + p}\dfrac{p}{100e^{-3p^2 + p}}$

$= p(6p - 1).$

When $p = 3$, $E = 51$: The demand is going down 51% per 1% increase in price at that price level. Thus, a large price decrease is advised.
(b) $E = 1$ when

$$p(6p - 1) = 1$$
$$6p^2 - p - 1 = 0$$
$$(3p + 1)(2p - 1) = 0$$
$$p = 1/2.$$

(We reject the solution $p = -1/3$ because we must have $p > 0$.) Revenue is maximized when $p = ¥0.50$.
(c) Demand would be $q = 100e^{-3/4+1/2} \approx 78$ paint-by-number sets per month.

11. (a) $E = -\dfrac{dq}{dp} \cdot \dfrac{p}{q}$

$$= -m \frac{p}{mp + b} = -\frac{mp}{mp + b}$$

(b) $E = 1$ when

$$-\frac{mp}{mp + b} = 1$$
$$-mp = mp + b$$
$$p = -\frac{b}{2m}$$

13. (a) $E = -\dfrac{dq}{dp} \cdot \dfrac{p}{q}$

$$= -(-r)\frac{k}{p^{r+1}} \frac{p}{k/p^r} = r$$

(b) E is independent of p.
(c) If $r = 1$ the revenue is not affected by the price. If $r > 1$ the revenue is always elastic, whereas if $r < 1$ the revenue is always inelastic. This is an unrealistic model because there should be a price at which the revenue is a maximum.

15. (a) We have two data points: $(p, q) = (2.00, 3000)$ and $(4.00, 0)$. The line through these two points is

$$q = -1500p + 6000.$$

(b) $E = -\dfrac{dq}{dp} \cdot \dfrac{p}{q}$

$$= -(-1500) \frac{p}{-1500p + 6000}$$
$$= \frac{1500p}{-1500p + 6000}$$

$E = 1$ when

$$\frac{1500p}{-1500p + 6000} = 1$$
$$1500p = -1500p + 6000$$
$$p = \$2 \text{ per hamburger.}$$

This gives a total weekly revenue of

$$R = pq$$
$$= 2(-1500 \cdot 2 + 6000) = \$6000.$$

17. $E = \dfrac{dq}{dx} \cdot \dfrac{x}{q}$

$$= 0.01(-0.0156x + 1.5)$$
$$\cdot \frac{x}{0.01(-0.0078x^2 + 1.5x + 4.1)}$$
$$= \frac{-0.0156x^2 + 1.5x}{-0.0078x^2 + 1.5x + 4.1}.$$

When $x = 20$, $E \approx 0.77$: At a family income level of \$20,000, the fraction of children attending a live theatrical performance is increasing by 0.77% per 1% increase in household income.

19. (a) $E = \dfrac{dq}{dx} \cdot \dfrac{x}{q}$

$$= \frac{0.3454}{x} \cdot \frac{x}{0.3454 \ln x - 3.047}$$
$$= \frac{0.3454}{0.3454 \ln x - 3.047}.$$

When $x = 60,000$, $E \approx 0.46$. The demand for computers is increasing by 0.46% per 1% increase in household income.
(b) E decreases as income increases because the denominator of E gets larger. **(c)** Unreliable; it predicts a likelihood greater than 1 at incomes of \$123,000 and above. ($0.3454 \ln x - 3.047 = 1$ when $x = e^{4.047/0.3454} \approx 123,000$) In a model appropriate for large incomes, one would expect the curve to level off at or below 1.
(d) E approaches 0 as x goes to infinity, so for very large x we have $E \approx 0$.

21. The income elasticity of demand is

$$\frac{dQ}{dY} \cdot \frac{Y}{Q} = a\beta P^\alpha Y^{\beta-1} \frac{Y}{aP^\alpha Y^\beta} = \beta.$$

An increase in income of x% will result in an increase in demand of βx%.

23. (a) The data points $(p, q) = (3.00, 407)$ and $(5.00, 223)$ give us the exponential function $q = 1000e^{-0.3p}$.

(b) $E = -\dfrac{dq}{dp} \cdot \dfrac{p}{q}$

$\quad = 300e^{-0.3p} \dfrac{p}{1000e^{-0.3p}} = 0.3p$

At $p = \$3$, $E = 0.9$; at $p = \$4$, $E = 1.2$; and at $p = \$5$, $E = 1.5$.

(c) $E = 1$ when $0.3p = 1$, so $p = \$3.33$.

(d) We first find the price that produces a demand of 200 pounds:

$\quad 1000e^{-0.3p} = 200$

$\quad e^{-0.3p} = 0.2$

$\quad p = -(\ln 0.2)/0.3 = \$5.36$.

Selling at a lower price would increase demand, but you cannot sell more than 200 pounds anyway so your revenue would go down. On the other hand, if you set the price higher than $5.36 the decrease in sales will outweigh the increase in price, which we know because the elasticity at $p = 5.36$ is $E \approx 1.6 > 1$. You should therefore set the price at $5.36 per pound.

25. the price is lowered

27. Start with $R = pq$ and differentiate with respect to p to obtain

$\quad \dfrac{dR}{dp} = q + p\dfrac{dq}{dp}.$

For a stationary point, $dR/dp = 0$, so

$\quad q + p\dfrac{dq}{dp} = 0.$

Rearranging gives

$\quad p\dfrac{dq}{dp} = -q,$ and hence

$\quad -\dfrac{dq}{dp} \cdot \dfrac{p}{q} = 1,$ or

$\quad E = 1,$

showing that stationary points of R correspond to points of unit elasticity.

29. The distinction is best illustrated by an example. Suppose that q is measured in weekly sales and p is the unit price in dollars. Then the quantity $-dq/dp$ measures the drop in weekly sales per $1 increase in price. The elasticity of demand E, on the other hand, measures the *percentage* drop in sales per *one percent* increase in price. Thus, $-dq/dp$ measures absolute change, while E measures fractional, or percentage, change.

Chapter 12 Review Exercises

1. $f(x) = 2x^3 - 6x + 1$ on $[-2, +\infty)$
End points: -2
Stationary points:
$$f'(x) = 6x^2 - 6$$
$f'(x) = 0$ when
$$6x^2 - 6 = 0$$
$$6(x^2 - 1) = 0$$
$$x = \pm 1$$
Singular points:
$f'(x)$ is defined for all x, so no singular points.
Thus, we have stationary points at $x = \pm 1$ and the endpoint $x = -2$. We test a point to the right of 1 as well:

x	-2	-1	1	2
$f(x)$	-3	5	-3	5

The graph must increase from $x = -2$ to $x = -1$, decrease to $x = 1$, and increase from then on.

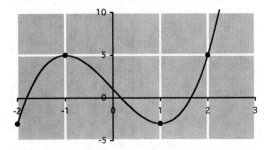

This gives absolute mins at $(-2, -3)$ and $(1, -3)$ and a relative max at $(-1, 5)$.

3. $g(x) = x^4 - 4x$ on $[-1, 1]$
End points: $-1, 1$
Stationary Points:
$$g'(x) = 4x^3 - 4$$
$$= 4(x^3 - 1)$$
$$= 4(x-1)(x^2 + x + 1)$$
$g'(x) = 0$ when
$$4(x-1)(x^2 + x + 1) = 0$$
$$x = 1$$
Singular points: None

x	$y = x^4 - 4x$
-1	5
1	-3

From the chart we see that g has an absolute maximum at $(-1, 5)$ and an absolute minimum at $(1, -3)$.

5. $g(x) = (x-1)^{2/3}$
End points: None.
Stationary points:
$$g'(x) = \tfrac{2}{3}(x-1)^{-1/3}$$
$g'(x)$ is never 0, so no stationary points.
Singular points:
$g'(x)$ is not defined at $x = 1$.
Thus, we have a single critical point at $x = 1$. We test a point on either side:

x	0	1	2
$g(x)$	1	0	1

The graph must decrease until $x = 1$ and then increase.

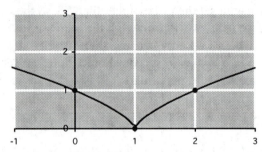

This gives an absolute min at $(1, 0)$.

7. $h(x) = \dfrac{1}{x} + \dfrac{1}{x^2}$
End points: The domain of h is all $x \neq 0$. Thus, there are no end-points of the domain.
Stationary points:
$$h'(x) = -\frac{1}{x^2} - \frac{2}{x^3}$$
$h'(x) = 0$ when
$$2x^2 = -x^3$$
$$x^3 + 2x^2 = 0$$
$$x^2(x + 2) = 0$$
$$x = -2 \;(x = 0 \text{ is not in the domain})$$
Singular points:
$h'(x)$ is defined for all x in the domain of h, so there are no singular points.
Thus, h has a stationary point at $x = -2$. We test points on either side of -2 and points to the right of $x = 0$:

472

x	−3	−2	−1	1	2
$h(x)$	−0.22	−0.25	0	2	0.75

The graph decreases to $x = -2$ then increases approaching the vertical asymptote at $x = 0$. On the other side of the asymptote it decreases.

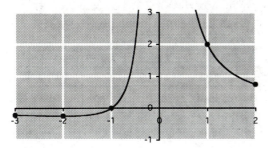

h has an absolute min at $(-2, -1/4)$.

9. Relative max at $x = 1$, point of inflection at $x = -1$.

11. Relative max at $x = -2$: the derivative goes from positive to negative, so the f must go from increasing to decreasing; relative min at $x = 1$: f goes from decreasing to increasing; point of inflection at $x = -1$: a min of f' gives a point of inflection of f.

13. One point of inflection, at $x = 0$. Note that f'' is not defined at $x = 0$, but does change from negative to positive there.

15. $s = \dfrac{2}{3t^2} - \dfrac{1}{t}$

$v = \dfrac{ds}{dt} = -\dfrac{4}{3t^3} + \dfrac{1}{t^2}$

$a = \dfrac{dv}{dt} = \dfrac{4}{t^4} - \dfrac{2}{t^3}$ m/sec^2

(b) At time $t = 1$, acceleration is

$a = \dfrac{4}{(1)^4} - \dfrac{2}{(1)^3} = 2$ m/sec^2

17. $f'(x) = 3x^2 - 12$

Stationary Points:

$f'(x) = 0$ when $3(x^2 - 4) = 0$, $x = \pm 2$; Singular Points: None; $f'(x)$ is always defined.

Possible points of inflection:

$f''(x) = 6x$

$f''(x) = 0$ when $x = 0$; $f''(x)$ is always defined.

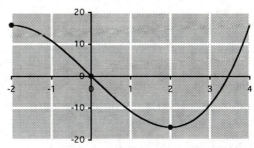

f has a relative max at $(-2, 16)$, an absolute min at $(2, -16)$, and a point of inflection at $(0, 0)$. It has no horizontal or vertical asymptotes.

19. The domain of f includes all numbers except 0.

Stationary points:

$f'(x) = \dfrac{2x \cdot x^3 - (x^2 - 3)(3x^2)}{x^6}$

$= \dfrac{-x^4 + 9x^2}{x^6} = \dfrac{-x^2 + 9}{x^4}$

$f'(x) = 0$ when $x = \pm 3$

Singular points: None; $f'(x)$ is defined for all x in the domain of f.

Inflection points:

$f''(x) = \dfrac{-2x \cdot x^4 - (-x^2 + 9)(4x^3)}{x^8}$

$= \dfrac{2x^5 - 36x^3}{x^8} = \dfrac{2(x^2 - 18)}{x^5}$

$f''(x) = 0$ when $x = \pm\sqrt{18} = \pm 3\sqrt{2}$; $f''(x)$ is defined for all x in the domain of f.

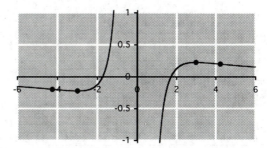

f has a relative min at $(-3, -2/9)$, a relative max at $(3, 2/9)$, and points of inflection at $(-3\sqrt{2}, -5\sqrt{2}/36)$ and $(3\sqrt{2}, 5\sqrt{2}/36)$. f has a vertical asymptote at $x = 0$ and a horizontal asymptote at $y = 0$.

473

21. The domain of g is $x \geq 0$.
Stationary points:

$$g'(x) = \sqrt{x} + (x - 3)\frac{1}{2\sqrt{x}} = \frac{3\sqrt{x}}{2} - \frac{3}{2\sqrt{x}}$$

$g'(x) = 0$ when

$$\sqrt{x} = \frac{1}{\sqrt{x}}, \; x = 1;$$

Singular points: $g'(x)$ is not defined when $x = 0$, so $x = 0$ is a singular point.

Inflection points: $g''(x) = \frac{3}{4\sqrt{x}} + \frac{3}{4x^{3/2}}$;

$g''(x)$ is never 0; $g''(x)$ is not defined when $x = 0$.

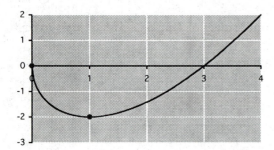

g has a relative max at $(0, 0)$ and an absolute min at $(1, -2)$.

23. Objective:
$$\text{Maximize } R = pq = p(-p^2 + 33p + 9)$$
$$= -p^3 + 33p^2 + 9p$$
End-points: $18, 28$
Stationary points:

$$\frac{dR}{dp} = -3p^2 + 66p + 9 = 0$$

when $p = \dfrac{-66 \pm \sqrt{66^2 - 4(-3)(9)}}{2(-3)}$

$p \approx -0.14$ or 22.14

We reject the negative value and obtain the following table

p	18	22.14	28
$R = -p^3 + 33p^2 + 9p$	5022	5522.61	4172

So the maximum revenue of $5522.61 occurs when $p = \$22.14$ per book

25. (a) Profit $P = R - C = pq - (9q + 100) =$
$p(-p^2 + 33p + 9) - 9(-p^2 + 33p + 9) - 100$
$= -p^3 + 42p^2 - 288p - 181$
(b) $P' = -3p^2 + 84p - 288$
$= -3(p^2 - 28p + 96)$
$= -3(p - 24)(p - 4)$
$P' = 0$ when $p = 4$ or 24. To see which, if either, gives us the maximum profit, we test points on either side:

p	0	4	24	30
$P(p)$	−181	−725	3275	1979

P decreases to a low at $p = 4$ then increases to a maximum at $p = 24$, after which it decreases again. So, the company should charge $24 per copy.
$P(24) = \$3275$
(c) For maximum revenue, the company should charge $22.14 per copy. At this price, the cost per book is decreasing with increasing price, while the revenue is not decreasing (its derivative is zero). Thus, the profit is increasing with increasing price, suggesting that the maximum profit will occur at a higher price. This is, in fact, what we just found.

27. Start by labeling the edges of the box:

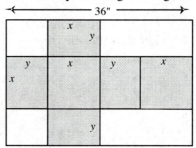

Objective: Maximize $V = xxy = x^2y$
subject to:
$$2x + 2y = 36$$
or $y = 18 - x$
Substitute into the objective
$$V = x^2(18-x) = 18x^2 - x^3$$

$$\frac{dV}{dx} = 36x - 3x^2 = 3x(12 - x)$$

Stationary point occurs at $x = 0, \; 12$
$x = 12$ in gives the maximum volume of
$$V = 18(12)^2 - (12)^3 = 864 \text{ cubic inches}$$

The height of the box is
$y = 18 - x = 18 - 12 = 6$ in.

29. (a) $E = -(-2p + 33)\, \dfrac{p}{-p^2 + 33p + 9}$

$= \dfrac{2p^2 - 33p}{-p^2 + 33p + 9}$

(b) $E(20) \approx 0.52$ and $E(25) \approx 2.03$. When the price is \$20, demand is dropping at a rate of 0.52% per 1% increase in the price; when the price is \$25, demand is dropping at a rate of 2.03% per 1% increase in the price.

(c) $E = 1$ when $2p^2 - 33p = -p^2 + 33p + 9$
$3p^2 - 66p - 9 = 0$
$p^2 - 22p - 3 = 0$
so $p \approx \$22.14$ per book (using the quadratic formula; the other solution is negative so we reject it).

31. (a) $E = -\dfrac{dq}{dp} \cdot \dfrac{p}{q}$

$= -(-2p+1)1000e^{-p^2+p} \cdot \dfrac{p}{1000e^{-p^2+p}}$

$= -(-2p+1)p = 2p^2 - p$

(b) $E(2) = 2(2)^2 - 2 = 6$
Interpretation: The demand is dropping at a rate of 6% per 1% increase in the price.

(c) For maximum revenue, $E = 1$:
$2p^2 - p = 1$
$2p^2 - p - 1 = 0$
$(2p+1)(p-1) = 0$
$p = \$1.00$ (reject the negative solution)
For the revenue,

$R = pq = p\,1000e^{-p^2+p}$
$R(1) = 1000e^{-1+1} = \$1000$ per month

33. (a) Weekly sales are growing fastest when the rate of change s' is a maximum. From the graph of s', we see that this occurred at about week 5 (see also the graph of s'' which becomes zero at that point).

(b) This point (a maximum in the graph of s') corresponds to a point of inflection on the graph of s. On the graph of s', that point is given by the

maximum, and by the t-intercept in the graph of s''.

(c) The graph appears to level of around $s = 10{,}500$; If weekly sales continue as predicted by the model, they will level off at around 10,500 books per week in the long term.

(d) The graph of s' appears to level off at $s' = 0$. If weekly sales continue as predicted by the model, the rate of change of sales approaches zero in the long term.

35. Let h be the distance between Marjory Duffin and John O'Hagan. Let y be Marjory Duffin's distance from the corner, and let x be John O'Hagan's distance. Then
$x^2 + y^2 = h^2$

$2x\dfrac{dx}{dt} + 2y\dfrac{dy}{dt} = 2h\dfrac{dh}{dt}$

(a) $2(2)(5) + 2(2)(5) = 2(2\sqrt{2})\dfrac{dh}{dt}$

$40 = 4\sqrt{2}\dfrac{dh}{dt}$

$\dfrac{dh}{dt} = \dfrac{10}{\sqrt{2}}$ ft/sec

(b) $2(1)(5) + 2(1)(5) = 2\sqrt{2}\dfrac{dh}{dt}$

$20 = 2\sqrt{2}\dfrac{dh}{dt}$

$\dfrac{dh}{dt} = \dfrac{10}{\sqrt{2}}$ ft/sec

(c) $2(h)(5) + 2(h)(5) = 2h\sqrt{2}\dfrac{dh}{dt}$

$20h = 2h\sqrt{2}\dfrac{dh}{dt}$

$\dfrac{dh}{dt} = \dfrac{10}{\sqrt{2}}$ ft/sec

(d) Since the answer to part (c) is independent of h, it also must hold as $h \to 0$, giving the same answer as parts (a) through (c).

Chapter 13 The Integral

13.1 The Indefinite Integral

1. $\int x^n \, dx = \dfrac{x^{n+1}}{n+1} + C; \; n = 5$

$\int x^5 \, dx = \dfrac{x^6}{6} + C$

3. $\int 6 \, dx = 6 \int 1 \, dx = 6x + C$

5. $\int x^n \, dx = \dfrac{x^{n+1}}{n+1} + C; \; n = 1$

$\int x \, dx = \dfrac{x^2}{2} + C$

7. $\int (x^2 - x) \, dx = \int x^2 \, dx - \int x \, dx$

$= \dfrac{x^3}{3} - \dfrac{x^2}{2} + C$

9. $\int (1 + x) \, dx = \int 1 \, dx + \int x \, dx$

$= x + \dfrac{x^2}{2} + C$

11. $\int x^n \, dx = \dfrac{x^{n+1}}{n+1} + C; \; n = -5$

$\int x^{-5} \, dx = \dfrac{x^{-4}}{(-4)} + C = -\dfrac{x^{-4}}{4} + C$

13. $\int (x^{2.3} + x^{-1.3}) \, dx$

$= \int x^{2.3} \, dx + \int x^{-1.3} \, dx$

$= \dfrac{x^{3.3}}{3.3} - \dfrac{x^{-0.3}}{0.3} + C$

15. $\int (u^2 - 1/u) \, du$

$= \int u^2 \, du - \int u^{-1} \, du$

$= \dfrac{u^3}{3} - \ln|u| + C$

17. $\int \sqrt[4]{x} \, dx = \int x^{1/4} \, dx = \dfrac{x^{5/4}}{5/4} + C = \dfrac{4x^{5/4}}{5} + C$

19. $\int (3x^4 - 2x^{-2} + x^{-5} + 4) \, dx$

$= 3 \int x^4 \, dx - 2 \int x^{-2} \, dx + \int x^{-5} \, dx + \int 4 \, dx$

$= \dfrac{3x^5}{5} + 2x^{-1} - \dfrac{x^{-4}}{4} + 4x + C$

21. $\int \left(\dfrac{2}{u} + \dfrac{u}{4} \right) du$

$= \int (2u^{-1} + (1/4)u) \, du$

$= 2\ln|u| + \dfrac{1}{4} \dfrac{u^2}{2} + C$

$= 2\ln|u| + u^2/8 + C$

23. $\int \left(\dfrac{1}{x} + \dfrac{2}{x^2} - \dfrac{1}{x^3} \right) dx$

$= \int (x^{-1} + 2x^{-2} - x^{-3}) \, dx$

$= \ln|x| + 2\dfrac{x^{-1}}{-1} - \dfrac{x^{-2}}{-2} + C$

$= \ln|x| - \dfrac{2}{x} + \dfrac{1}{2x^2} + C$

476

25. $\int (3x^{0.1} - x^{4.3} - 4.1)\, dx$

$= \dfrac{3x^{1.1}}{1.1} - \dfrac{x^{5.3}}{5.3} - 4.1x + C$

27. $\int \left(\dfrac{3}{x^{0.1}} - \dfrac{4}{x^{1.1}} \right) dx$

$= \int (3x^{-0.1} - 4x^{-1.1})\, dx$

$= 3\dfrac{x^{0.9}}{0.9} - 4\dfrac{x^{-0.1}}{-0.1} + C$

$= \dfrac{x^{0.9}}{0.3} + \dfrac{40}{x^{0.1}} + C$

29. $\int \left(5.1t - \dfrac{1.2}{t} + \dfrac{3}{t^{1.2}} \right) dt$

$= \int (5.1t - 1.2t^{-1} + 3t^{-1.2})\, dt$

$= \dfrac{5.1t^2}{2} - 1.2 \ln |t| + 3\dfrac{t^{-0.2}}{-0.2} + C$

$= 2.55t^2 - 1.2 \ln |t| - \dfrac{15}{t^{0.2}} + C$

31. $\int (2e^x + 5/x + 1/4)\, dx$

$= 2e^x + 5\ln|x| + \dfrac{x}{4} + C$

33. $\int \left(\dfrac{6.1}{x^{0.5}} + \dfrac{x^{0.5}}{6} - e^x \right) dx$

$= \int \left(6.1x^{-0.5} + \dfrac{x^{0.5}}{6} - e^x \right) dx$

$= 6.1 \dfrac{x^{0.5}}{0.5} + \dfrac{x^{1.5}}{6 \cdot 1.5} - e^x + C$

$= 12.2x^{0.5} + \dfrac{x^{1.5}}{9} - e^x + C$

35. $\int (2^x - 3^x)\, dx$

$= \dfrac{2^x}{\ln 2} - \dfrac{3^x}{\ln 3} + C$

37. $\int 100(1.1^x)\, dx$

$= \dfrac{100(1.1^x)}{\ln(1.1)} + C$

39. $\int \dfrac{x+2}{x^3}\, dx = \int \left(\dfrac{x}{x^3} + \dfrac{2}{x^3} \right) dx$

$= \int \left(\dfrac{1}{x^2} + \dfrac{2}{x^3} \right) dx$

$= \int (x^{-2} + 2x^{-3})\, dx$

$= -x^{-1} - x^{-2} = -\dfrac{1}{x} - \dfrac{1}{x^2} + C$

41. $f'(x) = x$, so

$f(x) = \int x\, dx = \dfrac{x^2}{2} + C.$

$f(0) = 1$, so

$\dfrac{0^2}{2} + C = 1$

$C = 1$

So, $f(x) = \dfrac{x^2}{2} + 1.$

43. $f'(x) = e^x - 1$, so

$f(x) = \int (e^x - 1)\, dx = e^x - x + C.$

$f(0) = 0$, so

$e^0 - 0 + C = 0$

$C = -1$

So, $f(x) = e^x - x - 1.$

45. $\int (3|x| - x)\, dx = 3(\tfrac{1}{2}x|x|) - x^2/2 + C$

$= \dfrac{3x|x| - x^2}{2} + C$

47. $C'(x) = 5 - \dfrac{x}{10,000}$

$$C(x) = \int \left(5 - \dfrac{x}{10,000}\right) dx$$

$$= 5x - \dfrac{x^2}{20,000} + K$$

$C(0) = 20,000$, so

$\quad 0 - 0 + K = 20,000$.

$\quad K = 20,000$

$$C(x) = 5x - \dfrac{x^2}{20,000} + 20,000.$$

49. $C'(x) = 5 + 2x + \dfrac{1}{x}$

$$C(x) = \int \left(5 + 2x + \dfrac{1}{x}\right) dx$$

$$= 5x + x^2 + \ln x + K$$

(Note that $x > 0$, so there is no need to write $|x|$.)

$\quad C(1) = 1000$, so $5 + 1 + \ln 1 + K = 1000$,

$K = 994$. $C(x) = 5x +$

$x^2 + \ln x + 994$.

51. (a) We are given

$\quad M'(t) = m(t) = 12t^2 - 20t + 10$

so $\quad M(t) = \int (12t^2 - 20t + 10) \, dt$

$$\quad = 12t^3/3 - 20t^2/2 + 10t + C$$

$$\quad = 4t^3 - 10t^2 + 10t + C$$

To find C we use the data from 2005 ($t = 0$):

$\quad M(0) = 1$

$\quad 4(0)^3 - 10(0)^2 + 10(0) + C = 1$

$\quad C = 1$

Thus $\quad M(t) = 4t^3 - 10t^2 + 10t + 1$

(b) Midway through 2008, $t = 3.5$, so membership was

$\quad M(3.5) = 4(3.5)^3 - 10(3.5)^2 + 10(3.5) +$

1

$\quad = 85$ million members

53. The rate of change of $I(t)$ is \$1200 per year;

$I'(t) = 1200$. So,

$I(t) = \int 1200 \, dt = 1200t + C$

$42,000 = I(0) = C$.

$I(t) = 42,000 + 1200t$

$I(5) = \$48,000$

55. We are given

$\quad S'(t) = s(t) = 12t^2 + 500t + 4700$

so $\quad S(t) = \int (12t^2 + 500t + 4700) \, dt$

$$\quad = 12t^3/3 + 500t^2/2 + 4700t + C$$

$$\quad = 4t^3 + 250t^2 + 4700t + C$$

Since $S(t)$ represents sales since 2000, it follows that $S(0) = 0$ (Sales since 2000 are zero in 2000.)

$\quad 4(0)^3 + 250(0)^2 + 4700(0) + C = 0$

$\quad C = 0$

Thus $\quad S(t) = 4t^3 + 250t^2 + 4700t$

At the end of 2005 (start of 2006), $t = 6$

$\quad S(6) = 4(6)^3 + 250(6)^2 + 4700(6)$

$\quad = 38,064$ million gallons

57. (a) $m = (100 - 65)/10 = 3.5$, so

$\quad H'(t) = 3.5t + 65$ billion dollars per year.

(b) $H(t) = \int (3.5t + 65) \, dt$

$$\quad = \dfrac{3.5}{2} t^2 + 65t + C$$

$H(0) = 700 = C$

Thus, $H(t) = 1.75t^2 + 65t + 700$ billion dollars

59. (a) From the quick example referred to in the hint:

Rate of change (velocity) of the percentage

$$v(t) = \int a(t) \, dt = \int -0.4 \, dt = -0.4t + C$$

To find C, use the fact that at time $t = 0$ $v = 1$ percentage point per year, so $v(0) = 1$, giving $C = 1$:

$\quad v(t) = -0.4t + 1$ points per year.

(b) From the quick example referred to in the hint:

Value of percentage

$$p(t) = \int v(t) \, dt = \int (-0.4t + 1) \, dt$$

$$= -0.2t^2 + t + C$$

To find C, use the fact that at time $t = 0$ $p = 13$ percentage points, so $p(0) = 13$, giving $C = 13$:

$$p(t) = -0.2t^2 + t + 13$$

In 2008, $t = 1$, so

$$p(1) = -0.2(1)^2 + 1 + 13 = 13.8\%$$

61. $v(t) = t^2 + 1$

(a) $s(t) = \int v(t) \, dt = \int (t^2 + 1) \, dt = \dfrac{t^3}{3} + t + C$

(b) $0 + 0 + C = 1$ so

$$s = \dfrac{t^3}{3} + t + 1$$

63. $a(t) = -32$

$$v(t) = \int a(t) \, dt = \int (-32) \, dt$$

$$= -32t + C$$

$v(0) = 0$ is given, so

$$0 = 0 + C$$
$$C = 0$$

and so $v(t) = -32t$.

After 10 seconds,

$$v(10) = -32(0) = -320$$

so the stone is traveling 320 ft/s downward.

65. (a) $v(t) = \int a(t) \, dt = \int (-32) \, dt = -32t + C$

At time $t = 0$ $v = 16$ ft/sec, so

$$16 = -32(0) + C$$
$$C = 16$$

giving

$$v(t) = -32t + 16$$

(b) $s(t) = \int v(t) \, dt = \int (-32t + 16) \, dt$

$$= -16t^2 + 16t + C$$

At time $t = 0$ $s = 185$ ft, so

$$185 = -16(0)^2 + 16(0) + C$$
$$C = 185$$

giving

$$s(t) = -16t^2 + 16t + 185$$

It reaches its zenith when $v(t) = 0$

$$-32t + 16 = 0$$
$$t = 16/32 = 0.5 \text{ sec}$$

Its height at that moment is

$$s(0.5) = -16(0.5)^2 + 16(0.5) + 185 = 189 \text{ feet,}$$

4 feet above the top of the tower.

67. (a) The ground speed is

$$v = 500 + 25 + 50t$$
$$= 525 + 50t.$$

Thus, $\dfrac{ds}{dt} = 525 + 50t$.

where s is the total distance traveled. Thus the total distance traveled is

$$s = \int [525 + 50t] \, dt$$

$$= 525t + 25t^2 + C$$

To obtain C, use the information that, at time $t = 0$, $s = 0$ as well (zero distance was traveled at time $t = 0$). Thus,

$$0 = 525(0) + 25(0)^2 + C$$

so $C = 0$,

and $s = 525t + 25t^2$

(b) At the end of a 1800-mile trip, $s = 1800$ and so, by (a)

$$1800 = 525t + 25t^2.$$

To obtain t we solve the quadratic

$$25t^2 + 525t - 1800 = 0$$

Dividing by 25 gives

$$t^2 + 21t - 72 = 0$$
$$(t + 24)(t - 3) = 0$$

So, rejecting the negative solution gives

$$t = 3 \text{ hours.}$$

(c) The negative solution indicates that, at time $t = -24$, the tail-wind would have been large and negative, causing the plane to be moving backwards through that position 24 hours prior to departure and arrive at the starting point of the flight at time 0!

69. From the formulas at the end of the section in the textbook,

$$v(t) = -32t + v_0.$$

The projectile has zero velocity when

$$v(t) = 0$$
$$0 = -32t + v_0$$
$$t = v_0/32.$$

This is when it reaches its highest point.

71. By Exercise 70, the ball reaches a maximum height of $v_0^2/64$ feet. Thus,

$$v_0^2/64 = 20$$
$$v_0 = (1280)^{1/2} \approx 35.78 \text{ ft/s}$$

73. (a) By Exercise 70, the chalk reaches a maximum height of $v_0^2/64$ feet. Thus,

$$v_0^2/64 = 100$$
$$v_0 = (6400)^{1/2} = 80 \text{ ft/s}$$

(b) $s(t) = -16t^2 + v_0 t + s_0$.
If we take the starting height as 0,

$$s_0 = 0, \text{ and so}$$
$$s(t) = -16t^2 + 100t.$$

The chalk strikes the ceiling when

$$s(t) = 100, \quad -16t^2 + 100t = 100$$
$$16t^2 - 100t + 100 = 0$$
$$4(4t - 5)(t - 5) = 0$$
$$t = 1.25 \text{ or } 5.$$

We take the first solution, which is the first time it strikes the ceiling. Now,

$$v(t) = -32t + v_0 = -32t + 100,$$

so the velocity when it strikes the ceiling is

$$v(1.25) = 60 \text{ ft/s}.$$

(c) Start with $s(0) = 100$ and $v(0) = -60$. So,

$$v(t) = -32t + v_0 = -32t - 60.$$

We have

$$s(t) = -16t^2 + v_0 t + s_0$$
$$= -16t^2 - 60t + 100.$$

Now we find when $s(t) = 0$:

$$-16t^2 - 60t + 100 = 0$$
$$-4(4t - 5)(t + 5) = 0$$
$$t = 1.25 \text{ or } -5$$

We take the positive solution and say that it takes 1.25 seconds to hit the ground.

75. Let v_0 be the speed at which Prof. Strong throws and let w_0 be the speed at which Prof. Weak throws. We have

$$v_0^2/64 = 2w_0^2/64$$

so

$$v_0 = w_0\sqrt{2}.$$

Thus, Prof. Strong throws $\sqrt{2} \approx 1.414$ times as fast as Prof. Weak.

77. The term *indefinite* refers to the arbitrary constant term in the indefinite integral; we do not obtain a definite value for C and hence the integral is "not definite."

79. Constant; since the derivative of a linear function is constant, linear functions are antiderivatives of constant functions.

81. No; there are infinitely many antiderivatives of a given function, each pair of them differing by a constant. Knowing the value of the function at a specific point suffices.

83. They differ by a constant,

$$G(x) - F(x) = \text{Constant}.$$

85. Antiderivative; marginal

87. $\int f(x)\, dx$ represents the total cost of manufacturing x items. The units of $\int f(x)\, dx$ are the product of the units of $f(x)$ and the units of x.

89. $\int [f(x) + g(x)]\, dx$ is, by definition, an antiderivative of $f(x) + g(x)$. Let $F(x)$ be an antiderivative of $f(x)$ and let $G(x)$ be an antiderivative of $g(x)$. Then, because the derivative of $F(x) + G(x)$ is $f(x) + g(x)$ (by the rule for sums of derivatives), $F(x) + G(x)$ is an antiderivative of $f(x) + g(x)$. In symbols,

$$\int [f(x) + g(x)]\, dx = F(x) + G(x) + C$$

$$= \int f(x)\, dx + \int g(x)\, dx$$

the sum of the indefinite integrals.

91. Answers will vary

$$\int x \cdot 1 \ dx \ = \ \int x \ dx \ = \ \frac{x^2}{2} \ + \ C,$$

whereas

$$\int x \ dx \cdot \int 1 \ dx \ = \ \left(\frac{x^2}{2} + D \right) \cdot (x + E),$$

which is not the same as $\frac{x^2}{2}$ + C, no matter what values we choose for the constants C, D and E.

93. If you take the <u>derivative</u> of the <u>indefinite integral</u> of $f(x)$, you obtain $f(x)$ back. On the other hand, if you take the <u>indefinite integral</u> of the <u>derivative</u> of $f(x)$, you obtain $f(x)+C$.

13.2 Substitution

1. $u = 3x-5$

$du/dx = 3$

$dx = \frac{1}{3}du$

$\int(3x-5)^3 \, dx = \int u^3 \, \frac{1}{3}du$

$= \frac{1}{3}\frac{u^4}{4} + C = u^4/12 + C$

$= (3x-5)^4/12 + C$

3. $\int(3x-5)^3 \, dx = \frac{(3x-5)^4}{(3)(4)} + C$

$= (3x-5)^4/12 + C$

5. $u = -x, \ du/dx = -1, \ dx = -du; \ \int e^{-x} \, dx$

$= -\int e^u \, du = -e^u + C = -e^{-x} + C$

7. $\int e^{-x} \, dx = \frac{1}{-1}e^{-x} + C = -e^{-x} + C$

9. $u = (x+1)^2, \ du/dx = 2(x+1), \ dx = \frac{du}{2(x+1)}$

$\int(x+1)e^{(x+1)^2} \, dx = \int(x+1)e^u \, \frac{du}{2(x+1)}$

$= \int\frac{1}{2}e^u \, du = \frac{1}{2}e^u + C = \frac{1}{2}e^{(x+1)^2} + C$

11. $u = 3x + 1, \ du = 3dx, \ dx = \frac{1}{3}du$

$\int(3x + 1)^5 \, dx = \int u^5 \, \frac{1}{3} \, du = \frac{u^6}{18} + C$

$= \frac{(3x + 1)^6}{18} + C$

13. $u = -2x + 2, \ du = -2 \, dx, \ dx = -\frac{1}{2} \, du$

$\int(-2x + 2)^{-2} \, dx = -\int u^{-2} \, \frac{1}{2} \, du = \frac{u^{-1}}{2} + C$

$= \frac{(-2x + 2)^{-1}}{2} + C$

15. $u = 3x - 4, \ du = 3 \, dx, \ dx = \frac{1}{3} \, du$

$\int 7.2\sqrt{3x - 4} \, dx = \int 7.2\sqrt{u} \, \frac{1}{3} \, du$

$= 2.4 \int u^{1/2} \, du = 2.4 \, \frac{u^{3/2}}{3/2} + C$

$= 1.6(3x - 4)^{3/2} + C$

17. $u = 0.6x + 2, \ du = 0.6 \, dx, \ dx = \frac{1}{0.6} \, du$

$\int 1.2e^{(0.6x+2)} \, dx = \int 1.2e^u \, \frac{1}{0.6} \, du = 2e^u + C$

$= 2e^{(0.6x+2)} + C$

19. $u = 3x^2 + 3, \ du = 6x \, dx, \ dx = \frac{1}{6x} \, du$

$\int x(3x^2 + 3)^3 \, dx = \int xu^3 \, \frac{1}{6x} \, du$

$= \frac{1}{6} \int u^3 \, du = \frac{u^4}{12} + C = \frac{(3x^2 + 3)^4}{24} + C$

21. $u = x^2 + 1, \ du = 2x \, dx, \ dx = \frac{1}{2x} \, du$

$\int x(x^2 + 1)^{1.3} \, dx = \int xu^{1.3} \, \frac{1}{2x} \, du$

$= \frac{1}{2} \int u^{1.3} \, du = \frac{u^{2.3}}{4.6} + C$

$= \frac{(x^2 + 1)^{2.3}}{4.6} + C$

23. $u = 3.1x - 2$, $du = 3.1\ dx$, $dx = \frac{1}{3.1}\ du$

$$\int (1 + 9.3e^{3.1x-2})\ dx = \int dx +$$

$$\int 9.3e^{3.1x-2}\ dx$$

$$= x + \int 9.3e^u \frac{1}{3.1}\ du = x + 3e^u + C$$

$$= x + 3e^{3.1x-2} + C$$

25. $u = 3x^2 - 1$, $du = 6x\ dx$, $dx = \frac{1}{6x}\ du$

$$\int 2x\sqrt{3x^2 - 1}\ dx = \int 2x\sqrt{u}\ \frac{1}{6x}\ du$$

$$= \frac{1}{3} \int u^{1/2}\ du = \frac{u^{3/2}}{9/2} + C$$

$$= \frac{2}{9}(3x^2 - 1)^{3/2} + C$$

27. $u = -x^2 + 1$, $du = -2x\ dx$, $dx = -\frac{1}{2x}\ du$

$$\int xe^{-x^2+1}\ dx = -\int xe^u \frac{1}{2x}\ du = -\frac{1}{2}\int e^u\ du$$

$$= -\frac{1}{2}e^u + C = -\frac{1}{2}e^{-x^2+1} + C$$

29. $u = -(x^2 + 2x)$, $du = -(2x + 2)\ dx$,

$$dx = -\frac{1}{2(x + 1)}\ du$$

$$\int (x + 1)e^{-(x^2+2x)}\ dx$$

$$= -\int (x + 1)e^u \frac{1}{2(x + 1)}\ du$$

$$= -\frac{1}{2}\int e^u\ du = -\frac{1}{2}e^u + C$$

$$= -\frac{1}{2}e^{-(x^2+2x)} + C$$

31. $u = x^2 + x + 1$, $du = (2x + 1)\ dx$,

$$dx = \frac{1}{2x + 1}\ du$$

$$\int \frac{-2x - 1}{(x^2 + x + 1)^3}\ dx = \int \frac{-2x - 1}{u^3} \cdot \frac{1}{2x + 1}\ du$$

$$= -\int u^{-3}\ du = -\frac{u^{-2}}{-2} + C$$

$$= \frac{(x^2 + x + 1)^{-2}}{2} + C$$

33. $u = 2x^3 + x^6 - 5$, $du = (6x^2 + 6x^5)\ dx$

$$dx = \frac{1}{6(x^2 + x^5)}\ du$$

$$\int \frac{x^2 + x^5}{\sqrt{2x^3 + x^6 - 5}}\ dx$$

$$= \int \frac{x^2 + x^5}{\sqrt{u}} \cdot \frac{1}{6(x^2 + x^5)}\ du = \frac{1}{6}\int u^{-1/2}\ du$$

$$= \frac{1}{6} \cdot \frac{u^{1/2}}{1/2} + C = \frac{1}{3}(2x^3 + x^6 - 5)^{1/2} + C$$

$$= \frac{1}{3}\sqrt{2x^3 + x^6 - 5} + C$$

35. $u = x - 2$, $du = dx$

$$\int x(x - 2)^5\ dx = \int xu^5\ dx$$

To remove the remaining x, solve for x in terms of u in the expression for u:

$u = x - 2$, so $x = u + 2$

The above integral is then

$$\int xu^5\ dx = \int (u + 2)u^5\ du$$

$$= \int (u^6 + 2u^5)\ du = \frac{1}{7}u^7 +$$

$$\frac{1}{3}u^6 + C = \frac{1}{7}(x - 2)^7 + \frac{1}{3}(x - 2)^6 + C$$

37. $u = x + 1$, $du = dx$

$$\int 2x\sqrt{x + 1}\ dx = \int 2x\sqrt{u}\ dx$$

To remove the remaining x, solve for x in terms of u in the expression for u:

$u = x + 1$, so $x = u - 1$

The above integral is then

$$\int 2x\sqrt{u} \, dx = \int 2(u - 1)\sqrt{u} \, du$$

$$= 2 \int (u^{3/2} - u^{1/2}) \, du$$

$$= 2 \frac{u^{5/2}}{5/2} - 2 \frac{u^{3/2}}{3/2} + C$$

$$= \frac{4}{5}(x + 1)^{5/2} - \frac{4}{3}(x + 1)^{3/2} + C$$

39. $u = 1 - e^{-0.05x}$, $du = 0.05e^{-0.05x} \, dx$

$$dx = \frac{1}{0.05e^{-0.05x}} \, du$$

$$\int \frac{e^{-0.05x}}{1 - e^{-0.05x}} \, dx$$

$$= \int \frac{e^{-0.05x}}{u} \cdot \frac{1}{0.05e^{-0.05x}} \, du$$

$$= 20 \int \frac{1}{u} \, du = 20\ln|u| + C$$

$$= 20\ln|1 - e^{-0.05x}| + C$$

41. $u = -\frac{1}{x}$, $du = \frac{1}{x^2} \, dx$, $dx = x^2 \, du$

$$\int \frac{3e^{-1/x}}{x^2} \, dx = \int \frac{3e^u}{x^2} x^2 \, du = \int 3e^u \, du$$

$$= 3e^u + C = 3e^{-1/x} + C$$

43. $\int \frac{e^x + e^{-x}}{2} \, dx = \int \frac{e^x}{2} + \frac{e^{-x}}{2} \, dx$

$$= \frac{e^x}{2} + \int \frac{e^{-x}}{2} \, dx$$

let $u = -x$, $du = -dx$, $dx = -du$

$$\frac{e^x}{2} + \int \frac{e^{-x}}{2} \, dx = \frac{e^x}{2} - \int \frac{e^u}{2} \, du$$

$$= \frac{e^x}{2} - \frac{e^u}{2} + C = \frac{e^x - e^{-x}}{2} + C$$

45. $u = e^x + e^{-x}$, $du = (e^x - e^{-x}) \, dx$

$$dx = \frac{1}{e^x - e^{-x}} \, du$$

$$\int \frac{e^x - e^{-x}}{e^x + e^{-x}} \, dx$$

$$= \int \frac{e^x - e^{-x}}{u} \cdot \frac{1}{e^x - e^{-x}} \, du$$

$$= \int \frac{1}{u} \, du = \ln|u| + C = \ln|e^x + e^{-x}| + C$$

47. $\int [(2x - 1)e^{2x^2 - 2x} + xe^{x^2}] \, dx$

$$= \int (2x - 1)e^{2x^2 - 2x} \, dx + \int xe^{x^2} \, dx \; .$$

For the first integral, let
$$u = 2x^2 - 2x, \; du = (4x - 2) \, dx,$$

$$dx = \frac{1}{2(2x - 1)} \, du$$

$$\int (2x - 1)e^{2x^2 - 2x} \, dx$$

$$= \int (2x - 1)e^u \frac{1}{2(2x - 1)} \, du$$

$$= \frac{1}{2} \int e^u \, du = \frac{1}{2} e^u + C = \frac{1}{2} e^{2x^2 - 2x} + C.$$

For the second integral, let
$$u = x^2, \; du = 2x \, dx, \; dx = \frac{1}{2x} \, du$$

$$\int xe^{x^2} \, dx = \int xe^u \frac{1}{2x} \, du = \frac{1}{2} \int e^u \, du$$

$$= \frac{1}{2} e^u + C = \frac{1}{2} e^{x^2} + C.$$

So, $\int [(2x - 1)e^{2x^2 - 2x} + xe^{x^2}] \, dx$

$$= \frac{1}{2} e^{2x^2 - 2x} + \frac{1}{2} e^{x^2} + C.$$

49. Let $u = ax + b$, $du = a \, dx$, $dx = \frac{1}{a} \, du$

$$\int (ax + b)^n \, dx = \int u^n \frac{1}{a} \, du$$

$$\frac{u^{n+1}}{a(n + 1)} + C = \frac{(ax + b)^{n+1}}{a(n + 1)} + C$$

(if $n \neq -1$)

484

51. Let $u = ax + b$, $du = a\,dx$, $dx = \dfrac{1}{a}\,du$

$$\int |ax + b|\ dx = \int |u|\,\frac{1}{a}\,du$$

$$= \frac{1}{a}\int |u|\,du = \frac{1}{a}\cdot\frac{1}{2}\,u|u| + C$$

(from Exercise 45 of Section 13.1)

$$= \frac{1}{2a}\,u|u| + C$$

$$= \frac{1}{2a}\,(ax + b)|ax + b| + C$$

53. $\int e^{ax+b}\ dx = \dfrac{1}{a}\,e^{ax+b} + C$

$$\int e^{-x}\ dx = \frac{1}{-1}e^{-x} + C = -e^{-x} + C$$

55. $\int e^{ax+b}\ dx = \dfrac{1}{a}\,e^{ax+b} + C$

$$\int e^{2x-1}\ dx = \tfrac{1}{2}\,e^{2x-1} + C$$

57. $\int (ax+b)^n\ dx = \dfrac{(ax+b)^{n+1}}{a(n+1)} + C$

$$\int (2x+4)^2\ dx = \frac{(2x+4)^3}{2(3)} + C = \frac{(2x+4)^3}{6} + C$$

59. $\int (ax+b)^{-1}\ dx = \dfrac{1}{a}\,\ln|ax+b| + C$

$$\int \frac{1}{5x-1}\ dx = \int (5x-1)^{-1}\ dx$$

$$= \tfrac{1}{5}\,\ln|5x - 1| + C$$

61. $\int (ax+b)^n\ dx = \dfrac{(ax+b)^{n+1}}{a(n+1)} + C$

$$\int (1.5x)^3\ dx = \frac{(1.5x)^4}{1.5(4)} = \tfrac{1}{6}\,(1.5x)^4 + C$$

63. $\int c^{ax+b}\ dx = \dfrac{1}{a\,\ln c}\,c^{ax+b} + C$

$$\int 1.5^{3x}\ dx = \frac{1.5^{3x}}{3\,\ln(1.5)} + C$$

65. By the shortcut,

$$\int |2x+4|\ dx = \frac{1}{2(2)}\,(2x+4)|2x+4| + C$$

$$= \tfrac{1}{4}(2x+4)|2x+4| + C$$

67. $\int c^{ax+b}\ dx = \dfrac{1}{a\,\ln c}\,c^{ax+b} + C$

$$\int (2^{3x+4} + 2^{-3x+4})\ dx = \frac{2^{3x+4}}{3\,\ln 2} + \frac{2^{-3x+4}}{-3\,\ln 2}$$

$$= \frac{2^{3x+4} - 2^{-3x+4}}{3\,\ln 2} + C$$

69. $f'(x) = x(x^2+1)^3$

So, $f(x) = \int x(x^2 + 1)^3\ dx$

$u = x^2 + 1$, $du = 2x\,dx$, $dx = \dfrac{1}{2x}\,du$

$$\int x(x^2 + 1)^3\ dx = \int xu^3\,\frac{1}{2x}\,du = \tfrac{1}{2}\int u^3\,du$$

$$= \tfrac{1}{8}\,u^4 + C = \tfrac{1}{8}\,(x^2 + 1)^4 + C.$$

$f(0) = 0$, so

$$\tfrac{1}{8}\,(0 + 1)^4 + C = 0$$

$$C = -\tfrac{1}{8}\,.$$

and so $f(x) = \tfrac{1}{8}\,(x^2 + 1)^4 - \tfrac{1}{8}$

71. $f'(x) = xe^{x^2-1}$

So, $f(x) = \int xe^{x^2-1}\ dx$

$u = x^2 - 1$, $du = 2x\, dx$, $dx = \dfrac{1}{2x}\, du$

$$\int xe^{x^2-1}\, dx = \int xe^u \dfrac{1}{2x}\, du = \dfrac{1}{2}\int e^u\, du$$

$$= \dfrac{1}{2}e^u + C = \dfrac{1}{2}e^{x^2-1} + C.$$

$f(1) = 1/2$, so

$$\dfrac{1}{2}e^{1-1} + C = 1/2$$

$$C = 0$$

and so $f(x) = \dfrac{1}{2}e^{x^2-1}$

73. $V(t) = \displaystyle\int (210 - 62e^{-0.05t})\, dt$

$$= 210t - \dfrac{62}{-0.05}e^{-0.05t} + C$$

by the shortcut

$$= 210t + 1240e^{-0.05t} + C$$

$V(0) = 0$ gives

$$0 = 210(0) + 1240e^{-0.05(0)} + C$$

$$0 = 1240 + C$$

$C = -1240$, so

$$V(t) = 210t + 1240e^{-0.05t} - 1240$$

75. $R(t) = \displaystyle\int (825 - 240e^{-0.4t})\, dt$

$$= 825t - \dfrac{240}{-0.4}e^{-0.4t} + C \text{ by the shortcut}$$

$$= 825t + 600e^{-0.4t} + C$$

$R(0) = 0$ gives

$$0 = 825(0) + 600e^{-0.4(0)} + C$$

$$0 = 600 + C$$

$C = -600$, so

$$R(t) = 825t + 600e^{-0.4t} - 600$$

The period 1996–2006 is represented by $[0, 10]$, so the total revenue earned is

$$R(10) = 825(10) + 600e^{-0.4(10)} - 600$$

$$\approx \$7660 \text{ million}$$

77. $C'(x) = 5 + 1/(x+1)^2$

So, $C(x) = \displaystyle\int [5 + (x+1)^{-2}]\, dx$

$$= \int 5\, dx + = \int (x+1)^{-2}\, dx$$

$$= 5x - (x+1)^{-1} + K.$$

(We used the shortcut formula

$$\int (ax+b)^n\, dx = \dfrac{(ax+b)^{n+1}}{a(n+1)} + K$$

to do the second integal)

$C(1) = 1000$, so

$$5 - 2^{-1} + K = 1000$$

$$K = 995.5.$$

Thus, $C(x) = 5x - \dfrac{1}{(x+1)} + 995.5$

79. Total number of articles is

$$N(t) = \int \dfrac{7e^{0.2t}}{5 + e^{0.2t}}\, dt$$

$u = 5 + e^{0.2t}$, $\dfrac{du}{dt} = 0.2e^{0.2t}$, $dt = \dfrac{du}{0.2e^{0.2t}}$

$$N(t) = \int \dfrac{7e^{0.2t}}{5 + e^{0.2t}}\, dt = \int \dfrac{7e^{0.2t}}{u} \cdot \dfrac{du}{0.2e^{0.2t}}$$

$$= \int 35 \cdot \dfrac{du}{u} = 35 \ln|u| + C$$

$$= 35 \ln(5 + e^{0.2t}) + C$$

At time $t = 0$, $N = 0$, so

$$0 = 35 \ln(5 + e^0) + C = 35 \ln 6 + C$$

$$C = -35 \ln 6 \approx -63$$

so $N(t) = 35 \ln(5 + e^{0.2t}) - 63$

(b) Since 2003 is represented by $t = 20$, we get

$$N(10) = 35 \ln(5 + e^{0.2(20)}) - 63 \approx 80$$

thousand articles

81. Using the substitution

$$u = 3 + e^{0.25t}$$

we get

Total sales

$$S(t) = \int \dfrac{900e^{0.25t}}{3 + e^{0.25t}}\, dt$$

$$= \int \dfrac{3600}{u}\, du$$

$= 3600 \ln(3 + e^{0.25t}) + C$

At time $t = 0$,

$0 = 3600 \ln(3 + 1) + C$,

So $C = -3600 \ln 4$

giving

$S(t) = 3600[\ln(3 + e^{0.25t}) - \ln 4]$

After 12 months,

$S(12) = 3600[\ln(3 + e^{0.25(12)}) - \ln 4]$

≈ 6310 sets

83. (a) $s = \int v(t)\, dt = \int [t(t^2 + 1)^4 + t]\, dt$

$= \int t(t^2 + 1)^4\, dt + \frac{1}{2} t^2$

For the integral on the left, take

$u = t^2 + 1,\ du = 2t\, dt,\ dt = \frac{1}{2t} du$

$\int t(t^2 + 1)^4\, dt = \int tu^4 \frac{1}{2t}\, du$

$= \frac{1}{2} \int u^4\, du = \frac{1}{10} u^5 + C$

$= \frac{1}{10} (t^2 + 1)^5 + C.$

So, $s = \frac{1}{10} (t^2 + 1)^5 + \frac{1}{2} t^2 + C$

(b) $s(0) = 1$ gives

$\frac{1}{10} (0 + 1)^5 + 0 + C = 1$

$C = \frac{9}{10}$

So, $s = \frac{1}{10} (t^2 + 1)^5 + \frac{1}{2} t^2 + \frac{9}{10}$

85. Total sales are given by

$S(t) = \int s(t)\, dt$

To integrate this expression we use the shortcut:

$S(t) = 4(t-2000)^3 + 250(t-2000)^2 + 4700(t-2000) + C$

$S(2003) = 0$, so

$0 = 4(3)^3 + 250(3)^2 + 4700(3) + C$

$C = -16,458$

$S(t) = 4(t-2000)^3 + 250(t-2000)^2$

$+ 4700(t-2000) - 16,458$ million

gallons.

Sales 2003–2008 are given by

$S(2008) = 4(8)^3 + 250(8)^2 + 4700(8) - 16,458 = 39,190$ million gallons

87. None; the substitution $u = x$ simply replaces the letter x throughout by the letter u, and thus does not change the integral at all. For instance, the integral $\int x(3x^2 + 1)\, dx$ becomes

$\int u(3u^2 + 1)\, du$ if we substitute $u = x$.

89. (D); To compute the integral, first break it up into a sum of two integrals:

$\int \frac{x}{x^2-1}\, dx + \int \frac{3x}{x^2+1}\, dx$, and then compute the

first using $u = x^2-1$ and the second using $u = x^2+1$.

91. The purpose of substitution is to introduce a new variable that is defined in terms of the variable of integration. One cannot say $u = u^2 + 1$, because u is not a new variable. Instead, define $w = u^2 + 1$ (or any letter different from u).

93. The integral $\int x(x^2 + 1)\, dx$ can be solved by the substitution $u = x^2 + 1$: $\int x(x^2 + 1)\, dx =$

$\frac{1}{2} \int u\, du = \frac{1}{4} u^2 + C = \frac{1}{4} (x^2 + 1)^2 + C.$

95. The substitution $u = -x$ leads to $\int e^{-u^2}\, du$,

which is just the original integral. The

substitution $u = x^2$ leads to $\int e^{-u} \frac{1}{2x}\, du =$

$$\int e^{-u} \frac{1}{2\sqrt{u}} \, du \quad \text{which is no easier to evaluate. The}$$

substitution $u = -x^2$ is similar.

13.3 The Definite Integral: Numerical and Graphical Approaches

1. $\Delta x = \frac{b-a}{n} = \frac{5-0}{5} = 1$

Left Sum = $[f(0) + f(1) + f(2) + f(3) + f(4)]\Delta x$
$\approx [14 + 6 + 2 + 2 + 6](1) = 30$

3. $\Delta x = \frac{b-a}{n} = \frac{9-1}{4} = 2$

Left Sum = $[f(1) + f(3) + f(5) + f(7)]\Delta x$
$\approx [0 + 4 + 4 + 3](2) = 22$

5. $\Delta x = \frac{b-a}{n} = \frac{3.5-1}{5} = 0.5$

Left Sum = $[f(1) + f(1.5) + f(2) + f(2.5) + f(3)]\Delta x$
$\approx [-1 + (-1) + (-2) + (-1) + 1](0.5) = -2$

7. $\Delta x = \frac{b-a}{n} = \frac{3-1}{3} = 1$

Left Sum = $[f(0) + f(1) + f(2)]\Delta x$
$\approx [0 + 0 + 0](1) = 0$

9. $f(x) = 4x - 1$, $\Delta x = \frac{b-a}{n} = \frac{2-0}{4} = 0.5$

Left Sum = $[f(0) + f(0.5) + f(1) + f(1.5)](0.5)$
$= 4$

11. $f(x) = x^2$, $\Delta x = \frac{b-a}{n} = \frac{2-(-2)}{4} = 1$

Left Sum = $[f(-2) + f(-1) + f(0) + f(1)](1) = 6$

13. $f(x) = 1/(1 + x)$, $\Delta x = \frac{b-a}{n} = \frac{1-0}{5} = 0.2$

Left Sum = $[f(0) + f(0.2) + f(0.4) + f(0.6) + f(0.8)](0.2) \approx 0.7456$

15. $f(x) = e^{-x}$, $\Delta x = \frac{b-a}{n} = \frac{10-0}{5} = 2$

Left Sum = $[f(0) + f(2) + f(4) + f(6) + f(8)](2)$
≈ 2.3129

17. $f(x) = e^{-x^2}$, $\Delta x = \frac{b-a}{n} = \frac{10-0}{4} = 2.5$

Left Sum = $[f(0) + f(2.5) + f(5) + f(7.5)](2.5)$
≈ 2.5048

19. The area is a square of height 1 and width 1, so has area $1 \times 1 = 1$.

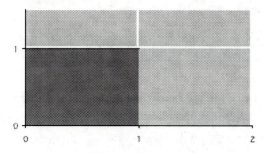

21. The area is a triangle of height 1 and base 1, so has area $\frac{1}{2}(1 \times 1) = \frac{1}{2}$.

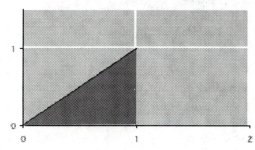

23. The area is a triangle of height $\frac{1}{2}$ and base 1, so has area $\frac{1}{2}\left(\frac{1}{2} \times 1\right) = \frac{1}{4}$.

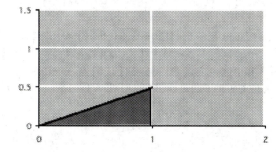

489

25. The area is a triangle of height 2 and base 2, so has area $\frac{1}{2}(2 \times 2) = 2$.

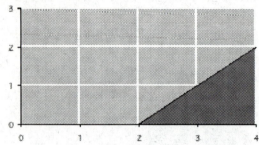

27. Since the are below the x axis is the same as the area above, the integral is 0.

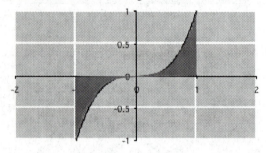

29. If counting grid squares, note that each grid square has an area of $1 \times 0.5 = 0.5$. Instead of counting grid squares, we average the left and right Riemann sums.
Left Sum $= (1 + 1 + 1.5 + 2)(1) = 5.5$
Right Sum $= (1 + 1.5 + 2 + 2)(1) = 5.5$
Total Change = Area under graph over $[1, 5]$ = Average of left- and right- sums $= (5.5+6.5)/2 = 6$

31. Left Sum $= (-1 - 0.5 + 0 + 0.5)(1) = -1$
Right Sum $= (-0.5 + 0 + 0.5 + 1)(1) = 1$
Total Change = Net area over $[2, 6]$ = Average of left- and right- sums $= (1+(-1))/2 = 0$

33. Note that each grid square has an area of $0.5 \times 0.5 = 0.25$. Note that the areas corresponding to $[-1, 0]$ and $[0, 1]$ cancel out, so we are left with the area above $[1, 2]$, which is 2 grid squares, or 0.5.

35. Technology formula:
TI 83/84 Plus: `4*√(1-x^2)`
Excel: `4*SQRT(1-x^2)`

$n = 10$: 3.3045; $n = 100$: 3.1604; $n = 1000$: 3.1436

37. Technology formula:
`2*x^1.2/(1+3.5*x^4.7)`
$n = 10$: 0.0275; $n = 100$: 0.0258; $n = 1000$: 0.0256

39. $C'(x) = 20 - \dfrac{x}{200}$

$$\int_0^5 C'(x)\,dx \approx [C'(0) + C'(1) + C'(2) + C'(3)$$
$$+ C'(4)](1) = \$99.95$$

41. The period given corresponds to $[0, 5]$

Total sales $= \displaystyle\int_0^5 s(t)\,dt$

$\approx [s(0) + s(1) + s(2) + s(3) + s(4)](1)$
$= 28{,}860$ million gallons ≈ 29 billion gallons

43. Total number of items $= \displaystyle\int_6^{14} n(t)\,dt$ million.

Using a Riemann sum with 4 subdivisions gives
$$\int_6^{14} n(t)\,dt \approx [n(6) + n(8) + n(10) +$$
$n(12)](2)$
$\approx [600 + 650 + 600 + 600](2)$
$= 4900$ million items

45. (a) Left sum $\approx (1 + 1.5 + 2 + 2.5 + 3 + 3.5 + 4.5 + 5)(2) = 46$,
or about 46,000 articles
Right Sum $\approx (1.5 + 2 + 2.5 + 3 + 3.5 + 4.5 + 5 + 5.5)(2) = 55$
or about 55,000 articles
(b) From the "Before we go on" discussion at the end of Example 4 we can estimate $\displaystyle\int_0^{16} r(t)\,dt$ as

490

$$\int_0^{16} r(t)\ dt \approx \text{Average of left- and right-sums}$$

$$= \frac{46+55.5}{2} = 50.5$$

Interpretation: Since $r(t)$ is given in thousands of articles per year, we conclude that a total of about 50,500 articles in *Physics Review* were written by researchers in Europe in the 16-year period beginning 1983.

47. The period 2005–2010 is represented by the interval $[0, 5]$. The total number of doctoral degrees awarded is given by the definite integral of its rate of change:

$$\text{Total change} = \int_0^5 f'(t)\ dt = \text{Area above}$$

interval $[0, 5]$
Left sum $= (2.25+2.25+2.5+2.25+2.5)(1) = 11.75$
Right Sum $= (2.25+2.5+2.25+2.5+2.5)(1) = 12.0$

$$\text{Average} = \frac{11.75+12}{2} = 11.875, \text{ or } 118,750$$

degrees

49. Left sum $= (4+5+5+6.5)(1) = 20.5$
Right Sum $= (5+5+6.5+4.5)(1) = 21$
Actual net income corresponds to the left sum, as it is the sum of the quarterly net incomes for 2006.

51. The total change in number of students $c(t)$ from China who took the GRE exams is given by the definite integral of its rate of change:

$$\text{Total change} = \int_2^4 c'(t)\ dt$$

$= \text{Area above interval } [2, 4]$
Left sum $= (35 + 32.5 + 30 + 20)(0.5) = 58.75$
Right Sum $= (32.5 + 30 + 20 + 17.5)(0.5) = 50$

$$\text{Average} = \frac{58.75 + 50}{2} = 54.375, \text{ or about } 54,000$$

students

53. $\Delta x = (4-0)/5 = 0.8$;

Left Riemann sum
$= [v(0) + v(0.8) + v(1.6) + v(2.4) + v(3.2)](0.8)$
$\qquad = [30 + 4.4 - 21.2 - 46.8 - 72.4](0.8)$
$\qquad = -84.8\text{ft}$
The answer represents the total change in position, so, after 4 seconds, the stone is about 84.8 ft. below where it started.

55. $v(t) = 40t^2$ ft/s
Height of rocket 2 seconds after launch

$$= \int_0^2 v(t)\ dt \approx [v(0) + v(0.2) + v(0.4) +$$

$$\dots + v(1.8)](0.2)$$
$$= 91.2 \text{ ft}$$

57. Here is a method of computing the Riemann sum on the Web Site: Go to
Web site $\to$ On Line Utilities $\to$ Numerical Integration Utility
Enter $-8.03*x^2 + 73x + 1060$ for $f(x)$, 1 and 9 for the left and right end-points, and 150 for the number of subdivisions. Then press "Left Sum" to obtain the left Riemann sum: $9452.9135 \approx 9450$ (rounded to 3 significant digits). Therefore, $\int_1^9 p(t)\ dt \approx 9450$. A total of about 9450 million barrels of oil was produced by Pemex from the start of 2001 the start of 2009.

59. Here is a method of computing the Riemann sum on the Web Site: Go to
Web site $\to$ On Line Utilities $\to$ Numerical Integration Utility
Enter $820*e^{(0.051*x)}$ for $f(x)$, 0 and 15 for the left and right end-points, and 100 for the number of subdivisions. Then press "Left Sum" to obtain the left Riemann sum: $18{,}403.454 \approx 18{,}400$ (rounded to the nearest 10). Therefore, $\int_0^{15} w(t)\ dt \approx 18{,}400$. A total of 18,400 wiretaps were authorized by U.S. state and federal courts during the 15-year period starting January 1990.

61. Yes. The Riemann sum gives an estimated area of $(0 + 15 + 18 + 8 + 7 + 16 + 20)(5)$ = 420 square feet.

63. (a) Graph:

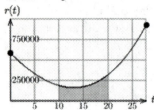

$r(t) = p(t)q(t)$ represents annual total expenditure on oil in the U.S., so $\int_{10}^{20} r(t) \, dt$

represents the total expenditure on oil in the U.S. from 1990 to 2000.

(b) Here is a method of computing the Riemann sum on the Web Site: Go to
Web site → On Line Utilities → Numerical Integration Utility
Enter
`(76x+5540)*(0.45x^2-12x+105)`
for $f(x)$, 10 and 20 for the left and right endpoints, and 200 for the number of subdivisions. Then press "Left Sum" to obtain the left Riemann sum: $2{,}010{,}295.5 \approx 2{,}010{,}000$ (rounded to 3 significant digits). Therefore, $\int_{10}^{20} r(t) \, dt \approx$

$2{,}010{,}000$. A total of $2{,}010{,}000$ million, or $2.01 trillion, was spent on oil in the U.S. from 1990 to 2000.

65. (a) $p(x) = \dfrac{1}{5.2\sqrt{2\pi}} e^{-(x-72.6)^2/54.08}$,

$\int_{60}^{100} p(x) \, dx \approx 0.994$, so approximately 99.4% of

students obtained between 60 and 100.

(b) $\int_{0}^{30} p(x) \, dx \approx 0$ (to at least 15 decimal places)

67. Stays the same: The graph is a horizontal line, and all Riemann sums give the exact area.

69. Increases: The left sum underestimates the function, by less as n increases.

71. The area under the curve and above the x axis equals the area above the curve and below the x axis.

73. Answers will vary. One example: Let r(t) be the rate of change of net income at time t. If r(t) is negative, then the net income is decreasing, so the change in net income, represented by the definite integral of r(t), is negative.

75. Answers may vary. For example:

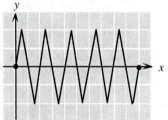

77. The total cost is $c(1) + c(2) + \ldots + c(60)$, which is represented by the Riemann sum

approximation of $\int_{1}^{61} c(t) \, dt$ with $n = 60$.

79. $[f(x_1) + f(x_2) + \ldots + f(x_n)]\Delta x = \displaystyle\sum_{k=1}^{n} f(x_k)\Delta x$

81. If increasing n by a factor of 10 does not change the value of the answer when rounded to three decimal places, then the answer is (likely) accurate to three decimal places.

13.4 The Definite Integral: Algebraic Approach and the Fundamental Theorem of Calculus

1. $\int_{-1}^{1} (x^2 + 2)\, dx = \left[\dfrac{x^3}{3} + 2x\right]_{-1}^{1}$

$= \dfrac{1}{3} + 2 - \left(\dfrac{-1}{3} - 2\right) = \dfrac{14}{3}$

3. $\int_{0}^{1} (12x^5 + 5x^4 - 6x^2 + 4)\, dx$

$= [2x^6 + x^5 - 2x^3 + 4x]_{0}^{1}$

$= 2 + 1 - 2 + 4 - (0) = 5$

5. $\int_{-2}^{2} (x^3 - 2x)\, dx = \left[\dfrac{x^4}{4} - x^2\right]_{-2}^{2}$

$= \dfrac{16}{4} - 4 - \left(\dfrac{16}{4} - 4\right) = 0$

7. $\int_{1}^{3} \left(\dfrac{2}{x^2} + 3x\right) dx = \int_{1}^{3} (2x^{-2} + 3x)\, dx$

$= \left[-2x^{-1} + \dfrac{3}{2}x^2\right]_{1}^{3}$

$= -\dfrac{2}{3} + \dfrac{27}{2} - \left(-2 + \dfrac{3}{2}\right) = \dfrac{40}{3}$

9. $\int_{0}^{1} (2.1x - 4.3x^{1.2})\, dx = \left[1.05x^2 - \dfrac{4.3}{2.2}x^{2.2}\right]_{0}^{1}$

$= 1.05 - \dfrac{4.3}{2.2} - (0) \approx -0.9045$

11. $\int_{0}^{1} 2e^x\, dx = [2e^x]_{0}^{1} = 2e - 2$

$= 2(e - 1)$

13. $\int_{0}^{1} \sqrt{x}\, dx = \int_{0}^{1} x^{1/2}\, dx = \left[\dfrac{2}{3}x^{3/2}\right]_{0}^{1}$

$= \dfrac{2}{3} - 0 = \dfrac{2}{3}$

15. $\int_{0}^{1} 2^x\, dx = \left[\dfrac{2^x}{\ln 2}\right]_{0}^{1} = \dfrac{2}{\ln 2} - \dfrac{1}{\ln 2} = \dfrac{1}{\ln 2}$

17. Let $u = 3x + 1$, $du = 3dx$, $dx = \dfrac{1}{3}du$;

when $x = 0$, $u = 1$; when $x = 1$, $u = 4$;

$\int_{0}^{1} 18(3x + 1)^5\, dx = \int_{1}^{4} 18u^5 \dfrac{1}{3}\, du = [u^6]_{1}^{4}$

$= 4^6 - 1 = 4095$

In Exercises 19–25, we use the shortcut integration formulas rather than substitution.

19. $\int_{-1}^{1} e^{2x-1}\, dx = \left[\dfrac{1}{2}e^{2x-1}\right]_{-1}^{1} = \dfrac{1}{2}e^1 - \dfrac{1}{2}e^{-3}$

$= \dfrac{1}{2}(e - e^{-3})$

21. $\int_{0}^{2} 2^{-x+1}\, dx = \left[-\dfrac{2^{-x+1}}{\ln 2}\right]_{0}^{2}$

$= -\dfrac{2^{-1}}{\ln 2} - \left(-\dfrac{2}{\ln 2}\right) = \dfrac{3}{2 \ln 2}$

23. $\int_{0}^{50} e^{-0.02x-1}\, dx = \left[-\dfrac{e^{-0.02x-1}}{0.02}\right]_{0}^{50}$

$= -50e^{-2} - (-50e^{-1}) = 50(e^{-1} - e^{-2})$

25. By the shortcut,

$\int_{0}^{4} |-3x+4|\, dx = \left[-\dfrac{1}{6}(-3x+4)|-3x+4|\right]_{0}^{4}$

$$= \left(-\frac{1}{6}(-8)(8)\right) - \left(-\frac{1}{6}(4)(4)\right)$$

$$= 80/6 = 40/3$$

27. $\displaystyle\int_{-1.1}^{1.1} e^{x+1}\, dx = [e^{x+1}]_{-1.1}^{1.1} = e^{2.1} - e^{-0.1}$

29. Let $u = 2x^2 + 1$, $du = 4x\, dx$, $dx = \dfrac{1}{4x}\, du$;

when $x = -\sqrt{2}$, $u = 5$; when $x = \sqrt{2}$, $u = 5$;

$$\int_{-\sqrt{2}}^{\sqrt{2}} 3x\sqrt{2x^2 + 1}\, dx = \int_0^0 3x\sqrt{u}\,\frac{1}{4x}\, du$$

$$= \frac{3}{4}\int_0^0 u^{1/2}\, du = \left[\frac{1}{2}u^{3/2}\right]_0^0 = 0 - 0 = 0$$

31. Let $u = x^2 + 2$, $du = 2x\, dx$, $dx = \dfrac{1}{2x}\, du$;

when $x = 0$, $u = 2$; when $x = 1$, $u = 3$;

$$\int_0^1 5xe^{x^2+2}\, dx = \int_2^3 5xe^u\,\frac{1}{2x}\, du = \frac{5}{2}\int_2^3 e^u\, du$$

$$= \left[\frac{5}{2}e^u\right]_2^3 = \frac{5}{2}(e^3 - e^2)$$

33. Let $u = x^3 - 1$, $du = 3x^2\, dx$, $dx = \dfrac{1}{3x^2}\, du$;

when $x = 2$, $u = 7$; when $x = 3$, $u = 26$;

$$\int_2^3 \frac{x^2}{x^3 - 1}\, dx = \int_7^{26} \frac{x^2}{u}\cdot\frac{1}{3x^2}\, du = \frac{1}{3}\int_7^{26}\frac{1}{u}\, du$$

$$= \left[\frac{1}{3}\ln|u|\right]_7^{26} = \frac{1}{3}(\ln 26 - \ln 7)\ \text{or}$$

$\dfrac{1}{3}\ln\!\left(\dfrac{26}{7}\right)$

35. Let $u = -x^2$, $du = -2x\, dx$, $dx = -\dfrac{1}{2x}\, du$;

when $x = 0$, $u = 0$; when $x = 1$, $u = -1$;

$$\int_0^1 x(1.1)^{-x^2}\, dx = -\int_0^{-1} x(1.1)^u\,\frac{1}{2x}\, du$$

$$= -\frac{1}{2}\int_0^{-1}(1.1)^u\, du = \left[-\frac{(1.1)^u}{2\ln 1.1}\right]_0^{-1}$$

$$= -\frac{(1.1)^{-1}}{2\ln 1.1} - \left(-\frac{1}{2\ln 1.1}\right) = \frac{0.1}{2.2\ln 1.1}$$

37. Let $u = 1/x$, $du = -1/x^2\, dx$, $dx = -x^2\, du$;

when $x = 1$, $u = 1$; when $x = 2$, $u = 1/2$;

$$\int_1^2 \frac{e^{1/x}}{x^2}\, dx = -\int_1^{1/2}\frac{e^u}{x^2}x^2\, du = -\int_1^{1/2} e^u\, du$$

$$= [-e^u]_1^{1/2} = -e^{1/2} - (-e) = e - e^{1/2}$$

39. Let $u = x + 1$, $du = dx$; when $x = 0$, $u = 1$; when $x = 2$, $u = 3$; $x = u - 1$;

$$\int_0^2 \frac{x}{x+1}\, dx = \int_1^3 \frac{x}{u}\, du = \int_1^3 \frac{u - 1}{u}\, du$$

$$= \int_1^3\left(1 - \frac{1}{u}\right)du = [u - \ln|u|]_1^3$$

$$= 3 - \ln 3 - (1 - \ln 1) = 2 - \ln 3$$

41. Let $u = x - 2$, $du = dx$; when $x = 1$, $u = -1$; when $x = 2$, $u = 0$; $x = u + 2$;

$$\int_1^2 x(x - 2)^5\, dx = \int_{-1}^0 xu^5\, du =$$

$$\int_{-1}^0 (u + 2)u^5\, du$$

$$= \int_{-1}^0 (u^6 + 2u^5)\, du = \left[\frac{u^7}{7} + \frac{u^6}{3}\right]_{-1}^0$$

$$= 0 + 0 - \left(\frac{-1}{7} + \frac{1}{3}\right) = -\frac{4}{21}$$

43. Let $u = 2x + 1$, $du = 2\,dx$, $dx = \frac{1}{2}\,du$;

when $x = 0$, $u = 1$; when $x = 1$, $u = 3$; $x = \frac{1}{2}$

$(u - 1)$;

$$\int_0^1 x\sqrt{2x + 1}\,dx = \int_1^3 x\sqrt{u}\,\frac{1}{2}\,du$$

$$= \frac{1}{2}\int_1^3 \frac{1}{2}(u - 1)u^{1/2}\,du = \frac{1}{4}$$

$$\int_1^3 (u^{3/2} - u^{1/2})\,du$$

$$= \left[\frac{1}{10}u^{5/2} - \frac{1}{6}u^{3/2}\right]_1^3$$

$$= \frac{3^{5/2}}{10} - \frac{3^{3/2}}{6} - \left(\frac{1}{10} - \frac{1}{6}\right)$$

$$= \frac{3^{5/2}}{10} - \frac{3^{3/2}}{6} + \frac{1}{15}$$

45. Graph:

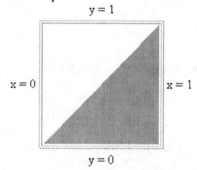

The line $y = x$ crosses the x axis at $x = 0$, so we can calculate the area as

$$\int_0^1 x\,dx = \left[\frac{x^2}{2}\right]_0^1 = \frac{1}{2}$$

47. Graph:

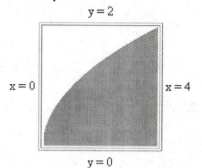

The curve $y = \sqrt{x}$ touches the x axis at $x = 0$ only, so we can calculate the area as

$$\int_0^4 \sqrt{x}\,dx = \int_0^4 x^{1/2}\,dx = \left[\frac{2}{3}x^{3/2}\right]_0^4$$

$$= \frac{16}{3} - 0 = \frac{16}{3}$$

49. The value of $|2x - 3|$ is never negative, so we can compute the area as

$$\int_0^3 |2x - 3|\,dx = \left[\frac{1}{4}(2x - 3)|2x - 3|\right]_0^3$$

$$= \frac{1}{4}(9) - \frac{1}{4}(-9) = \frac{9}{2}$$

51. Graph:

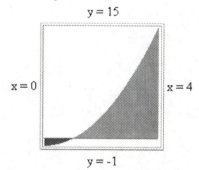

The curve $y = x^2 - 1$ crosses the x axis where $x^2 - 1 = 0$, so at $x = \pm 1$. We compute two integrals:

$$\int_0^1 (x^2 - 1)\,dx = \left[\frac{x^3}{3} - x\right]_0^1$$

$$= \frac{1}{3} - 1 - 0 = -\frac{2}{3}$$

and

$$\int_1^4 (x^2 - 1)\ dx = \left[\frac{x^3}{3} - x\right]_1^4$$

$$= \frac{64}{3} - 4 - \left(\frac{1}{3} - 1\right) = \frac{54}{3} \ .$$

The total area is the sum of the absolute values,

so $\frac{2}{3} + \frac{54}{3} = \frac{56}{3}$.

53. Graph:

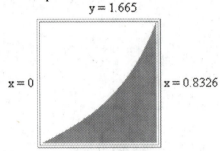

The curve $y = xe^{x^2}$ crosses the x axis at $x = 0$, so

we can calculate the area as $\int_0^{(\ln 2)^{1/2}} xe^{x^2}\ dx$.

Let $u = x^2$, $du = 2x\ dx$, $dx = \frac{1}{2x}\ du$; when $x = 0$, $u = 0$; when $x = (\ln 2)^{1/2}$, $u = \ln 2$.

$$\int_0^{(\ln 2)^{1/2}} xe^{x^2}\ dx = \int_0^{\ln 2} xe^u \frac{1}{2x}\ du$$

$$= \frac{1}{2}\int_0^{\ln 2} e^u\ du = \left[\frac{1}{2} e^u\right]_0^{\ln 2} = \frac{1}{2} e^{\ln 2} - \frac{1}{2}$$

$$= \frac{1}{2}\cdot 2 - \frac{1}{2} = \frac{1}{2}$$

55. $C'(x) = 5 + \dfrac{x^2}{1000}$

Change in cost $= \displaystyle\int_{10}^{100} C'(x)\ dx$

$$= \int_{10}^{100}\left(5 + \frac{x^2}{1000}\right) dx = \left[5x + \frac{x^3}{3000}\right]_{10}^{100}$$

$$= 500 + \frac{1000}{3} - \left(50 + \frac{1}{3}\right) = \$783$$

57. $v(t) = 60 - e^{-t/10}$ mph

Total distance traveled $= \displaystyle\int_1^6 v(t)\ dt$

$$= \int_1^6 (60 - e^{-t/10})\ dt = [60t + 10e^{-t/10}]_1^6$$

$$= 360 + 10e^{-6/10} - (60 + 10e^{-1/10})$$

$$\approx 296 \text{ miles}$$

59. The distance traveled is the displacement

$$\int_0^{10} |-10t + 40|\ dt =$$

$$\left[-\frac{1}{20}(-10t + 40)|-10t + 40|\right]_0^{10}$$

$$= \frac{1}{20}(60)^2 + \frac{1}{20}(40)^2 = 260 \text{ ft.}$$

61. The period given corresponds to $[0, 5]$

Total sales $= \displaystyle\int_0^5 (12t^2 + 500t + 4700)\ dt$

$$= [4t^3 + 250t^2 + 4700t]_0^5$$

$$= [4(5)^3 + 250(5)^2 + 4700(5)] -$$

$$[4(0)^3 + 250(0)^2 + 4700(0)]$$

$$= 30{,}250 \text{ million gallons} \approx 30 \text{ billion gallons}$$

63. The period given corresponds to $[1, 9]$

Total sales $= \displaystyle\int_1^9 (-8.03t^2 + 73t + 1060)\ dt$

$$= [-8.03t^3/3 + 73t^2/2 + 1060t]_1^9$$

$$= \left[-8.03(9)^3/3 + 73(9)^2/2 + 1060(9)\right] -$$
$$\left[-8.03(1)^3/3 + 73(1)^2/2 + 1060(1)\right]$$
$$\approx 9451 \text{ million barrels}$$

65. $\int\limits_{0}^{15} w(t) \ dt = \int\limits_{0}^{15} 820e^{0.051t} \ dt$

$$= \left[\frac{820}{0.051} e^{0.051t}\right]_{0}^{15}$$

$$= \frac{820}{0.051} (e^{0.051(15)} - e^{0.051(0)})$$

$$\approx 18{,}470 \text{ wiretaps}$$

A total of 18,470 wiretaps were authorized by U.S. state and federal courts during the 15-year period starting January 1990.

67. $V(t) = \int\limits_{0}^{t} (210 - 62e^{-0.05x}) \ dx$

$$= \left[210x - \frac{62}{-0.05}e^{-0.05x}\right]_{0}^{t} \quad \text{by the shortcut}$$

$$= 210t + 1240e^{-0.05t}$$
$$- [210(0) + 1240e^{-0.05(0)}]$$
$$= 210t + 1240e^{-0.05t} - 1240$$

69. $\int\limits_{0}^{10} (1 - e^{-t}) \ dt = [t + e^{-t}]_{0}^{10}$

$$= 10 + e^{-10} - (0 + e^{0}) \approx 9 \text{ gallons}$$

71. If $S(t)$ is the weekly sales rate, we have $S(t)$
$= 50(1 - 0.05)^t$ T-shirts per week after t weeks. The total sales over the coming year will be

$$\int\limits_{0}^{52} 50(0.95)^t \ dt = \left[\frac{50(0.95)^t}{\ln 0.95}\right]_{0}^{52}$$

$$= \frac{50}{\ln 0.95}[(0.95)^{52} - 1] \approx 907 \text{ T-shirts.}$$

73. TI-83/84 Plus:
On the home screen, enter

`fnInt(-0.065X^3+3.4X^2-`
`22X+3.6,X,8,10)`
and press [ENTER]
(You can find `fnInt` in the MATH menu (option 9).)
Web Site: Go to
Web site → On Line Utilities → Numerical Integration Utility
Enter `-0.065x^3+3.4x^2-22x+3.6` for $f(x)$, 8 and 10 for the left and right end-points.
Then press "Adaptive Quadrature"
Result: $68.326667 \approx 68$ ml.

75. TI-83/84 Plus:
On the home screen, enter
`fnInt(0.07X^4-`
`1.7X^3+9.9X^2+17X+390,X,1,9)`
and press [ENTER]
(You can find `fnInt` in the MATH menu (option 9).)
Web Site: Go to
Web site → On Line Utilities → Numerical Integration Utility
Enter `0.07x^4-1.7x^3+9.9x^2+17x+390` for $f(x)$, 1 and 9 for the left and right end-points.
Then press "Adaptive Quadrature."
Result: $4241.07 \approx 4200$ million items.
The interval $[1, 9]$ represents the start of Q1 in 2005 through the start of Q1 in 2007, and hence the two-year period 2005 through 2006. A total of around 4200 million items were listed on eBay during 2005 and 2006.

77. Change in cost $= \int\limits_{0}^{x} m(t) \ dt$

$$= C(x) - C(0) \text{ by the FTC,}$$
so

$$C(x) = C(0) + \int\limits_{0}^{x} m(t) \ dt \ .$$

$C(0)$ is the *fixed cost*.

79. Using the substitution
$u = 5 + e^{0.2t}$

497

we get

Total number of articles

$$N(t) = \int_0^{20} \frac{7e^{0.2t}}{5 + e^{0.2t}} \, dt$$

$$= \int_6^{5+e^4} \frac{35}{u} \, du$$

$$= 35[\ln(5+e^4) - \ln 6] \approx 80.355,$$

or 80,000 articles (to the nearest 1000 articles).

81. (a) $\dfrac{N}{1 + Ab^{-x}} = \dfrac{Nb^x}{(1 + Ab^{-x})b^x}$

$$= \frac{Nb^x}{b^x + A} = \frac{Nb^x}{A + b^x}$$

(b) Let $u = A + b^x$, $du = (\ln b)b^x \, dx$,

$$dx = \frac{1}{(\ln b)b^x} \, du.$$

$$\int \frac{N}{1 + Ab^{-x}} \, dx = \int \frac{Nb^x}{A + b^x} \, dx$$

$$= \int \frac{Nb^x}{u} \cdot \frac{1}{(\ln b)b^x} \, du = \frac{N}{\ln b} \int \frac{1}{u} \, du$$

$$= \frac{N \ln u}{\ln b} + C = \frac{N \ln(A + b^x)}{\ln b} + C$$

(c) Total number of graduates

$$= \int_6^{14} \left(220 + \frac{110}{1 + 3.8(1.27)^{-t}} \right) dt$$

$$= \left[220t + \frac{110 \ln(3.8 + 1.27^t)}{\ln 1.27} \right]_6^{14}$$

≈ 2401 thousand students.

83. $W = \displaystyle\int_{v_0}^{v_1} v \frac{d}{dv}(p) \, dv = \int_{v_0}^{v_1} v \frac{d}{dv}(mv) \, dv$

$$= \int_{v_0}^{v_1} v \cdot m \, dv = \left[\tfrac{1}{2} mv^2 \right]_{v_0}^{v_1}$$

$$= \tfrac{1}{2} mv_1^2 - \tfrac{1}{2} mv_0^2$$

85. They are related by the Fundamental Theorem of Calculus, which states (summarized briefly) that the definite integral of a suitably nice function can be calculated by evaluating the indefinite integral at the two endpoints and subtracting.

87. Computing its definite integral from a to b.

89. Calculate definite integrals by using antiderivatives.

91. An example is $v(t) = t - 5$.

93. An example is $f(x) = e^{-x}$.

95. By the FTC,

$$\int_a^x f(t) \, dt = G(x) - G(a)$$

where G is an antiderivative of f. Hence,

$$F(x) = G(x) - G(a).$$

Taking derivatives of both sides,

$$F'(x) = G'(x) + 0 = f(x),$$

as required. The result gives us a formula, in terms of area, for an antiderivative of any continuous function.

Chapter 13 Review Exercises

1. $\int (x^2 - 10x + 2)\, dx = \dfrac{x^3}{3} - 5x^2 + 2x + C$

3. $\int \left(\dfrac{4x^2}{5} - \dfrac{4}{5x^2} \right) dx = \int \left(\dfrac{4}{5}x^2 - \dfrac{4}{5}x^{-2} \right) dx$

$\quad = \dfrac{4}{5}\dfrac{x^3}{3} - \dfrac{4}{5}\dfrac{x^{-1}}{-1} + C$

$\quad = 4x^3/15 + 4/(5x) + C$

5. By the shortcut,

$\int e^{-2x+11}\, dx = \dfrac{1}{-2} e^{-2x+11} + C$

$\quad = -e^{-2x+11}/2 + C$

7. Let $u = x^2 + 4$, $du = 2x\, dx$, $dx = \dfrac{1}{2x} du$

$\int x(x^2 + 4)^{10}\, dx = \int xu^{10} \dfrac{1}{2x}\, du$

$\quad = \dfrac{1}{2} \int u^{10}\, du = \dfrac{1}{22} u^{11} + C$

$\quad = \dfrac{1}{22} (x^2 + 4)^{11} + C$

9. From the shortcut,

$\int 5e^{-2x}\, dx = -\dfrac{5}{2} e^{-2x} + C$

11. Put $u = x + 2$. $du/dx = 1$, $du = dx$.

$\int \dfrac{x + 1}{x + 2}\, dx = \int \dfrac{x + 1}{u}\, du = \int \dfrac{u - 1}{u}\, du$

(Solve for x in the equation $u = x + 2$)

$\quad = \int \left(1 - \dfrac{1}{u} \right) du = u - \ln |u| + C$

$\quad = (x + 2) - \ln|x + 2| + C$

(or $x - \ln|x + 2| + C$ if we incorporate the 2 in C).

13. $\Delta x = \dfrac{b-a}{n} = \dfrac{3-0}{6} = 0.5$

Left Sum $= [f(0) + f(0.5) + f(1) + f(1.5) + f(2) + f(2.5)]\Delta x$

$\quad \approx [-4 + 0 + 3 + 1 + 0 + 2](0.5) = 1$

15. $\Delta x = \dfrac{b-a}{n} = \dfrac{1-(-1)}{4} = 0.5$

Left Sum $= [f(-1) + f(-0.5) + f(0) + f(0.5)]\Delta x$

$\quad = [2 + 1.25 + 1 + 1.25](0.5) = 2.75$

17. $\Delta x = \dfrac{b-a}{n} = \dfrac{1-0}{5} = 0.2$

Left Sum $= [f(0) + f(0.2) + f(0.4) + f(0.6) + f(0.8)]\Delta x$

$\quad = [0 - 0.192 - 0.336 - 0.384 - 0.228](0.2)$

$\quad = -0.24$

19. Technology formulas:

TI-83/84 Plus: $e^{\wedge}(-x^2)$

Excel: EXP(-x^2)

$n = 10$: 0.7778

$n = 100$: 0.7500

$n = 1000$: 0.7471

21. Left Sum $= (0 + -0.5 + -1 + 0 + 0 + 1)(0.5) = -0.25$

Right Sum $= (-0.5 + -1 + 0 + 0 + 1 + 1)(0.5) = 0.25$

Total Change $=$ Area under graph over $[-1, 2] =$ Average of left- and right- sums $= (-0.25 + 0.25)/2 = 0$

23. $\displaystyle\int_0^1 (x - x^3)\, dx = \left[\dfrac{x^2}{2} - \dfrac{x^4}{4} \right]_0^1$

$\quad = \dfrac{1}{2} - \dfrac{1}{4} - (0) = \dfrac{1}{4}$

25. $\displaystyle\int_{-1}^1 (1 + e^x)\, dx = [x + e^x]_{-1}^1$

$\quad = (1 + e) - (-1 + e^{-1}) = 2 + e - e^{-1}$

27. Let $u = x^3 + 1$, $du = 3x^2\ dx$, $dx = \dfrac{1}{3x^2}$;

when $x = 0$, $u = 1$; when $x = 2$, $u = 9$

$$\int_0^2 x^2\sqrt{x^3 + 1}\ dx = \int_1^9 x\sqrt{u}\ \frac{1}{3x^2}\ du$$

$$= \frac{1}{3}\int_1^9 u^{1/2}\ du = \frac{2}{9}\ [u^{3/2}]_1^9$$

$$= \frac{2}{9}\ (27 - 1) = \frac{52}{9}$$

29. $u = 1 + 4e^{-2x}$, $du/dx = -8e^{-2x}$, $dx = -du/(8e^{-2x})$

$x = 0 \Rightarrow u = 1 + 4 = 5$

$x = \ln 2 \Rightarrow u = 1 + 4e^{-2\ln 2} = 1 + 1 = 2$

$$\int_0^{\ln 2} \frac{e^{-2x}}{1 + 4e^{-2x}}\ dx = -\int_5^2 \frac{e^{-2x}}{u\,8e^{-2x}}\ du$$

$$= \int_2^5 \frac{1}{8u}\ du = \frac{1}{8}\ [\ln u]_2^5 = \frac{1}{8}\ [\ln 5 - \ln 2]$$

31. $y = 4 - x^2$ crosses the x axis when $x = \pm 2$, so we can compute the area as

$$\int_{-2}^2 (4 - x^2)\ dx = \left[4x - \frac{x^3}{3}\right]_{-2}^2$$

$$= 8 - \frac{8}{3} - \left(-8 + \frac{8}{3}\right) = \frac{32}{3}$$

33. $y = xe^{-x^2}$ crosses the x axis at $x = 0$, so we

can compute the area as $\displaystyle\int_0^5 xe^{-x^2}\ dx$.

Let $u = -x^2$, $du = -2x\ dx$, $dx = -\dfrac{1}{2x}\ du$;

when $x = 0$, $u = 0$; when $x = 5$, $u = -25$. So the area is

$$\int_0^5 xe^{-x^2}\ dx = -\int_0^{-25} xe^{u}\ \frac{1}{2x}\ du$$

$$= \frac{1}{2}\int_0^{-25} e^{u}\ du = -\frac{1}{2}\ [e^{u}]_0^{-25}$$

$$= \frac{1 - e^{-25}}{2}.$$

35. (a) $N(t) = \displaystyle\int n(t)\ dt = \int (196 + t^2 - 0.16t^5)\ dt$

$$= 196t + t^3/3 - 0.16t^6/6 + C$$

When $t = 0$, $N = 0$, and substituting gives $C = 0$. So

$$N(t) = 196t + t^3/3 - 0.16t^6/6$$

$$= 196t + t^3/3 - 0.08t^6/3$$

(b) $N(6) = 196(6) + (6)^3/3 - 0.08(6)^6/3 = 3.84 \approx 4$ books.

37. (a) $a(t) = -32$, so

$$v(t) = \int(-32)\ dt = -32t + v_0, \text{ where } v_0 = $$

100, so

$$v(t) = -32t + 100. \text{ Then}$$

$$s(t) = \int(-32t + 100)\ dt$$

$$= -16t^2 + 100t + s_0, \text{ where } s_0 = 0, \text{ so}$$

$$s(t) = -16t^2 + 100t$$

(b) At its highest point the velocity is zero. From above,

$$v(t) = -32t + 100$$

This is zero when $t = 100/32 = 3.125$ seconds At that time, the height is

$$s(3.125) = -16(3.125)^2 + 100(3.125) = $$

156.25 ft

(c) The ball returns to Juan's hand when $s(t) = 0$:

$$0 = -16t^2 + 100t = t(-16t + 100)$$

which is zero when t $= 0$ or $t = 100/16 = 6.25$ seconds.

39. $\Delta t = (5-0)/10 = 0.5$

Left Sum $= (5 + 5 + 5 + 5 + 5 + 5 + 5 + 5 + 5 + 5)(0.5) = 25$

Hence, sales were approximately 25,000 copies.

41. The total number of books sold is given by the definite integral of the function shown over the interval $[0, 1.5]$:

$\Delta t = 0.25$

Left Sum $= (2 + 1 + 0 - 1 - 1 + 0)(0.25) = 0.25$

Right Sum $= (1 + 0 - 1 - 1 + 0 + 0)(0.5) = -0.25$

$$\text{Average} = \frac{0.25 + (-0.25)}{2} = 0$$

Total net sales amounted to 0.

43. The last 10 days of the period is represented by the interval $[0, 10]$

$\Delta t = (10-0)/5 = 2$

Left Sum $= (n(0) + n(2) + n(4) + n(6) + n(8))(2)$

$= (0 + 1968 + 3904 + 5856 + 7872)(2) = 39,200$ hits

45. Total cost $= \displaystyle\int_{5}^{7} \frac{1000(x+3)^2}{(8+(x+3)^3)^{3/2}} \, dx$

Put $u = 8 + (x+3)^3$

$du/dx = 3(x+3)^2$

$dx = du/[3(x+3)^2]$

$x = 5 \Rightarrow u = 8 + 8^3 = 520$

$x = 7 \Rightarrow u = 8 + 10^3 = 1008$

$\text{Cost} = \displaystyle\int_{520}^{1008} \frac{1000 \, du}{3u^{3/2}}$

$= \dfrac{2000}{3}[520^{-1/2} - 1008^{-1/2}]$

≈ 8.237 thousand dollars, or about \$8200.

47. If $S(t)$ is the weekly sales in week t, they were estimating that

$S(t) = 6400(2)^{t/2}$.

The total sales over the first five weeks would be

$\displaystyle\int_{0}^{5} S(t) \, dt = \int_{0}^{5} 6400(2)^{t/2} \, dt$

$= \dfrac{12,800}{\ln 2}[2^{t/2}]_{0}^{5} = \dfrac{12,800}{\ln 2}[2^{5/2} - 2^{0}]$

$\approx 86,000$ books.

49. Let $u = e^{0.55t} + 14.01$

$du = 0.55e^{0.55t} \, dt$

$dt = \dfrac{1}{0.55e^{0.55t}} \, du$;

when $t = 0$, $u = 15.01$; when $t = 5$, $u \approx 29.653$

$\displaystyle\int_{0}^{5} \left(6053 + \frac{4474e^{0.55t}}{e^{0.55t} + 14.01}\right) dt$

$= [6053t]_{0}^{5} + \displaystyle\int_{15.01}^{29.653} \frac{4474e^{0.55t}}{u} \cdot \frac{1}{0.55e^{0.55t}} \, du$

$= 30,265 + 8135 \displaystyle\int_{15.01}^{29.653} \frac{1}{u} \, du$

$= 30,265 + 8135[\ln u]_{15.01}^{29.653}$

$\approx 35,800$ books.

Chapter 14 Further Integration Techniques and Applications of the Integral

14.1 Integration by Parts

1.

D	I
+ 2x	e^x
− 2	e^x
+∫ 0 →	e^x

$$\int 2xe^x \, dx = 2xe^x - 2e^x + C = 2(x-1)e^x + C$$

3.

D	I
+ 3x − 1	e^{-x}
− 3	$-e^{-x}$
+∫ 0 →	e^{-x}

$$\int (3x-1)e^{-x} \, dx = -(3x-1)e^{-x} - 3e^{-x} + C$$

$$= -(3x+2)e^{-x} + C$$

5.

D	I
+ $x^2 - 1$	e^{2x}
− 2x	$\frac{1}{2}e^{2x}$
+ 2	$\frac{1}{4}e^{2x}$
−∫ 0 →	$\frac{1}{8}e^{2x}$

$$\int (x^2 - 1)e^{2x} \, dx$$

$$= \frac{1}{2}(x^2-1)e^{2x} - \frac{1}{2}xe^{2x} + \frac{1}{4}e^{2x} + C$$

$$= \frac{1}{4}(2x^2 - 2x - 1)e^{2x} + C$$

7.

D	I
+ $x^2 + 1$	e^{-2x+4}
− 2x	$-\frac{1}{2}e^{-2x+4}$
+ 2	$\frac{1}{4}e^{-2x+4}$
−∫ 0 →	$-\frac{1}{8}e^{-2x+4}$

$$\int (x^2 + 1)e^{-2x+4} \, dx$$

$$= -\frac{1}{2}(x^2+1)e^{-2x+4} - \frac{1}{2}xe^{-2x+4} - \frac{1}{4}e^{-2x+4} + C$$

$$= -\frac{1}{4}(2x^2 + 2x + 3)e^{-2x+4} + C$$

9.

D	I
+ 2 − x	2^x
− −1	$2^x/\ln 2$
+∫ 0 →	$2^x/(\ln 2)^2$

$$\int (2-x)2^x \, dx$$

$$= \frac{1}{\ln 2}(2-x)2^2 + \frac{1}{(\ln 2)^2}2^x + C$$

$$= \left[\frac{2-x}{\ln 2} + \frac{1}{(\ln 2)^2}\right]2^x + C$$

11.

	D	I
+	$x^2 - 1$	3^{-x}
−	$2x$	$-3^{-x}/\ln 3$
+	2	$3^{-x}/(\ln 3)^2$
$-\int$	$0 \quad \rightarrow$	$-3^{-x}/(\ln 3)^3$

$\int (x^2 - 1)3^{-x}\, dx$

$= -\dfrac{1}{\ln 3}(x^2 - 1)3^{-x} - \dfrac{2}{(\ln 3)^2}x3^{-x}$

$\quad - \dfrac{2}{(\ln 3)^3}3^{-x} + C$

$= -\left[\dfrac{x^2 - 1}{\ln 3} + \dfrac{2x}{(\ln 3)^2} + \dfrac{2}{(\ln 3)^3}\right]3^{-x} + C$

13.

	D	I
+	$x^2 - x$	e^{-x}
−	$2x - 1$	$-e^{-x}$
+	2	e^{-x}
$-\int$	$0 \quad \rightarrow$	$-e^{-x}$

$\int \dfrac{x^2 - x}{e^x}\, dx = \int (x^2 - x)e^{-x}\, dx$

$= -(x^2 - x)e^{-x} - (2x - 1)e^{-x} - 2e^{-x} + C$
$= -(x^2 + x + 1)e^{-x} + C$

15.

	D	I
+	x	$(x + 2)^6$
−	1	$(x + 2)^7/7$
$+\int$	$0 \quad \rightarrow$	$(x + 2)^8/56$

$\int x(x + 2)^6\, dx = \dfrac{1}{7}x(x + 2)^7 - \dfrac{1}{56}(x + 2)^8 + C$

17.

	D	I
+	x	$(x - 2)^{-3}$
−	1	$-(x - 2)^{-2}/2$
$+\int$	$0 \quad \rightarrow$	$(x - 2)^{-1}/2$

$\int \dfrac{x}{(x - 2)^3}\, dx = \int x(x - 2)^{-3}\, dx$

$= -\dfrac{1}{2}x(x - 2)^{-2} - \dfrac{1}{2}(x - 2)^{-1} + C$

$= -\dfrac{x}{2(x - 2)^2} - \dfrac{1}{2(x - 2)} + C$

19.

	D	I
+	$\ln x$	x^3
$-\int$	$1/x \quad \rightarrow$	$x^4/4$

$\int x^3 \ln x\, dx = \dfrac{1}{4}x^4 \ln x - \int \dfrac{1}{4}x^3\, dx$

$= \dfrac{1}{4}x^4 \ln x - \dfrac{1}{16}x^4 + C$

21.

	D	I
+	$\ln(2t)$	$t^2 + 1$
$-\int$	$1/t \quad \rightarrow$	$t^3/3 + t$

$\int (t^2 + 1)\ln(2t)\, dt = \left(\dfrac{1}{3}t^3 + t\right)\ln(2t) -$

$\int \left(\dfrac{1}{3}t^2 + 1\right)\, dt = \left(\dfrac{1}{3}t^3 + t\right)\ln(2t) - \dfrac{1}{9}t^3 - t$

$+ C$

23.

	D	I
+	$\ln t$	$t^{1/3}$
$-\int$	$1/t$ $\rightarrow$	$3t^{4/3}/4$

$$\int t^{1/3} \ln t \, dt = \frac{3}{4} t^{4/3} \ln t - \int \frac{3}{4} t^{1/3} \, dt$$

$$= \frac{3}{4} t^{4/3} \ln t - \frac{9}{16} t^{4/3} + C = \frac{3}{4} t^{4/3} \left(\ln t - \frac{3}{4} \right) + C$$

25.

	D	I
+	$\log_3 x$	1
$-\int$	$1/(x \ln 3)$ $\rightarrow$	x

$$\int \log_3 x \, dx = x \log_3 x - \int \frac{1}{\ln 3} \, dx$$

$$= x \log_3 x - \frac{x}{\ln 3} + C$$

27. $\int (xe^{2x} - 4e^{3x}) \, dx = \int xe^{2x} \, dx - \int 4e^{3x} \, dx$

$$= \int xe^{2x} \, dx - \frac{4}{3} e^{3x}$$

To evaluate the remaining integral we use integration by parts:

	D	I
+	x	e^{2x}
$-$	1	$\frac{1}{2} e^{2x}$
$+\int$	0 $\rightarrow$	$\frac{1}{4} e^{2x}$

$$\int xe^{2x} \, dx - \frac{4}{3} e^{3x} = \frac{1}{2} xe^{2x} - \frac{1}{4} e^{2x} - \frac{4}{3} e^{3x} + C$$

$$= \left(\frac{1}{2} x - \frac{1}{4} \right) e^{2x} - \frac{4}{3} e^{3x} + C$$

29. $\int (x^2 e^x - xe^{x^2}) \, dx = \int x^2 e^x \, dx - \int xe^{x^2} \, dx$.

To evaluate the first integral we use integration by parts:

	D	I
+	x^2	e^x
$-$	$2x$	e^x
$+$	2	e^x
$-\int$	0 $\rightarrow$	e^x

$$\int x^2 e^x \, dx = x^2 e^x - 2xe^x + 2e^x + C$$

$$= (x^2 - 2x + 2)e^x + C.$$

To evaluate the second integral we use substitution:

$$u = x^2, \, du = 2x \, dx, \, dx = \frac{1}{2x} \, du$$

$$\int xe^{x^2} \, dx = \int xe^u \frac{1}{2x} \, du = \frac{1}{2} \int e^u \, du = \frac{1}{2} e^u + C$$

$$= \frac{1}{2} e^{x^2} + C$$

Combining the two integrals we get

$$\int (x^2 e^x - xe^{x^2}) \, dx = (x^2 - 2x + 2)e^x - \frac{1}{2} e^{x^2} + C$$

31.

	D	I
+	$x + 1$	e^x
$-$	1	e^x
$+\int$	0 $\rightarrow$	e^x

$$\int_0^1 (x + 1)e^x \, dx = [(x + 1)e^x - e^x]_0^1$$

$$= [xe^x]_0^1 = e - 0 = e$$

33.

	D	I
+	x^2	$(x + 1)^{10}$
−	$2x$	$(x + 1)^{11}/11$
+	2	$(x + 1)^{12}/132$
−∫	0	$\to (x + 1)^{13}/1716$

$$\int_0^1 x^2(x + 1)^{10}\, dx =$$

$$\left[\frac{1}{11} x^2(x + 1)^{11} - \frac{1}{66} x(x + 1)^{12} + \frac{1}{858} (x + 1)^{13} \right]_0^1$$

$$= \frac{1}{11} 2^{11} - \frac{1}{66} 2^{12} + \frac{1}{858} 2^{13} - \frac{1}{858} = \frac{38{,}229}{286}$$

35.

	D	I
+	$3x - 4$	$(2x - 1)^{1/2}$
−	3	$\frac{1}{3}(2x - 1)^{3/2}$
+∫	0	$\to \frac{1}{15}(2x - 1)^{5/2}$

$$\int (3x-4)\sqrt{2x-1}\, dx$$

$$= \frac{1}{3}(3x-4)(2x-1)^{3/2} - \frac{1}{5}(2x-1)^{5/2} + C$$

37.

	D	I
+	$\ln(2x)$	x
−∫	$1/x$	$\to x^2/2$

$$\int_1^2 x \ln(2x)\, dx = \left[\frac{1}{2} x^2 \ln(2x) \right]_1^2 - \int_1^2 \frac{1}{2} x\, dx$$

$$= \left[\frac{1}{2} x^2 \ln(2x) \right]_1^2 - \left[\frac{1}{4} x^2 \right]_1^2$$

$$= 2 \ln 4 - \frac{1}{2} \ln 2 - \left(1 - \frac{1}{4} \right)$$

$$= 4 \ln 2 - \frac{1}{2} \ln 2 - \frac{3}{4} = \frac{7}{2} \ln 2 - \frac{3}{4}$$

39.

	D	I
+	$\ln(x + 1)$	x
−∫	$1/(x + 1)$	$\to x^2/2$

$$\int_0^1 x \ln(x + 1)\, dx$$

$$= \left[\frac{1}{2} x^2 \ln(x + 1) \right]_0^1 - \int_0^1 \frac{x^2}{2(x + 1)}\, dx$$

$$= \frac{1}{2} \ln 2 - \int_0^1 \frac{x^2}{2(x + 1)}\, dx$$

We evaluate this integral using a substitution:
$u = x + 1$, $du = dx$; $x = u - 1$; when $x = 0$, $u = 1$; when $x = 1$, $u = 2$.

$$\int_0^1 \frac{x^2}{2(x + 1)}\, dx = \int_1^2 \frac{x^2}{2u}\, du = \int_1^2 \frac{(u - 1)^2}{2u}\, du$$

$$= \frac{1}{2} \int_1^2 \frac{u^2 - 2u + 1}{u}\, du = \frac{1}{2} \int_1^2 \left(u - 2 + \frac{1}{u} \right) du$$

$$= \left[\frac{1}{4} u^2 - u + \frac{1}{2} \ln u \right]_1^2$$

$$= 1 - 2 + \frac{1}{2} \ln 2 - \left(\frac{1}{4} - 1 + 0 \right)$$

$$= -\frac{1}{4} + \frac{1}{2} \ln 2$$

Combining with our earlier calculation we get

$$\int_0^1 x \ln(x + 1)\, dx = \frac{1}{2} \ln 2 - \left(-\frac{1}{4} + \frac{1}{2} \ln 2 \right) = \frac{1}{4}$$

41. We calculate the area using $\int_0^{10} x e^{-x}\, dx$. To evaluate this integral we use integration by parts:

D	I
+ x	e^{-x}
− 1	$-e^{-x}$
+∫ 0 →	e^{-x}

$$\int_0^{10} xe^{-x}\,dx = [-xe^{-x} - e^{-x}]_0^{10}$$

$$= -10e^{-10} - e^{-10} - (0 - e^0) = 1 - 11e^{-10}$$

43. We calculate the area using $\int_1^2 (x+1)\ln x\,dx$

. To evaluate this integral we use integration by parts:

D	I
+ $\ln x$	$x + 1$
−∫ $1/x$ →	$x^2/2 + x$

$$\int_1^2 (x+1)\ln x\,dx$$

$$= \left[\left(\tfrac{1}{2}x^2 + x\right)\ln x\right]_1^2 - \int_1^2 \left(\tfrac{1}{2}x + 1\right)dx$$

$$= \left[\left(\tfrac{1}{2}x^2 + x\right)\ln x\right]_1^2 - \left[\tfrac{1}{4}x^2 + x\right]_1^2$$

$$= 4\ln 2 - 0 - \left[1 + 2 - \left(\tfrac{1}{4} + 1\right)\right] = 4\ln 2 - \tfrac{7}{4}$$

45.

D	I
+ x	$\|x-3\|$
− 1	$\tfrac{1}{2}(x-3)\|x-3\|$
+∫ 0 →	$\tfrac{1}{6}(x-3)^2\|x-3\|$

$$\int x\|x-3\|\,dx = \tfrac{1}{2}x(x-3)\|x-3\| - \tfrac{1}{6}(x-3)^2\|x-3\| + C$$

47.

D	I
+ $2x$	$\dfrac{\|x-3\|}{x-3}$
− 2	$\|x-3\|$
+∫ 0 →	$\tfrac{1}{2}(x-3)\|x-3\|$

$$\int 2x\,\frac{\|x-3\|}{x-3}\,dx = 2x\|x-3\| - (x-3)\|x-3\| + C$$

49.

D	I
+ $2x^2$	$\|-x+4\|$
− $4x$	$-\tfrac{1}{2}(-x+4)\|-x+4\|$
+ 4	$\tfrac{1}{6}(-x+4)^2\|-x+4\|$
−∫ 0 →	$-\tfrac{1}{24}(-x+4)^3\|-x+4\|$

$$\int 2x^2\|-x+4\|\,dx = -x^2(-x+4)\|-x+4\|$$

$$-\tfrac{2}{3}x(-x+4)^2\|-x+4\| - \tfrac{1}{6}(-x+4)^3\|-x+4\| + C$$

51.

D	I
+ x^2-2x+3	$\|x-4\|$
− $2x-2$	$\tfrac{1}{2}(x-4)\|x-4\|$
+ 2	$\tfrac{1}{6}(x-4)^2\|x-4\|$
−∫ 0 →	$\tfrac{1}{24}(x-4)^3\|x-4\|$

$$\int (x^2-2x+3)|x-4| \ dx \ = \frac{1}{2}(x^2-2x+3)(x-4)|x-4|$$

$$-\frac{1}{3}(x-1)(x-4)^2|x-4|+ \frac{1}{12}(x-4)^3|x-4|+C$$

53. We compute the displacement by integrating the velocity over the first two minutes, or 120 seconds: $\int_0^{120} 2000te^{-t/120} \ dt$. We evaluate this integral using integration by parts:

	D	I
+	t	$e^{-t/120}$
−	1	$-120e^{-t/120}$
$+\int$	0	$14{,}400e^{-t/120}$

$$\int_0^{120} 2000te^{-t/120} \ dt$$

$= 2000[-120te^{-t/120} - 14{,}400e^{-t/120}]_0^{120}$

$= 2000[-14{,}400e^{-1} - 14{,}400e^{-1} - (-14{,}400e^{0})] = 28{,}800{,}000(1 - 2e^{-1})$ ft $\approx$ 7,610,000 ft

55. We are given $C'(x) = 10 + \dfrac{\ln(x + 1)}{(x + 1)^2}$ and $C(0) = 5000$. So

$$C(x) = \int \left[10 + \frac{\ln(x + 1)}{(x + 1)^2}\right] dx$$

$$= 10x + \int (x + 1)^{-2} \ln(x + 1) \ dx$$

To evaluate this integral we use integration by parts:

	D	I
+	$\ln(x + 1)$	$(x + 1)^{-2}$
$-\int$	$1/(x + 1)$	$-(x + 1)^{-1}$

$C(x) = 10x - (x + 1)^{-1} \ln(x + 1) +$

$$\int (x + 1)^{-2} \ dx$$

$= 10x - (x + 1)^{-1} \ln(x + 1) - (x + 1)^{-1} + K$

To determine K we substitute $C(0) = 5000$

$5000 = -\ln 1 - 1 + K$

$K = 5001$

So,

$$C(x) \ = \ 10x - \frac{\ln(x + 1)}{x + 1} - \frac{1}{x + 1} + 5001$$

57. Rate of spending $= p(t)q(t)$

So total spending $= \int_0^8 p(t)q(t) \ dt$

$$= \int_0^8 1.2e^{0.12t}(-3.5t + 280) \ dt$$

	D	I
+	$-3.5t + 280$	$1.2e^{0.12t}$
−	-3.5	$10e^{0.12t}$
$+\int$	0	$\frac{250}{3}e^{0.12t}$

$$\int_0^8 1.2e^{0.12t}(-3.5t + 280) \ dt$$

$$= \left[e^{0.12t}(10[-3.5t+280] + 3.5\frac{250}{3}) \right]_0^8$$

$\approx$ 4252 billion dollars

59. Rate of consumption $= s(t)f(t)$

So total consumption $= \int_0^5 s(t)f(t) \ dt$

$$= \int_0^5 (600t+4600)(0.1+0.02t)^{1/2} \ dt$$

D	I
+ $600t$ $+4600$	$(0.1+0.02t)^{1/2}$
$-$ $\quad 600$	$\frac{100}{3}(0.1+0.02t)^{3/2}$
$+\int \quad 0 \quad \rightarrow$	$\frac{10000}{15}$ $(0.1+0.02t)^{5/2}$

$$\int_0^5 (600t+4600)(0.1+0.02t)^{1/2}\, dt$$

$$= \Big[(600t+4600)\tfrac{100}{3}(0.1+0.02t)^{3/2}$$

$$- 600\tfrac{10000}{15}(0.1+0.02t)^{5/2}\Big]_0^5$$

$$\approx 11{,}919 \text{ million gallons}$$

61. Rate of area = $s(t)a(t)$ thousand square feet per year

So total area = $\displaystyle\int_0^8 s(t)a(t)\, dt$

$$= \int_0^8 (-30t^2+240t+800)(40t+2000)\, dt$$

D	I
+ $40t+2000$	$-30t^2+240t+800$
$-$ $\quad 40$	$-10t^3+120t^2+800t$
$+\int \quad 0 \quad \rightarrow$	$-2.5t^4+40t^3$ $+400t^2$

$$\int_0^8 (-30t^2+240t+800)(40t+2000)\, dt$$

$$= \Big[(40t+2000)(-10t^3+120t^2+800t)$$

$$- 40(-2.5t^4+40t^3+400t^2)\Big]_0^8$$

$$= 19{,}353{,}600 \text{ thousand square feet, or about 19}$$
billion square feet.

63. Rate of revenue = $p(t)q(t)$

So total revenue to time $x = \displaystyle\int_0^x p(t)q(t)\, dt$

$$= \int_0^x 25e^{0.1t}(-8t^2+70t+1000)\, dt$$

D	I
+ $-8t^2+70t+1000$	$25e^{0.1t}$
$-$ $\quad -16t+70$	$250e^{0.1t}$
+ $\quad -16$	$2500e^{0.1t}$
$-\int \quad 0 \quad \rightarrow$	$25000e^{0.1t}$

$$\int_0^x 25e^{0.1t}(-8t^2+70t+1000)\, dt$$

$$= \Big[e^{0.1t}\{250[-8t^2+70t+1000]$$

$$- 2500[-16t+70] - 16(25{,}000)\}\Big]_0^x$$

$R(x) = e^{0.1x}(250[-8x^2+70x+1000] - 2500[-16x+70] - 400{,}000) + 325{,}000$

65. If $p(t)$ is the price in week t, we are told that $p(t) = 10 + 0.5t$. If $q(t)$ is the weekly sales, we are told that $q(t) = 50e^{-0.02t}$. The weekly revenue is therefore
$R(t) = p(t)q(t) = 50(10 + 0.5t)e^{-0.02t}$
and the total revenue over the coming year is

$$\int_0^{52} 50(10 + 0.5t)e^{-0.02t}\, dt$$

$$= \int_0^{52} 500e^{-0.02t}\, dt + 25\int_0^{52} te^{-0.02t}\, dt$$

We evaluate each integral:

$$\int_0^{52} 500e^{-0.02t}\, dt = [-25{,}000e^{-0.02t}]_0^{52}$$

$$= -25{,}000e^{-1.04} - (-25{,}000e^0)$$

$$= 25{,}000 - 25{,}000e^{-1.04}$$

We evaluate the second integral using integration by parts:

	D	I
+	t	$e^{-0.02t}$
−	1	$-50e^{-0.02t}$
+∫	0 →	$2500e^{-0.02t}$

$$\int_0^{52} te^{-0.02t}\, dt = [-50te^{-0.02t} - 2500e^{-0.02t}]_0^{52}$$

$$= -2600e^{-1.04} - 2500e^{-1.04} - (-2500e^0)$$

$$= 2500 - 5100e^{-1.04}$$

So, the total revenue is

$$25{,}000 - 25{,}000e^{-1.04} + 25(2500 - 5100e^{-1.04})$$

$$\approx \$33{,}598.$$

67. (a)

$$\begin{cases} p(x) & \text{if } x < a \\ q(x) & \text{if } x > a \end{cases}$$

$$= p(x) + \frac{1}{2}[q(x)-p(x)]\left[1+\frac{|x-a|}{x-a}\right]$$

$$\begin{cases} -0.1t + 3 & \text{if } 0 \le t \le 10 \\ -0.05t + 2.5 & \text{if } 10 \le t \le 20 \end{cases}$$

$$= -0.1t+3$$

$$+ \frac{1}{2}[(-0.05t+2.5)-(-0.1t+3)]\left[1+\frac{|t-10|}{t-10}\right]$$

$$= -0.1t+3 + [0.025t-0.25]\left[1+\frac{|t-10|}{t-10}\right]$$

$$= -0.075t+2.75 + [0.025t-0.25]\frac{|t-10|}{t-10}$$

(b) The total population change is

$$\int_0^{20} r(t)\, dt =$$

$$\int_0^{20} \left(-0.075t+2.75 + [0.025t-0.25]\frac{|t-10|}{t-10}\right) dt$$

$$= \int_0^{20} (-0.075t+2.75)\, dt$$

$$+ \int_0^{20} [0.025t-0.25]\frac{|t-10|}{t-10}\, dt$$

The first integral is

$$= \left[-0.0375t^2+2.75t\right]_0^{20} = 40$$

The second integral is done by parts:

	D	I
+	$0.025t-0.25$	$\dfrac{|t-10|}{t-10}$
−	0.025	$|t-10|$
+∫	0 →	$\frac{1}{2}(t-10)|t-10|$

$$\int_0^{20} [0.025t-0.25]\frac{|t-10|}{t-10}\, dt$$

$$= \left[(0.025t-0.25)|t-10|-0.0125(t-10)|t-10|\right]_0^{20}$$

$$= (2.5(10)-1.25) - (-2.5-1.25)$$

$$= 2.5$$

The sum of the integrals gives $40+2.5 = 42.5$ million people.

69. Answers will vary. Examples are xe^{x^2} and e^{x^2}

$$= 1 \cdot e^{x^2}$$

71. Answers will vary. Examples are Exercises 35 and 36, or more simply, integrals like

$$\int x(x+1)^5\, dx.$$

73. Substitution:
$$u = 3x^2 - x$$

75. Parts: Differentiate the first factor and integrate the second.

77. Substitution:
$$u = \ln(x+1)$$

79. Parts: Rewrite the integrand as
$$1 \cdot \ln(x^2) \quad \text{or as } 2 \ln x$$
and then integrate the first factor and differentiate the second.

81. $n+1$ times

83. If $f(x)$ is a polynomial of degree n, then $f^{(n+1)}(x) = 0$. Using integration by parts to evaluate $\displaystyle\int_0^b f(x)e^{-x}\, dx$ we get the following table:

	D	I
+	$f(x)$	e^{-x}
$-$	$f'(x)$	$-e^{-x}$
+	$f''(x)$	e^{-x}
	$\cdots$	
$\pm$	$f^{(n)}(x)$	$\pm e^{-x}$
$\mp \int$	$0 \quad \rightarrow$	$\mp e^{-x}$

So, $\displaystyle\int_0^b f(x)e^{-x}\, dx$

$= \left[-f(x)e^{-x} - f'(x)e^{-x} - \ldots - f^{(n)}(x)e^{-x} \right]_0^b$

$= -[f(b) + f'(b) + \ldots$
$\quad + f^{(n)}(b)]e^{-b} + [f(0) + f'(0) + \ldots + f^{(n)}(0)]e^0 = F(0) - F(b)e^{-b}$

14.2 Area Between Two Curves and Applications

1. Area $= \int_a^b [\text{Top} - \text{Bottom}]\ dx$

$= \int_0^2 [0 - (x^2-4)]\ dx$

$= \left[\dfrac{-x^3}{3} + 4x\right]_0^2$

$= -8/3 + 8 = 16/3$

3. Area $= \int_a^b [\text{Top} - \text{Bottom}]\ dx$

$= \int_1^2 [6x - (x^3-3x)]\ dx$

$= \int_1^2 [9x - x^3]\ dx$

$= \left[\dfrac{9x^2}{2} - \dfrac{x^4}{4}\right]_1^2$

$= (18-4) - (9/2 - 1/4) = 39/4$ or 9.75

5. Area $= \int_a^b [\text{Top} - \text{Bottom}]\ dx$

$= \int_0^2 [|x-1| - (-|x-1|)]\ dx$

$= \int_0^2 2|x-1|\ dx$

$= [(x-1)|x-1|]_0^2$

$= 1 - (-1) = 2$

7. Area $= \int_a^b [\text{Top} - \text{Bottom}]\ dx$

$= \int_0^2 [5-x^2 - (x^2-2x+1)]\ dx$

$+ \int_2^3 [(x^2-2x+1) - (5-x^2)]\ dx$

$= \int_0^2 [4+2x-2x^2]\ dx$

$+ \int_2^3 [2x^2-2x-4]\ dx$

$= \left[4x+x^2-\dfrac{2x^3}{3}\right]_0^2$

$+ \left[\dfrac{2x^3}{3}-x^2-4x\right]_2^3$

$= (8+4-16/3) + (54/3-9-12)$
$\quad - (16/3-4-8) = 31/3$

In these solutions we always take the integral of $f(x) - g(x)$ with $f(x) \geq g(x)$. Remember that, if you reverse the order, you will simply get the negative of that integral and should then take the absolute value.

9. We have $x^2 \geq 0 > -1$ for all x, so the two graphs do not cross. The area is

$\int_{-1}^1 [x^2 - (-1)]\ dx = \int_{-1}^1 (x^2 + 1)\ dx$

$= \left[\dfrac{1}{3} x^3 + x\right]_{-1}^1$

$= \dfrac{1}{3} + 1 - \left(-\dfrac{1}{3} - 1\right) = \dfrac{8}{3}$.

11. $-x = x$ when $x = 0$. The area is

$\int_0^2 [x - (-x)]\ dx = \int_0^2 2x\ dx = [x^2]_0^2 = 4 - 0 =$

4.

13. Area $= \int_a^b [\text{Top} - \text{Bottom}]\, dx$

$= \int_{-1}^1 [|x| - x^2]\, dx$

$= \left[\frac{1}{2}x|x| - \frac{x^3}{3} \right]_{-1}^1$

$= \left(\frac{1}{2} - \frac{1}{3} \right) - \left(-\frac{1}{2} + \frac{1}{3} \right) = \frac{1}{3}$

15. $x = x^2$ when $x^2 - x = 0$, $x(x-1) = 0$, $x = 0$ or $x = 1$. We calculate the area using two integrals:

$\int_{-1}^0 (x^2 - x)\, dx = \left[\frac{1}{3}x^3 - \frac{1}{2}x^2 \right]_{-1}^0$

$= 0 - \left(-\frac{1}{3} - \frac{1}{2} \right) = \frac{5}{6}$

and $\int_0^1 (x - x^2)\, dx = \left[\frac{1}{2}x^2 - \frac{1}{3}x^3 \right]_0^1$

$= \frac{1}{2} - \frac{1}{3} - (0) = \frac{1}{6}$.

The total area is therefore $\frac{5}{6} + \frac{1}{6} = 1$.

17. $x^2 - 2x = -x^2 + 4x - 4$ when

$2x^2 - 6x + 4 = 0$

$2(x-1)(x-2) = 0$

$x = 1, x = 2$

So, the area is

$\int_0^1 [(x^2 - 2x) - (-x^2 + 4x - 4)]\, dx$

$+ \int_1^2 [(-x^2 + 4x - 4) - (x^2 - 2x)]\, dx$

$= \int_0^1 (2x^2 - 6x + 4)\, dx + \int_1^2 (-2x^2 + 6x - 4)\, dx$

$= \left[\frac{2x^3}{3} - 3x^2 + 4x \right]_0^1 + \left[-\frac{2x^3}{3} + 3x^2 - 4x \right]_1^2$

$= (2/3 - 3 + 4) - 0 + (-16/3 + 12 - 8) - (-2/3 + 3 - 4)$

$= 2$

19. $2x^2 + 10x - 5 = -x^2 + 4x + 4$ when

$3x^2 + 6x - 9 = 0$

$3(x+3)(x-1) = 0$

$x = -3, x = 1$

So, the area is

$\int_{-3}^1 [(-x^2 + 4x + 4) - (2x^2 + 10x - 5)]\, dx$

$+ \int_1^2 [(2x^2 + 10x - 5) - (-x^2 + 4x + 4)]\, dx$

$= \int_{-3}^1 (-3x^2 - 6x + 9)\, dx$

$+ \int_1^2 (3x^2 + 6x - 9)\, dx$

$= \left[-x^3 - 3x^2 + 9x \right]_{-3}^1 +$

$\left[x^3 + 3x^2 - 9x \right]_1^2$

$= (-1 - 3 + 9) - (27 - 27 - 27)$

$+ (8 + 12 - 18) - (1 + 3 - 9)$

$= 39$

21. $e^x > x$ for all x. (Examine the graphs, or consider the fact that $e^x - x$ has its minimum value when its derivative, $e^x - 1$, is 0, which occurs when $x = 0$.) The area is $\int_0^1 (e^x - x)\, dx =$

$\left[e^x - \frac{1}{2}x^2 \right]_0^1 = e - \frac{1}{2} - (1) = e - \frac{3}{2} \approx 1.218$.

512

23. $(x - 1)^2 \geq 0 \geq -(x - 1)^2$, so the area is

$$\int_0^1 [(x - 1)^2 + (x - 1)^2] \, dx = \int_0^1 2(x - 1)^2 \, dx =$$

$$\left[\tfrac{2}{3}(x - 1)^3\right]_0^1 = 0 - \left(-\tfrac{2}{3}\right) = \tfrac{2}{3} \, .$$

25. $x = x^4$ when $x^4 - x = 0$, $x(x^3 - 1) = 0$, $x = 0$ or $x = 1$. So, the area is $\int_0^1 (x - x^4) \, dx =$

$$\left[\tfrac{1}{2}x^2 - \tfrac{1}{5}x^5\right]_0^1 = \tfrac{1}{2} - \tfrac{1}{5} = \tfrac{3}{10} \, .$$

27. $x^3 = x^4$ when $x^4 - x^3 = 0$, $x^3(x - 1) = 0$, $x = 0$ or $x = 1$. So, the area is $\int_0^1 (x^3 - x^4) \, dx =$

$$\left[\tfrac{1}{4}x^4 - \tfrac{1}{5}x^5\right]_0^1 = \tfrac{1}{4} - \tfrac{1}{5} - (0) = \tfrac{1}{20} \, .$$

29. $x^2 = x^4$ when $x^4 - x^2 = 0$, $x^2(x^2 - 1) = 0$, $x = 0$ or $x = \pm 1$.
So, the area is

$$\int_{-1}^0 (x^2 - x^4) \, dx + \int_0^1 (x^2 - x^4) \, dx$$

$$= \left[\tfrac{1}{3}x^3 - \tfrac{1}{5}x^5\right]_{-1}^0 + \left[\tfrac{1}{3}x^3 - \tfrac{1}{5}x^5\right]_0^1$$

$$= 0 - \left(-\tfrac{1}{3} + \tfrac{1}{5}\right) + \left(\tfrac{1}{3} - \tfrac{1}{5}\right) = \tfrac{4}{15} \, .$$

(In fact, since $x^2 \geq x^4$ on all of $[-1, 1]$, we could have used the single integral $\int_{-1}^1 (x^2 - x^4) \, dx$ to calculate this area.)

31. $x^2 - 2x = -x^2 + 4x - 4$ when
$2x^2 - 6x + 4 = 0$
$2(x-1)(x-2) = 0$
$x = 1, x = 2$

So, the area is $\int_1^2 [(-x^2 + 4x - 4) - (x^2 - 2x)] \, dx$

$$= \int_1^2 (-2x^2 + 6x - 4) \, dx = \left[-\frac{2x^3}{3} + 3x^2 - 4x\right]_1^2$$

$$= (-16/3 + 12 - 8) - (-2/3 + 3 - 4) = 1/3$$

33. $2x^2 + 10x - 5 = -x^2 + 4x + 4$ when
$3x^2 + 6x - 9 = 0$
$3(x+3)(x-1) = 0$
$x = -3, x = 1$
So, the area is

$$\int_{-3}^1 [(-x^2 + 4x + 4) - (2x^2 + 10x - 5)] \, dx$$

$$= \int_{-3}^1 (-3x^2 - 6x + 9) \, dx$$

$$= \left[-x^3 - 3x^2 + 9x\right]_{-3}^1$$

$$= (-1 - 3 + 9) - (27 - 27 - 27) = 32$$

35. Here are the graphs of $y = e^x$ and $y = 2$:

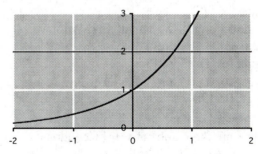

The two graphs intersect where $e^x = 2$, $x = \ln 2$. From the graph we can see that the area we want is the area between these two graphs for $0 \leq x \leq \ln 2$. So we compute $\int_0^{\ln 2} (2 - e^x) \, dx =$

$$[2x - e^x]_0^{\ln 2} = 2\ln 2 - e^{\ln 2} - (0 - 1) =$$
$$2\ln 2 - 2 + 1 = 2\ln 2 - 1.$$

37. Here are the graphs of $y = \ln x$ and $y = 2 - \ln x$:

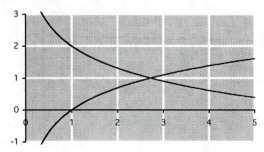

The two graphs intersect where $\ln x = 2 - \ln x$, $2 \ln x = 2$, $\ln x = 1$, $x = e$. From the graph we can see that the area we want is the area between these two graphs for $e \le x \le 4$. So we compute

$$\int_e^4 (2 \ln x - 2) \, dx = [2(x \ln x - x) - 2x]_e^4 \text{ (using)}$$

the antiderivative of $\ln x$ we derived in Section 14.1) $= [2x \ln x - 4x]_e^4 = 8 \ln 4 - 16 - (2e \ln e - 4e) = 8 \ln 4 + 2e - 16 \approx 0.5269$.

39. Formula for
Online Utilities
→ Numerical Integration Utility:
`abs(e^x-(2x+1))`
Left End-Point: -1, Right End-Point: 1 (Use "Adaptive Quadrature")
Formula for TI-83/84 Plus:
`fnInt(abs(e^x-(2x+1)),X,-1,1)`
Answer: 0.9138

41. Here are the graphs of $y = \ln x$ and $y = \frac{1}{2} x - \frac{1}{2}$:

The two graphs intersect at $x = 1$ and at a point somewhere between $x = 3$ and $x = 4$. We cannot solve the equation $\ln x = \frac{1}{2} x - \frac{1}{2}$ algebraically, but we can use technology to estimate the second intersection point. To four decimal places it is $x \approx 3.5129$. Therefore, the area is approximately

$$\int_1^{3.5129} \left(\ln x - \frac{1}{2} x + \frac{1}{2} \right) dx =$$

$$\left[x \ln x - x - \frac{1}{4} x^2 + \frac{1}{2} x \right]_1^{3.5129} =$$

$$3.5129 \ln(3.5129) - 3.5129 - \frac{1}{4}(3.5129)^2 +$$

$$\frac{1}{2} 3.5129 - \left(0 - 1 - \frac{1}{4} + \frac{1}{2} \right) \approx 0.3222.$$

43. Area $= \int_0^5 [R(t) - C(t)] \, dt$

$$= \int_0^5 [(100 + 10t) - (90 + 5t)] \, dt$$

$$= \int_0^5 (10 + 5t) \, dt = \left[10t + \frac{5}{2} t^2 \right]_0^5 = 112.5.$$

Since area under a curve represents total change, this area represents your total profit for the week, $112.50.

45. The area is

$$\int_2^6 [(-30t^2 + 240t + 800) - (-33t^2 + 240t + 700)] \, dt$$

$$= \int_2^6 (3t^2 + 100) \, dt = [t^3 + 100t]_2^6$$

$$= (216 + 600) - (8 + 200) = 608$$

Since the upper curve represents all homes and the lower curve represents homes for sale, the area between them represents the total number of housing starts (in thousands) not for sale purposes from the start of 2002 to the start of 2006. Thus, there were 608,000 housing starts from the start of 2002 to the start of 2006 not for sale purposes.

47. (a) June 2005 to the start of 2007 is represented by the interval $[0.5, 2]$
Graph: (The upper curve is Myspace.)

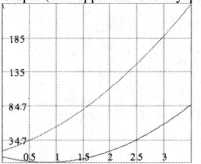

The rate at which people joined Myspace exceeded the rate at which people joined Facebook by

$m(t) - f(t)$
$= 10.5t^2 + 25t + 18.5 - (12t^2 - 20t + 10)$
$= -1.5t^2 + 45t + 8.5.$

Total from June 2005 to the start of 2007 is

$$= \int_{0.5}^{2} (-1.5t^2 + 45t + 8.5) \, dt$$

$$= \left[-0.5t^3 + \frac{45t^2}{2} + 8.5t \right]_{0.5}^{2}$$

$= (-4+90+17) - (-0.0625+5.625+4.25)$
$= 93.1875$ million members, or approximately 93 million members.

(b) The integral in part (a) corresponds to the area between the curves $y = f(t)$ and $y = m(t)$ over $[0.5, 2]$.

49. Area between graphs

$$= \int_{0}^{t} [820e^{0.051x} - 440e^{0.06x}] \, dx$$

(Notice that we need to change the variable of integration so as not to use the letter t for two different purposes.)

$$= \left[\frac{820}{0.051}e^{0.051x} - \frac{440}{0.06}e^{0.06x} \right]_{0}^{t}$$

$\approx 16078(e^{0.051t} - 1) - 7333.3(e^{0.06t} - 1)$
The total number of wiretaps authorized by federal courts from the start of 1990 up to time t was about $16078(e^{0.051t} - 1) - 7333.3(e^{0.06t} - 1)$.

51. Wrong: It could mean that the graphs of f and g cross, as shown in the caution at the start of this topic in the textbook.

53. The area between the export and import curves represents the U.S.'s accumulated trade deficit (that is, the total excess of imports over exports) from 1960 to 2007.

55. Up through day 4 the cost curve is above the revenue curve, so the enclosed area over $[0, 4]$ represents accumulated loss. Beyond day 4, the revenue curve is above he cost curve, so the enclosed area over $[4, 7]$ represents accumulated profit. Correct choice: (A).

57. The claim is wrong because the area under a curve can only represent income if the curve is a graph of income *per unit time*. The value of a stock price is not income per unit time—the income can only be realized when the stock is sold and it amounts to the current market price. The total net income (per share) from the given investment would be the stock price on the date of sale minus the purchase price of $50.

14.3 Averages and Moving Averages

1. Average $= \frac{1}{2} \int_{0}^{2} x^3 \, dx = \frac{1}{2} \left[\frac{1}{4} x^4 \right]_{0}^{2}$

$= \frac{1}{2} (4 - 0) = 2$

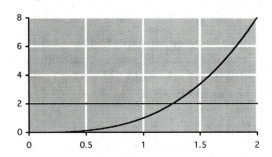

3. Average $= \frac{1}{2} \int_{0}^{2} (x^3 - x) \, dx =$

$\frac{1}{2} \left[\frac{1}{4} x^4 - \frac{1}{2} x^2 \right]_{0}^{2}$

$= \frac{1}{2} (4 - 2 - 0) = 1$

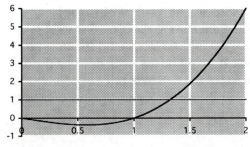

5. Average $= \frac{1}{2} \int_{0}^{2} e^{-x} \, dx = \frac{1}{2} \left[-e^{-x} \right]_{0}^{2}$

$= \frac{1}{2} (1 - e^{-2}) \approx 0.43$

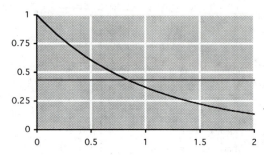

7. Average $= \frac{1}{4} \int_{0}^{4} |2x-5| \, dx =$

$\frac{1}{4} \frac{1}{4} \left[(2x-5)|2x-5| \right]_{0}^{4}$

$= \frac{1}{16} (9-(-25)) = 34/16 = 17/8$

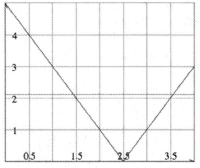

9. $\bar{r}(2) = (3 + 5 + 10)/3 = 6$, and so on.

x	0	1	2	3	4	5	6	7
$r(x)$	3	5	10	3	2	5	6	7
$\bar{r}(x)$			6	6	5	10/3	13/3	6

11. We must have $(1 + 2 + r(2))/3 = \bar{r}(2) = 3$, so $r(2) = 6$. Working from left to right we fill in the other missing values similarly.

x	0	1	2	3	4	5	6	7
$r(x)$	1	2	6	7	11	15	10	2
$\bar{r}(x)$			3	5	8	11	12	9

516

13. Moving average: $\bar{f}(x) = \frac{1}{5} \int\limits_{x-5}^{x} t^3 \, dt$

$= \frac{1}{5} \left[\frac{1}{4} t^4 \right]_{x-5}^{x} = \frac{1}{20} [x^4 - (x-5)^4]$

$= \frac{1}{20} (20x^3 - 150x^2 + 500x - 625)$

$= x^3 - \frac{15}{2} x^2 + 25x - \frac{125}{4}$

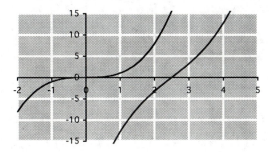

15. Moving average: $\bar{f}(x) = \frac{1}{5} \int\limits_{x-5}^{x} t^{2/3} \, dt$

$= \frac{1}{5} \left[\frac{3}{5} t^{5/3} \right]_{x-5}^{x} = \frac{3}{25} [x^{5/3} - (x-5)^{5/3}]$

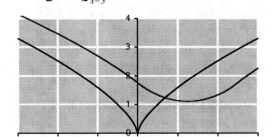

17. Moving average: $\bar{f}(x) = \frac{1}{5} \int\limits_{x-5}^{x} e^{0.5t} \, dt$

$= \frac{1}{5} [2e^{0.5t}]_{x-5}^{x} = \frac{2}{5} [e^{0.5x} - e^{0.5(x-5)}]$

$= \frac{2}{5} (1 - e^{-2.5}) e^{0.5x} \approx 0.367 e^{0.5x}$

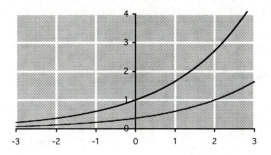

19. Moving average: $\bar{f}(x) = \frac{1}{5} \int\limits_{x-5}^{x} t^{1/2} \, dt$

$= \frac{1}{5} \left[\frac{2}{3} t^{3/2} \right]_{x-5}^{x} = \frac{2}{15} [x^{3/2} - (x-5)^{3/2}]$

(Note that the domain is $x \geq 5$.)

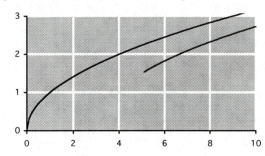

21. Moving average: $\bar{f}(x) = \frac{1}{5}$

$\int\limits_{x-5}^{x} \left[1 - \frac{|2t-1|}{2t-1} \right] dt$

$= \frac{1}{5} \left[t - \frac{1}{2} |2t-1| \right]_{x-5}^{x}$

$= \frac{1}{5} \left[x - \frac{1}{2} |2x-1| - (x-5) + \frac{1}{2} |2x-11| \right]$

$= \frac{1}{5} \left[5 - \frac{1}{2} |2x-1| + \frac{1}{2} |2x-11| \right]$

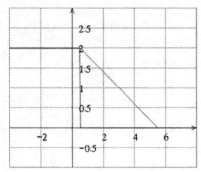

Tech formulas:

Function: `1-abs(2x-1)/(2x-1)`

Moving average:

`0.2(5-abs(2x-1)/2+abs(2x-11)/2)`

23. Moving average: $\bar{f}(x) = \frac{1}{5}\int_{x-5}^{x} (2-|t+1|+|t|)\, dt$

$$= \frac{1}{5}\left[2t - \frac{1}{2}(t+1)|t+1| + \frac{1}{2}t|t|\right]_{x-5}^{x}$$

$$= \frac{1}{5}\left[2x - \frac{1}{2}(x+1)|x+1| + \frac{1}{2}x|x| - 2(x-5) + \frac{1}{2}(x-4)|x-4| - \frac{1}{2}(x-5)|x-5|\right]$$

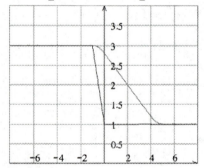

Tech formulas:

Function: `2-abs(x+1)+abs(x)`

Moving average:

`0.2(2x-(x+1)*abs(x+1)/2+x*abs(x)`
`/2-2(x-5)+(x-4)*abs(x-4)/2-(x-5)`
`*abs(x-5)/2)`

25. Plotting the moving average on a TI-83/84 Plus:

`Y`$_1$` = 10x/(1+5*abs(x))`
`Y`$_2$` = (1/3)fnInt(Y`$_1$`(T),T,X-3,X)`

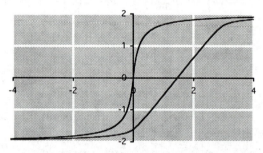

27. Plotting the moving average on a TI-83/84 Plus:

`Y`$_1$` = ln(1+x^2)`
`Y`$_2$` = (1/3)fnInt(Y`$_1$`(T),T,X-3,X)`

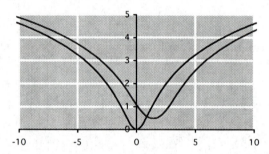

29. Plotting the moving average on a TI-83/84 Plus:

`Y`$_1$` = abs(x)-abs(x-1)+abs(x-2)-`
`abs(x-3)+abs(x-4)`
`Y`$_2$` = (1/3)fnInt(Y`$_1$`(T),T,X-3,X)`

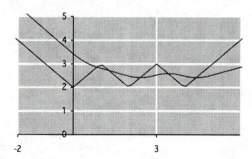

31. Plotting the moving average on a TI-83/84
Plus:

```
Y₁ = abs(x)/x-abs(x-1)/(x-1)
+abs(x-2)/(x-2)-abs(x-3)/(x-3)
Y₂ = (1/3)fnInt(Y₁(T),T,X-3,X)
```

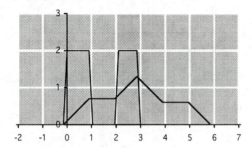

33. Plotting the moving average on a TI-83/84
Plus:

```
Y₁ = abs(x)/x+abs(x-1)/(x-1)
+abs(x-2)/(x-2)+abs(x-3)/(x-3)
Y₂ = (1/3)fnInt(Y₁(T),T,X-3,X)
```

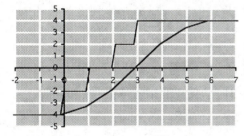

35. The average of $6718.8 million is obtained by adding all the profit figures (their sum is 33,594) and dividing by the number of data points (5).

37. Average

$$= \frac{1}{7} \int_0^7 (0.13t + 1.5) \, dt = \frac{1}{7} [0.065t^2 + 1.5t]_0^7$$

$$= \$1.955 \text{ million}$$

39. Average

$$= \frac{1}{3} \int_0^3 (12t^2 - 20t + 10) \, dt$$

$$= \frac{1}{3} [4t^3 - 10t^2 + 10t]_0^3$$

$$= 16 \text{ million members per year}$$

41. Average

$$= \frac{1}{10} \int_0^{10} 97.2(1.20)^t \, dt$$

$$= \frac{1}{10} \left[\frac{97.2(1.20)^t}{\ln(1.2)} \right]_0^{10}$$

$$\approx 277 \text{ tons}$$

43. The amount you have in the account at time t is $A(t) = 10,000e^{0.08t}$, $0 \le t \le 1$. The average amount over the first year is

$$\int_0^1 10,000e^{0.08t} \, dt = [125,000e^{0.08t}]_0^1$$

$$= \$10,410.88.$$

45. The amount in the account begins at $3000 at the beginning of the month and then declines linearly to 0 by the end of the month. So, the amount in the account during the month is

$$A(t) = 3000 - 3000t, \; 0 \le t \le 1.$$

The average over one month is therefore

$$\int_0^1 (3000 - 3000t) \, dt = [3000t - 1500t^2]_0^1$$

$$= \$1500.$$

Since the average over each month is $1500, the average over several months is also $1500.

47. TI-83/84 Plus:
On the home screen, enter
```
(1/13)*fnInt(0.07X^4-1.7X^3
+9.9X^2+17X+390,X,1,14)
```
and press [ENTER]
(You can find `fnInt` in the MATH menu (option 9).)
Web Site: Go to
Web site → On Line Utilities → Numerical Integration Utility
Enter
```
(0.07x^4-
1.7x^3+9.9x^2+17x+390)/13
```
for $f(x)$, 1 and 14 for the left and right end-points. Then press "Adaptive Quadrature."
Result: 537 million items per quarter.

519

49.

Year t	1999	2000	2001	2002	2003	2004	2005	2006	2007	2008
Stock Price	36	44	41	36	38	49	56	71	92	62
Moving average (rounded)				39	40	41	45	54	67	70

Each 4-year moving average is computed by averaging that year's figure with that of the preceding three years:

1999: $(36+44+41+36)/4 = 39.25 \approx 39$

1999: $(44+41+36+37)/4 = 39.75 \approx 40$

and so on.

The moving average continued up at a lower rate.

51. (a) To obtain the moving averages from January to June, use the fact that the data repeats every 12 months. Graph:

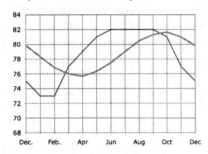

(b) The 12-month moving average is constant and equal to the year-long average of approximately 79°.

53. (a) Graph:

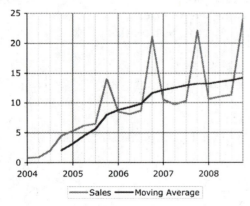

(b)

The moving average figures in Q1 2008 and Q4 2008 are 13.175 and 14.2 respectively. The average rate of change of the moving average over the interval from the first quarter to the fourth is therefore

$$\frac{14.2 - 13.175}{4 - 1} \approx 0.3 \text{ million iPods per quarter.}$$

55. (a) Average $= \frac{1}{8} \int_0^8 (12t^2 + 500t + 4700)\, dt$

$= \frac{1}{8}\left[4t^3 + 250t^2 + 4700t\right]_0^8$

$= 6956 \approx 7000$ million gallons per year

(b) Moving average

$= \frac{1}{2} \int_{t-2}^{t} (12x^2 + 500x + 4700)\, dx$

$= \frac{1}{2}\left[4x^3 + 250x^2 + 4700x\right]_{t-2}^{t}$

$=$

$\frac{1}{2}\left[4[t^3-(t-2)^3] + 250[t^2-(t-2)^2] + 9400\right]$

(c) The function is quadratic because the t^3 terms cancel.

57. (a) The line through $(t, s) = (0, 240)$ and $(25, 600)$ is

$s = 14.4t + 240.$

(b) $\bar{s}(t) = \frac{1}{4} \int_{t-4}^{t} (14.4x + 240)\, dx$

$$= \frac{1}{4} [7.2x^2 + 240x]_{t-4}^{t}$$

$$= \frac{1}{4} \{7.2[t^2 - (t-4)^2] + 960\}$$

$$= \frac{1}{4} (57.6t + 844.4) = 14.4t + 211.2$$

(c) The slope of the moving average is the same as the slope of the original function (because the original is linear).

59. $\bar{f}(x) = \frac{1}{a} \int\limits_{x-a}^{x} (mt + b) \, dt = \frac{1}{a} \left[\frac{m}{2} t^2 + bt \right]_{x-a}^{x}$

$$= \frac{1}{a} \left[\frac{m}{2} [x^2 - (x-a)^2] + bx - b(x-a) \right]$$

$$= \frac{1}{a} \left[max - \frac{ma^2}{2} + ab \right] = mx + b - \frac{ma}{2}$$

61. The moving average "blurs" the effects of short-term oscillations in the price, and shows the longer-term trend of the stock price.

63. They repeat every six months.

65. The area above the x-axis equals the area below the x-axis. Example: $y = x$ on $[-1,1]$.

67. This need not be the case; for instance, the function $f(x) = x^2$ on $[0, 1]$ has average value $1/3$, whereas the value midway between the maximum and minimum is $1/2$.

69. (C): A shorter term moving average most closely approximates the original function, since it averages the function over a shorter period, and continuous functions change by only a small amount over a small period.

14.4 Applications to Business and Economics: Consumers' and Producers' Surplus and Continuous Income Streams

1. $q = 5 - p/2$, so $\bar{q} = 5 - 5/2 = 5/2$. The consumers' surplus is

$$\int_0^{5/2} (10 - 2q - 5)\, dq = \int_0^{5/2} (5 - 2q)\, dq$$

$$= [5q - q^2]_0^{5/2} = 5\left(\frac{5}{2}\right) - \left(\frac{5}{2}\right)^2 - (0) =$$

6.25.

3. $q = (100 - p)^2/9$, so $\bar{q} = (100 - 76)^2/9 = 64$. The consumers' surplus is

$$\int_0^{64} (100 - 3\sqrt{q} - 76)\, dq = \int_0^{64} (24 - 3\sqrt{q})\, dq$$

$$= [24q - 2q^{3/2}]_0^{64} = 24(64) - 2(64)^{3/2} - (0)$$

$$= \$512.$$

5. $q = -\frac{1}{2}\ln(p/500)$, so

$$\bar{q} = -\frac{1}{2}\ln(1/5) = \frac{1}{2}\ln 5.$$

The consumers' surplus is

$$\int_0^{(\ln 5)/2} (500e^{-2q} - 100)\, dq$$

$$= [-250e^{-2q} - 100q]_0^{(\ln 5)/2}$$

$$= -250(1/5) - 50\ln 5 - (-250) = \$119.53.$$

7. $\bar{q} = 100 - 2(20) = 60$; $p = 50 - q/2$. The consumers' surplus is

$$\int_0^{60} \left(50 - \frac{1}{2}q - 20\right) dq = \int_0^{60} \left(30 - \frac{1}{2}q\right) dq$$

$$= \left[30q - \frac{1}{4}q^2\right]_0^{60} = 30(60) - \frac{1}{4}(60)^2 - (0)$$

$$= \$900.$$

9. $\bar{q} = 100 - 0.25(10)^2 = 75$; $p = 2\sqrt{100 - q}$. The consumers' surplus is

$$\int_0^{75} (2\sqrt{100 - q} - 10)\, dq$$

$$= \left[-\frac{4}{3}(100 - q)^{3/2} - 10q\right]_0^{75}$$

$$= -\frac{4}{3}(100 - 75)^{3/2} - 10(75)$$

$$\qquad - \left(-\frac{4}{3}(100)^{3/2} - 0\right)$$

$$= \$416.67.$$

11. $\bar{q} = 500e^{-0.5} - 50$; $p = -2\ln\left(\frac{1}{500}q + \frac{1}{10}\right)$.

The consumers' surplus is

$$\int_0^{500e^{-0.5}-50} \left[-2\ln\left(\frac{1}{500}q + \frac{1}{10}\right) - 1\right] dq$$

$$= -2 \int_0^{500e^{-0.5}-50} \ln\left(\frac{1}{500}q + \frac{1}{10}\right) dq -$$

$$[q]_0^{500e^{-0.5}-50}$$

$$= -2 \int_0^{500e^{-0.5}-50} \ln\left(\frac{1}{500}q + \frac{1}{10}\right) dq$$

$$\qquad - 500e^{-0.5} + 50.$$

We evaluate the remaining integral using substitution (and the integral of $\ln u$ we found by integration by parts):

Let $u = \frac{1}{500}q + \frac{1}{10}$; $du = \frac{1}{500}dq$, $dq = 500\, du$. When $q = 0$, $u = 0.1$; when $q = 500e^{-0.5} - 50$,

$$u = e^{-0.5}.$$

$$\int_{0}^{500e^{-0.5}-50} \ln\left(\tfrac{1}{500}\, q + \tfrac{1}{10}\right)\, dq = \int_{0.1}^{e^{-0.5}} 500 \ln u\, du$$

$$= 500[u \ln u - u]_{0.1}^{e^{-0.5}}$$

$= 500[e^{-0.5} \ln(e^{-0.5}) - e^{-0.5} - (0.1 \ln(0.1) - 0.1)] \approx -289.769.$

The consumers' surplus is therefore
$-2(-289.769) - 500e^{-0.5} + 50 = \$326.27.$

13. $q = p/2 - 5$, so $\bar{q} = 20/2 - 5 = 5$. The producers' surplus is

$$\int_{0}^{5} [20 - (10 + 2q)]\, dq = \int_{0}^{5} (10 - 2q)\, dq$$

$$= [10q - q^2]_{0}^{5} = 50 - 25 - (0) = \$25.$$

15. $q = (p/2 - 5)^3$, so $\bar{q} = (12/2 - 5)^3 = 1$. The producers' surplus is

$$\int_{0}^{1} [12 - (10 + 2q^{1/3})]\, dq =$$

$$\int_{0}^{1} (2 - 2q^{1/3})\, dq$$

$$= \left[2q - \tfrac{3}{2} q^{4/3}\right]_{0}^{1} = 2 - \tfrac{3}{2} - (0) = \$0.50.$$

17. $q = 2 \ln(p/500)$, so
$\bar{q} = 2 \ln(1000/500) = 2 \ln 2$.
The producers' surplus is

$$\int_{0}^{2 \ln 2} (1000 - 500e^{0.5q})\, dq$$

$$= [1000q - 1000e^{0.5q}]_{0}^{2 \ln 2}$$

$= 2000 \ln 2 - 2000 - (-1000)$

$= 2000 \ln 2 - 1000 = \$386.29.$

19. $\bar{q} = 2(40) - 50 = 30; p = q/2 + 25$. The producers' surplus is

$$\int_{0}^{30} \left[40 - \left(\tfrac{1}{2} q + 25\right)\right] dq$$

$$= \int_{0}^{30} \left(15 - \tfrac{1}{2} q\right) dq = \left[15q - \tfrac{1}{4} q^2\right]_{0}^{30}$$

$= 450 - 225 - (0) = \$225.$

21. $\bar{q} = 0.25(10)^2 - 10 = 15$
$p = \sqrt{4q + 40} = 2\sqrt{q + 10}$.
The producers' surplus is

$$\int_{0}^{15} (10 - 2\sqrt{q + 10})\, dq$$

$$= \left[10q - \tfrac{4}{3}(q + 10)^{3/2}\right]_{0}^{15}$$

$$= 150 - \tfrac{500}{3} - \left(-\tfrac{4}{3} \cdot 10^{3/2}\right) = \$25.50.$$

23. $\bar{q} = 500e^{0.05(10)} - 50 = 500e^{0.5} - 50$
$p = 20 \ln\left(\tfrac{1}{500}\, q + \tfrac{1}{10}\right)$.
The producers' surplus is

$$\int_{0}^{500e^{0.5}-50} \left[10 - 20 \ln\left(\tfrac{1}{500}\, q + \tfrac{1}{10}\right)\right] dq$$

$$= \left[10q - (20q + 1000) \ln\left(\tfrac{1}{500}\, q + \tfrac{1}{10}\right)\right.$$

$$\left. + (20q + 1000)\right]_{0}^{500e^{0.5}-50}$$

(using the substitution $u = \tfrac{1}{500}\, q + \tfrac{1}{10}$ as in Exercise 11)

$= 5000e^{0.5} - 500 - 10{,}000e^{0.5} \ln e^{0.5}$
$\quad + 10{,}000e^{0.5} -$

$\left(-1000 \ln\left(\tfrac{1}{10}\right) + 1000\right)$

$= \$12{,}684.63.$

25. $TV = \int_0^{10} 30{,}000 \; dt = [30{,}000t]_0^{10} =$

$300{,}000$; $FV = \int_0^{10} 30{,}000 e^{0.07(10-t)} \; dt =$

$\left[-\frac{30{,}000}{0.07} e^{0.07(10-t)} \right]_0^{10} = \$434{,}465.45$

27. $TV = \int_0^{10} (30{,}000 + 1000t) \; dt$

$= [30{,}000t + 500t^2]_0^{10} = \$350{,}000;$

$FV = \int_0^{10} (30{,}000 + 1000t) e^{0.07(10-t)} \; dt =$

$\left[-\frac{1}{0.07}(30{,}000 + 1000t) \; e^{0.07(10-t)} - \frac{1000}{0.07^2} e^{0.07(10-t)} \right]_0^{10}$

(using integration by parts) $= \$498{,}496.61$

29. $TV = \int_0^{10} 30{,}000 e^{0.05t} \; dt =$

$[600{,}000 e^{0.05t}]_0^{10}$

$= \$389{,}232.76;$

$FV = \int_0^{10} 30{,}000 e^{0.05t} e^{0.07(10-t)} \; dt$

$= \int_0^{10} 30{,}000 e^{0.7} e^{-0.02t} \; dt$

$= [-1{,}500{,}000 e^{0.7} e^{-0.02t}]_0^{10} = \$547{,}547.16$

31. $TV = \int_0^5 20{,}000 \; dt = [20{,}000t]_0^5 =$

$100{,}000$; $PV = \int_0^5 20{,}000 e^{-0.08t} \; dt = [-$

$250{,}000 e^{-0.08t}]_0^5$

$= \$82{,}419.99$

33. $TV = \int_0^5 (20{,}000 + 1000t) \; dt$

$= [20{,}000t + 500t^2]_0^5 = \$112{,}500;$

$PV = \int_0^5 (20{,}000 + 1000t) e^{-0.08t} \; dt$

$= [-(250{,}000 + 12{,}500t) e^{-0.08t} - 156{,}250 e^{-0.08t}]_0^5$ (by integration by parts)

$= \$92{,}037.48$

35. $TV = \int_0^5 20{,}000 e^{0.03t} \; dt$

$= \left[\frac{20{,}000}{0.03} e^{0.03t} \right]_0^5$

$= \$107{,}889.50;$

$PV = \int_0^5 20{,}000 e^{0.03t} e^{-0.08t} \; dt$

$= \int_0^5 20{,}000 e^{-0.05t} \; dt = [-400{,}000 e^{-0.05t}]_0^5$

$= \$88{,}479.69$

37. To find the equilibrium tuition set demand equal to supply:

$20{,}000 - 2p = 7500 + 0.5p$

so $\bar{p} = \$5000.$

The equilibrium supply is thus

$\bar{q} = 20{,}000 - 2(5{,}000) = 10{,}000.$

To find the consumers' surplus we solve for

$p = 10{,}000 - q/2$ and compute

$CS = \int_0^{10{,}000} \left(10{,}000 - \frac{q}{2} - 5000 \right) dq$

$= \int_0^{10{,}000} \left(5000 - \frac{q}{2} \right) dq = \left[5000q - \frac{q^2}{4} \right]_0^{10{,}000}$

$= \$25{,}000{,}000.$

To find the producers' surplus we solve for
$$p = 2q - 15{,}000$$
and compute

$$PS = \int_0^{10{,}000} [5000 - (2q - 15{,}000)] \, dq$$

$$= \int_0^{10{,}000} (20{,}0000 - 2q) \, dq$$

$$= [20{,}000q - q^2]_0^{10{,}000} = \$100{,}000{,}000.$$

The total social gain is \$125 million.

39. $\bar{q} = b - m\bar{p}$, $p = \dfrac{1}{m}(b - q)$

$$CS = \int_0^{b-m\bar{p}} \left[\frac{1}{m}(b - q) - \bar{p}\right] dq$$

$$= \left[\frac{b}{m} q - \frac{1}{2m} q^2 - \bar{p}q\right]_0^{b-m\bar{p}}$$

$$= \frac{b}{m}(b - m\bar{p}) - \frac{1}{2m}(b - m\bar{p})^2$$
$$\qquad - \bar{p}(b - m\bar{p}) - (0)$$

$$= \frac{1}{2m}(b - m\bar{p})[2b - (b - m\bar{p}) - 2m\bar{p}]$$

$$= \frac{1}{2m}(b - m\bar{p})^2$$

41. Total revenue $= \displaystyle\int_2^8 (\, 0.6t^2 - 2t + 30) \, dt$

$$= \left[\frac{0.6}{3}t^3 - t^2 + 30t\right]_2^8 = 220.8 \approx 220$$

This gives a total revenue of about €220 billion

43. Total revenue $= \displaystyle\int_0^6 162e^{0.11t} \, dt$

$$= \left[\frac{162}{0.11}e^{0.11t}\right]_0^6 \approx 1380 \text{ (to the nearest 10)}$$

This gives a total revenue of about \$1380 million.

45. Total revenue

$$= \int_2^8 (0.6t^2 - 2t + 30)e^{0.04(8-t)} \, dt$$

$$=$$
$$\left[-\frac{(0.6t^2 - 2t + 30)e^{0.04(8-t)}}{0.04} - \frac{(1.2t - 2)e^{0.04(8-t)}}{0.04^2} - \frac{1.2e^{0.04(8-t)}}{0.04^3}\right]_2^8$$

$$\approx 250 \text{ (to the nearest 10)}$$

This gives a total revenue of about €250 billion.

47. Total revenue $= \displaystyle\int_0^6 162e^{0.11t}e^{-0.05(6-t)} \, dt$

$$= \int_0^6 162e^{-0.3}e^{0.16t} \, dt$$

$$= \left[\frac{162}{0.16}e^{-0.3}e^{0.16t}\right]_0^6 \approx 1210$$

This gives a total revenue of about \$1210 billion.

49. $R(t) = 12 \times 700 = \$8400/\text{year}.$

$$FV = \int_0^{45} 8400e^{0.06(45-t)} \, dt$$

$$= [-140{,}000e^{0.06(45-t)}]_0^{45} = \$1{,}943{,}162.44$$

51. $R(t) = 12 \times 700e^{0.03t} = 8400e^{0.03t}$

$$FV = \int_0^{45} 8400e^{0.03t}e^{0.06(45-t)} \, dt$$

$$= \int_0^{45} 8400e^{2.7}e^{-0.03t} \, dt = [-280{,}000e^{2.7}e^{-}$$

$$^{0.03t}]_0^{45}$$

$$= \$3{,}086{,}245.73$$

53. $R(t) = 3375.$ $PV = \displaystyle\int_0^{30} 3375e^{-0.04t} \, dt$

$$= [-84{,}375e^{-0.04t}]_0^{30} = \$58{,}961.74$$

55. $R(t) = 100{,}000 + 5000t$

$$PV = \int_{0}^{20}(100{,}000 + 5000t)e^{-0.05t}\, dt$$

$$= [-20(100{,}000 + 5000t)e^{-0.05t} -$$
$$400(5000)e^{-0.05t}]_{0}^{20}$$

$$= \$1{,}792{,}723.35$$

57. Total

59. She is correct, provided there is a positive rate of return, in which case the future value (which includes interest) is greater than the total value (which does not).

61. $PV < TV < FV$

14.5 Improper Integrals and Applications

1. $\displaystyle\int_1^{+\infty} x\,dx = \lim_{M\to+\infty}\int_1^M x\,dx = \lim_{M\to+\infty}\left[\tfrac{1}{2}x^2\right]_1^M$

$\displaystyle = \lim_{M\to+\infty}\left(\tfrac{1}{2}M^2 - \tfrac{1}{2}\right) = +\infty;\ \text{diverges}$

3. $\displaystyle\int_{-2}^{+\infty} e^{-0.5x}\,dx = \lim_{M\to+\infty}\int_{-2}^M e^{-0.5x}\,dx$

$\displaystyle = \lim_{M\to+\infty}\left[-2e^{-0.5x}\right]_{-2}^M$

$\displaystyle = \lim_{M\to+\infty}(-2e^{-0.5M} + 2e) = 2e;\ \text{converges}$

5. $\displaystyle\int_{-\infty}^2 e^x\,dx = \lim_{M\to-\infty}\int_M^2 e^x\,dx = \lim_{M\to-\infty}\left[e^x\right]_M^2$

$\displaystyle = \lim_{M\to-\infty}(e^2 - e^M) = e^2;\ \text{converges}$

7. $\displaystyle\int_{-\infty}^{-2} \frac{1}{x^2}\,dx = \lim_{M\to-\infty}\int_M^{-2}\frac{1}{x^2}\,dx = \lim_{M\to-\infty}\left[-x^-\right.$

$\left.{}^1\right]_M^{-2}$

$\displaystyle = \lim_{M\to-\infty}\left(\tfrac{1}{2} + \tfrac{1}{M}\right) = \tfrac{1}{2}\ ;\ \text{converges}$

9. $\displaystyle\int_0^{+\infty} x^2 e^{-6x}\,dx = \lim_{M\to+\infty}\int_0^M x^2 e^{-6x}\,dx$

$\displaystyle = \lim_{M\to+\infty}$

$\displaystyle\left[-\tfrac{1}{6}x^2 e^{-6x} - \tfrac{2}{36}xe^{-6x} - \tfrac{2}{216}e^{-6x}\right]_0^M = \lim_{M\to+\infty}$

$\displaystyle\left(-\tfrac{1}{6}M^2 e^{-6M} - \tfrac{2}{36}Me^{-6M}\right.$

$\displaystyle\left. -\tfrac{2}{216}e^{-6M} + \tfrac{2}{216}\right)$

$\displaystyle = \tfrac{2}{216} = \tfrac{1}{108}\ ;\ \text{converges}$

11. $\displaystyle\int_0^5 \frac{2}{x^{1/3}}\,dx = \lim_{r\to0^+}\int_r^5 \frac{2}{x^{1/3}}\,dx = \lim_{r\to0^+}\left[3x^{2/3}\right]_r^5$

$\displaystyle = \lim_{r\to0^+}(3\times5^{2/3} - 3r^{2/3}) = 3\times5^{2/3};\ \text{converges}$

13. $\displaystyle\int_{-1}^2 \frac{3}{(x+1)^2}\,dx = \lim_{r\to-1^+}\int_r^2 \frac{3}{(x+1)^2}\,dx$

$\displaystyle = \lim_{r\to-1^+}\left[-3(x+1)^{-1}\right]_r^2$

$\displaystyle = \lim_{r\to-1^+}\left(-1 + \frac{3}{r+1}\right) = +\infty;\ \text{diverges}$

15. $\displaystyle\int_{-1}^2 \frac{3x}{x^2-1}\,dx = \lim_{r\to-1^+}\int_r^0 \frac{3x}{x^2-1}\,dx$

$\displaystyle + \lim_{r\to1^-}\int_0^r \frac{3x}{x^2-1}\,dx + \lim_{r\to1^+}\int_r^2 \frac{3x}{x^2-1}\,dx.$

Now, $\displaystyle\int \frac{3x}{x^2-1}\,dx = \tfrac{3}{2}\ln|x^2 - 1| + C$

by substitution, so

$\displaystyle\lim_{r\to-1^+}\int_r^0 \frac{3x}{x^2-1}\,dx = \lim_{r\to-1^+}$

$\displaystyle\left[\tfrac{3}{2}\ln|x^2 - 1|\right]_r^0$

$\displaystyle = \lim_{r\to-1^+}\left(-\tfrac{3}{2}\ln|r^2 - 1|\right) = +\infty$

Since this one part diverges, the whole integral diverges. (In fact, all three parts diverge.)

17. $\displaystyle\int_{-2}^{2} \frac{1}{(x+1)^{1/5}} \, dx$

$= \displaystyle\lim_{r \to -1^-} \int_{-2}^{r} \frac{1}{(x+1)^{1/5}} \, dx + \lim_{r \to -1^+}$

$\displaystyle\int_{r}^{2} \frac{1}{(x+1)^{1/5}} \, dx$

$= \displaystyle\lim_{r \to -1^-} \left[\frac{5}{4}(x+1)^{4/5} \right]_{-2}^{r} + \lim_{r \to -1^+}$

$\left[\frac{5}{4}(x+1)^{4/5} \right]_{r}^{2}$

$= \displaystyle\lim_{r \to -1^-} \left[\frac{5}{4}(r+1)^{4/5} - \frac{5}{4} \right]$

$\qquad + \displaystyle\lim_{r \to -1^+} \left[\frac{5}{4} 3^{4/5} - \frac{5}{4}(r+1)^{4/5} \right]$

$= \frac{5}{4}(3^{4/5} - 1)$; converges

19. $\displaystyle\int_{-1}^{1} \frac{2x}{x^2 - 1} \, dx$

$= \displaystyle\lim_{r \to -1^+} \int_{r}^{0} \frac{2x}{x^2 - 1} \, dx + \lim_{r \to 1^-} \int_{0}^{r} \frac{2x}{x^2 - 1} \, dx$

$= \displaystyle\lim_{r \to -1^+} [\ln|x^2 - 1|]_{r}^{0} + \lim_{r \to 1^-} [\ln|x^2 - 1|]_{0}^{r}$

(use the substitution $u = x^2 - 1$)

$= \displaystyle\lim_{r \to -1^+} (-\ln|r^2 - 1|) + \lim_{r \to 1^-} \ln|r^2 - 1|$

$= +\infty - \infty$; diverges

(Note that the infinities don't cancel. For convergence we need each part of the integral to converge on its own.)

21. $\displaystyle\int_{-\infty}^{+\infty} xe^{-x^2} \, dx$

$= \displaystyle\lim_{M \to -\infty} \int_{M}^{0} xe^{-x^2} \, dx + \lim_{M \to +\infty} \int_{0}^{M} xe^{-x^2} \, dx$

$= \displaystyle\lim_{M \to -\infty} \left[-\frac{1}{2}e^{-x^2} \right]_{M}^{0} + \lim_{M \to +\infty}$

$\left[-\frac{1}{2}e^{-x^2} \right]_{0}^{M}$

(use the substitution $u = -x^2$)

$= \displaystyle\lim_{M \to -\infty} \left(-\frac{1}{2} + \frac{1}{2}e^{-M^2} \right)$

$\qquad + \displaystyle\lim_{M \to +\infty} \left(-\frac{1}{2}e^{-M^2} + \frac{1}{2} \right)$

$= -\frac{1}{2} + \frac{1}{2} = 0$; converges

23. $\displaystyle\int_{0}^{+\infty} \frac{1}{x \ln x} \, dx$

$= \displaystyle\lim_{r \to 0^+} \int_{r}^{1/2} \frac{1}{x \ln x} \, dx + \lim_{r \to 1^-} \int_{1/2}^{r} \frac{1}{x \ln x} \, dx$

$\qquad + \displaystyle\lim_{r \to 1^+} \int_{r}^{2} \frac{1}{x \ln x} \, dx + \lim_{M \to +\infty} \int_{2}^{M} \frac{1}{x \ln x} \, dx$

$\displaystyle\lim_{r \to 0^+} \int_{r}^{1/2} \frac{1}{x \ln x} \, dx = \lim_{r \to 0^+} [\ln |\ln x|]_{r}^{1/2}$

(use the substitution $u = \ln x$)

$= \displaystyle\lim_{r \to 0^+} [\ln |\ln(1/2)| - \ln |\ln r|] = -\infty$

Without checking the remaining parts of the integral we can say that the whole integral diverges.

25. $\displaystyle\int_0^{+\infty} \frac{2x}{x^2 - 1}\, dx$

$\displaystyle = \lim_{r \to 1^-} \int_0^r \frac{2x}{x^2 - 1}\, dx + \lim_{r \to 1^+} \int_r^2 \frac{2x}{x^2 - 1}\, dx$

$\displaystyle \qquad + \lim_{M \to +\infty} \int_2^M \frac{2x}{x^2 - 1}\, dx$

$\displaystyle = \lim_{r \to 1^-} [\ln|x^2 - 1|]_0^r + \lim_{r \to 1^+} [\ln|x^2 - 1|]_r^2$

$\displaystyle \qquad + \lim_{M \to +\infty} [\ln|x^2 - 1|]_2^M$

$\displaystyle = \lim_{r \to 1^-} \ln|r^2 - 1| + \lim_{r \to 1^+} (\ln 3 - \ln|r^2 - 1|)$

$\displaystyle \qquad + \lim_{M \to +\infty} (\ln|M^2 - 1| - \ln 3)$

$= -\infty + \infty + \infty;$ diverges

27. $0.9, 0.99, 0.999, \ldots$ Converges to 1.

29. $7.602, 95.38, 993.1, \ldots$ Diverges

31. $1.368, 1.800, 1.937, 1.980, 1.994, 1.998,$
$1.999, 2.000, \ldots$
Converges to 2

33. $9.000, 99.00, 999.0, \ldots$
Diverges to $+\infty$

35. Total number of homes $= \displaystyle\int_0^{+\infty} 1.05 e^{-0.376t}\, dt$

$\displaystyle = \lim_{M \to +\infty} \int_0^M 1.05 e^{-0.376t}\, dt$

$\displaystyle = \lim_{M \to +\infty} \left[\frac{1.05}{-0.376} e^{-0.376t} \right]_0^M$

$\displaystyle = \lim_{M \to +\infty} \left[\frac{1.05}{-0.376} [e^{-0.376M} - 1] \right]$

$\displaystyle = \frac{1.05}{0.376} \approx 2.79$ million homes

37. Annual sales $= S(t) = 415(0.97)^t$ billion cigarettes per year.

Total sales $= \displaystyle\int_0^{+\infty} 415(0.97)^t\, dt$

$\displaystyle = \lim_{M \to +\infty} \int_0^M 415(0.97)^t\, dt$

$\displaystyle = \lim_{M \to +\infty} \left[\frac{415}{\ln 0.97} (0.97)^t \right]_0^M$

$\displaystyle = \lim_{M \to +\infty} \left[\frac{415}{\ln 0.97} (0.97)^M - \frac{415}{\ln 0.97} \right]$

$\displaystyle = -\frac{415}{\ln 0.97} \approx 13600$ billion cigarettes

39. Annual sales $= S(t) = 200(0.90)^t$.

Total sales $= \displaystyle\int_0^{+\infty} 200(0.90)^t\, dt$

$\displaystyle = \lim_{M \to +\infty} \int_0^M 200(0.90)^t\, dt$

$\displaystyle = \lim_{M \to +\infty} \left[\frac{200}{\ln 0.90} (0.90)^t \right]_0^M$

$\displaystyle = \lim_{M \to +\infty} \left[\frac{200}{\ln 0.90} (0.90)^M - \frac{200}{\ln 0.90} \right]$

$\displaystyle = -\frac{200}{\ln 0.90} \approx 1900.$

No, you will not sell more than about 2000 of them.

41. $\displaystyle \lim_{t \to +\infty} N(t) = 2.8 + \int_1^{+\infty} 0.20 t^{-0.93}\, dt$

$\displaystyle = 2.8 + \lim_{M \to +\infty} \int_1^M 0.20 t^{-0.93}\, dt$

$\displaystyle \approx 2.8 + \lim_{M \to +\infty} [2.857 t^{0.07}]_1^M$

$\displaystyle = 2.8 + \lim_{M \to +\infty} [2.857 M^{0.07} - 2.857] = +\infty$

43. (a) The revenue per cell phone user is $P(t) = 350 e^{-0.1t}$; multiplying by the number of users

529

gives the annual revenue as $R(t) = 350e^{-0.1t}(39t + 68)$ million dollars per year.

(b) Total revenue $= \int_0^{+\infty} 350e^{-0.1t}(39t + 68)\, dt$

$= \lim_{M \to +\infty} \int_0^M 350e^{-0.1t}(39t + 68)\, dt$

$= \lim_{M \to +\infty} [-3500e^{-0.1t}(39t + 68)$

$\quad - 1,365,000e^{-0.1t}]_0^M$

$= \lim_{M \to +\infty} [-3500e^{-0.1M}(39M + 68)$

$\quad - 1,365,000e^{-0.1M} + 3500(68) +$

$1,365,000]$

$= 3500(68) + 1,365,000$

$= \$1,603,000$ million.

45. The total investment is

$\int_0^{+\infty}(0.2t^2 + 3.5t + 60)e^{-0.03t}\, dt$

$= \lim_{M \to +\infty} \int_0^M (0.2t^2 + 3.5t + 60)e^{-0.03t}\, dt$

Tech formula:
```
(0.2x^2 + 3.5x + 60)*e^(-0.03x)
```
Following the technology note in the textbook, we compute

$\int_0^M (0.2t^2 + 3.5t + 60)e^{-0.03t}\, dt$

where $M = 10, 100, 1000, 10000,..$ This gives integrals of approximately
715, 13560, 20704, 20704, ...
so we conclude that

$\int_0^{+\infty}(0.2t^2 + 3.5t + 60)e^{-0.03t}\, dt \approx 20{,}704,$

or about \$20,700 million.

47. $\int_0^{+\infty} N(t)\, dt = \int_0^{+\infty} \frac{80(7)^t}{20 + 7^t}\, dt$

$= \lim_{M \to +\infty} \int_0^M \frac{80(7)^t}{20 + 7^t}\, dt$

$= \lim_{M \to +\infty} \left[\frac{80}{\ln 7} \ln(20 + 7^t)\right]_0^M$

(use the substitution $u = 20 + 7^t$)

$= \lim_{M \to +\infty} \left[\frac{80}{\ln 7} \ln(20 + 7^M) - \frac{80}{\ln 7} \ln(21)\right]$

$= +\infty.$

$\int_0^{+\infty} N(t)\, dt$ diverges, indicating that there is no bound to the expected total future online sales of mousse.

$\int_{-\infty}^0 N(t)\, dt = \int_{-\infty}^0 \frac{80(7)^t}{20 + 7^t}\, dt$

$= \lim_{M \to -\infty} \int_M^0 \frac{80(7)^t}{20 + 7^t}\, dt$

$= \lim_{M \to -\infty} \left[\frac{80}{\ln 7} \ln(20 + 7^t)\right]_M^0$

$= \lim_{M \to -\infty} \left[\frac{80}{\ln 7} \ln(21) - \frac{80}{\ln 7} \ln(20 + 7^M)\right]$

$= \frac{80}{\ln 7} \ln(21) - \frac{80}{\ln 7} \ln(20))$

≈ 2.006

$\int_{-\infty}^0 N(t)\, dt$ converges to approximately 2.006,

indicating that total online sales of mousse prior to the current year amounted to approximately 2 million gallons.

49. 1

51. 0.1587

53. The value per bottle is $P(t) = 85e^{0.4t}$. The annual sales rate is $Q(t) = 500e^{-t}$. The annual income is

$$R(t) = P(t)Q(t) = 42{,}500e^{-0.6t}$$

The total income is

$$\int_0^{+\infty} 42{,}500e^{-0.6t}\,dt = \lim_{M \to +\infty}$$

$$\int_0^M 42{,}500e^{-0.6t}\,dt$$

$$= \lim_{M \to +\infty} [-70{,}833e^{-0.6t}]_0^M$$

$$= \lim_{M \to +\infty} [-70{,}833e^{-0.6M} + 70{,}833]$$

$$= \$70{,}833$$

55. (a) $\displaystyle\int_{0.2}^{+\infty} \frac{1}{5.6997k^{1.081}}\,dk$

$$= \lim_{M \to +\infty} \int_{0.2}^M \frac{1}{5.6997k^{1.081}}\,dk$$

$$= \lim_{M \to +\infty} [-2.166k^{-0.081}]_{0.2}^M$$

$$= \lim_{M \to +\infty} [-2.166M^{-0.081} + 2.166(0.2)^{-0.081}]$$

$$\approx 2.468 \text{ meteors on average}$$

(b) $\displaystyle\int_0^1 \frac{1}{5.6997k^{1.081}}\,dk = \lim_{r \to 0} \int_r^1 \frac{1}{5.6997k^{1.081}}\,dk$

$$= \lim_{r \to 0} [-2.166k^{-0.081}]_r^1$$

$$= \lim_{r \to 0} [-2.166 + 2.166r^{-0.081}]$$

$$= +\infty; \text{ the integral diverges.}$$

We can interpret this as saying that the number of impacts by meteors smaller than 1 megaton is very large. (This makes sense because, for example, this number includes meteors no larger than a grain of dust.)

57. (a) $\displaystyle\Gamma(1) = \int_0^{+\infty} e^{-t}\,dt = \lim_{M \to +\infty} \int_0^M e^{-t}\,dt$

$$= \lim_{M \to +\infty} [-e^{-t}]_0^M = \lim_{M \to +\infty} [-e^{-M} + 1]$$

$$= 1$$

$$\Gamma(2) = \int_0^{+\infty} te^{-t}\,dt = \lim_{M \to +\infty} \int_0^M te^{-t}\,dt$$

$$= \lim_{M \to +\infty} [-te^{-t} - e^{-t}]_0^M$$

$$= \lim_{M \to +\infty} [-Me^{-M} - e^{-M} + 1]$$

$$= 1$$

(b) $\displaystyle\Gamma(n+1) = \int_0^{+\infty} t^n e^{-t}\,dt = \lim_{M \to +\infty} \int_0^M t^n e^{-t}\,dt$

$$= \lim_{M \to +\infty} \left([-t^n e^{-t}]_0^M + \int_0^M n t^{n-1} e^{-t}\,dt \right)$$

$$= \lim_{M \to +\infty} [-M^n e^{-M} + 0] + n \int_0^{+\infty} t^{n-1} e^{-t}\,dt$$

$$= n\Gamma(n)$$

(c) If n is a positive integer, then applying part (b) several times we get

$$\Gamma(n) = (n-1)\Gamma(n-1)$$
$$= (n-1)(n-2)\Gamma(n-2)$$
$$= \cdots$$
$$= (n-1)(n-2)\cdots 1\Gamma(1)$$
$$= (n-1)!$$

by part (a).

59. The integral does not converge, so the number given by the FTC is meaningless.

61. Yes; the integrals converge to 0, and the FTC also gives 0.

63. (a) Not improper. $|x|/x$ is not defined at zero, but $\lim_{x \to 0^-} |x|/x = -1$ and $\lim_{x \to 0^+} |x|/x = 1$. Since these limits are finite, the integral is not improper.

(b) Improper, since $x^{-1/3}$ has infinite left- and right limits at 0.

(c) Improper, since $(x-2)/(x^2-4x+4) = 1/(x-2)$, which has an infinite left limit at 2.

65. In all cases, you need to rewrite the improper integral as a limit and use technology to evaluate the integral of which you are taking the limit. Evaluate for several values of the endpoint approaching the limit. In the case of an integral in which one of the limits of integration is infinite, you may have to instruct the calculator or computer to use more subdivisions as you approach $+\infty$.

67. Answers will vary.

14.6 Differential Equations and Applications

1. $y = \int (x^2 + \sqrt{x})\, dx = \dfrac{x^3}{3} + \dfrac{2x^{3/2}}{3} + C$

3. $y\, dy = x\, dx;$ $\int y\, dy = \int x\, dx$; $\dfrac{y^2}{2} = \dfrac{x^2}{2} + C$

5. $\dfrac{1}{y}\, dy = x\, dx;$ $\int \dfrac{1}{y}\, dy = \int x\, dx$; $\ln |y| = \dfrac{x^2}{2} +$
$C;$ $|y| = e^{x^2/2+C} = e^C e^{x^2/2}$; $y = Ae^{x^2/2}$ (where
$A = \pm e^C$)

7. $\dfrac{1}{y^2}\, dy = (x+1)\, dx;$ $\int \dfrac{1}{y^2}\, dy = \int (x+1)\, dx$;
$-\dfrac{1}{y} = \dfrac{1}{2}(x+1)^2 + K = \dfrac{(x+1)^2 + C}{2}$;
$y = -\dfrac{2}{(x+1)^2 + C}$

9. $y\, dy = \dfrac{\ln x}{x}\, dx;$ $\int y\, dy = \int \dfrac{\ln x}{x}\, dx$;
$\dfrac{y^2}{2} = (\ln x)^2 + C$ (substitute $u = \ln x$);
$y = \pm\sqrt{(\ln x)^2 + C}$

11. $y = \int (x^3 - 2x)\, dx = \dfrac{x^4}{4} - x^2 + C;$
$1 = 0 - 0 + C;$ $C = 1;$ $y = \dfrac{x^4}{4} - x^2 + 1$

13. $y^2\, dy = x^2\, dx;$ $\int y^2\, dy = \int x^2\, dx$;
$\dfrac{y^3}{3} = \dfrac{x^3}{3} + C;$ $y^3 = x^3 + K;$ $8 = 0 + K;$ $K = 8;$
$y^3 = x^3 + 8;$ $y = (x^3 + 8)^{1/3}$

15. $\dfrac{1}{y}\, dy = \dfrac{1}{x}\, dx;$ $\int \dfrac{1}{y}\, dy = \int \dfrac{1}{x}\, dx$;
$\ln |y| = \ln |x| + C;$ $|y| = e^{\ln|x|+C} = e^C e^{\ln|x|} = e^C |x|;$
$y = Ax;$ $2 = A \times 1;$ $A = 2;$ $y = 2x$

17. $\dfrac{1}{y+1}\, dy = x\, dx;$ $\int \dfrac{1}{y+1}\, dy = \int x\, dx$;
$\ln |y+1| = \dfrac{x^2}{2} + C;$ $\ln 1 = 0 + C;$ $C = 0;$
$\ln |y+1| = \dfrac{x^2}{2}$; $|y+1| = e^{x^2/2}$; $y+1 = e^{x^2/2}$
(note that $y(0) + 1 = 1 > 0$); $y = e^{x^2/2} - 1$

19. $\dfrac{1}{y^2}\, dy = \dfrac{x}{x^2+1}\, dx;$ $\int \dfrac{1}{y^2}\, dy = \int \dfrac{x}{x^2+1}\, dx$;
$-\dfrac{1}{y} = \dfrac{1}{2}\ln(x^2+1) + C;$ $1 = 0 + C;$ $C = 1;$
$-\dfrac{1}{y} = \dfrac{1}{2}\ln(x^2+1) + 1;$ $y = -\dfrac{2}{\ln(x^2+1)+2}$

21. With $s(t) =$ monthly sales after t months,
$\dfrac{ds}{dt} = -0.05s;$ $s = 1000$ when $t = 0.$ $\dfrac{1}{s}\, ds =$
$-0.05\, dt;$ $\int \dfrac{1}{s}\, ds = \int (-0.05)\, dt$; $\ln |s| =$
$-0.05t + C;$ $|s| = e^{-0.05t+C} = e^C e^{-0.05t};$ $s = Ae^{-0.05t};$ $1000 = A \times 1;$ $A = 1000;$ $s = 1000e^{-0.05t}$
quarts per month.

23. (a) By Example 5,
$$H(t) = T_a + (T_0 - T_a)e^{-kt}$$
$$= 75 + (200 - 75)e^{-0.05t}$$
$$= 75 + 125e^{-0.05t}$$

(b) The coffee will have cooled to $80°$ when
$$80 = 75 + 125e^{-0.05t}$$
$$125e^{-0.05t} = 5$$
$$e^{-0.05t} = 0.04$$
$$-0.05t = \ln 0.04$$
$$t = -(\ln 0.04)/0.05 \approx 64.4 \text{ minutes}$$

25. $\dfrac{dH}{dt} = -k(H - 75); \dfrac{1}{H - 75} dH = -k\, dt;$

$\displaystyle\int \dfrac{1}{H - 75} dH = \int(-k)\, dt\; 0$

$\ln(H - 75) = -kt + C$

$H(t) = 75 + Ae^{-kt}.$

$H(0) = 190,$ so

$190 = 75 + A;\; A = 115.$

$H(10) = 150,$ so

$150 = 75 + 115e^{-10k}$

$k = -\dfrac{1}{10}\ln\left(\dfrac{150 - 75}{115}\right) \approx 0.04274.$

$H(t) = 75 + 115e^{-0.04274t}$ degrees Fahrenheit
after t minutes.

27. With $S(t) = $ total sales after t months, $\dfrac{dS}{dt} =$

$0.1(100{,}000 - S);\; S(0) = 0.\; \dfrac{1}{100{,}000 - S}\, dS$

$= 0.1\; dt;\; \displaystyle\int \dfrac{1}{100{,}000 - S}\, dS = \int 0.1\; dt\;;$

$-\ln(100{,}000 - S) = 0.1t + C;\; S(t) = 100{,}000$
$- Ae^{-0.1t}.\; 0 = 100{,}000 - A;\; A = 100{,}000.\; S(t)$
$= 100{,}000 - 100{,}000e^{-0.1t} = 100{,}000(1 - e^{-0.1t})$ monitors after t months.

29. $-\dfrac{p}{q}\dfrac{dq}{dp} = 0.05p - 1.5;\; \dfrac{1}{q}\, dq =$

$\left(-0.05 + \dfrac{1.5}{p}\right) dp;\; \displaystyle\int \dfrac{1}{q}\, dq =$

$\displaystyle\int\left(-0.05 + \dfrac{1.5}{p}\right) dp;\; \ln q = -0.05p + 1.5\ln p$

$+ C;\; q = Ae^{-0.05p}p^{1.5}.\; q(20) = 20,$ so $20 = Ae^{-1}(20)^{1.5};\; A = e(20)^{-0.5} \approx 0.6078.\; q = 0.6078e^{-0.05p}p^{1.5}.$

31. $p = 1$ and $f(t) = e^{-t}$

$y = e^{-pt}\displaystyle\int f(t)e^{pt}\, dt$

$= e^{-t}\displaystyle\int e^{-t}e^{t}\, dt$

$= e^{-t}\displaystyle\int 1\, dt$

$y = e^{-t}(t + C)$

Initial condition:

$1 = e^{0}(0 + C)$

$1 = C$

So the solution is

$y = e^{-t}(t + 1)$

33. Rewrite the given equation in the form

$\dfrac{dy}{dt} + py = f(t)$

by dividing both sides by 2:

$\dfrac{dy}{dt} - \dfrac{1}{2}y = t$

$p = -\dfrac{1}{2}$ and $f(t) = t$

$y = e^{-pt}\displaystyle\int f(t)e^{pt}\, dt$

$= e^{t/2}\displaystyle\int te^{-t/2}\, dt$

$= e^{t/2}\left[-2te^{-t/2} - 4e^{-t/2} + C\right]$

(using integration by parts)

Initial condition:

$1 = [-4 + C]$

$C = 5$

So the solution is

$y = e^{t/2}\left[-2te^{-t/2} - 4e^{-t/2} + 5\right]$

35. $L\dfrac{di}{dt} + Ri = V(t)$ with $L = R = 1$ gives

$\dfrac{di}{dt} + i = 5\left[1 + \dfrac{|t-1|}{t-1}\right]$

$p = 1$ and $f(t) = 5\left[1 + \dfrac{|t-1|}{t-1}\right]$

$i = e^{-pt}\displaystyle\int f(t)e^{pt}\, dt$

$= 5e^{-t}\displaystyle\int\left[1 + \dfrac{|t-1|}{t-1}\right]e^{t}\, dt$

$- 5e^{-t}\left(\left[1 + \dfrac{|t-1|}{t-1}\right](e^{t} - e) + C\right)$

(using integration by parts)

534

Initial condition: No current flowing at time $t = 0$:

$i = 0$ when $t = 0$

$$0 = 5\left([1 + (-1)](1-e) + C \right)$$

$C = 0$

So the solution is

$$i = 5e^{-t}(e^t - e)\left[1 + \frac{|t-1|}{t-1} \right]$$

Technology formula for graph:

`5*e^(-x)*(e^x-e)*(1+abs(x-1)/(x-1))`

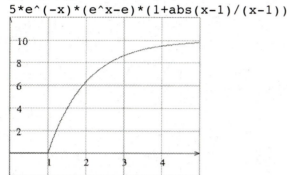

37. (a) $\dfrac{dp}{dt} = k[D(p) - S(p)]$

$\quad = k(20{,}000 - 1000p)$

(b) $\dfrac{1}{20{,}000 - 1000p}\, dp = k\, dt;$

$\displaystyle\int \dfrac{1}{20{,}000 - 1000p}\, dp = \int k\, dt\;;$

$-\ln(20{,}000 - 1000p) = kt + C;$

$p(t) = 20 - Ae^{-kt}.$

(c) $p(0) = 10$ and $p(1) = 12$, so $10 = 20 - A;$

$A = 10;\; 12 = 20 - 10e^{-k};\; k = -\ln\left(\dfrac{20 - 12}{10}\right) \approx$

$0.2231;\; p(t) = 20 - 10e^{-0.2231t}$ dollars after t months.

39. If $y = \dfrac{CL}{e^{-aLt} + C}$ then $\dfrac{dy}{dt} = \dfrac{-aCL^2e^{-aLt}}{(e^{-aLt} + C)^2}$ and

$ay(L - y) = a\left(\dfrac{CL}{e^{-aLt} + C}\right)\left(\dfrac{-Le^{-aLt}}{e^{-aLt} + C}\right) =$

$\dfrac{-aCL^2e^{-aLt}}{(e^{-aLt} + C)^2}$ also, so this y satisfies the

differential equation.

41. $a = 1/4$ and $L = 2$, so $S = \dfrac{2C}{e^{-0.5t} + C}$ for

some C. $S = 0.001$ when $t = 0$, so

$0.001 = \dfrac{2C}{1 + C};\; C = \dfrac{0.001}{2 - 0.001} = \dfrac{1}{1999}.$

$S = \dfrac{2/1999}{e^{-0.5t} + 1/1999}.$

Graph:

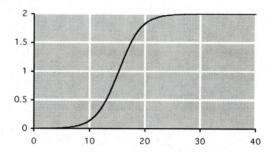

It will take about 27 months to saturate the market.

43. (a) $\dfrac{1}{y \ln(y/b)}\, dy = -a\, dt;\; \displaystyle\int \dfrac{1}{y \ln(y/b)}\, dy =$

$\displaystyle\int (-a)\, dt\;;\; \ln[\ln(y/b)] = -at + C$ [use the

substitution $u = \ln(y/b)$]; $y = be^{Ae^{-at}}$, for some constant A.

(b) $5 = 10e^A,\; A = \ln 0.5 \approx -0.69315;$

$y = 10e^{-0.69315e^{-t}}$. Graph:

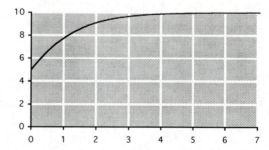

45. A general solution gives all possible solutions to the equation, using at least one arbitrary constant. A particular solution is one specific function that satisfies the equation. We obtain a particular solution by substituting specific values for any arbitrary constants in the general solution.

47. Example: $\dfrac{d^2y}{dx^2} = 1$ has general solution

$y = \dfrac{1}{2}x^2 + Cx + D$ (integrate twice).

49. Differentiate to get the differential equation

$\dfrac{dy}{dx} = -4e^{-x} + 3$.

Chapter 14 Review Exercises

1.

	D	I
+	$x^2 + 2$	e^x
−	$2x$	e^x
+	2	e^x
−∫	0 →	e^x

$\int (x^2 + 2)e^x \, dx = (x^2 + 2)e^x - 2xe^x + 2e^x + C$

$= (x^2 - 2x + 4)e^x + C$

3.

	D	I
+	$\ln(2x)$	x^2
−∫	$1/x$ →	$x^3/3$

$\int x^2 \ln(2x) \, dx = \frac{1}{3} x^3 \ln(2x) - \int \frac{1}{3} x^2 \, dx$

$= \frac{1}{3} x^3 \ln(2x) - \frac{1}{9} x^3 + C$

5.

	D	I
+	$2x$	$|2x+1|$
−	2	$\frac{1}{4}(2x+1)|2x+1|$
+∫	0 →	$\frac{1}{24}(2x+1)^2|2x+1|$

$\int 2x|2x+1| \, dx$

$= \frac{1}{2}x(2x+1)|2x+1| - \frac{1}{12}(2x+1)^2|2x+1| + C$

7.

	D	I
+	$5x$	$\dfrac{|-x+3|}{-x+3}$
−	5	$-|-x+3|$
+∫	0 →	$\frac{1}{2}(-x+3)|-x+3|$

$\int 5x \dfrac{|-x+3|}{-x+3} \, dx$

$= -5x|-x+3| - \frac{5}{2}(-x+3)|-x+3| + C$

9.

	D	I
+	$x^3 + 1$	e^{-x}
−	$3x^2$	$-e^{-x}$
+	$6x$	e^{-x}
−	6	$-e^{-x}$
+∫	0 →	e^{-x}

$\int_{-2}^{2} (x^3 + 1)e^{-x} \, dx$

$= [-(x^3 + 1)e^{-x} - 3x^2 e^{-x} - 6xe^{-x} - 6e^{-x}]_{-2}^{2}$

$= [-(x^3 + 3x^2 + 6x + 7)e^{-x}]_{-2}^{2}$

$= -39e^{-2} - e^2 \approx -12.67$

11. $x^3 = 1 - x^3$ when $x^3 = 1/2$, $x = 1/2^{1/3}$.

Area $= \displaystyle\int_{0}^{1/2^{1/3}} [(1 - x^3) - x^3] \, dx +$

$\displaystyle\int_{1/2^{1/3}}^{1} [x^3 - (1 - x^3)] \, dx$

$$= \int_0^{1/2^{1/3}} (1 - 2x^3)\, dx + \int_{1/2^{1/3}}^1 (2x^3 - 1)\, dx$$

$$= \left[x - \tfrac{1}{2} x^4 \right]_0^{1/2^{1/3}} + \left[\tfrac{1}{2} x^4 - x \right]_{1/2^{1/3}}^1$$

$$= \frac{1}{2^{1/3}} - \frac{1}{4 \cdot 2^{1/3}} + \frac{1}{2} - 1 - \frac{1}{4 \cdot 2^{1/3}} + \frac{1}{2^{1/3}}$$

$$= \frac{3}{2 \cdot 2^{1/3}} - \frac{1}{2} \approx 0.6906$$

13. $1 - x^2 = x^2$ when $x = \pm 1/\sqrt{2}$, so

$$\text{Area} = \int_{-1/\sqrt{2}}^{1/\sqrt{2}} [(1 - x^2) - x^2]\, dx$$

$$= \int_{-1/\sqrt{2}}^{1/\sqrt{2}} (1 - 2x^2)\, dx$$

$$= \left[x - \tfrac{2}{3} x^3 \right]_{-1/\sqrt{2}}^{1/\sqrt{2}}$$

$$= \frac{1}{\sqrt{2}} - \frac{2}{6\sqrt{2}} + \frac{1}{\sqrt{2}} - \frac{2}{6\sqrt{2}}$$

$$= \frac{4}{3\sqrt{2}} = \frac{2\sqrt{2}}{3}$$

15. $\bar{f} = \dfrac{1}{2-(-2)} \displaystyle\int_{-2}^2 x^3 - 1\, dx = \dfrac{1}{4}\left[\dfrac{x^4}{4} - x \right]_{-2}^2$

$$= \frac{1}{4}[2 - 6] = -1$$

17. Average $= \dfrac{1}{1-0} \displaystyle\int_0^1 x^2 e^x\, dx$

$$= [x^2 e^x - 2xe^x + 2e^x]_0^1$$

(using integration by parts)

$$= e - 2$$

19. $\bar{f} = \dfrac{1}{2} \displaystyle\int_{x-2}^x (3t + 1)\, dt = \dfrac{1}{2}\left[\dfrac{3t^2}{2} + t \right]_{x-2}^x$

$$= \frac{1}{2}\left[\frac{3x^2}{2} + x - \left(\frac{3(x-2)^2}{2} + x - 2 \right) \right]$$

$$= 3x - 2$$

21. $\dfrac{1}{2} \displaystyle\int_{x-2}^x t^{4/3}\, dt = \left[\dfrac{3}{14} t^{7/3} \right]_{x-2}^x$

$$= \frac{3}{14} [x^{7/3} - (x - 2)^{7/3}]$$

23. $q = 100 - 2p$, so $\bar{q} = 100 - 20 = 80$. The consumers' surplus is

$$CS = \int_0^{80} \left(50 - \tfrac{1}{2} q - 10 \right) dq$$

$$= \int_0^{80} \left(40 - \tfrac{1}{2} q \right) dq$$

$$= [40q - \tfrac{1}{4} q^2]_0^{80} = 3200 - 1600 = \$1600.$$

25. $q = 2p - 100$, so $\bar{q} = 200 - 100 = 100$. The producers' surplus is

$$PS = \int_0^{100} \left(100 - 50 - \tfrac{1}{2} q \right) dq$$

$$= \int_0^{100} \left(50 - \tfrac{1}{2} q \right) dq$$

$$= [50q - \tfrac{1}{4} q^2]_0^{100} = 5000 - 2500 = \$2500.$$

27. $\displaystyle\int_1^{+\infty} \frac{1}{x^5}\, dx = \lim_{M \to +\infty} \int_1^M \frac{1}{x^5}\, dx$

$$= \lim_{M \to +\infty} \left[-\tfrac{1}{4} x^{-4} \right]_1^M$$

$$= \lim_{M \to +\infty} \left[-\tfrac{1}{4} M^{-4} + \tfrac{1}{4} \right] = \frac{1}{4}$$

29. $\displaystyle\int_{-1}^{1} \frac{x}{(x^2-1)^{5/3}}\, dx$

$\displaystyle = \lim_{r \to -1^+} \int_{r}^{0} \frac{x}{(x^2-1)^{5/3}}\, dx$

$\displaystyle + \lim_{r \to 1^-} \int_{0}^{r} \frac{x}{(x^2-1)^{5/3}}\, dx$

$\displaystyle = \lim_{r \to -1^+}\left[\frac{-3}{4(x^2-1)^{2/3}}\right]_{r}^{0}$

$\displaystyle + \lim_{r \to 1^-}\left[\frac{-3}{4(x^2-1)^{2/3}}\right]_{0}^{r}$

(use the substitution $u = x^2 - 1$)

$= +\infty - \infty$; diverges

(Note that the infinities don't cancel. For convergence we need each part of the integral to converge on its own.)

31. $\displaystyle\int_{0}^{+\infty} 2xe^{-x^2}\, dx = \lim_{M \to +\infty}\int_{0}^{M} 2xe^{-x^2}\, dx$

$\displaystyle = \lim_{M \to +\infty}\left[-e^{-x^2}\right]_{0}^{M}$ (use $u = -x^2$)

$\displaystyle = \lim_{M \to +\infty}\left[-e^{-M^2} - (-1)\right] = 1$

33. $\dfrac{1}{y^2}\, dy = x^2\, dx;\ \displaystyle\int\frac{1}{y^2}\, dy = \int x^2\, dx\ ;$

$-\dfrac{1}{y} = \dfrac{x^3}{3} + K = \dfrac{x^3+C}{3}\ ;\ y = -\dfrac{3}{x^3+C}$

35. $y\, dy = \dfrac{1}{x}\, dx;\ \displaystyle\int y\, dy = \int\frac{1}{x}\, dx\ ;\ \dfrac{y^2}{2} = \ln |x| +$

$C;\ \dfrac{1}{2} = 0 + C;\ \dfrac{y^2}{2} = \ln |x| + \dfrac{1}{2}\ ;\ y =$

$\sqrt{2 \ln |x| + 1}$ (note that $y(1) > 0$)

37. Rate of spending $= p(t)q(t)$

So total spending $= \displaystyle\int_{0}^{5} p(t)q(t)\, dt$

$= \displaystyle\int_{0}^{5} 9e^{0.09t}(45t + 200)\, dt$

	D	I
$+$	$45t + 200$	$9e^{0.09t}$
$-$	45	$100e^{0.09t}$
$+\int$	0	$\dfrac{10{,}000}{9}e^{0.09t}$

$\displaystyle\int_{0}^{5} 9e^{0.09t}(45t + 200)\, dt =$

$\left[e^{0.09t}\left(100[45t+200] - 45\dfrac{10{,}000}{9}\right)\right]_{0}^{5}$

$\approx 18{,}238 \approx 18{,}200$ dollars

39. The monthly cost is $2000e^{0.01t}$ dollars after t months, so the average amount over two years is

$\dfrac{2000}{24-0}\displaystyle\int_{0}^{24} e^{0.01t}\, dt = 2000\left[100e^{0.01t}\right]_{0}^{24}$

$= \dfrac{200{,}000}{24}(e^{0.24}-1) \approx \$2260$

(b) $\dfrac{2000}{4}\displaystyle\int_{t-4}^{t} e^{0.01s}\, ds$

$= 500\left[100e^{0.01s}\right]_{t-4}^{t}$

$= 50{,}000(e^{0.01t} - e^{0.01(t-4)})$

$= 50{,}000e^{0.01t}(1 - e^{-0.04})$

$\approx 1960.53e^{0.01t}$

41. (a) For equilibrium price,

$20{,}000(28-p)^{1/3} = 40{,}000(p-19)^{1/3}$

gives, by cubing both sides,

$28-p = 8(p-19)$

$9p = 180$, so

$\bar{p} = 20$ and

$\bar{q} = 20{,}000(28-20)^{1/3} = 40{,}000$

(b) Consumers' surplus:

Solve the demand equation for p:

$$q = 20{,}000(28-p)^{1/3}$$

$$(28-p)^{1/3} = \frac{q}{20{,}000} = \frac{q}{2\times10^4}$$

$$28-p = \frac{q^3}{8\times10^{12}}$$

$$p = 28-\frac{q^3}{8\times10^{12}}$$

$$CS = \int_0^{40{,}000}\left(28-\frac{q^3}{8\times10^{12}} - 20\right)dq$$

$$= \left[28q - \frac{q^4}{32\times10^{12}} - 20q\right]_0^{40{,}000}$$

$$= \$240{,}000$$

Producers' surplus:

Solve the supply equation for p:

$$q = 40{,}000(p-19)^{1/3}$$

$$(p-19)^{1/3} = \frac{q}{40{,}000} = \frac{q}{4\times10^4}$$

$$p-19 = \frac{q^3}{64\times10^{12}}$$

$$p = 19+\frac{q^3}{64\times10^{12}}$$

$$PS = \int_0^{40{,}000}\left(20 - 19 - \frac{q^3}{64\times10^{12}}\right)dq$$

$$= \left[q - \frac{q^4}{256\times10^{12}}\right]_0^{40{,}000}$$

$$= \$30{,}000$$

43. The price follows the function $p(t) = 40 - 2t$ while the quantity sold per week follows $q(t) = 5000e^{-0.1t}$, where t is measured in weeks. The revenue per week is therefore $R(t) = p(t)q(t)$ and the total revenue over the next 8 weeks is

$$\int_0^8 5000(40 - 2t)e^{-0.1t}\,dt$$

$$= 5000[-(400 - 20t)e^{-0.1t} + 200e^{-0.1t}]_0^8$$

(using integration by parts)

$$= 5000[(20t - 200)e^{-0.1t}]_0^8$$

$$= 5000[-40e^{-0.8} + 200] \approx \$910{,}000$$

45. (a) The rate at which money is deposited is $100{,}000 + 10{,}000(12t)$ dollars per month after t years; converting to dollars per year gives $R(t) = 1{,}200{,}000 + 1{,}440{,}000t$ dollars per year after t years. The total deposited over two years is

$$\int_0^2 (1{,}200{,}000 + 1{,}440{,}000t)e^{0.06(2-t)}\,dt$$

$$= \left[-\frac{1{,}200{,}000 + 1{,}440{,}000t}{0.06}e^{0.06(2-t)}\right.$$

$$\left.-\frac{1{,}440{,}000}{0.06^2}e^{0.06(2-t)}\right]_0^2$$

$$\approx \$5{,}549{,}000$$

(b) The principal is given by

$$\int_0^2 (1{,}200{,}000 + 1{,}440{,}000t)\,dt$$

$$= [1{,}200{,}000t + 720{,}000t^2]_0^2 = \$5{,}280{,}000,$$

so the interest is the remaining \$269,000.

47. The revenue stream is $R(t) = 50e^{0.1t}$ million dollars per year. The present value if the next year's revenue is

$$\int_0^1 50e^{0.1t}e^{-0.06t}\,dt$$

$$= \int_0^1 50e^{0.04t}\,dt = [1250e^{0.04t}]_0^1$$

$$\approx \$51 \text{ million}$$

49. The money $y(t)$ in the account satisfies the differential equation $\dfrac{dy}{dt} = 0.0001y^2$. We solve this equation:

$$\frac{1}{y^2}\,dy = 0.0001\,dt$$

$$\int \frac{1}{y^2}\, dy = \int 0.0001 \; dt$$

$$-\frac{1}{y} = 0.0001t + C$$

$$y(0) = 10,000$$

so $C = -1/10,000 = -0.0001$

$$y = \frac{1}{0.0001 - 0.0001t} = \frac{10,000}{1 - t}$$

The amount in the account would approach infinity one year after the deposit.

Chapter 15 Functions of Several Variables

15.1 Functions of Several Variables from the Numerical, Algebraic, and Graphical Viewpoints

1. (a) $f(0, 0) = 0^2 + 0^2 - 0 + 1 = 1$
(b) $f(1, 0) = 1^2 + 0^2 - 1 + 1 = 1$
(c) $f(0, -1) = 0^2 + (-1)^2 - 0 + 1 = 2$
(d) $f(a, 2) = a^2 + 2^2 - a + 1 = a^2 - a + 5$
(e) $f(y, x) = y^2 + x^2 - y + 1$
(f) $f(x + h, y + k) =$
$(x + h)^2 + (y + k)^2 - (x + h) + 1$

3. (a) $f(0, 0) = 0 + 0 - 0 = 0$
(b) $f(1, 0) = 0.2 + 0 - 0 = 0.2$
(c) $f(0, -1) = 0 + 0.1(-1) - 0 = -0.1$
(d) $f(a, 2) = 0.2a + 0.1(2) - 0.01a(2) =$
$0.18a + 0.2$
(e) $f(y, x) = 0.2y + 0.1x - 0.01xy =$
$0.1x + 0.2y - 0.01xy$
(f) $f(x + h, y + k) =$
$0.2(x + h) + 0.1(y + k) - 0.01(x + h)(y + k)$

5. (a) $g(0, 0, 0) = e^{0+0+0} = 1$
(b) $g(1, 0, 0) = e^{1+0+0} = e$
(c) $g(0, 1, 0) = e^{0+1+0} = e$
(d) $g(z, x, y) = e^{z+x+y} = e^{x+y+z}$
(e) $g(x + h, y + k, z + l) = e^{x+h+y+k+z+l}$

7. (a) $g(0, 0, 0) = \dfrac{0}{0 + 0 + 0}$ does not exist

(b) $g(1, 0, 0) = \dfrac{0}{1 + 0 + 0} = 0$

(c) $g(0, 1, 0) = \dfrac{0}{0 + 1 + 0} = 0$

(d) $g(z, x, y) = \dfrac{zxy}{z^2 + x^2 + y^2} = \dfrac{xyz}{x^2 + y^2 + z^2}$

(e) $g(x + h, y + k, z + l) =$
$\dfrac{(x + h)(y + k)(z + l)}{(x + h)^2 + (y + k)^2 + (z + l)^2}$

9. (a) f increases by 2.3 units for every 1 unit of increase in x.
(b) f decreases by 1.4 units for every 1 unit of increase in y.
(c) f decreases by 2.5 units for every 1 unit increase in z.

11. Neither, because of the y^2 term

13. Linear

15. Linear

17. Interaction, because $g(x, y, z) = \frac{1}{4}x - \frac{3}{4}y + \frac{1}{4}yz$

19. (a) $f(20, 10) = 107$
(b) $f(40, 20) = -14$
(c) $f(10, 20) - f(20, 10) = -6 - 107 = -113$

21.

	$x \to$			
	10	**20**	**30**	**40**
y **10**	52	107	162	217
$\downarrow$ **20**	94	194	294	394
30	136	281	426	571
40	178	368	558	748

25.

18
4
0.0965
47,040

27.

6.9078
1.5193
5.4366
0

29. Let z = annual sales of Z (in millions of dollars), x = annual sales of X, and y = annual sales of Y. A linear model has the form $z = ax + by + c$. We are told that $a = -2.1$ and $b = 0.4$, and that $z(6, 6) = 6$. Thus, $6 = -2.1(6) + 0.4(6) + c$, so $c = 16.2$. The model is $z = -2.1x + 0.4y + 16.2$.

31.

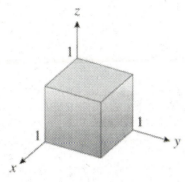

33.

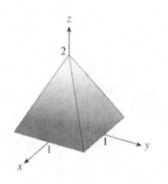

35.

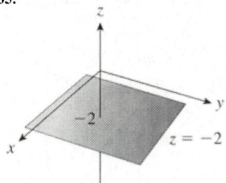

37.

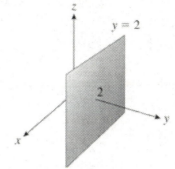

39.

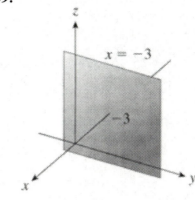

41. (H): The plane with z intercept 1, x intercept 1/3, and y intercept $-1/2$.

43. (B): An inverted paraboloid with apex at 1 on the z axis.

45. (F): A hemisphere lying below the xy plane.

47. (C): A surface of revolution having the z axis as a vertical asymptote.

49. $c = 0$: $2x^2 + 2y^2 = 0$ gives $x^2 + y^2 = 0$. This is the circle centered at the origin of radius 0 (that is, the single point $(0, 0)$).
$c = 2$: $2x^2 + 2y^2 = 2$, or $x^2 + y^2 = 1$. This is the circle centered at the origin of radius 1.
$c = 18$: $2x^2 + 2y^2 = 18$, or $x^2 + y^2 = 9$. This is the circle centered at the origin of radius 3.
Sketch:

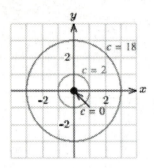

51. $c = -2$: $y + 2x^2 = -2$ gives
$\quad y = -2x^2 - 2$ (parabola)
$c = 0$: $y + 2x^2 = 0$ gives
$\quad y = -2x^2$ (parabola)
$c = 2$: $y + 2x^2 = 2$ gives
$\quad y = -2x^2 + 2$ (parabola)
Sketch:

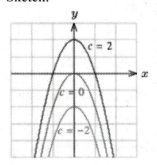

53. $c = -1$: $2xy - 1 = -1$ gives $xy = 0$ (the union of the x- and y-axes)
$c = 0$: $2xy - 1 = 0$ gives $y = 1/(2x)$
(hyperbola)
$c = 1$: $2xy - 1 = 2$ gives $y = 3/(2x)$
(hyperbola)
Sketch:

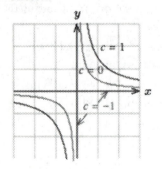

55. The plane with x, y, and z intercepts all 1.

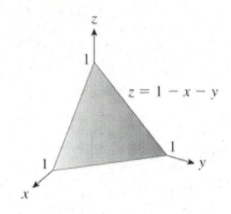

57. The plane with x intercept 1, y intercept 2, and z intercept -2.

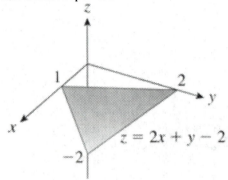

59. The plane with x intercept -2 and z intercept 2, parallel to the y axis.

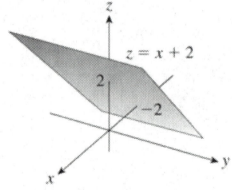

61. The plane passing through the origin, the line $z = x$ in the xz plane, and the line $z = y$ in the yz plane.

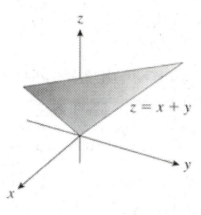

63. The cross section at $z = 1$ is the circle $x^2 + y^2 = 1/2$ and the cross section at $z = 2$ is the circle $x^2 + y^2 = 1$.

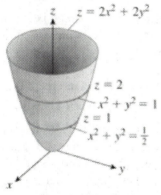

65. The cross section at $x = 0$ is the parabola $z = 2y^2$ and the cross section at $z = 1$ is the ellipse $x^2 + 2y^2 = 1$.

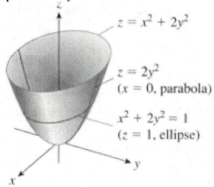

67. The graph is a cone with vertex at 2 on the z axis. The cross section at $y = 0$ is $z = 2 + |x|$ and the cross section at $z = 3$ is the circle $x^2 + y^2 = 1$.

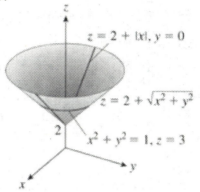

69. The graph is an inverted cone with apex at the origin. The cross section at $z = -4$ is the circle $x^2 + y^2 = 4$ and the cross section at $y = 1$ is a part of the hyperbola $z^2 - 4x^2 = 4$.

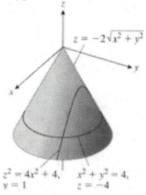

71. The cross sections $x = c$ are all copies of the parabola $z = y^2$.

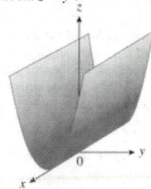

73. The cross sections $x = c$ are all copies of the hyperbola $z = 1/y$.

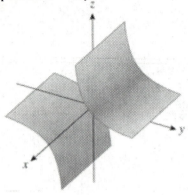

75. The cross sections $z = c$ are circles. The cross section $y = 0$ is the bell-shaped curve $z = e^{-x^2}$.

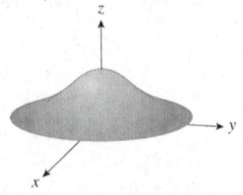

77. (a) The marginal cost of cars is $6000 per car, the coefficient of x. The marginal cost of trucks is $4000 per truck, the coefficient of y.
(b) The graph is a plane with x intercept -40, y intercept -60, and z intercept $240,000$.
(c) The slice $x = 10$ is the straight line with equation $z = 300,000 + 4000y$. It describes the cost function for the manufacture of trucks if car production is held fixed at 10 cars per week.
(d) The level curve $z = 480,000$ is the straight line $6000x + 4000y = 240,000$. It describes the number of cars and trucks you can manufacture to maintain weekly costs at $480,000.

79. $C(x, y) = 10 + 0.03x + 0.04y$ where C is the cost in dollars, x is the number of video clips sold per month, and y is the number of audio clips sold per month.

81. (a) 2003 is represented by $t = 20$. We are told that $x = 38$. Therefore,
$y = 82 - 0.78t - 1.02x$
$\quad = 82 - 0.78(20) - 1.02(38) = 27.64 \approx$ 28%
in 2003, 28% of the articles were written by researchers in the U.S.
(b) 1983 is represented by $t = 0$, We are told that $y = 61$. Therefore
$$y = 82 - 0.78t - 1.02x$$
$$61 = 82 - 0.78(0) - 1.02x$$
$$1.02x = 82 - 61 = 21$$
$$x = 21/1.02 \approx 21\%$$
In 1983, 21% of the articles were written by researchers in Europe.
(c) Since y is measured in percentage points and t in years, the coefficient -0.78 is measured in percentage points per year.

83. The graph is a plane with x_1 intercept 0.3, x_2 intercept 33, and x_3 intercept 0.66. The slices by $x_1 = constant$ are straight lines that are parallel to each other. Thus, the rate of change of General Motors' share as a function of Ford's share does not depend on Chrysler's share. Specifically, GM's share decreases by 0.02 percentage points per 1 percentage-point increase in Ford's market share, regardless of Chrysler's share.

85. (a) Since x and y must be in thousands, $x = 1300$ and $y = 781$. So
$N(1300, 750) = 27 -$
$0.08(1300)+0.08(781)+0.0002(1300)(781)$
$\quad \approx 189$ thousand prisoners
(b) $y = 300$:
$\quad N(x, 300) = 27 - 0.08x + 0.08(300) +$
$0.0002x(300)$
$\quad = 27+24-0.08x+0.06x$
$\quad = -0.02x + 51$
$y = 500$:
$\quad N(x, 500) = 27 - 0.08x + 0.08(500) +$
$0.0002x(500)$
$\quad = 27+40-0.08x+0.10x$
$\quad = 0.02x + 67$
Since units of the slope are units of N (thousands of prisoners in federal prisons) per unit of x

(thousands of prisoners in state prisons), we interpret the results as follows:
When there are 300,000 prisoners in local jails, the number in federal prisons decreases by 20 per 1000 additional prisoners in state prisons.
When there are 500,000 prisoners in local jails, the number in federal prisons increases by 20 per 1000 additional prisoners in state prisons.

87. (a) The slices $x = constant$ and $y = constant$ are straight lines. **(b)** No. Even though the slices $x = constant$ and $y = constant$ are straight lines, the level curves are not, and so the surface is not a plane. **(c)** The slice $x = 10$ has a slope of 3800. The slice $x = 20$ has a slope of 3600. Manufacturing more cars lowers the marginal cost of manufacturing trucks.

89. (a) $R(12,000, 5000, 5000) = \$9980$
(b) $R(z) = R(5000, 5000, z) =$
$10,000 - 0.01(5000) - 0.02(5000) - 0.01z + 0.00001(5000)z = 9850 + 0.04z$

91. $f(x, t) = Ax + Bt + C$
Substituting the given data:
2000 ($t = 0$): $160 = 27A + C$
2004 ($t = 4$): $250 = 42A + 4B + C$
2008 ($t = 8$): $370 = 62A + 8B + C$
This is a system of 3 linear equations in 3 unknowns. We can eliminate C by subtracting the first equation from the second and the second from the third:
$90 = 15A + 4B$
$120 = 20A + 4B$
from which we get $A = 6$, $B = 0$.
Substituting this into the first original equation gives $C = -2$.
So our model is $f(x, t) = 6x - 2$
Given $t = 6$ and $x = 52.5$ gives Walmart's earnings as
$f(52.5, 6) = 6(52.5) - 2 = \$313$ billion

93. $U(11, 10) - U(10, 10) \approx 5.75$. This means that, if your company now has 10 copies of Macro Publish and 10 copies of Turbo Publish, then the purchase of one additional copy of Macro Publish will result in a productivity increase of approximately 5.75 pages per day.

95. (a) $(a, b, c) = (3, 1/4, 1/\pi)$ and $(a, b, c) = (1/\pi, 3, 1/4)$ both work. In fact, if we take any positive values for a and b we can take $c = \dfrac{3}{4\pi ab}$
. **(b)** $V(a, a, a) = \frac{4}{3}\pi a^3 = 1$ gives $a = \left(\dfrac{3}{4\pi}\right)^{1/3}$.
The resulting ellipsoid is a sphere with radius a.

97. $P(100, 500,000) = 1000(100^{0.5})(500,000^{0.5}) \approx 7,000,000$

99. (a) $100 = K(1000)^a(1,000,000)^{1-a}$, $10 = K(1000)^a(10,000)^{1-a}$
(b) Taking logs of both sides of the first equation we get
$\log 100$
$= \log K + a \log 1000 + (1 - a) \log 1,000,000;$
$2 = \log K + 3a + 6(1 - a);$
$\log K - 3a = -4.$
From the second equation we get $\log K - a = -3$ similarly.
(c) Solving we get $\log K = -2.5$ and $a = 0.5$, so $K = 10^{-2.5} \approx 0.003162$
(d) $P(500, 1,000,000) = 0.003162(500^{0.5})(1,000,000^{0.5}) = 71$ pianos (to the nearest piano)

101. (a) We first need to convert n into years:
5 days = 5/365 years. So, $B(1.5\times10^{14}, 5/365)$
$= \dfrac{1.5\times10^{14}\times5}{5.1\times10^{14}\times365} \approx 4 \times 10^{-3}$ grams per square meter.
(b) The total weight of sulfates in the Earth's atmosphere.

103. (a) The value of N would be doubled.
(b) $N(R, f_p, n_e, f_l, f_i, L) = R f_p n_e f_l f_i L$, where here L is the average lifetime of an intelligent civilization. **(c)** Take the logarithm of both sides, since this would yield the linear function $\ln N = \ln R + \ln f_p + \ln n_e + \ln f_l + \ln f_i + \ln f_c + \ln L$.

105. They are reciprocals of each other.

107. For example, $f(x, y) = x^2 + y^2$.

109. For example, $f(x, y, z) = xyz$

111. For example, take $f(x, y) = x + y$. Then setting $y = 3$ gives $f(x, 3) = x + 3$. This can be viewed as a function of the single variable x. Choosing other values for y gives other functions of x.

113. If $f = ax + by + c$, then fixing $y = k$ gives $f = ax + (bk + c)$, a linear function with slope a and intercept $bk + c$. The slope is independent of the choice of k.

115. That CDs cost more than cassettes.

117. Plane

119. (B) Traveling in the direction B results in the shortest trip to nearby isotherms, and hence the fastest rate of increase in temperature.

121. Agree: Any slice through a plane is a straight line.

123. The graph of a function of three or more variables lives in four-dimensional (or higher) space, which makes it difficult to draw and visualize.

125. We need one dimension for each of the variables plus one dimension for the value of the function.

15.2 Partial Derivatives

1. $f_x(x, y) = -40; f_y(x, y) = 20; f_x(1, -1) = -40; f_y(1, -1) = 20$

3. $f_x(x, y) = 6x + 1; f_y(x, y) = -3y^2; f_x(1, -1) = 7; f_y(1, -1) = -3$

5. $f_x(x, y) = -40 + 10y; f_y(x, y) = 20 + 10x; f_x(1, -1) = -50; f_y(1, -1) = 30$

7. $f_x(x, y) = 6xy; f_y(x, y) = 3x^2; f_x(1, -1) = -6; f_y(1, -1) = 3$

9. $f_x(x, y) = 2xy^3 - 3x^2y^2 - y; f_y(x, y) = 3x^2y^2 - 2x^3y - x; f_x(1, -1) = -4; f_y(1, -1) = 4$

11. $f_x(x, y) = 6y(2xy + 1)^2; \quad f_y(x, y) = 6x(2xy + 1)^2; f_x(1, -1) = -6; f_y(1, -1) = 6$

13. $f_x(x, y) = e^{x+y}; f_y(x, y) = e^{x+y}; f_x(1, -1) = 1; f_y(1, -1) = 1$

15. $f_x(x, y) = 3x^{-0.4}y^{0.4}; \quad f_y(x, y) = 2x^{0.6}y^{-0.6}; f_x(1, -1)$ is undefined; $f_y(1, -1)$ is undefined

17. $f_x(x, y) = 0.2ye^{0.2xy}; \quad f_y(x, y) = 0.2xe^{0.2xy}; f_x(1, -1) = -0.2e^{-0.2}; f_y(1, -1) = 0.2e^{-0.2}$

19. $f_{xx}(x, y) = 0; f_{yy}(x, y) = 0; f_{xy}(x, y) = f_{yx}(x, y) = 0; f_{xx}(1, -1) = 0; f_{yy}(1, -1) = 0; f_{xy}(1, -1) = f_{yx}(1, -1) = 0$

21. $f_{xx}(x, y) = 0; f_{yy}(x, y) = 0; f_{xy}(x, y) = f_{yx}(x, y) = 10; f_{xx}(1, -1) = 0; f_{yy}(1, -1) = 0; f_{xy}(1, -1) = f_{yx}(1, -1) = 10$

23. $f_{xx}(x, y) = 6y; f_{yy}(x, y) = 0; f_{xy}(x, y) = f_{yx}(x, y) = 6x; f_{xx}(1, -1) = -6; f_{yy}(1, -1) = 0; f_{xy}(1, -1) = f_{yx}(1, -1) = 6$

25. $f_{xx}(x, y) = e^{x+y}; f_{yy}(x, y) = e^{x+y}; f_{xy}(x, y) = f_{yx}(x, y) = e^{x+y}; f_{xx}(1, -1) = 1; f_{yy}(1, -1) = 1; f_{xy}(1, -1) = f_{yx}(1, -1) = 1$

27. $f_{xx}(x, y) = -1.2x^{-14}y^{0.4}; \quad f_{yy}(x, y) = -1.2x^{0.6}y^{-1.6}; f_{xy}(x, y) = f_{yx}(x, y) = 1.2x^{-0.4}y^{-0.6}; f_{xx}(1, -1)$ is undefined; $f_{yy}(1, -1)$ is undefined; $f_{xy}(1, -1)$ and $f_{yx}(1, -1)$ are undefined

29. $f_x(x, y, z) = yz; f_y(x, y, z) = xz; f_z(x, y, z) = xy; \quad f_x(0, -1, 1) = -1; f_y(0, -1, 1) = 0; f_z(0, -1, 1) = 0$

31. $f_x(x, y, z) = \dfrac{4}{(x + y + z^2)^2}; f_y(x, y, z) = \dfrac{4}{(x + y + z^2)^2}; f_z(x, y, z) = \dfrac{8z}{(x + y + z^2)^2}; f_x(0, -1, 1)$ is undefined; $f_y(0, -1, 1)$ is undefined; $f_z(0, -1, 1)$ is undefined

33. $f_x(x, y, z) = e^{yz} + yze^{xz}; f_y(x, y, z) = xze^{yz} + e^{xz}; f_z(x, y, z) = xy(e^{yz} + e^{xz}); \quad f_x(0, -1, 1) = e^{-1} - 1; f_y(0, -1, 1) = 1; f_z(0, -1, 1) = 0$

35. $f_x(x, y, z) = 0.1x^{-0.9}y^{0.4}z^{0.5}; f_y(x, y, z) = 0.4x^{0.1}y^{-0.6}z^{0.5}; f_z(x, y, z) = 0.5x^{0.1}y^{0.4}z^{-0.5}; f_x(0, -1, 1)$ is undefined; $f_y(0, -1, 1)$ is undefined, $f_z(0, -1, 1)$ is undefined

37. $f_x(x, y, z) = yze^{xyz}, f_y(x, y, z) = xze^{xyz}, f_z(x, y, z) = xye^{xyz}; f_x(0, -1, 1) = -1; f_y(0, -1, 1) = f_z(0, -1, 1) = 0$

39. $f_x(x, y, z) = 0; f_y(x, y, z) = -\dfrac{600z}{y^{0.7}(1 + y^{0.3})^2}; f_z(x, y, z) = \dfrac{2000}{1 + y^{0.3}}; f_x(0, -1, 1)$ is undefined (because $f(0, -1, 1)$ is); $f_y(0, -1, 1)$ is undefined; $f_z(0, -1, 1)$ is undefined

41. $\partial C/\partial x = 6000$: The marginal cost to manufacture each car is $6000. $\partial C/\partial y = 4000$: The marginal cost to manufacture each truck is $4000.

43. $y = 82 - 0.78t - 1.02x;$ $\frac{\partial y}{\partial t} = -0.78.$ Since units of $\partial y/\partial t$ are units of y (percentage of articles written by researchers in the U.S.) per unit of t (years), we conclude that the number of articles written by researchers in the U.S. was decreasing at a rate of 0.78 percentage points per year. $\frac{\partial y}{\partial x} = -1.02.$ Since units of $\partial y/\partial x$ are units of y (percentage of articles written by researchers in the U.S.) per unit of y (percentage of articles written by researchers in Europe), we conclude that the number of articles written by researchers in the U.S. was decreasing at a rate of 1.02 percentage points per one percentage point increase in articles written in Europe.

45. $C_x(x, y) = 6000 - 20y;$ $C_x(10, 20) = \$5600$ per car

47. (a) $\partial M/\partial c = -3.8,$ $\partial M/\partial f = 2.2.$ For every 1 point increase in the percentage of Chrysler owners who remain loyal, the percentage of Mazda owners who remain loyal decreases by 3.8 points. For every 1 point increase in the percentage of Ford owners who remain loyal, the percentage of Mazda owners who remain loyal increases by 2.2 points.
(b) $M(0.56, 0.56, 0.72, 0.50, 0.43) \approx 16\%$

49. (a) For a Hispanic family, $x = 0.$ $z(20, 0) = \$36,600$
For a white family, $x = 1.$ $z(20, 1) = \$52,900$
(b) The rate of change of median income is
$z_t(t, x) = 270 + 140x$
For a Hispanic family in 2000, this is
$z_t(20, 0) = \$270$ per year
For a white family in 2000, the rate of change is
$z_t(20, 1) = \$410$ per year
(c) Since the median income was increasing faster for white families, the gap was widening: The median income of a white family was rising faster.
(d) Looking at the solution to part (b), we see that the coefficient 140 of xt is the difference between the rates of change of white income and black income. Thus, it is the rate at which the income gap is widening..

51. The marginal cost of a car is $C_x(x, y) = \$6000 + 1000e^{-0.01(x+y)}$ per car. The marginal cost of a truck is $C_y(x, y) = \$4000 + 1000e^{-0.01(x+y)}$ per truck. Both marginal costs decrease as production rises.

53. $\overline{C}(x, y) = \dfrac{200,000 + 6000x + 4000y - 100,000e^{-0.01(x+y)}}{x + y};$

$\overline{C}_x(x, y) = \dfrac{(6000 + 1000e^{-0.1(x+y)})(x + y) - (200,000 + 6000x + 4000y - 100,000e^{-0.01(x+y)})}{(x + y)^2} =$

$\dfrac{-200,000 + 2000y + (1000x + 1000y + 100,000)e^{-0.01(x+y)}}{(x + y)^2};$

$\overline{C}_x(50, 50) = -\$2.64$ per car. This means that at a production level of 50 cars and 50 trucks per week, the average cost per vehicle is decreasing by \$2.64 for each additional car manufactured.

$\overline{C}_y(x, y) = \dfrac{(4000 + 1000e^{-0.01(x+y)})(x + y) - (200,000 + 6000x + 4000y - 100,000e^{-0.01(x+y)})}{(x + y)^2} =$

$\dfrac{-200,000 - 2000x + (1000x + 1000y + 100,000)e^{-0.01(x+y)}}{(x + y)^2};$

$\overline{C}_y(50, 50) = -\$22.64$ per truck. This means that at a production level of 50 cars and 50 trucks per week, the average cost per vehicle is decreasing by \$22.64 for each additional truck manufactured.

55. No. Your revenue function is $R(x, y) =$ $15,000x + 10,000y - 5000\sqrt{x+y}$, so your marginal revenue from the sale of cars is

$R_x(x, y) = \$15,000 - \dfrac{2500}{\sqrt{x+y}}$ per car and your

marginal revenue from the sale of trucks is

$R_y(x, y) = \$10,000 - \dfrac{2500}{\sqrt{x+y}}$ per truck. These

increase with increasing x and y. In other words, you will earn more revenue per vehicle with increasing sales, and so the rental company will pay more for each additional vehicle it buys.

57. $P_z(x, y, z) = 0.016x^{0.4}y^{0.2}z^{-0.6}$;
$P_z(10, 100,000, 1,000,000) \approx 0.0001010$
papers/\$, or approximately 1 paper per \$10,000 increase in the subsidy.

59. (a) $U_x(x, y) = 4.8x^{-0.2}y^{0.2} + 1$; $U_y(x, y) =$ $1.2x^{0.8}y^{-0.8}$; $U_x(10, 5) = 5.18$, $U_y(10, 5) =$ 2.09. This means that, if 10 copies of Macro Publish and 5 copies of Turbo Publish are purchased, the company's daily productivity is increasing at a rate of 5.18 pages per day for each additional copy of Macro purchased and by 2.09 pages per day for each additional copy of Turbo purchased. **(b)** $\dfrac{U_x(10, 5)}{U_y(10, 5)} \approx 2.48$ is the ratio of the usefulness of one additional copy of Macro to one of Turbo. Thus, with 10 copies of Macro and 5 copies of Turbo, the company can expect approximately 2.48 times the productivity per additional copy of Macro compared to Turbo.

61. $F_y(x, y, z) =$
$-\dfrac{2KQq(y - b)}{[(x - a)^2 + (y - b)^2 + (z - c)^2]^2}$. With
$(a, b, c) = (0, 0, 0), K = 9\times10^9, Q = 10$, and $q = 5, F_y(2, 3, 3) \approx -6\times10^9$ N/sec.

63. (a) $A_P(P, r, t) = (1 + r)^t$; $A_r(P, r, t) =$ $tP(1 + r)^{t-1}$; $A_t(P, r, t) = P(1 + r)^t \ln(1 + r)$; $A_P(100, 0.1, 10) = 2.59$; $A_r(100, 0.1, 10) =$ 2,357.95; $A_t(100, 0.1, 10) = 24.72$. Thus, for a \$100 investment at 10% interest, after 10 years

the accumulated amount is increasing at a rate of \$2.59 per \$1 of principal, at a rate of \$2,357.95 per increase of 1 in r (note that this would correspond to an increase in the interest rate of 100%), and at a rate of \$24.72 per year. **(b)** $A_P(100, 0.1, t)$ tells you the rate at which the accumulated amount in an account bearing 10% interest with a principal of \$100 is growing per \$1 increase in the principal, t years after the investment.

65. (a) $P_x = Ka\left(\dfrac{y}{x}\right)^b$ and $P_y = Kb\left(\dfrac{x}{y}\right)^a$. They

are equal precisely when $\dfrac{a}{b} = \left(\dfrac{x}{y}\right)^b\left(\dfrac{x}{y}\right)^a$.

Substituting $b = 1-a$ now gives $\dfrac{a}{b} = \dfrac{x}{y}$.
(b) The given information implies that $P_x(100,$ $200) = P_y(100, 200)$. By part (a), this occurs precisely when $a/b = x/y = 100/200 = 1/2$. But $b = 1 - a$, so $a/(1 - a) = 1/2$, giving $a = 1/3$ and $b = 2/3$.

67. $u_t(r, t) =$
$-\dfrac{1}{4\pi Dt^2} e^{-r^2/(4Dt)} + \dfrac{r^2}{16\pi D^2 t^3} e^{-r^2/(4Dt)}$.
Taking $D = 1$, $u_t(1, 3) \approx -0.0075$, so the concentration is decreasing at 0.0075 parts of nutrient per part of water/sec.

69. f is increasing at a rate of s units per unit of x, f is increasing at a rate of t units per unit of y, and the value of f is r when $x = a$ and $y = b$

71. The marginal cost of building an additional orbicus; zonars per unit.

73. Answers will vary. One example is $f(x, y) =$ $-2x + 3y$. Others are $f(x, y) = -2x + 3y + 9$ and $f(x, y) = xy - 3x + 2y + 10$.

75. (a) b is the z-intercept of the plane, m is the slope of the intersection of the plane with the xz-plane, n is the slope of the intersection of the plane with the yz-plane.

(b) Write $z = b + rx + sy$. We are told that $\partial z/\partial x = m$, so $r = m$. Similarly, $s = n$. Thus, $z = b + mx + ny$. We are also told that the plane passes through (h, k, l). Substituting gives $l = b + mh + nk$. This gives b as $l - mh - nk$. Substituting in the equation for z therefore gives $z = l - mh - nk + mx + ny = l + m(x - h) + n(y - k)$, as required.

15.3 Maxima and Minima

1. P: relative minimum; Q: none of the above; R: relative maximum

3. P: saddle point; Q: relative maximum; R: none of the above

5. Relative minimum

7. Neither

9. Saddle point

11. $f_x = 2x$; $f_y = 2y$; $f_{xx} = 2$; $f_{yy} = 2$; $f_{xy} = 0$. $f_x = 0$ when $x = 0$; $f_y = 0$ when $y = 0$, so $(0, 0)$ is the only critical point. $H = 4$ and $f_{xx} > 0$, so f has a relative minimum at $(0, 0, 1)$.

13. $g_x = -2x - 1$; $g_y = -2y + 1$; $g_{xx} = -2$; $g_{yy} = -2$; $g_{xy} = 0$. $g_x = 0$ when $x = -1/2$; $g_y = 0$ when $y = 1/2$, so $(-1/2, 1/2)$ is the only critical point.
$H = 4$ and $g_{xx} < 0$, so g has a relative maximum at $(-1/2, 1/2, 3/2)$.

15. $k_x = 2x - 3y$; $k_y = -3x + 2y$
Critical points:
  $2x - 3y = 0$
  $-3x + 2y = 0$
Solution: $(0, 0)$, which is the only critical point.
$k_{xx} = 2$; $k_{xy} = -3$; $k_{yy} = 2$
$H = (2)(2) - (-3)^2 = -5$
Since $H < 0$, k has a saddle point at $(0, 0, 0)$.

17. $f_x = 2x + 2y - 2$; $f_y = 2x + 4y + 4$
Critical points:
  $2x + 2y - 2 = 0 \Rightarrow x + y = 1$
  $2x + 4y + 4 = 0 \Rightarrow x + 2y = -2$
Solution: $(4, -3)$, which is the only critical point.
$f_{xx} = 2$; $f_{xy} = 2$; $f_{yy} = 4$
$H = (2)(4) - 2^2 = 4$
Since $H > 0$ and $f_{xx} > 0$, f has a minimum at $(4, -3, -10)$.

(The corresponding z coordinate is
  $f(4, -3) = -10$.)

19. $g_x = -2x - 2y - 3$; $g_y = -2x - 6y - 2$
Critical points:
  $-2x - 2y - 3 = 0 \Rightarrow -2x - 2y = 3$
  $-2x - 6y - 2 \Rightarrow -2x - 6y = 2$
Solution: $(-7/4, 1/4)$, which is the only critical point.
$g_{xx} = -2$; $g_{xy} = -2$; $g_{yy} = -6$
$H = (-2)(-6) - (-2)^2 = 8$
Since $H > 0$ and $g_{xx} < 0$ g has a maximum at $(-7/4, 1/4, 19/8)$.
(The corresponding z coordinate is
  $f(-7/4, 1/4) = 19/8$.)

21. $h_x = 2xy - 4x$; $h_y = x^2 - 8y$; $h_{xx} = 2y - 4$; $h_{yy} = -8$; $h_{xy} = 2x$. $h_x = 0$ when $x = 0$ or $y = 2$; $h_y = 0$ when $x^2 = 8y$. The two possibilities are $x = 0$, so $y = 0$, or $y = 2$, so $x^2 = 16$ or $x = \pm 4$. This gives three critical points: $(0, 0)$, $(-4, 2)$, and $(4, 2)$. $H(x, y) = -8(2y - 4) - 4x^2 = 32 - 16y - 4x^2$; $H(0, 0) = 32$ and $h_{xx}(0, 0) = -4 < 0$; $H(-4, 2) = -64 = H(4, 2)$. Hence h has a relative maximum at $(0, 0, 0)$ and saddle points at $(\pm 4, 2, -16)$.

23. $f_x = 2x + 2y^2$; $f_y = 4xy + 4y$
Critical points:
  $2x + 2y^2 = 0 \Rightarrow x = -y^2$
  $4xy + 4y = 0 \Rightarrow 4y(x + 1) = 0 \Rightarrow y = 0$
  or $x = -1$
$y = 0$ gives, using the first equation, $x = 0$
$x = -1$ gives, using the first equation, $y^2 = 1$, so $y = \pm 1$
Critical points: $(0, 0, 0)$, $(-1, -1, 1)$, $(-1, 1, 1)$
(We get the z-coordinate by substituting for x and y in the original function.)
  $f_{xx} = 2$; $f_{xy} = 4y$; $f_{yy} = 4x + 4$
$(0, 0, 0)$: $H = (2)(4) - 0^2 = 8$ and $f_{xx} > 0$, giving a minimum at $(0, 0, 0)$
$(-1, \pm 1, 1)$: $H = (2)(0) - (\pm 4)^2 = -16$, giving saddle points at $(-1, \pm 1, 1)$

25. $s_x = 2xe^{x^2+y^2}$; $s_y = 2ye^{x^2+y^2}$; $s_{xx} = (2 + 4x^2)e^{x^2+y^2}$; $s_{yy} = (2 + 4y^2)e^{x^2+y^2}$; $s_{xy} = 4xye^{x^2+y^2}$. $s_x = 0$ when $x = 0$; $s_y = 0$ when $y = 0$; so the only critical point is $(0, 0)$. $H(0, 0) = 4 - 0 = 4$ and $s_{xx}(0, 0) = 2 > 0$, so s has a relative minimum at $(0, 0, 1)$.

27. $t_x = 4x^3 + 8y^2$; $t_y = 16xy + 8y^3$; $t_{xx} = 12x^2$; $t_{yy} = 16x + 24y^2$; $t_{xy} = 16y$. $t_x = 0$ when $x^3 = -2y^2$ (notice that $x \le 0$ in this case); $t_y = 0$ when $y = 0$ or $y^2 = -2x$; if $y = 0$ then $x = 0$; if $y^2 = -2x$ then $x^3 = 4x$, so $x = 0$ (and $y = 0$) or $x = -2$ and $y = \pm 2$. This gives three critical points: $(0, 0)$ and $(-2, \pm 2)$. $H(-2, \pm 2) = 3072 \pm 32 > 0$ and $t_{xx}(-2, \pm 2) = 48 > 0$. $H(0, 0) = 0$ so the second derivative test is inconclusive. We can see from the graph of t that the origin is not a max, min, or saddle point. Or we can look at the slice along $x = y$ (suggested by the graph) where $t = 3x^4 + 8x^3$; this function increases as x approaches 0 and then increases again as x becomes larger than 0. So, t has two relative minima at $(-2, \pm 2, -16)$ and $(0, 0)$ is a critical point that is not a relative extremum or saddle point.

29. $f_x = 2x$; $f_y = 1 - e^y$; $f_{xx} = 2$; $f_{yy} = -e^y$; $f_{xy} = 0$. $f_x = 0$ when $x = 0$; $f_y = 0$ when $y = 0$. This gives one critical point, $(0, 0)$. $H(0, 0) = -2$, so f has a saddle point at $(0, 0, -1)$ and no other critical points.

31. $f_x = -(2x + 2)e^{-(x^2+y^2+2x)}$;
$f_y = -2ye^{-(x^2+y^2+2x)}$;
$f_{xx} = (4x^2 + 8x + 2)e^{-(x^2+y^2+2x)}$;
$f_{yy} = (4y^2 - 2)e^{-(x^2+y^2+2x)}$;
$f_{xy} = 2y(2x + 2)e^{-(x^2+y^2+2x)}$.
$f_x = 0$ when $x = -1$; $f_y = 0$ when $y = 0$. This gives one critical point, $(-1, 0)$. $H(-1, 0) = 4e^2 > 0$ and $f_{xx}(-1, 0) = -2e < 0$, so f has a relative maximum at $(-1, 0, e)$.

33. $f_x = y - \dfrac{2}{x^2}$; $f_y = x - \dfrac{2}{y^2}$; $f_{xx} = \dfrac{4}{x^3}$; $f_{yy} = \dfrac{4}{y^3}$; $f_{xy} = 1$. $f_x = 0$ when $y = \dfrac{2}{x^2}$; $f_y = 0$ when $x = \dfrac{2}{y^2} = \dfrac{1}{2}x^4$, $x = 2^{1/3}$ ($x = 0$ is excluded because it is not in the domain of f). The corresponding value of y is $y = 2^{1/3}$. This gives one critical point, $(2^{1/3}, 2^{1/3})$. $H(2^{1/3}, 2^{1/3}) = 3$ and $f_{xx}(2^{1/3}, 2^{1/3}) = 2 > 0$, so f has a relative minimum at $(2^{1/3}, 2^{1/3}, 3(2^{2/3}))$.

35. $g_x = 2x - \dfrac{2}{x^2 y}$; $g_y = 2y - \dfrac{2}{xy^2}$; $g_{xx} = 2 + \dfrac{4}{x^3 y}$; $g_{yy} = 2 + \dfrac{4}{xy^3}$; $g_{xy} = \dfrac{2}{x^2 y^2}$. $g_x = 0$ when $y = \dfrac{1}{x^3}$; $g_y = 0$ when $x = \dfrac{1}{y^3} = x^9$, $x = \pm 1$ ($x = 0$ is excluded because it is not in the domain of f). This gives two critical points, $(1, 1)$ and $(-1, -1)$. $H(1, 1) = 32$ and $g_{xx}(1, 1) = 6 > 0$; $H(-1, -1) = 32$ and $g_{xx}(-1, -1) = 6 > 0$. So, g has relative minima at $(1, 1, 4)$ and $(-1, -1, 4)$.

37. f has an absolute minimum at $(0, 0, 1)$: $x^2 + y^2 + 1 \ge 1$ for all x and y because $x^2 + y^2 \ge 0$.

39. The relative maximum at $(0, 0, 0)$ is not absolute. For example,
$h(10, 10) = 400 > h(0, 0)$.

41. $M_c = 8 + 8c - 20f$; $M_f = -40 - 20c + 80f$; $M_{cc} = 8$; $M_{ff} = 80$; $M_{cf} = -20$. $M_c = 0$ and $M_f = 0$ when $(c, f) = (2/3, 2/3)$; $H(2/3, 2/3) = 240$ and $M_{cc}(2/3, 2/3) = 8 > 0$, so M has a minimum at $(2/3, 2/3, 1/3)$. Thus, at least 1/3 of all Mazda owners would choose another new Mazda, and this lowest loyalty occurs when 2/3 of Chrysler and Ford owners remain loyal to their brands.

43. The subsidy is $S(x, y) = 500x + 100y$, so the net cost is $N(x, y) = C(x, y) - S(x, y) = 4000 + 100x^2 + 50y^2 - 500x - 100y$. $N_x = 200x - 500$; $N_y = 100y - 100$. $N_{xx} = 200$; $N_{yy} = 100$; $N_{xy} = 0$. $N_x = 0$ when $x = 2.5$; $N_y = 0$ when $y =$

1. $H = 20{,}000$ and $N_{xx} = 100 > 0$, so N has a maximum at $(2.5,\ 1)$. The firm should remove 2.5 pounds of sulfur and 1 pound of lead per day.

45. The total revenue is $R = p_1 q_1 + p_2 q_2 =$
$100{,}000 p_1 - 100 p_1^2 + 10 p_1 p_2 + 150{,}000 p_2 +$
$10 p_1 p_2 - 100 p_2^2 = 100{,}000 p_1 + 150{,}000 p_2 -$
$100 p_1^2 + 20 p_1 p_2 - 100 p_2^2$. For convenience, write R_1 for $\partial R / \partial p_1$ and R_2 for $\partial R / \partial p_2$. Then $R_1 = 100{,}000 - 200 p_1 + 20 p_2$; $R_2 = 150{,}000 + 20 p_1 - 200 p_2$; $R_{11} = -200$; $R_{22} = -200$; $R_{12} = 20$.
$R_1 = 0$ and $R_2 = 0$ when $(p_1,\ p_2) = (580.81, 808.08)$. $H = 39{,}600$ and $R_{11} = -200 < 0$, so R has a maximum at this critical point. You should charge \$580.81 for the Ultra Mini and \$808.08 for the Big Stack.

47. Let $l =$ length, $w =$ width, and $h =$ height. We are told that $l + w + h \le 62$; for the largest possible volume we will want $l + w + h = 62$, so $l = 62 - w - h$. The volume is $V = lwh = (62 - w - h)wh = 62wh - w^2 h - wh^2$. $V_w = 62h - 2wh - h^2$; $V_h = 62w - w^2 - 2wh$; $V_{ww} = -2h$; $V_{hh} = -2w$; $V_{wh} = 62 - 2w - 2h$. $V_w = 0$ when $62 - 2w - h = 0$ ($h = 0$ is not in the domain); $V_h = 0$ when $62 - w - 2h = 0$; this occurs when $w = h = 62/3 \approx 20.67$. $H \approx 1280$ and $V_{ww} < 0$, so V has a maximum at this critical point. The largest volume bag has dimensions $l = w = h \approx 20.67$ in and volume ≈ 8827 cubic inches.

49. Let $l =$ length, $w =$ width, and $h =$ height. We are told that $l + 2(w + h) \le 108$; for the largest possible volume we will want $l + 2(w + h) = 108$, so $l = 108 - 2(w + h)$. The volume is $V = lwh = (108 - 2w - 2h)wh = 108wh - 2w^2 h - 2wh^2$. $V_w = 108h - 4wh - 2h^2$; $V_h = 108w - 2w^2 - 4wh$; $V_{ww} = -4h$; $V_{hh} = -4w$; $V_{wh} = 108 - 4w - 4h$. $V_w = 0$ when $108 - 4w - 2h = 0$; $V_h = 0$ when $108 - 2w - 4h = 0$; this occurs when $w = h = 18$. $H = 3888$ and

$V_{ww} < 0$, so V has a maximum at this critical point. The corresponding length is $l = 108 - 2(w + h) = 36$. So, the largest volume package has dimensions 18 in $\times$ 18 in $\times$ 36 in and volume $= 11{,}664$ cubic inches.

51.

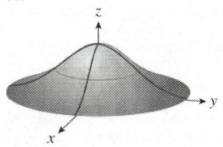

53. The function below has a relative max at $(0,\ 0,\ -1)$, has the unit circle $x^2 + y^2 = 1$ as a vertical asymptote, and takes on values larger than -1 as $x^2 + y^2$ gets large. The particular function graphed is $f(x,\ y) = -\dfrac{1}{|x^2 + y^2 - 1|}$.

Continues up indefinitely

Continues down indefinitely
Function not defined on circle

55. H must be positive.

57. No. In order for there to be a relative maximum at $(a,\ b)$, *all* vertical planes through $(a,\ b)$ should yield a curve with a relative maximum at $(a,\ b)$. It could happen that a slice by another vertical plane through $(a,\ b)$ (such as $x - a = y - b$) does not yield a curve with a relative maximum at $(a,\ b)$. An example is $f(x,\ y) = x^2 + y^2 - \sqrt{xy}$, at the point $(0,\ 0)$. Look at the slices through $x = 0$, $y = 0$ and $y = x$.

59. $\overline{C}_x = \dfrac{\partial}{\partial x}\left(\dfrac{C}{x+y}\right) = \dfrac{(x+y)C_x - C}{(x+y)^2}$. If this

is zero, then $(x+y)C_x = C$, or $C_x = \dfrac{C}{x+y} = \overline{C}$.

Similarly, if $\overline{C}_y = 0$ then $C_y = \overline{C}$. This is reasonable because if the average cost is decreasing with increasing x, then the average cost is greater than the marginal cost C_x. Similarly, if the average cost is increasing with increasing x, then the average cost is less than the marginal cost C_x. Thus, if the average cost is stationary with increasing x, then the average cost equals the marginal cost C_x. (The situation is similar for the case of increasing y.)

61. The equation of the tangent plane at the point (a, b) is $z = f(a, b) + f_x(a, b)(x - a) + f_y(a, b)(y - b)$. If f has a relative extremum at (a, b), then $f_x(a, b) = 0 = f_y(a, b)$. Substituting these into the equation of the tangent plane gives $z = f(a, b)$, a constant. But the graph of $z = constant$ is a plane parallel to the xy-plane.

15.4 Constrained Maxima and Minima and Applications

1. The objective function is $1 - x^2 - y^2 - z^2$. Substituting the constraint equation $z = 2y$ gives the objective function as

$$h(x, y) = 1 - x^2 - y^2 - (2y)^2$$
$$= 1 - x^2 - 5y^2$$

$$h_x = -2x, \; h_y = -10y$$

Critical points:

$$-2x = 0$$
$$-10y = 0$$

$(x, y) = (0, 0)$ is the only critical point.

$$h_{xx} = -2, \; h_{xy} = 0, \; h_{yy} = -10$$
$$H = h_{xx}h_{yy} - (h_{xy})^2 = 20 > 0$$

Since $h_{xx} < 0$, the critical point is a local maximum. That it is an absolute maximum can be seen by considering the graph of $h(x, y)$. The corresponding value of the objective function is

$$h(0, 0) = 1 - (0)^2 - 5(0)^2 = 1$$

When $x = 0$ and $y = 0$, $z = 2y = 0$ as well, so f has an absolute maximum of 1 at the point $(0, 0, 0)$.

3. The objective function is $1 - x^2 - x - y^2 + y - z^2 + z$. Substituting the constraint equation $y = 3x$ gives the objective function as

$$h(x, z) = 1 - x^2 - x - 9x^2 + 3x - z^2 + z$$
$$= 1 - 10x^2 + 2x - z^2 + z$$

$$h_x = -20x + 2, \; h_z = -2z + 1$$

Critical points:

$$-20x + 2 = 0$$
$$\Rightarrow x = 1/10$$
$$-2z + 1 = 0$$
$$\Rightarrow z = 1/2$$

$(x, z) = (1/10, 1/2)$ is the only critical point.

$$h_{xx} = -20, \; h_{xz} = 0, \; h_{zz} = -2$$
$$H = h_{xx}h_{zz} - (h_{xz})^2 = 40 > 0$$

Since $h_{xx} < 0$, the critical point is a local maximum. That it is an absolute maximum can be seen by considering the graph of $h(x, y) = 1 - $

$10x^2 + 2x - z^2 + z$. The corresponding value of the objective function is

$$h(1/10, \; 1/2)$$
$$= 1 - 10/100 + 2/10 - 1/4 + 1/2$$
$$= 1.35$$

When $x = 1/10$ and $z = 1/2$, $y = 3x = 3/10$, so f has an absolute maximum of 1.35 at the point $(1/10, 3/10, 1/2)$.

5. The objective function is $xy + 4xz + 2yz$. Solving the constraint equation $xyz = 1$ for z gives

$$z = \frac{1}{xy}$$

$$S = xy + 4x\frac{1}{xy} + 2y\frac{1}{xy}$$
$$= xy + \frac{4}{y} + \frac{2}{x}$$

$$S_x = y - \frac{2}{x^2}, \; S_y = x - \frac{4}{y^2}$$

Critical points:

$$y - \frac{2}{x^2} = 0$$

$$x - \frac{4}{y^2} = 0$$

Substituting the first equation in the second gives

$$x - \frac{4}{(2/x^2)^2} = 0$$
$$x - x^4 = 0$$
$$x(1 - x^3) = 0$$
$$x = 1$$

(We reject $x = 0$ because $x > 0$). Substituting $x = 1$ into the equation $y - \frac{2}{x^2} = 0$ gives

$$y = 2.$$

$(x, y) = (1, 2)$ is the only critical point.

$$S_{xx} = 4/x^3, \; S_{xy} = 1 \; S_{yy} = 8/y^3$$
$$S_{xx}(1, 2) = 4/1^3 = 4$$
$$S_{xy}(1, 2) = 1$$
$$S_{yy}(1, 2) = 8/2^3 = 1$$
$$H = S_{xx}S_{yy} - (S_{xy})^2 = 4 - 1 = 3 > 0$$

Since $S_{xx}(1, 2) > 0$, the critical point is a local minimum. That it is an absolute minimum can be seen by considering the graph of $S(x, y) = xy + $

$\dfrac{4}{y} + \dfrac{2}{x}$ for $x > 0$, $y > 0$. The corresponding value of the objective function is

$$S(1, 2) = (1)(2) + \frac{4}{2} + \frac{2}{1} = 6$$

When $x = 1$ and $y = 2$, $z = 1/xy = 1/2$, so S has an absolute minimum of 6 at $(1, 2, 1/2)$.

7. The constraint is $x + 2y = 40$, or $x + 2y - 40 = 0$.

$f(x, y) = xy$

$g(x, y) = x + 2y - 4$

$L(x, y) = xy - \lambda(x + 2y - 40)$

(1) $L_x = 0 \Rightarrow y - \lambda = 0 \Rightarrow y = \lambda$

(2) $L_y = 0 \Rightarrow x - 2\lambda = 0 \Rightarrow x = 2\lambda$

(3) Constraint $\Rightarrow x + 2y = 40$

Substitute Equation (1) into (2) to obtain

$x = 2y$

Substitute in Equation (3) to obtain

$2y + 2y = 40$

$4y = 40$

$y = 10$

The corresponding value of x is

$x = 2(10) = 20$

The corresponding value of the objective is

$f(20, 10) = (20)(10) = 200$

9. The constraint is $x^2 + y^2 = 8$, or $x^2 + y^2 - 8 = 0$.

$f(x, y) = 4xy$

$g(x, y) = x^2 + y^2 - 8$

$L(x, y) = 4xy - \lambda(x^2 + y^2 - 8)$

(1) $L_x = 0 \Rightarrow 4y - 2\lambda x = 0 \Rightarrow 2y = \lambda x$

(2) $L_y = 0 \Rightarrow 4x - 2\lambda y = 0 \Rightarrow 2x = \lambda y$

(3) Constraint $\Rightarrow x^2 + y^2 = 8$

Divide Equation (1) by x to obtain

$\lambda = \dfrac{2y}{x}$

Substitute in Equation (2) to obtain

$2x = \dfrac{2y}{x} y = \dfrac{2y^2}{x}$

or $2x^2 = 2y^2$,

giving

$y = \pm x$

Substituting into the constraint gives

$x^2 + x^2 = 8$

or $2x^2 = 8$, giving

$x^2 = 4$,

$x = \pm 2$, so $y = \pm 2$ as well.

Thus we have 4 critical points: $(-2, -2)$, $(-2, 2)$, $(2, -2)$ and $(2, 2)$

Substitute each of these into the objective function:

$f(-2, -2) = 4(-2)(-2) = 16$

$f(-2, 2) = 4(-2)(2) = -16$

$f(2, -2) = 4(2)(-2) = -16$

$f(2, 2) = 4(2)(2) = 16$

The first and last of these give a maximum of 16.

11. The constraint is $x + 2y = 10$, or $x + 2y - 10 = 0$.

$f(x, y) = x^2 + y^2$

$g(x, y) = x + 2y - 10$

$L(x, y) = x^2 + y^2 - \lambda(x + 2y - 10)$

(1) $L_x = 0 \Rightarrow 2x - \lambda = 0 \Rightarrow 2x = \lambda$

(2) $L_y = 0 \Rightarrow 2y - 2\lambda = 0 \Rightarrow y = \lambda$

(3) Constraint $\Rightarrow x + 2y = 10$

Substitute Equation (2) in Equation (1) to obtain

$2x = y$

Substituting into the constraint gives

$x + 4x = 10$

$5x = 10$

$x = 2$

so $y = 2x = 4$

Thus we have one critical point: $(2, 4)$

The corresponding value of the objective function is

$$f(2, 10) = 2^2 + 4^2 = 20$$

13. The constraint is $z = 2y$, or $z - 2y = 0$.

$f(x, y, z) = 1 - x^2 - y^2 - z^2$

$g(x, y, z) = z - 2y$

$L(x, y, z) = 1 - x^2 - y^2 - z^2 - \lambda(z - 2y)$

(1) $L_x = 0 \Rightarrow -2x = 0$

(2) $L_y = 0 \Rightarrow -2y + 2\lambda = 0 \Rightarrow y = \lambda$

(3) $L_z = 0 \Rightarrow -2z - \lambda = 0 \Rightarrow -2z = \lambda$

(4) Constraint $\Rightarrow z = 2y$

The first equation tells us that $x = 0$.

Substituting the third equation in the second gives

$y = -2z$, or

$y + 2z = 0$

Combining this with the constraint equation

$z - 2y = 0$

gives a system of 2 equations in 2 unknowns, whose solution is

$y = z = 0$

Thus the only critical point is $(x, y, z) = (0, 0, 0)$

and the corresponding value of the objective is

$f(0, 0, 0) = 1 - 0^2 - 0^2 - 0^2 = 1$

That this is an absolute maximum is seen from the fact that $f(x, y, z) = 1 - x^2 - y^2 - z^2$ can never be larger than 1.

15. The constraint is $3x = y$, or $3x - y = 0$.

$f(x, y, z) = 1 - x^2 - x - y^2 + y - z^2 + z$

$g(x, y, z) = 3x - y$

$L(x, y, z) = 1 - x^2 - x - y^2 + y - z^2 + z - \lambda(3x - y)$

(1) $L_x = 0 \Rightarrow -2x - 1 - 3\lambda = 0 \Rightarrow -2x - 1 = 3\lambda$

(2) $L_y = 0 \Rightarrow -2y + 1 + \lambda = 0 \Rightarrow -2y + 1 = -\lambda$

(3) $L_z = 0 \Rightarrow -2z + 1 = 0$

(4) Constraint $\Rightarrow 3x = y$

Equation (3) tells us that $z = 1/2$.

Substituting (2) in (1) gives

$-2x - 1 = 6y - 3$ or

$2x + 6y = 2$, that is

$x + 3y = 1$

Combining this with the constraint equation

$3x - y = 0$

gives a system of 2 equations in 2 unknowns, whose solution is

$x = 1/10, \ y = 3/10$

Thus the only critical point is $(x, y, z) = (1/10, 3/10, 1/2)$

and the corresponding value of the objective is

$f(1/10, 3/10, 1/2)$

$= 1 - 1/100 - 1/10 - 9/100 + 3/10 - 1/4 + 1/2 = 1.35$

17. The constraint is $xyz = 1$, or $xyz - 1 = 0$.

$f(x, y, z) = xy + 4xz + 2yz$

$g(x, y, z) = xyz - 1$

$L(x, y, z) = xy + 4xz + 2yz - \lambda(xyz - 1)$

(1) $L_x = 0 \Rightarrow y + 4z = \lambda yz$

(2) $L_y = 0 \Rightarrow x + 2z = \lambda xz$

(3) $L_z = 0 \Rightarrow 4x + 2y = \lambda xy$

(4) Constraint $\Rightarrow xyz = 1$

Solve Equation (1) for λ:

$\lambda = \dfrac{1}{z} + \dfrac{4}{y}$

Substituting in (2) gives

$x + 2z = (\dfrac{1}{z} + \dfrac{4}{y})xz$

$x + 2z = x + \dfrac{4xz}{y}$

$2z = \dfrac{4xz}{y}$

$1 = \dfrac{2x}{y}$

$y = 2x$

Substituting the expression for λ in (3) gives

$4x + 2y = (\dfrac{1}{z} + \dfrac{4}{y})xy$

$4x + 2y = \dfrac{xy}{z} + 4x$

$2y = \dfrac{xy}{z}$

$2 = \dfrac{x}{z}$

$z = x/2$

Substituting the expressions we obtained for y and z in the constraint equation gives:

$x(2x)(x/2) = 1$

$x^3 = 1$

$x = 1$

The corresponding values of y and z are

$y = 2x = 2$

$z = x/2 = 1/2$

Therefore, the only critical point is $(1, 2, 1/2)$ and the corresponding value of the objective is

$f(1, 2, 1/2) = (1)(2) + 4(1)(1/2) + 2(2)(1/2) = 6$

Therefore, the minimum value of the objective function is 6, and occurs at the point $(1, 2, 1/2)$.

19. Let x be the length of the east and west sides, and let y be the length of the north and south sides. The problem translates to:

Maximize $A = xy$ subject to $8x + 4y = 80$
The constraint is $8x + 4y = 80$, or $8x + 4y - 80 = 0$.

$f(x, y) = xy$
$g(x, y) = 8x + 4y - 80$
$L(x, y) = xy - \lambda(8x+4y-80)$
(1) $L_x = 0 \Rightarrow y - 8\lambda = 0 \Rightarrow y = 8\lambda$
(2) $L_y = 0 \Rightarrow x - 4\lambda = 0 \Rightarrow x = 4\lambda$
(3) Constraint $\Rightarrow 8x + 4y = 80$

Solve Equation (1) for λ to obtain
$\lambda = y/8$
Substitute in Equation (2) to obtain
$x = 4y/8 = y/2$
or $2x - y = 0$
The constraint equation is
$8x + 4y = 80$
or $2x + y = 20$
So we have a system of two linear equations in two unknowns:
$2x - y = 0$
$2x + y = 20$
The solution is $(x, y) = (5, 10)$
The corresponding value of the objective function is
$A = xy = 5 \times 10 = 50$ sq. ft.

21. The problem translates to:
Maximize $R = pq$ subject to $q = 200{,}000 - 10{,}000p$
The constraint is $q = 200{,}000 - 10{,}000p$, or $q - 200{,}000 + 10{,}000p = 0$.

$f(p, q) = pq$
$g(p, q) = q - 200{,}000 + 10{,}000p$
$L(p, q) = pq - \lambda(q-200{,}000+10{,}000p)$
(1) $L_p = 0 \Rightarrow q-10{,}000\lambda = 0 \Rightarrow q = 10{,}000\lambda$
(2) $L_q = 0 \Rightarrow p-\lambda = 0 \Rightarrow p = \lambda$
(3) Constraint $\Rightarrow q = 200{,}000 - 10{,}000p$

Substitute Equation (2) in Equation (2) to obtain
$q = 10{,}000p$
The constraint equation is
$q = 200{,}000 - 10{,}000p$
Equating these two expressions for q gives
$10{,}000p = 200{,}000 - 10{,}000p$
$20{,}000p = 200{,}000$
$p = \$10$

We are not asked for any further information.

23. We want to maximize $f(x, y, z) = xyz$ subject to $x^2 + y^2 + z^2 - 1 = 0$. Using Lagrange multipliers, we need to solve the system $yz = 2\lambda x$, $xz = 2\lambda y$, $xy = 2\lambda z$, and $x^2 + y^2 + z^2 - 1 = 0$. We solve the first equation for $\lambda = \dfrac{yz}{2x}$ and substitute in the second to find $xz = \dfrac{y^2 z}{x}$ or $x^2 z = y^2 z$, giving $x^2 = y^2$ (assuming that $z \neq 0$, but $z = 0$ makes $xyz = 0$ and we can easily find larger values). Substituting the expression for λ into the third equation gives $x^2 = z^2$ similarly. From the last equation we then get $3x^2 - 1 = 0$, or $x = \pm 1/\sqrt{3}$. The corresponding values of y and z are also $\pm 1/\sqrt{3}$, so we get eight points to check: $(\pm 1/\sqrt{3}, \pm 1/\sqrt{3}, \pm 1/\sqrt{3})$ with all eight choices of signs. Checking, we see that the largest value of xyz occurs when all of the signs or exactly two of them are positive, that is, at the points $(1/\sqrt{3}, 1/\sqrt{3}, 1/\sqrt{3})$, $(-1/\sqrt{3}, -1/\sqrt{3}, 1/\sqrt{3})$, $(1/\sqrt{3}, -1/\sqrt{3}, -1/\sqrt{3})$, and
$(-1/\sqrt{3}, 1/\sqrt{3}, -1/\sqrt{3})$.

25. Minimize $d(x, y, z) = x^2 + y^2 + z^2$ subject to $x^2 + y - 1 - z = 0$. Using Lagrange multipliers, we need to solve $2x = 2\lambda x$, $2y = \lambda$, $2z = -\lambda$, and $x^2 + y - 1 - z = 0$. From the first equation, either $x = 0$ or $\lambda = 1$. If $x = 0$, substitute $y = \lambda/2$ and $z = -\lambda/2$ in the last equation to get $\lambda/2 - 1 + \lambda/2 = 0$, $\lambda = 1$. This gives the point $(0, 1/2, -1/2)$. On the other hand, if $\lambda = 1$, then $y = 1/2$, $z = -1/2$, and $x^2 + 1/2 - 1 + 1/2 = 0$, so $x = 0$, giving the same point. Since the distance may get arbitrarily large, but can get no less than 0, there must be a minimum distance and it must be at the point we found, $(0, 1/2, -1/2)$.

27. Minimize $d(x, y, z) = (x + 1)^2 + (y - 1)^2 + (z - 3)^2$ subject to $-2x + 2y + z - 5 = 0$. Using Lagrange multipliers, we need to solve $2(x + 1) = -2\lambda$, $2(y - 1) = 2\lambda$, $2(z - 3) = \lambda$, and $-2x + 2y + z - 5 = 0$. Solve the first three equations for $x = -\lambda - 1$, $y = \lambda + 1$, and $z = \lambda/2 + 3$. Substitute these in the last equation: $-2(-\lambda - 1) + 2(\lambda + 1) + (\lambda/2 + 3) - 5 = 0$, $\frac{9}{2}\lambda + 2 = 0$, $\lambda = -4/9$. This gives $x = -5/9$, $y = 5/9$, and $z = 25/9$. Thus, the point on the plane closest to $(-1, 1, 3)$ is $(-5/9, 5/9, 25/9)$.

29. Let the length, width, and height be l, w, and h, respectively. We wish to minimize $C = 20 \times 2 \times lw + 10(2lh + 2wh) = 40lw + 20lh + 20wh$ subject to $lwh = 2$. Solve the constraint for $l = \frac{2}{wh}$ and substitute to get $C(w, h) = \frac{80}{h} + \frac{40}{w} + 20wh$, $w > 0$ and $h > 0$. $C_w = -\frac{40}{w^2} + 20h = 0$ when $h = \frac{2}{w^2}$; $C_h = -\frac{80}{h^2} + 20w = 0$ when $w = \frac{4}{h^2} = w^4$, giving $w = 1$ (recall that $w > 0$). Thus, $h = 2$ and $l = 1$. Making either of h or w small or both large makes C large, so the point we found must be a minimum. Thus, the dimensions of the box of least cost are $l \times w \times h = 1 \times 1 \times 2$.

31. Let $l = $ length, $w = $ width, and $h = $ height. We want to maximize the volume $V = lwh$. We are told that $l + 2(w + h) \leq 108$; for the largest possible volume we will want $l + 2(w + h) = 108$. In the solution for Exercise 41 in Section 15.4 we showed one way to solve this problem. Here we use the alternative, which is Lagrange multipliers. So, we need to solve $wh = \lambda$, $lh = 2\lambda$, $lw = 2\lambda$, and $l + 2w + 2h - 108 = 0$. Substitute $\lambda = wh$ in the second equation to get $lh = 2wh$, so $l = 2w$ (we certainly must have $h > 0$ for a maximum volume). If we substitute $\lambda = wh$ in the third equation we get $lw = 2wh$, so $l = 2h$, hence $w = h = \frac{1}{2}l$. Substituting now in the last equation gives $3l = 108$, so $l = 36$, $w = 18$, and $h = 18$. Thus, the largest volume package has dimensions $18 \text{ in} \times 18 \text{ in} \times 36 \text{ in}$ and volume $= 11{,}664$ cubic inches.

33. Let L be the cost of lightweight cardboard and let H be the cost of heavy-duty cardboard.

Minimize $C = 2Hlw + 2Llh + 2Lwh$ subject to $lwh = 2$. Solve the constraint for $l = \frac{2}{wh}$ and substitute to get $C(w, h) = \frac{4H}{h} + \frac{4L}{w} + 2Lwh$. $C_w = -\frac{4L}{w^2} + 2Lh = 0$ when $h = \frac{2}{w^2}$; $C_h = -\frac{4H}{h^2} + 2Lw = 0$ when $w = \frac{2H}{Lh^2} = \frac{H}{2L}w^4$, so $w = \left(\frac{2L}{H}\right)^{1/3}$. Substituting in the expressions for l and h we find $h = 2^{1/3}\left(\frac{H}{L}\right)^{2/3}$ and $l = w = \left(\frac{2L}{H}\right)^{1/3}$. Thus, the box of least cost has dimensions $\left(\frac{2L}{H}\right)^{1/3} \times \left(\frac{2L}{H}\right)^{1/3} \times 2^{1/3}\left(\frac{H}{L}\right)^{2/3}$.

35. If (x, y, z) is one corner of the box (with $x > 0$, $y > 0$, and $z > 0$), the box will have dimensions $2x \times 2y \times z$. We need to maximize $V = 4xyz$ subject to $x^2 + y^2 + z - 1 = 0$. Using Lagrange multipliers we must solve $4yz = 2\lambda x$, $4xz = 2\lambda y$, $4xy = \lambda$, and $x^2 + y^2 + z - 1 = 0$. Solve the first equation for $\lambda = 2\frac{yz}{x}$ and substitute in the second and third equations: $4xz = 4\frac{y^2z}{x}$, $x^2 = y^2$; $4xy = 2\frac{yz}{x}$, $2x^2 = z$. Substituting $y^2 = x^2$ and $z = 2x^2$ in the last equation gives $x^2 + x^2 + 2x^2 = 1$, $x^2 = \frac{1}{4}$, $x = \frac{1}{2}$; $y = \frac{1}{2}$; $z = \frac{1}{2}$. The dimensions of the box are $1 \times 1 \times \frac{1}{2}$.

37. The objective is to maximize the productivity:
$$q = 50n^{0.6}r^{0.4}$$
subject to
$$150n + 60r = 1500 \quad (n, r \geq 0)$$
So, $g(n, r) = 150n + 60r - 1500$,
$$L(n, r) = 50n^{0.6}r^{0.4} - \lambda(150n + 60r - 1500)$$
and the system we need to solve is:
$$L_n = 0: \quad 30n^{-0.4}r^{0.4} = 150\lambda$$
$$L_r = 0: \quad 20n^{0.6}r^{-0.6} = 60\lambda$$
$$g = 0: \quad 150n + 60r = 1500$$
Rewrite the first two equations as

561

$$30\left(\frac{r}{n}\right)^{0.4} = 150\lambda \qquad 20\left(\frac{n}{r}\right)^{0.6} = 60\lambda$$

Dividing the first by the second:

$$\frac{3}{2}\left(\frac{r}{n}\right)^{0.4}\left(\frac{r}{n}\right)^{0.6} = \frac{5}{2}$$

that is, $3\dfrac{r}{n} = 5$,

giving $r = \dfrac{5}{3}n$

Substituting this result into the constraint equation gives

$$150n + 100n = 1500$$
$$250n = 1500$$

so $n = 6$ laborers, $r = \dfrac{5}{3}n = 10$ robots

for a productivity of

$$q = 50(6)^{0.6}(10)^{0.4} \approx 368 \text{ pairs}$$

of socks per day.

39. Method 1: Solve $g(x, y, z) = 0$ for one of the variables and substitute in $f(x, y, z)$. Then find the maximum value of the resulting function of two variables. Advantage (Answers may vary): We can use the second derivative test to check whether the resulting critical points are maxima, minima, saddle points, or none of these. Disadvantage (Answers may vary): We may not be able to solve $g(x, y, z) = 0$ for one of the variables.

Method 2: Use the method of Lagrange Multipliers. Advantage (Answers may vary): We do not need to solve the constraint equation for one of the variables. Disadvantage (Answers may vary): The method does not tell us whether the critical points obtained are maxima, minima, saddle points, or none of these.

41. If the only constraint is an equality constraint, and if it is impossible to eliminate one of the variables in the objective function by substitution (solving the constraint equation for a variable or some other method).

43. Answers may vary: Maximize $f(x, y) = 1 - x^2 - y^2$ subject to $x = y$.

45. Yes. There may be relative extrema at points on the boundary of the domain of the function. The partial derivatives of the function need not be 0 at such points.

47. If the solution were located in the interior of one of the line segments making up the boundary of the domain of f, then the derivative of a certain function would be 0. This function is obtained by substituting the linear equation $C(x, y) = 0$ in the linear objective function. But because the result would again be a linear function, it is either constant, or its derivative is a nonzero constant. In either event, extrema lie on the boundary of that line segment; that is, at one of the corners of the domain.

15.5 Double Integrals and Applications

1. $\displaystyle\int_0^1 \int_0^1 (x - 2y)\, dx\, dy = \int_0^1 \left[\tfrac{1}{2}x^2 - 2xy\right]_{x=0}^1 dy = \int_0^1 \left(\tfrac{1}{2} - 2y\right) dy = \left[\tfrac{1}{2}y - y^2\right]_{y=0}^1 = -\tfrac{1}{2}$

3. $\displaystyle\int_0^1 \int_0^2 (ye^x - x - y)\, dx\, dy = \int_0^1 \left[ye^x - \tfrac{1}{2}x^2 - xy\right]_{x=0}^2 dy = \int_0^1 (e^2 y - 2 - 3y)\, dy =$

$\left[\tfrac{1}{2}e^2 y^2 - 2y - \tfrac{3}{2}y^2\right]_{y=0}^1 = \tfrac{1}{2}e^2 - \tfrac{7}{2}$

5. $\displaystyle\int_0^2 \int_0^3 e^{x+y}\, dx\, dy = \int_0^2 [e^{x+y}]_{x=0}^3 \, dy = \int_0^2 (e^{3+y} - e^y)\, dy = [e^{3+y} - e^y]_{y=0}^2 = e^5 - e^2 - e^3 + 1 =$

$(e^3 - 1)(e^2 - 1)$. This may also be found by writing $\displaystyle\int_0^2 \int_0^3 e^{x+y}\, dx\, dy = \int_0^2 \int_0^3 e^x e^y\, dx\, dy =$

$\left(\displaystyle\int_0^3 e^x\, dx\right)\left(\int_0^2 e^y\, dy\right)$ as in Exercise 57.

7. $\displaystyle\int_0^1 \int_0^{2-y} x\, dx\, dy = \int_0^1 \left[\tfrac{1}{2}x^2\right]_{x=0}^{2-y} dy = \int_0^1 \tfrac{1}{2}(2 - y)^2\, dy = \left[-\tfrac{1}{6}(2 - y)^3\right]_{y=0}^1 = -\tfrac{1}{6} + \tfrac{8}{6} = \tfrac{7}{6}$

9. $\displaystyle\int_{-1}^1 \int_{y-1}^{y+1} e^{x+y}\, dx\, dy = \int_{-1}^1 [e^{x+y}]_{x=y-1}^{y+1}\, dy = \int_{-1}^1 (e^{2y+1} - e^{2y-1})\, dy = \left[\tfrac{1}{2}e^{2y+1} - \tfrac{1}{2}e^{2y-1}\right]_{y=-1}^1 =$

$\tfrac{1}{2}(e^3 - e - e^{-1} + e^{-3})$

11. $\displaystyle\int_0^1 \int_{-x^2}^{x^2} x\, dy\, dx = \int_0^1 [xy]_{y=-x^2}^{x^2}\, dx = \int_0^1 2x^3\, dx = \left[\tfrac{1}{2}x^4\right]_{x=0}^1 = \tfrac{1}{2}$

13. $\displaystyle\int_0^1 \int_0^x e^{x^2}\, dy\, dx = \int_0^1 [ye^{x^2}]_{y=0}^x\, dx = \int_0^1 xe^{x^2}\, dx = \left[\tfrac{1}{2}e^{x^2}\right]_{x=0}^1 = \tfrac{1}{2}e - \tfrac{1}{2} = \tfrac{1}{2}(e-1)$

15. $\displaystyle\int_0^2 \int_{1-x}^{8-x} (x+y)^{1/3}\, dy\, dx = \int_0^2 \left[\tfrac{3}{4}(x+y)^{4/3}\right]_{y=1-x}^{8-x}\, dx = \int_0^2 \left(\tfrac{3}{4}(16) - \tfrac{3}{4}\right)\, dx = \int_0^2 \tfrac{45}{4}\, dx = \left[\tfrac{45}{4}x\right]_{x=0}^2 =$
$\dfrac{45}{2}$

17. $\displaystyle\int_{-1}^1 \int_0^{1-x^2} 2\, dy\, dx = \int_{-1}^1 [2y]_{y=0}^{1-x^2}\, dx = \int_{-1}^1 2(1-x^2)\, dx = \left[2x - \tfrac{2}{3}x^3\right]_{x=-1}^1 = 2 - \tfrac{2}{3} + 2 - \tfrac{2}{3} = \tfrac{8}{3}$

19. $\displaystyle\int_{-1}^1 \int_0^{1-y^2} (1+y)\, dx\, dy = \int_{-1}^1 [x(1+y)]_{x=0}^{1-y^2}\, dy = \int_{-1}^1 (1-y^2)(1+y)\, dy = \int_{-1}^1 (1+y-y^2-y^3)\, dy =$
$\left[y + \tfrac{1}{2}y^2 - \tfrac{1}{3}y^3 - \tfrac{1}{4}y^4\right]_{y=-1}^1 = 1 + \tfrac{1}{2} - \tfrac{1}{3} - \tfrac{1}{4} + 1 - \tfrac{1}{2} - \tfrac{1}{3} + \tfrac{1}{4} = \tfrac{4}{3}$

21. $\displaystyle\int_{-1}^1 \int_{-\sqrt{1-y^2}}^{\sqrt{1-y^2}} xy^2\, dx\, dy = \int_{-1}^1 \left[\tfrac{1}{2}x^2y^2\right]_{x=-\sqrt{1-y^2}}^{\sqrt{1-y^2}}\, dy = \int_{-1}^1 0\, dy = 0$

23. $\displaystyle\int_{-1}^0 \int_{-x-1}^{x+1} (x^2+y^2)\, dy\, dx + \int_0^1 \int_{x-1}^{-x+1} (x^2+y^2)\, dy\, dx = \int_{-1}^0 \left[x^2y + \tfrac{1}{3}y^3\right]_{y=-x-1}^{x+1}\, dx +$
$\displaystyle\int_0^1 \left[x^2y + \tfrac{1}{3}y^3\right]_{y=x-1}^{-x+1}\, dx = \int_{-1}^0 \left(2x^2(x+1) + \tfrac{2}{3}(x+1)^3\right)\, dx + \int_0^1 \left(2x^2(-x+1) + \tfrac{2}{3}(-x+1)^3\right)\, dx =$
$\displaystyle\int_{-1}^0 \left(\tfrac{8}{3}x^3 + 4x^2 + 2x + \tfrac{2}{3}\right)\, dx + \int_0^1 \left(-\tfrac{8}{3}x^3 + 4x^2 - 2x + \tfrac{2}{3}\right)\, dx = \left[\tfrac{2}{3}x^4 + \tfrac{4}{3}x^3 + x^2 + \tfrac{2}{3}x\right]_{x=-1}^0 +$
$\left[-\tfrac{2}{3}x^4 + \tfrac{4}{3}x^3 - x^2 + \tfrac{2}{3}x\right]_{x=0}^1 = -\tfrac{2}{3} + \tfrac{4}{3} - 1 + \tfrac{2}{3} - \tfrac{2}{3} + \tfrac{4}{3} - 1 + \tfrac{2}{3} = \tfrac{2}{3}$

25. $\displaystyle\int_0^2\int_0^{2-x} y\,dy\,dx = \int_0^2\left[\tfrac{1}{2}y^2\right]_{y=0}^{2-x}dx = \int_0^2\tfrac{1}{2}(2-x)^2\,dx = \left[-\tfrac{1}{6}(2-x)^3\right]_{x=0}^2 = \tfrac{8}{6} = \tfrac{4}{3}$. $A = \tfrac{1}{2}(2)(2) = 2$

by the geometric formula for the area of a triangle. So, the average value $= \dfrac{4/3}{2} = \dfrac{2}{3}$.

27. $\displaystyle\int_0^1\int_{y-1}^{1-y} e^y\,dx\,dy = \int_0^1\left[xe^y\right]_{x=y-1}^{1-y}dy = \int_0^1 2(1-y)e^y\,dy = \left[2(1-y)e^y + 2e^y\right]_{y=0}^1 = 2e - 4 = 2(e-$

$2)$. $A = \tfrac{1}{2}(2)(1) = 1$ by the formula for the area of a triangle. So, the average value $= 2(e-2)$.

29. $\displaystyle\int_{-2}^0\int_{x/2+1}^{x+2} x^2\,dy\,dx + \int_0^2\int_{-x/2+1}^{-x+2} x^2\,dy\,dx = \tfrac{4}{3}$ as calculated in Exercise 24. $A = 2\cdot\tfrac{1}{2}(2)(1) = 2$, since A

consists of the area of two triangles. So, the average value $= \dfrac{4/3}{2} = \dfrac{2}{3}$.

31. $\displaystyle\int_0^1\int_0^{1-x} f(x,y)\,dy\,dx$

35. $\displaystyle\int_1^4\int_1^{2\sqrt{y}} f(x,y)\,dx\,dy$

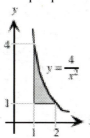

33. $\displaystyle\int_0^{\sqrt{2}}\int_{x^2-1}^1 f(x,y)\,dy\,dx$

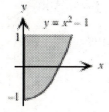

37. $\displaystyle V = \int_0^1\int_0^2 (1-x^2)\,dy\,dx = \int_0^1\left[(1-x^2)y\right]_{y=0}^2 dx = \int_0^1 2(1-x^2)\,dx = \left[2x - \tfrac{2}{3}x^3\right]_{x=0}^1 = 2 - \tfrac{2}{3} = \tfrac{4}{3}$

565

39. The top face of the tetrahedron is the graph of $z = 1 - x - y$. So, the volume is $V =$

$$\int_0^1 \int_0^{1-x} (1 - x - y)\, dy\, dx = \int_0^1 \left[(1 - x)y - \tfrac{1}{2}y^2 \right]_{y=0}^{1-x} dx = \int_0^1 \left[(1 - x)^2 - \tfrac{1}{2}(1 - x)^2 \right] dx =$$

$$\int_0^1 \tfrac{1}{2}(1 - x)^2\, dx = \left[-\tfrac{1}{6}(1 - x)^3 \right]_{x=0}^1 = \tfrac{1}{6}.$$

41. $\displaystyle\int_{45}^{55} \int_8^{12} 10{,}000x^{0.3}y^{0.7}\, dy\, dx = 10{,}000 \int_{45}^{55}\left[\frac{1}{1.7}x^{0.3}y^{1.7} \right]_{y=8}^{12} dx = \frac{10{,}000(12^{1.7} - 8^{1.7})}{1.7} \int_{45}^{55} x^{0.3}\, dx =$

$\dfrac{10{,}000(12^{1.7} - 8^{1.7})}{(1.7)(1.3)} [x^{1.3}]_{x=45}^{55} = \dfrac{10{,}000(12^{1.7} - 8^{1.7})(55^{1.3} - 45^{1.3})}{(1.7)(1.3)} \approx 6{,}471{,}000.$ $A = (12 - 8)(55 -$

$45) = 40.$ So, the average is approximately $\dfrac{6{,}471{,}000}{40} \approx 162{,}000$ gadgets.

43. $R(p, q) = pq$ with $40 \le p \le 50$ and $8000 - p^2 \le q \le 10{,}000 - p^2$. To find the average we first compute

$$\int_{40}^{50} \int_{8000-p^2}^{10{,}000-p^2} pq\, dq\, dp = \tfrac{1}{2}\int_{40}^{50} [pq^2]_{q=8000-p^2}^{10{,}000-p^2} dp =$$

$$\tfrac{1}{2}\int_{40}^{50} [p(10{,}000 - p^2)^2 - p(8000 - p^2)^2]\, dp =$$

$$\tfrac{1}{12}[-(10{,}000 - p^2)^3 + (8000 - p^2)^3]_{p=40}^{50} =$$

$6{,}255{,}000{,}000.$ Next we compute $A =$

$$\int_{40}^{50} \int_{8000-p^2}^{10{,}000-p^2} dq\, dp = \int_{40}^{50} [q]_{q=8000-p^2}^{10{,}000-p^2} dp =$$

$$\int_{40}^{50} [10{,}000 - p^2 - (8000 - p^2)]\, dp =$$

$$\int_{40}^{50} 2000\, dp = [2000p]_{p=40}^{50} = 100{,}000 -$$

$80{,}000 = 20{,}000.$ The average revenue is thus $\dfrac{6{,}255{,}000{,}000}{20{,}000} = \$312{,}750.$

45. $R(p, q) = pq$ with $15{,}000/q \le p \le 20{,}000/q$ and $500 \le q \le 1000$. To find the average we first compute

$$\int_{500}^{1000} \int_{15{,}000/q}^{20{,}000/q} pq\, dp\, dq = \tfrac{1}{2}\int_{500}^{1000} [p^2 q]_{p=15{,}000/q}^{20{,}000/q} dq =$$

$$\tfrac{1}{2}\int_{500}^{1000} \left(\frac{20{,}000^2}{q} - \frac{15{,}000^2}{q} \right) dq =$$

$$\frac{175{,}000{,}000}{2} \int_{500}^{1000} \frac{1}{q}\, dq = \frac{175{,}000{,}000}{2} [\ln$$

$q]_{q=500}^{1000} = \dfrac{175{,}000{,}000}{2}(\ln 1000 - \ln 500) =$

$\dfrac{175{,}000{,}000}{2}\ln 2.$ Next we compute $A =$

$$\int_{500}^{1000} \int_{15{,}000/q}^{20{,}000/q} dp\, dq = \int_{500}^{1000} [p]_{p=15{,}000/q}^{20{,}000/q} dq =$$

$$5000 \int_{500}^{1000} \frac{1}{q}\, dq = 5000[\ln q]_{q=500}^{1000} =$$

$5000(\ln 1000 - \ln 500) = 5000 \ln 2$. The average revenue is thus $\dfrac{175{,}000{,}000 \ln 2}{2 \cdot 5000 \ln 2} =$ $\$17{,}500$.

47. Total population $= \displaystyle\int_0^{20} \int_0^{30} e^{-0.1(x+y)}\, dy\, dx =$

$-10 \displaystyle\int_0^{20} [e^{-0.1(x+y)}]_{y=0}^{30}\, dx =$

$-10 \displaystyle\int_0^{20} (e^{-0.1x-3} - e^{-0.1x})\, dx =$

$-10(e^{-3} - 1) \displaystyle\int_0^{20} e^{-0.1x}\, dx = 100(e^{-3} - 1)[e^{-}$

$^{0.1x}]_{x=0}^{20} = 100(e^{-3} - 1)(e^{-2} - 1) \approx 82.16$ hundred people, or 8216 people.

49. $\displaystyle\int_0^1 \int_0^1 (x^2 + 2y^2)\, dy\, dx =$

$\displaystyle\int_0^1 \left[x^2 y + \tfrac{2}{3} y^3 \right]_{y=0}^{1} dx = \displaystyle\int_0^1 \left(x^2 + \tfrac{2}{3} \right) dx =$

$\left[\tfrac{1}{3} x^3 + \tfrac{2}{3} x \right]_{x=0}^{1} = 1$. $A = 1$, the area of the square. Hence, the average temperature is $1/1 = 1$ degree.

51. The area between the curves $y = r(x)$ and $y = s(x)$ and the vertical lines $x = a$ and $x = b$ is given by $\displaystyle\int_a^b \int_{r(x)}^{s(x)} dy\, dx$ assuming that $r(x) \le s(x)$ for $a \le x \le b$.

53. The first step in calculating an integral of the form $\displaystyle\int_a^b \int_{r(x)}^{s(x)} f(x,\, y)\, dy\, dx$ is to evaluate the integral

$\displaystyle\int_{r(x)}^{s(x)} f(x,\, y)\, dy$, obtained by holding x constant and integrating with respect to y.

55. Paintings per picasso per dali

57. The left-hand side is $\displaystyle\int_a^b \int_c^d f(x)g(y)\, dx\, dy =$

$\displaystyle\int_a^b \left(g(y) \int_c^d f(x)\, dx \right) dy$ (because $g(y)$ is treated as a constant in the inner integral) $=$

$\left(\displaystyle\int_c^d f(x)\, dx \right) \int_a^b g(y)\, dy$ (because $\displaystyle\int_c^d f(x)\, dx$ is a constant and can therefore be taken outside the integral with respect to y). For example,

$\displaystyle\int_0^1 \int_1^2 ye^x\, dx\, dy = \tfrac{1}{2}(e^2 - e)$ if we compute it as

an iterated integral or as $\displaystyle\int_1^2 e^x\, dx \int_0^1 y\, dy$.

Chapter 15 Review Exercises

1. $f(x, y, z) = \dfrac{x}{y+xz} + x^2y$

$f(0, 1, 1) = \dfrac{0}{1+(0)(1)} + 0^2(1) = 0$

$f(2, 1, 1) = \dfrac{2}{1+(2)(1)} + 2^2(1) = 14/3$

$f(-1, 1, -1) = \dfrac{-1}{1+(-1)(-1)} + (-1)^2(1) = 1/2$

$f(z, z, z) = \dfrac{z}{z+z^2} + z^2z = \dfrac{1}{1+z} + z^3$

$f(x+h, y+k, z+l)$
$= \dfrac{x+h}{y+k+(x+h)(z+l)} + (x+h)^2(y+k)$

3. Decreases by 0.32 units; increases by 12.5 units

5. Reading left to right, starting at the top: $4, 0, 0,$ $3, 0, 1, 2, 0, 2$

7. Answers may vary. Two examples are $f(x, y) = 3(x - y)/2$ and $f(x, y) = 3(x - y)^3/8$.

9. $f_x = 2x + y, f_y = x, f_{yy} = 0$

11. $f(x, y) = 4x + 5y - 6xy$
$f_x(x, y) = 4 - 6y$
$f_{zz}(x, y) = 4$
$f_{xx}(1, 0) = f_{xx}(3, 2) = 4$
Therefore $f_{xx}(1, 0) - f_{xx}(3, 2) = 0$

13. $\dfrac{\partial f}{\partial x} = \dfrac{-x^2 + y^2 + z^2}{(x^2 + y^2 + z^2)^2}$

$\dfrac{\partial f}{\partial y} = -\dfrac{2xy}{(x^2 + y^2 + z^2)^2}$

$\dfrac{\partial f}{\partial z} = -\dfrac{2xz}{(x^2 + y^2 + z^2)^2}$

$\left.\dfrac{\partial f}{\partial x}\right|_{(0,1,0)} = 1$

15. $f_x = 2(x - 1); f_y = 2(2y - 3); f_{xx} = 2; f_{yy} = 4; f_{xy} = 0. f_x = 0$ when $x = 1; f_y = 0$ when $y =$

$3/2,$ so $(1, 3/2)$ is the only critical point. $H = 8$ and $f_{xx} > 0,$ so f has an absolute minimum at $(1, 3/2)$.

17. $h_x = ye^{xy}; h_y = xe^{xy}; h_{xx} = y^2e^{xy}; h_{yy} = x^2e^{xy};$ $h_{xy} = (xy + 1)e^{xy}. h_x = 0$ when $y = 0; h_y = 0$ when $x = 0,$ so $(0, 0)$ is the only critical point. $H(0, 0) = -1 < 0,$ so h has a saddle point at $(0, 0).$

19. Note that the domain of f contains every point except $(0, 0). f_x = \dfrac{2x}{x^2 + y^2} - 2x; f_y = \dfrac{2y}{x^2 + y^2} - 2y; f_{xx} = \dfrac{-2x^2 + 2y^2}{(x^2 + y^2)^2} - 2; f_{yy} = \dfrac{2x^2 - 2y^2}{(x^2 + y^2)^2} - 2;$

$f_{xy} = \dfrac{-4xy}{(x^2 + y^2)^2}. f_x = 0$ if either $x = 0$ or $x^2 + y^2 = 1; f_y = 0$ if either $y = 0$ or $x^2 + y^2 = 1;$ since $(0, 0)$ is not in the domain, the critical points are all the points on the circle $x^2 + y^2 = 1.$ At such a point we can substitute $y^2 = 1 - x^2$ to get $f_{xx} =$
$(2 - 4x^2) - 2 = -4x^2, f_{yy} = 4x^2 - 4;$ and $f_{xy} = \pm 4x(1 - x^2)^{1/2}.$ Hence $H = -4x^2(4x^2 - 4) - 16x^2(1 - x^2) = 0.$ To resolve what happens along the circle we graph the function:

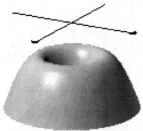

f has an absolute maximum at each point on the circle $x^2 + y^2 = 1.$

21. The objective function is $V = xyz.$ Solving the constraint equation $x + y + z = 1$ for z gives
$z = 1 - x - y$
$V = xy(1 - x - y)$
$\quad = xy - x^2y - xy^2$
$V_x = y - 2xy - y^2, \quad V_y = x - x^2 - 2xy$
Critical points:
$y - 2xy - y^2 = 0$

568

$x - x^2 - 2xy = 0$

Dividing the first equation by y gives

$1 - 2x - y = 0$

$2x + y = 1$

Dividing the second equation by x gives

$1 - x - 2y = 0$

$x + 2y = 1$

The system of equations

$2x + y = 1$

$x + 2y = 1$

has solution $(x, y) = (1/3, 1/3)$. The corresponding value of z is

$z = 1 - x - y = 1 - 2/3 = 1/3$

$V_{xx} = -2y$, $V_{xy} = 1 - 2x - 2y$, $V_{yy} = -2x$

Evaluating these at $(1/3, 1/3)$ gives

$V_{xx} = -2/3$, $V_{xy} = -1/3$, $V_{yy} = -2/3$

$H = V_{zz}V_{yy} - (V_{xy})^2 = 4/9 - 1/9 > 0$

$V_{xx} < 0$

so V does have a relative maximum at $(1/3, 1/3, 1/3)$. The value of V at that point is $V = xyz = 1/27$. That it is the absolute maximum can be seen as interpreting the problems as maximizing the volume of a box whose dimensions add up to 1.

23. We minimize the square of the distance of (x, y, z) to the origin: $d(x, y, z) = x^2 + y^2 + z^2$ subject to $z = \sqrt{x^2 + 2(y-3)^2}$, which we rewrite as $x^2 + 2(y - 3)^2 - z^2 = 0$ $(z \geq 0)$. Solving the second equation for z^2 and substituting in the formula for d gives

$d = x^2 + y^2 + x^2 + 2(y - 3)^2$

$= 2x^2 + 3y^2 - 12y + 18$

$d_x = 4x$, $d_y = 6y - 12$

The critical points are

$4x = 0$ or $x = 0$

$6y - 12 = 0$, or $y = 2$

The corresponding value of z is

$z = \sqrt{x^2 + 2(y-3)^2} = \sqrt{0^2 + 2(2-3)^2} = \sqrt{2}$.

The critical point is therefore $(0, 2, \sqrt{2}\,)$. Since there must be at least one point on the given surface closest to the origin, the critical point $(0, 2, \sqrt{2}\,)$ must be that point.

25. The constraint is $xy = 2$, or $xy - 2 = 0$.

$f(x, y) = x^2 + y^2$

$g(x, y) = xy - 2$

$L(x, y) = x^2 + y^2 - \lambda(xy - 2)$

(1) $L_x = 0 \Rightarrow 2x = \lambda y \Rightarrow \lambda = 2x/y$

(2) $L_y = 0 \Rightarrow 2y = \lambda x$

(3) Constraint $\Rightarrow xy = 2$

Substitute Equation (1) in Equation (2) to obtain

$2x^2/y = 2y \Rightarrow x^2 = y^2 \Rightarrow x = \pm y$

Substituting into the constraint gives

$x^2 = 2$ (we must reject $y = -x$ here)

$x = \pm\sqrt{2}$

so $y = 2/x = \pm\sqrt{2}$

Thus we have two critical points: $(\sqrt{2}, \sqrt{2})$ and $(-\sqrt{2}, -\sqrt{2})$

The corresponding value of the objective function is

$f(\pm\sqrt{2}, \pm\sqrt{2}) = 2 + 2 = 4$

27. The constraint is $xyz = 1$, or $xyz - 1 = 0$.

$f(x, y, z) = xy + x^2z^2 + 4yz$

$g(x, y, z) = xyz - 1$

$L(x, y, z) = xy + x^2z^2 + 4yz - \lambda(xyz - 1)$

(1) $L_x = 0 \Rightarrow y + 2xz^2 = \lambda yz$

(2) $L_y = 0 \Rightarrow x + 4z = \lambda xz$

(3) $L_z = 0 \Rightarrow 2x^2z + 4y = \lambda xy$

(4) Constraint $\Rightarrow xyz = 1$

Solve Equation (1) for λ:

$\lambda = \dfrac{1}{z} + \dfrac{2xz}{y}$

Substituting in (2) gives

$x + 4z = (\dfrac{1}{z} + \dfrac{2xz}{y})xz$

$x + 4z = x + \dfrac{2x^2z^2}{y}$

$4z = \dfrac{2x^2z^2}{y}$

$2 = \dfrac{x^2z}{y}$

$y = x^2z/2$

Substituting the expression for λ in (3) gives

$$2x^2z + 4y = \left(\frac{1}{z} + \frac{2xz}{y}\right)xy$$

$$2x^2z + 4y = \frac{xy}{z} + 2x^2z$$

$$4y = \frac{xy}{z}$$

$$4 = \frac{x}{z}$$

$$z = x/4$$

Substituting the expressions we obtained for y and z in the constraint equation gives:

$$x[2x^2(x/4)/2](x/4) = 1$$

$$x^5 = 32$$

$$x = 2$$

The corresponding values of y and z are

$$z = x/4 = 2/4 = 1/2$$

$$y = x^2z/2 = 1$$

Therefore, the only critical point is $(2, 1, 1/2)$ and the corresponding value of the objective is

$$f(2, 1, 1/2) = (2)(1) + (2)^2(1/2)^2 + 4(1)(1/2)$$
$$= 5$$

Therefore, the minimum value of the objective function is 5, and occurs at the point $(2, 1, 1/2)$.

29. $\displaystyle\int_0^1 \int_0^2 2xy \, dx \, dy = \int_0^1 [x^2y]_{x=0}^2 \, dy = \int_0^1 4y \, dy =$

$$[2y^2]_{y=0}^1 = 2$$

31. $\displaystyle\int_0^2 \int_0^{2x} \frac{1}{x^2 + 1} \, dy \, dx = \int_0^2 \left[\frac{y}{x^2 + 1}\right]_{y=0}^{2x} \, dy =$

$$\int_0^2 \frac{2x}{x^2 + 1} \, dy = [\ln(x^2 + 1)]_{x=0}^2 = \ln 5$$

33. $\displaystyle\int_0^1 \int_y^{2-y} (x^2 - y^2) \, dx \, dy =$

$$\int_0^1 \left[\frac{1}{3}x^3 - xy^2\right]_{x=y}^{2-y} \, dy$$

$$= \int_0^1 \left[\frac{1}{3}(2 - y)^3 - (2 - y)y^2 - \frac{1}{3}y^3 + y^3\right] \, dy$$

$$= \int_0^1 \left(\frac{8}{3} - 4y + \frac{4}{3}y^3\right) \, dy$$

$$= \left[\frac{8}{3}y - 2y^2 + \frac{1}{3}y^4\right]_{y=0}^1$$

$$= 1$$

35. (a) $h(x, y) = 5000 - 0.8x - 0.6y$ hits per day (x = number of new customers at JungleBooks.com, y = number of new customers at FarmerBooks.com)
(b) Set $x = 100$ and $h = 4770$:
$4770 = 5000 - 0.8(100) - 0.6y$
so $y = 250$ new customers.
(c) $h(x, y, z) = 5000 - 0.8x - 0.6y + 0.0001z$
(z = number of new Internet shoppers)
(d) Set $x = y = 100$ and $h = 5000$:
$5000 = 5000 - 0.8(100) - 0.6(100) + 0.0001z$
so $z = 1.4$ million new Internet shoppers.

37. (a) $h(2000, 3000) = 2320$ hits per day
(b) $h_y = 0.08 + 0.00003x$ hits (daily) per dollar spent on television advertising per month; this increases with increasing x.
(c) Set $h_y = 1/5 = 0.2$: $0.2 = 0.08 + 0.00003x$,
$x = \$4000$ per month

39. (A) because $h_x(x, 0) = 0.05$.

41. (a) $P(10, 1000) \approx 15,800$ additional orders per day

(b) Minimize $C(x, y) = 150x + y$ subject to $1000x^{0.9}y^{0.1} = 15,000$.

Using Lagrange multipliers we need to solve the system

$$150 = 900\lambda x^{-0.1}y^{0.1}$$
$$1 = 100\lambda x^{0.9}y^{-0.9}$$
$$1000x^{0.9}y^{0.1} - 15,000 = 0$$

From the first equation we get

$$\lambda = \frac{x^{0.1}}{6y^{0.1}}$$

Substituting in the second equation gives

$$1 = \frac{100x}{6y}$$

so $y = \frac{50}{3} x$

Substituting in the last equation gives

$$1000\left(\frac{50}{3}\right)^{0.1} x = 15,000$$

so $x = 15\left(\frac{3}{50}\right)^{0.1} \approx 11$

43. $\displaystyle\int_{1200}^{1500} \int_{1800}^{2000}(3x + 10y) \, dy \, dx =$

$\displaystyle\int_{1200}^{1500} [3xy + 5y^2]_{y=1800}^{2000} \, dx$

$= \displaystyle\int_{1200}^{1500} (600x + 3,800,000) \, dx$

$= [300x^2 + 3,800,000x]_{x=1200}^{1500}$

$= 1,383,000,000.$

$A = (1500 - 1200)(2000 - 1800) = 60,000$,
so the average profit is $1,383,000,000/60,000 =$ $23,050.

Chapter 16 Trigonometric Models

16.1 Trigonometric Functions, Models, and Regression

1.

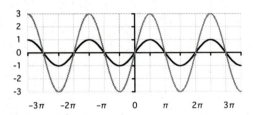

3.

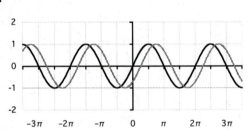

5.

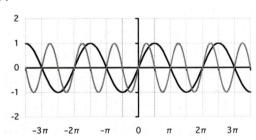

7.

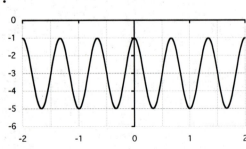

9.

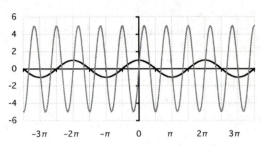

11.

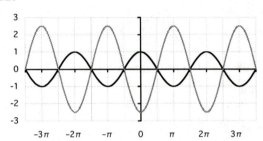

13. From the graph, $C = 1$, $A = 1$, $\alpha = 0$, and $P = 1$, so $\omega = 2\pi$. Thus, $f(x) = \sin(2\pi x) + 1$.

15. From the graph, $C = 0$, $A = 1.5$, $\alpha = 0.25$, and $P = 0.5$, so $\omega = 4\pi$. Thus, $f(x) = 1.5\sin[4\pi(x - 0.25)]$.

17. From the graph, $C = -50$, $A = 50$, $\alpha = 5$, and $P = 20$, so $\omega = \pi/10$. Thus, $f(x) = 50\sin[\pi(x - 5)/10] - 50$.

19. From the graph, $C = 0$, $A = 1$, $\alpha = 0$, and $P = 1$, so $\omega = 2\pi$. Thus, $f(x) = \cos(2\pi x)$.

21. From the graph, $C = 0$, $A = 1.5$, $\alpha = 0.375$, and $P = 0.5$, so $\omega = 4\pi$. Thus, $f(x) = 1.5\cos[4\pi(x - 0.375)]$. Alternatively, $\alpha = -0.125$, so $f(x) = 1.5\cos[4\pi(x + 0.125)]$.

23. From the graph, $C = 40$, $A = 40$, $\alpha = 10$, and $P = 20$, so $\omega = \pi/10$. Thus, $f(x) = 40\cos[\pi(x - 10)/10] + 40$.

25. $f(t) = 4.2\sin(\pi/2 - 2\pi t) + 3$

572

27. $g(x) = 4 - 1.3\sin[\pi/2 - 2.3(x - 4)]$

29. $\sin^2 x + \cos^2 x = 1$; $(\sin^2 x + \cos^2 x)/\cos^2 x = 1/\cos^2 x$; $\tan^2 x + 1 = \sec^2 x$.

31. $\sin(\pi/3) = \sin(\pi/6 + \pi/6) = \sin(\pi/6)\cos(\pi/6) + \cos(\pi/6)\sin(\pi/6) = (1/2)(\sqrt{3}/2) + (\sqrt{3}/2)(1/2) = \sqrt{3}/2$

33. $\sin(t + \pi/2) = \sin t \cos(\pi/2) + \cos t \sin(\pi/2) = \cos t$ because $\cos(\pi/2) = 0$ and $\sin(\pi/2) = 1$.

35. $\sin(\pi - x) = \sin \pi \cos x - \cos \pi \sin x = \sin x$ because $\sin \pi = 0$ and $\cos \pi = -1$.

37. $\tan(x + \pi) = \dfrac{\sin(x + \pi)}{\cos(x + \pi)} = \dfrac{\sin x \cos \pi + \cos x \sin \pi}{\cos x \cos \pi - \sin x \sin \pi} = \dfrac{-\sin x}{-\cos x} = \tan x$

39. (a) $P = 2\pi/0.602 \approx 10.4$ years.
(b) Maximum: $58.8 + 57.7 = 116.5 \approx 117$; minimum: $58.8 - 57.7 = 1.1 \approx 1$
(c) $1.43 + P/4 + P = 1.43 + 13.05 \approx 14.5$ years, or midway through 2011

41. (a)

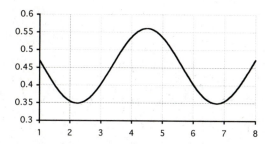

Maximum sales occurred when $t \approx 4.5$ (during the first quarter of 1996). Minimum sales occurred when $t \approx 2.2$ (during the third quarter of 1995) and $t \approx 6.8$ (during the third quarter of 1996). **(b)** Maximum quarterly revenues were $0.561 billion; minimum quarterly revenues were $0.349 billion. **(c)** The maximum and minimum values are $C \pm A$, so the maximum is $0.455 + 0.106 = 0.561$ and the minimum is $0.455 - 0.106 = 0.349$.

43. Amplitude = 0.106, vertical offset = 0.455, phase shift = $-1.61/1.39 \approx -1.16$, angular frequency = 1.39, period = 4.52. In 1995 and 1996, quarterly revenue from the sale of computers at Computer City fluctuated in cycles of 4.52 quarters about a baseline of $0.455 billion. Every cycle, quarterly revenue peaked at $0.561 billion ($0.106 above the baseline) and dipped to a low of $0.349 billion. Revenue peaked in the middle of the first quarter of 1996 (at $t = -1.16 + (5/4)\times 4.52 = 4.49$).

45. $C = 12.5$, $A = 7.5$, period = 52 weeks, so $\omega = \pi/26$, $\alpha = 52/4 = 13$. So, $P(t) = 7.5\sin[\pi(t - 13)/26] + 12.5$.

47. Since the high point is 82 and the low point 75, the baseline is their average:
$C = 78.5$
The amplitude is therefore:
$A = 82-78.5 = 3.5$
The period (low point to low point) is $P = 12$ months, so
$\omega = 2\pi/12 = \pi/6$
The phase shift is the number of months to the high point:
$\beta = 7$
Thus, the model is
$T(t) = A \cos[\omega(t-\beta)] + C = 3.5\cos[\pi(t-7)/6] + 78.5$

49. $C = (4.5+7)/2 = 5.75$, $A = (7-4.5)/2 = 1.25$, $P = 4$, so $\omega = \pi/2$, $\alpha + P/4 =$ distance to first high $= 2$. So, $\alpha = 2-1 = 1$
$n(t) = 1.25\sin[\pi(t-1)/2] + 5.75$

51. $C = (4.5+7)/2 = 5.75$ $A = (7-4.5)/2 = 1.25$, $P = 4$, so
$\omega = \pi/2$, $\beta =$ distance to first high $= 2$
$n(t) = 1.25\cos[\pi(t-2)/2] + 5.75$

53. $C = 10$, $A = 5$, period = 13.5 hours, so $\omega = 2\pi/13.5$, $\alpha = 5 - 13.5/4 = 1.625$. So, $d(t) = 5\sin[2\pi(t - 1.625)/13.5] + 10$.

573

55. (a) $C = 7.5$, $A = 2.5$, period $= 1$ year, so
$\omega = 2\pi$, $\alpha = 0.75$. So, $u(t) =$
$2.5\sin[2\pi(t - 0.75)] + 7.5$. **(b)** $c(t) =$
$1.04^t\{2.5\sin[2\pi(t-0.75)] + 7.5\}$.

57. Graph:

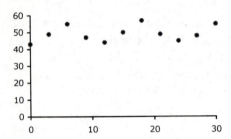

(a) Rough estimates: $C = 50$, $A = 8$, $P = 12$, $\beta = 6$
(b) $f(t) = 5.882\cos[2\pi(t-5.696)/ 12.263] + 49.238$
Graph:

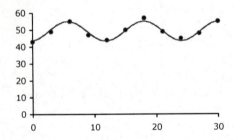

(c) Domestic air travel on U.S. air carriers shows a pattern that repeats itself every $\underline{12}$ months, from a low of $\underline{43}$ to a high of $\underline{55}$ billion revenue passenger miles. (Rounding answers to the nearest whole number.)

59. (a)

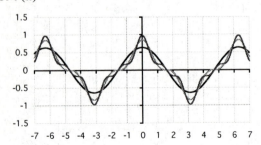

(b) $y_{11} = \dfrac{2}{\pi} \cos x + \dfrac{2}{3\pi} \cos 3x + \dfrac{2}{5\pi} \cos 5x +$
$\dfrac{2}{7\pi} \cos 7x + \dfrac{2}{9\pi} \cos 9x + \dfrac{2}{11\pi} \cos 11x$
Graph of all four functions:

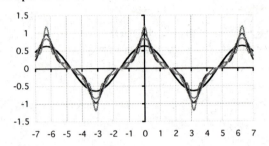

Graph of y_{11} alone:

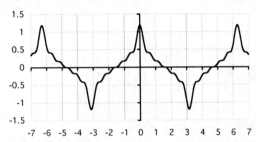

(c) Multiply the amplitudes by 3 and change ω to
$1/2$: $y_{11} = \dfrac{6}{\pi} \cos \dfrac{x}{2} + \dfrac{6}{3\pi} \cos \dfrac{3x}{2} + \dfrac{6}{5\pi} \cos \dfrac{5x}{2}$
$+ \dfrac{6}{7\pi} \cos \dfrac{7x}{2} + \dfrac{6}{9\pi} \cos \dfrac{9x}{2} + \dfrac{6}{11\pi} \cos \dfrac{11x}{2}$

61.

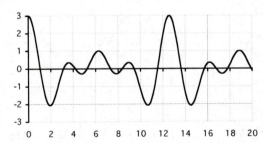

The period is approximately 12.6 units.

63. Lows: $B - A$; Highs: $B + A$

65. He is correct. The other trig functions can be obtained from the sine function by first using the formula $\cos x = \sin(x + \pi/2)$ to obtain cosine,

574

and then using the formulas $\tan x = \dfrac{\sin x}{\cos x}$, $\cot x$ $= \dfrac{\cos x}{\sin x}$, $\sec x = \dfrac{1}{\cos x}$, $\csc x = \dfrac{1}{\sin x}$ to obtain the rest.

67. The largest B can be is A. Otherwise, if B is larger than A, the low figure for sales would have the negative value of $A - B$.

16.2 Derivatives of Trigonometric Functions and Applications

1. $f'(x) = \cos x + \sin x$

3. $g'(x) = \cos x \tan x + \sin x \sec^2 x = \sin x (1 + \sec^2 x)$

5. $h'(x) = -2 \csc x \cot x - \sec x \tan x + 3$

7. $r'(x) = \cos x - x \sin x + 2x$

9. $s'(x) = (2x - 1)\tan x + (x^2 - x + 1)\sec^2 x$

11. $t'(x) = -[\csc^2 x (1 + \sec x) + \cot x \sec x \tan x]/(1 + \sec x)^2$

13. $k'(x) = -2 \cos x \sin x$

15. $j'(x) = 2 \sec^2 x \tan x$

17. $f'(x) = \cos(3x-5)(3) = 3\cos(3x-5)$

19. $f'(x) = -\sin(-2x+5)(-2) = 2\sin(-2x+5)$

21. $p'(x) = \pi \cos\left[\dfrac{\pi}{5}(x - 4)\right]$

23. $u'(x) = -(2x - 1)\sin(x^2 - x)$

25. $v'(x) = (2.2x^{1.2} + 1.2)\sec(x^{2.2} + 1.2x - 1) \cdot \tan(x^{2.2} + 1.2x - 1)$

27. $w'(x) = \sec x \tan x \tan(x^2 - 1) + 2x \sec x \sec^2(x^2 - 1)$

29. $y'(x) = e^x[-\sin(e^x)] + e^x \cos x - e^x \sin x = e^x[-\sin(e^x) + \cos x - \sin x]$

31. $z'(x) = \dfrac{\sec x \tan x + \sec^2 x}{\sec x + \tan x} = \dfrac{\sec x (\tan x + \sec x)}{\sec x + \tan x} = \sec x$

33. $\dfrac{d}{dx} \sec x = \dfrac{d}{dx}\left[\dfrac{1}{\cos x}\right] = \dfrac{\sin x}{\cos^2 x} =$

$\dfrac{1}{\cos x} \dfrac{\sin x}{\cos x} = \sec x \tan x$

35. $\dfrac{d}{dx} \csc x = \dfrac{d}{dx}\left[\dfrac{1}{\sin x}\right] = -\dfrac{\cos x}{\sin^2 x} =$

$-\dfrac{1}{\sin x} \dfrac{\cos x}{\sin x} = -\csc x \cot x$

37. $\dfrac{d}{dx}[e^{-2x} \sin(3\pi x)] =$

$-2e^{-2x} \sin(3\pi x) + 3\pi e^{-2x} \cos(3\pi x) =$

$e^{-2x}[-2\sin(3\pi x) + 3\pi\cos(3\pi x)]$

39. $\dfrac{d}{dx} [\sin(3x)]^{0.5} = 0.5[\sin(3x)]^{-0.5} \, 3\cos(3x) =$

$1.5[\sin(3x)]^{-0.5} \, \cos(3x)$

41. $\dfrac{d}{dx} \sec\left(\dfrac{x^3}{x^2 - 1}\right) =$

$\dfrac{3x^2(x^2 - 1) - 2x^4}{(x^2 - 1)^2} \, \sec\left(\dfrac{x^3}{x^2 - 1}\right) \tan\left(\dfrac{x^3}{x^2 - 1}\right) =$

$\dfrac{x^4 - 3x^2}{(x^2 - 1)^2} \, \sec\left(\dfrac{x^3}{x^2 - 1}\right) \tan\left(\dfrac{x^3}{x^2 - 1}\right)$

43. $\dfrac{d}{dx} [\ln |x| \cot(2x - 1)] = \dfrac{\cot(2x - 1)}{x} -$

$2 \ln |x| \csc^2(2x - 1)$

45. (a) Not differentiable at 0:

$\lim\limits_{h \to 0^+} \dfrac{|\sin h| - |\sin 0|}{h} = \lim\limits_{h \to 0^+} \dfrac{\sin h}{h} = 1$ but

$\lim\limits_{h \to 0^-} \dfrac{|\sin h| - |\sin 0|}{h} = \lim\limits_{h \to 0^-} \dfrac{-\sin h}{h} = -1.$

(b) $\sin(1) \approx 0.84 > 0$ so $|\sin x| = \sin x$ for x near 1; hence $f'(1)$ exists, and $f'(1) \approx$

$\dfrac{\sin 1.0001 - \sin 0.9999}{0.0002} \approx 0.5403.$

47. 0: Write $f(x) = (\sin^2 x)/x$.

x	$f(x)$
-0.1	-0.0997
-0.01	-0.01
-0.001	-0.001
-0.0001	-0.0001
0	
0.0001	0.0001
0.001	0.001
0.01	0.01
0.1	0.0997

By L'Hospital's rule: $\displaystyle\lim_{x \to 0} \frac{\sin^2 x}{x} =$

$\displaystyle\lim_{x \to 0} \frac{2 \sin x \cos x}{1} = 0.$

49. 2: Write $f(x) = (\sin 2x)/x$.

x	$f(x)$
-0.1	1.98669331
-0.01	1.99986667
-0.001	1.99999867
-0.0001	1.99999999
0	
0.0001	1.99999999
0.001	1.99999867
0.01	1.99986667
0.1	1.98669331

By L'Hospital's rule: $\displaystyle\lim_{x \to 0} \frac{\sin 2x}{x} =$

$\displaystyle\lim_{x \to 0} \frac{2 \cos 2x}{1} = 2.$

51. Does not exist: Write $f(x) = (\cos x - 1)/x^3$.

x	$f(x)$
-0.1	4.99583472
-0.01	49.9995833
-0.001	499.999958
-0.0001	4999.99997
0	
0.0001	-5000
0.001	-499.99996
0.01	-49.999583
0.1	-4.9958347

By L'Hospital's rule: $\displaystyle\lim_{x \to 0} \frac{\cos x - 1}{x^3} =$

$\displaystyle\lim_{x \to 0} \frac{-\sin x}{3x^2} = \lim_{x \to 0} \frac{-\cos x}{6x}$ does not exist.

53. $1 = (\sec^2 y)\dfrac{dy}{dx}$, so $\dfrac{dy}{dx} = 1/\sec^2 y$

55. $1 + \dfrac{dy}{dx} + \left(y + x\dfrac{dy}{dx}\right)\cos(xy) = 0$,

$[1 + x\cos(xy)]\dfrac{dy}{dx} = -[1 + y\cos(xy)]$, so $\dfrac{dy}{dx} =$

$-[1 + y\cos(xy)]/[1 + x\cos(xy)]$

57. $c'(t) = 7\pi\cos[2\pi(t - 0.75)]$; $c'(0.75) \approx$ 21.99 per *year* $\approx \$0.42$ per week.

59. $N'(t) = 34.7354\cos[0.602(t - 1.43)]$, $N'(6) \approx -32.12$. On January 1, 2003, the number of sunspots was decreasing at a rate of 32.12 sunspots per year.

61. $c'(t) = 1.035^t[\ln(1.035)(0.8\sin(2\pi t) + 10.2) + 1.6\pi\cos(2\pi t)]$; $c'(1) \approx \$5.57$ per year, or $\$0.11$ per week.

63. (a) $d(t) = 5\cos(2\pi t/13.5) + 10$ **(b)** $d'(t) = -(10\pi/13.5)\sin(2\pi t/13.5)$; $d'(7) \approx 0.270$. At noon, the tide was rising at a rate of 0.270 feet per hour.

65. $V(t) = 110|\sin(100\pi t)|$

$\dfrac{dV}{dt} = 110\,\dfrac{|\sin(100\pi t)|}{\sin(100\pi t)}(100\pi)\cos(100\pi t)$

577

$$= 11{,}000\pi \, \frac{|\sin(100\pi t)|}{\sin(100\pi t)} \cos(100\pi t)$$

Graph:

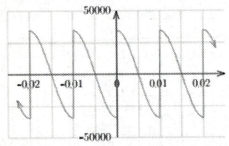

The sudden jumps in the graph are due to the non-differentiability of V at the times $0, \pm 0.01, \pm 0.02,$ .... The derivative is negative immediately to the left and positive immediately to the right of these points.

67. (a) $p(0) = 1.2\cos(5\pi(0) + \pi) = -1.2$ cm; 1.2 cm above the rest position
(b) Velocity $= p'(t) = -6\pi\sin(5\pi t + \pi)$ cm/sec
$t = 0$: $p'(0) = -6\pi\sin(5\pi(0) + \pi) = 0$ cm/sec; not moving
$t = 0.1$: $p'(0.1) = -6\pi\sin(5\pi(0.1) + \pi) = 6\pi \approx$ 18.85 cm/sec; moving downward at 18.85 cm/sec
(c) Period $= P = 2\pi/\omega = 2\pi/(5\pi) = 0.4$ sec
Frequency $= 1/P = 2.5$ cycles per second

69. (a) Velocity $= p'(t) = 1.2e^{-0.1t}[-0.1\cos(5\pi t + \pi) - 5\pi\sin(5\pi t + \pi)]$ cm/sec
$t = 0$: $p'(0) = 1.2e^{-0.1(0)}[-0.1\cos(5\pi(0) + \pi) - 5\pi\sin(5\pi(0) + \pi)] = 0.12$ cm/sec; moving downward at 0.12 cm/sec
$t = 0.1$; $p'(0.1) = 1.2e^{-0.1(0.1)}[-0.1\cos(5\pi(0.1) + \pi) - 5\pi\sin(5\pi(0.1) + \pi)] \approx 18.66$ cm/sec; moving downward at 18.66 cm/sec
(b) Graphs:

p:

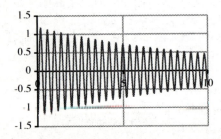

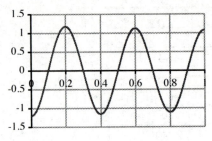

p':

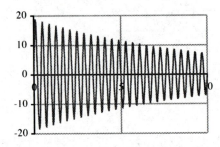

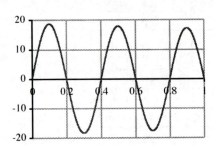

By zooming in on the graph of p', we can see that the slope of p is greatest when $t \approx 0.1$

71. (a) (C): From what we are told we must have $\alpha = -500$ and $P = 40$, so $\omega = 2\pi/40$. **(b)** $A'(t) = -(\pi/20)\sin[2\pi(t + 500)/40]$, $A'(-150) \approx 0.157$. The tilt was increasing at a rate of 0.157 degrees per thousand years.
73. -6; 6

75. Answers will vary. Examples: $f(x) = \sin x$; $f(x) = \cos x$

77. Answers will vary. Examples: $f(x) = e^{-x}$; $f(x) = -2e^{-x}$

79. The graph of $\cos x$ slopes down over the interval $(0, \pi)$, so that its derivative is negative

over that interval. The function $-\sin x$, and not
$\sin x$, has this property.

81. The velocity is $p'(t) = A\omega\cos(\omega t+d)$ which
is a maximum when its derivative,
$p''(t) = -A\omega^2\sin(\omega t+d)$, is zero. But this occurs
when $\sin(\omega t+d) = 0$, so that $p(t)$ is zero as well,
meaning that the stock is at yesterday's close.

83. The derivative of $\sin x$ is $\cos x$. When $x = 0$,
this is $\cos(0) = 1$. Thus, the tangent to the graph
of $\sin x$ at the point $(0, 0)$ has slope 1, which
means it slopes upward at $45°$.

16.3 Integrals of Trigonometric Functions and Applications

1. $\int(\sin x - 2 \cos x)\,dx = -\cos x - 2 \sin x + C$

3. $\int(2 \cos x - 4.3 \sin x - 9.33)\,dx = 2 \sin x + 4.3 \cos x - 9.33x + C$

5. $\int\left(3.4 \sec^2 x + \dfrac{\cos x}{1.3} - 3.2e^x\right) = 3.4 \tan x + \dfrac{\sin x}{1.3} - 3.2e^x + C$

7. $\int 7.6 \cos(3x - 4)\,dx = \dfrac{7.6}{3} \sin(3x - 4) + C$

9. $\int x \sin(3x^2 - 4)\,dx = -\dfrac{1}{6} \cos(3x^2 - 4) + C$: Substitute $u = 3x^2 - 4$.

11. $\int(4x + 2)\sin(x^2 + x)\,dx = -2 \cos(x^2 + x) + C$: Substitute $u = x^2 + x$.

13. $\int(x + x^2)\sec^2(3x^2 + 2x^3)\,dx = \dfrac{1}{6} \tan(3x^2 + 2x^3) + C$: Substitute $u = 3x^2 + 2x^3$.

15. $\int x^2 \tan(2x^3)\,dx = -\dfrac{1}{6} \ln|\cos(2x^3)| + C$: Substitute $u = 2x^3$.

17. $\int 6 \sec(2x - 4)\,dx = 3 \ln|\sec(2x - 4) + \tan(2x - 4)| + C$: Substitute $u = 2x - 4$.

19. $\int e^{2x} \cos(e^{2x} + 1)\,dx = \dfrac{1}{2} \sin(e^{2x} + 1) + C$: Substitute $u = e^{2x} + 1$.

21. $\int_{-\pi}^{0} \sin x\,dx = [-\cos x]_{-\pi}^{0} = -\cos(0) + \cos(-\pi) = -1 - 1 = -2$

23. $\int_{0}^{\pi/3} \tan x\,dx = -[\ln|\cos x|]_{0}^{\pi/3} = -\ln|\cos(\pi/3)| + \ln|\cos(0)| = -\ln(1/2) + \ln(1) = \ln(2)$

25. $\int_{1}^{\sqrt{\pi+1}} x \cos(x^2 - 1)\,dx = \int_{0}^{\pi} \dfrac{1}{2} \cos u\,du$ (substitute $u = x^2 - 1$) $= \dfrac{1}{2} [\sin u]_{0}^{\pi} = \dfrac{1}{2} [\sin(\pi) - \sin(0)] = 0$

27. $\int_{1/\pi}^{2/\pi} \dfrac{\sin(1/x)}{x^2}\,dx = -\int_{\pi}^{\pi/2} \sin u\,du$ (substitute $u = 1/x$) $= [\cos u]_{\pi}^{\pi/2} = \cos(\pi/2) - \cos(\pi) = 1$

29. $\int \cos(ax + b)\,dx = \int \dfrac{1}{a} \cos u\,du$ (substitute $u = ax + b$) $= \dfrac{1}{a} \sin u + C = \dfrac{1}{a} \sin(ax + b) + C$

31. $\int \cot x\,dx = \int \dfrac{\cos x}{\sin x}\,dx = \int \dfrac{1}{u}\,du$ (substitute $u = \sin x$) $= \ln|u| + C = \ln|\sin x| + C$

33. $\int \sin(4x)\,dx = -\dfrac{1}{4} \cos(4x) + C$

35. $\int \cos(-x + 1) \, dx = -\sin(-x + 1) + C$

37. $\int \sin(-1.1x - 1) \, dx$

$\qquad = \frac{1}{1.1} \cos(-1.1x - 1) + C$

39. $\int \cot(-4x) \, dx = -\frac{1}{4} \ln |\sin(-4x)| + C$

41. $\displaystyle\int_{-\pi/2}^{\pi/2} \sin x \, dx = 0$ because, by symmetry, there

is as much area above the x axis as below.

43. $\displaystyle\int_{0}^{2\pi} (1 + \sin x) \, dx = 2\pi$ because, by symmetry,

the average value of $1 + \sin x$ over $[0, 2\pi]$ is 1, hence the area under the curve is the same as the area under the line of height 1, which is 2π.

45.

	D	I
+	x	$\sin x$
−	1	$-\cos x$
$+\int$	$0 \quad \rightarrow$	$-\sin x$

$\int x \sin x \, dx = -x \cos x + \sin x + C$

47.

	D	I
+	x^2	$\cos(2x)$
−	$2x$	$\frac{1}{2} \sin(2x)$
+	2	$-\frac{1}{4} \cos(2x)$
$-\int$	$0 \quad \rightarrow$	$-\frac{1}{8} \sin(2x)$

$\int x^2 \cos(2x) \, dx = \left(\frac{x^2}{2} - \frac{1}{4}\right)\sin(2x) + \frac{x}{2} \cos(2x)$
$+ C$

49.

	D	I
+	$\sin x$	e^{-x}
−	$\cos x$	$-e^{-x}$
$+\int$	$-\sin x \quad \rightarrow$	e^{-x}

$\int e^{-x} \sin x \, dx = -e^{-x} \sin x - e^{-x} \cos x -$

$\int e^{-x} \sin x \, dx$, so $2 \int e^{-x} \sin x \, dx = -e^{-x} \sin x -$

$e^{-x} \cos x + C$; $\int e^{-x} \sin x \, dx = -\frac{1}{2} e^{-x} \sin x -$

$\frac{1}{2} e^{-x} \cos x + C$

51.

	D	I
+	x^2	$\sin x$
−	$2x$	$-\cos x$
+	2	$-\sin x$
$-\int$	$0 \quad \rightarrow$	$\cos x$

$$\int_0^\pi x^2 \sin x \, dx = [-x^2 \cos x + 2x \sin x + 2 \cos$$

$$x]_0^\pi = \pi^2 - 4$$

53. Average $= \dfrac{1}{\pi} \int_0^\pi \sin x \, dx = \dfrac{1}{\pi} [-\cos x]_0^\pi =$

$\dfrac{2}{\pi}$

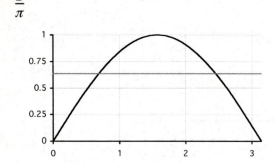

55. $\displaystyle\int_0^{+\infty} \sin x \, dx = \lim_{M \to +\infty} \int_0^M \sin x \, dx =$

$\displaystyle\lim_{M \to +\infty} [-\cos x]_0^M = \lim_{M \to +\infty} (1 - \cos M)$

diverges

57.

	D	I
+	$\cos x$	e^{-x}
$-$	$-\sin x$	$-e^{-x}$
$+\int$	$-\cos x$ $\rightarrow$	e^{-x}

$\displaystyle\int e^{-x} \cos x \, dx = -e^{-x} \cos x + e^{-x} \sin x -$

$\displaystyle\int e^{-x} \cos x \, dx$, so $2 \displaystyle\int e^{-x} \cos x \, dx = -e^{-x} \cos x +$

$e^{-x} \sin x + C$; $\displaystyle\int e^{-x} \cos x \, dx = -\dfrac{1}{2} e^{-x} \cos x +$

$\dfrac{1}{2} e^{-x} \sin x + C$; $\displaystyle\int_0^{+\infty} e^{-x} \cos x \, dx =$

$\displaystyle\lim_{M \to +\infty} \int_0^M e^{-x} \cos x \, dx = \lim_{M \to +\infty} [-\dfrac{1}{2} e^{-x} \cos x +$

$\dfrac{1}{2} e^{-x} \sin x]_0^M = \lim_{M \to +\infty} (-\dfrac{1}{2} e^{-M} \cos M +$

$\dfrac{1}{2} e^{-M} \sin M + \dfrac{1}{2}) = \dfrac{1}{2}$; converges

59. $C(t) = \displaystyle\int \left\{ 0.04 - 0.1 \sin\left[\dfrac{\pi}{26}(t - 25)\right] \right\} dt =$

$0.04t + \dfrac{2.6}{\pi} \cos\left[\dfrac{\pi}{26}(t - 25)\right] + K$; $C(12) =$

1.50 gives $K = 1.02$, so $C(t) = 0.04t +$

$\dfrac{2.6}{\pi} \cos\left[\dfrac{\pi}{26}(t - 25)\right] + 1.02$

61. Position $= \displaystyle\int_0^{10} 3\pi \cos\left[\dfrac{\pi}{2}(t - 1)\right] dt =$

$\left[6 \sin\left[\dfrac{\pi}{2}(t - 1)\right] \right]_0^{10} = 6 \sin \dfrac{9\pi}{2} - 6 \sin\left(-\dfrac{\pi}{2}\right) =$

12 feet

63. Average $=$

$\dfrac{1}{2} \displaystyle\int_5^7 \{57.7 \sin[0.602(t - 1.43)] + 58.8\} \, dt =$

$\dfrac{1}{2} [-95.847 \cos[0.602(t - 1.43)] + 58.8t]_5^7 \approx$

79 sunspots

65. $P(t) = 7.5 \sin[\pi(t - 13)/26] + 12.5$ (see Exercise 45 in Section 16.1). Average $=$

$\dfrac{1}{13} \displaystyle\int_0^{13} \{7.5 \sin[\pi(t - 13)/26] + 12.5\} \, dt =$

$\dfrac{1}{13} [-62.07 \cos[\pi(t - 13)/26] + 12.5t]_0^{13} \approx$

7.7%

67. (a) Average voltage over [0, 1/6] is

$$6 \int_{0}^{1/6} 165 \cos(120\pi t) \, dt = 2.63[\sin(120\pi t)]_{0}^{1/6} =$$

0. In one second the voltage goes through 60 periods of the cosine wave, hence reaches its maximum 60 times; hence the electricity has a frequency of 60 cycles per second.

(b)

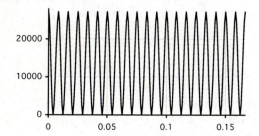

(c) We can use technology to estimate the average $\overline{S} \approx 13{,}612.5$, hence the RMS voltage is approximately $\sqrt{13{,}612.5} \approx 116.673$ volts. Or, we can notice that the graph in (b) appears to be a sinusoid with an average value of $\overline{S} = 165^2/2 = 13{,}612.5$, so that the RMS voltage is $165/\sqrt{2} \approx 116.673$ volts.

69. $TV = \int_{0}^{1} [50{,}000 + 2000\pi \sin(2\pi t)] \, dt =$

$[50{,}000t - 1000 \cos(2\pi t)]_{0}^{1} = \$50{,}000$

71. The integral over a whole number of periods is always zero, by symmetry.

73. 1, because the average value of $2 \cos x$ will be approximately 0 over a large interval.

75. Integrate twice to get $s = -\dfrac{K}{\omega^2} \sin(\omega t - \alpha)$
$+ Lt + M$ for constants L and M.

Chapter 16 Review Exercises

1. $C = 1$, $A = 2$, $\alpha = 0$, and $P = 2\pi$, so $\omega = 1$. Thus, $f(x) = 2 \sin x + 1$.

3. $C = 2$, $A = 2$, $\alpha = 1$, and $P = 2$, so $\omega = \pi$. Thus, $f(x) = 2 \sin[\pi(x - 1)] + 2$. We could also take $\alpha = -1$, getting $f(x) = 2 \sin[\pi(x + 1)] + 2$

In each of Exercises 5–8, substitute $\sin z = \cos(z - \pi/2)$ in the answer from the corresponding one of Exercises 1–4, and simplify as necessary.

5. $f(x) = 2 \cos(x - \pi/2) + 1$

7. $f(x) = 2 \cos[\pi(x - 3/2)] + 2 = 2 \cos[\pi(x + 1/2)] + 2$

9. $-2x \sin(x^2 - 1)$

11. $2e^x \sec^2(2e^x - 1)$

13. $4x \sin(x^2) \cos(x^2)$

15. $\int 4 \cos(2x - 1) \, dx = 2 \sin(2x - 1) + C$

(substitute $u = 2x - 1$)

17. $\int 4x \sec^2(2x^2 - 1) \, dx = \tan(2x^2 - 1) + C$

(substitute $u = 2x^2 - 1$)

19. $\int x \tan(x^2 + 1) \, dx = -\frac{1}{2}\ln|(\cos(x^2 + 1)| + C$

(substitute $u = x^2 + 1$)

21. $\int_{\ln(\pi/2)}^{\ln(\pi)} e^x \sin(e^x) \, dx = \int_{\pi/2}^{\pi} \sin u \, du$ (substitute $u = e^x$) $= [-\cos u]_{\pi/2}^{\pi} = -\cos \pi + \cos(\pi/2) = 1$

23.

	D	I
$+$	x^2	$\sin x$
$-$	$2x$	$-\cos x$
$+$	2	$-\sin x$
$-\int$	0	$\cos x$

$\int x^2 \sin x \, dx = -x^2 \cos x + 2x \sin x + 2 \cos x + C$

25. $C = 10{,}500$, $A = 1500$, $\alpha = 52/2 = 26$, and $P = 52$, so $\omega = 2\pi/52$. Thus, $s(t) = 1500 \cdot \sin[(2\pi/52)(t - 26)] + 10{,}500 \approx 1500 \cdot \sin(0.12083t - 3.14159) + 10{,}500$

27. $R'(t) = (-0.05)20{,}000e^{-0.05t} \sin[(\pi/6)(t - 2)] + 20{,}000e^{-0.05t}(\pi/6) \cos[(\pi/6)(t - 2)]$

$= 20{,}000e^{-0.05t}\{-0.05\sin[(\pi/6)(t - 2)] + (\pi/6) \cos[(\pi/6)(t - 2)]\}$ dollars/month

$R'(20) = 20{,}000e^{-1}\{-0.05\sin[(\pi/6)(18)] + (\pi/6) \cos[(\pi/6)(18)]\}$

$\approx -\$3852/\text{month}$;

decreasing at a rate of $3852 per month

29. Using technology to evaluate $\int_0^{20} R(t) \, dt$:

```
100000+20000*e^(-0.05*x)
        *sin((pi/6)*(x-2))
```

Answer: $2,029,700

31. Total consumption

$= \int_0^t \{150 + 50 \sin[(\pi/2)(x - 1)]\} \, dx$

$= [150x - (100/\pi) \cos[(\pi/2)(x - 1)]]_0^t$